Tectonic–Magmatic Evolution and Mineralization Effect in the Southern Central Asian Orogenic Belt

Tectonic–Magmatic Evolution and Mineralization Effect in the Southern Central Asian Orogenic Belt

Editors

Jiafu Chen

Nan Ju

Zhonghai Zhao

MDPI

Basel • Beijing • Wuhan • Barcelona • Belgrade • Novi Sad • Cluj • Manchester

Editors
Jiafu Chen
Department of Geology
Northeastern University
Shenyang
China

Nan Ju
Shenyang Center
China Geological Survey
Shenyang
China

Zhonghai Zhao
College of Mining
Liaoning Technical University
Fuxin
China

Editorial Office
MDPI AG
Grosspeteranlage 5
4052 Basel, Switzerland

This is a reprint of articles from the Special Issue published online in the open access journal *Minerals* (ISSN 2075-163X) (available at: https://www.mdpi.com/journal/minerals/special_issues/IZ97DTFYB1).

For citation purposes, cite each article independently as indicated on the article page online and as indicated below:

Lastname, A.A.; Lastname, B.B. Article Title. *Journal Name* **Year**, *Volume Number*, Page Range.

ISBN 978-3-7258-2561-5 (Hbk)
ISBN 978-3-7258-2562-2 (PDF)
doi.org/10.3390/books978-3-7258-2562-2

Contents

Preface

Orogenic research, including the temporal and spatial features of orogens, has long been a major task of the international community because it deals with the fundamental principles of the formation and growth of continents. The Central Asian Orogenic Belt (CAOB), also known as the Central Asian Orogenic System or the Altaid Tectonic Collage, occupies a broad area between the Siberian, Baltic, North China, and Tarim cratons. It is the largest accretionary orogen and one of the most important sites of Phanerozoic crustal growth in the world. It is the result of long-lived multi-stage tectonic evolution, including Proterozoic to Paleozoic accretion and collision, Mesozoic intracontinental modification, and Cenozoic rapid deformation and uplift. The accretionary and collisional orogenesis of its early history generated a huge orogenic collage consisting of diverse tectonic units including island arcs, ophiolites, accretionary prisms, seamounts, oceanic plateaus, and micro-continents. These incorporated orogenic components preserved valuable detailed information on orogenic processes and continental crust growth, which make the CAOB a key region for the understanding of continental evolution, mantle–crust interaction, and associated mineralization.

Abundant high-quality data (such as zircon U-Pb ages, whole-rock geochemistry, and Sr-Nd-Pb-Hf-O isotopes) have recently been obtained from various orogens in the CAOB, with the rapid advancement of testing technology. It has been accepted that the accretionary orogenic process is very complex, and more in-depth work is needed. The south CAOB is constituted by several key terranes, microcontinents, arcs, and suture zones, including the Kazakhstan–West Junggar orocline system, Tianshan orogenic belt, Beishan orogenic belt, Xing–Meng orogenic belt, and the northern margin of the North China Craton. The architecture of these orogens remains controversial, in particular, in terms of their crustal composition, deformation, metamorphism, sedimentation, kinematics, and ore deposit genesis. The final closure time and subduction process of the Paleo–Asian Ocean still requires further discussion and more evidence. Furthermore, the south CAOB has experienced superimposition of different orogenic domains and has been overprinted by younger tectonism, the mechanism of which remains to be understood. Thus, we assembled papers for this Special Issue of *Minerals*, which provide new geologic, geophysical, petrological, geochronological, and other data and examine advances on tectonic evolution and the mineralization effect of the south domain of the CAOB.

We appreciate the efforts made by all authors, reviewers, and editors.

Jiafu Chen, Nan Ju, and Zhonghai Zhao
Editors

minerals

Editorial

Editorial for Special Issue "Tectonic–Magmatic Evolution and Mineralization Effect in the Southern Central Asian Orogenic Belt"

Jiafu Chen [1,*], Nan Ju [2,3] and Zhonghai Zhao [4]

1 Department of Geology, Northeastern University, Shenyang 110819, China
2 Shenyang Center of China Geological Survey, Shenyang 110034, China; junan@mail.cgs.gov.cn
3 Northeast Geological S&T Innovation Center of China Geological Survey, Shenyang 110034, China
4 College of Mining, Liaoning Technical University, Fuxin 123000, China; zhaozhonghai@lntu.edu.cn
* Correspondence: chenjiafu@mail.neu.edu.cn

The Central Asian Orogenic Belt (CAOB), one of the world's largest orogens, extending from the Ural Mountains in the west to the Russian and the Chinese Far East, is the result of long-lived multi-stage tectonic evolution, including Proterozoic to Paleozoic accretion and collision, Mesozoic intracontinental modification, and Cenozoic rapid deformation and uplift [1–3]. The accretionary and collisional orogenesis of its early history generated a huge orogenic collage consisting of diverse tectonic units, including island arcs, ophiolites, accretionary prisms, seamounts, oceanic plateaus, and micro-continents. These incorporated orogenic components have preserved valuable detailed information on the orogenic process and continental crust growth, which makes the CAOB a key region for understanding continental evolution, mantle–crust interaction, and the associated mineralization.

However, the architecture of the CAOB remains controversial, particularly in terms of its crustal composition, deformation, metamorphism, sedimentation, and kinematics. Furthermore, the CAOB has experienced the superimposition of different orogenic domains and has been overprinted by younger tectonism, the mechanism of which is still not understood [3]. We have assembled papers for this Special Issue of Minerals that provide new geologic, geophysical, petrological, and geochronological data and study advances in tectonic evolution and the mineralization effect in the south domain of CAOB. For this Special Issue, we have collected 15 papers, encompassing the archipelago system of the West Junggar orogenic belt, Tianshan orogenic belt, Beishan orogenic belt, and Xing-Meng orogenic belt and the northern margin of the North China Craton.

Citation: Chen, J.; Ju, N.; Zhao, Z. Editorial for Special Issue "Tectonic–Magmatic Evolution and Mineralization Effect in the Southern Central Asian Orogenic Belt". *Minerals* **2024**, *14*, 994. https://doi.org/10.3390/min14100994

Received: 26 September 2024
Accepted: 28 September 2024
Published: 30 September 2024

Liu et al. (Contribution 1) report zircon U-Pb chronological and geochemical research on the volcano-sedimentary strata of the Middle Devonian Kulumudi Formation in West Junggar. The zircon U-Pb age of the lithic crystal tuff is 386 ± 2 Ma, and the maximum depositional age of the sandstone is 341 ± 3 Ma, indicating that this rock unit should be redefined as belonging to the Lower Carboniferous Jiangbasitao Formation. Geochemical data and Zircon Hf isotopic data indicate that the Kulumudi Formation and Jiangbasitao Formation were both formed in a juvenile arc setting with ocean–continent subduction. It is further concluded that the southward subduction of the ocean basin represented by the Darbut–Karamay ophiolitic mélange beneath the newly accreted arc crustal segments produced a juvenile arc.

Wang et al. (Contribution 2) conducted zircon U-Pb dating, whole-rock geochemical analysis, and molybdenite Re-Os isotope chronological analysis to constrain the magma origins and the timing of mineralization for the Yuhaixi Mo (Cu) deposit in the eastern Dananhu–Tousuquan island arc, Eastern Tianshan. Zircon U-Pb dating suggests that the emplaced ages of the Yuhaixi monzonitic granite, diorite, and granite are 359 ± 2 Ma, 299 ± 2 Ma, and 307 ± 2 Ma, respectively. The Re-Os dating of molybdenite yields a well-constrained ^{187}Re-^{187}Os isochron age of 354 ± 7 Ma. Geochemical data suggest that the Yuhaixi Mo (Cu) deposit was likely sourced from the juvenile lower crust and was formed in an arc setting.

Wu et al. (Contribution 3) present a systematic study of granodiorites and bimodal volcanic rocks from the north Huitongshan–Zhangfangshan ophiolitic belt in the Southern Beishan orogenic belt. The bimodal volcanic rocks consist of rhyolite and basaltic andesite. The Early Permian granodiorites (289 $\pm$ 2 Ma) and rhyolite (272 $\pm$ 2Ma) have geochemical features similar to those of typical A_2-type granitoids. The authors conclude that the Early Permian Zhangfangshan granodiorites and Baishantang bimodal volcanic rocks formed in a post-collision extensional setting and the Huitongshan–Zhangfangshan ocean had been closed before the Early Permian.

Chen et al. (Contribution 4) report new geochronological and geochemical results for the Early Devonian plutons from the Luotuoquan complex in the eastern segment of the Southern Beishan orogenic belt, including monzogranite (405 Ma) and syenogranite (399 Ma). Geochemical data enable the authors to state that these A-type granites might occur in a post-collisional extension setting. They further point out that the closure of the Beishan Ocean occurred before the Early Devonian.

Wang et al. (Contribution 5) present petrographic, geochronologic, and whole-rock elemental analyses and Sr–Nd–Hf–O isotopic composition analyses for Paleozoic Huashigou granitoids in the South Qilian along the northeastern margin of the Tibetan Plateau. They conclude that the Late Devonian (369 $\pm$4 Ma) and Middle Permian (262 $\pm$1 Ma) granitic rocks were formed in the orogenic–post-orogenic extensional setting after the closure of the Paleo-Tethys Ocean and were derived from the partial melting of crustal materials at different depths.

Wang et al. (Contribution 6) present a geochronological and geochemical study on Early Cretaceous volcanic rocks (129 $\pm$ 1 Ma) from the Longjiang formation in the middle of the Da Hinggan mountains, which belongs to the Zhalantun Duobaoshan island arc. Geochemical data enable the authors to state that these rocks were formed in a superimposed post-collisional and continental arc environment, possibly associated with the Mongol–Okhotsk Ocean closure and the oblique subduction of the Pacific Plate.

Wang et al. (Contribution 7) report new geochronological, geochemical, and isotopic results for the monzogranites (209 $\pm$ 1 Ma) and rhyolitic porphyries (170 $\pm$ 1 Ma) associated with the Wunugetu deposit, situated in the southern Erguna block. They conclude that the Wunugetu deposit was formed in an active continental margin setting and was influenced by the Late Triassic–Middle Jurassic southeastward subduction of the Mongol-Okhotsk Ocean.

Zhao et al. (Contribution 8) conducted skarn mineralogy, fluid inclusion (FI), and H-O isotopic analyses for the Da'anhe deposit in the eastern section of the Xingmeng orogenic belt. The data enable the authors to state that the Da'anhe deposit is a calcareous reduced skarn Au deposit related to the metasomatism of gabbro-dioritic magma and marble in the middle-deep part of the crust. They further point out that the distinctive nature of the ore-forming magma (source, reducing conditions, and high-water content: ΔFMQ = 0.17, $\log(fO_2) = -15.11$, water content = 6.80 wt.%) was key to the formation of the Da'anhe skarn gold deposit.

Na et al. (Contribution 9) studied the geochronology, petrology, and geochemistry of the Hadayang schists (360 $\pm$ 2 Ma and 355 $\pm$ 3Ma) in the eastern Xing'an block. Geochemical data indicate that the magma source of the schists was a mantle that consisted of both spinel and garnet lherzolite, with a partial melting degree of 1–5% and that it had undergone olivine, orthopyroxene, and plagioclase fractional crystallization. They conclude that Late Devonian–Early Carboniferous basic–intermediate magmatic rocks in the eastern Xing'an block were formed in an intracontinental extension tectonic setting similar to that of the North American Basin and Range basalt.

Zhang et al. (Contribution 10) conducted molybdenite Re-Os dating, H-O-S-Pb isotopic compositions, and fluid inclusion (FI) analyses on the Lower Urgen molybdenum deposit, which is situated in the northern Great Xing'an Range. Systematic data indicate that the Lower Urgen molybdenum deposit is coeval with Late Jurassic–Early Cretaceous molybdenum mineralization events in NE China and that the hydrothermal process can be divided into three stages. Isotopic data suggest that the ore-forming fluids were initially of magmatic origin, with the increasing incorporation of meteoric water, and that the ore-forming materials originated from granitic magmas.

Pan et al. (Contribution 11) conducted a systematic investigation into sedimentary rocks from the Heilongjiang Complex, NE China. They identified two episodes of orogenic events including crustal thickening, magmatic flare-up, and landward migration. The Permian crustal thickening event was caused by a shallowing subducting slab angle and thus triggered magmatic flare-ups. The shallow slab subduction enhanced plate coupling and caused the westward subduction of the Mudanjiang Oceanic Plate, generating the second flare-up in this region and the crustal thickening event.

Li et al. (Contribution 12) reported new geochronological and geochemical results for the Middle Silurian–Middle Devonian magmatic rocks in Yitong County, Jilin Province, NE China, including tonalite, basalt, andesite, and metamorphic olivine-bearing basalt. Geochemical data enable the authors to state that the Middle–Late Silurian magmatic rocks were formed in an active continental margin environment influenced by the southward subduction of the Paleo-Asian Ocean, while the Middle Devonian volcanic rocks formed in an extensional environment after the arc–continent collision.

Chen et al. (Contribution 13) conducted a systematic investigation into Permian granitic rocks in the Chifeng area, located at the junction of the Xing'an–Mongolian orogenic belt and the northern margin of the North China Craton. They conclude that the Permian tectono-magmatic evolution of the Paleo-Asian Ocean can be divided into three events: (1) Early Permian granite formed in a subduction setting; (2) Middle Permian granite formed in a syn-collision setting; and (3) Late Permian–Early Triassic granite formed in an extensional setting induced by the slab break-off responsible for the closing of the PAO.

Yu et al. (Contribution 14) present in situ pyrite trace element and sulfur isotope composition measurements from the Beiwagou Pb-Zn Deposit, Liaodong Peninsula, Northeast China. They suggest that most mineralization events occurred due to the compaction and lithification during diagenesis, which led the sulfate to undergo thermochemical sulfate reduction, a loss of permeability, and vertical dewatering.

Tang et al. (Contribution 15) conducted geochronological, geochemical, and apatite fission track thermochronology studies on sandstone from the Upper Triassic Dajianggang Formation at the northeastern margin of the North China Plate, and they conclude that the main source of the Dajianggang Formation was the Late Triassic magmatic arc of the Changchun–Yanji suture zone, possibly mixed with materials from the northern margin of the North China Craton in a convergent tectonic setting. They further point out that the orogenic process in the study area ceased during the Late Triassic period and that the remote effect of Pacific subduction did not impact the study area until 30 Ma.

Funding: This work is supported by National Natural Science Foundation of China (42473076, 42102087, U2244201, 42372108), the China Postdoctoral Science Foundation (2022M712966), and the funding project of the Northeast Geological Science and Technology Innovation Center of the China Geological Survey (QCJJ2023-27).

Conflicts of Interest: The authors declare no conflicts of interest.

List of Contributions:

1. Liu, B.; Hou, L.X.; Xu, Y.; Ju, N.; Ma, J.X.; Xie, Z.H.; Hong, Y.B.H. Zircon U-Pb-Hf Isotopes and Whole-Rock Geochemistry of the "Kulumudi Formation" from the Laofengkou Area (West Junggar): Implications of the Construction of a Juvenile Arc in the Junggar–Balkhash Ocean. *Minerals* **2024**, *14*, 14. https://doi.org/10.3390/min14010014.
2. Wang, D.; Xue, C.; Zhao, Y.; Li, C.; Xi, B.; Yang, Y.; Tian, Q.; Kang, X.; Wu, X. Ore Genesis and the Magmatism of the Yuhaixi Mo(Cu) Deposit in Eastern Tianshan, NW China: Constraints from Geology, Geochemistry, Zircon U-Pb and Molybdenite Re-Os Dating. *Minerals* **2023**, *13*, 1368. https://doi.org/10.3390/min13111368.

3. Wu, L.; Zhai, X.; Wang, E.; Chen, W.; Song, G.; Zheng, F.; Zhao, J.; Wang, J.; Wang, H. Early Permian Post-Collision Extensional Setting in the Southern Beishan Orogenic Belt: Evidence from the Zhangfangshan Granodiorite and the Baishantang Bimodal Volcanic Rocks. *Minerals* **2023**, *13*, 1468. https://doi.org/10.3390/min13121468.
4. Chen, W.; Zhang, M.; Tao, G.; Li, X.; Yu, Q.; Fan, X.; Zhang, J. Ocean–Continent Conversion in Beishan Orogenic Belt: Evidence from Geochemical and Zircon U-Pb-Hf Isotopic Data of Luotuoquan A-Type Granite. *Minerals* **2023**, *13*, 1411. https://doi.org/10.3390/min13111411.
5. Wang, Y.; Chen, W.; Wang, J.; Jia, Z.; Tang, Q.; Di, P. Petrogenesis and Tectonic Evolution of Huashigou Granitoids in the South Qilian Orogen, NW China: Constraints from Geochronology, Geochemistry, and Sr–Nd–Hf–O Isotopes. *Minerals* **2024**, *14*, 71. https://doi.org/10.3390/min14010071.
6. Wang, S.C.; Hao, Y.J.; Shi, L.; Tang, Z.; Zhu, S. Geochronology, Geochemical Characterization and Tectonic Background of Volcanic Rocks of the Longjiang Formation in the Lengjimanda Plate Area, Middle DaHinggan Mountains. *Minerals* **2024**, *14*, 719. https://doi.org/10.3390/min14070719.
7. Wang, Q.; Yang, Y.; Fu, Q.; Zhang, Z.; Guo, X.; Wu, T.; Chai, L.; Zhou, Y.; An, Y. Origin and Geological Implications of Monzogranites and Rhyolitic Porphyries in the Wunugetu Porphyry Copper–Molybdenum Deposit, Northeast China: Evidence from Zircon U-Pb-Hf Isotopes and Whole-Rock Geochemistry. *Minerals* **2024**, *14*, 310. https://doi.org/10.3390/min14030310.
8. Zhao, C.; Sun, F.; Sun, J.; Wang, J.; Han, J.; Chu, X.; Bai, C.; Yu, D.; Xu, Z.; Yi, L.; et al. Genesis and Prospecting Potential of the Da'anhe Skarn Au Deposit in the Central of the Lesser Xing'an Range, NE China: Evidence from Skarn Mineralogy, Fluid Inclusions and H-O Isotopes. *Minerals* **2024**, *14*, 214. https://doi.org/10.3390/min14030214.
9. Na, F.; Song, W.; Liu, Y.; Fu, J.; Wang, Y.; Sun, W. Geochemistry, Chronology and Tectonic Implications of the Hadayang Schists in the Northern Great Xing'an Range, Northeast China. *Minerals* **2023**, *13*, 1379. https://doi.org/10.3390/min13111379.
10. Zhang, G.; Xie, W.; Wen, S.; Zeng, Q.; Zhou, L.; Wang, H.; Zhang, K.; Tang, T.; Ma, P. Ore Genesis of the Lower Urgen Porphyry Molybdenum Deposit in the Northern Great Xing'an Range, Northeast China: Constraints from Molybdenite Re-Os Dating, Fluid Inclusions, and H-O-S-Pb Isotopes. *Minerals* **2023**, *13*, 1189. https://doi.org/10.3390/min13091189.
11. Pan, Y.; Xu, M.; Liu, K.; Wang, M. Continental Arc Flare-Ups and Crustal Thickening Events in NE China: Insights from Detrital Zircon U-Pb Dating and Trace Elements from the Heilongjiang Complex. *Minerals* **2023**, *13*, 1121. https://doi.org/10.3390/min13091121.
12. Li, B.; Chen, J.; Yang, F.; Liu, M.; Zang, Y.; Zhang, C. Middle Silurian–Middle Devonian Magmatic Rocks in the Eastern Segment of the Northern Margin of the North China Craton: Implications for Regional Tectonics. *Minerals* **2024**, *14*, 641. https://doi.org/10.3390/min14070641.
13. Chen, J.; Tian, D.; Li, B.; Shi, Y.; Gao, Z.; Tian, Y.; Li, W.; Zhang, C.; Wang, Y. Permian Granitic Plutons from the Northern Margin of the North China Craton: Implications for the Tectonic Evolution of the Central Asian Orogenic Belt. *Minerals* **2023**, *13*, 1554. https://doi.org/10.3390/min13121554.
14. Yu, Q.; Wang, Z.; Sun, Q.; Wang, K. In Situ Trace Element and Sulfur Isotope Composition of Pyrite from the Beiwagou Pb-Zn Deposit, Liaodong Peninsula, Northeast China: Implications for Ore Genesis. *Minerals* **2023**, *13*, 1176. https://doi.org/10.3390/min13091176.
15. Tang, S.; Li, S.; Zhang, X.; Zhang, D.; Wang, H.; Nie, L.; Zhou, X.; Li, M. Late Triassic Tectonic Setting in Northeastern Margin of North China Craton: Insight into Sedimentary and Apatite Fission Tracks. *Minerals* **2023**, *13*, 975. https://doi.org/10.3390/min13070975.

References

1. Windley, B.F.; Alexeiev, D.; Xiao, W.J.; Kröner, A.; Badarch, G. Tectonic models for accretion of the Central Asian Orogenic Belt. *J. Geol. Soc.* **2007**, *164*, 31–47. [CrossRef]
2. Xiao, W.; Windley, B.F.; Sun, S.; Li, J.; Huang, B.; Han, C.; Yuan, C.; Sun, M.; Chen, H. A Tale of Amalgamation of Three Permo-Triassic Collage Systems in Central Asia: Oroclines, Sutures, and Terminal Accretion. *Annu. Rev. Earth Planet. Sci.* **2015**, *43*, 477–507. [CrossRef]
3. Xiao, W.J.; Windley, B.F.; Han, C.M.; Liu, W.; Wan, B.; Zhang, J.E.; Ao, S.J.; Zhang, Z.Y.; Song, D.F. Late Paleozoic to early Triassic multiple roll-back and oroclinal bending of the Mongolia collage in Central Asia. *Earth-Sci. Rev.* **2018**, *186*, 94–128. [CrossRef]

minerals

MDPI

Article

Zircon U-Pb-Hf Isotopes and Whole-Rock Geochemistry of the "Kulumudi Formation" from the Laofengkou Area (West Junggar): Implications of the Construction of a Juvenile Arc in the Junggar–Balkhash Ocean

Bo Liu [1,2], Lin-Xiao Hou [1], Yan Xu [3], Nan Ju [4,5,*], Jing-Xuan Ma [1], Zhi-Hao Xie [1] and Yang-Bai-He Hong [6]

1 Key Laboratory of Ministry of Education on Safe Mining of Deep Metal Mines, Department of Geology, School of Resources and Civil Engineering, Northeastern University, Shenyang 110819, China; liubo@mail.neu.edu.cn (B.L.); 502023290037@smail.nju.edu.cn (L.-X.H.); 2201068@stu.neu.edu.cn (J.-X.M.)
2 State Key Laboratory of Continental Dynamics, Northwest University, Xi'an 710069, China
3 Ministry of Education, Key Laboratory of Orogenic Belts and Crustal Evolution, School of Earth and Space Sciences, Peking University, Beijing 100871, China; yan_xu@pku.edu.cn
4 Shenyang Center of China Geological Survey, Shenyang 110034, China
5 Northeast Geological S&T Innovation Center of China Geological Survey, Shenyang 110034, China
6 Department of Geological Sciences, Stockholm University, 10691 Stockholm, Sweden; yangbaihe.hong@geo.su.se
* Correspondence: junan@mail.cgs.gov.cn

Abstract: The properties of ancient magmatic arcs are crucial for understanding the tectonic evolution of the Central Asian Orogenic Belt. The Middle Devonian Kulumudi Formation in the Laofengkou area of West Junggar lacks accurate chronological data constraints, which hampers the knowledge of the nature of the Late Paleozoic magmatic arcs in the West Junggar and circum-Balkhash areas. In this contribution, samples of pyroclastic rocks and sedimentary rocks were collected from the volcano–sedimentary strata of the Kulumudi Formation. Petrography, zircon U-Pb-Hf isotopic analysis and whole-rock geochemistry were carried out to constrain the age and the tectonic setting of the Kulumudi Formation. The zircon U-Pb age of the lithic crystal tuff from the Kulumudi Formation on the northeast side of the Alemale Mountains was 386 ± 2 Ma, accurately indicating that this rock unit formed during the Middle Devonian. However, the fine sandstone near the Huojierte Mongolian Township, originally assigned as the "Kulumudi Formation", yielded a maximum depositional age of 341 ± 3 Ma. Combined with the stratigraphic contact, this rock unit was redefined to belong to the Lower Carboniferous Jiangbasitao Formation. According to the whole-rock geochemistry study, the lithic crystal tuff of the Kulumudi Formation was characterized as medium potassium–calc–alkaline series rock, which is relatively enriched in light rare earth elements and large ion lithophile elements (i.e., Rb, Ba, K) and depleted in high-field-strength elements (i.e., Nb, Ta, Ti), showing similar geochemical characteristics to the volcanic arc rocks. By contrast, the fine sandstone from the Jiangbasitao Formation had Al_2O_3/SiO_2 (0.25–0.29) and K_2O/Na_2O (1.29–1.72) ratios close to those derived from the continental arc and active continental margin and was characterized as part of the continental arc field in the La-Th-Sc and Th-Sc-Zr/10 tectonic discrimination diagrams. Zircon Hf isotope analysis showed that the $\varepsilon_{Hf}(t)$ values of the Kulumudi Formation were +5.6–+12.8, and those of the Jiangbasitao Formation were +11.43–+15.48, both of which show highly positive juvenile characteristics. The above data indicate that the Kulumudi Formation and Jiangbasitao Formation both formed in a juvenile arc setting with ocean–continent subduction. Combined with the previous work, it was concluded that the southward subduction of the ocean basin represented by the Darbut–Karamay ophiolitic mélanges beneath the newly accreted arc crustal segments produced a juvenile arc with positive Hf isotope characteristics.

Keywords: Kulumudi Formation; West Junggar; Junggar–Balkhash Ocean; Central Asian Orogenic Belt; Hf isotope

Citation: Liu, B.; Hou, L.-X.; Xu, Y.; Ju, N.; Ma, J.-X.; Xie, Z.-H.; Hong, Y.-B.-H. Zircon U-Pb-Hf Isotopes and Whole-Rock Geochemistry of the "Kulumudi Formation" from the Laofengkou Area (West Junggar): Implications of the Construction of a Juvenile Arc in the Junggar–Balkhash Ocean. *Minerals* **2024**, *14*, 14. https://doi.org/10.3390/min14010014

Academic Editor: Jean-Michel Lafon

Received: 24 October 2023
Revised: 11 December 2023
Accepted: 16 December 2023
Published: 21 December 2023

1. Introduction

The Central Asian Orogenic Belt (CAOB), or the Altaids [1], is located between the Siberian Craton, the European Craton and the Tarim–North China Craton (Figure 1a). It is characterized by the significant accretion of arcs (both intra-oceanic and continental), accretionary complexes, seamounts and oceanic plateaus since the Neoproterozoic, which formed during the long-term evolution of the Paleo-Asian Ocean [2,3]. The CAOB is a complex collage that consists of three major oroclines, including the Kazakhstan Orocline [4], the Mongolian Orocline [5] and the NE China Orocline [6]. Among these, the formation of the Kazakhstan Orocline has been attributed to the large-angle bending of the Devonian–Carboniferous magmatic arcs during the Late Paleozoic consumption of the Junggar–Balkhash Ocean (the southwestern branch of the Paleo-Asian Ocean) [2,4,7–10]. These two Devonian–Carboniferous magmatic arcs are difficult to reconstruct using a traditional study of igneous rocks because of the long-term subduction erosion [11]. However, the arc-derived sedimentary rocks retained on the surface might have recorded information about the provenances [11,12]. By using whole-rock geochemical compositions and detrital zircon U-Pb-Hf isotope characteristics of these arc-related sedimentary rocks, we can trace the properties of ancient magmatic arcs. Moreover, compared with the well-accepted Early-to-Middle-Devonian outer arc belt (OAB) with a Precambrian continental basement, the basement nature of the Late-Devonian-to-Carboniferous inner arc belt (IAB) remains debated [9,10,13].

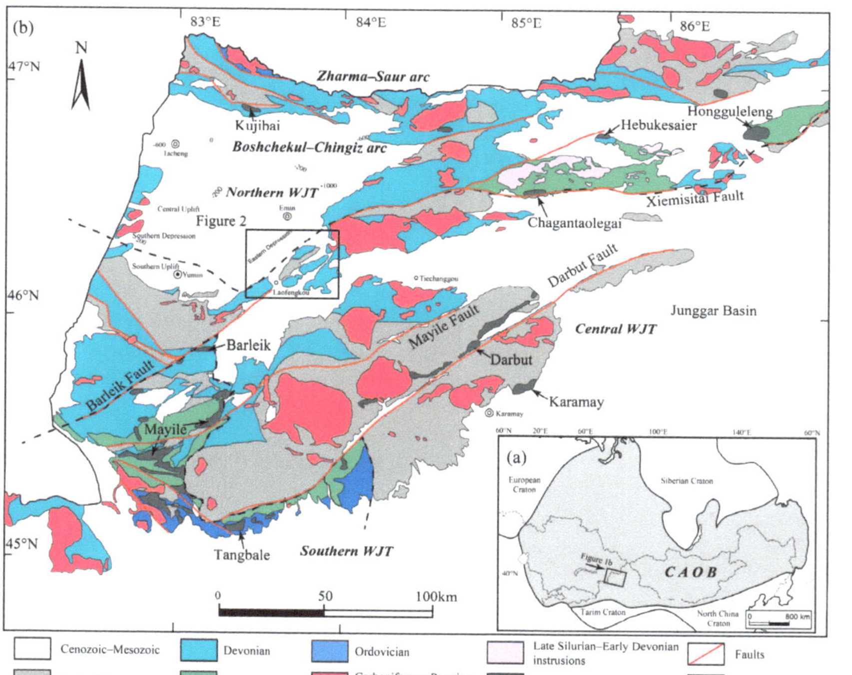

Figure 1. (**a**) Simplified tectonic map of the Central Asian Orogenic Belt (modified after [14]). (**b**) Geological map of the West Junggar terrane (modified after [15]).

The West Junggar Terrane (WJT) in the southwestern CAOB is located on the northern limb of the Kazakhstan Orocline (Figure 1a) [9,10]. The WJT is characterized by widespread Paleozoic accretionary complexes and Carboniferous–Permian intrusive rocks (Figure 1b), and its formation was closely associated with the subduction and closure of the Junggar–Balkhash Ocean [2,7,16–18]. Current research on the tectonic evolution of the WJT focuses on the Ediacaran–Early Paleozoic [15,17,19,20] and Carboniferous–Permian periods [14,18,21]. It is generally accepted that the Ediacaran–Early Paleozoic evolution of the WJT was dominated by a single or multiple intra-oceanic arc(s) [15,17,19,20,22,23]. Previous work showed that the WJT was a place of active convergence during the Early Carboniferous [24,25], while the Late Carboniferous–Permian tectonic setting was suggested to be a post-collisional [25–27] or continuous normal/ridge-related subduction system [7,28,29]. However, relatively little research has been focused on the Devonian tectonic evolution of the WJT [9,30]. Therefore, a timely study of the Devonian tectonic setting is the key to understanding the whole Paleozoic tectonic evolution of the Junggar–Balkhash Ocean.

The Laofengkou area is located in the central part of the WJT, where the Late Paleozoic volcano–sedimentary strata and the Carboniferous–Permian intrusive rocks are extensively outcropping (Figure 2), but the accurate sedimentary age of the Middle Devonian Kulumudi Formation is still unclear. In this contribution, samples of pyroclastic rocks and sedimentary rocks were collected from the Middle Devonian Kulumudi Formation (Figure 2). Petrography, zircon U-Pb-Hf isotopic and whole-rock geochemistry analyses were carried out to accurately constrain the age and tectonic setting of the Middle Devonian Kulumudi Formation, which could provide new insights into the Late Paleozoic evolution of West Junggar. Combined with published data from the West Junggar and circum-Balkhash areas, the nature of the Late Paleozoic magmatic arcs in the tectonic frame of the Kazakhstan Orocline is discussed.

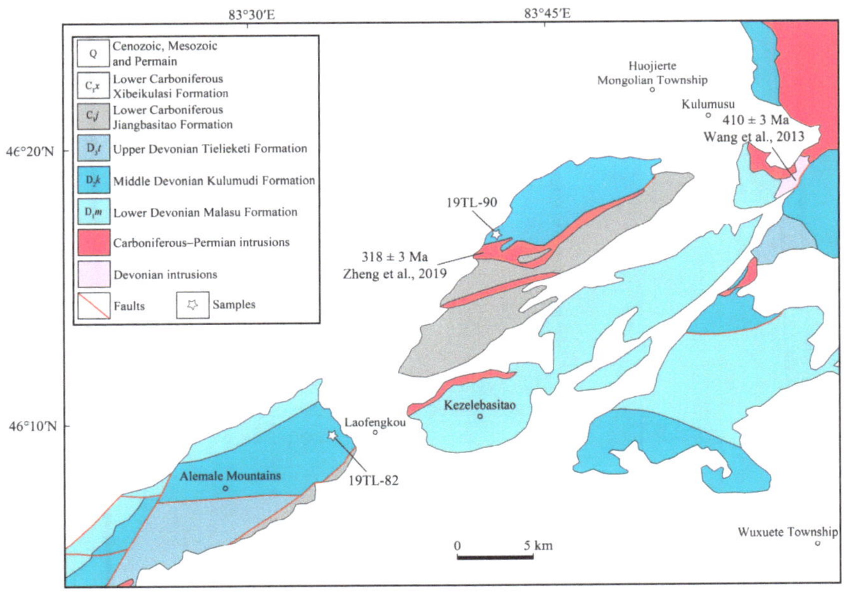

Figure 2. Geological map of the Laofengkou area (modified after [31]). Published age data are from [22,32].

2. Geological Setting and Sampling

2.1. Tectonic Units of the WJT

The WJT is located at the junction of the Siberian and Kazakhstan–Yili plates (Figure 1a), sandwiched between the North Tianshan and the Irtysh–Zaisan suture zones. According to juvenile Sr-Nd-Hf isotopic data, the WJT has been interpreted to lack of ancient Precambrian continental basement [16,33,34]. Taking the Chagantaolegai ophiolite mélange [35,36] and the Barleik–Mayile–Tangbale ophiolite mélanges as the tectonic boundary (Figure 1b; [21,37]), the WJT can be divided into the northern, central and southern parts.

The northern WJT is composed of the Zharma–Saur arc and Boshchekul–Chingiz arc from north to south, which are separated by the EW-trending Kujibai–Hebukesaier-Hongguleleng ophiolitic mélanges [36,38]. The Zharma–Saur arc (346–321 Ma [24]) formed during the southward subduction of the Irtysh–Zaisan Ocean [2,24,39,40], while the Boshchekul–Chingiz arc is supposed to be related to the northward subduction of the Junggar–Balkhash Ocean [2,41] or the southward subduction of the Irtysh–Zaisan Ocean [42,43] during the Late Silurian–Early Devonian period.

The central WJT is dominated by the Late Paleozoic accretionary complexes (Figure 1b) and Late Carboniferous–Early Permian granitoids [26,27]. Based on the recent identification of Late Devonian diorite (~368 Ma) with typical juvenile Nd isotopes, the central WJT was suggested to be the eastern extension of the northern limb of the IAB [9]. The Devonian–Early Carboniferous Karamay and Darbut ophiolitic mélanges [44–47] occur in the accretionary complexes and represent the youngest ophiolite mélanges in the WJT. The Paleozoic volcano–sedimentary sequences of the central WJT can be divided into the Mayile Mountains area and Karamay area. In the Mayile Mountains area, the Devonian strata are dominated by the Kulumudi and Tielieketi formations, comprising a set of clastic and volcanic rocks under neritic to marine environments [21,31,37,48–50]. The zircon U-Pb ages of the tuffs from the Tielieketi Formation range from ~369 to 363 Ma [21,50], and the maximum depositional age of this formation is ~356 Ma [21]. The Lower Carboniferous strata include the Heishantou and Jiangbasitao formations, and the zircon U-Pb ages of tuffs from the latter are ~336 to 339 Ma [51]. The Paleozoic strata exposed in the Karamay area include the Carboniferous and Permian volcano–sedimentary rocks. The Lower Carboniferous strata consist of the Tailegula, Baogutu and Xibeikulasi formations, and there are different understandings of the stratigraphic sequences among them [52–55]. A limited outcrop of the Upper Carboniferous Molaoba Formation shows typical fluvial sedimentary characteristics [21]. Upwards, Permian volcano–sedimentary sequences are a set of intermediate-acid terrestrial volcanic rocks containing fossils of Angara flora [56,57].

The southern WJT contains the Ediacaran–Silurian arc and accretionary complexes. The records of intra-oceanic arc magmatism are mainly distributed in the Barleik (540–502 Ma [22,37,58]), Mayile (530–515 Ma [20,23]) and Tangbale (572–549 Ma [22,23]) areas. The accretionary complexes are mainly composed of the Barleik, Mayile and Tangbale SSZ-type ophiolite mélanges from north to south and the subduction-related metamorphic rocks (504–492 Ma [17]). Previous studies have found that the oldest ophiolitic mélanges of the southern WJT formed at 572 to 512 Ma [20,59–61], suggesting that the initial intra-oceanic subduction of the Junggar–Balkhash Ocean occurred in the Ediacaran [15]. Recently, the occurrences of quartz diorites with adakitic geochemical features (394–390 Ma [10]) were interpreted as supporting evidence of an eastern extension of the Kazakhstan OAB in the southern WJT.

2.2. Geology of the Study Area

The Laofengkou area is located in the northeast of the Mayile Mountains in the central WJT and is dominated by the Devonian–Early Carboniferous strata and granitoid intrusions (Figure 2). The Middle Devonian Kulumudi Formation is composed of littoral–shallow marine terrigenous sedimentary rocks, pyroclastic rocks with a small amount of limestone and acid volcanic rocks [31]. The zircon U-Pb age of the tuff for the Kulumudi Formation near the Laofengkou area (Tiechanggou area, Figure 1b) is 379 ± 5 Ma [48]. In terms of the Laofengkou area, the Middle Devonian Kulumudi Formation exposed northeast of the Alemale Mountains consists of tuffaceous siltstone and lithic crystal tuff (Figure 3a). Meanwhile, the Kulumudi Formation is composed of fine sandstone interbedded with sandstone southwest of the Huojierte Mongolian Township (Figure 3d).

Figure 3. Photomicrographs and field photographs of the dark green lithic crystal tuff (**a–c**) and gray-green fine sandstone (**d–f**). Pl—plagioclase; Cpx—clinopyroxene; Deb—debris.

2.3. Sample Description

Samples were collected from the above locations (Figure 2), including the lithic crystal tuff and fine sandstone. The dark green lithic crystal tuff (19TL-82, 46°7′48″ N, 83°35′57″ E) had a tuffaceous texture and massive structure (Figure 3a), which was mainly composed of crystal debris (65%), lithic debris (25%) and volcanic ashes (10%) (Figure 3b,c). The crystal debris consisted of plagioclase (0.2–0.7 mm) and clinopyroxene (~0.2 mm). Locally, this sample had undergone clayization. The gray-green fine sandstone (19TL-90, 46°17′20″ N, 83°42′22″ E) showed a fine-grained texture and massive structure (Figure 3d). It developed plenty of xenomorphic carbonate minerals and a few plagioclases with altered edges and polysynthetic twins (Figure 3e,f).

3. Analytical Methods

3.1. Zircon U-Pb and Lu-Hf Isotope Analysis

Representative zircon grains were selected from the samples according to their color, shape and integrity; mounted in epoxy resin; and polished. The zircon grains with obvious oscillating zoning under cathodoluminescence (CL) images were considered. Zircon U-Pb dating and trace element analyses were undertaken using laser ablation inductively coupled plasma mass spectrometry (LA-ICP-MS) at Beijing Createch Testing Technology Co., Ltd., Beijing, China, using a RESOlution 193 nm laser ablation system attached to an Agilent 7500 ICP-MS instrument. The laser spot was 24 μm in diameter, and the frequency was 10 Hz. The energy density was about 10 J/cm^2. GJ-1 standard zircons were used during the experiments. Offline data calculations and concordia diagrams were processed using the ICPMSDataCal program [62] and Isoplot 4.0, respectively. The zircon U–Pb data and rare earth element compositions are summarized in Table S1 and Table S2, respectively.

In situ zircon Hf isotope analysis was carried out using a RESOlution SE 193 laser-ablation system attached to a Nu Plasma II multiple collector ICP-MS instrument at State Key Laboratory of Continental Dynamics, Northwest University, Xi'an, China. Zircon GJ-1 was used as the standard sample. The laser spot was 43 μm in diameter and the energy density was 6 J/cm^2. The detailed experimental process and related parameters are described by [63,64]. A two-stage model was selected for the depleted-mantle-based Hf model ages (T_{DM}) calculation. The zircon Hf data are summarized in Table S3.

3.2. *Whole-Rock Major and Trace Element Analysis*

The whole-rock major and trace element compositions were determined at Beijing Createch Testing Technology Co., Ltd., Beijing, China. Major elements were analyzed using a Shimadzu X-ray fluorescence spectrometer with an accuracy better than 5%. The trace elements were determined using an Analyticjena PQMS elite ICP-MS with an accuracy better than 5%. The analytical data are presented in Table S4.

4. Analytical Results

4.1. *Zircon U-Pb and Lu-Hf Isotope Analysis*

4.1.1. Zircon U-Pb Ages

Zircons from the lithic crystal tuff (19TL-82) were mainly euhedral and prismatic with lengths of 50–150 μm and showed obvious oscillatory zoning (Figure 4a), which is typical of tuff origin [65]. Based on REE patterns (Table S2), zircons shown by the solid lines exhibited positive Ce and slightly negative Eu anomalies, which is typical of igneous origin (Figure 5a). This is consistent with their Th/U ratios (0.31–0.97, Table S1). Zircons shown by the dotted line were relatively rich in La (Figure 5a), which might have been subjected to hydrothermal metasomatism or metamorphism at a later stage [66], and were excluded from consideration. This sample yielded 30 valid zircon ages ranging from 380 ± 4 Ma to 421 ± 7 Ma ($^{206}Pb/^{238}U$ age, Table S1), of which the older age of 421 ± 7 Ma might have been derived from an inherited zircon. The weighted mean $^{206}Pb/^{238}U$ age of the remaining 29 zircons was 386 ± 2 Ma (MSWD = 1.20, Figure 6a,b), representing that the lithic crystal tuff formed during the Middle Devonian.

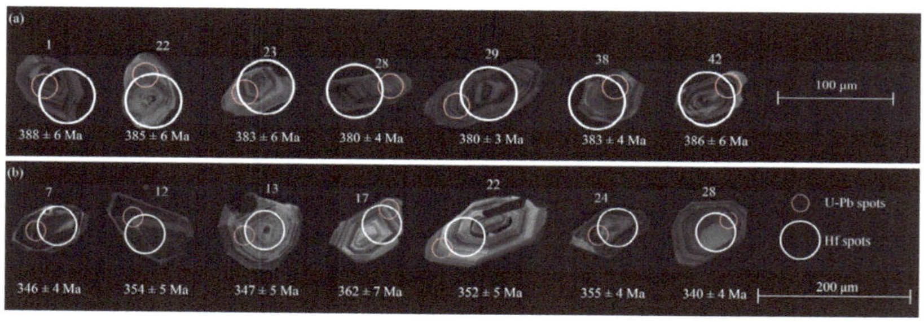

Figure 4. Cathodoluminescence (CL) images of representative zircons of the lithic crystal tuff (**a**) and fine sandstone (**b**).

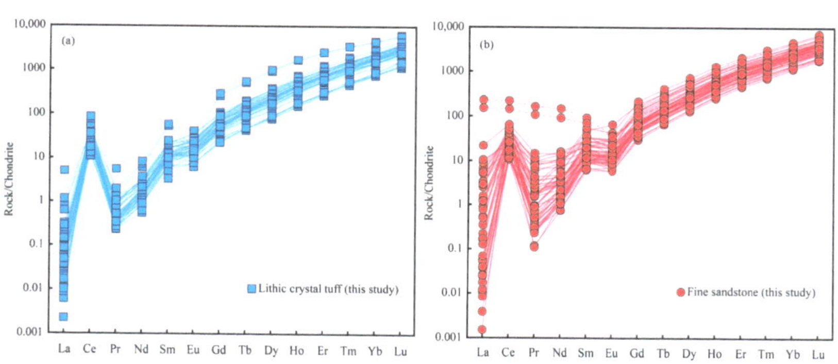

Figure 5. Chondrite-normalized REE patterns of the zircons of the lithic crystal tuff (**a**) and fine sandstone (**b**). Chondrite normalization values are from [67].

Detrital zircons from the fine sandstone (19TL-90) were mainly euhedral and short prismatic, with clear oscillating zoning and 50–200 μm in length (Figure 4b). Zircons shown by the dotted lines were also not considered (Figure 5b, Table S2). The sample yielded 48 usable zircon ages with a variation from 340 ± 4 Ma to 382 ± 5 Ma, with Th/U ratios ranging from 0.40 to 1.11 (Table S1). The Early Carboniferous ages comprised the largest population (41 of the 48 results, Table S1) and exhibited a single peak at 350 Ma (Figure 6c,d). The maximum depositional age of this sandstone was 341 ± 3 Ma ($^{206}Pb/^{238}U$ age, n = 5) using the weighted mean age of the youngest cluster of two or more grain ages (n ≥ 2) overlapping in age at 1σ (YC1σ (2+)).

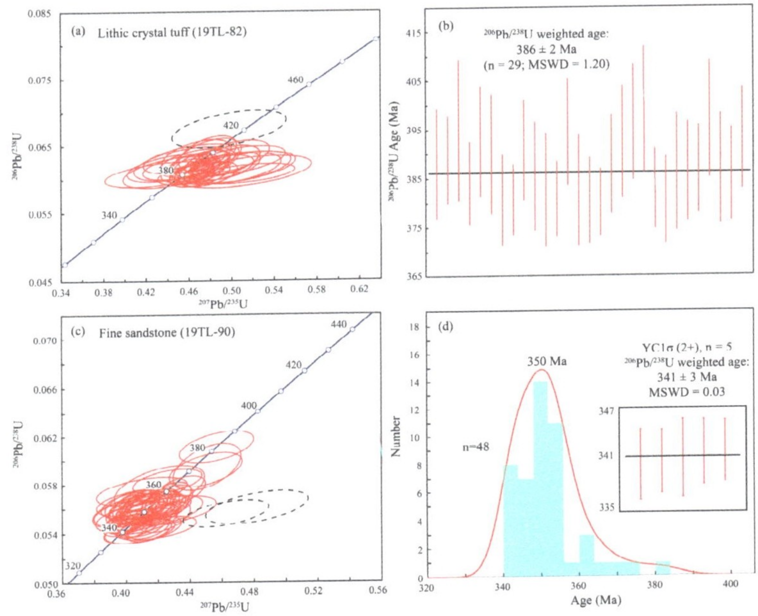

Figure 6. Zircon U-Pb concordia diagrams and weighted average age diagrams of the lithic crystal tuff (**a**,**b**) and fine sandstone (**c**,**d**). The red solid circles represent valid zircon ages that were used for further discussion, wheras the black dotted ones were not considered during the calculation.

4.1.2. Hf Isotopes

Nineteen zircon grains from the lithic crystal tuff that were selected for U–Pb dating were analyzed for in situ Lu–Hf isotopic compositions. They had initial $^{176}Hf/^{177}Hf$ ratios from 0.282685 to 0.282895 (Table S3) and exhibited $\varepsilon_{Hf}(t)$ values from +5.6 to +12.8 (Figure 7a). All zircon grains were characterized by young Hf model ages of 564–1032 Ma (Table S3).

Twenty-eight detrital zircon grains from the fine sandstone had initial $^{176}Hf/^{177}Hf$ ratios from 0.282931 to 0.282991 (Table S3) and their $\varepsilon_{Hf}(t)$ values ranged from +11.4 to +15.5, consistent with the Early Carboniferous sedimentary rocks in the adjacent area (Figure 7b). These detrital zircons had comparable young Hf model ages of 366–626 Ma (Table S3).

4.2. Whole-Rock Major and Trace Element Analysis

The whole-rock geochemical analysis was carried out for the lithic crystal tuff (19TL-80/81) and fine sandstone (19TL-88/89). To avoid the influence of loss on ignition (LOI), all contents for major oxides were recalculated after deducting the LOI. Whole-rock geochemical data are shown in Table S4.

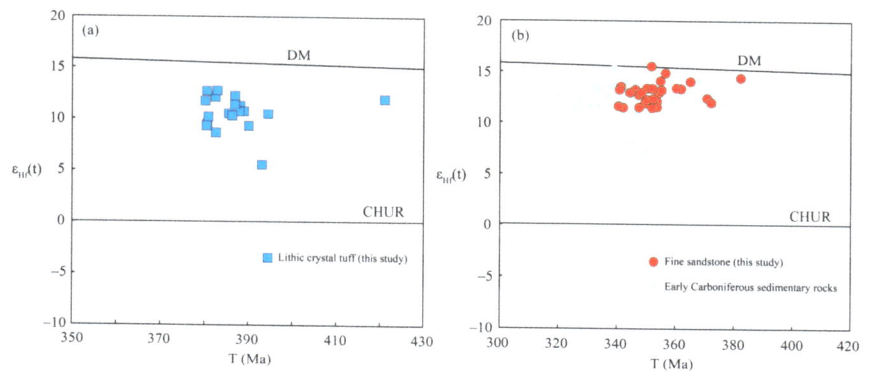

Figure 7. The $\varepsilon_{Hf}(t)$ vs. T/(Ma) diagrams of the lithic crystal tuff (**a**) and fine sandstone (**b**). Data for the Early Carboniferous sedimentary rocks are from [68].

4.2.1. Major Elements

The lithic crystal tuff samples (19TL-80/81) had high SiO_2 contents (62.32–63.29 wt.%) and medium Na_2O (3.93–4.23 wt.%) and K_2O (1.18–1.36 wt.%) contents. Their MgO contents ranged from 2.38 wt.% to 2.73 wt.%, with a relatively high Mg# value of 56. They were plotted in the andesite and dacite fields (Figure 8a) and medium-K calc-alkaline series field (Figure 8b). The SiO_2/Al_2O_3 ratios (3.50–3.97) of the fine sandstone samples (19TL-88/89) were comparatively low. These fine sandstone samples had medium Na_2O/K_2O and relatively high Fe_2O_3/K_2O ratios, and thus, were plotted in the litharenite fields (Figure 8c).

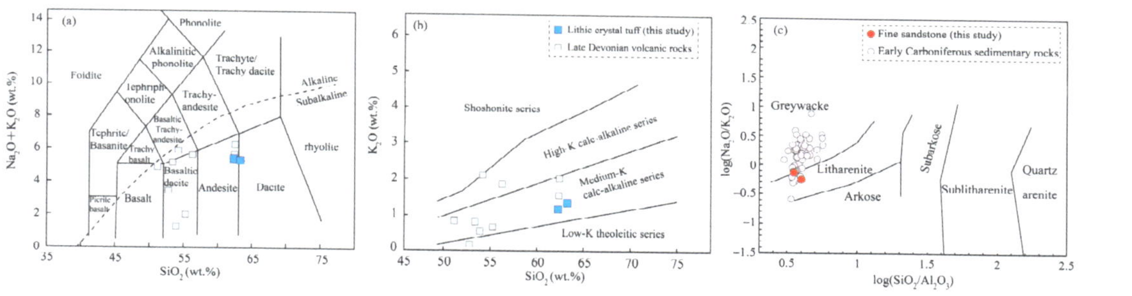

Figure 8. TAS classification diagram ((**a**), from [69]), SiO_2 vs. K_2O diagram of the lithic crystal tuff ((**b**), from [70]); lg[w(Na_2O)/(K_2O)] vs. lg[w(SiO_2)/w(Al_2O_3)] diagram ((**c**), from [71]) of the fine sandstone. Data for the Late Devonian volcanic rocks and the Early Carboniferous clastic rocks are from [50] and [68,72], respectively.

4.2.2. Trace Elements

The lithic crystal tuff samples showed enrichment of light rare earth elements (LREEs) and depletion of heavy rare earth elements (HREEs) with δEu values of 0.94–0.97 (Figure 9a), implying the absence of obvious Eu anomalies. They were rich in large-ion lithophile elements (LILEs, i.e., Rb, Ba, K) and depleted in high-field-strength elements (HFSEs, i.e., Nb, Ta, Ti), showing similar geochemical characteristics to those of volcanic arc rocks (Figure 9b). The fine sandstone samples had enriched LREEs and relatively flat HREEs (Figure 9c). The δEu values ranged from 0.66 to 0.74, indicating an obvious Eu negative anomaly, similar to that of open PAAS (δEu = 0.65 [73]). The samples were also deficient in HFSEs (i.e., Nb, Ta, Ti) (Figure 9d) and exhibited low ferromagnesian trace element concentrations (Sc, 8.0–8.9; Co, 9.8–11.9; Ni, 4.5–6.5).

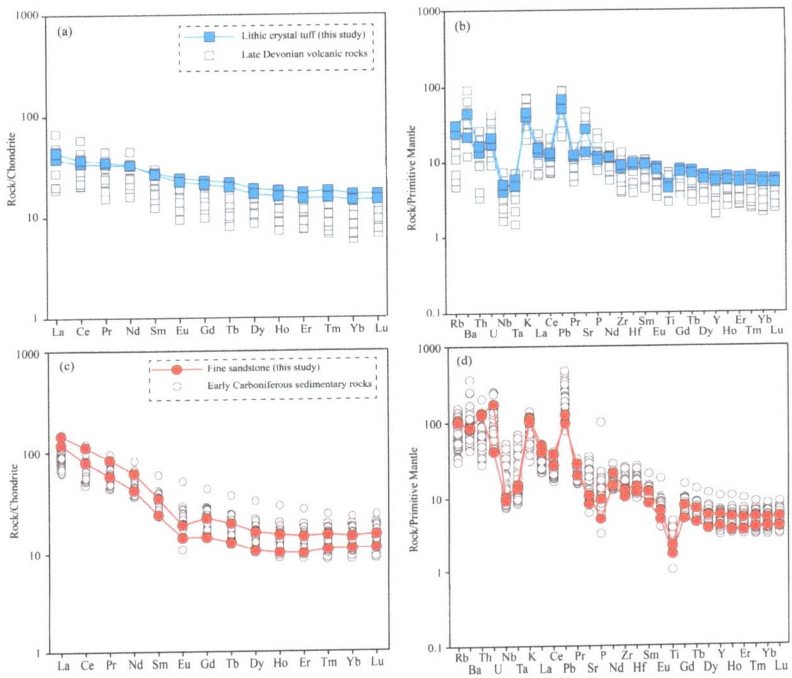

Figure 9. Chondrite-normalized REE patterns (**a**,**c**) and primitive-mantle-normalized spidergrams (**b**,**d**) of the lithic crystal tuff and fine sandstone. Data for Later Devonian volcanic rocks and Early Carboniferous sedimentary rocks are from [50] and [68,72], respectively.

5. Discussion

5.1. Sedimentary Age of the Kulumudi Formation

Previous work on the stratigraphy of the Middle Devonian Kulumudi Formation was mainly based on the lithostratigraphic correlation and paleontological fossils [31,74]. Among these fossils, occurrences of *Barradeophylum* sp., *Pachyyfavosites* sp., *Paraspirifer* sp., *Psilophyton bella* (*Tchirkoua*) *Aneurophyton* aff. Germanicak Etw., etc., indicate typical Middle Devonian characteristics [31,74]. In this study, the zircon U-Pb dating result for the lithic crystal tuff from the Middle Devonian Kulumudi Formation was 386 ± 2 Ma (Figure 6b), indicating that the stratum northeast of the Alemale Mountains belongs to the Middle Devonian, which is consistent with previous results [31,74].

Differently, based on the detrital zircon ages, the maximum depositional age of detrital zircons from the fine sandstone was 341 ± 3 Ma (Figure 6c). This result shows that the stratum originally assigned as the "Kulumudi Formation" southwest of the Huojierte Mongolian Township was formed during the Early Carboniferous. Indeed, the result is similar to the age of the adjacent Lower Carboniferous Jiangbasitao Formation (336 ± 3 Ma, 339 ± 2 Ma [51]). Importantly, the originally assigned "Kulumudi Formation" and the Lower Carboniferous Jiangbasitao Formation were invaded by the Late Carboniferous intrusive rocks (318 ± 3 Ma [22]) together (Figure 2). The Jiangbasitao Formation is a set of terrigenous sedimentary rocks mainly composed of the conglomerate, sandy conglomerate, carbonaceous shale, argillaceous siltstone and fine sandstone [31], which is consistent with the rocks observed southwest of the Huojierte Mongolian Township. Thus, our new chronological data, together with the above lithologic correlation, indicate that parts of the "Kulumudi Formation" should be redefined as the Lower Carboniferous Jiangbasitao Formation.

5.2. *Tectonic Setting*

5.2.1. The Middle Devonian Kulumudi Formation

The lithic crystal tuff samples (~386 Ma) of the Middle Devonian Kulumudi Formation were plotted in the andesite and dacite fields in the TAS classification diagram (Figure 8a), and thus, they belonged to the medium potassium calc-alkaline series (Figure 8b). These samples were rich in LREEs and LILEs and depleted in HREEs and HFSEs (Figure 9a,b), indicating geochemical characteristics similar to the volcanic arc rocks [75]. Furthermore, the samples had relatively low Nb/La ratios (0.32–0.34) and high La/Ta (35.53–53.82) and Th/Nb ratios (0.38–0.39), confirming that they were formed in a subduction-related setting [76]. Their high Ba/La values (16.47–29.58) also imply a strong effect on the magma source area from a subduction zone or related fluids [77]. Furthermore, the $\varepsilon_{Hf}(t)$ values of the zircons from the lithic crystal tuff show highly positive characteristics, ranging from +5.6 to +12.8 (Figure 7a), indicating the feature of a juvenile arc. Considering the whole-rock geochemistry and Hf isotope analysis, the lithic crystal tuff from the Middle Devonian Kulumudi Formation is suggested to be formed in a juvenile arc setting.

5.2.2. The Early Carboniferous Jiangbasitao Formation

The chemical index of alteration (CIA) and index of compositional variability (ICV) can be used to assess the weathering intensity and source composition of the sedimentary rocks. The relatively low CIA values (63.06–63.97) for the fine sandstone from the Jiangbasitao Formation suggest a low degree of chemical weathering. The high ICV values ranging from 1.40–1.45 indicate an immature source. Their δEu values ranged from 0.66 to 0.74 and clear Nb-Ta negative anomalies (Figure 9d), similar to many subduction-related compositionally intermediate lavas [75]. The contents and ratios of major and trace elements in clastic rocks can distinguish different tectonic settings, such as oceanic arc, continental arc, active continental margin and passive continental margin [78,79]. The contents of TiO_2 (0.38–0.50 wt.%) in fine sandstone samples from the Jiangbasitao Formation were similar to those in active continental margin and continental arc [80]. The Al_2O_3/SiO_2 (0.25–0.29) and K_2O/Na_2O (1.29–1.72) ratios were close to those derived from the continental arc and active continental margin, respectively [80]. Moreover, all samples were plotted in the continental arc field in the La-Th-Sc and Th-Sc-Zr/10 tectonic discrimination diagrams (Figure 10a,b). The highly positive $\varepsilon_{Hf}(t)$ values (+11.4 to +15.5) of the detrital zircons from the fine sandstone suggest a juvenile arc (Figure 7b), which is consistent with the interpretation that the Early Carboniferous sedimentary rocks from the Darbut area were derived from a juvenile arc setting (Figure 7b). Thus, combined with published data from adjacent areas [68,72]), the new whole-rock geochemistry and Hf isotope analysis suggest an Early Carboniferous juvenile arc in the central WJT rather than a continental arc with Precambrian basement [13].

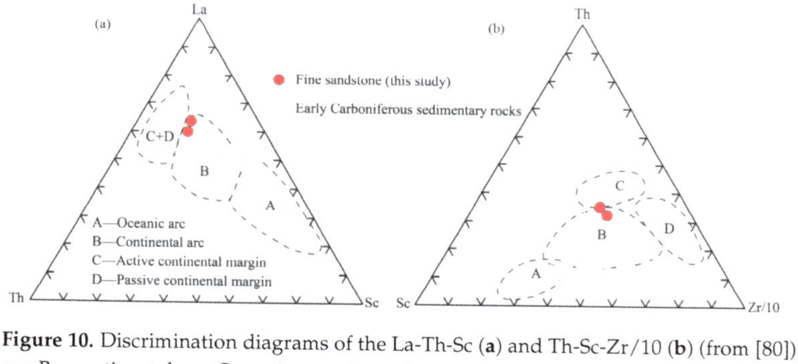

Figure 10. Discrimination diagrams of the La-Th-Sc (**a**) and Th-Sc-Zr/10 (**b**) (from [80]). A—oceanic arc; B—continental arc; C—active continental margin; D—passive continental margin. Data for Early Carboniferous sedimentary rocks are from [68,72].

5.3. Tectonic Evolution of West Junggar during the Middle Devonian–Early Carboniferous

During the Ediacaran–Cambrian period, the southern WJT was suggested to be a southward intra-oceanic subduction setting in the Tangbale area, forming the Tangbale ophiolitic mélange (572–531 Ma [15,20,59–61]). With the northward migration of the subduction zone, the Mayile ophiolite mélange (516–517 Ma [20,81]) and Barleik ophiolite mélange (512 ± 7 Ma [81]) gradually formed. These Ediacaran–Early Paleozoic SSZ-type ophiolitic mélanges, combined with the intra-oceanic arc igneous rocks [23] and subduction-related metamorphic rocks [17], clearly indicate an intra-oceanic arc setting for the southern WJT [15]. Meanwhile, the northward subduction of the oceanic crust of the Junggar–Balkhash Ocean formed the Middle Silurian to Early Devonian Boshchekul–Chingiz arc [41].

There are different interpretations of the ocean basin represented by the Darbut–Karamay ophiolite mélanges during the Devonian and Early Carboniferous. Some researchers interpret the basin as a residual ocean basin trapped by the Junggar–Balkhash Ocean under the constraint of the Kazakhstan Orocline [46], based on the fact that the Lower Carboniferous continuous volcano–sedimentary strata are developed on both sides of the Darbut ophiolite mélange [45,82]. Alternatively, the opening of the Devonian Karamay back-arc ocean basin [41] was suggested to result from the continuous lithosphere extension during the Early–Middle Devonian. Regardless of the residual ocean basin or back-arc ocean basin model, a consensus has been reached about the existence of an Early Devonian–Early Carboniferous oceanic basin in the central WJT. Importantly, previous studies found that the Middle Devonian Kulumudi Formation unconformably overlies the Ediacaran–Early Paleozoic accretionary complexes in the southern and central WJT [20,49,83], indicating that the Ediacaran–Early Paleozoic intra-oceanic subduction had ended before the Middle Devonian [15]. Such a pre-Middle Devonian lateral accretion of juvenile oceanic arc crust was further evidenced by recent dating results of detrital zircons from the southern WJT [12]. A major sedimentary transition from Middle–Late Silurian bathyal facies to Devonian overall littoral–terrestrial facies [49] and the absence of Late Silurian arc-related volcanism also support a pre-Late Silurian termination of the intra-oceanic subduction–accretion processes.

Since the Early Devonian, the oceanic crust represented by the Darbut–Karamay ophiolitic mélanges has subducted southward, forming the active continental margin in the southern and central WJT. Early–Middle Devonian adakitic magmatism (394–390 Ma) with high positive $\varepsilon_{Hf}(t)$ values of +5.6 to +10.3 was newly recognized in the southern WJT and their formation was contributed to the partial melting of a juvenile and slightly thickened lower crust as the eastern extension of the Kazakhstan Early–Middle Devonian OAB [10]. Similarly, Liu and Han [9] suggested that the tectonic setting of the central WJT (northern IAB) evolved to a juvenile Late Devonian IAB built on Ediacaran–Early Paleozoic accreted terranes (intra-oceanic arcs and accretionary complexes), contrasting with the Yili (southern IAB) and Balkhash (central IAB) arcs built on a Precambrian continental basement. Combined with the above-published results, the positive $\varepsilon_{Hf}(t)$ values of the studied Middle Devonian tuff (+5.6–+12.8) and Early Carboniferous sandstone (+11.4–+15.5) show typical juvenile arc characteristics, indicating that the active continental margin of central WJT lacked ancient continental basement during the Middle Devonian–Early Carboniferous subduction [9,10] The occurrences of only two ancient zircons for pyroxene diorite (408 Ma [13]) from the central WJT may not indicate a Precambrian basement. Thus, it was concluded that the southward subduction of the ocean basin represented by the Darbut–Karamay ophiolitic mélanges beneath the newly accreted arc crustal segments resulted in a juvenile arc with positive Hf isotope characteristics.

6. Conclusions

(1) The lithic crystal tuff from the Kulumudi Formation northeast of the Alemale Mountains in the Laofengkou area yielded a zircon U-Pb age of 386 ± 2 Ma and was formed during the Middle Devonian. The maximum depositional age for the fine sandstone originally assigned into the "Kulumudi Formation" southwest of the Huo-

jierte Mongolian Township was 341 ± 3 Ma, belonging to the Lower Carboniferous Jiangbasitao Formation.
(2) According to the whole-rock geochemistry and Hf isotope analysis, the Middle Devonian Kulumudi Formation and Lower Carboniferous Jiangbasitao Formation formed in a juvenile arc setting related to the ocean–continent subduction.
(3) During the Middle Devonian to Early Carboniferous, the ocean basin represented by the Darbut–Karamay ophiolitic mélanges subducted southward beneath the newly accreted arc crustal segments, forming a juvenile arc with positive Hf isotopic characteristics.

Supplementary Materials: The following supporting information can be downloaded at: https://www.mdpi.com/article/10.3390/min14010014/s1, Table S1: Zircon U-Pb data of pyroclastic rocks and sedimentary rocks from the Laofengkou area, West Junggar; Table S2: Rare earth element compositions ($\times 10^{-6}$) of zircons for pyroclastic rocks and sedimentary rocks from the Laofengkou area, West Junggar; Table S3: Zircon in situ Lu-Hf isotope compositions for pyroclastic rocks and sedimentary rocks from the Laofengkou area, West Junggar; Table S4: Whole-rock major (wt.%) and trace element ($\times 10^{-6}$) compositions for pyroclastic rocks and sedimentary rocks from the Laofengkou area, West Junggar.

Author Contributions: Conceptualization, B.L.; methodology, B.L.; software, L.-X.H. and Z.-H.X.; formal analysis, L.-X.H. and J.-X.M.; investigation, Y.-B.-H.H.; writing—original draft preparation, B.L. and L.-X.H.; writing—review and editing, B.L. and Y.X.; funding acquisition, N.J. All authors have read and agreed to the published version of the manuscript.

Funding: This study was supported by the National Key R&D Program of China (2023YFB2603602); National Natural Science Foundation of China (41802236, 42102087, U2244201); China Postdoctoral Science Foundation (2022M712966); Second Tibetan Plateau Scientific Expedition and Research (2021QZKK0304); "the Fundamental Research Funds for the Central Universities" (N2201014); "the Fundamental Research Funds for the Educational Department of Liaoning Provience" (JYTMS20230617); and Opening Foundation of State Key Laboratory of Continental Dynamics, Northwest University (20LCD07).

Data Availability Statement: The authors confirm that all data are presented in the paper.

Acknowledgments: Three anonymous reviewers are thanked for their constructive comments. We are particularly grateful to Zhi-An Bao for help with the zircon Hf isotopic analyses at Northwest University. We thank Jia-Wei Li and Chen He for their help in the field.

Conflicts of Interest: The authors declare no conflict of interest.

References

1. Şengör, A.M.C.; Natal'in, B.A.; Burtman, V.S. Evolution of the Altaid tectonic collage and Palaeozoic crustal growth in Eurasia. *Nature* **1993**, *364*, 299–307. [CrossRef]
2. Windley, B.F.; Alexeiev, D.; Xiao, W.J.; Kröner, A.; Badarch, G. Tectonic models for accretion of the Central Asian Orogenic Belt. *J. Geol. Soc.* **2007**, *164*, 31–47. [CrossRef]
3. Kröner, A.; Hegner, E.; Lehmann, B.; Heinhorst, J.; Wingate, M.T.D.; Liu, D.Y.; Ermelov, P. Palaeozoic arc magmatism in the Central Asian Orogenic Belt of Kazakhstan: SHRIMP zircon ages and whole-rock Nd isotopic systematics. *J. Asian Earth Sci.* **2008**, *32*, 118–130. [CrossRef]
4. Li, P.; Sun, M.; Rosenbaum, G.; Yuan, C.; Safonova, I.; Cai, K.; Jiang, Y.; Zhang, Y. Geometry, kinematics and tectonic models of the Kazakhstan Orocline, Central Asian Orogenic Belt. *J. Asian Earth Sci.* **2018**, *153*, 42–56. [CrossRef]
5. Xiao, W.; Windley, B.F.; Sun, S.; Li, J.; Huang, B.; Han, C.; Yuan, C.; Sun, M.; Chen, H. A Tale of Amalgamation of Three Permo-Triassic Collage Systems in Central Asia: Oroclines, Sutures, and Terminal Accretion. *Annu. Rev. Earth Planet. Sci.* **2015**, *43*, 477–507. [CrossRef]
6. Liu, Y.; Xiao, W.; Ma, Y.; Li, S.; Peskov, A.Y.; Chen, Z.; Zhou, T.; Guan, Q. Oroclines in the Central Asian Orogenic Belt. *Natl. Sci. Rev.* **2023**, *10*, nwac243. [CrossRef]
7. Xiao, W.; Han, C.; Yuan, C.; Sun, M.; Lin, S.; Chen, H.; Li, Z.; Li, J.; Sun, S. Middle Cambrian to Permian subduction-related accretionary orogenesis of Northern Xinjiang, NW China: Implications for the tectonic evolution of central Asia. *J. Asian Earth Sci.* **2008**, *32*, 102–117. [CrossRef]

8. Choulet, F.; Cluzel, D.; Faure, M.; Lin, W.; Wang, B.; Chen, Y.; Wu, F.; Ji, W. New constraints on the pre-Permian continental crust growth of Central Asia (West Junggar, China) by U-Pb and Hf isotopic data from detrital zircon: Pre-Permian continental crust growth of Central Asia. *Terra Nova* **2012**, *24*, 189–198. [CrossRef]
9. Liu, B.; Han, B.F. Identification of lateral inhomogeneity of arc basement by reconstructing the Late Devonian arc belt in the southwestern Central Asian Orogenic Belt. *J. Geodyn.* **2019**, *132*, 101668. [CrossRef]
10. Ren, R.; Liu, B.; Han, B.F.; Guan, S.W. Early-Middle Devonian adakitic magmatism generated by slab retreat in southern West Junggar, NW China: Implications for tectonic correlation with central and East Kazakhstan. *Int. Geol. Rev.* **2023**, *65*, 2315–2331. [CrossRef]
11. Safonova, I.; Perfilova, A. Survived and disappeared intra-oceanic arcs of the Paleo-Asian Ocean: Evidence from Kazakhstan. *Natl. Sci. Rev.* **2023**, *10*, nwac215. [CrossRef] [PubMed]
12. Zhang, P.; Wang, G.C.; Shen, T.; Polat, A.; Zhu, C. Paleozoic convergence processes in the southwestern Central Asian Orogenic Belt: Insights from U–Pb dating of detrital zircons from West Junggar, northwestern China. *Geosci. Front.* **2021**, *12*, 531–548. [CrossRef]
13. Wu, C.; Hong, T.; Xu, X.-W.; Cao, M.-J.; Li, H.; Zhang, G.-L.; You, J.; Ke, Q.; Dong, L.-H. Tectonic evolution of the Paleozoic Barluk continental arc, West Junggar, NW China. *J. Asian Earth Sci.* **2018**, *160*, 48–66. [CrossRef]
14. Han, B.F.; Guo, Z.J.; He, G.Q. Timing of major suture zones in North Xinjiang, China: Constraints from stitching plutons. *Acta Petrol. Sin.* **2010**, *26*, 2233–2246. (In Chinese with English abstract)
15. Liu, B.; Han, B.F.; Xu, Z.; Ren, R.; Chen, J.F. The Ediacaran to Early Paleozoic evolution of the Junggar– Balkhash Ocean: A synthesis of the ophiolitic mélanges in the southern West Junggar terrane, NW China. *Geol. J.* **2020**, *55*, 1689–1707. [CrossRef]
16. Chen, B.; Arakawa, Y. Elemental and Nd-Sr isotopic geochemistry of granitoids from the West Junggar foldbelt (NW China), with implications for Phanerozoic continental growth. *Geochim. Cosmochim. Acta* **2005**, *69*, 1307–1320. [CrossRef]
17. Liu, B.; Han, B.-F.; Xu, Z.; Ren, R.; Zhang, J.-R.; Zhou, J.; Su, L.; Li, Q.-L. The Cambrian initiation of intra-oceanic subduction in the southern Paleo-Asian Ocean: Further evidence from the Barleik subduction-related metamorphic complex in the West Junggar region, NW China. *J. Asian Earth Sci.* **2016**, *123*, 1–21. [CrossRef]
18. Liu, B.; Han, B.F.; Ren, R.; Chen, J.F.; Wang, Z.Z.; Zheng, B. Petrogenesis and tectonic implications of the Early Carboniferous to the Late Permian Barleik plutons in the West Junggar (NW China). *Lithos* **2017**, *272–273*, 232–248. [CrossRef]
19. Xu, Z.; Han, B.-F.; Ren, R.; Zhou, Y.-Z.; Zhang, L.; Chen, J.-F.; Su, L.; Li, X.-H.; Liu, D.-Y. Ultramafic–mafic mélange, island arc and post-collisional intrusions in the Mayile Mountain, West Junggar, China: Implications for Paleozoic intra-oceanic subduction–accretion process. *Lithos* **2012**, *132–133*, 141–161. [CrossRef]
20. Ren, R.; Han, B.-F.; Xu, Z.; Zhou, Y.-Z.; Liu, B.; Zhang, L.; Chen, J.-F.; Su, L.; Li, J.; Li, X.-H.; et al. When did the subduction first initiate in the southern Paleo-Asian Ocean: New constraints from a Cambrian intra-oceanic arc system in West Junggar, NW China. *Earth Planet. Sci. Lett.* **2014**, *388*, 222–236. [CrossRef]
21. Liu, B.; Han, B.F.; Chen, J.F.; Ren, R.; Zheng, B.; Wang, Z.Z.; Feng, L.X. Closure time of the Junggar-Balkhash Ocean: Constraints from Late Paleozoic volcano-sedimentary sequences in the Barleik Mountains, West Junggar, NW China. *Tectonics* **2017**, *36*, 2823–2845. [CrossRef]
22. Zheng, B.; Han, B.F.; Liu, B.; Wang, Z.Z. Ediacaran to Paleozoic magmatism in West Junggar Orogenic Belt, NW China, and implications for evolution of Central Asian Orogenic Belt. *Lithos* **2019**, *338–339*, 111–127. [CrossRef]
23. Liao, W.; Han, B.F.; Xu, Y.; Li, A. Ediacaran initial subduction and Cambrian slab rollback of the Junggar Ocean: New evidence from igneous tectonic blocks and gabbro enclave in Early Palaeozoic accretionary complexes, southern West Junggar, NW China. *Geol. Mag.* **2021**, *158*, 1811–1829. [CrossRef]
24. Chen, J.F.; Han, B.F.; Ji, J.Q.; Zhang, L.; Xu, Z.; He, G.Q.; Wang, T. Zircon U–Pb ages and tectonic implications of Paleozoic plutons in northern West Junggar, North Xinjiang, China. *Lithos* **2010**, *115*, 137–152. [CrossRef]
25. Gao, R.; Xiao, L.; Pirajno, F.; Wang, G.C.; He, X.X.; Yang, G.; Yan, S.W. Carboniferous–Permian extensive magmatism in the West Junggar, Xinjiang, northwestern China: Its geochemistry, geochronology, and petrogenesis. *Lithos* **2014**, *204*, 125–143. [CrossRef]
26. Han, B.F.; Ji, J.Q.; Song, B.; Chen, L.H.; Zhang, L. Late Paleozoic vertical growth of continental crust around the Junggar Basin, Xinjiang, China (part I): Timing of postcollisional plutonism. *Acta Petrol. Sin.* **2006**, *22*, 1077–1086. (In Chinese with English abstract)
27. Hong, Y.B.H.; Liu, B.; Han, B.F.; Ma, J.X.; Chen, J.D. Late Carboniferous closure of the Junggar-Balkhash Ocean: Insights from the early Permian post-accretionary magmatism in the Barleik Mountains of West Junggar, NW China. *Lithos* **2022**, *432–433*, 106900. [CrossRef]
28. Geng, H.; Sun, M.; Yuan, C.; Xiao, W.; Xian, W.; Zhao, G.; Zhang, L.; Wong, K.; Wu, F. Geochemical, Sr–Nd and zircon U–Pb–Hf isotopic studies of Late Carboniferous magmatism in the West Junggar, Xinjiang: Implications for ridge subduction? *Chem. Geol.* **2009**, *266*, 364–389. [CrossRef]
29. Tang, G.J.; Wang, Q.; Wyman, D.A.; Li, Z.X.; Zhao, Z.H.; Jia, X.H.; Jiang, Z.Q. Ridge subduction and crustal growth in the Central Asian Orogenic Belt: Evidence from Late Carboniferous adakites and high-Mg diorites in the western Junggar region, northern Xinjiang (west China). *Chem. Geol.* **2010**, *277*, 281–300. [CrossRef]
30. Zhang, J.E.; Chen, Y.C.; Xiao, W.J.; Wakabayashi, J.; Windley, B.F.; Yin, J. Sub-parallel ridge-trench interaction and an alternative model for the Silurian-Devonian archipelago in Western Junggar and North-Central Tianshan in NW China. *Earth-Sci. Rev.* **2021**, *217*, 103648 [CrossRef]

31. XBGMR (Xinjiang Bureau of Geology and Mineral Resources, Xinjiang, China). *Regional Geological Map of Tuoli Area (L-44-XXIV)*; Map Scale 1:200 000; XBGMR: Urumqi, China, 1975. (In Chinese)
32. Wang, J.R.; Jia, Z.L.; Li, T.D.; Ma, J.L.; Zhao, L.; He, Y.B.; Zhang, W.; Liu, K.X. Discovery of Early Devonian adakite in West Junggar, Xinjiang: Implications for geotectonics and Cu mineralization. *Acta Petrol. Sin.* **2013**, *29*, 840–852. (In Chinese with English abstract)
33. Han, B.F.; He, G.Q.; Wang, S.G. Postcollisional mantle-derived magmatism, underplating and implications for basement of the Junggar Basin. *Sci. China Ser. D Earth Sci.* **1999**, *42*, 113–119. [CrossRef]
34. Zheng, B.; Han, B.F.; Wang, Z.Z.; Liu, B.; Feng, L.X. An example of Phanerozoic continental crustal growth: The West Junggar Orogenic Belt, Northwest China. *Lithos* **2020**, *376*, 105745. [CrossRef]
35. Zhao, L.; He, G.Q. Geochronology and geochemistry of the Cambrian (~518 Ma) Chagantaolegai ophiolite in northern West Junggar (NW China): Constraints on spatiotemporal characteristics of the Chingiz–Tarbagatai megazone. *Int. Geol. Rev.* **2014**, *56*, 1181–1196. [CrossRef]
36. Yang, Y.Q.; Zhao, L.; Zheng, R.G.; Xu, Q.Q. Evolution of the early Paleozoic Hongguleleng–Balkybey Ocean: Evidence from the Hebukesaier ophiolitic mélange in the northern West Junggar, NW China. *Lithos* **2019**, *324–325*, 519–536. [CrossRef]
37. Xu, Z.; Han, B.F.; Ren, R.; Zhou, Y.Z.; Su, L. Palaeozoic multiphase magmatism at Barleik Mountain, southern West Junggar, Northwest China: Implications for tectonic evolution of the West Junggar. *Int. Geol. Rev.* **2013**, *55*, 633–656. [CrossRef]
38. Yang, Y.Q.; Zhao, L.; Zhang, J.; Niu, M.L. Middle Paleozoic archipelago amalgamation and tectonic transform in the northern West Junggar, NW China: Constraints from magmatism and deformation. *Gondwana Res.* **2021**, *98*, 147–165. [CrossRef]
39. Yin, J.; Chen, W.; Yuan, C.; Yu, S.; Xiao, W.; Long, X.; Li, J.; Sun, J. Petrogenesis of Early Carboniferous adakitic dikes, Sawur region, northern West Junggar, NW China: Implications for geodynamic evolution. *Gondwana Res.* **2015**, *27*, 1630–1645. [CrossRef]
40. Xu, Y.; Han, B.F.; Liao, W.; Li, A. The Serpukhovian–Bashkirian Amalgamation of Laurussia and the Siberian Continent and Implications for Assembly of Pangea. *Tectonics* **2022**, *41*, e2022TC007218. [CrossRef]
41. Chen, J.F.; Han, B.F.; Zhang, L.; Xu, Z.; Liu, J.L.; Qu, W.J.; Li, C.; Yang, J.H.; Yang, Y.H. Middle Paleozoic initial amalgamation and crustal growth in the West Junggar (NW China): Constraints from geochronology, geochemistry and Sr–Nd–Hf–Os isotopes of calc-alkaline and alkaline intrusions in the Xiemisitai-Saier Mountains. *J. Asian Earth Sci.* **2015**, *113*, 90–109. [CrossRef]
42. Shen, P.; Shen, Y.C.; Li, X.H.; Pan, H.D.; Zhu, H.P.; Meng, L.; Dai, H.W. Northwestern Junggar Basin, Xiemisitai Mountains, China: A geochemical and geochronological approach. *Lithos* **2012**, *140–141*, 103–118. [CrossRef]
43. Yin, J.Y.; Chen, W.; Xiao, W.J.; Yuan, C.; Windley, B.F.; Yu, S.; Cai, K.D. Late Silurian–early Devonian adakitic granodiorite, A-type and I-type granites in NW Junggar, NW China: Partial melting of mafic lower crust and implications for slab roll-back. *Gondwana Res.* **2017**, *43*, 55–73. [CrossRef]
44. Gu, P.Y.; Li, Y.J.; Zhang, B.; Tong, L.L.; Wang, J.N. LA-ICP-MS zircon U-Pb dating of gabbro in the Darbut ophiolite, western Junggar, China. *Acta Petrol. Silica* **2009**, *25*, 1364–1372. (In Chinese with English abstract)
45. Chen, S.; Guo, Z.J. Time constraints, tectonic setting of Dalabute ophiolitic complex and its significance for Late Paleozoic tectonic evolution in West Junggar. *Acta Petrol. Sin.* **2010**, *26*, 2336–2344. (In Chinese with English abstract)
46. Chen, S.; Pe-Piper, G.; Piper, D.J.W.; Guo, Z. Ophiolitic mélanges in crustal-scale fault zones: Implications for the Late Palaeozoic tectonic evolution in West Junggar, China: Fault-zone ophiolite, West Junggar. *Tectonics* **2014**, *33*, 2419–2443. [CrossRef]
47. Zhao, L.; He, G.Q. LA-ICP-MS zircon age of dacite in Tailegula Formation and their constraints on emplacement of darbut ophiolite in West Junggar. *Acta Sci. Nat. Univ. Pekin.* **2016**, *52*, 871–880. (In Chinese with English abstract)
48. Bai, J.K.; Chen, J.L.; Tang, Z.; Zhang, Y. Redefinition of the Middle Devonian Kulumudi Formation in the south of Tiechanggou town, West Junggar, Xinjiang and its geological implications. *Northwestern Geol.* **2015**, *48*, 72–80. (In Chinese with English abstract)
49. Bai, J.K.; Chen, J.L.; Yan, Z.; Tang, Z.; Xu, X.Y.; Li, J.L. The timing of opening and closure of the Mayile oceanic basin: Evidence from the angular unconformity between the Middle Devonian and its underlying geological body in the southern West Junggar. *Acta Petrol. Sin.* **2015**, *31*, 133–142. (In Chinese with English abstract)
50. Xiang, K.P.; Li, Y.J.; Li, Z.; Wang, R.; Yang, G.X.; Duan, F.H.; Li, G.Y. LA ICP-MS zircon age and geochemistry of the Aladeyikesai Formation volcanic rocks in the Halaalate Mountain of West Junggar, Xinjiang, and their tectonic significance. *Acta Geol. Sin.* **2015**, *89*, 843–855. (In Chinese with English abstract) [CrossRef]
51. Fan, T.T.; Zhou, X.H.; Liu, Y.Q.; Li, W.; Zheng, Z.Y.; Hu, T.; Liang, H. Crystal Tuff Zircon LA-ICP-MS U-Pb Ages from the Lower Jiangbasitao Formation in the East Daheishan Area Xinjiang and Their Geological Implications. *Acta Sedimentol. Sin.* **2011**, *29*, 312–320. (In Chinese with English abstract)
52. Guo, L.S.; Liu, Y.L.; Wang, Z.H.; Song, D.; Xu, F.J.; Su, L. The zircon U-Pb LA-ICP-MS geochronology of volcanic rocksin Baogutu areas, western Junggar. *Acta Petrol. Sin.* **2010**, *26*, 471–477. (In Chinese with English abstract)
53. Li, Y.J.; Tong, L.L.; Zhang, B.; Liu, J.; Zhang, T.J.; Wang, J.N. On the old and new relationship between Xibeikulasi formation and Baogutu formation of the Carboniferous system, West Junggar. *Xinjiang Geol.* **2010**, *28*, 130–136. (In Chinese with English abstract)
54. Gong, Y.M.; Zong, R.W. Paleozoic Stratigraphic Regionalization and Paleogeographic Evolution in Western Junggar, Northwestern China. *Earth Sci.-J. China Univ. Geosci.* **2015**, *40*, 461–484. (In Chinese with English abstract)
55. An, F.; Wei, S.N.; Fang, Z.K.; Zheng, B.; Deng, Y. Petrology and zircon U-Pb geochronology of tuffaceous conglomerate in the Xibeikulasi Formation, West Junggar, and its implication for sedimentary sequence. *Acta Geol. Sin.* **2019**, *93*, 1954–1967. (In Chinese with English abstract)

56. Dou, Y.W.; Sun, Z.H. The Age of Harjiawu Formation and Karagang Formation in North Xinjiang. *Geol. Rev.* **1985**, *31*, 489–494. (In Chinese with English abstract)
57. Li, J.W.; Liu, B.; Han, B.F.; Hong, Y.B.H. Identification of a Late Carboniferous Angular Unconformity in the Jiayier Mountains of the West Junggar and its Constraints on the Closure Time of the Junggar-Balkhash Ocean. *Geotecton. Metallog.* **2023**, *47*, 284–307. (In Chinese with English abstract) [CrossRef]
58. Wen, Z.G.; Wang, H.Q.; Du, Y.; Liu, S.B.; Wang, J.W. The formation age and tectonic significance of Yemuqi granite in WestJunggar, Xinjiang. *Geol. Bull. China* **2019**, *38*, 1297–1305. (In Chinese with English abstract)
59. Kwon, S.T.; Tilton, G.R.; Coleman, R.G.; Feng, Y. Isotopic studies bearing on the tectonics of the West Junggar Region, Xinjiang, China. *Tectonics* **1989**, *8*, 719–727. [CrossRef]
60. Jian, P.; Liu, D.Y.; Shi, Y.R.; Zhang, F.Q. SHRIMP dating of SSZ ophiolites from Northern Xinjiang Province, China: Implications for generation of oceanic crust in the central Asian orogenic belt. In *Structural and Tectonic Correlation across the Central Asia Orogenic Collage: North-Eastern Segment. Guidebook and Abstract Volume of the Siberian Workshop IGCP-480*; Sklyarov, E.V., Ed.; IEC SB RAS: Irkutsk, Russia, 2005; Volume 246.
61. Yang, G.; Li, Y.; Santosh, M.; Gu, P.; Yang, B.; Zhang, B.; Wang, H.; Zhong, X.; Tong, L. A Neoproterozoic seamount in the Paleoasian Ocean: Evidence from zircon U–Pb geochronology and geochemistry of the Mayile ophiolitic mélange in West Junggar, NW China. *Lithos* **2012**, *140–141*, 53–65. [CrossRef]
62. Liu, Y.S.; Hu, Z.C.; Gao, S.; Günther, D.; Xu, J.; Gao, C.G.; Chen, H.H. In situ analysis of major and trace elements of anhydrous minerals by LA-ICP-MS without applying an internal standard. *Chem. Geol.* **2008**, *257*, 34–43. [CrossRef]
63. Yuan, H.L.; Gao, S.; Dai, M.N.; Zong, C.L.; Günther, D.; Fontaine, G.H.; Liu, X.M.; Diwu, C.R. Simultaneous determinations of U–Pb age, Hf isotopes and trace element compositions of zircon by excimer laser-ablation quadrupole and multiple-collector ICP-MS. *Chem. Geol.* **2008**, *247*, 100–118. [CrossRef]
64. Bao, Z.A.; Chen, K.Y.; Kang, L.; Zong, C.L.; Nie, X.J.; Lv, N.; Liang, P.; Yuan, H.L. Comparison of synthetic zircon, high-temperature and high-pressure sintered zircon and fast hot-pressing sintered zircon for in situ hafnium isotope analysis by LA-MC-ICP-MS. *J. Anal. At. Spectrom.* **2023**, *38*, 2313–2323. [CrossRef]
65. Lan, Z.W.; Huyskens, M.H.; Le Hir, G.; Mitchell, R.N.; Yin, Q.Z.; Zhang, G.Y.; Li, X.H. Massive volcanism may have foreshortened the Marinoan snowball Earth. *Geophys. Res. Lett.* **2022**, *49*, e2021GL097156. [CrossRef]
66. Hoskin, P.W.O. Trace-element composition of hydrothermal zircon and the alteration of Hadean zircon from the Jack Hills, Australia. *Geochim. Cosmochim. Acta* **2005**, *69*, 637–648. [CrossRef]
67. Sun, S.S.; McDonough, W.F. Chemical and isotopic systematics of oceanic basalts: Implications for mantle composition and processes. *Geol. Soc. Lond. Spec. Publ.* **1989**, *42*, 313–345. [CrossRef]
68. Weng, K.; Dong, Y.P.; Xu, X.Y.; Ma, Z.P.; Chen, B. Carboniferous sedimentary provenance and tectonic setting in the Darbut region of Western Junggar (NW China): Evidence from mineralogy, geochemistry and detrital zircon U–Pb dating. *J. Geol. Soc.* **2021**, *178*, jgs2020-132. [CrossRef]
69. Le Bas, M.J.; Le Maitre, R.W.; Streckeisen, A.; Zanettin, B. A Chemical Classification of Volcanic Rocks Based on the Total Alkali-Silica Diagram. *J. Petrol.* **1986**, *27*, 745–750. [CrossRef]
70. Peccerillo, A.; Taylor, S.R. Geochemistry of eocene calc-alkaline volcanic rocks from the Kastamonu area, Northern Turkey. *Contrib. Mineral. Petrol.* **1976**, *58*, 63–81. [CrossRef]
71. Pettijohn, F.J.; Potter, P.E.; Siever, R. *Sand and Sandstone*; Springer: Berlin/Heidelberg, Germany, 1972. [CrossRef]
72. Tao, H.F.; Sun, S.; Wang, Q.C.; Yang, X.F.; Jiang, L. Petrography and geochemistry of lower carboniferous greywacke and mudstones in Northeast Junggar, China: Implications for provenance, source weathering, and tectonic setting. *J. Asian Earth Sci.* **2014**, *87*, 11–25. [CrossRef]
73. Taylor, S.T.; McLennan, S.M. *The Continental Crust: Composition and Evolution*; Blackwell: Oxford, UK, 1985; Volume 312.
74. Wei, R.Z.; Dong, A.G.; Li, S.; Wang, R.J.; Chai, J.Z. Discovery and significance of Early Devonian fossils in the Mayileshan area, western Junggar, Xinjiang, China. *Geol. Bull. China* **2011**, *30*, 101–105. (In Chinese with English abstract)
75. Stolz, A.J.; Jochum, K.P.; Spettel, B.; Hofmann, A.W. Fluid- and melt-related enrichment in the subarc mantle: Evidence from Nb/Ta variations in island-arc basalts. *Geology* **1996**, *24*, 587–590. [CrossRef]
76. Condie, K.C. Geochemistry and Tectonic Setting of Early Proterozoic Supracrustal Rocks in the Southwestern United States. *J. Geol.* **1986**, *94*, 845–864. [CrossRef]
77. Edwards CM, H.; Morris, J.D.; Thirlwall, M.F. Separating mantle from slab signatures in arc lavas using B/Be and radiogenic isotope systematics. *Nature* **1993**, *362*, 530–533. [CrossRef]
78. Bhatia, M.R. Plate Tectonics and Geochemical Composition of Sandstones: A Reply. *J. Geol.* **1985**, *93*, 85–87. [CrossRef]
79. Bhatia, M.R.; Crook, K.A.W. Trace element characteristics of graywackes and tectonic setting discrimination of sedimentary basins. *Contrib. Mineral. Petrol.* **1986**, *92*, 181–193. [CrossRef]
80. Bhatia, M.R. Plate Tectonics and Geochemical Composition of Sandstones. *J. Geol.* **1983**, *91*, 611–627. [CrossRef]
81. Zhao, W.P. Paleozoic stratigraphic division in the south of western Junggar and some problems about Barleik ophiolites. In Proceedings of the International Workshop on the Large Balkash–Western Junggar Copper–Gold Province, Karamay, China, 22–27 August 2011; pp. 32–39. (In Chinese with English abstract)

82. Chen, S.; Guo, Z.J.; Pe-Piper, G.; Zhu, B.B. Late Paleozoic peperites in West Junggar, China, and how they constrain regional tectonic and palaeoenvironmental setting. *Gondwana Res.* **2013**, *23*, 666–681. [CrossRef]
83. Zhang, J.E.; Xiao, W.J.; Luo, J.; Chen, Y.C.; Windley, B.F.; Song, D.F.; Han, C.M.; Safonova, I. Collision of the Tacheng block with the Mayile-Barleik-Tangbale accretionary complex in Western Junggar, NW China: Implication for Early-Middle Paleozoic architecture of the western Altaids. *J. Asian Earth Sci.* **2018**, *159*, 259–278. [CrossRef]

minerals

MDPI

Article

Ore Genesis and the Magmatism of the Yuhaixi Mo(Cu) Deposit in Eastern Tianshan, NW China: Constraints from Geology, Geochemistry, Zircon U-Pb and Molybdenite Re-Os Dating

Di Wang [1,2], Chunji Xue [1,3,*], Yun Zhao [1], Chao Li [2,*], Binbin Xi [4], Yang Yang [4], Qinglei Tian [4], Xunshan Kang [4] and Xing Wu [4]

1 State Key Laboratory of Geological Processes and Mineral Resources, China University of Geosciences, Beijing 100083, China; shadywangdi@gmail.com (D.W.); yun.zhao@cugb.edu.cn (Y.Z.)
2 National Research Center for Geoanalysis, Beijing 100037, China
3 The National 305 Project Office of Xinjiang, Urumqi 830000, China
4 No. 704 Geological Party, Xinjiang Geological Exploration Bureau for Nonferrous Metals, Hami 839005, China
* Correspondence: chunji.xue@cugb.edu.cn (C.X.); re-os@163.com (C.L.)

Abstract: The Yuhaixi Mo(Cu) deposit is a new discovery in the eastern section of the Dananhu-Tousuquan island arc, Eastern Tianshan. However, the genesis of the Yuhaixi Mo(Cu) deposit is still not fully understood. The Yuhaixi intrusion is composed of monzonitic granites, diorites, granites, and gabbro dikes, among which disseminated or veinlet Mo and Cu mineralization is mainly hosted by the monzonitic granites. The LA-ICP-MS zircon U-Pb dating yields emplacement ages of 359.4 ± 1.6 Ma for the monzonitic granite, 298.8 ± 1.8 Ma for the diorite, and 307.0 ± 2.3 Ma for the granite. The Re-Os dating of molybdenite hosted by monzonitic granite yields a well-constrained ^{187}Re-^{187}Os isochron age of 354.1 ± 6.8 Ma (MSWD = 1.7) with a weighted average age of 344.5 ± 3.1 Ma. The Mo mineralization is closely associated with the Yuhaixi monzonitic granite. The Yuhaixi monzonitic granite rocks are characterized by high silica (SiO_2 > 70 wt.%), low MgO (0.23–0.36), Ni, Cr contents, and they are enriched in light rare earth elements (LREEs) and large ion lithophile elements (LILEs: e.g., K, Ba, Pb and Sr), and depleted in heavy rare earth elements (HREEs) and high field-strength elements (HFSEs: e.g., Nb, Ta and Ti). They are weak peraluminous and have high εHf(t) (11.37–17.59) and εNd(t) (1.36–7.75) values, and varied initial $^{87}Sr/^{86}Sr$ (0.7037–0.7128) values. The Yuhaixi post-ore granites exhibit similar geochemical and isotopic signatures to the Yuhaixi monzonitic granite. These characteristics suggest that the Yuhaixi felsic rocks are likely sourced from the partial melting of the juvenile lower crust. The Yuhaixi diorite has low SiO_2, and K_2O contents, relatively high Na_2O, MgO ($Mg^{\#}$ = 45–53) contents, and depletions in HFSE (e.g., Nb, Ta, and Ti). These geochemical features, coupled with isotopic data such as low initial $^{87}Sr/^{86}Sr$ (≤0.7043), high εNd(t) (2.5 to 3.0) and εHf(t) (≥11.6) values, and young Hf model ages, suggest that their parental magmas possibly originated from the partial melting of the depleted lithospheric mantle that was metasomatized by hydrous melts or fluids from the subducting oceanic plate. Integrating our new results with previous works on the Dananhu-Tousuquan island arc belt, we suggest that the Yuhaixi Mo(Cu)deposit is likely sourced from the juvenile lower crust, which was formed in an arc setting, where the bipolar subduction of the North Tianshan oceanic slab forms the Dananhu Tousuquan belt to the north and the Aqishan-Yamansu belt to the south. The eastern section of the Dananhu-Tousuquan island arc is a promising target for late Paleozoic porphyry Mo(Cu) deposits.

Keywords: LA-ICP-MS zircon U-Pb dating; geochemistry; Sr-Nd-Hf isotope; molybdenite Re-Os dating; Yuhaixi Mo deposit; Eastern Tianshan

Citation: Wang, D.; Xue, C.; Zhao, Y.; Li, C.; Xi, B.; Yang, Y.; Tian, Q.; Kang, X.; Wu, X. Ore Genesis and the Magmatism of the Yuhaixi Mo(Cu) Deposit in Eastern Tianshan, NW China: Constraints from Geology, Geochemistry, Zircon U-Pb and Molybdenite Re-Os Dating. *Minerals* **2023**, *13*, 1368. https://doi.org/10.3390/min13111368

Academic Editor: Alain Chauvet

Received: 22 August 2023
Revised: 15 October 2023
Accepted: 21 October 2023
Published: 26 October 2023

1. Introduction

Porphyry ore deposits are the primary source of Cu, Mo, and Au [1–3]. Porphyry-type deposits commonly form in the young continental and oceanic arcs [4], and they are usually associated with magmatic–hydrothermal fluids exsolved from relatively oxidized intermediate-felsic magmas with high oxidation states [1,4]. However, porphyry Mo(Cu) deposits in the margins of North China Craton mainly occurred in a post-collision extension setting and the ore-forming magmas stemmed from an old lower crust source [5–14]. Therefore, the origin and evolution processes of ore-forming magmas and their tectonic setting are significant for understanding the formation of economic porphyry deposits [1,2,4].

Recently, numerous early Paleozoic porphyry Cu-Mo deposits have been discovered in the Eastern Tianshan, southern margin of the Central Asian Orogenic Belt (CAOB). These porphyry-type deposits are closely related to the Paleozoic intermediate to felsic magmatic hydrothermal activity in the Eastern Tianshan [15–17]. The Dananhu-Tousuquan arc belt (DTA) is located in the northern part of the Eastern Tianshan, which hosts Tuwu-Yandong, Fuxing, Linglong, Chihu, Yuhai porphyry deposits. Previous studies are mainly focused on these porphyry deposits in the middle section, which have revealed that most of the porphyry Cu-Mo deposits were likely related to the subduction of the North Tianshan Oceanic plate in the late Paleozoic [4–10]. However, the genesis of the porphyry Mo(Cu) deposits in the eastern section of the DTA has not been fully understood.

The Yuhaixi porphyry Mo(Cu) deposit is a new discovery in the eastern section of the DTA, which provides a good opportunity for investigating the generation mechanisms for the porphyry Mo deposits in this section [10–14]. In this study, zircon LA-ICP-MS U-Pb dating and trace elements, zircon Hf isotope compositions, whole-rock geochemical data, and molybdenite Re-Os isotope dating were carried out to constrain the magma origins and the timing of mineralization in the Yuhaixi Mo(Cu) deposit. This study provides new insights on the evolution of the DTA and its potential for porphyry Mo(Cu) mineralization in Eastern Tianshan.

2. Geological Settings

2.1. Regional Geology

The CAOB, which is placed between the European and Siberian Cratons in the north and the North China and Tarim Cratons in the south, is the largest Phanerozoic accretionary orogenic belt in the world [8]. The west–east trending Eastern Tianshan orogenic belt is situated in the southern margin of the CAOB. The Eastern Tianshan is a Paleozoic island arc system, which is located between the Jungar Basin and the Tarim Craton (Figure 1), characterized by a series of complicated tectonic history, which has witnessed multiple stages of subduction and collisional events [11–17]. It can be divided into three major tectonic zones from the north to the south: the Bogeda-Haerlike belt, the Jueluotage belt, and the Central Tianshan block (Figure 1c). The Eastern Tianshan belt withstands a series of E–W trending faults, including the Kanggur, the Yamansu, the Dacaotan, and the Aqikekuduke faults [12]. The Dananhu-Tousuquan island arc belt mainly consists of Devonian to Carboniferous sedimentary–volcanic rocks. The Devonian Dananhu Formation consists of basaltic to andesitic volcanic rocks, which were overlain by the Carboniferous Gandun Formation sedimentary rocks and Qie'shan Group basaltic–andesitic volcanic/pyroclastic rocks and sedimentary rocks. Most of the Cu-Mo deposits in the Eastern Tianshan porphyry are hosted in the middle section of the Dananhu-Tousuquan island arc belt [15,18–25]. The Kangguer shear zone mainly comprises volcanic–sedimentary rocks of the Carboniferous Gandun and Wutongwozi formations, and volcanic and pyroclastic rocks of the Carboniferous Yamansu Formation [15,16]. Most rocks were metamorphosed and ductile deformed in the Kangguer shear zone [11,15]. The Aqishan-Yamansu belt comprises bimodal volcanic rocks of the Carboniferous Yamansu Formation, as well as clastic rocks, andesitic tuff, and the Permian Kula Formation marine and terrestrial clastic rocks [15,25–29].

2.2. Ore Deposit Geology

The Yuhaixi Mo(Cu) deposit is located in the eastern section of the Dananhu-Tousuquan arc belt, approximately 110 km southeast of Hami City, Xinjiang (Figure 1b). The main lithostratigraphic units in the area include granulite, hornblende schist of the lower Carboniferous

Yanchi Formation, and glutenite of the Tertiary Putaogou Formation (Figure 2A; [5]). The intrusive complex at Yuhaixi is intermediate to felsic in composition (Figure 2B). In the middle and eastern area, various granitoid intrusions occur as stocks and dikes that cover an area of approximately 3 km^2. These intrusions are composed of monzonitic granite, granite, diorite (Figure 2B). The monzonitic granites host most of the Mo and Cu mineralization at Yuhaixi. Postmineralization plutons comprising granite and diorite intruded the mineralized monzonitic granite. The gabbro dikes mark the final stage of magmatic activity, having intruded the monzonitic granite and the Yanchi Formation (Figure 2B). In the study area, the monzonitic granite and granite generally exhibit low or moderate degrees of alteration. The monzonitic granite is primarily composed of plagioclase (~25%), K-feldspar (~35%), quartz (~25%), and biotite (~5%), with accessory apatite and zircon (Figure 4A,D). The diorite is light gray to gray-white in color and has a microgranular or aphanitic groundmass. It mainly consists of plagioclase (~45%), hornblende (~20%), quartz (~15%), and biotite (~15%), with minor accessory minerals (Figure 4B,E). The granite is coarse-grained and shows typical equigranular texture. It consists of plagioclase (~35%), quartz (~30%), and biotite (~5%), with negligible magnetite, apatite, and zircon (Figure 4C,F). The Yuhaixi intrusive complex has been affected by subordinate faults associated with the regional-scale Kanggur fault [27–37]. The early NE-trending fault dips 65° to 85° and is inferred to be the predominant structure in the Yuhaixi area (Figure 2A). Geologic surveying defined that the ore zone consists of 6 orebodies, which is based on the drill hole ZK3201 and ZK3601. Each individual orebody is about 2–10 m thick (Figure 3A,B). The monzonitic granite have a genetic relationship with disseminated molybdenite and chalcopyrite (Figure 4G–I) and associated hydrothermal alterations (Figure 4I–K).

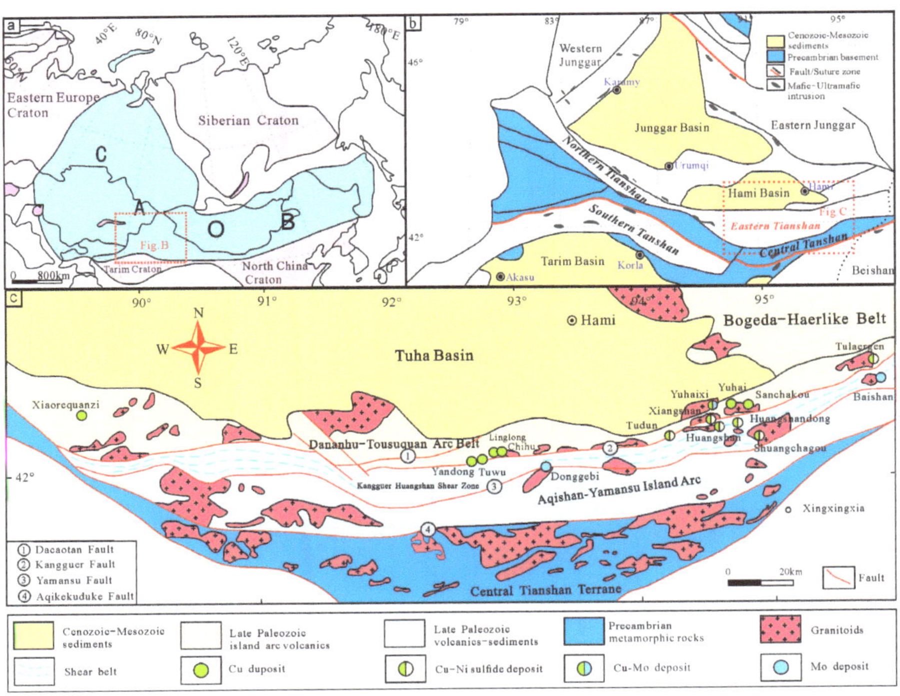

Figure 1. (**a**) Location of Central Asian Orogenic Belt (CAOB) [4]; (**b**) geological map of NW China showing the mains tectonic units; (**c**) geological map of Eastern Tianshan showing major tectonic units, faults and copper deposits (modified from [12]).

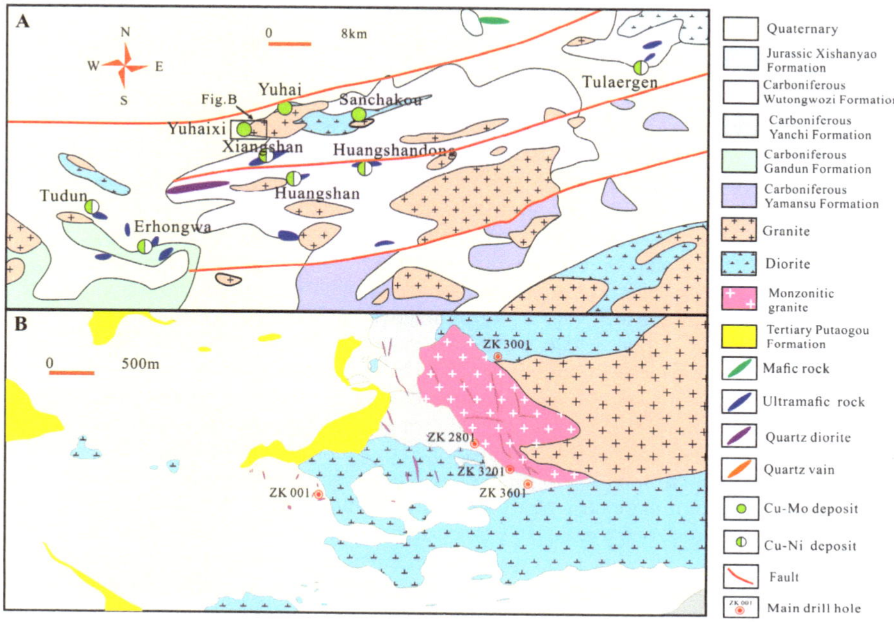

Figure 2. (**A**) Geological map of the Yuhaixi region showing the distribution of Cu and Cu–Ni deposits. (**B**) Simplified geological map of the Yuhaixi Mo(Cu) deposit (modified from [30]).

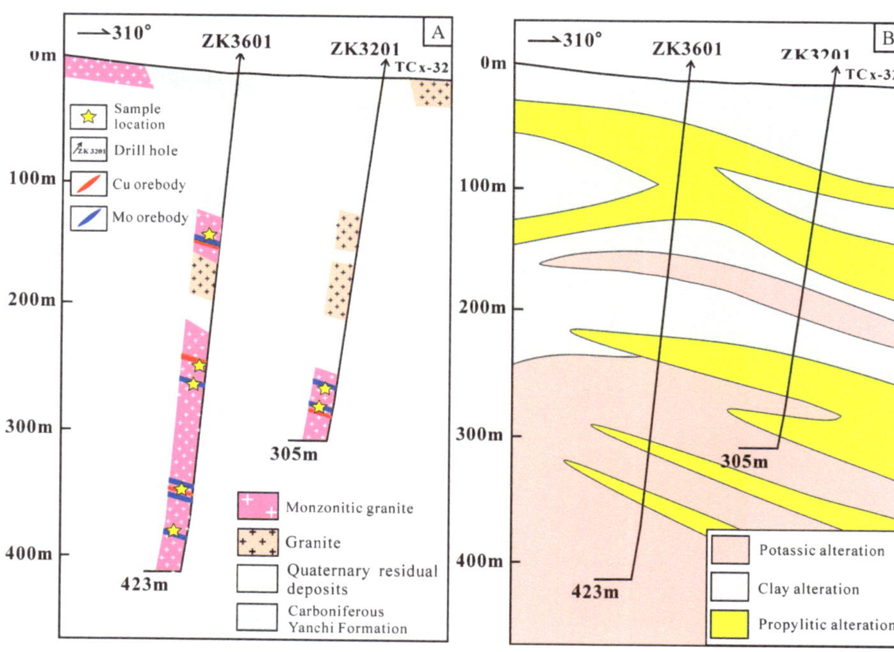

Figure 3. Geologic cross section of the ZK3601 and ZK3201 exploration lines across the Yuhaixi porphyry Mo deposit with lithology and mineralization (**A**) and alteration (**B**).

Figure 4. Field, hand specimen, and microscope photos of magmatic intrusions in the Yuhaixi area. (**A–C**) Hand specimen of the monzonitic granite (**A**), diorite (**B**), granite (**C**). (**D**) Photomicrograph of the monzonitic granite, showing K-felspar, quartz, and muscovite phenocrysts under cross-polarized light. (**E**) Photomicrograph of the diorite, showing plagioclase, quartz, and hornblende phenocrysts under cross-polarized light. (**F**) Photomicrograph of the granite, showing biotite, quartz, and plagioclase phenocrysts under cross-polarized light. (**G**) Photomicrograph of the anhedral chalcopyrite and euhedral molybdenite assemblages under reflected light. Abbreviations: Group; Qtz, quartz; Pl, plagioclase; Bt, biotite; Mus, muscovite; Kfs, K-feldspar; Ccp, chalcopyrite; Mo, molybdenite. (**H**,**I**) The molybdenite in the monzonitic granite hand specimen, with potassic alteration. (**J**) The Carboniferous Yanchi Formation rocks and the monzonitic granite in the exploratory trench. (**K**) Propylitic alteration in the exploratory trench. (**L**) The gabbro dike intruded into early monzonitic granite.

3. Materials and Methods

3.1. Sampling

Three representative intrusive samples (YHX-ZK-1, ZK001-412, 721-6) from the Yuhaixi outcrops (granite) and drill holes (monzonitic granite, diorite) were collected for zircon LA-ICP-MS U-Pb dating, trace element and Hf isotope analyses. Nine samples, including monzonitic granite (Sample No. YHX-ZK-14, YHX-ZK2, YHX-ZK3), diorite (Sample No. ZK001-25, ZK001-350, ZK001-396), and granite (Sample No. 721-1, 721-3, 721-5), were chosen for major and trace element analysis and Sr-Nd isotope analyses. In addition, nine molybdenite samples from Mo orebodies of the Yuhaixi deposit were collected from the disseminated ores for Re-Os isotope analyses.

3.2. Zircon U-Pb Dating and Trace Elements

Three samples, including one monzonitic granite (sample YHX-ZK-1), one diorite (sample ZK001-1), and one granite (721-6), were chosen for the zircon LA-ICP-MS U-Pb dating and trace element analysis.

The zircon grains were separated by conventional density and magnetic techniques and then carefully hand-picked under a binocular microscope. Subsequently, they were mounted on epoxy resin discs and polished to expose the crystal cross-sections at the Langfang Regional Geological Survey. The selection of potential target sites for the U–Pb dating of all the mounted zircons were based on photomicrographs of both transmitted and reflected light, as well as cathodoluminescence (CL) images. Zircon U–Pb dating and trace element analyses were conducted using an Agilent 7500 a inductively coupled plasma mass spectrometer (ICP-MS) simultaneously coupled with a GeoLas 2005 at the Tianjin Institute of Geology and Mineral Resources. The analytical procedures were described in [38]. Laser ablation was operated at a constant energy of 60 mJ, with a repetition rate of 4 Hz and a spot diameter of 32 μm. Zircons 91500 and GJ-1, NIST SRM 610 [30] were used as external standards. Zircon 91500 was analyzed twice for every six analyses in order to calibrate the isotope fractionation. NIST SRM 610 was analyzed once every eight analyses to make the instrumental drift and mass discrimination correction of the trace element analysis. Individual errors in analyses were cited at the 1σ level, and the weighted mean $^{206}Pb/^{238}U$ ages were quoted at the 95% confidence level. The ICPMSDataCal software GeoLas 2005 were used for the adjustment of background and ablation signals, time drift correction [38]. Concordia diagrams and weighted mean calculations were determined using Isoplot 3.0 [39]. Zircon Ce anomalies were calculated using the method of the lattice strain model [17].

3.3. Zircon Hf Isotopes

In situ Hf isotope analyses were conducted using a Neptune MC-ICP-MS and New Wave UP 213 ultraviolet LA-MC-ICP-MS at the National Research Center for Geoanalysis, Beijing, China. The analysis was undertaken on the adjacent spots used for the LA–ICP–MS zircon U–Pb dating to match the Hf isotope data with the U–Pb ages. The ablated samples wrapped in helium were transported from the laser ablation chamber to the ICP-MS torch via a mixing chamber of Argon. Based on the zircon size, the stationary beam spot size was set to either 40 or 50 μm. During testing, GJ-1 international standard [30] zircon samples were used as a reference for the instrumental mass bias correction. The weighted average of the $^{176}Hf/^{177}Hf$ of the GJ-1 zircon samples was 0.281017 ± 0.000007 (2 SD, n = 10), which is consistent with the values reported by [18]. More detailed operating conditions for the MC-ICP-MS instrument, the laser ablation system, and the analytical method are given in [39,40].

3.4. Whole-Rock Major and Trace Elements

Whole-rock major and trace elements analyses were performed at the National Research Center for Geoanalysis, Beijing, China. The samples were powdered to approximately 200 mesh before testing. The major elements were determined using a Philips PW 2404 X-ray fluorescence (XRF) spectrometer with a rhodium X-ray source. The trace elements and the rare earth elements were determined using an Element-I plasma mass spectrometer (Finnigan-MAT Ltd., Bremen, German); two national geological standard reference samples, GSR-3 and GSR-15, were used for the analytical quality control purpose. The analytical precision for the major elements is better than 1% and for the trace elements is better than 5%, and the analytical procedures were described by [19].

3.5. Whole-Rock Sr-Nd Isotopes

Sr–Nd isotopic analyses were carried out using a Neptune multi-collector ICP-MS at the National Research Center for Geoanalysis, Beijing, China, using analytical procedures described by [40]. The Separation of the Sr and REE were performed using cation columns, and the Nd fractions were further separated using HDEHP-coated Kef columns. The measured $^{87}Sr/^{86}Sr$ and $^{143}Nd/^{144}Nd$ ratios were normalized to $^{86}Sr/^{88}Sr$ = 0.1193 and $^{146}Nd/^{144}Nd$ = 0.7217, respectively. The reported $^{87}Sr/^{86}Sr$ and $^{143}Nd/^{144}Nd$ ratios were adjusted to the NBS SRM 987 standard $^{87}Sr/^{86}Sr$ = 0.71027 and the Shin Etsu JNdi-1 standard $^{143}Nd/^{144}Nd$ = 0.512116 [11], respectively.

3.6. Re-Os Isotopic Analyses

Nine molybdenite samples were collected from the Yuhaixi porphyry Mo(Cu) deposits for Re-Os isotope analyses. Among them, five molybdenite samples came from drill holes ZK3601. Four molybdenite samples came from drill hole ZK3201. The photographs and photomicrographs of the representative samples are shown in Figure 4. Molybdenite occurs as disseminations in monzonitic granite or quartz veinlets and molybdenite is cogenetic with chalcopyrite (Figure 4H). The molybdenite was magnetically separated and then handpicked under a binocular microscope at the Langfang Regional Geological Survey. Fresh, nonoxidized molybdenite powders (<0.1 mm in size and purity >99%) were used for Re-Os isotope analyses. ^{187}Re and ^{187}Os concentrations of molybdenite were measured using a TJA PQ ExCell ICP–MS at the Re-Os Laboratory of National Research Center of Geoanalysis, Chinese Academy of Geological Sciences, Beijing. The chemical separation of the Re and Os and the analytical procedure were in accordance with [18]. The weighed molybdenite samples were loaded into a Carius tube through a thin-neck long funnel. The mixed ^{190}Os and ^{185}Re spike solutions and 2 mL HCl and 4 mL HNO_4 were loaded, while the bottom part of the tube was frozen at −50 °C to −80 °C in an ethanol–liquid nitrogen slush, and the top was sealed using an oxygen–propane torch. The tube was then placed in a stainless-steel jacket and heated for 24 h at 230 °C. After 24 h and upon cooling, the sample-bearing tubes were opened to transfer the supernatants out and the Os was separated using the method of direct distillation from the Carius tube for 50 min and trapped in 2 mL of water that was used for the ICP–MS (X-Series) determination of the Os isotope ratio. After that, the residual Re-bearing solution was saved in a 150 mL Teflon beaker for Re separation. The residual Re-bearing solution was heated to near-dryness. The acetone phase was transferred to a 150 mL Teflon beaker that contained 1 mL of water. Finally, the solution was picked up in 2% HNO_3, which was used for the ICP–MS(X-Series) determination of the Re isotope ratio.

The average blanks for the method were ca. 3 pg Re and ca. 0.5 pg Os. The working conditions of the instrument were controlled by the reference material JDC, which produced a measured value of 139.8 ± 2.0 Ma, which is consistent with the recommended value of 139.6 ± 3.8 Ma [41]. The analytical uncertainty in the Re-Os model ages includes 1.02% uncertainty (at 95% confidence level) for the ^{187}Re decay constant. The Re-Os model ages were calculated following the equation: $t = [\ln(1 + {}^{187}Os/{}^{187}Re)]/\lambda$, where λ is the decay constant of ^{187}Re ($\lambda\ {}^{187}Re = 1.666 \times 10^{-11}$ year^{-1}; [41]) and denotes the age. The Re-Os isochron ages were calculated using the least-squares method [42], employing the program ISOPLOT 3.0 [23].

4. Results

4.1. Zircon U-Pb Dating and Zircon Trace Elements

The Yuhaixi monzonitic granite sample (YHX-ZK-1), diorite sample (ZK001-412), and granite sample (721-6) were selected for LA-ICP-MS zircon U-Pb dating, and the analytical data are listed in Table 1. Most of the zircon grains are euhedral–subhedral and show prismatic forms (100–200 µm long) with aspect ratios of 3:1 to 3:2 and characterized by clear oscillatory growth zoning in the CL images (Figure 5). All samples have varying

U(58–453 ppm) and Th (30–417 ppm) contents with high Th/U ratios (>0.3), which is consistent with a magmatic origin [21].

Table 1. Whole-rock geochemical data of the studied intrusive rocks in the Yuhaixi Mo(Cu) deposit (major elements: wt.%; trace elements: ppm).

Rock Type	**Monzonitic Granite**			**Diorite**			**Granite**		
Sample No.	**YHX-ZK-1**	**YHX-ZK-2**	**YHX-ZK-3**	**ZK001-25**	**ZK001-350**	**ZK001-396**	**721-1**	**721-3**	**721-5**
SiO_2	74.2	76.57	74.64	53.55	54.58	54.16	74.16	73.79	73.75
TiO_2	0.17	0.18	0.14	0.95	0.94	0.8	0.2	0.18	0.18
Al_2O_3	13.57	12.42	12.9	17.98	17.33	17.59	13.77	14.15	14.17
TFe_2O_3	1.23	1.31	1	8.69	8.37	7.9	1.49	1.4	1.39
MnO	0.03	0.06	0.03	0.17	0.16	0.14	0.08	0.06	0.06
MgO	0.28	0.36	0.23	4.06	4.84	5.06	0.45	0.39	0.39
CaO	1.23	1.12	1.9	7.97	7.8	7.61	1.65	1.63	1.64
Na_2O	3.33	3.39	2.77	3.86	4.11	4.11	4.46	4.57	4.55
K_2O	4.97	3.75	4.53	1.15	0.9	1.14	2.65	2.7	2.81
P_2O_5	0.04	0.05	0.04	0.24	0.16	0.25	0.07	0.07	0.07
LOI	0.63	0.73	1.82	1.12	0.8	1.02	0.5	0.47	0.5
TOTAL	99.68	99.94	99.98	99.74	99.99	99.77	99.48	99.41	99.52
$Na_2O + K_2O$	8.3	7.14	7.29	5.01	5.01	5.25	7.11	7.28	7.37
$Mg^{\#}$	28.87	32.88	29.08	45.44	50.76	53.31	35	33.18	33.34
A/CNK	1.04	1.05	1.06	0.81	0.79	0.81	1.04	1.05	1.05
A/NK	1.25	1.2	1.29	2.37	2.24	2.2	1.35	1.35	1.27
Li	5.44	6.67	1.31	8.67	10.66	8.71	13.79	16.67	12.95
Be	0.89	2.39	0.94	0.97	1.02	0.96	1.05	1.52	1.04
Sc	2.15	3.53	1.14	19.92	24.87	23.37	1.38	2.6	1.38
V	10.66	12.16	9.27	201.07	241.97	116.9	8.35	8.39	7.22
Cr	19.13	0.78	10.53	22.33	67.75	129.44	5.68	8.87	4.1
Co	1.97	1.01	0.56	22.83	30.73	29.04	1.01	1.03	0.88
Ni	0.11	0.73	0.86	18.48	46.27	81.17	0.26	1.81	1.52
Cu	114.22	37.2	10.2	65.43	132.91	55.89	2.01	1.23	2.88
Zn	15.83	74.89	27.69	103.43	102.49	105.55	68.29	44.17	49.97
Ga	14.88	15.67	11.72	22.98	23.43	24.02	15.15	16.25	15.12
Rb	64.57	75.8	64.65	25.2	17.69	25.25	45.6	62.97	45.83
Sr	188.36	168.2	137.76	833.76	784.8	940.51	379.1	400.17	371.33
Y	7.83	14.27	5.77	28.58	25.65	24.64	19.57	18.59	19.64
Zr	94.78	114.16	89.03	364.46	160.19	23.13	120.58	110.57	114.79
Nb	5.35	12.47	5.34	4.52	5.26	4.15	5.4	6.28	4.93
Cs	0.24	0.4	0.4	0.51	0.5	0.57	0.39	1.39	0.38
Ba	853	592	598	369	314	436	1248	1277	1333
La	27.76	27.03	17.31	18.18	19.09	15.35	20.76	35.15	27.57
Ce	51.7	59	38.94	45.57	47.45	41.31	45.29	68.09	53.25
Pr	5.86	6.19	3.93	6.63	6.78	6.23	4.71	8.11	6.57
Nd	21.54	22.74	14.66	30.25	30.26	28.08	18.11	30.56	25.29
Sm	3.39	4.1	2.56	6.66	6.09	5.78	3.35	5.4	4.51
Eu	1.15	0.86	0.82	1.91	1.57	1.58	1.4	1.54	1.51
Gd	2.94	3.4	2.14	5.46	5.05	4.69	2.99	4.43	3.96
Tb	0.36	0.49	0.27	0.9	0.83	0.79	0.49	0.64	0.6
Dy	1.64	2.58	1.16	5.67	4.84	4.65	2.98	3.42	3.47
Ho	0.3	0.52	0.22	1.2	0.99	0.94	0.67	0.67	0.71
Er	0.81	1.43	0.6	3.3	2.77	2.57	1.97	1.87	1.97
Tm	0.11	0.22	0.08	0.49	0.41	0.37	0.3	0.28	0.29
Yb	0.67	1.45	0.53	3.12	2.71	2.42	1.99	1.73	1.95
Lu	0.11	0.24	0.08	0.48	0.44	0.39	0.33	0.28	0.32
Hf	2.77	3.64	2.5	8.98	4.93	1.27	3.67	3.46	3.54

Table 1. *Cont.*

Rock Type	Monzonitic Granite			Diorite			Granite		
Sample No.	YHX-ZK-1	YHX-ZK-2	YHX-ZK-3	ZK001-25	ZK001-350	ZK001-396	721-1	721-3	721-5
Ta	0.3	1.34	0.29	0.32	0.34	0.25	0.31	0.48	0.3
Tl	0.21	0.24	0.19	0.09	0.06	0.08	0.18	0.2	0.16
Pb	15.31	11.3	11.69	6.2	5.45	8.02	8.42	9.44	8.78
Bi	0.05	0.03	0.03	0.12	0.1	0.26	0.02	0.04	0.02
Th	5.77	8.18	5.8	3.06	3.35	1.18	5.21	9.97	7.37
U	1.13	1.95	0.99	2.19	1.46	0.83	1.19	1.82	1.46
$(La/Yb)_N$	27.9	12.55	21.98	3.92	4.74	4.28	7.02	13.71	9.54
$(Tb/Yb)_N$	2.37	1.5	2.24	1.29	1.36	1.45	1.1	1.63	1.36
$(La/Sm)_N$	5.15	4.15	4.26	1.72	1.97	1.67	3.9	4.09	3.84
Th/Ce	0.11	0.14	0.15	0.07	0.07	0.03	0.12	0.15	0.14
Th/U	5.12	4.2	5.84	1.4	2.29	1.42	4.39	5.48	5.05
Ce/Pb	3.38	5.22	3.33	7.35	8.7	5.15	5.38	7.22	6.06
ΣREE	118.34	130.24	83.29	129.82	129.29	115.14	105.33	162.16	131.97
Eu	1.09	0.68	1.05	0.94	0.84	0.9	1.33	0.94	1.07
Sr/Y	24.05	11.78	23.88	29.17	30.59	38.17	19.38	21.53	18.91

Note: $Mg^{\#} = 100 \times (MgO/40.3044)/(MgO/40.3044 + 0.8998 \times Fe_2O_3{}^T/71.8440)$.

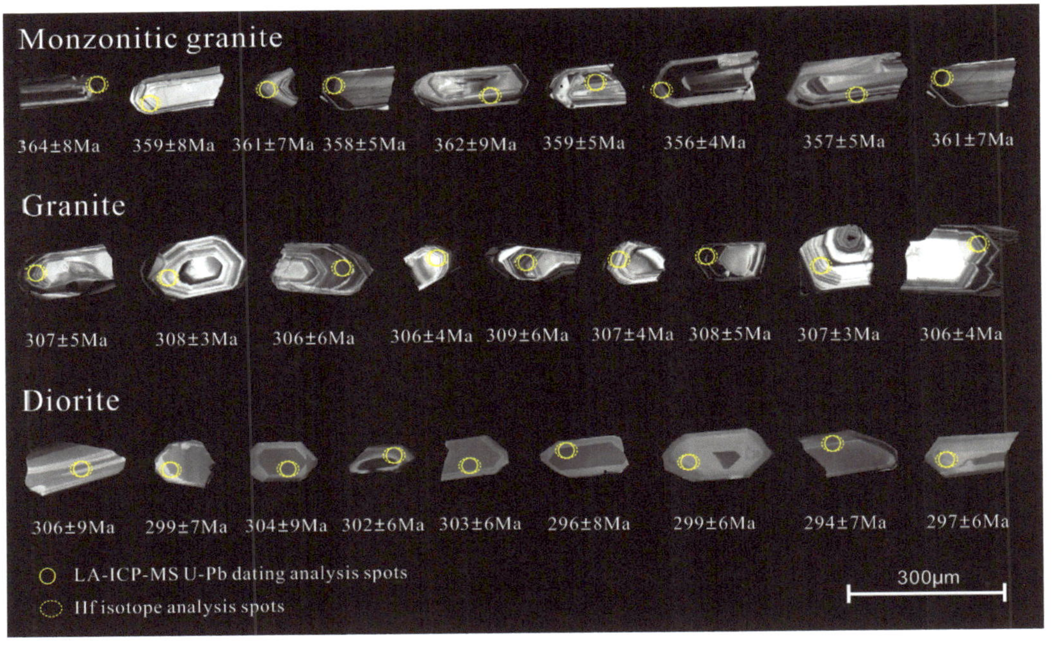

Figure 5. Cathodoluminescence images of representative zircon grains of the Yuhaixi showing the inner structures and analyzed spots.

Except for two discordant spot (02, 05), the remaining 17 analyses from the monzonitic granite sample (YHX-ZK-1) yielded concordant $^{206}Pb/^{238}U$ ages varying from 352 to 362 Ma (Figure 6), with a weighted mean age of 359.4 ± 1.6 Ma (MSWD = 0.89), which is interpreted as the best estimate of emplacement age for the Yuhaixi monzonitic granite.

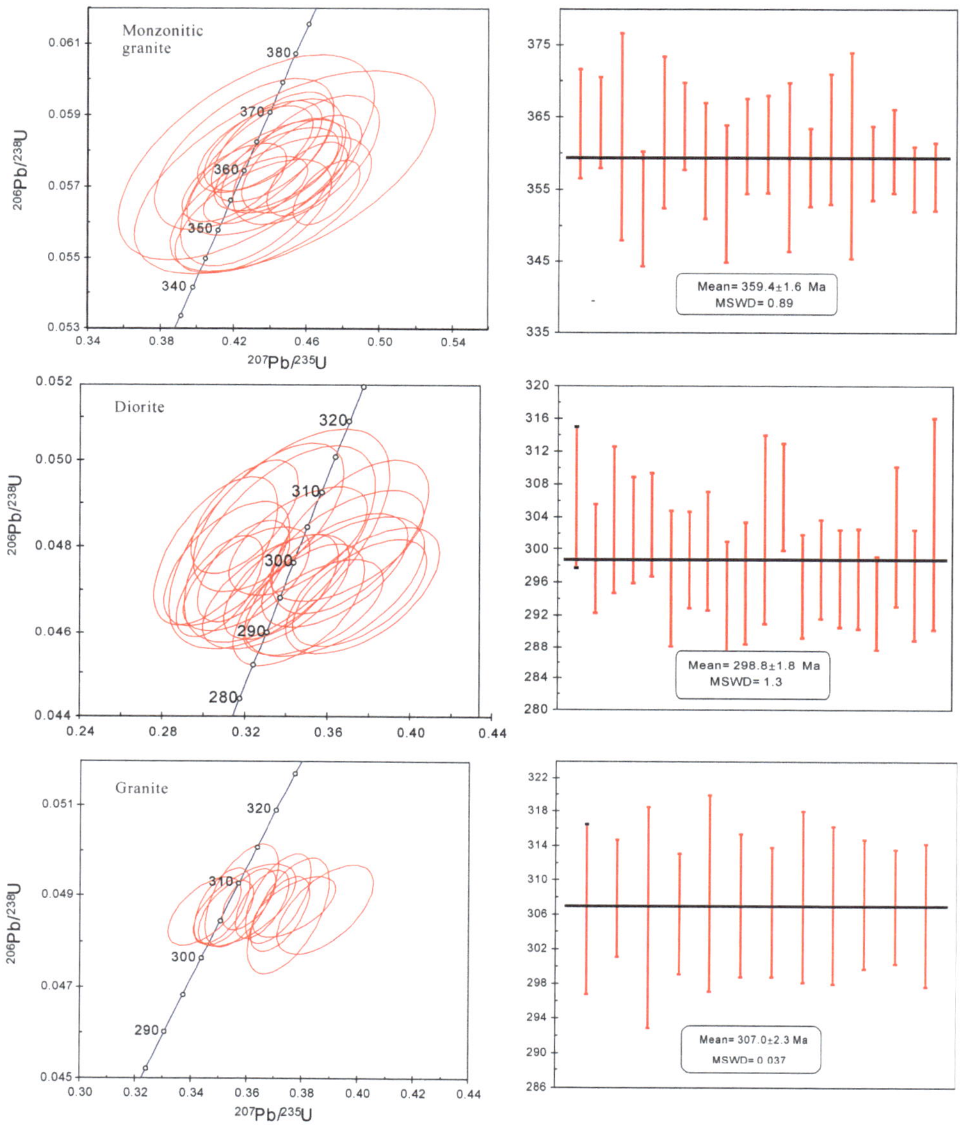

Figure 6. U-Pb concordia diagrams and weighted average ages of the zircons from monzonitic granite, diorite, and granite in the Yuhaixi porphyry Mo(Cu) deposit.

Except for three discordant spots (07, 09, 11), the remaining 12 analytical spots from the granite sample (721-6) had $^{206}Pb/^{238}U$ ages ranging from 306 to 309 Ma (Figure 6), with a weighted mean age of 307.0 ± 2.3 Ma (MSWD = 0.037).

Except for three discordant spots (03, 07 and 16), the remaining 19 analytical spots from the diorite sample (ZK001-412) had $^{206}Pb/^{238}U$ ages ranging from 296 to 306 Ma. (Figure 6), with a weighted mean age of 298.8 ± 1.8 Ma (MSWD = 1.3) (Table 2).

Table 2. LA–ICP–MS zircon U–Pb data for the studied intrusive rocks in the Yuhaixi Mo(Cu) deposit.

Spot	Element ($\times 10^{-6}$)			Isotope Ration						Apparent Age (Ma)					
	Pb	Th	U	$^{207}Pb/^{206}Pb$	1σ	$^{207}Pb/^{235}U$	1σ	$^{206}Pb/^{238}U$	1σ	$^{207}Pb/^{206}Pb$	1σ	$^{207}Pb/^{235}U$	1σ	$^{206}Pb/^{238}U$	1σ
							Monzonitic granite								
YHX-1-1	97	849	1156	0.0547	0.0034	0.4414	0.0273	0.0581	0.0012	398	173	371	19	364	8
YHX-1-2	23	174	302	0.0539	0.0033	0.4362	0.0264	0.0581	0.001	369	134	368	19	364	6
YHX-1-3	26	108	348	0.0525	0.0071	0.4269	0.0573	0.0578	0.0023	309	281	361	41	362	14
YHX-1-4	20	137	265	0.0543	0.0043	0.4309	0.0348	0.0562	0.0013	383	180	364	25	352	8
YHX-1-5	14	120	173	0.0558	0.006	0.4258	0.0401	0.0579	0.0017	456	275	360	29	363	11
YHX-1-6	74	835	840	0.0532	0.0035	0.4319	0.0259	0.058	0.001	345	145	365	18	364	6
YHX-1-7	7	41	89	0.0568	0.006	0.4393	0.0366	0.0573	0.0013	487	236	370	26	359	8
YHX-1-8	8	46	102	0.0548	0.0056	0.4404	0.0449	0.0565	0.0016	406	227	371	32	354	10
YHX-1-9	13	69	182	0.0561	0.0038	0.4465	0.027	0.0576	0.0011	457	155	375	19	361	7
YHX-1-10	90	535	1167	0.0559	0.0033	0.4573	0.0267	0.0576	0.0011	450	127	382	19	361	7
YHX-1-11	20	144	245	0.0546	0.0059	0.436	0.0369	0.0571	0.0019	398	210	367	26	358	12
YHX-1-12	16	95	218	0.0531	0.0032	0.4203	0.0234	0.0571	0.0009	345	144	356	17	358	5
YHX-1-13	18	106	232	0.059	0.005	0.4598	0.0328	0.0578	0.0015	565	183	384	23	362	9
YHX-1-14	12	68	157	0.0571	0.0092	0.448	0.0673	0.0574	0.0023	494	356	376	47	360	14
YHX-1-15	104	1050	1184	0.0577	0.0022	0.4602	0.0181	0.0572	0.0008	517	83	384	13	359	5
YHX-1-16	14	92	182	0.0564	0.0038	0.4436	0.0279	0.0575	0.001	478	146	373	20	360	6
YHX-1-17	235	1618	3001	0.0594	0.002	0.471	0.0151	0.0568	0.0007	583	74	392	10	356	4
YHX-1-18	25	241	307	0.0543	0.003	0.4322	0.0234	0.0569	0.0008	389	122	365	17	357	5
								Diorite							
zk001-1	4	30	60	0.0498	0.0057	0.3316	0.0297	0.0487	0.0014	187	244	291	23	306	9
zk001-2	5	40	85	0.0559	0.0052	0.3636	0.0288	0.0475	0.0011	450	207	315	21	299	7
zk001-3	5	47	72	0.0584	0.0059	0.3712	0.0311	0.0482	0.0015	543	222	321	23	304	9
zk001-4	7	72	108	0.055	0.0048	0.3695	0.0304	0.048	0.0011	413	196	319	23	302	6
zk001-5	8	79	128	0.052	0.0046	0.3458	0.0274	0.0481	0.001	283	199	302	21	303	6
zk001-6	6	69	93	0.0604	0.0077	0.3636	0.0356	0.0471	0.0013	617	275	315	26	296	8
zk001-7	9	85	133	0.0502	0.0037	0.3193	0.0181	0.0474	0.001	206	168	281	14	299	6
zk001-8	8	72	128	0.0499	0.0053	0.3111	0.0249	0.0476	0.0012	191	230	275	19	300	7
zk001-9	5	49	87	0.0568	0.0055	0.3463	0.026	0.0466	0.0012	483	213	302	20	294	7
zk001-10	4	28	58	0.0597	0.0061	0.3772	0.0304	0.047	0.0012	591	222	325	22	296	7
zk001-11	5	47	69	0.0506	0.0055	0.3367	0.0377	0.048	0.0019	233	224	295	29	303	11
zk001-12	10	91	149	0.0481	0.004	0.3198	0.0212	0.0487	0.0011	106	181	282	16	306	7
zk001-13	8	85	118	0.0581	0.0045	0.3758	0.0265	0.0469	0.001	600	170	324	20	296	6

Table 2. *Cont.*

Spot	Element ($\times 10^{-6}$)			Isotope Ration						Apparent Age (Ma)					
	Pb	Th	U	$^{207}Pb/^{206}Pb$	1σ	$^{207}Pb/^{235}U$	1σ	$^{206}Pb/^{238}U$	1σ	$^{207}Pb/^{206}Pb$	1σ	$^{207}Pb/^{235}U$	1σ	$^{206}Pb/^{238}U$	1σ
zk001-14	9	86	127	0.0489	0.0045	0.3209	0.025	0.0473	0.001	143	200	283	19	298	6
zk001-15	6	57	84	0.0485	0.0038	0.3079	0.017	0.0471	0.001	124	174	273	13	297	6
zk001-16	8	90	117	0.0505	0.0037	0.3377	0.0227	0.0471	0.001	217	177	295	17	296	6
zk001-17	6	46	101	0.0554	0.0049	0.3534	0.0275	0.0466	0.0009	428	198	307	21	293	6
zk001-18	11	102	153	0.0461	0.004	0.3058	0.0245	0.0479	0.0014	400	-202	271	19	302	9
zk001-19	5	41	83	0.053	0.0054	0.3352	0.029	0.0469	0.0011	328	230	294	22	296	7
zk001-20	4	36	67	0.0533	0.0082	0.3365	0.0437	0.0482	0.0021	343	315	295	33	303	13
Granite															
721-6-01	6	106	87	0.0584	0.0021	0.3884	0.0141	0.0487	0.0008	546	84	333	10	307	5
721-6-02	26	475	350	0.0534	0.001	0.3608	0.0079	0.0489	0.0005	346	44	313	6	308	3
721-6-03	30	369	453	0.0559	0.0014	0.3752	0.012	0.0486	0.001	456	56	323	9	306	6
721-6-05	20	177	327	0.0525	0.001	0.3526	0.008	0.0486	0.0006	306	46	307	6	306	4
721-6-07	16	131	258	0.0529	0.0012	0.3582	0.0108	0.049	0.0009	324	52	311	8	309	6
721-6-08	16	248	216	0.0522	0.0011	0.3511	0.0082	0.0488	0.0007	295	50	306	6	307	4
721-6-11	17	198	269	0.0513	0.0015	0.3449	0.0108	0.0487	0.0006	254	67	301	8	306	4
721-6-13	12	178	170	0.0553	0.0014	0.3714	0.0096	0.0489	0.0008	433	57	321	7	308	5
721-6-16	5	63	62	0.0536	0.0023	0.3594	0.0151	0.0488	0.0007	367	96	312	11	307	5
721-6-21	21	204	338	0.0542	0.001	0.3656	0.0078	0.0488	0.0006	376	38	316	6	307	4
721-6-22	12	193	162	0.0559	0.0017	0.3772	0.0122	0.0488	0.0005	450	64	325	9	307	3
721-6-23	33	417	478	0.0553	0.0012	0.3721	0.0087	0.0486	0.0007	433	51	321	6	306	4

The calculated results of zircon Ce^{4+}/Ce^{3+} and Ti-in-zircon thermometer have been used to estimate the magma temperatures and oxidation states [17], with the detailed calculation procedures being presented by [17]. In this study, the zircon Ce^{4+}/Ce^{3+} values of the monzonitic granite, diorite, and granite were calculated to be ~6–248 (avg. 60), 11–182 (avg. 54), and 6–241 (avg. 85), respectively (Table 3). Ti-in-zircon temperatures were calculated to be 590–954 °C (avg. 744 °C) for the monzonitic granite, 698–775 °C (avg. 728 °C), for the diorite and 653–917 °C (avg. 712 °C) for the granite (Table 3). The zircon REE patterns are commonly featured by HREE enrichments with positive Ce anomalies and negative Eu anomalies (Figure 7).

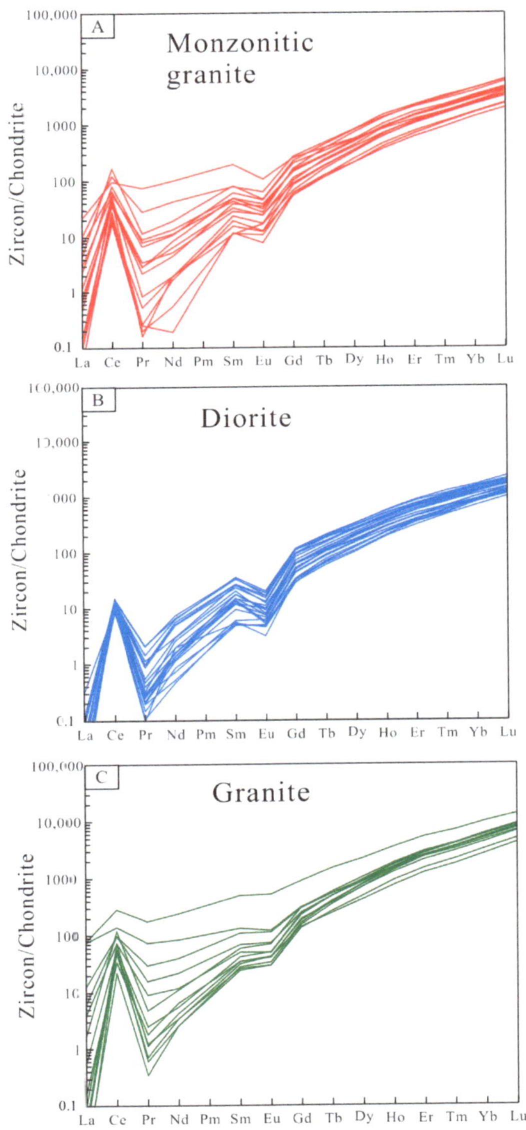

Figure 7. Chondrite-normalized REE patterns for zircon grains from the Yuhai intrusions. Chondrite values are from [20].

Table 3. Trace element abundance (in ppm), Eu anomalies, and Ce^{4+}/Ce^{3+} in zircon and Ti-in-zircon temperature.

Analysis	Ti	Y	La	Ce	Pr	Nd	Sm	Eu	Gd	Tb	Dy	Ho	Er	Tm	Yb	Lu	Eu/Eu*	T
YHX-ZK-1	38	1860	9.55	101	11	63.5	34.9	7.57	63	15.5	154	56.7	257	55	525	114	0.52	732
YHX-ZK-2	8.1	3041	3.15	99.1	3.4	25.5	15.3	4.62	70	22.1	261	108	466	105	918	200	0.56	745
YHX-ZK-3	5.2	2093	0.04	32.2	0.3	2.73	7.59	2.57	39	12.6	159	65.3	298	64.9	606	130	0.45	747
YHX-ZK-4	5.9	1501	0.2	33.4	0.4	3.21	5.53	1.82	23	8.64	112	49.2	245	54.7	521	125	0.41	748
YHX-ZK-5	27	3836	12.5	137	15	88.1	57.1	11.1	118	33.3	356	138	601	128	1137	252	0.49	677
YHX-ZK-6	6.1	1496	0.35	32.9	0.4	3.96	6.36	1.77	29	10.4	128	50.8	236	52.5	526	108	0.47	715
YHX-ZK-7	6.2	1879	0.24	28.6	0.4	5.12	9.43	2.32	41	13.2	167	64	299	63.8	597	135	0.35	631
YHX-ZK-8	13	2846	1.38	136	1.4	11.6	15.8	3.42	59	20.4	257	97.9	431	90.9	824	187	0.37	815
YHX-ZK-9	12	2797	0.69	65.9	1	6.91	8.41	2	43	17.2	232	94.2	452	97.8	925	206	0.41	745
YHX-ZK-10	6	1867	0.14	43	0.1	1.15	4.61	1.29	26	11	138	62.7	310	68.3	629	148	0.36	637
YHX-ZK-11	17	2089	2.31	53.1	0.8	6.62	11.9	3.34	53	15.7	175	74.6	350	72.1	694	151	0.70	731
YHX-ZK-12	1.6	1329	0.02	22.7	0	0.98	3.79	0.95	22	8.39	111	44.5	223	50.2	473	103	0.52	884
YHX-ZK-13	22	2186	0.92	46.7	1.1	8.24	9.48	2.81	38	12.9	160	65.9	305	68.2	674	140	0.42	955
YHX-ZK-14	24	1112	0.54	28.1	0.2	2.7	3.87	1.32	22	6.98	98	40.1	178	38.9	357	83	0.48	693
YHX-ZK-15	17	5436	17.6	178	20	124	76.3	16.8	177	49.1	527	185	800	168	1441	298	0.41	692
YHX-ZK-16	3.3	2710	0.51	38	0.7	7.62	11.2	3.02	58	19.3	227	85.2	377	81.3	747	159	0.64	626
zk001-412-1	9.1	956	0.02	9.89	0.1	3.49	5.43	1.25	26	7.73	87	31.7	146	30.8	277	57	0.47	799
zk001-412-2	8.4	515	0.11	8.09	0	1.22	1.07	0.34	7.6	3.4	41	16.8	76.2	18.7	182	38	0.43	741
zk001-412-3	6.2	820	0.01	12.3	0	0.85	2.37	0.57	17	5.93	69	27.9	124	27.9	259	54	0.40	799
zk001-412-4	6.7	907	0.01	11.5	0.1	1.83	4.09	0.6	19	6.5	78	30.1	132	29.2	275	60	0.35	787
zk001-412-5	5.9	809	0.04	12.1	0	0.75	2.68	0.4	13	5.12	66	25.9	123	26.4	256	56	0.40	778
zk001-412-6	8.8	574	0.01	10.1	0	0.71	1.9	0.52	12	4.17	51	18.8	85.9	18.9	177	37	0.30	749
zk001-412-7	9.4	446	0.01	7.37	0	0.53	1.01	0.36	7.9	2.86	34	14.3	66.8	15.7	149	33	0.30	782
zk001-412-8	9.8	658	0.01	7.69	0	1.61	3.52	0.74	16	5.21	58	21.9	96.2	20.4	189	39	0.23	757
zk001-412-9	8.7	1111	0	12.1	0.3	4.6	6.9	1.26	31	9.23	101	36.7	163	34.3	308	63	0.22	745
zk001-412-10	14	1237	0.01	11.2	0.2	3.91	7.2	1.52	31	10.1	112	40.9	182	37.6	336	69	0.37	783
zk001-412-11	8.6	649	0.01	8.97	0	0.98	2.79	0.43	12	4.38	52	21.5	98.7	22.1	216	48	0.42	790
zk001-412-12	11	1265	0	12.5	0.3	4.06	6.62	1.42	31	9.51	110	42	191	41.8	354	77	0.32	794
zk001-412-13	5.8	552	0.01	8.09	0	0.29	1.13	0.24	9.1	3.27	42	17.5	82.8	19.2	189	41	0.28	766
zk001-412-14	6.7	1109	0.02	11.7	0.1	3.15	5.28	1.16	29	8.81	103	37.7	161	36	324	65	0.26	766
7216-1	6.4	1636	0.05	26.9	0.2	3.83	6.68	3.07	36	11.3	134	52.8	256	56.2	555	131	0.55	752
7216-2	9.3	3942	0.51	95	0.6	6.55	11.3	5.1	77	26.8	330	131	610	129	1234	276	0.48	789
7216-3	6.3	3785	1.19	58.2	1.1	7.02	9.66	3.63	58	22.6	299	125	600	130	1255	272	0.46	733
7216-4	5.3	3706	0.16	46.2	0.3	3.23	8.03	3.7	61	22.6	297	121	577	127	1200	265	0.45	739
7216-5	5.6	3438	0	43.7	0.1	1.51	5.39	2.45	42	18	253	111	558	127	1250	279	0.44	666
7216-6	2.4	3024	0	17.6	0	1.5	4.7	2.16	39	16.2	224	97.3	495	112	1110	252	0.45	776
7216-7	8.2	2765	0.02	57	0.1	2.48	6.69	2.97	49	17.3	223	90.1	431	94.1	890	208	0.50	771
7216-8	9.4	3352	1.86	77.4	1.9	13	13.1	5.35	64	22.6	273	109	517	113	1072	241	0.45	747
7216-9	11	3309	3.84	77.9	3.6	23.4	20.8	8.18	76	25	285	109	489	103	973	215	0.56	772
7216-10	48	7715	23.6	227	21	141	95.1	37.6	228	69.6	704	248	1096	220	2021	419	0.63	695
7216-11	6.1	1957	0.02	42	0.2	2.01	5.07	2.16	35	12.2	155	64.5	308	68.6	672	156	0.57	742
7216-12	11	3451	22.8	111	8.8	49.4	25.3	8.75	79	25.3	291	111	506	110	1044	235	0.49	712
7216-13	7.9	1906	0.08	37.7	0.2	2.07	5.13	2.62	33	11.5	148	62.1	301	69.1	693	158	0.51	730
7216-14	3.4	720	0	11.6	0.1	1.63	2.86	1.43	14	4.39	55	22.6	113	27	300	77	0.46	753
7216-15	5.8	1816	0.02	34.9	0.1	1.93	4.62	2.52	32	11.4	143	59.5	281	64	620	144	0.48	731

[1] $(Eu/Eu^*) = Eu/(Sm \times Gd)^{1/2}$ [18]. [2] Ti-in-zircon temperatures are calculated using the equation proposed by [22]: $\log(\text{ppm Ti-in-zircon}) = (5.711 \pm 0.072) - (4800 \pm 86)/T(K) - \log\alpha SiO_2 + \log\alpha TiO_2$, where $\alpha SiO_2 = 1$, $\alpha TiO_2 = 0.6$ are used in the calculation.

4.2. Whole-Rock Major and Trace Elements

The representative whole-rock geochemical data for the Yuhaixi intrusive samples are presented in Table 1. The Yuhaixi monzonitic granite, granite and diorite samples plot inside the granite, granite and monzodiorite fields, respectively on the $Na_2O + K_2O$ vs. SiO_2 diagram (Figure 8a; [37]). The Yuhaixi monzonitic granites are characterized by high SiO_2 (74.20~76.57 wt.%) and total alkali contents ($K_2O + Na_2O$ = 7.14–8.30 wt.%), but low MgO (0.23~0.36 wt.%), TiO_2 (0.14–0.17 wt.%), and P_2O_5 (0.04–0.05 wt.%), with low $Mg^{\#}$ = 28–32 [100 × molecular $Mg^{2+}/(Mg^{2+} + Fe^{2+})$], which belong to the high-K calc-alkaline series (Figure 8b). The Yuhaixi post-ore granite exhibited similar major element compositions, but they show relatively high Na_2O and low K_2O contents relative to the monzonitic granites. These rocks are characterized by high SiO_2 and $K_2O + Na_2O$ (7.11–7.37 wt.%) contents, which suggests that they are high-K calc-alkaline series rocks (Figure 8b). The Yuhaixi diorite is chemically different from the Yuhaixi monzonitic granite and granite, and has SiO_2, Al_2O_3, Na_2O, K_2O, TFe_2O_3, TiO_2, and P_2O_5 contents of 53.55–54.58 wt.%, 17.33–17.98 wt.%, 3.86–4.11 wt.%, 0.90–1.15 wt.%, 7.90–8.69 wt.%, 0.80–0.95 wt.%, and 0.16–0.25 wt.%, respectively. The Yuhaixi diorite belongs to the calc-alkaline series (Figure 8b). In the A/NK vs. A/CNK diagram, the Yuhaixi monzonitic granite and granite samples are metaluminous to weak peraluminous (Figure 8c), with an aluminum saturation index ($Al_2O_3/(Na_2O + K_2O + CaO)$) ranging from 0.66–1.07.

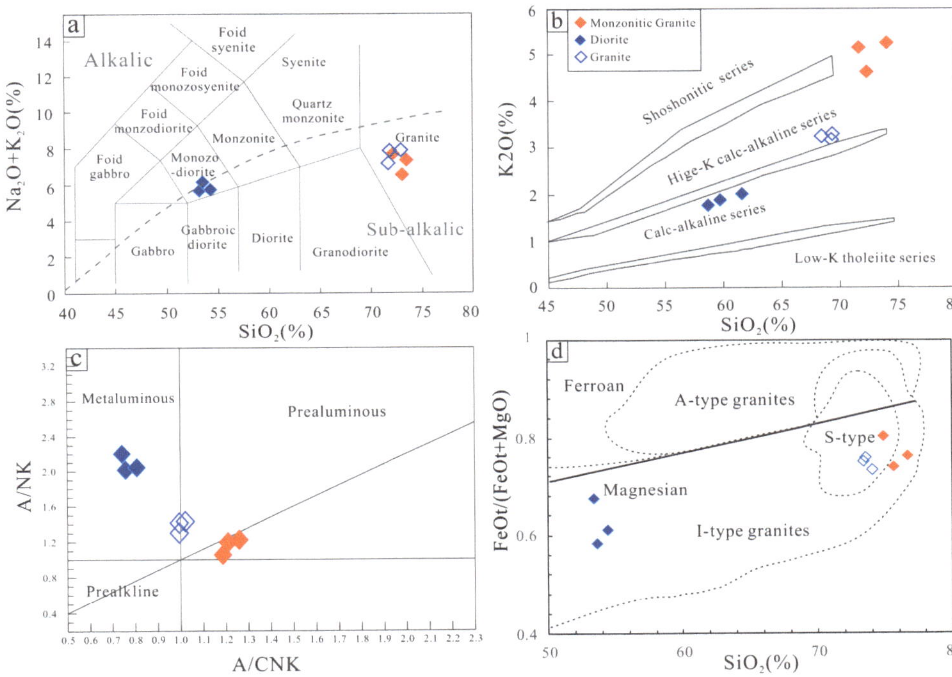

Figure 8. Classification and series diagrams of intrusions in the Yuhaixi porphyry Mo(Cu) deposit. (**a**) $Na_2O + K_2O$ vs. SiO_2 plot diagram [43]. (**b**) K_2O vs. SiO_2 diagram [44]. (**c**) A/NK vs. A/CNK plot diagram [45]. (**d**) FeOt/(FeO+MgO) vs. SiO_2 diagram.

As for the chondrite-normalized REE patterns, the Yuhaixi diorite samples are moderately fractionated ($(La/Yb)_N$ = 4.19–7.64), with light rare earth element (LREE) enrichment and heavy rare earth element (HREE) depletion in the absence of clear Eu anomalies (Eu/Eu* = 0.92–1.27) (Figure 9A). The overall rare earth content is low (ΣREE = 32.1~162.16) in Yuhaixi, among which the rare earth content of granite is the lowest (ΣREE = 32.1~78.49). The granite samples show a slight positive Eu anomaly (δEu = 0.94~1.33). The diorite samples show weaker negative Eu anomalies (δEu = 0.84–0.94) (Figure 9A). In the primitive mantle-normalized trace element spider diagram (Figure 9A; [30]), different magmatic rocks show different characteristics, which are generally characterized by enrichment of large-ion lithophile elements (LILE: Rb, Ba, K, U) and relative depletion of high-field-strength elements (HFSE: Nb, Ta, Zr). The monzonite granite is enriched in Rb, Ba, U, K, Pb, and depleted of Nb, Ta, Sr, Ti, P. Compared with granite, the LILE enrichment and the HFSE depletion in monzonitic granite and diorite are more significant (Figure 9A).

4.3. Zircon Hf Isotopes

Zircon Hf isotope analysis results are listed in Table 4. The monzonitic granite, diorite, and granite yielded εHf(t) values of 11.37–17.59, 11.59–13.46, and 11.73–14.76, respectively. The zircon Hf single- and two-stage model ages are 280–538 Ma and 240–637 Ma for the monzonitic granite, 399–472 Ma and 458–578 Ma for the diorite, and 354–478 Ma and 377–572 Ma for the granite, respectively (Table 4).

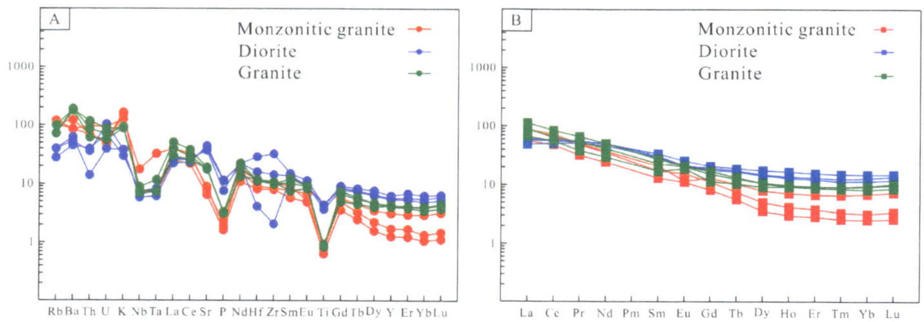

Figure 9. (**A**) primitive mantle-normalized trace element abundance spider diagram and (**B**) chondrite-normalized REE of the intrusions in the Yuhaixi porphyry Mo(Cu) deposit. (normalization values are from [20]). The N-MORB, E-MORB, and OIB patterns are from [20].

4.4. Whole-Rock Sr-Nd Isotopes

The initial $^{87}Sr/^{86}Sr$ (Isr) isotope ratios and εNd(t) were calculated using the $^{87}Sr/^{86}Sr$ ratios, Nd values, and zircon U-Pb ages analyzed in this study (Table 5). All rock samples from the Yuhaixi Mo(Cu) deposit (e.g., monzonitic granite, diorite and granite) are characterized by low $(^{87}Sr/^{86}Sr)i$ values with obviously varied ranges (e.g., 0.7043–0.7128, 0.7037–0.7041, and 0.7042–0.7043, respectively). They show positive and narrow ranges of εNd(t) (1.36–3.4, 5.16–7.75, and 4.62–6.44, respectively), corresponding to T_{DM2} ages of 832–1002 Ma, 430–642 Ma, and 545–693 Ma, respectively (Table 5).

4.5. Re-Os Isotopic Ages

The Re-Os isotopic data for nine molybdenite samples from the Yuhaixi porphyry Mo deposit are listed in Table 6 and are plotted in an isochron diagram in Figure 10. The concentrations of ^{187}Re and ^{187}Os ranged from 93.9 to 322.9 μg/g and from 539.5 to 1889.2 ng/g, respectively. Nine samples yielded model ages ranging from 337.9 ± 6.7 to 350.2 ± 6.1 Ma and a well-constrained ^{187}Re–^{187}Os isochron age of 354.1 ± 6.8 Ma, with MSWD = 1.7 and an initial ^{187}Os of 24 ± 17ng/g (Figure 10A). The data also yields a weighted average age of 344.5 ± 3.1 Ma (MSWD = 2.1) (Figure 10B). These ages are concordant within the errors, indicating that the Yuhaixi Mo(Cu) deposit was formed in the Carboniferous.

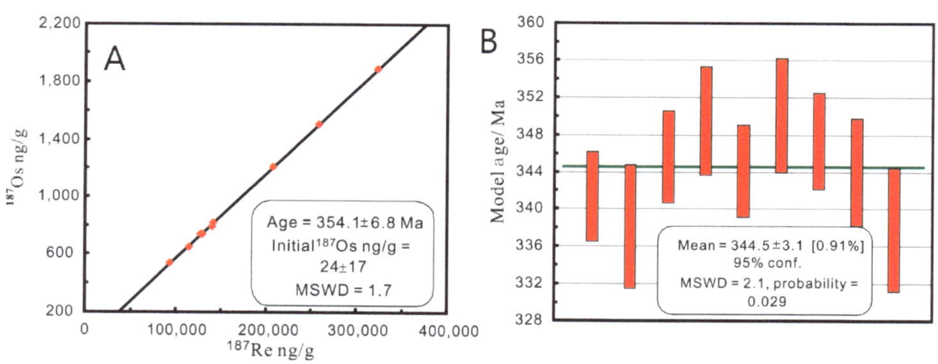

Figure 10. Re-Os isochron diagram and weighted average model age diagram for four molybdenite samples in the Yuhaixi porphyry Mo(Cu) deposit.

Table 4. In situ zircon Hf isotopic data on the studied intrusive rocks in the Yuhaixi Mo(Cu) deposit.

Sample No.	Age (Ma)	$^{176}Hf/^{177}Hf$	1σ	$^{176}Lu/^{177}Hf$	1σ	$^{176}Yb/^{177}Hf$	1σ	eHf (0)	eHf (T)	T_{DM}	T_{DMC}	fs
						Monzonitic granite						
YHX-ZK-1	364.07	0.283072	0.000014	0.004346	0.000038	0.125695	0.000845	10.61	17.59	279.05	240.34	−0.87
YHX-ZK-2	364.28	0.282921	0.000016	0.002443	0.000049	0.068981	0.001428	5.27	12.71	487.64	553.83	−0.93
YHX-ZK-3	362.36	0.282939	0.000013	0.002261	0.000021	0.062658	0.00063	5.91	13.34	458.91	511.47	−0.93
YHX-ZK-4	352.34	0.282972	0.000017	0.002645	0.000092	0.076706	0.002935	7.09	14.23	414.25	446.94	−0.92
YHX-ZK-5	362.89	0.282947	0.000014	0.001811	0.000013	0.051511	0.000516	6.18	13.74	442.01	486.6	−0.95
YHX-ZK-6	363.7	0.282929	0.000015	0.002913	0.000044	0.085132	0.001176	5.56	12.87	481.81	542.82	−0.91
YHX-ZK-7	358.97	0.282965	0.000014	0.001719	0.000046	0.048263	0.001292	6.83	14.32	414.48	445.98	−0.95
YHX-ZK-8	354.4	0.282978	0.000016	0.003775	0.000091	0.106083	0.002773	7.3	14.22	418.43	449.15	−0.89
YHX-ZK-9	361	0.282947	0.000013	0.001313	0.000006	0.035821	0.000127	6.21	13.84	435.14	478.51	−0.96
YHX-ZK-10	361.26	0.282985	0.000016	0.003241	0.000026	0.088301	0.000827	7.54	14.72	401.89	422.3	−0.9
YHX-ZK-11	358.08	0.282945	0.000016	0.002979	0.000054	0.083992	0.001507	6.12	13.3	459.23	511.2	−0.91
YHX-ZK-12	358.06	0.282903	0.000013	0.002089	0.000038	0.056193	0.001068	4.62	12.01	509.8	593.52	−0.94
YHX-ZK-13	361.99	0.28288	0.000014	0.001761	0.000034	0.046745	0.000888	3.82	11.37	538.24	637.96	−0.95
YHX-ZK-14	359.72	0.282956	0.000014	0.001655	0.000009	0.046342	0.000231	6.52	14.05	426.25	464.2	−0.95
YHX-ZK-15	358.69	0.282968	0.000016	0.003117	0.000047	0.087435	0.001308	6.93	14.08	426.66	461.17	−0.91
YHX-ZK-16	360.35	0.282957	0.000016	0.001848	0.000006	0.048001	0.000285	6.53	14.03	427.98	466	−0.94
YHX-ZK-17	356.41	0.283007	0.000014	0.003289	0.000063	0.097506	0.001856	8.3	15.37	369.99	376.79	−0.9
YHX-ZK-18	356.81	0.282927	0.000018	0.003289	0.000045	0.096394	0.001396	5.49	12.57	490.31	557.13	−0.9
						Diorite						
zk001-412-1	306.4	0.282943	0.000014	0.000731	0.00001	0.02065	0.000259	6.04	12.64	434.94	513.32	−0.98
zk001-412-2	298.94	0.282967	0.000014	0.000688	0.000001	0.01916	0.000048	6.9	13.34	400.38	462.42	−0.98
zk001-412-3	303.67	0.282969	0.000016	0.000888	0.000002	0.026311	0.000066	6.95	13.46	400.35	458.66	−0.97
zk001-412-4	302.43	0.282938	0.000016	0.001176	0.000012	0.033611	0.000286	5.87	12.29	447.19	532.71	−0.96
zk001-412-6	303.05	0.282944	0.000012	0.000827	0.000005	0.023606	0.00012	6.09	12.59	434.22	513.67	−0.98
zk001-412-7	296.45	0.282927	0.000012	0.000631	0.000004	0.018343	0.000149	5.5	11.9	455.47	553.11	−0.98
zk001-412-8	298.79	0.282937	0.000014	0.000837	0.000016	0.024897	0.000592	5.85	12.26	443.87	531.68	−0.97
zk001-412-9	299.82	0.282946	0.000014	0.000701	0.000003	0.020066	0.000103	6.16	12.62	430.05	509.63	−0.98
zk001-412-10	293.73	0.282967	0.000011	0.000607	0.000005	0.017496	0.000137	6.91	13.26	398.9	463.59	−0.98
zk001-412-11	295.9	0.282925	0.000013	0.000574	0.000001	0.016515	0.000039	5.41	11.81	458.27	558.37	−0.98
zk001-412-12	302.54	0.282914	0.000015	0.000449	0.000011	0.012607	0.000408	5.02	11.59	472.15	577.62	−0.99
zk001-412-13	306.43	0.28294	0.000013	0.00094	0.000005	0.027755	0.000198	5.95	12.51	440.87	521.65	−0.97
zk001-412-14	295.58	0.282937	0.000012	0.00096	0.000001	0.02861	0.000066	5.83	12.14	446.22	536.6	−0.97
zk001-412-15	297.65	0.282955	0.000012	0.000885	0.000003	0.025982	0.000121	6.48	12.86	419.08	492.32	−0.97
zk001-412-16	296.55	0.282953	0.000013	0.000649	0.000003	0.018522	0.000146	6.41	12.81	419.37	494.69	−0.98
zk001-412-17	296.48	0.282952	0.000014	0.001076	0.000004	0.032282	0.000155	6.38	12.69	425.56	502.38	−0.97
zk001-412-18	293.48	0.282934	0.000013	0.000679	0.000002	0.019199	0.000088	5.73	12.05	446.96	540.9	−0.98
zk001-412-19	301.64	0.282966	0.000017	0.000677	0.000003	0.018615	0.000083	6.87	13.38	401.16	462.08	−0.98
zk001-412-20	295.72	0.282931	0.000011	0.000607	0.000005	0.016322	0.000143	5.62	12.01	450.39	545.54	−0.98
zk001-412-21	303.2	0.282938	0.000015	0.000835	0.000009	0.024164	0.000315	5.87	12.38	442.95	527.54	−0.97

Table 4. *Cont.*

Sample No.	Age (Ma)	$^{176}Hf/^{177}Hf$	1σ	$^{176}Lu/^{177}Hf$	1σ	$^{176}Yb/^{177}Hf$	1σ	eHf (0)	eHf (T)	T_{DM}	T_{DMC}	fs
						Granite						
7216-1	306.63	0.282925	0.000034	0.00212	0.000015	0.052126	0.000742	5.41	11.73	477.66	571.8	−0.94
7216-2	307.95	0.28298	0.000027	0.004211	0.000022	0.108632	0.000672	7.36	13.28	421.22	473.44	−0.87
7216-3	305.7	0.283025	0.000023	0.004513	0.000057	0.114445	0.001314	8.94	14.76	354.81	376.83	−0.86
7216-4	306.15	0.282992	0.000019	0.004137	0.000074	0.106852	0.002601	7.78	13.68	401.84	446.26	−0.88
7216-5	308.54	0.282968	0.000014	0.003326	0.000032	0.088744	0.001145	6.94	13.06	428.42	488.23	−0.9
7216-6	307.07	0.282947	0.000021	0.002676	0.000013	0.07104	0.000482	6.19	12.41	452.27	528.74	−0.92
7216-7	306.32	0.282989	0.000019	0.004237	0.000041	0.110844	0.001475	7.69	13.58	406.88	453.14	−0.87
7216-8	308.07	0.282945	0.000019	0.0022	0.000039	0.05607	0.001128	6.11	12.44	449.83	527.37	−0.93
7216-9	307.16	0.282962	0.00002	0.001427	0.000076	0.033621	0.001796	6.71	13.18	416.07	479.25	−0.96

Note: $\varepsilon Hf(0) = [(^{176}Hf/^{177}Hf)_S/(^{176}Hf/^{177}Hf)_{CHUR,0} - 1] \cdot 10{,}000$; $\varepsilon Hf(t) = \{[(^{176}Hf/^{177}Hf)_S - (^{176}Lu/^{177}Hf)_S \cdot (e^{\lambda t} - 1)]/[(^{176}Hf/^{177}Hf)_{CHUR,0} - (^{176}Lu/^{177}Hf)_{CHUR,0} \cdot (e^{\lambda t} - 1)] - 1\} \cdot 10{,}000$; $T_{DM1} = 1/\lambda \cdot \ln\{1 + [(^{176}Hf/^{177}Hf)_S - (^{176}Hf/^{177}Hf)_{DM}]/[(^{176}Lu/^{177}Hf)_S - (^{176}Lu/^{177}Hf)_{DM}]$; $T_{DM2} = T_{DM1} - (T_{DM1} - t) \cdot (f_{cc} - f_s) \cdot (f_{cc} - f_{DM})$; $f_{Lu/Hf} = [(^{176}Lu/^{177}Hf)_S/(^{176}Lu/^{177}Hf)_{CHUR,0}] - 1$, where $(^{176}Hf/^{177}Hf)_S$ and $(^{176}Lu/^{177}Hf)_S$ are the measured values of the samples, s = sample, and t = crystallization time of zircon; $(^{176}Lu/^{177}Hf)_{CHUR,0} = 0.0332$ and $(^{176}Hf/^{177}Hf)_{CHUR,0} = 0.282772$ [23]; $(^{176}Lu/^{177}Hf)DM = 0.0384$ and $(^{176}Hf/^{177}Hf)DM = 0.28325$ [24]; $f_{cc} = -0.55$ and $f_{DM} = 0.16$; and $\lambda = 1.867 \times 10^{-12}/yr^{-1}$ [46] were used in the calculation.

Table 5. Sr–Nd isotopic compositions of the studied intrusive rocks in the Yuhaixi Mo(Cu) deposit.

Sample No.	Rock Type	Age (Ma)	Rb	Sr	$^{87}Rb/^{86}Sr$	$^{87}Sr/^{86}Sr$	I_{Sr}	Sm	Nd	$^{147}Sm/^{144}Nd$	$^{143}Nd/^{144}Nd$	Sm/Nd	I_{Nd}	εNd(0)	εNd(t)	T_{DM}
YHX-ZK-1	MG	359.4	30.9563	107	0.8371	0.711	0.7062	4.545	13.95	0.197	0.5127	0.326	0.5122	1.3655	1.3597	3992
YHX-ZK-2	MG	359.4	28.9275	68.65	1.2193	0.711	0.7043	1.948	6.905	0.171	0.5127	0.282	0.5123	1.4045	2.6103	1555
S3201-20	GNG	364	58.6	289	0.5866	0.706	0.7032	1.64	10.4	0.095	0.5126			−0.68	4.03	705
S3201-26	GNG	364	69.9	203	0.9964	0.709	0.7037	2.56	14.9	0.104	0.5126			−0.76	3.56	765
ZK001-25	DI	298.8	11.6175	657.9	0.0511	0.704	0.7037	1.996	17.47	0.069	0.5128	0.114	0.5127	2.887	7.7512	386
ZK001-350	DI	298.8	8.715	648.6	0.0389	0.704	0.7037	1.968	19.86	0.06	0.5128	0.099	0.5126	2.4969	7.7096	382
ZK001-396	DI	298.8	11.9025	749.4	0.046	0.704	0.7041	3.966	17.23	0.139	0.5128	0.23	0.5125	2.9651	5.1618	739
721-1	GR	307	14.9663	212.8	0.2035	0.705	0.7043	1.443	9.591	0.091	0.5128	0.151	0.5126	2.2823	6.435	493
721-3	GR	307	14.2313	194.2	0.212	0.705	0.7043	2.673	12.85	0.126	0.5127	0.208	0.5125	1.8337	4.6214	727
721-5	GR	307	14.5125	209	0.2009	0.705	0.7043	1.739	12.24	0.086	0.5127	0.142	0.5126	1.9507	6.3017	493

Abbreviation: MG, monzonitic granite; GNG, gneissic granite; DI, diorite; GR, granite. $(^{87}Sr/^{86}Sr)_i = (^{87}Sr/^{86}Sr)_s - (^{87}Rb/^{86}Sr)_s \times (e^{\lambda t} - 1)$; $^{87}Sr/^{86}Sr = (Rb/Sr) \times 2.8956$; $\lambda_{Rb\text{-}Sr} = 1.42 \times 10^{-11}/a$; $(^{143}Nd/^{144}Nd)_i = (^{143}Nd/^{144}Nd)_s - (^{147}Sm/^{144}Nd)_s \times (e^{\lambda t-1})$; $^{147}Sm/^{144}Nd = (Sm/Nd) \times 0.60456$; $\lambda_{Sm\text{-}Nd} = 6.54 \times 10^{-12}/a$; $\varepsilon Nd(t) = 10{,}000\ [(^{143}Nd/^{144}Nd)i/(^{143}Nd/^{144}Nd)_{CHUR(t)} - 1]$; $(^{143}Nd/^{144}Nd)_{CHUR(t)} = (^{143}Nd/^{144}Nd)_{CHUR(0)} - (^{147}Sm/^{144}Nd)_{CHUR} \times (e^{\lambda t} - 1)$; $(^{143}Nd/^{144}Nd)_{CHUR(0)} = 0.512638$; $(^{147}Sm/^{144}Nd)_{CHUR} = 0.1967$; $T_{DM} = 1/\lambda \times \ln\{1 + [(^{143}Nd/^{144}Nd)_S - (^{143}Nd/^{144}Nd)_{DM}]/[(^{147}Sm/^{144}Nd)_S - (^{147}Sm/^{144}Nd)_{DM}]\}$; $(^{147}Sm/^{144}Nd)_{DM} = 0.21357$; $(^{143}Nd/^{144}Nd)_{DM} = 0.51315$; $(^{147}Sm/^{144}Nd)_{crust} = 0.118$.

Table 6. Molybdenite Re-Os isotopic data for the Yuhaixi Mo(Cu) deposit.

Sample No.	Weight (g)	Ores Type	Occurrence	Re/μg·g^{-1}		Os/ng·g$_{-1}$		^{187}Re/μg·g^{-1}		^{187}Os/ng·g^{-1}		Model Age (Ma)	
				Measured	2σ	Measured	2σ	Measured	2σ	Measured	2σ	Measured	2σ
Mo-01	0.00306	Mo mineralized MG	Dissemination	206.7	1.8	0.0072	0.2	129.9	1.1	741	4.2	341.4	4.9
Mo-02	0.00516	Mo mineralized MG	Dissemination	223.4	3.4	0.0022	0.1	140.4	2.2	793.4	5.8	338.2	6.7
Mo-03	0.00141	Mo mineralized MG	Dissemination	226	1.8	0.008	0.1	142.1	1.2	820.5	5.7	345.7	5.0
Mo-04	0.00311	Mo mineralized MG	Dissemination	411.5	5.0	0.0071	0.2	258.6	3.1	1510.8	9.3	349.6	5.9
Mo-05	0.00205	Mo mineralized MG	Dissemination	204.1	1.8	0.0087	0.3	128.3	1.1	737.7	4.5	344.2	5.0
Mo-06	0.00051	Mo mineralized MG	Dissemination	513.7	5.4	0.0087	0.4	322.9	3.4	1889.2	18.7	350.2	6.1
Mo-07	0.0031	Ccp-Mo-Qz veinlet	Veinlet	330.7	3.2	0.0072	0.2	207.8	2.0	1206.3	7.2	347.4	5.2
Mo-08	0.00509	Ccp-Mo-Qz veinlet	Veinlet	149.3	1.9	0.0023	0.1	93.9	1.2	539.5	3.3	344.1	5.8
Mo-09	0.00511	Ccp-Mo-Qz veinlet	Veinlet	182.6	2.9	0.0021	0.1	114.8	1.8	648	4.0	337.9	6.7

Abbreviation: MG, monzonitic granite. Decay constant: $\lambda^{187}Re = 1.666 \times 10^{-11}$ year^{-1} [46]. Uncertainty in the Re and Os concentrations includes errors associated with the weighing of the sample and diluent, the calibration error of the diluent, the mass spectrometry analytical error, and the measurement error of the isotope ratios for the test sample; the confidence level is 95%. Uncertainty in the Re-Os model ages includes the uncertainty of the ^{187}Re decay constant, with a confidence level of 95%. Uncertainties for ages are absolute (2σ).

5. Discussion

5.1. Timing of Magmatism and Mineralization of Yuhaixi Mo(Cu) Deposit

Multiple magmatic activities were documented in the eastern segment of the Dananhu-Tousuquan island arc belt [25–29]. Our study reveals that at least two stages (~359 Ma and ~307–299 Ma) of magmatic activities occurred in the Yuhaixi area. Zircon U-Pb dating showed that Yuhaixi monzonitic granite, granite, and diorite were emplaced at ~359 Ma, ~307 Ma, and ~299 Ma, respectively. However, other intrusive rocks near the Yuhaixi orefield were reported to form at ~443–430 Ma, earlier than that of Yuhaixi intrusive rocks. For example, Wang et al. (2016) obtained the age of the rocks in the Yuhai Mo(Cu) deposit to be 441.6 ± 2.5 Ma, 430.3 ± 2.6 Ma for the diorite and granodiorite, respectively [5]; Wang et al. (2015) obtained the age of the Sanchakou pluton to be 443 Ma; Wang et al. (2016a) obtained the age of the felsic intrusion in the Sanchakou mining area to be 440–426 Ma [30]; and Wang et al. (2018) obtained the age of the Yuhai quartz diorite to be 443.5 ± 4.1 Ma [12]. Overall, at least three stages of magmatic activities were identified and recorded in the eastern segment of the Dananhu-Tousuquan island arc belt, namely ~430–443 Ma, ~359 Ma, and ~307–299 Ma.

The molybdenite Re-Os dating shows that the Yuhaixi Mo(Cu) deposit was formed at 354 ± 6.8 Ma (Figure 10), which is approximately coeval with the emplaced ages of the Yuhaixi monzonitic granite (359.4 ± 1.6 Ma). Previous studies have shown that the Yuhai molybdenite age [5] and the Sanchakou molybdenite age are concentrated in 370–350 Ma (Figure 11 [30]), which are consistent with the molybdenite Re-Os age of Yuhaixi Mo(Cu) deposit. So, the Yuhaixi Mo(Cu) deposit is suggested to form at ~354–360 Ma. The Yuhaixi granite and diorite rocks are post-ore intrusive plutons. Field investigations revealed that Yuhaixi Mo(Cu) deposits are characterized by disseminated or veinlet ores, and Mo mineralization mainly occurs in the potassic alteration zone (Figure 4I). It is highly likely that the monzonitic granite contributed to the generation of the Yuhaixi deposit.

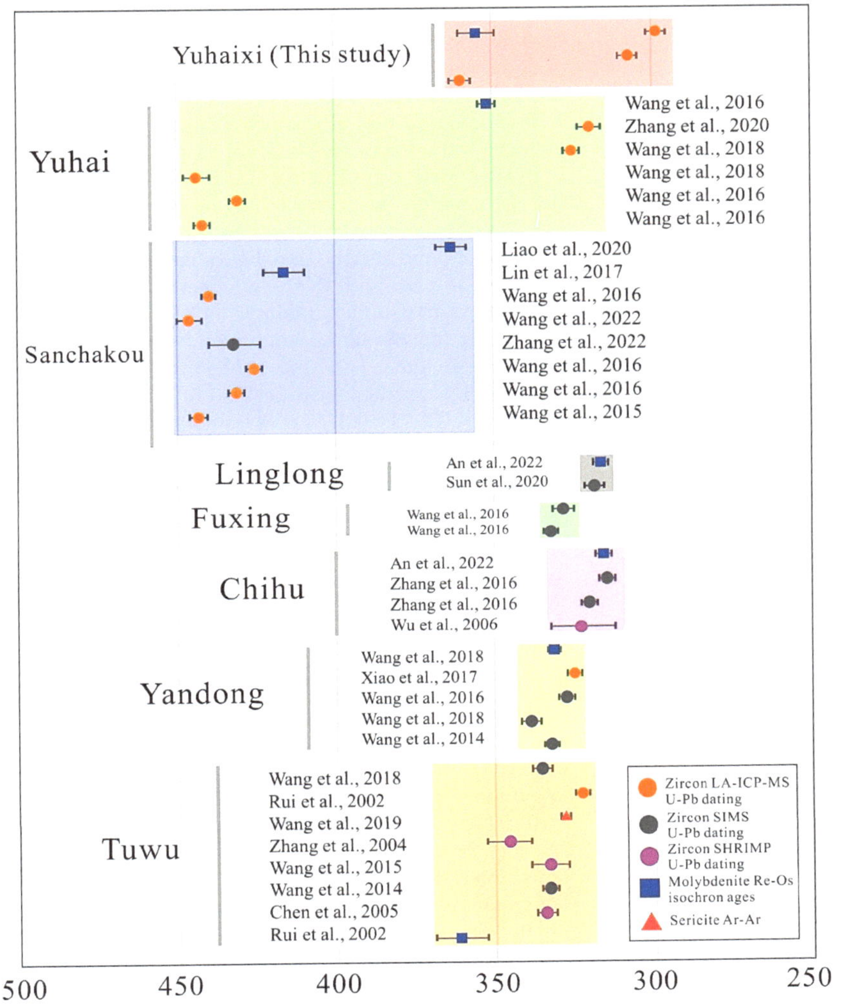

Figure 11. Summary of geochronological data of major porphyry Cu deposits in the Eastern-Tianshan. Error bars are 2σ [5–7,9,12,47–58].

5.2. Petrogenesis

The Yuhaixi monzonitic granite is characterized by high SiO_2 and $K_2O + Na_2O$ contents, low Al_2O_3 content, and the depletion of aluminum-rich minerals (e.g., muscovite, tourmaline, and garnet). They have a weak peraluminous character (1 < A/CNK < 1.1), and they exhibit low Sr (<188 ppm) and Y contents with relatively low Sr/Y ratios (<25). These features suggest that the Yuhaixi intrusion can be classified as the I- or A-type granite [21]. In addition, the Yuhaixi monzonitic granite shows low FeOt/(FeOt + MgO) ratios, Zr content, and $10^4 \times$ Ga/Al ratios (2.1–2.4), excluding the possibility of A-type granite (Figure 8d). Therefore, the Yuhaixi monzonitic granite likely belongs to I-type granites [59]. The Yuhaixi post-ore granite is a high silica granite and exhibits similar geochemical signatures to the causative monzonitic granite. The Yuhaixi post-ore granite also has relatively low A/CNK (<1.1) and Ga/Al ratios, which is consistent with the features of I-type granites.

Generally, the $Mg^{\#}$ values of magmatic rocks formed by the partial melting of the basaltic lower crust are less than 40 regardless of the degree of melting [43,44,60]. If mantle material was involved in the origin of ore-forming felsic rocks, $Mg^{\#}$ values of these rocks will be higher than 40 [60]. Thus, the $Mg^{\#}$ values can be used as an important indicator to track the addition of mantle-derived magma. Overall, the Yuhaixi felsic rocks both show high silica and low MgO, Cr, Ni contents, which is in favor of the crust origin. They both are enriched in LREE and LILE and depleted in HFSE elements (e.g., Nb, Ta) in the primitive mantle-normalized diagram (Figure 9). The Yuhaixi mozonitic granite samples have significant variable $^{87}Sr/^{86}Sr$ ratios, ranging from 0.7041–0.7127, and high εHf(t) (11.37–17.59) and εNd(t) (1.36–3.4) values, suggesting that it may have formed by melting the juvenile crust with the involvement of crustal components [43–45,60]. In addition, the Yuhaixi monzonitic granite is high silica granite and is lithologically homogeneous, arguing against the significant upper crustal contamination and fractional crystallization of basaltic magmas [40]. The studied monzonitic granite is characterized by high K_2O (3.8–5.0 wt.%) and K_2O/Na_2O (1.1–1.6) ratios and shows relatively low Rb/Sr (0.34–0.47) and high K/Rb (396–615) ratios without obvious Eu-negative anomaly, indicating that it is less evolved than high-K granites. Thus, we proposed that Yuhaixi monzonitic granite likely originated from the partial melting of the juvenile lower crust with the involvement of hydrous melts or fluids sourced from continental crustal components (e.g., subducting sediments). However, Yuhaixi post-ore granite shows high Na_2O (4.5–4.6 wt.%) contents and low K_2O/Na_2O (~0.6) ratios and it has low initial $^{87}Sr/^{86}Sr$ ratios (0.7042–0.7043) but high εHf(t) (11.7–14.8) and εNd(t) (4.6–6.4) values, which suggests that Yuhaixi granite likely originated from the partial melting of the juvenile crust with the addition of ocean crustal components (e.g., subducting oceanic slab) [45].

Compared to the Yuhaixi felsic rocks, Yuhaixi diorites have relatively low silica (53–54 wt.%) and high MgO (4.1–5.1 wt.%), Na_2O (3.9–4.1 wt.%), and Al_2O_3 (17.3–18 wt.%) contents with high Cr, V, and Ni contents. Their high MgO ($Mg^{\#} > 45$) and high transition metal (e.g., Cr, Ni, V) contents suggest the Yuhaixi high-Al diorites may be derived from the depleted lithospheric mantle source [20,23,25]. They have relatively low initial $^{87}Sr/^{86}Sr$ ratios, and high εNd(t) (5.16–7.75) and εHf(t) values (11.59–13.46), further supporting the depleted mantle origin. However, the Yuhaixi diorite samples are enriched in LREEs and LILEs, depleted in HREEs and HFSEs, and exhibit high Sr contents (785–941 ppm), Ba/La, Ba/Th ratios, but low Th/Yb, and Th/Nb ratios. Plank and Langmuir (1998) pointed out that the magma formed from the source area metasomatized by the slab dehydration fluid usually has high Sr, Ba content and Ba/Th ratio (>170) [45]. Therefore, the Yuhaixi diorite is likely derived from partial melting of the metasomatized mantle wedge (Figure 12) [31].

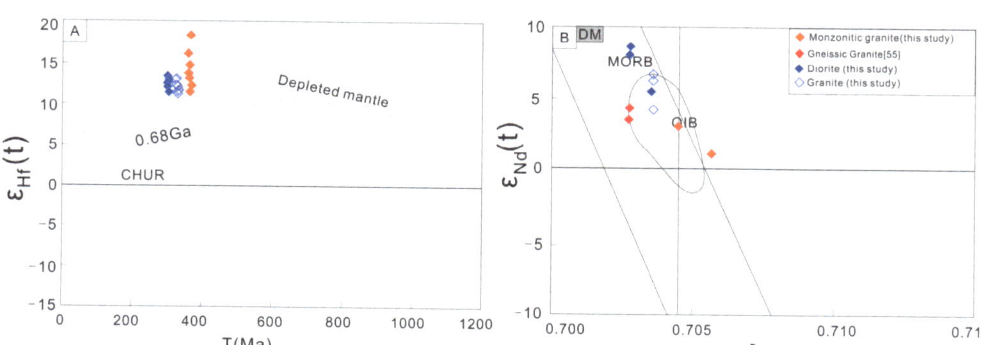

Figure 12. (**A**) Zircon εHf (t) vs. U-Pb ages (T) diagram for the Yuhaixi intrusions. (**B**) εNd(t) vs. I_{Sr} diagram for the Yuhaixi intrusions.

In a word, the Yuhaixi felsic intrusive rocks are high silica ($SiO_2 > 70$ wt.%) and are characterized by high total alkali and low MgO, Cr, Ni contents and high εNd(t) and εHf(t) values. These characteristics suggest that their melts are mainly derived from the partial melting of the juvenile lower crust. However, the post-ore diorite has a feature of high Al_2O_3, Na_2O contents with depleted isotopic signatures, indicating that diorite melts likely originated from the partial melting of depleted lithospheric mantle metasomatized by fluids or hydrous melts from the subducting slabs [5,9,12,21,23].

5.3. Implication for Tectonic Evolution and Porphyry Mineralization in Eastern Tianshan

The tectonic evolution of the Dananhu-Tousuquan arc belt has been widely addressed in previous studies [5,12,27–29], and a growing number of Paleozoic arc-related magmatic rocks have been reported in the Dananhu-Tousuquan island arc belt [11–16,25–30]. As reflected by the tectonic discrimination diagrams (Figure 13), Yuhaixi intrusive rocks including felsic and intermediate rocks all fall within the volcanic arc granitoid field, suggesting that they were formed in a subduction tectonic setting during Paleozoic era. Stratigraphic [29] and tectonic [30] studies indicated that the early Paleozoic Dananhu-Tousuquan island arcs were formed by the N-dipping subduction of the North Tianshan oceanic plate [5,12,27,29]. In the early Paleozoic, the partial melting of the low-angle, young subducted North Tianshan oceanic slab probably generated the Sanchakou adakite rocks (ca. 443 Ma–430 Ma; Unpublished data) (Figure 14A). The Carboniferous bipolar subduction of the North Tianshan Ocean formed the major Dananhu-Tousuquan island arc belt to the north and the Aqishan-Yamansu arc belt to the south (Figure 14B; [9,27,34–36,40]). During this period, asthenosphere upwelling triggered by the roll back or retraction of the subduction plate resulted in the partial melting of the juvenile lower crust to generate the Yuhaixi monzonitic granite (~360 Ma; Figure 14B). Meanwhile, with the emplacement of these magmas, the arc-related porphyry Mo mineralization was formed in the Dananhu-Tousuquan island arc belt (Figure 14B). The Eastern Tianshan bimodal magmatism, e.g., the Baiyanggou gabbro (295.8 ± 2.8 Ma) and rhyolite (293 ± 1.7 Ma) [61,62], the Cheguluquan rhyolite (294.5 ± 3.6 Ma) and basalt (293.6 ± 2.3 Ma) [63], and the youngest Hongshishan ophiolite in this belt (~310 Ma), suggest that the Eastern Tianshan tectonic setting was dominated by extension, which implies that the collision between the Dananhu-Tousuquan and Aqishan-Yamansu belts likely occurred during the late Carboniferous to early Permian periods [38,41]. This idea is also favored by the occurrence of diorite (298.0 ± 1.8 Ma) and granite (307.0 ± 2.3 Ma) in the Yuhaixi area. Moreover, many extension-related magmatic Cu-Ni sulfide deposits that occurred in the post-collisional setting were documented and reported in ca. 300–275 Ma [64], which also support the view of the collision between the Dananhu-Tousuquan and Aqishan-Yamansu belts [65–67].

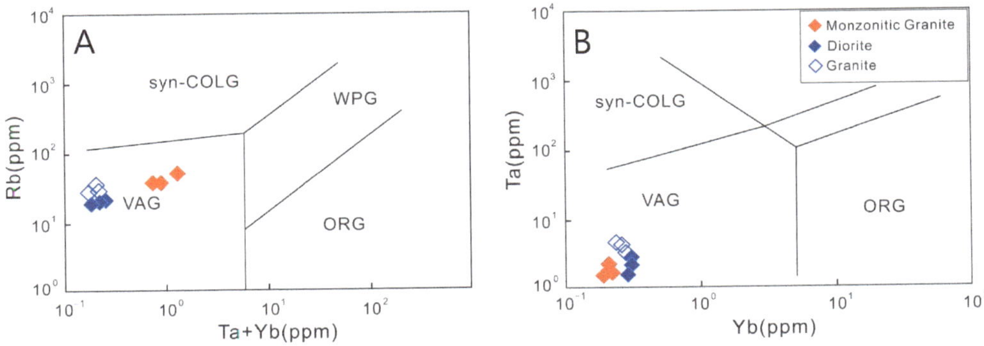

Figure 13. Tectonic discrimination diagrams for the Yuhaixi intrusions. (**A**) Rb vs. Ta + Yb diagram [33]; (**B**) Ta vs. Yb diagram [33]. Syn-COLG = Syn-collision granites; WPG = Within plate granites; ORG = Ocean ridge granites; VAG = Volcanic arc granites.

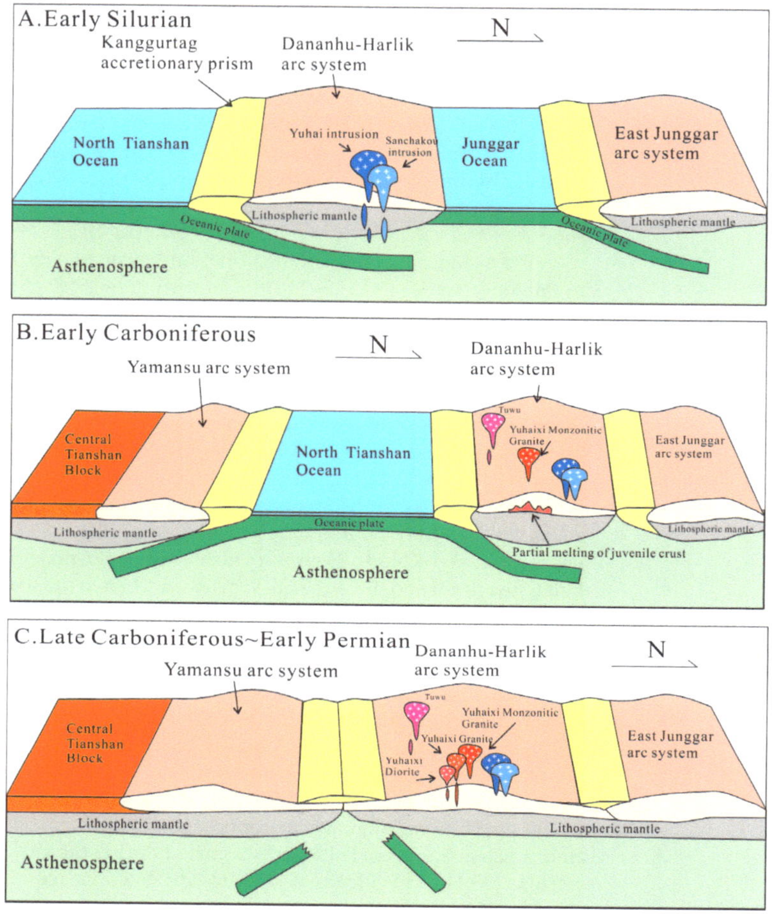

Figure 14. Schematic cartoons illustrating the tectono–magmatic–metallogenic evolution model of the Yuhaixi porphyry Mo(Cu) deposit in eastern Tianshan. (**A**) N-dipping subduction of the North Tianshan ocean plate gave rise to the Bogeda-Haerlike and Dananhu-Tousuquan island arcs in the early Paleozoic period. (**B**) Bipolar subduction of the North Tianshan ocean plate gave rise to the Dananhu-Tousuquan and Aqishan-Yamansu arcs in the Carboniferous period, forming Yuhaixi monzonitic granite and granite. (**C**) The Permian period post-collisional extension forming Yuhaixi diorite.

6. Conclusions

(1) Zircon LA-ICP-MS U-Pb dating suggests that the emplaced ages of the Yuhaixi monzonitic granite, diorite, and granite are 359.4 ± 1.6 Ma, 298.8 ± 1.8 Ma, and 307.0 ± 2.3 Ma, respectively.

(2) The Re-Os dating of molybdenite hosted by Yuhaixi monzonitic granite yields a well-constrained ^{187}Re–^{187}Os isochron age of 354.1 ± 6.8 Ma (MSWD = 1.7).

(3) Whole-rock geochemical characteristics and Sr-Nd-Hf isotopic compositions indicate that the Yuhaixi monzonitic granite and granite were formed via the partial melting of the juvenile crust. The post-ore diorite was formed via the partial melting of the metasomatized mantle wedge.

(4) The eastern section of the Dananhu-Tousuquan island arc is a promising target for late Paleozoic porphyry Mo(Cu) deposits.

Author Contributions: Conceptualization, D.W. and C.X.; investigation, D.W., Y.Z., C.X., B.X., Y.Y., Q.T., X.K. and X.W.; methodology, D.W., C.L. and C.X.; data curation, D.W. and C.L.; writing—original draft preparation, D.W.; writing—review and editing, D.W., Y.Z. and C.X.; supervision, C.X.; funding acquisition, C.X., C.L. and Y.Z.; Resources, X.K. All authors have read and agreed to the published version of the manuscript.

Funding: This research was funded by the National Key R&D Program of China (2017YFC0601202 and 2020YFA0714800) and Open Research Project from the State Key Laboratory of Geological Processes and Mineral Resources (GPMR202307). Chunji Xue acknowledges funding under the Xinjiang Uygur Autonomous Region Introduced Project "Tianchi talent".

Data Availability Statement: All data are presented in the paper.

Acknowledgments: The authors would like to thank the managers and geological staff of No. 1 Geological Team of Xinjiang for their support with the fieldwork. The authors are also deeply grateful for the reviews and constructive suggestions of the anonymous reviewers.

Conflicts of Interest: The authors declare no conflict of interest.

References

1. Sillitoe, R.H. Porphyry copper systems. *Econ. Geol.* **2010**, *105*, 3–41. [CrossRef]
2. Sinclair, W.D. Porphyry deposits. Mineral Deposits of Canada: A Synthesis of Major Deposit-Types, District Metallogeny, the Evolution of Geological Provinces, and Exploration Methods. In *Geological Association of Canada, Mineral Deposits Division, Special Publication No. 5*; Geological Association of Canada: St. John's, NL, Canada, 2007; pp. 223–243.
3. Gao, J.; Klemd, R.; Zhu, M.; Wang, X.-S.; Li, J.; Wan, B.; Xiao, W.; Zeng, Q.; Shen, P.; Sun, J.; et al. Large-scale porphyry-type mineralization in the Central Asian metallogenic domain: A review. *J. Asian Earth Sci.* **2018**, *165*, 7–36. [CrossRef]
4. Yan, D.-P.; Qiu, L. Geology of China and adjacent regions: An introduction. *J. Asian Earth Sci.* **2020**, *203*, 104533. [CrossRef]
5. Wang, Y.; Zhang, F.; Liu, J. The genesis of the ores and intrusions at the Yuhai Cu-Mo deposit in eastern Tianshan, NW China: Constraints from geology, geochronology, geochemistry, and Hf isotope systematics. *Ore Geol. Rev.* **2016**, *77*, 312–331. [CrossRef]
6. Xiao, B.; Chen, H.; Hollings, P.; Han, J.; Wang, Y.; Yang, J.; Cai, K. Magmatic evolution of the Tuwu–Yandong porphyry Cu belt, NW China: Constraints from geochronology, geochemistry and Sr–Nd–Hf isotopes. *Gondwana Res.* **2017**, *43*, 74–91. [CrossRef]
7. Wang, Y.H.; Zhang, F.F.; Liu, J.J.; Que, C.Y. Genesis of the Fuxing porphyry Cu deposit in Eastern Tianshan, China: Evidence from fluid inclusions and C–H–O–S–Pb isotope systematics. *Ore Geol. Rev.* **2016**, *79*, 46–61. [CrossRef]
8. Zhang, H.; Qiu, L.; Yan, D.-P.; Zhao, Z.; Cai, K.; Zhang, J.; Chen, S.; Li, Y.; Song, Y.; Zheng, Y.; et al. Late-Permian subduction-to-collision transition and closure of Paleo-Asian Ocean in eastern Central Asian Orogenic Belt: Evidence from borehole cores in the Songliao Basin, Northeast China. *Gondwana Res.* **2023**, *122*, 74–92. [CrossRef]
9. Zhang, F.; Wang, Y.; Liu, J. Petrogenesis of Late Carboniferous granitoids in the Chihu area of Eastern Tianshan, Northwest China, and tectonic implications: Geochronological, geochemical, and zircon Hf–O isotopic constraints. *Int. Geol. Rev.* **2016**, *58*, 949–966. [CrossRef]
10. Wang, Y.; Chen, H.; Baker, M.J.; Han, J.; Xiao, B.; Yang, J.; Jourdan, F. Multiple mineralization events of the Paleozoic Tuwu porphyry copper deposit, Eastern Tianshan: Evidence from geology, fluid inclusions, sulfur isotopes, and geochronology. *Miner. Depos.* **2018**, *54*, 1053–1076. [CrossRef]
11. Xiao, B.; Chen, H.; Wang, Y.; Han, J.; Xu, C.; Yang, J. Chlorite and epidote chemistry of the Yandong Cu deposit, NW China: Metallogenic and exploration implications for Paleozoic porphyry Cu systems in the Eastern Tianshan. *Ore Geol. Rev.* **2018**, *100*, 168–182. [CrossRef]
12. Wang, Y.F.; Chen, H.Y.; Han, J.S. Paleozoic tectonic evolution of the Dananhu-Tousuquan island arc belt, Eastern Tianshan: Constraints from the mliaoagmatism of the Yuhai porphyry Cu deposit, Xinjiang, NW China. *J. Asian Earth Sci.* **2018**, *153*, 282–306. [CrossRef]
13. Xiao, W.; Windley, B.F.; Allen, M.B.; Han, C. Paleozoic multiple accretionary and collisional tectonics of the Chinese Tianshan orogenic collage. *Gondwana Res.* **2013**, *23*, 1316–1341. [CrossRef]
14. Sun, H.; Li, H.; Danišík, M.; Xia, Q.; Jiang, C.; Wu, P.; Yang, H.; Fan, Q.; Zhu, D. U–Pb and Re-Os geochronology and geochemistry of the Donggebi Mo deposit, Eastern Tianshan, NW China: Insights into mineralization and tectonic setting. *Ore Geol. Rev.* **2017**, *86*, 584–599. [CrossRef]
15. Zhang, S.; Chen, H.; Hollings, P.; Zhao, L.; Gong, L. Tectonic and magmatic evolution of the Aqishan-Yamansu belt: A Paleozoic arc-related basin in the Eastern Tianshan (NW China). *GSA Bull.* **2020**, *133*, 1320–1344. [CrossRef]
16. Hou, G.S.; Tang, H.F.; Liu, C.Q.; Wang, Y.B. Geochronological and geochemical study on the wallrock of Tuwu-Yandong porphyry copper deposits, eastern Tianshan mountains. *Acta Petrol. Sin.* **2005**, *21*, 1729–1736.
17. Ballard, R.J.; Palin, M.J.; Campbell, I.H. Relative oxidation states of magmas inferred from Ce (IV)/Ce (III) in zircon: Application to porphyry copper deposits of northern Chile. *Contrib. Mineral. Petrol.* **2002**, *144*, 347–364. [CrossRef]
18. Elhlou, S.; Belousova, E.; Griffin, W.; Pearson, N.; O'Reilly, S. Trace element and isotopic composition of GJ-red zircon standard by laser ablation. *Geochim. Cosmochim. Acta* **2006**, *70*, A158. [CrossRef]

19. Liang, Q.; Jing, H.; Gregoire, D.C. Determination of trace elements in granites by inductively coupled plasma mass spectrometry. *Talanta* **2000**, *51*, 507–513. [CrossRef]
20. Sun, S.S.; McDonough, W.F. Chemical and isotopic systematics of oceanic basalts: Implications for mantle composition and processes. In *Magmatism in the Ocean Basin*; Geological Society Special Publication: London, UK, 1989; Volume 42, pp. 313–345.
21. Hoskin, P.W.O.; Schaltegger, U. The composition of zircon and igneous and metamorphic petrogenesis. *Rev. Mineral. Geochem.* **2003**, *53*, 27–62. [CrossRef]
22. Ferry, J.M.; Watson, E.B. New thermodynamic models and revised calibrations for the Ti-in-zircon and Zr-in-rutile thermometers. *Contrib. Mineral. Petrol.* **2007**, *154*, 429–437. [CrossRef]
23. Blichert-Toft, J.; Albarède, F. The Lu-Hf isotope geochemistry of chondrites and the evolution of the mantle-crust system. *Earth Planet. Sci. Lett.* **1997**, *148*, 243–258.
24. Griffin, W.L.; Pearson, N.J.; Belousova, E.; Jackson, S.V.; Van Achterbergh, E.; O'Reilly, S.Y.; Shee, S.R. The Hf isotope composition of cratonic mantle: LAM-MC-ICPMS analysis of zircon megacrysts in kimberlites. *Geochim. Cosmochim. Acta* **2000**, *64*, 133–147. [CrossRef]
25. Chen, F.W.; Li, H.Q.; Chen, Y.C.; Wang, D.H.; Wang, J.L.; Liu, D.Q.; Zhou, R.H. Zircon SHRIMP U–Pb dating and its geological significance of mineralization in Tuwu-Yandong porphyry copper mine, East Tianshan Mountain. *Acta Geol. Sin.* **2005**, *79*, 256–261.
26. Wu, H.; Li, H.Q.; Chen, F.W.; Lu, Y.F.D.; Mei, Y.P. Zircon SHRIMP U–Pb dating of plagiogranite porphyry in the Chihu molybdenum-copper district, Hami, East Tianshan. *Geol. Bull. China* **2006**, *25*, 549–552.
27. Du, L.; Yuan, C.; Li, X.-P.; Zhang, Y.; Huang, Z.; Long, X. Petrogenesis and Geodynamic Implications of the Carboniferous Granitoids in the Dananhu Belt, Eastern Tianshan Orogenic Belt. *J. Earth Sci.* **2019**, *30*, 1243–1252. [CrossRef]
28. Rui, Z.Y. Discussion on metallogenic epoch of Tuwu and Yandong porphyry copper deposits in eastern Tianshan Mountains, Xinjiang. *Miner. Depos.* **2002**, *21*, 16–22.
29. Zhang, L.; Qin, K.; Xiao, W. Multiple mineralization events in the eastern Tianshan district, NW China: Isotopic geochronology and geological significance. *J. Southeast Asian Earth Sci.* **2008**, *32*, 236–246. [CrossRef]
30. Wang, Y.-H.; Zhang, F.-F. Petrogenesis of early Silurian intrusions in the Sanchakou area of Eastern Tianshan, Northwest China, and tectonic implications: Geochronological, geochemical, and Hf isotopic evidence. *Int. Geol. Rev.* **2016**, *58*, 1294–1310. [CrossRef]
31. Wood, D.A. The application of a Th–Hf–Ta diagram to problems of tectonomagmatic classification and to establishing the nature of crustal contamination of basaltic lavas of the British Tertiary Volcanic Province. *Earth Planet. Sci. Lett.* **1980**, *50*, 11–30. [CrossRef]
32. Du, L.; Zhu, H.; Yuan, C.; Zhang, Y.; Huang, Z.; Li, X.-P.; Long, X. Paleozoic crustal evolution and tectonic switching in the Northeastern Tianshan: Insights from zircon Hf isotopes of granitoids. *J. Geol. Soc.* **2020**, *178*, jgs2020-035. [CrossRef]
33. Pearce, J.A.; Harris, N.B.W.; Tindle, A.G. Trace Element Discrimination Diagrams for the Tectonic Interpretation of Granitic Rocks. *J. Pet.* **1984**, *25*, 956–983. [CrossRef]
34. Harris, N.B.W.; Pearce, J.A.; Tindle, A.G. Geochemical characteristics of collision-zone magmatism. In *Geological Society*; Special Publications: London, UK, 1986; Volume 19, pp. 67–81.
35. Zhang, Z.; Ding, H.; Dong, X.; Tian, Z. Partial melting of subduction zones. *Acta Petrol. Sin.* **2020**, *36*, 2589–2615.
36. Tang, G.; Wang, Q.; Wyman, D.A.; Li, Z.-X.; Zhao, Z.-H.; Jia, X.-H.; Jiang, Z.-Q. Ridge subduction and crustal growth in the Central Asian Orogenic Belt: Evidence from Late Carboniferous adakites and high-Mg diorites in the western Junggar region, northern Xinjiang (west China). *Chem. Geol.* **2010**, *277*, 281–300. [CrossRef]
37. Li, J.Y. Late Neoproterozoic and Paleozoic tectonic framework and evolution of Eastern Xinjiang, NW China. *Geol. Rev.* **2004**, *50*, 304–322, (In Chinese with English Abstract).
38. Liu, Y.; Hu, Z.; Zong, K.; Gao, C.; Gao, S.; Xu, J.; Chen, H. Reappraisement and refinement of zircon U–Pb isotope and trace element analyses by LA–ICP–MS. *Chin. Sci. Bull.* **2010**, *55*, 1535–1546. [CrossRef]
39. Ludwig, K.R. *Isoplot/Ex, Version 3: A Geochronological Toolkit for Microsoft Excel: Berkley*; Berkeley Geochronology Center: Berkeley, CA, USA, 2003.
40. Tang, G.-J.; Wang, Q.; Wyman, D.A.; Li, Z.-X.; Zhao, Z.-H.; Yang, Y.-H. Late Carboniferous high εNd(t)–εHf(t) granitoids, enclaves and dikes in western Junggar, NW China: Ridge-subduction-related magmatism and crustal growth. *Lithos* **2012**, *140–141*, 86–102. [CrossRef]
41. Andao, D.; Hongliao, H.; Ningwan, Y.; Xiaoqiu, Z.; Yali, S.; Dezhong, S.; Shaozhen, C.; Wenjun, Q. A Study of the Rhenium Osmium Geochronometry of Molybdenites1. *Acta Geol. Sin. Engl. Ed.* **1995**, *8*, 171–181. [CrossRef]
42. Qu, W.J. Highly precise Re-Os dating of molybdenite by ICP-MS with Carius tube sample digestion. *Rock Miner. Anal.* **2003**, *22*, 254–257.
43. Wang, Q.; Xu, J.-F.; Jian, P.; Bao, Z.-W.; Zhao, Z.-H.; Li, C.-F.; Xiong, X.-L.; Ma, J.-L. Petrogenesis of Adakitic Porphyries in an Extensional Tectonic Setting, Dexing, South China: Implications for the Genesis of Porphyry Copper Mineralization. *J. Pet.* **2005**, *47*, 119–144. [CrossRef]
44. Liu, S.-A.; Li, S.; He, Y.; Huang, F. Geochemical contrasts between early Cretaceous ore-bearing and ore-barren high-Mg adakites in central-eastern China: Implications for petrogenesis and Cu–Au mineralization. *Geochim. Cosmochim. Acta* **2010**, *74*, 7160–7178. [CrossRef]

45. Plank, T.; Langmuir, C.H. The chemical composition of subducting sediment and its consequences for the crust and mantle. *Chem. Geol.* **1998**, *145*, 325–394. [CrossRef]
46. Smoliar, M.I.; Walker, R.J.; Morgan, J.W. Re-Os ages of group IIA, IIIA, IVA, and IVB iron meteorites. *Science* **1996**, *271*, 1099–1102. [CrossRef]
47. Zhang, W.; Zhang, F.F.; Wang, Y.H.; Xue, C.J.; Wang, J.P.; Sun, M.; Wang, K. Origin of Paleozoic granitoids in the Yuhai Cu–Modeposit, Eastern Tianshan, NW China and implications for regional metallogeny. *Ore Geol. Rev.* **2020**, *121*, 103465. [CrossRef]
48. Liao, K.L.; Lyu, C.L. The geological significance of molybdenite Re-Os isotopic dating result of Sanchakouxi Cu deposit in East Tianshan, Xinjiang. *Miner. Resour. Geol.* **2020**, *34*, 471–476, (In Chinese with English Abstract).
49. Lin, T.; Deng, Y.F.; Qu, W.J.; Zhou, T.F.; Yuan, F.; Deng, G. The genesis, petrology, chronology and geochemical evidence of the Sanchakou copper deposit in the East Tianshan area, Xinjiang. *J. Mineral. Petrol.* **2017**, *37*, 47–61, (In Chinese with English Abstract).
50. Wang, Y.; Chen, H.; Falloon, T.J.; Han, J. The Paleozoic-Mesozoic magmatic evolution of the Eastern Tianshan, NW China: Constraints from geochronology and geochemistry of the Sanchakou intrusive complex. *Gondwana Res.* **2022**, *103*, 1–22. [CrossRef]
51. Zhang, F.F.; Wang, Y.H.; Liu, J.J.; Xue, C.J.; Wang, J.P.; Zhang, W.; Li, Y.Y. Paleozoic magmatism and mineralization potential of the Sanchakou copper deposit, eastern Tianshan, Northwest China: Insights from geochronology, mineral chemistry, and isotopes. *Econ. Geol.* **2022**, *117*, 165–194. [CrossRef]
52. Wang, C.; Chen, B.; Ma, X.H.; Yan, X.L. Petrogenesis of Early and late Paleozoic Plutons in Sanchakou Area of East Tianshan and Their Implications for Evolution of Kangur Suture Zone. *J. Earth Sci. Environ.* **2015**, *37*, 5.
53. An, W.; Xue, C.; Zhao, Y.; Chao, L. Two Periods of Porphyry Cu Mineralization and Metallogenic Implications in the Tuwu–Yandong Belt (NW China), Based on Re-Os Systematics of Molybdenite. *Minerals* **2022**, *12*, 1127. [CrossRef]
54. Sun, M.; Wang, Y.H.; Zhang, F.F.; Lin, S.Y.; Xue, C.J.; Liu, J.J.; Zhu, D.C.; Wang, K.; Zhang, W. Petrogenesis of Late Carboniferous intrusions in the Linglong area of Eastern Tianshan, NW China, and tectonic implications: Geochronological, geochemical, and zircon Hf–O isotopic constraints. *Ore Geol. Rev.* **2020**, *120*, 103462. [CrossRef]
55. Wang, Y.; Chen, H.; Xiao, B.; Han, J.; Fang, J.; Yang, J.; Jourdan, F. Overprinting mineralization in the Paleozoic Yandong porphyry copper deposit, Eastern Tianshan, NW China—Evidence from geology, fluid inclusions and geochronology. *Ore Geol. Rev.* **2018**, *100*, 148–167. [CrossRef]
56. Shen, P.; Pan, H.; Dong, L. Yandong porphyry Cu deposit, Xinjiang, China—geology, geochemistry and SIMS U–Pb zircon geochronology of host porphyries and associated alteration and mineralization. *J. Asian Earth Sci.* **2014**, *80*, 197–217. [CrossRef]
57. Zhang, L.; Qin, K.Z.; Ying, J.F.; Xia, B. The relationship between ore-forming processes and adakitic rock in Tuwu-Yandong porphyry copper metallogenic belt, eastern Tianshan mountains. *Acta Petrol. Sin.* **2004**, *20*, 259–268.
58. Wang, Y.H.; Xue, C.J.; Liu, J.J.; Wang, J.P.; Yang, J.T.; Zhang, F.F.; Zhao, Z.N.; Zhao, Y.J.; Liu, B. Early Carboniferous adakitic rocks in the area of the Tuwu deposit, eastern Tianshan, NW China: Slab melting and implications for porphyry copper mineralization. *J. Asian Earth Sci.* **2015**, *103*, 332–349. [CrossRef]
59. Whalen, J.B.; Currie, K.L.; Chappell, B.W. A-type granites: Geochemical characteristics, discrimination and petrogenesis. *Contrib. Mineral. Petr.* **1987**, *95*, 407–419. [CrossRef]
60. Richards, J.P.; Kerrich, R. Special Paper: Adakite-Like Rocks: Their Diverse Origins and Questionable Role in Metallogenesis. *Econ. Geol.* **2007**, *102*, 537–576. [CrossRef]
61. Chen, X.; Shu, L.; Santosh, M.; Zhao, X. Island arc-type bimodal magmatism in the eastern Tianshan Belt, Northwest China: Geochemistry, zircon U-Pb geochronology and implications for the Paleozoic crustal evolution in Central Asia. *Lithos* **2013**, *168–169*, 48–66. [CrossRef]
62. Chen, X.; Shu, L.; Santosh, M. Late Paleozoic post-collisional magmatism in the Eastern Tianshan Belt, Northwest China: New insights from geochemistry, geochronology and petrology of bimodal volcanic rocks. *Lithos* **2011**, *127*, 581–598. [CrossRef]
63. Yuan, C.; Sun, M.; Wilde, S.; Xiao, W.; Xu, Y.; Long, X.; Zhao, G. Post collisional plutons in the Balikun area, East Chinese Tianshan: Evolving magmatism in response to extension and slab break-off. *Lithos* **2010**, *119*, 269–288. [CrossRef]
64. Qiu, M.; Zhao, Y.; Xue, C.; Chen, J.; Ma, X.; Wang, L.; Yu, L.; You, Y. genesis of the Ni–Cu sulfide mineralization of the carboniferous Haibaotan intrusion, Eastern Tianshan, Central Asian Orogenic Belt. *Int. J. Earth Sci.* **2022**, *112*, 1–26. [CrossRef]
65. Rickwood, P.C. Boundary lines within petrologic diagrams which use oxides of major and minor elements. *Lithos* **1989**, *22*, 247–263. [CrossRef]
66. Irvine, T.N.; Baragar, W.R.A. A Guide to the Chemical Classification of the Common Volcanic Rocks. *Can. J. Earth Sci.* **1971**, *8*, 523–548. [CrossRef]
67. Huang, B.; Chen, S.; Li, C.; Tian, Q.; Wang, C.; Wu, J.; Chen, M.; Han, J.; Wang, Y. Geochemical Features and Geological Significance of Yuhaixi Plutons in Eastern Tianshan, Xinjiang. *Earth Sci.* **2018**, *43*, 2943–2965.

minerals

MDPI

Article

Early Permian Post-Collision Extensional Setting in the Southern Beishan Orogenic Belt: Evidence from the Zhangfangshan Granodiorite and the Baishantang Bimodal Volcanic Rocks

Lei Wu [1], Xinwei Zhai [1,*], Erteng Wang [1], Wanfeng Chen [2], Gaorui Song [1], Feifei Zheng [1], Jiaolong Zhao [1], Jinrong Wang [1] and Haidong Wang [1]

[1] Key Laboratory of Mineral Resources in Western China (Gansu Province), School of Earth Sciences, Lanzhou University, Lanzhou 730000, China; wul20@lzu.edu.cn (L.W.); wanget19@lzu.edu.cn (E.W.); songgr21@lzu.edu.cn (G.S.); zhengff2023@lzu.edu.cn (F.Z.); jlz@lzu.edu.cn (J.Z.); jrwang@lzu.edu.cn (J.W.); whaidong2023@lzu.edu.cn (H.W.)

[2] College of Geography and Tourism, Hunan University of Arts and Science, Changde 430700, China; epwanfeng@163.com

* Correspondence: zhaixw926@lzu.edu.cn

Abstract: Outcrops of late Paleozoic magmatic rocks are common in the Southern Beishan Orogenic Belt (SBOB), Southern Central Asian Orogenic Belt (CAOB), which is a key object for the understanding of regional tectonism and defining the final closure time of the Paleo-Asian Ocean (PAO). We present zircon U-Pb chronology and whole-rock geochemistry data for late Paleozoic granodiorites and bimodal volcanic rocks from the Shuangyingshan-Huaniushan unit in the north Huitongshan-Zhangfangshan ophiolitic belt in the SBOB. The Zhangfangshan granodiorites (LA-ICP-MS, Ca. 288 Ma) are A2-type granite enriched in Rb, Th, Pb and LREEs and depleted in Nb, Ta, Ti, Sr, Ba and HREEs. They have varying MgO and $^{T}Fe_2O_3$ contents with high $Mg^{\#}$ (38.56~48.97) values; the Lu/Yb ratios (0.14~0.15) of these granodiorites are similar to mantle-derived magma. A clear plagioclase zoning structure and acicular apatite occur in mineral assemblages derived from magma mixing between mafic and felsic magmas. The Baishantang bimodal volcanic rocks (272 Ma) consist of rhyolite and basaltic andesite. Baishantang rhyolites are A2-type felsic rock enriched in Rb, Th, Pb and LREEs and depleted in Nb, Ta, Ti, Sr, Ba and HREEs, with negative $\varepsilon_{Nd}(t)$ and $\varepsilon_{Hf}(t)$ (-5.2~-4.8 and -2.2~-1.9, respectively). Rhyolites originated from the partial melting of the crust, influenced by mantle material. Basaltic andesites belong to calc-alkaline series and have an enrichment of Rb, Ba, Th, U, Pb and LREEs, are weakly enriched in Zr-Hf, and are depleted in Nb, Ta, Ti and HREEs. The Nd-Hf isotopes of these basaltic andesites are not coupled with negative $\varepsilon_{Nd}(t)$ (-2.8~-0.4) and positive $\varepsilon_{Hf}(t)$ (1.8~5.5) values. These characteristics indicate that they originated from the partial melting of the mantle mixed with sediment-derived melts. In combination with previous studies, our findings show that the early Permian Zhangfangshan granodiorites and Baishantang bimodal volcanic rocks formed in a post-collision extensional setting, and the Huitongshan-Zhangfangshan ocean had been closed before early Permian.

Keywords: early Permian; post-collisional extension environment; granodiorite; bimodal volcanic rocks; Beishan Orogenic Belt

Citation: Wu, L.; Zhai, X.; Wang, E.; Chen, W.; Song, G.; Zheng, F.; Zhao, J.; Wang, J.; Wang, H. Early Permian Post-Collision Extensional Setting in the Southern Beishan Orogenic Belt: Evidence from the Zhangfangshan Granodiorite and the Baishantang Bimodal Volcanic Rocks. *Minerals* **2023**, *13*, 1468. https://doi.org/10.3390/min13121468

Academic Editors: Alexey V. Ivanov, Jiafu Chen, Nan Ju and Zhonghai Zhao

Received: 2 October 2023
Revised: 14 November 2023
Accepted: 16 November 2023
Published: 22 November 2023

1. Introduction

The Central Asian Orogenic Belt (CAOB), is the largest Phanerozoic accretionary orogenic belt in the world, situated between the European, Sibrian, Tarim and North China cratons (Figure 1a) [1–5]. The CAOB evolved from the Neoproterozoic to the late Paleozoic–early Mesozoic through multiple accretion of arc/back-arc systems, ophiolites and microcontinental fragments during the closure of the Paleo-Asian Ocean (PAO) [5–8].

Voluminous magmatic and metamorphic rocks are widely exposed in CAOB and associated with accretionary, intra-continental, collisional and post-collisional processes [9–14], and the reference therein. These magmatic and metamorphic rocks provide good clues for the study of the closure time of PAO and evolution process of orogenic belt.

The Beishan Orogenic Belt (BOB), located in the southern margin of the CAOB, is an important part of the CAOB (Figure 1). It is a key area in the study of the tectonic evolution of the southern CAOB and the closure time of the PAO [15]. Four magmatic events occurred in the BOB [16]: Mesoproterozoic magmatic events representing the formation of an ancient basement in the Southern Beishan Orogenic Belt (SBOB) [17,18]; Neoproterozoic magmatism which represents the reworked ancient crust during the evolution of the Rodinia supercontinent [19,20]; Cambrian-Silurian arc magmatism related to the subduction of the Beishan ocean/PAO [21–23]; Ordovician–early late Paleozoic magmatic events correlating with the opening and closure of the branch of PAO [16,24], and the reference therein. Although a large number of late Paleozoic igneous rocks are distributed in the SBOB (Figure 1b), the formation environment of these magmatic rocks is still controversial (subduction, syn-collision or post collision) [12–14,25–30], which hampers the understanding of the final closure time of the PAO [12–14,25–30] and the tectonic evolution of the SBOB. The study of Carboniferous-Permian magmatism, especially magmatic rocks intruded into the ophiolites, provides an ideal target to investigate the late Paleozoic tectonic evolution of the SBOB and the final closure time of the PAO.

In this paper, we focus on the early Permian Zhangfangshan granodiorite intruded into the Zhangfangshan ophiolites and Baishantang bimodal volcanic rocks in the SBOB (Figure 1b) using zircon U-Pb isotopic dating, whole-rock major oxides, trace elements and Nd-Hf isotopic analysis in order to discuss their petrogenesis and the implications for the closure time of PAO and the tectonic evolution of the SBOB.

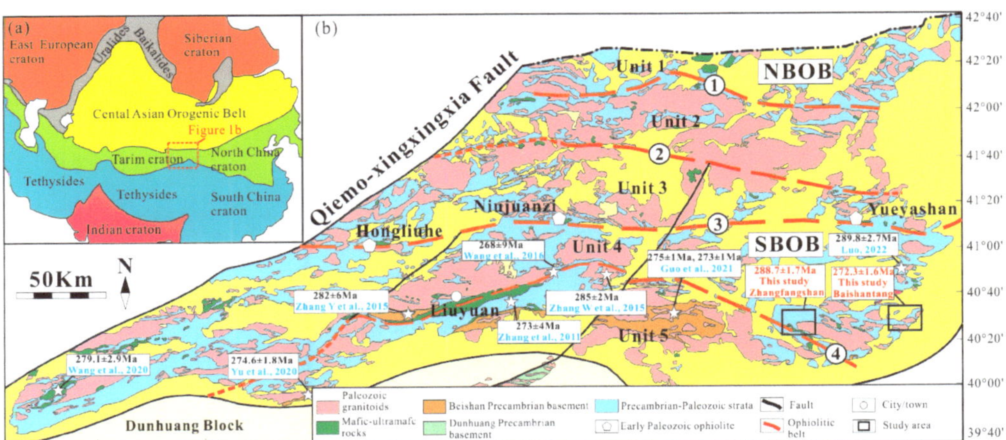

Figure 1. (**a**) Simplified tectonic map of the Central Asian Orogenic Belt showing the location of the Beishan Orogenic Belt (BOB) (modified after [31]). (**b**) Simplified geological map of the BOB (modified after [17]). (**b**) Early Permian magmatism in the SBOB (age data are taken from [13,26,32–37]). (1): Hongshishan ophiolitic belt; (2): Jijitaizi-Xiaohuangshan ophiolitic belt; (3): Hongliuhe-Niujuanzi-Yueyashan ophiolitic belt; (4): Huitongshan-Zhangfangshan ophiolitic belt; Unit 1: Queershan unit; Unit 2: Heiyingshan-Hanshan unit; Unit 3: Mazongshan unit; Unit 4: Shuangyingshan-Huaniushan unit; Unit 5: Shibanshan unit.

2. Geological Setting

The BOB is located at the southern margin of the CAOB and juxtaposed against the Dunhuang block to the south [15,17,30]. From north to south, there are four important ophiolite belts in the BOB, namely, Hongshishan, Jijitaizi-Xiaohuangshan, Hongliuhe-Niujuanzi-Yueyashan (HNY) and Huitongshan-Zhangfangshan (HZ) ophiolitic belts (Figure 1b) [30,38–41]. The BOB is sub-divided into Queershan, Heiyingshan-Hanshan, Mazongshan, Shuangyingshan-Huaniushan (SH) and Shibanshan units from north to south by these ophiolitic belts (Figure 1b) [15].

The study regions are located at the SH unit bounded by the HNY ophiolitic belt to the north and the HZ ophiolitic belt to the south. Neoproterozoic–late Paleozoic strata and intrusive rocks are widely exposed in this unit. Neoproterozoic-Cambrian strata are mainly composed of metasedimentary-sedimentary rocks [42]. Ordovician-Devonian strata are characterized by volcano-sedimentary formations. Carboniferous clastic and carbonate strata are sparsely distributed in this unit. Permian clastic and volcanic rocks are mainly exposed on the southern margin of the SH unit [43,44].

The Zhangfangshan granodiorite and the Baishantang bimodal volcanic rocks crop out in the east of the SH unit. The Zhangfangshan granodiorite is located in the north of the Zhangfangshan ophiolitic belt and west of the Baishantang area, intruded into the Ordovician-Silurian strata, ophiolites and Carboniferous strata while making unconformable contact with Jurassic strata (Figures 2a and 3a,b).

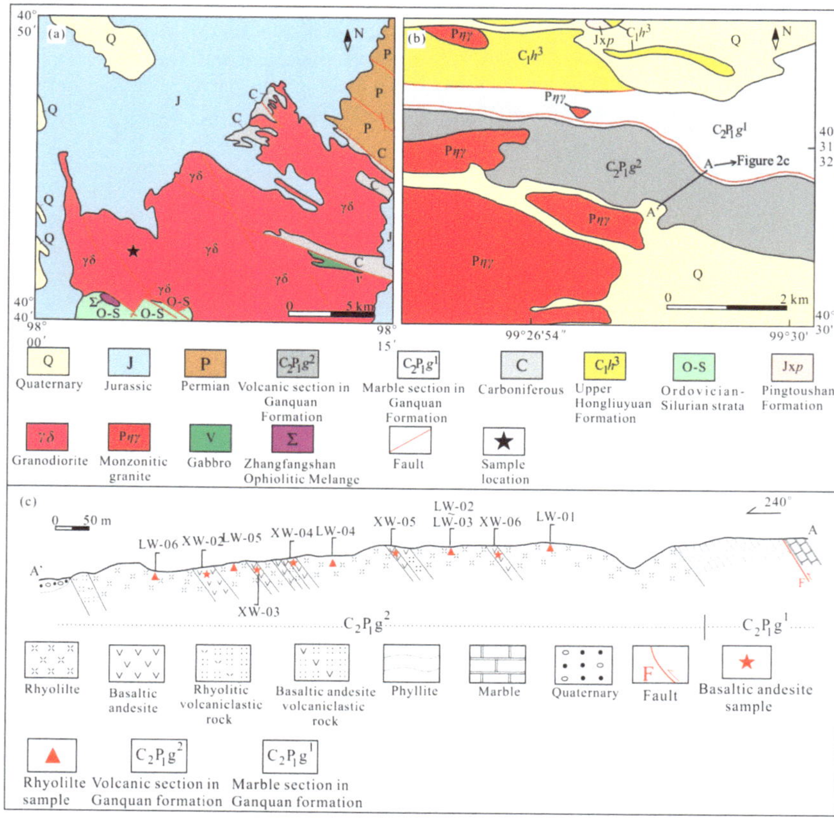

Figure 2. Geological maps of (**a**) the Zhangfangshan granodiorite and (**b**) the Baishantang bimodal volcanic rocks (after [45,46]); (**c**) profile of the Baishantang bimodal volcanic rocks.

Figure 3. Field photos and photomicrographs of the Zhangfangshan granodiorite and Baishantang bimodal volcanic rocks. (**a**,**b**) Zhangfangshan granodiorite intruded into Zhangfangshan ophiolitic mélange (ZOM); (**c**) Photo of the Zhangfangshan granodiorite sample. Hornblende can be clearly seen in the rock; (**d–f**) Photomicrographs of the Zhangfangshan granodiorite. Tiny acicular apartite and zonal structure of plagioclase are detected (**d**,**f**). The mineral assemblages of granodiorite (**e**). Quartz crystal as inclusions occur in plagioclase (**f**); (**g**,**h**) Field photos of the Baishantang bimodal volcanic rocks. Rhyolite and basaltic andesite are interbedded (**g**), and the distribution area of rhyolite is larger than that of basaltic andesite (**h**); (**i**) Plane-polarized photograph of basaltic andesite; (**j**) Cross-polarized photograph of basaltic andesite, where epidotization occurs in plagioclase; (**k**) Plane-polarized photograph of rhyolite; (**l**) Cross-polarized photograph of rhyolite, where epidotization occurs in plagioclase. Ap: apatite; Hbl: hornblende; Bt: biotite; Pl: plagioclase; Qtz: quartz; Chl: chlorite; Ep: Epidote.

The Baishantang bimodal volcanic rocks are exposed in the volcanic section in Ganquan Formation. The Ganquan Formation is exposed in the south of the Hongliuyuan Formation and separated from it by faults (Figure 2b). The Ganquan Formation can be divided into marble and volcanic sections. The marble section is mainly composed of marble with rhyolite, phyllite, mica quartz schist and metabasaltic and is separated from the volcanic section by a fault to the south. The volcanic section is mainly composed of rhyolite and basaltic andesite, with a small amount of rhyolitic and basaltic andesite pyroclastic rocks. Rhyolites have larger distribution areas than the basaltic andesites, and they are interbedded to form bimodal volcanic rocks (Figures 2c and 3g,h). Due to intense weathering, only few complete rhyolite and basaltic andesite outcrops can be seen in the study area.

3. Sample Description

3.1. Zhangfangshan Granodiorite

The granodiorite is grayish white, medium- to coarse-grained and massive structure. Its mineral composition includes plagioclase (40~45 vol%), quartz (35~40 vol%), biotite (10~15 vol%) and hornblende (5~10 vol%). Accessory minerals include apatite, zircon and others (Figure 3d–f). One sample was collected for zircon U-Pb dating, and six samples were collected for the whole-rock major oxides and trace elements study.

3.2. Baishantang Bimodal Volcanic Rocks

The rhyolite is light red to light pink in color, massive and porphyritic in texture, with up to 20 vol% phenocrysts, mainly comprising plagioclase and a minor amount of quartz (Figure 3k,l). The directional arrangement of phenocrysts can be observed in Figure 3k,l. The groundmass is usually a fine-grained assemblage of feldspar, quartz and minor Fe-Ti oxides. Plagioclase has a certain degree of kaolinization and epidotization.

The basaltic andesite is dark grey with a typical porphyritic texture (Figure 3i,j). The phenocrysts are mainly plagioclase (~15 vol%), while the matrix consists of plagioclase microcrystals and accessory minerals. Plagioclase has a certain degree of kaolinization and epidotization. One rhyolite was collected for zircon U-Pb dating; six rhyolite samples and five basaltic andesite samples were collected for whole-rock major oxides and trace elements study. Three rhyolite samples and three basaltic andesite samples were collected for the whole rock Nd-Hf isotopes analysis.

4. Analytical Methods

4.1. Zircon U-Pb Dating

Zircon cathodoluminescence (CL) images of the Zhangfangshan granodiorite and the Baishantang rhyolite were taken at the Key Laboratory of Mineral Resources of Western China, Lanzhou University, Lanzhou, China, and Langfang Chenxin Geological Service Co., Hebei, China, respectively. After separation using the conventional heavy liquid and magnetic techniques, zircon grains were purified through hand-picking under a binocular microscope, embedded in an epoxy and then polished to expose half of the zircon grains [47]. The zircon grains with internal ring structures, no clear inclusions and few fractures in the CL were chosen for U-Pb dating. Zircon LA-ICP-MS U-Pb isotope analyses were performed at the Key Laboratory of Mineral Resources of Western China, Lanzhou University, using ICP-MS Agilent 7500a (Agilent, Santa Clara, CA, USA). The zircon standard 91500 [48] was used as the age reference. The glass NIST 610 and Si were used as an internal standard for Pb, Th and U [49]. The Glitter (ver. 4.0) [50] and Isoplot R [51] programs were employed to process the data. The analytical procedures are described in detail by Yuan et al. [52].

4.2. Whole-Rock Major Oxides and Trace Elements Analyses

All analyses of major oxides and trace elements for the Zhangfangshan granodiorite and Baishantang bimodal volcanic rocks were carried out at the Key Laboratory of Mineral Resources of Western China, Lanzhou University. Major oxides were determined through ICP optical emission spectroscopy (ICP-OES) (manufactured by Agilent, Santa Clara, CA, USA) using a Leeman Prodigy system with analytical precision greater than 2%. Loss on ignition (LOI) was obtained by heating approximately 0.5 g of the dried-sample powder at 1000 °C for 2 h. Trace elements were determined through inductively coupled plasma mass spectrometry (ICP-MS) using an Agilent 7700× system. The relative standard deviation was less than 10% for the determination of trace elements, including rare earth elements (REEs). The US Geological Survey reference materials AGV-2 and BCR-2 were used as standards. Analytical procedures of major oxides and trace elements have been described by Li et al. [53] and Gao et al. [54], respectively.

4.3. Whole-Rock Nd-Hf Isotope Analyses

Sm-Nd and Lu-Hf isotopic analyses were performed on Nu Instruments Nu Plasma II MC-ICP-MS (Nu Instrument, Wrexham, Wales, UK) at the Key Laboratory of Mineral Resources of Western China, Lanzhou University. The US Geological Survey reference materials BCR-2 and AGV-2 were used as standards. JNDI and Alfa Hf standard solution were periodically analyzed to correct instrumental drift. Mass fractionation corrections for Nd and Hf isotopic ratios were based on 0.7219 for $^{146}Nd/^{144}Nd$ and 0.7325 for $^{179}Hf/^{177}Hf$, respectively. Standard BCR-2 yielded average values of 0.512642 ± 0.00003 (2σ, n = 5) for $^{143}Nd/^{144}Nd$ and 0.282871 ± 0.00004 (2σ, n = 2) for $^{176}Hf/^{177}Hf$. Standard AGV-2 yielded 0.512742 ± 0.00003 (2σ, n = 5) for $^{143}Nd/^{144}Nd$ and 0.282950 ± 0.00004 (2σ, n = 1) for $^{176}Hf/^{177}Hf$. Analytical procedures were described by Bao et al. [55].

5. Analytical Results

The zircon U-Pb isotopic data, whole-rock major oxides, trace element data and Nd-Hf isotopic data are listed in Supplementary Table S1, Table S2 and Table S3, respectively.

5.1. Zircon U-Pb Geochronology

Fifteen zircons from the granodiorite samples are euhedral to subhedral and columnar in shape, with length-to-width ratios of 2:1~3:1. The zircons display well-preserved concentric oscillatory zoning (Figure 4a) and high Th/U ratios (0.44~0.65) (Table S1), indicating a typical magmatic origin [56]. All fifteen analyses yielded similar apparent $^{206}Pb/^{238}U$ ages of 297~284 Ma, with a weighted mean age of 288.7 ± 1.7 Ma (MSWD = 1.1, n = 15) (Figure 5a).

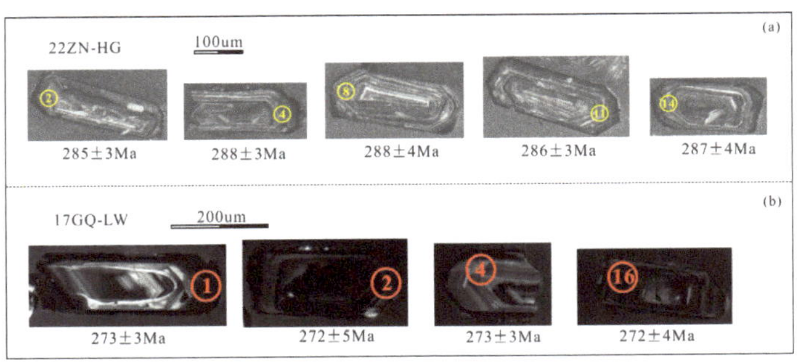

Figure 4. Representative zircons cathodoluminescence images for zircons in the Zhangfangshan granodiorite (**a**) and the Baishantang rhyolite (**b**). Yellow and red numbers represent the U-Pb analytical sites of the Zhangfangshan granodiorite and the Baishantang rhyolite, respectively.

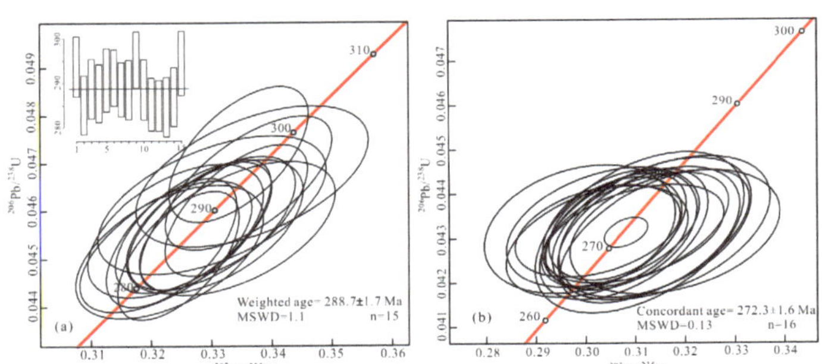

Figure 5. U-Pb dating results of zircons for the Zhangfangshan granodiorite (**a**) and the Baishantang rhyolilte (**b**).

Eighteen zircons from the Baishantang rhyolite are euhedral to subhedral and display well-preserved concentric oscillatory zoning (Figure 4b). Their Th/U ratios are 0.46~1.30 (Table S1), indicating a typical magmatic origin. Sixteen analyses yield close apparent $^{206}Pb/^{238}U$ ages of 271~273 Ma and a concordant age of 272.3 ± 1.6 Ma (MSWD = 0.13, n = 16) (Figure 5b). The other two analyses yield old $^{206}Pb/^{238}U$ ages (both of GQ-LW-03 and GQ-LW-07 are 395 Ma) (Table S1).

5.2. Whole-Rock Major Oxides and Trace Elements

5.2.1. The Zhangfangshan Granodiorite

The composition of major oxides is as follows: SiO_2 (64.78~69.21 wt.%), Al_2O_3 (13.55~14.58 wt.%), $^{T}Fe_2O_3$ (2.04~3.47 wt.%), MgO (0.65~1.68 wt.%), Na_2O (3.92~7.22 wt.%), K_2O (3.20~4.03 wt.%) and TiO_2 (0.33~0.40 wt.%). The Na_2O + K_2O values are 7.12~10.74 wt.%, $Mg^{\#}$ values are 38.56~48.97 and A/CNK ratios are less than 1 (0.62~0.90). The samples' plots are in the granite field and are high-K calc-alkaline series and metaluminous (Figure 6). Chondrite-normalized REE patterns of Zhangfangshan granodiorites (Figure 7a) show a weakly fractionated REE pattern ($(La/Yb)_N$ = 3.39~5.27), enrichments of LREEs (LREE/HREE = 4.08~6.14) and negative Eu anomalies (δEu = 0.27~0.48). Primitive mantle-normalized trace elements' patterns (Figure 7b) show enrichments of Rb, Th and Pb; weak enrichments of Zr and Hf; and a depletion of Nb, Ta, Ti, Sr and Ba.

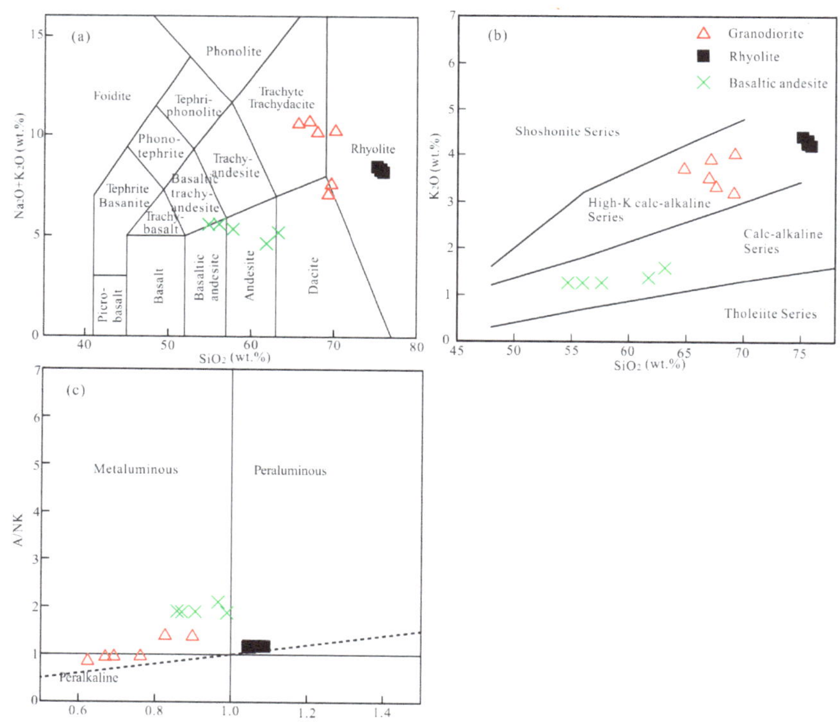

Figure 6. Geochemical classification diagrams for the Zhangfangshan granodiorites and Baishantang bimodal volcanic rocks. (**a**) TAS diagram [57], (**b**) K_2O versus SiO_2 diagram [58], (**c**) A/NK versus A/CNK diagram [59].

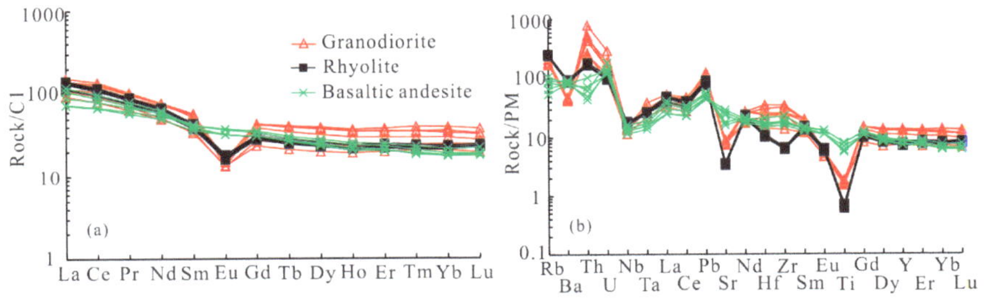

Figure 7. Chondrite (C1)-normalized REE patterns (**a**) and Primitive Mantle (PM)-normalized multi-element diagrams (**b**) for the Zhangfangshan granodiorites and Baishantang bimodal volcanic rocks. C1 and PM values are from Sun and McDonough [60].

5.2.2. Baishantang Bimodal Volcanic Rocks

Baishantang rhyolites have high SiO_2 (75.12~75.87 wt.%), Na_2O (4.00~4.07 wt.%), K_2O (4.21~4.41 wt.%) and $Na_2O + K_2O$ (8.21~8.48 wt.%) contents; moderate Al_2O_3 (13.14~13.53 wt.%) contents; and low ${}^{T}Fe_2O_3$ (1.41~1.74 wt.%), MgO (0.26~0.37 wt.%) and TiO_2 (0.13~0.14 wt.%) contents; their $Mg^{\#}$ values are 23.72~33.58. In the TAS diagram (Figure 6a), all samples are plotted in the rhyolite field in the TAS diagram (Figure 6a) and belong to the high-K calc-alkaline series in the K_2O-SiO_2 diagram (Figure 6b). The A/CNK ratios of all rhyolites are 1.04~1.09 and plotted in the peraluminous (Figure 6c). Rhyolite samples display LREE-enriched REE patterns ($(La/Yb)_N$ = 5.18~6.28, LREEs/HREEs = 5.71~6.55) and have obvious negative Eu anomalies (δEu = 0.48~0.52) (Figure 7a). They are enriched in Rb, Th and Pb; depleted in HREEs and Ba, Nb, Ta, Ti and Sr; and weakly depleted in Zr and Hf (Figure 7b).

Baishantang basaltic andesites have moderate SiO_2 (54.66~63.14 wt.%) contents; high Al_2O_3 (14.36~16.13 wt.%), MgO (3.47~4.26 wt.%), Na_2O (3.25~4.32 wt.%), ${}^{T}Fe_2O_3$ (7.97~10.71 wt.%) and TiO_2 (1.28~1.87 wt.%) contents; and low P_2O_5 (0.40~0.55 wt.%) and K_2O (1.26~1.59 wt.%) contents. The $Mg^{\#}$ values are 44.08~46.30, and the $Na_2O + K_2O$ values are 4.63~5.58 wt.%. The samples belong to calc-alkaline series (Figure 6b). Basaltic andesite samples display LREE-enriched REE patterns ($(La/Yb)_N$ = 3.85~6.49, LREEs/HREEs = 4.09~6.06) and have no obvious negative Eu anomalies (δEu = 0.92~0.99) (Figure 7a). They are enriched in Rb, Ba, Th, U and Pb; weakly enriched in Zr and Hf; and depleted in Nb, Ta and Ti (Figure 7b).

5.3. Whole-Rock Nd-Hf Isotope of the Baishantang Bimodal Volcanic Rocks

Whole-rock Nd-Hf isotopic compositions of Baishantang rocks are plotted in Figure 8. The initial Nd and Hf isotope ratios were calculated using the zircon U-Pb ages (272 Ma). The ${}^{176}Hf/{}^{177}Hf$ ratios of 0.282674~0.282682 and ${}^{143}Nd/{}^{144}Nd$ ratios of 0.512252~0.512264 were measured for three rhyolite samples, which correspond to initial ${}^{176}Hf/{}^{177}Hf$ ratios of 0.282542~0.282549 and initial ${}^{143}Nd/{}^{144}Nd$ ratios of 0.512023~0.512041. They had negative $\varepsilon_{Hf}(t)$ (−2.2~−1.9) and $\varepsilon_{Nd}(t)$ (−5.2~−4.8) values, with two-stage Hf and Nd model ages (T_{DM2}) between 1427~1411 Ma and 1460~1433 Ma, respectively. The ${}^{176}Hf/{}^{177}Hf$ ratios of 0.282721~0.282831 and ${}^{143}Nd/{}^{144}Nd$ ratios of 0.512381~0.512522 were measured for three basaltic andesite samples, which correspond to initial ${}^{176}Hf/{}^{177}Hf$ ratios of 0.282655~0.282760 and initial ${}^{143}Nd/{}^{144}Nd$ ratios of 0.512146~0.512205. Basalitic andesites have a negative $\varepsilon_{Nd}(t)$ (−2.8~−0.4) but positive $\varepsilon_{Hf}(t)$ (1.8~5.5).

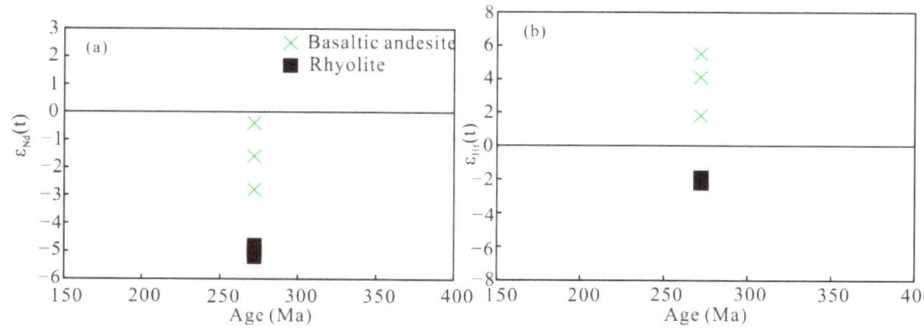

Figure 8. (**a**) $\varepsilon_{Nd}(t)$ values of the Baishantang bimodal volcanic rocks; (**b**) $\varepsilon_{Hf}(t)$ values of the Baishantang bimodal volcanic rocks.

6. Discussion

6.1. Early Permian Magmatism in the SBOB

In this research, we studied granodiorite and obtained new zircon U-Pb ages of 288.7 ± 1.7 Ma. Zircon grains from this pluton are euhedral-subhedral with clear concentric oscillatory patterns, without overgrowth rim and inherited cores (Figure 4a). Meanwhile, these zircons have relatively high Th/U ratios (0.44~0.65), indicative of a magmatic origin. In the field, the Zhangfangshan granodiorite intruded the Zhangfangshan ophiolite (362 Ma) [61], suggesting that the Zhangfangshan granodiorite was formed in the early Permian.

The Baishantang basaltic andesite and rhyolite are spatially associated with each other, and their SiO_2 contents have obvious discontinuity, which clearly suggests that they are likely coeval and, thus, constitute a bimodal volcanic suit. The timing of the Baishantang bimodal volcanic magmatism can be determined by the eruption age of the rhyolite [62]. The basaltic andesite and rhyolite outcrops are in the volcanic section of the Ganquan Formation. The Ganquan Formation is widely distributed in the SBOB and consists of marble and volcanic sections. Through studies, it has been determined that the Ganquan Formation is in contact with the Upper Carboniferous Shibanshan and Lower Permian Shuangbaotang Formations [63]. The fossils in the marble section and the zircon U-Pb isotopic age in the volcanic section indicate that the Ganquan Formation was formed in the late Carboniferous–early Permian while the volcanic section was formed in the early Permian [63]. All the dated zircons in this study exhibit well-developed zoning and have relatively high Th/U ratios (0.46~1.30), indicative of a magmatic origin. Sixteen analyses yielded a concordant age of 272.3 ± 1.6 Ma; this result is consistent with the previous study and is interpreted as the crystallization age of the Baishantang rhyolite. $^{206}Pb/^{207}Pb$ ages of 395 Ma are yield by GQ-LW-3 and GQ-LW-7, clearly older than other ages, and represent the captured zircon ages. Therefore, the Baishantang bimodal volcanic rocks were formed in early Permian.

There are other early Permian magmatisms in the SBOB. Granitic plutons with an age of 290~270 Ma were reported in the Shuwojing, Western Shuwojing, Bandaoshan, Wohushan, Liuyuan and Yinwaxia, including syenogranite, biotite monzogranite, biotite granite, monzogranite and granodiorite [13,27,33,35]. Meanwhile, the ultramafic-mafic rocks and mafic-felsic dykes are also widely distributed in the SBOB, such as Podong, Cihai, Yitiaoshan, Gubaoquan, Yinwaxia and Niujuanzi [26,31,32,34,64–66]. These mafic-felsic rocks, combined with the Zhangfangshan granodiorite and the Baishantang bimodal volcanic rocks studied in this paper, indicate that the early Permian magmatic activities are widespread in the SBOB.

6.2. Petrogensis

6.2.1. The Zhangfangshan Granodiorite

Generally, granites are categorized into I-, S-, A- and M-type granite based on geochemical and mineralogical features [67–69]. Early Permian granodiorites (Ca. 288 Ma) have high K_2O contents (3.20~4.03 wt.%), unlike M-type granites with low K_2O contents (<1 wt.%). Zhangfangshan granodiorites are metaluminous (A/CNK < 1.1), and Al-bearing minerals are lacking (e.g., garnet and andalusite), indicating that Zhangfangshan granodiorites are not S-type granites. Therefore, the Zhangfangshan granodiorites are I- or A-type granites. All granodiorite samples display high $K_2O + Na_2O$ (7.12~10.74 wt.%) and FeO_t contents (2.04~3.47 wt.%), low Rb contents (109.56~134.60 ppm), high 10,000*Ga/Al (2.8~3.3) and Zr + Nb + Ce + Y (368.17~544.83 ppm) values, demonstrating that the Zhangfangshan granodiorites show an affinity to A-type granites [69–74]. In the discrimination diagrams for I-S-A-type granite by Whalen et al. [70] (Figure 9a,b), all samples are plotted in the A-type granite field rather than the I and S-type field. A high calculated zircon saturation temperature is the most typical feature of A-type granite [59,75]. The calculated zircon saturation temperatures of Zhangfangshan granodiorite samples are 895~1004 °C [76], higher than the average temperatures of A- and I-type granites (Figure 9c). In summary, the studied granodiorites can be regarded as A-type granites. Based on different geochemical characteristics, A-type granites can be further divided into A1- and A2-type granites [77]. All samples have high Y/Nb (3.47~5.99), Rb/Nb (10.73~14.51) and Ce/Nb (5.99~8.36) ratios and are plotted within the A2 field in the Nb-Y-Ce diagram (Figure 9d). In conclusion, Zhangfangshan granodiorites are defined as A2-type granites.

Various models have been proposed to illustrate the formation of A-type granites, including the fractionation-crystallization process of mantle-derived alkaline magma [78,79], magma mixing between mafic and felsic magma, and partial melting of the lower continental crust caused by mantle magma underplating [71].

The Zhangfangshan granodiorites have high SiO_2 (64.78~69.21 wt.%) and K_2O (3.20~4.03 wt.%) contents, are enriched in LREEs and have a depletion of HFSEs as well as having negative Sr and Ti anomalies, indicating that the rocks are difficult to be formed by a high differentiation of mantle-derived magma [71]. The Y/Ho ratios (28.40~30.45) of granodiorites are close to those of chondrite (28), and the Zr/Hf ratios (34.7~38.9) of the samples belong to "CHARAC field (Zr/Hf ratios of 26~46)" [71], indicating that the fractional crystallization is limited. Meanwhile, if the felsic rocks were derived from the fractional crystallization of basaltic magma, voluminous mafic rocks should be exposed in the area [80]. However, there are no coeval intermediate-basic rocks exposed in the region. In conclusion, the Zhangfangshan granodiorites were not formed by the fractional crystallization of mantle-derived magmas.

The Ce/Pb (2.56~4.53) and La/Nb (2.60~3.61) values of granodiorites were found to be close to those of the crust (Ce/Pb < 15, La/Nb = 2.2) [81], indicating that the granodiorite had crustal affinity. However, samples also have high MgO contents and $Mg^{\#}$ values (>40, average 46.58), including Na_2O contents greater than or close to those of K_2O. These features are different from those of melts formed by the partial melting of the basaltic lower crust. Meanwhile, the Lu/Yb ratios (0.14~0.15) of these samples are similar to the ratio of the mantle-derived magma (0.15), suggesting a contribution from the mantle. Zhanfangshan granodiorite formation is not solely derived from the partial melting of the lower crust, and the addition of high-Mg mantle-derived mafic magma is required. Due to the strong Quaternary cover, no mafic enclaves have been found in the Zhangfangshan pluton, but we can still find evidence of magma mixing: (a) the obvious zonal structure of plagioclase, the existence of acicular apatite, and the presence of felsic inclusion in plagioclase crystals. These petrological features are special mineral phenomena in rocks formed from magma mixing [82]. (b) Previous studies have suggested that increasing Fe + Mg contents would be expected due to the basification of relatively felsic magma [71]. Zhangfangshan granodiorites display a positive correlation of FeO_t + MgO with TiO_2 (Figure 10a). The samples also have variations in MgO and TFe_2O_3 contents, which favors magma mixing between mantle-derived mafic magma and crust-derived felsic melts [71,83]. (c) The Zhangfangshan granodiorite samples show magma-mixing trends in the 1/V-Rb/V diagram (Figure 10b). Therefore, potential magma mixing may occur in the Zhanfangshan granodiorite formation process.

Granodiorites have the characteristics of depletions of Ba, Eu, Sr, Ti, Nb and Ta and an obvious plagioclase ring structure, which indicate the crystallization differentiation of the granodiorite and fractionation of plagioclase and Ti-bearing phases.

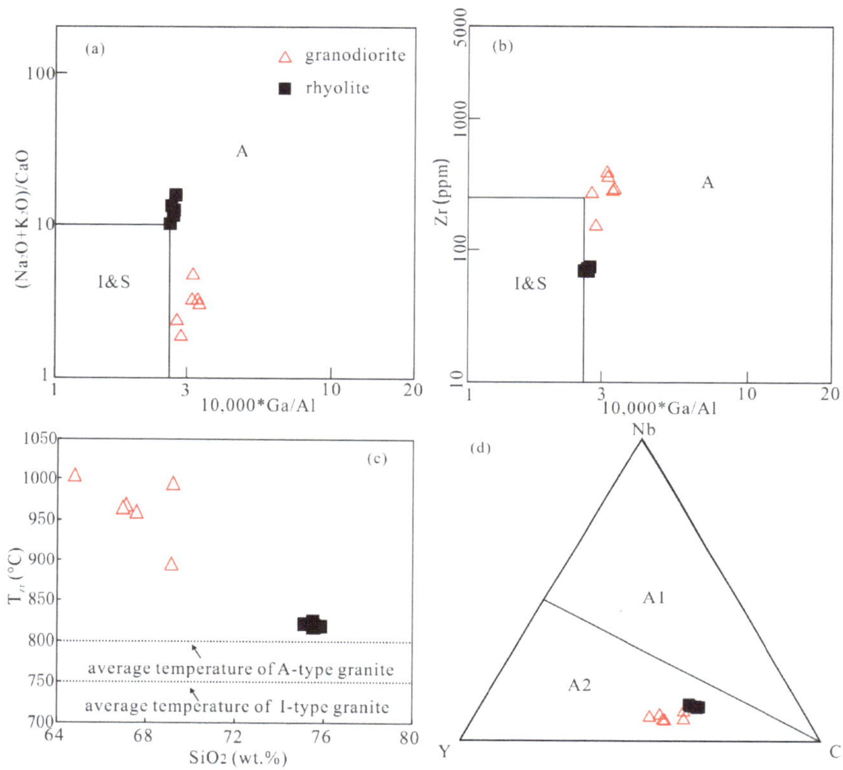

Figure 9. (**a**) ($Na_2O + K_2O$)/CaO and (**b**) Zr vs. 10,000*Ga/Al diagram [79]; (**c**) T_{Zr} (°C) vs. SiO_2 diagram; (**d**) Nb-Y-Ce diagram [77]. I: I-type felsic rock; S: S-type felsic rock; A: A-type felsic rock; A1: A1-type felsic rock; A2: A2-type felsic rock; T_{Zr} (°C): zircon saturation temperature; average temperature of A and I- type granite after [73,84].

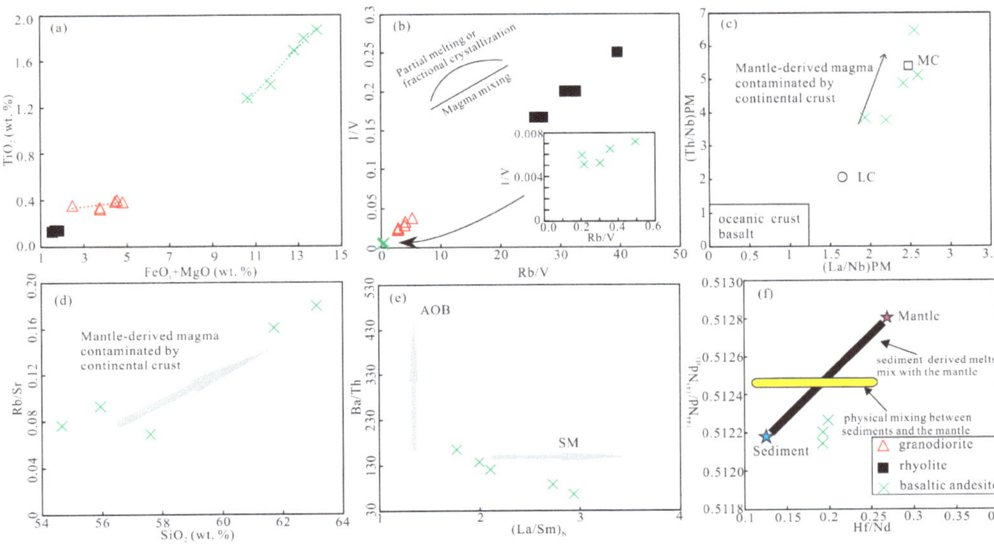

Figure 10. (**a**) TiO_2 vs. FeO_t + MgO diagram [82]; (**b**) 1/V vs. Rb/V diagram [85]; (**c**) $(Th/Nb)_{PM}$ vs. $(La/Nb)_{PM}$ diagram [86]; (**d**) Rb/Sr vs. SiO_2 diagram [87]; (**e**) Ba/Th vs. $(La/Sm)_N$ diagram [88]; (**f**) $^{144}Nd/^{143}Nd(i)$ vs. Hf/Nd diagram [88]. C: crust; M: mantle; Cm: crust-mantle mixed; LC: lower crust; MC: middle crust; AOB: altered oceanic basalt; SM: sediment metasomatism.

6.2.2. The Baishantang Bimodal Volcanic Rocks

The Baishantang rhyolites have the characteristics of being weakly peraluminous (A/CNK are 1.04~1.09) and having no aluminum-rich minerals, indicating that the rhyolites are I- or A-type granites rather than S-type granites. Further, characteristics such as high 10,000*Ga/Al and FeO_t/MgO values (2.60~2.73 and 3.53~5.73); low contents of MgO (0.26~0.37 wt.%) and CaO (0.53~0.82 wt.%); enrichment of Rb, Th and Pb; and depletion of Nb, Ta, Ti, Sr and Ba pertain to an A-type signature. Although the Zr + Nb + Ce + Y values (173.40~356.60 ppm and most of them are less than 350 ppm) are different from values of typical A-type granite (Zr + Nb + Ce + Y > 350 ppm), Baishantang rhyolites also have low Rb contents (154.00~163.50 ppm), high FeO_t contents (1.27~1.57 wt.%) and zircon saturation temperatures (817~824 °C) (Figure 9c), which are different from I-type but similar to A-type felsic rocks. Meanwhile, the almost constant Rb/Sr values of rhyolites (2.13~2.26) show that the Baishantang rhyolites do not belong to highly fractionated felsic rock [89]. In the discrimination diagrams of I-S-A-type granite (Figure 9a,b), all samples are plotted in the A-type field. Therefore, Baishantang rhyolites are A-type felsic rocks. All of these samples have relatively high Y/Nb (2.58~2.69), Ce/Nb (4.87~5.59) and Rb/Nb (11.88~13.69) ratios [71], plotting into the A2 field in the Nb-Y-Ce diagram (Figure 9d). In conclusion, the Baishantang rhyolites are defined as A2-type felsic rocks.

As A2-type felsic rocks, the Baishantang rhyolites have high SiO_2 and K_2O contents, have low MgO contents and $Mg^{\#}$ values, are enriched in LREEs, and are depleted in HFSEs, Sr and P. These characteristics indicate that rhyolites are difficult to form through the high differentiation of mantle-derived magma. The Y/Ho ratios (25.51~26.69) of the samples are close to those of chondrite, indicating that their fractional crystallization is limited [71]. Meanwhile, the exposed area of the rhyolite is much larger than that of the basaltic andesite. Therefore, the Baishantang rhyolite is probably derived from the partial melting of crustal or magma mixing rather than the crystallization differentiation of mantle-derived magma. A-type felsic rocks formed through magma mixing between crust- and mantle-derived magma generally exhibit a wide range of geochemical and isotopic compositions [83]. However, the samples have stable and high SiO_2 contents and low MgO contents and $Mg^{\#}$ values; the Cr (1.00~8.00 ppm) and Ni (1.20~4.40 ppm) contents are also obviously lower than those of mantle-derived magma (Cr = 300~500 ppm, Ni = 300~400 ppm) [60]. These features illustrate that the Baishantang rhyolite could not be formed through magma mixing. Although the samples show a positive correlation in the 1/V-Rb/V diagram (Figure 10b), the rhyolites have homogeneous and negative $\varepsilon_{Nd}(t)$ and $\varepsilon_{Hf}(t)$ values (Figure 8), and no special mineral phenomenon can be observed in the photomicrographs, implying that magma mixing is not the genesis of the rhyolites. The Ce/Pb (4.01~4.99), Lu/Yb (0.15~0.16), Nd/Th (1.94~2.14) and Nb/U (5.46~6.30) ratios of Baishantang rhyolites are different from those of mantle-derived magma but close to the average ratios of the crust (Ce/Pb < 15, 0.16 < Lu/Yb < 0.18, Nd/Th = 3, Nb/U = 9.7) [60,81], indicating that the rhyolite has crustal affinity. The Baishantang rhyolites have negative $\varepsilon_{Nd}(t)$ and $\varepsilon_{Hf}(t)$ values, and the two-stage model ages (T_{DM2}) of Nd and Hf are 1460~1433 Ma and 1411~1427 Ma, which are larger than the crystallization age of rhyolites. Studies have shown that if the $\varepsilon_{Nd}(t)$ ($\varepsilon_{Hf}(t)$) of intermediate-felsic rocks is negative, and the two-stage model age is much larger than its crystallization age, then the rocks originated from the partial melting of ancient crust, and the zircon age represents the reconstruction time of the crust [90–92]. There are Mesoproterozoic crustal materials in the SBOB, such as the Jiujing biotite plagioclase gneiss (1408 $\pm$ 4 Ma) [18] and the Gubaoquan paragneiss (detrital zircon age ~1.4 Ga) [93]. The two-stage model ages of the rhyolites are consistent with the previous results, suggesting that the Baishantang rhyolite is related to the partial melting of ancient continental crust [94]. The rhyolite samples show depletions in Eu, Ba, Nb, Ta, Sr and Ti, which could be due to the fractionation of plagioclase and Ti-bearing phases [26].

The Baishantang basaltic andesite has high contents of MgO, Al_2O_3, Na_2O, $^{T}Fe_2O_3$ and TiO_2, has low contents of P_2O_5 and K_2O, and belongs to the calc-alkaline series (Figure 6b). These high $Mg^{\#}$ values (44.08~46.30) are in contrast to the magma produced through partial melting of the basic lower crust (in which the $Mg^{\#}$ value is lower than 45). The Lu/Yb ratios (0.15) are similar to the mantle value (Lu/Yb = 0.15) [60], indicating the mantle's origin. However, these rocks display low Ce/Pb ratios (4.51~8.75) and Nd/Th ratios (3.33~6.95), which are close to the values of the continental crust (Ce/Pb < 15, Nd/Th $\approx$ 3) [95], and they are depleted in HFSEs (Nb, Ta, Ti) while being enriched in LILEs (Rb, Ba, Th, U and Pb), implying the influence of crustal material on magma. The Nb/Ta and high $(Th/Nb)_{PM}$ ratios are considered to be important parameters for crustal contamination [96,97]. The $(Th/Nb)_{PM}$ values of these samples are 3.78~6.47, larger than 1 [98], and the Nb/Ta values (14.15~14.56) for basaltic andesites are between those of the mantle (17.5 $\pm$ 2.5) and the crust (11.4) [60,92]. Meanwhile, all basaltic andesites have extremely heterogeneous isotopic signatures (Figure 8) and are plotted along the mantle-derived magma contaminated by the continental crust trend as shown in the $(La/Nb)_{PM}$–$(Th/Nb)_{PM}$ and Rb/Sr-SiO_2 diagrams (Figure 10c,d). All these features indicate that the magmas were modified by crustal materials during magma emplacement. Baishantang basaltic andesites have negative $\varepsilon_{Nd}(t)$ but positive $\varepsilon_{Hf}(t)$ values, which seems to indicate an opposite mantle source region (enriched mantle for $\varepsilon_{Nd}(t)$ but depleted mantle for εHf(t)). According to the formula for the crust Nd-Hf isotope ($\varepsilon_{Hf}(t) = 1.34 \times \varepsilon_{Nd}(t) + 2.82$) [98], the actual $\varepsilon_{Hf}(t)$ values are obviously larger than the values coupled with the whole rock (−0.93~2.28), and Nd-Hf isotopes are decoupled. Several mechanisms that can lead to Nd-Hf isotopic decoupling: (1) deeply subducted oceanic basalt [99]; (2) the lower crust that has undergone an ancient anataxis with residual phases of garnet and zircon may be an optional reservoir [88,100,101]; (3) mixing of components with different geochemical features [88,102–104]. The magmatic rock formed by the melting of basaltic crust has lower $Mg^{\#}$ ($Mg^{\#}$ < 45) values [105] and obvious differentiation between MREE and HREE during anatexis [102]. The features of Baishantang basaltic andesites, such as high $Mg^{\#}$ values and weak differentiation between MREEs and HREEs ($(Dy/Yb)_N$ = 1.28~1.40), contradict the idea that decoupling was caused by deeply subducted ocean basalt or the lower crust undergoing an ancient anatexis event.

Nd-Hf isotopic decoupling may result from the mixing of components with different geochemical features. As discussed above, the crustal material played an important role in the basaltic andesite formation (Figure 10c,d). Basaltic andesites have high $(La/Sm)_N$ ratios and low Ba/Th values (Figure 10e), which are thought to reflect sediment involvement. The influence of sediments on magma can be classified as sediment-derived melt mixing and physical mixing. The $(^{143}Nd/^{144}Nd)$–(Hf/Nd) diagram (Figure 10f) can distinguish between these two models [106]. All samples are distributed along an oblique line rather than a horizontal line (Figure 10f), indicating that sediment-derived melt mixing, rather than physical mixing between sediments, and mantle-derived magma play an important role in basaltic andesites formation. The partial melting of the mantle mixed with sediment-derived melts or bulk subducted sediments could generate magmas with Nd-Hf isotopic decoupling [88]. Therefore, the Nd-Hf isotopic decoupling of basaltic andesite is a result of the mixing of sediment-derived melts and mantle magma.

The Dy/Yb values of basaltic andesites are 1.92~2.09. These values are between the spinel lherzolite mantle melts (<1.5) and garnet lherzolite mantle melts (>2.5), indicating that the magma comes from the partial melting of the two [107]. The transformation depth of the spinel and garnet is 70~80 km [108], and it is speculated that the formation depth of basaltic andesite magma is about 70~80 km.

6.3. Tectonic Implications

The late Paleozoic tectonic setting of the SBOB and the closure time of the PAO are the focus of this research. Paleozoic magmatic rocks are widely distributed in the SBOB, which record the subduction, closure and post-collision evolution history of the PAO. The Xiaoquandong 395 Ma A-type granite and Hongliuhe 404 Ma A-type granite intruded into the Hongliuhe ophiolite in the SH unit, indicating that the HNY ocean closed before the early Devonian [44,109,110]. Magmatism in the SH and Shibanshan units after early Devonian indicate the evolution process of the back-arc basin represented by the HZ ophiolitic belt [44]. Late Devonian–early Carboniferous arc magmatism occurred in the Tanshanzidong, Huaniushan and Shuangfengshan in the SH unit [44], but it did not occur in the Shibanshan unit, indicating the northward subduction of the HZ ocean during this time [94].

Late Carboniferous to early Permian magmatic rocks are widely distributed in the Shibanshan and SH units, constituting the main magmatic events during the late Paleozoic. Qiaowan 304 Ma granite [27], Liuyuan and Shuwojing 285~275 Ma granite [12,13] and Yinwaxia 273 Ma rhyolite [111], and other acidic rocks, are the main types of late Carboniferous–early Permian magmatic rocks in the SBOB. These rocks have the geochemical characteristics of calc-alkaline to high K calc-alkaline and belong to highly differentiated I- or A-type felsic rocks. Except for these felsic rocks, the Podong, Cihai and Niujuanzi 282~280 Ma tholeiite-Calc-alkaline series diabase and other mafic rocks are also widely distributed in the SBOB [64–66]. These felsic and mafic rocks were formed in the extensional tectonic setting [12,13,64–66,111]. The A1-type felsic rocks were formed in the non-orogenic but the A2-type in syn-collision or post-collision tectonic environments [77]. The Zhangfangshan granodiorite and Baishantang rhyolite are A2-type felsic rocks, and all samples are plotted in the POG field in the Nb-Y and Ta-Yb diagrams (Figure 11a,b), suggesting the post-collision tectonic setting. Baishantang basaltic andesites have high Zr (180~258 ppm) concentrations and Zr/Y (5.37~6.53) ratios, similar to basalts formed in the within-plate extensional setting (Zr > 70 ppm and Zr/Y > 3) [112]. All basaltic andesite samples are plotted in the WPB field in the Zr/Y-Zr and Ti-Zr diagrams (Figure 11c,d), suggesting a within-plate extensional environment. The formation age and tectonic setting of the granodiorites and the bimodal volcanic rocks in this study are similar to late Carboniferous–early Permian magmatic rocks, showing that the late Carboniferous–early Permian magmatism in the SBOB were formed in the same tectonic environment. Meanwhile, intermediate-basic dikes in 282~280 Ma are widely distributed in the SBOB, indicating an extensional tectonic background caused by lithospheric delamination [25,26]. Based on the regional stratigraphic, the late Carboniferous to early middle Permian in the SBOB is a continuous transgressive sedimentary sequence, and there is no unconformity interface reflecting the continental crust of the SBOB, which was further stretched on the basis of a post-collisional extension in the late Carboniferous [63,113–115]. Therefore, combined with magmatism and stratigraphic characteristics in the SBOB, the HZ ocean has been closed, and the SBOB was in a post-collisional extensional setting during early Permian (Figure 12).

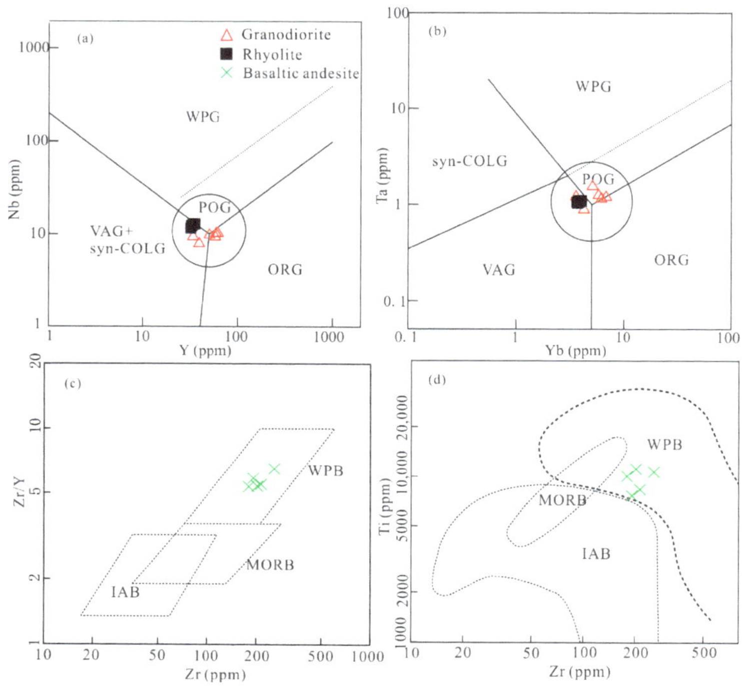

Figure 11. Tectonic setting discrimination diagrams for the Zhangfangshan granodiorite and the Baishantang bimodal volcanic rocks: (**a**) Nb vs. Y diagram [116]; (**b**) Ta vs. Yb diagram [116]; (**c**) Zr/Y vs. Zr diagram [117]; (**d**) Ti vs. Zr diagram [117]. MORB: mid-ocean ridge basalt; IAB: island-arc basalt; WPB: within-plate basalt; VAG: volcanic arc granite; POG: post-orogenic granite; ORG: ocean ridge granite; WPG: within-plate granite; syn-COLG: syn-collisional granite.

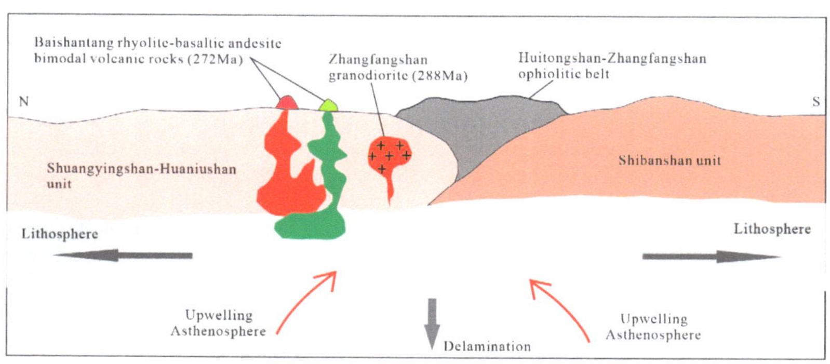

Figure 12. Schematic illustration of the proposed genetic model for the early Permian in the Southern Beishan Orogenic Belt.

7. Conclusions

A petrographic, geochronological, whole-rock major-oxides, trace-element and Nd-Hf isotopic study of the Zhangfangshan granodiorite and the Baishantang bimodal volcanic rocks in the SBOB led to the following conclusions.

1. The Zhangfangshan granodiorites have $^{206}Pb/^{238}U$ ages of 288.7 ± 1.7 Ma, and the Baishantang rhyolites have $^{206}Pb/^{238}U$ ages of 272.3 ± 1.6 Ma.
2. The Zhangfangshan granodiorite belong to the A2-type granite and are generated from magma mixing between mafic and felsic magmas.
3. The Baishantang rhyolites belong to the A2-type granite and are generated from the partial melting of crust. The Baishantang basaltic andesites belong to the calc-alkaline series, which originated from the partial melting of the mantle and mixing via sediment-derived melts.
4. The Zhangfangshan granodiorites and the Baishantang bimodal volcanic rocks were formed in the post-collision extensional setting. The SBOB was in the post-collision period, and the HZ ocean was closed before the early Permian.

Supplementary Materials: The following supporting information can be downloaded at: https://www.mdpi.com/article/10.3390/min13121468/s1, Table S1: Results of U-Pb dating using zircons from the Zhangfangshan granodiorites and the Baishantang bimodal volcanic rocks; Table S2: Whole-rock major (wt.%) and trace-element (ppm) analyses of the Zhangfangshan granodiorites and the Baishantang bimodal volcanic rocks; Table S3: Whole-rock Nd-Hf isotopic data of the Baishantang bimodal volcanic rocks.

Author Contributions: Methodology, L.W., X.Z., E.W. and J.Z.; Investigation, L.W., E.W., W.C., G.S., F.Z. and H.W.; Resources, L.W.; Data curation, L.W., E.W., W.C., G.S., F.Z. and H.W.; Writing—original draft, L.W.; Writing—review and editing, X.Z., J.Z. and J.W.; Project administration, J.W.; Funding acquisition, X.Z. and J.W. All authors have read and agreed to the published version of the manuscript.

Funding: This research was supported by the China Atomic Energy Authority (CAEA) through the Geological Disposal Program and the National Second Expedition to the Tibetan Plateau (2019QZKK0901).

Data Availability Statement: The original contributions presented in the study are included in the article/Supplementary Material.

Acknowledgments: We would like to thank Conghui Xiong and Xiaoli Yan for their valuable help with the analysis, Honggang Zhao for his help on the field. We sincerely thank the editors and all reviewers for their valuable feedback that we have used to improve the quality of our manuscript.

Conflicts of Interest: The authors declare no conflict of interest.

References

1. Briggs, S.M.; Yin, A.; Manning, C.E.; Chen, Z.L.; Wang, X.F.; Grove, M. Late Paleozoic tectonic history of the Ertix Fault in the Chinese Altai and its implications for the development of the Central Asian Orogenic System. *Geol. Soc. Am. Bull.* **2007**, *119*, 944–960. [CrossRef]
2. Briggs, S.M.; Yin, A.; Manning, C.E.; Chen, Z.L.; Wang, X.F. Tectonic development of the southern Chinese Altai Range as determined by structural geology, thermobarometry, 40Ar/39Ar thermochronology, and Th/Pb ion microprobe monazite geochronology. *Geol. Soc. Am. Bull.* **2009**, *121*, 1381–1393. [CrossRef]
3. Şengör, A.M.C.; Natal'in, B.A.; Burtman, V.S. Evolution of the Altaid tectonic collage and Palaeozoic crustal growth in Eurasia. *Nature* **1993**, *364*, 299–307. [CrossRef]
4. Jahn, B.M.; Wu, F.Y.; Chen, B. Granitoids of the Central Asian Orogenic Belt and continental growth in the Phanerozoic. *Geol. Soc. Am. Spec. Pap.* **2000**, *350*, 181–193.
5. Windley, B.F.; Alexeiev, D.; Xiao, W.; Kroner, A.; Badarch, G. Tectonic models for accretion of the Central Asian Orogenic Belt. *J. Geol. Soc. Lond.* **2007**, *164*, 31–47. [CrossRef]
6. Xiao, W.J.; Pirajno, F.; Seltmann, R. Geodynamicsand metallogeny of the Altaid orogen. *J. Asian Earth Sci.* **2008**, *32*, 77–81. [CrossRef]
7. Khain, E.V.; Bibikova, E.V.; Kröner, A.; Zhuravlev, D.Z.; Sklyarov, E.V.; Fedotova, A.A.; Kravchenko-Berezhnoy, I.R. The most ancient ophiolite of the Central Asian fold belt: U-Pb and Pb-Pb zircon ages for the Dunzhugur Complex, Eastern Sayan, Siberia, and geodynamic implications. *Earth Planet. Sci. Lett.* **2002**, *199*, 311–325. [CrossRef]

8. Chen, X.H.; Dong, S.W.; Shi, W.; Ding, W.C.; Zhang, Y.P.; Li, B.; Shao, Z.G.; Wang, Y. Construction of the continental Asia in Phanerozoic: A review. *Acta Geol. Sin.* **2022**, *96*, 26–51. (In Chinese with English abstract) [CrossRef]
9. Xiao, W.J.; Song, D.F.; Windley, B.F.; Li, J.L.; Han, C.M.; Wan, B.; Zhang, J.E.; Ao, S.J.; Zhang, Z.Y. Accretionary processes and metallogenies of the Central Asian Orogenic Belt. Advances and perspectives. Science China. *Earth Sci.* **2020**, *63*, 329–361. [CrossRef]
10. Huang, H.; Wang, T.; Tong, Y.; Qin, Q.; Ma, X.X.; Yin, J.Y. Rejuvenation of ancient micro-continents during accretionary orogenesis: Insights from the Yili Block and adjacent regions of the SW Central Asian Orogenic Belt. *Earth-Sci. Rev.* **2020**, *208*, 103255. [CrossRef]
11. Şengör, A.M.C.; Natal'in, B.A.; Sunal, G.; Van der Voo, R. The tectonics of the Altaids: Crustal growth during the construction of the continental lithosphere of central asia between ~750 and ~130 Ma ago. *Annu. Rev. Earth Planet. Sci.* **2018**, *46*, 439–494. [CrossRef]
12. Li, S.; Wilde, S.A.; Wang, T. Early Permian post-collisional high-K granitoids from Liuyuan area in southern Beishan orogen, NW China: Petrogenesis and tectonic implications. *Lithos* **2013**, *179*, 99–119. [CrossRef]
13. Zhang, W.; Pease, V.; Meng, Q.P.; Zheng, R.G.; Thomsen, T.B.; Wohlgemuth-Ueberwasser, C.; Wu, T.R. Timing, petrogenesis, and setting of granites from the southern Beishan late Palaeozoic granitic belt, Northwest China and implications for their tectonic evolution. *Int. Geol. Rev.* **2015**, *57*, 1975–1991. [CrossRef]
14. Zheng, R.G.; Li, J.Y.; Zhang, J.; Xiao, W.J. A prolonged subduction-accretion in the southern Central Asian Orogenic Belt: Insights from anatomy and tectonic affinity for the Beishan complex. *Gondwana Res.* **2021**, *95*, 88–112. [CrossRef]
15. Xiao, W.J.; Mao, Q.G.; Windley, B.F.; Han, C.M.; Qu, J.F.; Zhang, J.E.; Ao, S.J.; Guo, Q.Q.; Cleven, N.R.; Lin, S.F.; et al. Paleozoic multiple accretionary and collisional processes of the Beishan orogenic collage. *Am. J. Sci.* **2010**, *310*, 1553–1594. [CrossRef]
16. Li, J.; Wu, C.; Chen, X.H.; Zuza, A.V.; Haproff, P.J.; Yin, A.; Shao, Z.G. Tectonic evolution of the Beishan orogen in central Asia: Subduction, accretion, and continent-continent collision during the closure of the Paleo-Asian Ocean. *Geol. Soc. Am. Bull.* **2023**, *135*, 5–6. [CrossRef]
17. Yuan, Y.; Zong, K.Q.; Cawood, P.A.; Cheng, H.; Yu, Y.Y.; Guo, J.L.; Liu, Y.S.; Hu, Z.C.; Zhang, W.; Li, M. Implication of Mesoproterozoic (~1.4 Ga) magmatism within microcontinents along the southern Central Asian Orogenic Belt. *Precambrian Res.* **2019**, *327*, 314–326. [CrossRef]
18. He, Z.Y.; Sun, L.X.; Mao, L.J.; Zong, K.Q.; Zhang, Z.M. Zircon U-Pb and Hf isotopic study of gneiss and granodiorite from the southern Beishan orogenic collage: Mesoproterozoic magmatism and crustal growth. *Chin. Sci. Bull.* **2015**, *60*, 389–399. (In Chinese with English abstract) [CrossRef]
19. Wang, B.R.; Yang, X.S.; Li, S.C.; Teng, C.; Yang, X.J.; Huang, F.Y.; Zhang, X.F.; Cao, J.; Zhou, Y.; Zhang, H.C.; et al. Geochronology, geochemistry, and tectonic implications of early Neoproterozoic granitic rocks from the eastern Beishan Orogenic Belt, southern Central Asian Orogenic Belt. *Precambrian Res.* **2021**, *352*, 106016. [CrossRef]
20. Soldner, J.; Yuan, C.; Schulmann, K.; Štípská, P.; Jiang, Y.; Zhang, Y.; Wang, X. Grenville an evolution of the Beishan Orogen, NW China: Implications for development of an active Rodinian margin. *Geol. Soc. Am. Bull.* **2020**, *132*, 1657–1680. [CrossRef]
21. Lv, C.L.; Yang, F.Q.; Luo, J.S.; Ma, W.F.; Liu, W.X.; Shi, W.M. Isotopic age of rock mass in Gongpoquan mining area in Beishan of Gansu and its geological significance. *Miner. Resour. Geol.* **2021**, *35*, 83–89. (In Chinese with English abstract)
22. Song, D.F.; Xiao, W.J.; Han, C.M.; Tian, Z.H. Geochronological and geochemical study of gneiss schist complexes and associated granitoids, Beishan Orogen, southern Altaids. *Int. Geol. Rev.* **2013**, *55*, 1705–1727. [CrossRef]
23. Yuan, Y.; Zong, K.Q.; He, Z.Y.; Klemd, R.; Jiang, H.Y.; Zhang, W.; Liu, Y.S.; Hu, Z.C.; Zhang, Z.M. Geochemical evidence for Paleozoic crustal growth and tectonic conversion in the Northern Beishan Orogenic Belt, southern Central Asian Orogenic Belt. *Lithos* **2018**, *302–303*, 189–202. [CrossRef]
24. Zhao, K.Q.; Ma, S.M.; Xi, M.J.; Yang, J.Z.; Cai, Y.W.; Gong, J.J. The LA-ICP-MS Zircon U-Pb ages and geochemical characteristics of Late Paleozoic intermediate-acidic intrusive complexes in Shibanjing area, Beishan Mountains, Inner Mongolia, and their geological significance. *Dizhi Lunping* **2020**, *66*, 69–87. (In Chinese with English abstract)
25. Zheng, R.G.; Li, J.Y.; Zhang, J.; Xiao, W.J.; Wang, Q.J. Permian oceanic slab subduction in the southmost of Central Asian Orogenic Belt: Evidence from adakite and high-Mg diorite in the southern Beishan. *Lithos* **2020**, *358*, 105406. [CrossRef]
26. Zhang, Y.Y.; Yuan, C.; Sun, M.; Long, X.P.; Xia, X.P.; Wang, X.Y.; Huang, Z.Y. Permian doleritic dikes in the Beishan Orogenic Belt, NW China: Asthenosphere–lithosphere interaction in response to slab break-off. *Lithos* **2015**, *233*, 174–192. [CrossRef]
27. Zhang, W.; Wu, T.R.; Zheng, R.G.; Feng, J.C.; Luo, H.L.; He, Y.K.; Xu, C. Post-collisional Southeastern Beishan granites: Geochemistry, geochronology, Sr–Nd–Hf isotopes and their implications for tectonic evolution. *J. Asian Earth Sci.* **2012**, *58*, 51–63. [CrossRef]
28. Su, B.X.; Qin, K.Z.; Patrick, A.S.; Liu, P.P.; Tang, D.M.; Malaviarachchi, S.P.K.; Xiao, Q.H.; Sun, H.; Dai, Y.C.; Hu, Y. Geochemistry and geochronology of acidic rocks in the Beishan region, NW China: Petrogenesis and tectonic implications. *J. Asian Earth Sci.* **2011**, *41*, 31–43. [CrossRef]
29. Chen, S.; Guo, Z.J.; Qi, J.F.; Zhang, Y.Y.; Georgia, P.; David, J.W.P. Early Permian volcano-sedimentary successions, Beishan, NW China: Pepe rites demonstrate an evolving rift basin. *J. Volcanol. Geotherm. Res.* **2016**, *309*, 31–44. [CrossRef]
30. Zuo, G.C.; Zhang, S.L.; He, G.Q.; Zhang, Y. Plate tectonic characteristics during the early Paleozoic in Beishan near the Sino-Mongolian border region, China. *Tectonophysics* **1991**, *188*, 385–392. [CrossRef]

31. Xiao, W.J.; Windley, B.F.; Allen, M.B.; Han, C.M. Paleozoic multiple accretionary and collisional tectonics of the Chinese Tianshan orogenic collage. *Gondwana Res.* **2013**, *23*, 1316–1341. [CrossRef]
32. Wang, Y.H.; Zhang, F.F.; Li, B.C.; Xue, C.J.; Liu, J.J.; Zhao, Y.; Zhang, W. Geology and genesis of the Cihai mafic intrusions in Beishan Terrane, Xinjiang, Northwest China: Implication for iron mineralization and tectonic setting. *Ore Geol. Rev.* **2020**, *121*, 103573. [CrossRef]
33. Yu, S.; Zhao, B.B.; Jia, X.; Zhai, X.W.; Ren, J.D. Geochemistry, Geochronology Characteristics and Tectonic Significance of Yitiaoshan Diorite in the Southern Margin of Beishan Orogenic Belt. *Northwest. Geol.* **2022**, *55*, 267–279. (In Chinese with English abstract)
34. Luo, X.Q. Geochemistry and Geochronology of Wohushan Early Permian Syenogranite in the Southern Belt of the Beishan mountains. *Gansu Geol.* **2022**, *31*, 7–16, (In Chinese with English abstract).
35. Guo, X.G.; Chen, S.Y.; Gou, R.; Liu, X.; Wang, Q.; Pu, Q.L. Geochemistry, Chronology and Genesis of Marine Basalts in HouhongquanArea, Beishan, Gansu and Inner Mongolia. *Earth Sci.* **2021**, *46*, 3945–3964. (In Chinese with English abstract)
36. Zhang, W.; Feng, J.C.; Zheng, R.G.; Wu, T.R.; Luo, H.L.; He, Y.K.; Jing, X. LA-ICP MS zircon U–Pb ages of the granites from the south of Yin'aoxia and their tectonic significances. *Acta Petrol. Sin.* **2011**, *27*, 1649–16661. (In Chinese with English abstract)
37. Wang, J.T.; Dong, Y.P.; Zeng, Z.C.; Yang, Z.; Sun, S.S.; Zhang, F.F.; Zhou, B.; Sun, J.P. Geochronology, geochemistry and geological significance of the Huangcaotan pluton in the southern Beishan Orogenic Belt. *Geoscience* **2016**, *30*, 937–949. (In Chinese with English abstract)
38. Ao, S.J.; Xiao, W.J.; Han, C.M.; Li, X.H.; Qu, J.F.; Zhang, J.E.; Guo, Q.Q.; Tian, Z.H. Cambrian to early Silurian ophiolite and accretionary processes in the Beishan collage, NW China: Implications for the architecture of the Southern Altaids. *Geol. Mag.* **2012**, *149*, 606–625. [CrossRef]
39. Tian, Z.; Xiao, W.; Windley, B.F.; Lin, L.N.; Han, C.; Zhang, J.E.; Wan, B.; Ao, S.; Song, D.; Feng, J. Structure, age, and tectonic development of the Huoshishan-Niujuanzi ophiolitic mélange, Beishan, southernmost Altaids. *Gondwana Res.* **2014**, *25*, 820–841. [CrossRef]
40. Su, B.X.; Qin, K.Z.; Santosh, M.; Sun, H.; Tang, D. The Early Permian mafic-ultramafic complexes in the Beishan Terrane, NW China: Alaskan-type intrusives or rift cumulates? *J. Asian Earth Sci.* **2013**, *66*, 175–187. [CrossRef]
41. Hou, Q.Y.; Wang, Z.; Liu, J.B.; Wang, J.; Li, D.P. Geochemistry Characteristics and SHRIMP Dating of Yueyashan Ophiolite in Beishan Orogen. *Geoscience* **2012**, *26*, 1008–1018. (In Chinese with English abstract)
42. Zuo, G.C.; Zhang, S.L.; He, G.Q.; Zhang, Y. Early Paleozoic Plate Tectonics in Beishan Area. *Sci. Geol. Sin.* **1990**, *25*, 305–314+411. (In Chinese with English abstract)
43. Niu, Y.Z.; Shi, G.R.; Wang, J.Q.; Liu, C.Y.; Zhou, J.L.; Lu, J.C.; Song, B.; Xu, W. The closing of the southern branch of the Paleo-Asian Ocean: Constraints from sedimentary records in the southern Beishan Region of the Central Asian Orogenic Belt, NW China. *Mar. Pet. Geol.* **2021**, *124*, 104791. [CrossRef]
44. Wang, H.T. Tectono-Magmatism and Its Geological Significance in the Beishan Area of the Southern Part of the Central Asian Orogenic Belt. Ph.D. Thesis, Lanzhou University, Lanzhou, China, 2019. (In Chinese with English abstract)
45. BGGP. *Hongliudaquan Regional Geological Survey Report (1:200,000)*; Gansu Bureau of Geology: Gansu, China, 1971. (In Chinese with English abstract)
46. BGGP. *Shibanquan Regional Geological Survey Report (1:50,000)*; Gansu Bureau of Geology: Gansu, China, 2013. (In Chinese with English abstract)
47. Song, B.; Zhang, Y.H.; Wan, Y.S.; Jian, P. Mount Making and Procedure of the SHRIMP Dating. *Geol. Rev.* **2002**, *48*, 26–30. (In Chinese with English abstract)
48. Wiedenbeck, M.; Hanchar, J.M.; Peck, W.H.; Sylvester, P.; Vley, J.; Whitehouse, M.; Kronz, A.; Morishita, Y.; Nasdala, L.; Fiebig, J.; et al. Further characterisation of the 91500 zircon crystal. *Geostand. Geoanal. Res.* **2004**, *28*, 9–39. [CrossRef]
49. Pearce, N.J.G.; Perkins, W.T.; Westgate, J.A.; Gorton, M.P.; Chenery, S.P. A Compilation of New and Published Major and Trace Element Data for NIST SRM 610 and NIST SRM 612 Glass Reference Materials. *Geostand. Newsl.* **1997**, *21*, 115–144. [CrossRef]
50. Jackson, S.E.; Pearson, N.J.; Griffin, W.L.; Belousova, E.A. The application of laser ablation-inductively coupled plasma-mass spectrometry to in situ U–Pb zircon geochronology. *Chem. Geol.* **2004**, *211*, 47–69. [CrossRef]
51. Vermeesch, P. IsoplotR: A free and open toolbox for geochronology. *Geosci. Front.* **2018**, *9*, 1479–1493. [CrossRef]
52. Yuan, H.L.; Gao, S.; Liu, X.M.; Li, H.M.; Gunther, D.; Wu, F.Y. Accurate U–Pb age and trace element determinations of zircon by laser ablation-inductively coupled plasma-mass spectrometry. *Geostand. Geoanal. Res.* **2004**, *28*, 353–370. [CrossRef]
53. Li, X.H.; Liu, Y.; Tu, X.L.; Hu, G.Q.; Zeng, W. Precise determination of chemical compositions in silicate rocks using ICP-AES and ICP-MS: A comparative study of sample digestion techniques of alkali fusion and acid dissolution. *Geochimica* **2002**, *31*, 289–294.
54. Gao, J.J.; Liu, J.H.; Li, X.G.; Yan, Q.S.; Wang, X.J.; Wang, H.M. The determination of 52 elements in marine geological samples by an inductively coupled plasma optical emission spectrometry and an inductively coupled plasma mass spectrometry with a high-pressure closed digestion method. *Acta Oceanol. Sin.* **2017**, *36*, 113–121. [CrossRef]
55. Bao, Z.; Zong, C.; Fang, L.; Yuan, H.; Chen, K.; Dai, M. Determination of Hf-Sr-Nd isotopic ratios by MC-ICP-MS using rapid acid digestion after flux-free fusion in geological materials. *Acta Geochim.* **2018**, *37*, 244–256. [CrossRef]
56. Wu, Y.B.; Zheng, Y.F. Genetic mineralogy of zircon and its constraints on U-Pb dating interpretation. *Chin. Sci. Bull.* **2004**, *49*, 1589–1604, (In Chinese with English abstract). [CrossRef]

57. Winchester, J.A.; Floyd, P.A. Geochemical discrimination of different magma series and their differentiation products using immobile elements. *Chem. Geol.* **1977**, *20*, 325–343. [CrossRef]
58. Rickwood, P.C. Boundary lines within petrologic diagrams which use oxides of major and minor elements. *Lithos* **1989**, *22*, 247–263. [CrossRef]
59. Maniar, P.D.; Piccoli, P.M. Tectonic discrimination of granitoids. *Geol. Soc. Am. Bull.* **1989**, *101*, 35–643. [CrossRef]
60. Sun, W.; McDonough, W. Chemical and isotopic systematics of oceanic basalts: Implications for mantle composition and processes. *Geol. Soc. Lond. Spec. Publ.* **1989**, *42*, 313–345. [CrossRef]
61. Yu, J.Y.; Li, X.M.; Wang, G.Q.; Wu, P.; Yan, Q.J. Zircon U-Pb ages of Huitongshan and Zhangfangshan ophiolite in Beishan of Gansu-Inner Mongolia border area and their significance. *Geol. Bull. China* **2012**, *31*, 2038–2045. (In Chinese with English abstract)
62. Yang, S.H.; Miao, L.C.; Zhang, F.C.; Meng, Q.R.; Zhu, M.S.; Munkhtsengel, B.; Chinmedtseren, A. Zircon age and geochemistry of the Tost bimodal volcanic rocks: Constraints on the Early Carboniferous tectonic evolution of the South Mongolia. *J. Asian Earth Sci.* **2016**, *120*, 29–42. [CrossRef]
63. Niu, Y.Z.; Lu, J.C.; Liu, C.Y.; Xu, W.; Shi, J.Z.; Song, B. Geochronology and distribution of the Upper Carboniferous—Lower Permian Ganquan Formation in the Beishan region, northwestern China and its tectonic implication. *Geol. Rev.* **2018**, *64*, 806–827. (In Chinese with English abstract)
64. Zheng, R.G.; Wu, T.R.; Zhang, W.; Meng, Q.P.; Zhang, Z.Y. Geochronology and geochemistry of late Paleozoic magmatic rocks in the Yinwaxia area, Beishan: Implications for rift magmatism in the southern Central Asian Orogenic Belt. *J. Asian Earth Sci.* **2014**, *91*, 39–55. [CrossRef]
65. Xue, S.C.; Li, C.; Qin, K.Z.; Tang, D.M. A non-plume model for the Permian protracted (266–286 Ma) basaltic magmatism in the Beishan–Tianshan region, Xinjiang, Western China. *Lithos* **2016**, *256–257*, 243–249. [CrossRef]
66. Chen, B.; Qin, K.Z.; Tang, D.M.; Mao, Y.J.; Feng, H.Y.; Xue, S.C.; Yao, Z.S. Lithological, chronological and geochemical characteristics of Cihai iron deposit, Eastern Xinjiang: Constraints on genesis of mafic-ultramafic and syenite intrusions and mineralization. *Acta Petrol. Sin.* **2015**, *31*, 2156–2174.
67. Whalen, J.B. Geochemistry of an island–arc plutonic suite: The Uasilau-Yau Yau intrusive complex, New Britain, PNG. *J. Petrol.* **1985**, *26*, 603–632. [CrossRef]
68. Chappell, B.W.; White, A.J.R. I-and S-type granites in the Lachlan fold belt. Earth Environ. *Earth Environ. Sci. Trans. R. Soc. Edinb.* **1992**, *83*, 1–26. [CrossRef]
69. Eby, G.N. The A-type granitoids: A review of their occurrence and chemical characteristics and speculations on their petrogenesis. *Lithos* **1990**, *26*, 115–134. [CrossRef]
70. Whalen, J.B.; Currie, K.L.; Chappell, B.W. A-type granites: Geochemical characteristics, discrimination and petrogenesis. *Contrib. Mineral. Petrol.* **1987**, *95*, 407–419. [CrossRef]
71. Quan, Y.K.; Mu, M.S.; Yang, D.B.; Yan, X.Y.; Wang, A.Q.; Hao, L.R.; Wang, F. Geochronology and in-situ apatite geochemistry of late Paleoproterozoic A-type granites in the Jiao-Liao-Ji Belt, North China Craton: Implications for petrogenesis and tectonic evolution. *Geochemistry* **2023**, 126009. [CrossRef]
72. King, P.L.; White, A.J.R.; Chappell, B.W.; Allen, C.M. Characterization and origin of aluminous A-type granites from the Lachlan Fold Belt, southeastern Australia. *J. Petrol.* **1997**, *3*, 371–391. [CrossRef]
73. Liu, Z.Y.; Yan, D.D.; Zhao, H.; Gong, T.T.; Xiong, F.H. Petrogenesis and Geological implications of Early Paleozoic Gouli granite in East Kunlun: Constraints from U-Pb Geochronology and Petrogeochemistry. *Mineral. Petrol.* **2022**, *42*, 7–19. (In Chinese with English abstract)
74. Wang, Q.; Zhao, Z.H.; Xiong, X.L. The ascertainment of Late-Yanshanian A-type granite in Tongbai-Dabie orogenic belt. *Acta Petrol. Mineral.* **2000**, *19*, 297–306+315. (In Chinese with English abstract)
75. Wu, F.Y.; Liu, X.C.; Ji, W.Q.; Wang, J.M.; Yang, L. Highly fractionated granites: Recognition and research. *Sci. China Earth Sci.* **2017**, *60*, 1201–1219. (In Chinese with English abstract) [CrossRef]
76. Watson, E.B.; Harrison, T.M. Zircon saturation revisited: Temperature and composition effects in a variety of crustal magma types. *Earth Planet. Sci. Lett.* **1983**, *64*, 295–304. [CrossRef]
77. Eby, G.N. Chemical subdivision of the A-type granitoids: Petrogenetic and tectonic implications. *Geology* **1992**, *20*, 641–644. [CrossRef]
78. Turner, S.P.; Foden, J.D.; Morrison, R.S. Derivation of some A-type magmas by fractionation of basaltic magma: An example from the Padthaway Ridge, South Australia. *Lithos* **1992**, *28*, 151–179. [CrossRef]
79. Mushkin, A.; Navon, O.; Halicz, L.; Hartmann, G.; Stein, M. The petrogenesis of A-type magmas from the Amram Massif, southern Israel. *J. Petrol.* **2003**, *44*, 815–832. [CrossRef]
80. Mu, M.S.; Yang, D.B.; Yang, H.T.; Wang, A.Q.; Hao, L.R.; Wang, F. Petrogenesis of late Paleoproterozoic post-collisional magmatism in southern North China Craton: Insights from geochemistry and Nd-Hf isotopic compositions of A-type granites. *Precambrian Res.* **2022**, *383*, 106887. [CrossRef]
81. Chapman, J.B.; Ducea, M.N.; DeCelles, P.G.; Profeta, I. Tracking changes in crustal thickness during orogenic evolution with Sr/Y: An example from the North American Cordillera. *Geology* **2015**, *43*, 919–922. [CrossRef]
82. Zhu, Y.; Lai, S.; Qin, J.; Zhu, R.; Zhao, S.W.; Liu, M.; Zhang, F.Y.; Zhang, Z.Z.; Yang, H. Magma mixing for the genesis of Neoproterozoic Mopanshan granitoids in the western Yangtze Block, South China. *J. Asian Earth Sci.* **2022**, *231*, 105227. [CrossRef]

83. Griffin, W.L.; Wang, X.; Jackson, S.E.; Pearson, N.J.; O'Reilly, S.Y.; Xu, X.S.; Zhou, X.M. Zircon chemistry and magma mixing, SE China: In-situ analysis of Hf isotopes, Tonglu and Pingtan igneous complexes. *Lithos* **2002**, *61*, 237–269. [CrossRef]
84. Wang, E.T.; Zhai, X.W.; Chen, W.F.; Ma, Z.; Wu, L.; Guo, Z.A.; Wang, Y.; Song, G.R.; Wang, J.R. Late Paleozoictec tonics of Southern Central Asian orogenic belt: Evidence from magmatic rocks in the northern Alxa, Northwest China. *Front. Earth Sci.* **2022**, *10*, 1045122. [CrossRef]
85. Hofmann, A.W.; Jochum, K.P.; Seufert, M.; White, W.M. Nb and Pb in oceanic basalts: New constraints on mantle evolution. *Earth Planet. Sci. Lett.* **1986**, *79*, 33–45. [CrossRef]
86. Saunders, A.D.; Storey, M.; Kent, R.W.; Norry, M.J. Consequences of plume lithosphere interactions. *Geol. Soc. Lond. Spec. Publ.* **1992**, *68*, 41–60. [CrossRef]
87. Dong, M.; Lang, X.H.; Deng, Y.L.; Wang, X.H. Geochronology and Geochemistry Implications for Early Eocene Rongma Gabbros in Southern Margin of the Lhasa Terrane, Tibet. *Earth Sci.* **2022**, *47*, 1349–1370.
88. Li, H.; Li, Y.J.; Xu, X.Y.; Yang, G.X.; Wang, Z.P.; Xu, Q.; Wang, L.J. Petrogenesis and tectonic implications of alkali basalts in Kalamaili area, east Junggar, Xinjiang (NW China): Constraints from petrology, geochronology and geochemistry. *Acta Geol. Sin.* **2021**, *95*, 3282–3300.
89. Nowell, G.; Pearson, D.; Bell, D.; Carlson, R.; Smith, C.; Kempton, P.; Noble, S. Hf isotope systematics of kimberlites and their megacrysts: New constraints on their source regions. *J. Petrol.* **2004**, *45*, 1583–1612. [CrossRef]
90. Schmitz, M.D.; Vervoort, J.D.; Bowring, S.A.; Patchett, P.J. Decoupling of the Lu- Hf and Sm-Nd isotope systems during the evolution of granulitic lower crust beneath Southern Africa. *Geology* **2004**, *32*, 405. [CrossRef]
91. Sun, P.; Wang, Q.; Hao, L.L.; Dan, W.; Qu, Q.; Jiang, Z.Q.; Tang, G.J. A mélange contribution to arc magmas recorded by Nd–Hf isotopic decoupling: An example from the southern Qiangtang Block, central Tibet. *J. Asian Earth Sci.* **2021**, *221*, 104931. [CrossRef]
92. Wu, F.Y.; Li, X.H.; Zheng, Y.F.; Gao, S. Lu-Hf isotopic systematics and their application in petrology. *Acta Petrol. Sin.* **2007**, *23*, 185–220.
93. Yu, J.C.; Mo, X.X.; Yu, X.H.; Zhu, D.C.; Li, Y.C.; Huang, X.F. Petrogenesis and geological implications of the Late Triassic potassic-ultrapotassic rocks in Changdu Block, northern segment of the Sanjiang area. *Acta Petrol. Sin.* **2014**, *30*, 3334–3344. (In Chinese with English abstract)
94. Marini, J.C.; Chauvel, C.; Maury, R.C. Hf Isotope Compositions of Northern Luzon Arc Lavas Suggest Involvement of Pelagic Sediments in Their Source. *Contrib. Mineral. Petrol.* **2005**, *149*, 216–232. [CrossRef]
95. Yogodzinski, G.; Vervoort, J.; Brown, S.T.; Gerseny, M. Subduction Controls of Hf and Nd Isotopes in Lavas of the Aleutian Island Arc. *Earth Planet. Sci. Lett.* **2010**, *300*, 226–238. [CrossRef]
96. Liu, X.C.; Chen, B.L.; Jahn, B.M.; Wu, G.G.; Liu, Y.S. Early Paleozoic (ca. 465 Ma) eclogites from Beishan (NW China) and their bearing on the tectonic evolution of the southern Central Asian Orogenic Belt. *J. Asian Earth Sci.* **2010**, *42*, 715–731. [CrossRef]
97. Li, R.W.; Zhang, X.; Shi, Q.; Chen, W.F.; An, Y.; Huang, Y.S.; Liu, Y.X.; Wang, J.R. Geochemistry and Nd–Hf isotopes of the Early Permian volcanic rocks in Hangwula of northern Alxa area and their tectonic significance. *Geol. Bull. China* **2020**, *39*, 647–658. (In Chinese with English abstract)
98. Nielsen, S.G.; Marschall, H.R. Geochemical evidence for me'lange melting in global arcs. *Sci. Adv.* **2017**, *3*, e1602402. [CrossRef]
99. Zhang, G.Z.; Xin, H.T.; Duan, L.F.; Niu, W.C.; Tian, J.; Zhang, Y. Geochemical characteristics and tectonic implications of the end Early Permian high magnesium gabbro from northern Beishan orogenic belt, Inner Mongolia. *Earth Sci.* **2020**, *47*, 3258–3269. (In Chinese with English abstract)
100. McKenzie, D.; O'Nions, R.K. Partial Melt Distribution From Inversion of Rare Earth Element Concentrations. *J. Petrol.* **1991**, *32*, 1021–1091. [CrossRef]
101. Miller, C.F.; McDowell, S.M.; Mapes, R.W. Hot and cold granites? Implications of zircon saturation temperatures and preservation inheritance. *Geology* **2003**, *31*, 529–532. [CrossRef]
102. Schiano, P.; Monzier, M.; Eissen, J.P.; Martin, H.; Koga, K.T. Simple mixing as the major control of the evolution of volcanic suites in the Ecuadorian Andes. *Contrib. Mineral. Petrol.* **2010**, *160*, 297–312. [CrossRef]
103. Zhu, D.C.; Mo, X.X.; Pan, G.T.; Zhao, Z.D.; Dong, G.C.; Shi, Y.R.; Liao, Z.L.; Wang, L.Q.; Zhou, C.Y. Petrogenesis of the Earliest Early Cretaceous Mafic Rocks from the Cona Area of the Eastern Tethyan Himalaya in South Tibet: Interaction between the Incubating Kerguelen Plume and the Eastern Greater India Lithosphere? *Lithos* **2008**, *100*, 147–173. [CrossRef]
104. Xie, F.Q.; Wu, J.H.; Sun, Y.H.; Wang, L.D.; Wu, J.Z.; Jia, W.J. Permian to Triassic tectonic evolution of the Alxa Tectonic Belt, NW China: Constraints from petrogenesis and geochronology of felsic intrusions. *Lithos* **2021**, *384*, 105980. [CrossRef]
105. Rudnick, R.L.; Gao, S. Composition of the Continental Crust. *Treatise Geochem.* **2003**, 3, 1–64.
106. Condie, K.C. Episodic continental growth models: After thoughts and extensions. *Tectonophysics* **2000**, *322*, 153–162. [CrossRef]
107. Collins, W.J.; Belousova, E.A.; Kemp, A.I.S.; Murphy, J.B. Twocontrasting Phanerozoic orogenic systems revealed by hafnium isotope data. *Nat. Geosci.* **2011**, *4*, 333–337. [CrossRef]
108. Bu, T.; Wang, G.Q.; Huang, B.T.; Dong, Z.C.; Guo, L. Neoproterozoic A-type granites in northern Beishan Orogenic Belt: Early response of the Rodinia supercontinent break-up. *Acta Petrol. Sin.* **2022**, *38*, 2988–3002. [CrossRef]
109. Wang, H.T.; Ren, W.X.; Zhao, Q.X.; Liu, Z.R.; Wang, T.C.; Zhao, H.X. Geochemical Characteristics and Tectonic Significance of A-typeGranite in the South Margin of Zhaobishan, Beishan Area. *Northwest. Geol.* **2016**, *49*, 39–49. (In Chinese with English abstract)

110. Zhang, Y.Y.; Guo, Z.J. Accurate constraint on formation and emplacement age of Hongliuhe ophiolite, boundary region between Xinjiang and Gansu Provinces and its tectonic implications. *Acta Petrol. Sin.* **2008**, *24*, 803–809. (In Chinese with English abstract)
111. Zheng, R.G.; Wang, Y.P.; Zhang, Z.Y.; Zhang, W.; Meng, Q.P.; Wu, T.R. Geochronology and Geochemistry of Yinwaxia Acidic Volcanic Rocks in the Southern Beishan: New Evidence for Permian Continental Rifting. *Geotecton. Metallog.* **2016**, *40*, 1031–1048. (In Chinese with English abstract) [CrossRef]
112. Xia, L.Q.; Xia, Z.C.; Xu, X.Y.; Li, X.M.; Ma, Z.P. The discrimination between continental basalt and island arc basalt based on geochemical method. *Acta Petrol. Mineral.* **2007**, *26*, 77–89. (In Chinese with English abstract)
113. Zuo, G.C.; He, G.Q. *Plate Tectonics and Metallogenic Regularities in Beishan Region*; Peking University Press: Beijing, China, 1990; pp. 1–226. (In Chinese with English abstract)
114. Zhang, D.D.; Wang, J.Q.; Niu, Y.Z.; Zhang, Y.X.; Shi, J.Z.; Liu, Z.L. Provenances of the Lower-Middle Permian in Beishan region, Central Asian Orogenic Belt and its paleo geographic implication. *Acta Geol. Sin.* **2023**, *97*, 307–324.
115. Niu, Y.Z.; Liu, C.Y.; Shi, G.R.; Lu, J.C.; Xu, W.; Shi, J.Z. Unconformity-bounded upper Paleozoic mega sequences in the Beishan region (NW China) and implications for the timing of the Paleo-Asian Ocean closure. *J. Asian Earth Sci.* **2018**, *167*, 11–32. [CrossRef]
116. Pearce, J.A.; Harris, N.B.W.; Tindle, A.G. Trace element discrimination diagrams for the tectonic interpretation of granitic rocks. *J. Petrol.* **1984**, *25*, 956–983. [CrossRef]
117. Pearce, J.A.; Norry, M.J. Petrogenetic implications of Ti, Zr, Y, and Nb variations in volcanic rocks. *Contrib. Mineral. Petrol.* **1979**, *69*, 33–47. [CrossRef]

minerals

MDPI

Article

Ocean–Continent Conversion in Beishan Orogenic Belt: Evidence from Geochemical and Zircon U-Pb-Hf Isotopic Data of Luotuoquan A-Type Granite

Wenliang Chen [1], Minjie Zhang [2,3,*], Guanghuo Tao [1], Xiaofeng Li [1], Qian Yu [2,3], Xiaojie Fan [2,3] and Jingwei Zhang [2,3]

1 Hebei Institute of Geological Survey, Shijiazhuang 050000, China; hebeiddycwl@163.com (W.C.); hebeiddytgh@163.com (G.T.); hebeiddylxf@163.com (X.L.)

2 Hebei Key Laboratory of Strategic Critical Mineral Resources, Hebei GEO University, Shijiazhuang 050031, China; yuqian_hgu@163.com (Q.Y.); fxj18233385502@163.com (X.F.); zhangjinglv@163.com (J.Z.)

3 College of Resources, Hebei GEO University, Shijiazhuang 050031, China

* Correspondence: mj-zhang@hgu.edu.cn

Abstract: Devonian magmatism is one of the most important tectonothermal events in the Central Asian Orogenic Belt (CAOB). However, little is known regarding the petrogenesis and geodynamic setting of the widely distributed Devonian granitoids in the eastern Southern Beishan Orogenic Belt (SBOB). Early-Devonian granitic magmatism has been recognized from the Luotuoquan area, and the granites were emplaced between 404.9 Ma and 399.4 Ma according to LA-ICPMS zircon U–Pb dating. Geochemically, the granites exhibit high SiO_2 and Al_2O_3 contents and are enriched in light rare earth elements as well as Rb, Th, Nd, Zr, and Hf, while being depleted in heavy rare earth elements and Ba, U, Sr, and Ti, with distinct rare earth element fractionation and pronounced negative Eu anomalies. According to the comprehensive analysis, they closely resemble the features typically associated with A-type granites. The zircons $\varepsilon_{Hf}(t)$ values are within the range of +0.90–+5.19 (averaged 3.23) for the monzogranite and syenogranite, whereas their T_{DM2} values fall between 1.05 and 1.34 Ga, suggesting that the magma source of the monzogranite–syenogranite originated from the partial melting of the Mesoproterozoic crust dominated by metagreywackes. Furthermore, the monzogranite and syenogranite exhibit high temperatures (average 782 °C), thin crustal thickness (average 30 km), and A-type characteristics, suggesting their formation in post-collision extensional settings. We propose the closure of the Beishan Ocean occurred before the early Devonian, followed by a transition in the Southern Beishan Orogenic Belt from a compressional to an extensional setting.

Keywords: zircon U–Pb dating; geochemistry; A-type granite; Luotuoquan complex; Beishan Orogenic Belt

Citation: Chen, W.; Zhang, M.; Tao, G.; Li, X.; Yu, Q.; Fan, X.; Zhang, J. Ocean–Continent Conversion in Beishan Orogenic Belt: Evidence from Geochemical and Zircon U-Pb-Hf Isotopic Data of Luotuoquan A-Type Granite. *Minerals* **2023**, *13*, 1411. https://doi.org/10.3390/min13111411

Academic Editors: Simon Paul Johnson, Jean-Michel Lafon, Clemente Recio, Jiafu Chen, Nan Ju and Zhonghai Zhao

Received: 5 September 2023
Revised: 26 October 2023
Accepted: 1 November 2023
Published: 4 November 2023

1. Introduction

The Central Asian Orogenic Belt (CAOB) is formed by the sequential accretion of multiple microcontinents, island arcs, seamounts, oceanic plateaus, and accretionary complexes from the early Neoproterozoic to the late Paleozoic [1–3]. The Beishan Orogenic Belt (BOB) connects the Tarim, Kazakhstan, and North China cratons [4–6] (Figure 1a) and is considered a critical southern Central Asian Orogenic Belt segment, vital for understanding crustal growth and tectonic evolution [5–16]. The Southern Beishan Orogenic Belt (SBOB), primarily consisting of the Shuangyingshan and Huaniushan units (Figure 1), is thought to record the collision and subduction among ancient microcontinents with Mesoproterozoic to Neoproterozoic basements [17–21]. However, the tectonic and thermal evolution of the SBOB remains unclear, particularly the timing of the Beishan Ocean closure during the early Paleozoic. Current views suggest the closure occurred in the later Silurian–early Devonian [22–25], Carboniferous [25], early Permian [26], or after the late Permian [27–29]. Thus, the Paleozoic crustal growth and geodynamics of the SBOB are rather controversial compared to those of the Mesozoic [11]. Further investigation of the tectonic and crustal development of the SBOB during the early–late Paleozoic is needed.

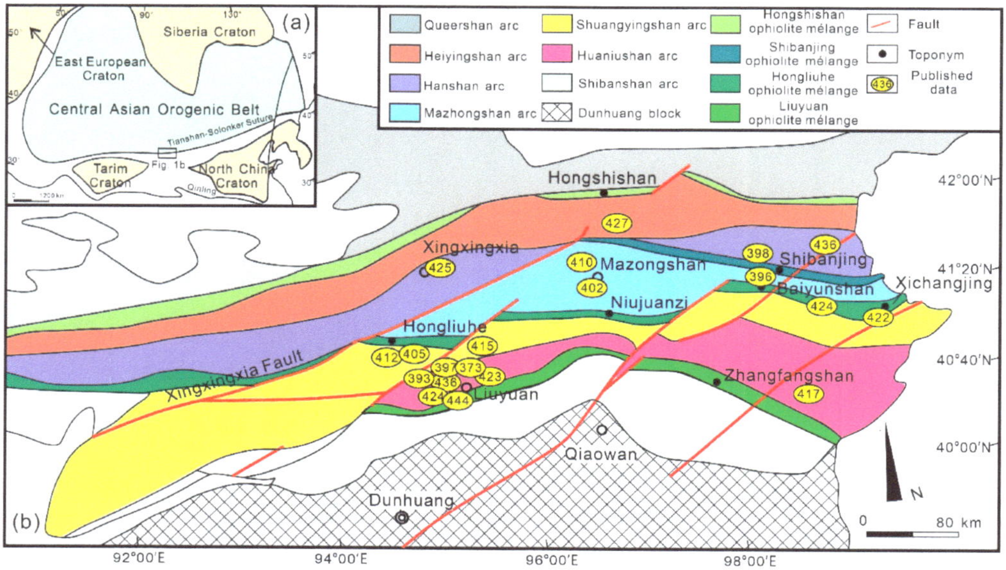

Figure 1. (**a**) Tectonic setting of the BOB (modified after [5]). (**b**) Simplified tectonic map of the Beishan orogenic collage and its adjacent area showing the tectonic subdivisions (modified after [5,13]). The zircon U-Pb age data of granitoids in (**b**) are from previous studies [4,6,10,30–35].

Herein, we present comprehensive whole-rock geochemical, in situ zircon U–Pb geochronology, and Hf isotopic data of A-type granites from the Luotuoquan complex. In addition to previously published findings, we discuss the genetic mechanism and tectonic settings of the early Paleozoic granitoids of the SBOB, elucidating their significant implications for understanding the tectonic evolution during this period in the BOB while also providing valuable insights into Beishan Ocean closure.

2. Geological Setting

The BOB comprises a complex assemblage of blocks, magmatic arcs, and ophiolitic mélanges that were formed through the subduction–accretion process of the Paleo-Asian Ocean [5]. Based on the spatial and temporal distributions of the ophiolitic mélanges and rock associations, the BOB is divided into several arcs [5], comprising (from north to south) the Queershan, Heiyingshan, Hanshan, Mazongshan, Shuangyingshan, Huaniushan, and Shibanshan arcs, which are separated by the Hongshishan, Shibanjing, Hongliuhe, and Liuyuan ophiolitic mélanges, respectively (Figure 1b).

The SBOB is situated within the Hongliuhe and Liuyuan ophiolitic mélanges [10] (Figure 1b) and comprises the Shuangyingshan and Huaniushan Units. In addition, the southern Beishan has recorded the subduction and collision of several microcontinental fragments and is characterized by early Paleozoic volcanic–sedimentary formations and intrusive plutons generated during early Paleozoic subduction [20,22,36,37]. Recent geochronological studies in the region reveal three stages of pluton emplacement in the Mid-Ordovician to late Silurian [6,23,38–40], early Devonian [4,6,40–42], and Late Devonian to early Carboniferous [6,38].

The Luotuoquan complex is situated in the eastern segment of the SBOB (Figure 1b) and intrudes into Silurian granitoids (Figure 2). It primarily consists of monzogranite, syenogranite, and minor basic rock. Field observations reveal that the syenogranite exhibits weak mylonitization and has intruded into the monzogranite (Figure 2). The Devonian samples are derived from the Luotuoquan granite complex, comprising five monzogranite samples (PM03-2, PM03-26, PM03-41, PM12-20, and PM17-4 and three syenogranite samples (D5195, D8373, and PM06-34). The monzogranite samples display medium- to coarse-grained massive rocks, primarily composed of quartz (20–25 vol%), K-feldspar (35–40 vol%), plagioclase (30–35 vol%), and orientated biotite and/or muscovite (5 vol%) (Figure 3a,b). The syenogranite samples exhibit well-developed mylonitization and a granoblastic texture. They predominantly consist of K-feldspar (55–60 vol%), plagioclase (10–15 vol%, An = 5–10), and quartz (20–25 vol%) with minor biotite (5 vol%) (Figure 3c,d).

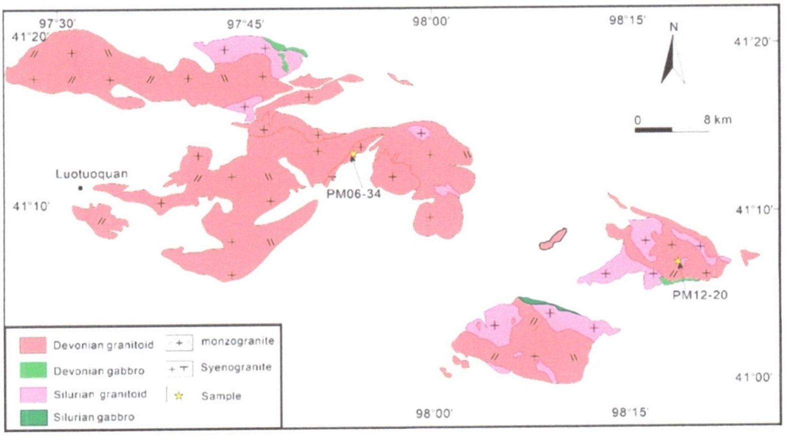

Figure 2. Geological map of the mafic–ultramafic intrusions and granite plutons in the Luotuoquan area.

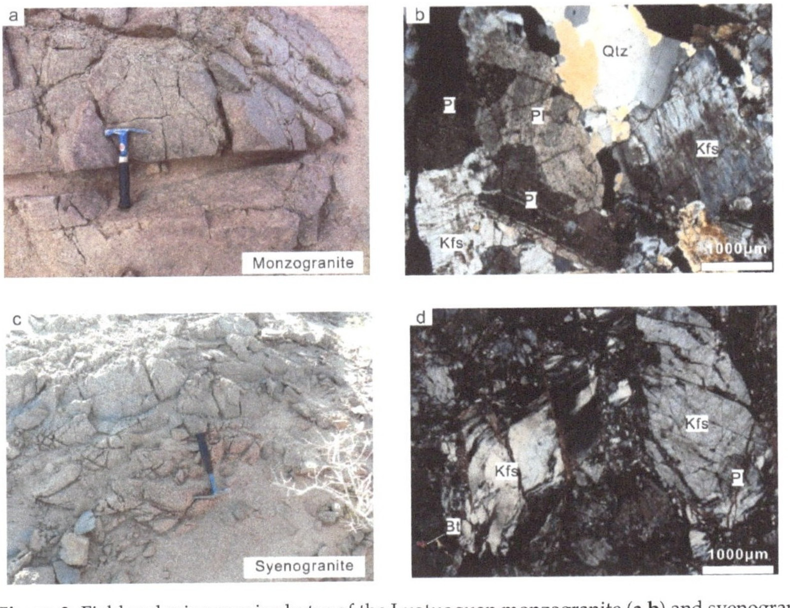

Figure 3. Field and microscopic photos of the Luotuoquan monzogranite (**a**,**b**) and syenogranite (**c**,**d**). Qtz-quartz; Pl-plagioclase; Kfs-potassium feldspar; Bt-biotite.

3. Analytic Methods

3.1. Zircon Dating and CL Imaging

Zircons were separated from the granite samples PM06-34 and PM12-20 of the Luotuoquan complex for laser ablation inductively coupled plasma mass spectrometry (LA-ICP-MS) U–Pb dating. Zircon grains were extracted using standard density and magnetic separation techniques. The selected zircon grains were handpicked under a stereoscopic microscope and mounted in epoxy resin before being polished to dissect the crystals in half for analysis. Cathodoluminescence (CL) and reflected-light photomicrographic analysis of the prepared sample targets were utilized to image the morphology and internal structure of the zircons to aid in selecting zircon grains for U–Pb dating. Zircon U–Pb dating analyses were conducted on a quadrupole inductively coupled plasma mass spectrometer (ICP-MS) (THERMO-ICAPRQ) coupled to a 193-nm ArF Excimer laser (Resolution-LR, Applied Spectra, West Sacramento, CA, USA) at Hebei Key Laboratory of Strategic Critical Mineral Resources. The laser spot size was set to 29 μm, the laser energy density was 3 J/cm^{2,} and the repetition rate was 8 Hz. Each analysis comprised a 10 s blank, a 40 s sampling ablation, and a 20 s sample-chamber flushing after the ablation. The ablated material was carried into the ICP-MS by the high-purity helium gas stream with a flux of 0.4 L/min. The whole laser path was fluxed with argon (0.9 L/min) to increase energy stability. A zircon 91,500 standard was used for external age calibration, and a zircon GJ–1 standard was used as a secondary standard to supervise the deviation of age calculation. Calibrations for trace element concentration were carried out using NIST SRM610 as an external standard and Si as the internal standard. ICPMSDataCal (Ver. 4.6) [43] and Isoplot 3.0 [44] programs were used for data reduction.

3.2. In Situ Lu-Hf Isotopes

In situ zircon Lu–Hf isotopic analyses were performed using a Neptune Plus MC–ICP–MS (Thermo Fisher Scientific, Brunswick, Germany) equipped with a Geolas 2005 excimer ArF laser ablation system (LambdaPhysik, Göttingen, Germany). All data on zircon were acquired in a single-spot ablation mode at a spot size of 44 μm. The energy density of laser ablation used in this study was ~7.0 J/cm^{-2}. Each measurement consisted of a 20-s acquisition of the background signal, followed by a 50-s acquisition of ablation signals. Instrumental conditions and data acquisition were as described by Wu et al. (2006) [45]. Zircon 91,500 was used as the reference standard. The chondritic ratios of ^{176}Hf/^{177}Hf = 0.282772 and ^{176}Lu/^{177}Hf = 0.0332 were used in our calculation of ε_{Hf}(t) values [46]. Single-stage model ages (T_{DM1}) were calculated by reference to depleted mantle with a present day ^{176}Hf/^{177}Hf ratio of 0.28325 and ^{176}Lu/^{177}Hf ratio of 0.0384 [47]. The two-stage Hf model age (T_{DM2}), also interpreted as crust formation age, was calculated by projecting the zircon ^{176}Hf/^{177}Hf (t) back to the depleted mantle model growth curve, assuming a mean crustal value for Lu/Hf (^{176}Lu/^{177}Hf = 0.015) [48].

3.3. Whole-Rock Geochemical Analysis

Whole-rock geochemical analyses were performed at the Institute of Regional Geology Survey of Hebei Province. Fresh chips of whole-rock samples were powdered to 200 mesh using a tungsten carbide ball mill. Major and trace elements were analyzed by X-ray fluorescence (Axios X; PANalytical B.V.) and inductively coupled plasma mass spectrometry (XSeries II; Thermo Fisher Scientific), respectively. The analytical precision is generally better than 2% for major elements. For trace element analyses, sample powders were digested using HF + HNO_3 mixture in high-pressure Teflon bombs at 190 °C for 48 h or longer. The analytical precision is generally better than 5% for trace elements.

4. Result

4.1. Zircon U-Pb Dating and Lu-Hf Isotope Compositions

The zircons from monzogranite (PM12-20) and syenogranite (PM06-34) are similar in crystal morphology with sizes ranging from 80 to 200 μm and aspects ratios of 2:1–3:1.

The CL images reveal oscillatory zoning and rare inherited cores (Figure 4). The Th/U ratios are >0.1, indicating magmatic origin (Table 1). Sixteen analyses from sample PM12-34 (monzogranite) were concordant and yielded a weighted mean age of 404.9 ± 1.4 Ma (Figure 4a). Sixteen spots were analyzed for dating from sample PM06-12, apart from six spots (RZ7, RZ9-11, RZ13, and RZ15), which represent the age of the inherited core, the remaining ten analyses yielded a weighted mean age of 399.4 ± 5.1 Ma (Figure 4b).

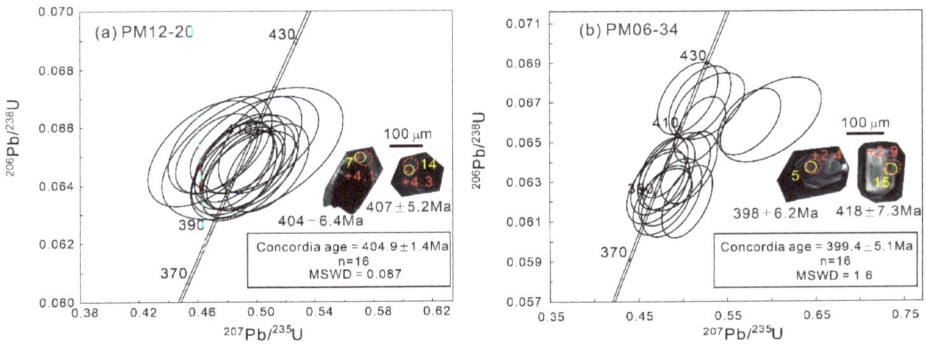

Figure 4. U–Pb concordia diagrams of zircons from Luotuoquan monzogranite (**a**) and syenogranite (**b**). The yellow line circle represents the spot of LA-ICP-MS analysis for U–Pb dating. The red dashed line circle represents the spot of LA-MC-ICP-MS analysis for Lu–Hf isotope compositions.

The analytical results are presented in Table 2, including $\varepsilon_{Hf}(t)$ values and model ages calculated using $^{206}Pb/^{238}U$ ages. The 32 spots from the Luotuoquan granites exhibit initial $^{176}Hf/^{177}Hf$ ratios ranging from 0.282205 to 0.282415 and positive $\varepsilon_{Hf}(t)$ values of +0.90–+5.19. Furthermore, the T_{DM2} model ages range from 1.05 to 1.34 Ga.

4.2. Major and Trace Elements

The whole-rock major and trace element compositions of the five monzogranite samples and three syenogranite samples were analyzed, and the results are presented in Table 3.

4.2.1. Major Elements

The samples possess high SiO_2 (71.04–76.00 wt.%), K_2O (4.02–5.59 wt.%), and $K_2O + Na_2O$ (7.04–8.62 wt.%); however, they possess low Al_2O_3 (12.10–14.30 wt.%), MgO (0.07–0.63 wt.%), and CaO (0.76–2.53 wt.%). In the SiO_2 versus $K_2O + Na_2O$ diagram (Figure 5a), the samples are plotted within the granitoid fields, which is consistent with the petrographic observations. They further display low A/CNK values (molar $Al_2O_3/(CaO + Na_2O + K_2O)$) ranging from 0.85 to 1.08 and exhibit peraluminous and high K characteristics (Figure 5).

4.2.2. Trace Elements

In the chondrite-normalized rare earth element patterns, the samples exhibit relative enrichment of light rare earth elements (LREE)($[La/Yb]_N$ = 2.63–16.87) and significant negative Eu anomalies (Eu/Eu* = 0.23–0.45), while the REE abundances range from 161.2 to 459.7 ppm (Figure 6a). All samples are depleted in Ba, U, Sr, P, and Ti and enriched in Rb, Th, Nd, Zr, and Hf relative to the primitive mantle (Figure 6b).

Table 1. LA-ICP-MS zircon U–Pb dating results of the Luotuoquan granites.

No	Pb	U	Isotopic Ratios						Isotopic Age (Ma)					
	(ppm)		$^{207}Pb/^{206}Pb$	1σ	$^{207}Pb/^{235}U$	1σ	$^{206}Pb/^{238}U$	1σ	$^{207}Pb/^{206}Pb$	1σ	$^{207}Pb/^{235}U$	1σ	$^{206}Pb/^{238}U$	1σ
PM12-20: monzogranite														
RZ1	75	294	0.0529	0.0028	0.4754	0.0234	0.0654	0.0011	324	120	395	16.1	409	7.2
RZ2	101	457	0.0556	0.0026	0.4958	0.0213	0.0645	0.0009	439	104	409	14.5	403	6.1
RZ3	84	420	0.0539	0.0020	0.4851	0.0179	0.0647	0.0010	365	88	402	12.3	404	6.5
RZ4	69	360	0.0540	0.0021	0.4845	0.0178	0.0648	0.0009	372	87	401	12.2	405	6.3
RZ5	120	529	0.0554	0.0020	0.5026	0.0188	0.0654	0.0011	428	81.5	413	12.7	408	6.6
RZ6	72	296	0.0551	0.0028	0.4922	0.0244	0.0646	0.0012	417	115	406	16.6	403	7.1
RZ7	74	300	0.0557	0.0026	0.4939	0.0214	0.0647	0.0010	443	99.1	408	14.6	404	6.4
RZ8	110	486	0.0549	0.0020	0.492	0.0172	0.0645	0.0009	406	81.5	406	11.7	403	6.2
RZ9	60.6	330	0.0510	0.0028	0.4651	0.0257	0.0651	0.0012	239	130	388	17.8	406	7.1
RZ10	233	900	0.0559	0.0022	0.5074	0.0188	0.0646	0.0010	456	82.4	417	12.7	403	6.3
RZ11	56.3	231	0.0548	0.0026	0.4878	0.0211	0.0645	0.0010	467	106	403	14.4	403	6.4
RZ12	34.4	149	0.0551	0.0048	0.4974	0.0408	0.0651	0.0015	413	193	410	27.7	406	9.4
RZ13	127	530	0.0544	0.0019	0.4882	0.0171	0.0644	0.0010	387	81.5	404	11.7	402	6.8
RZ14	86	388	0.0545	0.0022	0.4912	0.0189	0.0651	0.0009	391	90.7	406	12.8	407	5.2
RZ15	76	326	0.0550	0.0032	0.4988	0.0289	0.0654	0.0011	413	131	411	19.6	408	7.3
RZ16	82	360	0.0524	0.0021	0.4729	0.0184	0.0651	0.0009	306	92.6	393	12.7	407	6.1
PM06-34: syenogranite														
RZ1	48.3	213	0.0554	0.0025	0.4770	0.0221	0.062	0.0011	428	104	396	15.2	387	6.4
RZ2	51.2	227	0.0549	0.0029	0.4719	0.0234	0.0627	0.0012	406	117	392	16.1	392	7.3
RZ3	46.1	192	0.0567	0.0031	0.4925	0.0255	0.0635	0.0013	480	114	407	17.4	397	8.1
RZ4	114	422	0.0546	0.0022	0.4798	0.0198	0.0623	0.0010	394	90.7	398	13.6	390	6.1
RZ5	123	585	0.0537	0.0021	0.4785	0.0182	0.0637	0.0010	367	88.9	397	12.5	398	6.2
RZ6	92	305	0.0574	0.0026	0.5068	0.0212	0.0637	0.0012	506	100	416	14.3	398	7.2
RZ7	77	337	0.0565	0.0028	0.5297	0.0266	0.0668	0.0011	472	105	432	17.7	417	7.0
RZ8	61.6	384	0.0585	0.0022	0.5046	0.0190	0.0620	0.0012	550	83.3	415	12.8	388	7.1
RZ9	85	398	0.0623	0.0026	0.5750	0.0235	0.0660	0.0011	687	88.9	461	15.1	412	7.2
RZ10	98	447	0.0570	0.0022	0.5192	0.0193	0.0656	0.0011	500	83.3	425	12.9	410	6.1
RZ11	54.2	278	0.0547	0.0025	0.5056	0.0227	0.0672	0.0012	398	99	415	15.3	419	7.5
RZ12	52.6	241	0.0547	0.0025	0.4720	0.0204	0.0630	0.0011	398	102	393	14.1	394	7.2
RZ13	39.9	168	0.0654	0.0038	0.5996	0.0377	0.0661	0.0014	787	122	477	23.9	412	8.1
RZ14	75	267	0.0557	0.0027	0.4814	0.0214	0.0634	0.0012	439	107	399	14.7	396	7.2
RZ15	95	424	0.0557	0.0023	0.5192	0.0233	0.0670	0.0012	439	94.4	425	15.6	418	7.3
RZ16	45.2	220	0.0564	0.0027	0.4873	0.0226	0.0622	0.0013	478	103	403	15.4	389	8.0

Table 2. In situ zircon Hf isotopic results of the Luotuoquan granites.

No	Age (Ma)	176Yb/177Hf	176Lu/177Hf	176Hf/177Hf	2σ	εHf(0)	εHf(t)	2σ	TDM1 (Ma)	f(Lu/Hf)	TDM2 (Ma)
PM12-20: monzogranite											
RZ1	409	0.014856	0.000530	0.282242	0.000015	−18.7	3.72	0.51	894	−0.98	1160
RZ2	403	0.031534	0.001173	0.28240	0.000017	−13.2	2.96	0.59	927	−0.96	1204
RZ3	404	0.025478	0.000856	0.28238	0.000018	−13.9	3.82	0.64	876	−0.97	1150
RZ4	405	0.041885	0.001418	0.282381	0.000014	−13.8	3.39	0.49	895	−0.96	1178
RZ5	408	0.030106	0.001031	0.282337	0.000015	−15.4	3.51	0.51	887	−0.97	1173
RZ6	403	0.040490	0.001374	0.282318	0.000015	−16.1	3.71	0.54	873	−0.96	1156
RZ7	404	0.048124	0.001711	0.282350	0.000017	−14.9	4.12	0.59	884	−0.95	1131
RZ8	403	0.034757	0.001174	0.282334	0.000014	−15.5	2.32	0.51	934	−0.96	1244
RZ9	406	0.028390	0.000962	0.282279	0.000014	−17.4	3.12	0.51	902	−0.97	1196
RZ10	403	0.032415	0.001107	0.282360	0.000015	−14.6	2.66	0.54	914	−0.97	1223
RZ11	403	0.033578	0.001125	0.282319	0.000014	−16.0	3.30	0.50	889	−0.97	1182
RZ12	406	0.054288	0.001987	0.282317	0.000014	−16.1	3.88	0.49	879	−0.94	1148
RZ13	402	0.008480	0.000266	0.282205	0.000016	−20.1	3.58	0.55	907	−0.99	1164
RZ14	407	0.029075	0.000976	0.282415	0.000014	−12.6	4.34	0.49	873	−0.97	1119
RZ15	408	0.054699	0.001828	0.282399	0.000015	−13.2	3.74	0.54	875	−0.94	1158
RZ16	407	0.029610	0.001011	0.282348	0.000016	−15.0	4.57	0.58	847	−0.97	1105
PM06-34: syenogranite											
RZ1	387	0.032029	0.001097	0.282317	0.000018	−16.1	2.96	0.64	890	−0.97	1192
RZ2	392	0.011210	0.000352	0.282254	0.000018	−18.3	5.19	0.64	814	−0.99	1054
RZ3	397	0.045344	0.001502	0.282343	0.000015	−15.2	2.52	0.54	914	−0.95	1227
RZ4	390	0.028869	0.000986	0.282321	0.000019	−15.9	4.66	0.67	831	−0.97	1086
RZ5	398	0.031534	0.001173	0.282400	0.000015	−13.2	2.39	0.51	922	−0.96	1236
RZ6	398	0.025478	0.000856	0.282380	0.000017	−13.9	2.31	0.59	928	−0.97	1241
RZ7	417	0.041885	0.001418	0.282381	0.000016	−13.8	0.95	0.58	998	−0.96	1341
RZ8	388	0.030106	0.001031	0.282337	0.000014	−15.4	2.92	0.51	893	−0.97	1195
RZ9	412	0.040490	0.001374	0.282318	0.000014	−16.1	1.81	0.51	958	−0.96	1283
RZ10	410	0.048124	0.001711	0.282350	0.000014	−14.9	0.90	0.50	993	−0.95	1339
RZ11	419	0.034757	0.001174	0.282334	0.000014	−15.5	4.30	0.49	862	−0.96	1131
RZ12	394	0.028390	0.000962	0.282279	0.000014	−17.4	2.11	0.50	927	−0.97	1250
RZ13	412	0.032415	0.001107	0.282360	0.000013	−14.6	2.98	0.46	910	−0.97	1209
RZ14	396	0.033578	0.001125	0.282319	0.000013	−16.0	4.68	0.47	833	−0.97	1090
RZ15	418	0.054288	0.001987	0.282317	0.000017	−16.1	2.91	0.61	922	−0.94	1218
RZ16	389	0.008480	0.000266	0.282205	0.000016	−20.1	3.12	0.58	884	−0.99	1183

Table 3. Major and trace element compositions of the Luotuoquan granites.

Samples	Monzogranirte					Syenogranite		
	PM03-2	PM03-26	PM03-41	PM12-20	PM17-4	D5195	D8373	PM06-34
SiO_2	71.04	71.80	74.95	72.80	74.10	71.60	76.00	72.30
TiO_2	0.29	0.32	0.14	0.29	0.12	0.49	0.18	0.17
Al_2O_3	14.20	14.30	13.20	13.60	14.10	13.30	12.10	12.10
TFe_2O_3	2.41	2.61	1.11	2.57	1.08	3.63	1.70	1.86
MnO	0.06	0.06	0.01	0.05	0.01	0.03	0.01	0.03
MgO	0.55	0.63	0.27	0.41	0.30	0.62	0.07	0.63
CaO	2.16	1.98	0.76	1.33	1.29	2.00	1.03	2.53
Na_2O	3.45	3.49	3.03	3.10	3.51	3.01	2.46	2.69
K_2O	4.29	4.02	5.59	4.93	4.87	4.03	4.89	4.89
P_2O_5	0.12	0.12	0.03	0.09	0.04	0.11	0.04	0.29
LoI	1.52	0.77	0.76	0.92	0.51	1.23	1.43	2.48
Total	100.07	100.07	99.85	100.03	99.99	100.06	99.93	99.99
Rb	211.0	132.0	167.0	177.0	215.0	112.0	114.0	134.0
Ba	406.0	444.0	398.0	345.0	466.0	556.0	499.0	755.0
Th	28.60	11.50	11.30	20.70	16.70	29.20	21.20	24.80
U	4.81	2.41	2.23	2.95	1.76	2.10	1.30	1.90
Ta	1.10	0.76	0.90	1.08	0.80	0.81	0.40	0.72
Nb	14.00	9.53	9.56	11.50	10.10	15.10	5.13	12.40
Sr	90.50	50.10	60.40	65.80	84.50	86.00	59.00	79.00
Zr	333.0	202.0	245.0	259.0	236.0	365.0	147.0	275.0
Hf	10.90	6.98	8.40	8.90	8.01	12.50	6.10	9.50
Y	55.52	57.88	58.64	43.61	44.51	47.40	43.40	37.70
Ga	19.10	18.20	14.30	17.00	20.60	25.40	21.70	17.20
La	40.28	25.93	19.75	47.56	42.23	93.60	48.60	68.20
Ce	93.30	56.65	45.19	108.9	83.71	189.0	89.60	140.0
Pr	10.97	7.51	5.87	14.30	11.20	25.60	12.90	19.20
Nd	43.77	31.14	24.06	55.80	44.47	94.90	50.90	74.00
Sm	9.62	7.69	6.25	11.46	9.12	17.20	11.50	14.10
Eu	1.13	0.96	0.86	1.37	1.00	1.22	1.18	1.20
Gd	7.84	6.18	5.11	9.07	7.45	15.10	9.35	11.50
Tb	1.61	1.44	1.33	1.62	1.42	2.11	1.53	1.80
Dy	9.84	9.66	9.57	8.49	8.51	9.62	8.13	9.20
Ho	1.87	1.89	1.89	1.50	1.53	1.64	1.50	1.70
Er	4.68	4.72	4.75	3.75	3.81	4.56	4.35	4.00
Tm	0.84	0.93	0.96	0.63	0.69	0.66	0.73	0.60
Yb	4.87	5.52	5.38	3.57	4.11	3.98	4.67	3.70
Lu	0.85	0.96	0.89	0.66	0.75	0.54	0.64	0.50
∑REE	231.5	161.2	131.9	268.7	220.0	459.7	245.6	349.7
LREE/HREE	6.14	4.15	3.41	8.17	6.78	11.00	6.95	9.57
δEu	0.39	0.41	0.45	0.40	0.36	0.23	0.34	0.27

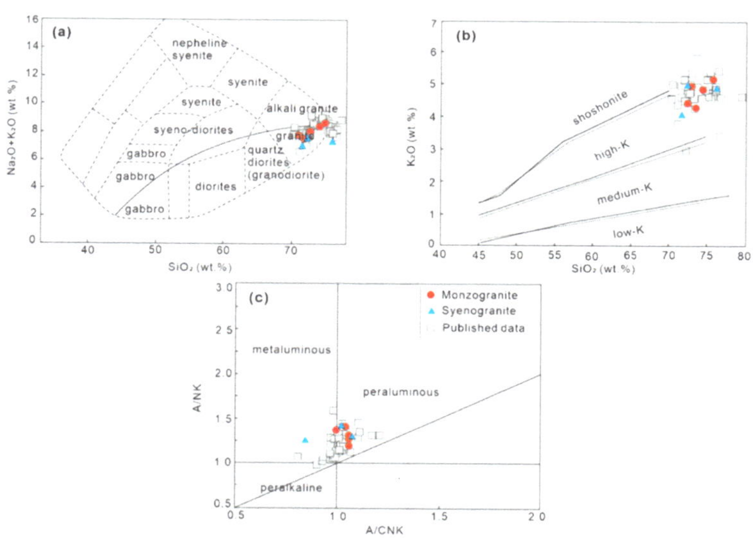

Figure 5. (**a**) Total alkalis (Na_2O + K_2O) vs. SiO_2 diagram [49]; (**b**) K_2O vs. SiO_2 diagram [50]; (**c**) A/NK vs. A/CNK diagram [51]. Published data of the early Devonian granitoids in the Beishan area are from [4,6,30,31,42,52,53].

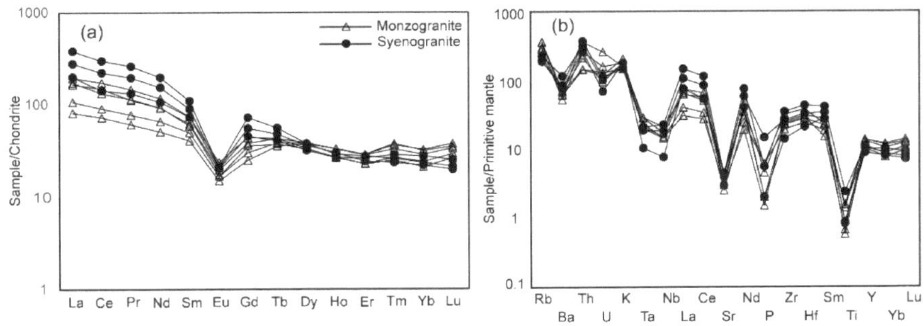

Figure 6. Chondrite-normalized REE element patterns (**a**) and primitive mantle-normalized trace element spider diagrams (**b**) for the Luotuoquan granites. Normalization values are from [54].

5. Discussion

5.1. Devonian Magmatism in the Southern Beishan Orogenic Belt

The LA–ICP–MS zircon U–Pb ages obtained in this study (404–399 Ma) are consistent with the timing of pluton emplacement in the SBOB, including the Liuyuan monzogranite (397 ± 7 Ma) [40], Shuangfengshan A-type granites (415 ± 3 Ma) [4], Huitongshan K-feldspar granite (397 ± 3 Ma) [42], and Shijinpo granitoid (404 ± 2 Ma) [53]. Previous studies have established that the Silurian–Devonian was a crucial period of magmatism in the BOB (Figure 1a), with age peaks of 440, 420, and 400 Ma. The Luotuoquan granites, as products of Devonian magmatism, offer valuable insights into the geodynamic evolution of the BOB during the mid-Paleozoic.

5.2. Petrogenesis and Magma Sources

The syenogranite samples in this study have undergone mylonitization, as evidenced by petrographic observation (Figure 2d). Therefore, it is imperative to evaluate the impact of alteration on both major and trace elements before discussing their petrogenesis [55]. Our samples demonstrate differentiated correlations between "immobile alteration" elements (e.g., Zr) and the other trace elements (Figure 7), where Nb, Ta, Th, and Hf display a stronger correlation with Zr than Rb and Ba, suggesting that this alteration has less effect on these elements. Consequently, the subsequent discussion will primarily focus on more immobile elements such as Nb, Ta, Zr, and REEs.

The Luotuoquan monzogranite and syenogranite exhibit similar geochemical characteristics and are discussed together. They have high SiO_2 (71.04–76.00 wt.%) and K_2O + Na_2O (7.04–8.62 wt.%) but low Al_2O_3 (mean of 13.36 wt.%), MgO (mean of 0.03 wt.%), and CaO (mean of 1.64 wt.%), indicating A-type granite geochemical characteristics. Nevertheless, highly fractionated I-type and S-type granites also share similarities with A-type granites. Highly fractionated S-type granites tend to exhibit higher P_2O_5 (mean of 0.14 wt.%) and lower Na_2O (mean of 2.81 wt.%) than A-type granites [56]. The monzogranite–syenogranite displays low P_2O_5 (mean of 0.11 wt.%) and high Na_2O (mean of 3.09 wt.%), indicating it does not belong to the highly fractionated S-type granite. Compared with highly fractionated I-type granite, A-type granite shows iron enrichment and magnesium depletion with higher $Fe_2O_3^T$/MgO ratios. The $Fe_2O_3^T$/MgO value of Luotuoquan granite ranges from 2.95 to 24.28 (mean of 6.95), which is significantly greater than that of highly fractionated I-type granite (2.27) and S-type granite (2.38) [57,58]. Additionally, typical A-type granites are enriched in trace elements such as Th, Nb, Ta, Zr, Hf, Ga, and Y while being depleted in Sr, Ti, P, Cr, Co, Ni, V, etc., with obvious negative Eu anomaly [57,59]. All samples exhibit high Th, Zr, K, Ga, Y, and Yb but low Sr, P, Eu, and Ti, which also implies that Luotuoquan granite had A-type geochemical features. Previous studies suggested that A-type granitoids share similar geochemistry characteristics of high 10,000× Ga/Al values (>2.6) and Zr + Nb + Ce + Y (>350 ppm) [57]. The discrimination diagrams (Figure 8) demonstrate that the majority, if not all, of the samples are plotted within the A-type granite field. We, therefore, conclude that the Luotuoquan monzogranite–syenogranite pluton are typical A-type intrusions.

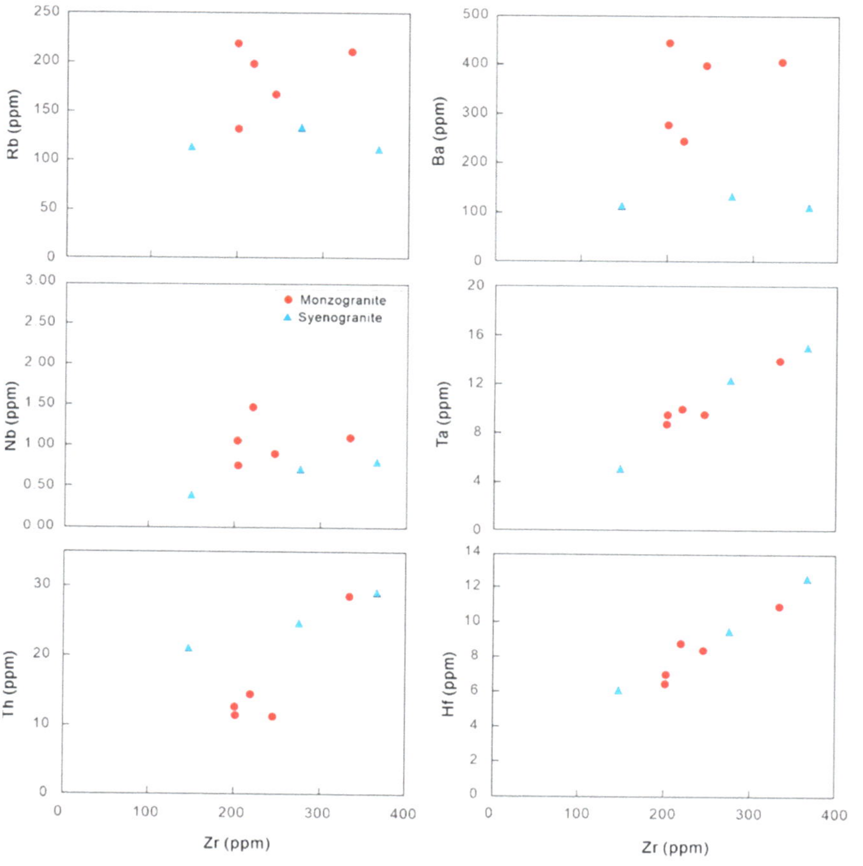

Figure 7. Trace elements vs. Zr for Luotuoquan A-type granites.

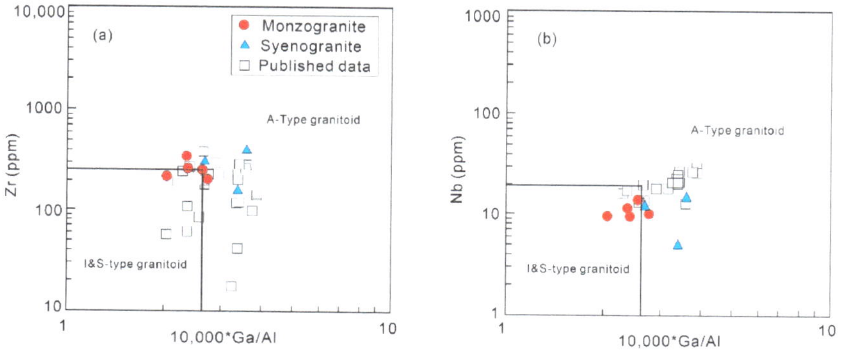

Figure 8. *Cont.*

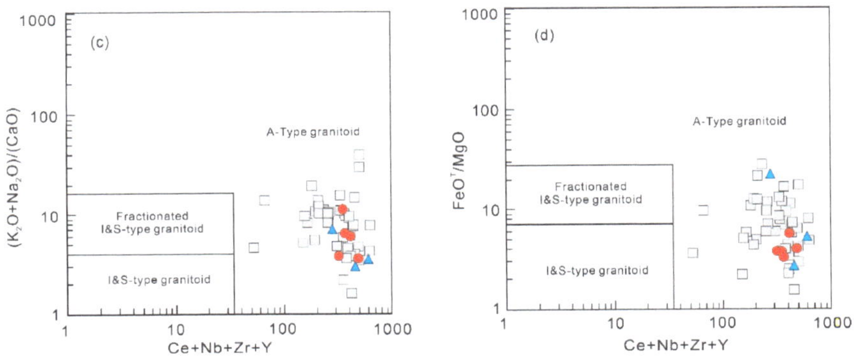

Figure 8. Discrimination diagrams for early Devonian granitoids from BOB. Zr (**a**) and Nb (**b**) vs. 10,000 ∗ Ga/Al and (K_2O + Na_2O)/CaO (**c**) and FeO^T/MgO (**d**) vs. Ce + Nb + Zr + Y diagrams [57]. Published data are from the same references as in Figure 5.

Various petrogenetic models have been proposed for A-type granites: (1) high crystallization differentiation of mantle-derived basaltic magma [60–63]; (2) mixing of mantle-derived and crustal materials [64–67]; (3) Partial melting of lower crustal material: partial melting of granulite facies remnants after granitic magma extraction [55,56,59,68], and partial melting of calc-alkaline tonalite-granodiorite [69]; (4) the melting of lower crustal rocks due to heating by mantle magma underplating [58,70,71]. Mafic rocks contemporaneous with A-type granite are rare in the study area, and no mafic microgranular enclaves are found in the granites, indicating little crust–mantle mixing. Experimental petrology shows residual granulite facies in the lower crust are low in K, Si and high in Ca, Al, and Mg [69], which cannot explain the production of Luotuoquan A-type granite rich in Si, alkali, low in Al and Mg by partial melting.

The CaO/Na_2O ratios distinguish between pelite-derived melts ($CaO/Na_2O < 0.5$) and melts derived from greywackes or igneous sources (CaO/Na_2O = 0.3–1.5) [72]. High-temperature melts generally exhibit lower Al_2O_3/TiO_2 ratios than low-temperature melts [73]. The monzogranite–syenogranite is characterized by high CaO/Na_2O ratios (mean of 0.53) and low Al_2O_3/TiO_2 ratios (mean of 63.73), suggesting a source of metagreywackes or metamorphic igneous rock. The samples are all plotted within the field of partial melts derived from metagreywackes in the molar Al_2O_3/(MgO + FeO^T) vs. molar CaO/(MgO + FeO^T) and molar K_2O/Na_2O vs. molar CaO/(MgO + FeO^T) diagrams (Figure 9). Therefore, the granitic magma source is likely related to partial melts from metagreywackes.

The Hf isotope analysis of U-Pb dated zircon grains can trace the original magma sources and distinguish between the reworking of continental crust and the remelting of juvenile crust [74,75]. In this study, the early Devonian Luotuoquan monzogranite–syenogranite has considerably positive zircon $\varepsilon_{Hf}(t)$ of +0.9–+5.2 (Figure 10) and slightly young two-stage Hf model ages of 1.05–1.34 Ga (Figure 10). The SBOB is thought to have developed abundant Mesoproterozoic to Neoproterozoic basement complex, which is a suite of metamorphosed clastic rock [17,18,20,37]. The significantly positive $\varepsilon_{Hf}(t)$ values and relatively young two-stage Hf model ages suggest that the granitic rocks originated from either the depleted mantle or through partial melting of recently accreted juvenile crustal material within the depleted mantle. Relevant mafic rocks are rarely contemporaneous, and the absence of mafic microgranular enclaves suggests limited direct involvement of newly derived mantle magma in granite formation. Therefore, the early Devonian granitoids from the SBOB primarily originated from partial melting of the overlying Mesoproterozoic crust facilitated by the underplating of mantle-derived magma.

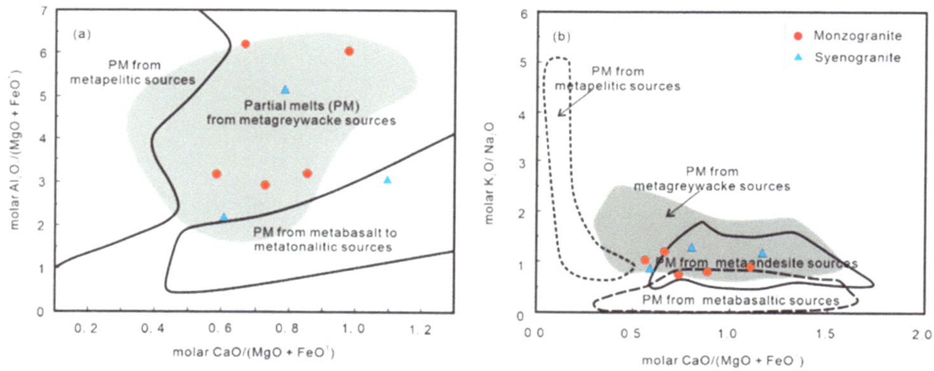

Figure 9. Discrimination diagrams of the source area of the Luotuoquan monzogranite and syenogranite. (**a**) molar Al_2O_3/(MgO + FeO^T) vs. molar CaO/(MgO + FeO^T) [76]; (**b**) molar K_2O/Na_2O vs. molar CaO/(MgO + FeO^T) [77].

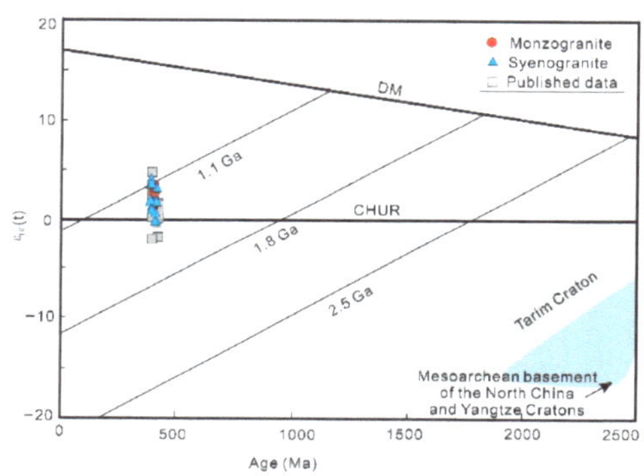

Figure 10. Zircon ε_{Hf}(t) vs. Age (Ma) diagrams of granitoids from the SBOB (base map after [75]). Published data of granitoids from [6,42]. The data on Tarim, North China, and Yangtze blocks were from [78].

5.3. Geodynamic Evolution

Recent research suggests that the BOB is an accretionary orogen with multiple island arcs and mélange belts that experienced subduction–collision during the Paleozoic [5,12,28,30,37]. The zircon U–Pb age of the ophiolite complexes in the central part of the BOB is 536–519 Ma [7,79–81], and the lower Cambrian Shuangyingshan Formation unconformable overlies the Neoproterozoic Xichangjing Formation [38], indicating the Beishan Ocean was formed before the Cambrian. Previous studies have proposed that the magmatic effects of slab windows vary dramatically during ridge subduction, forming adakitic magmas due to slab melting and A-type magmas resulting from asthenosphere upwelling through the slab window [54,82]. Widespread early Paleozoic magmatic rocks (464–424 Ma) [6,10,16,32,37,40,83,84] and ophiolitic mélange (462–420 Ma) [9,13,39,85] in the BOB indicate slab subduction setting during the Ordovician to Devonian.

The A-type granite has garnered significant attention due to its distinctive tectonic background. The Luotuoquan A-type granites fall within the volcanic arc and within-plate granites on tectonic discrimination diagrams (Figure 11a,b) rather than pertaining to ocean ridge granites. The studied granites are generally plotted within the A2 field (Figure 11c,d). The A2-type granitoids represent magmas sourced from the underplated crust or continental crust that has experienced a cycle of island-arc magmatism or continent-continent collision [86]. We propose that their potential origins were associated with island-arc magmatism. The monzogranite in the Hongliuhe ophiolite has a weighted age of 412.4 ± 2.9 Ma, indicating that the closure of the oceanic basin was completed prior to this event [41]. The granites from the Liuyuan area, with ages ranging from 436-423 Ma [40], 415 Ma [42], and 397 Ma [4,40], represent post-collision background products that are possibly associated with subduction plate detachment. Moreover, recent discoveries have revealed the presence of early Devonian post-collision granites in the middle of BOB. U–Pb dating has determined that these granites range in age from 402 to 387 Ma [15,31,35,52]. The aforementioned evidence suggests that the formation of Luotuoquan A-type granites can be attributed to a post-collision tectonic setting.

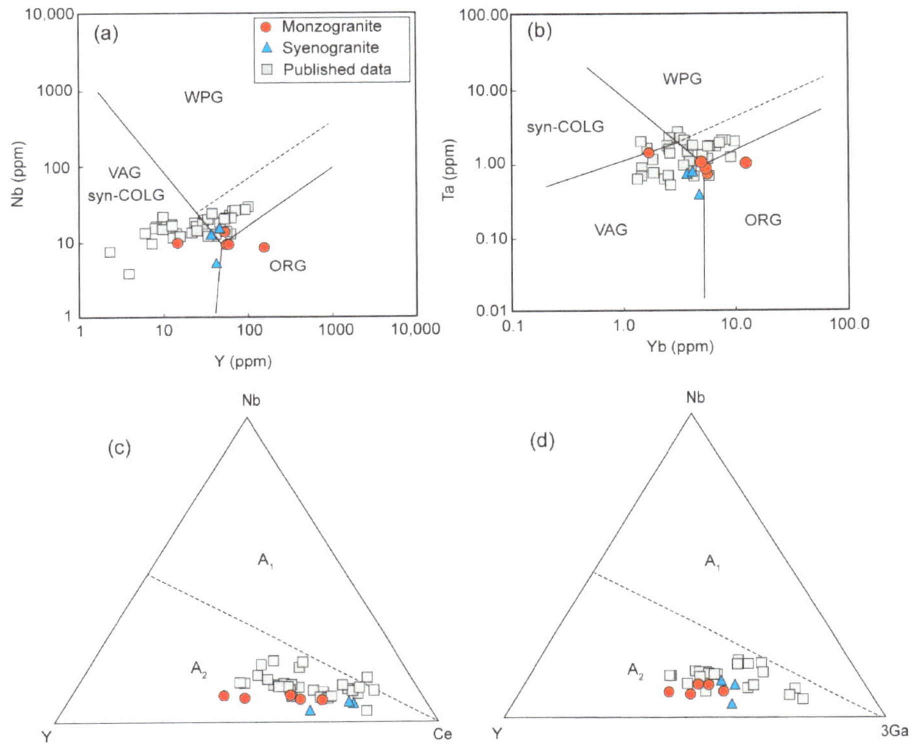

Figure 11. Nb vs. Y (**a**), Ta vs. Yb (**b**), Nb-Y-Ce (**c**), Nb-Y-3Ga (**d**) tectonic discriminant diagrams of the early Devonian granitoids from BOB ((**a**,**b**) modified after [56], (**c**,**d**) modified after [86]). Syn-COLG-syn-collision granites; VAG-volcanic arc granites; WPG-within plate granites; ORG-ocean ridge granites; post-COLG-post collision granites. Published data are from the same references as in Figure 5.

Whole-rock geochemical data of 460–390 Ma granites from the SBOB were collected to calculate zircon saturation temperatures ($\ln D_{Zr} = (10{,}108 \pm 32)/T(K) - (1.16 \pm 0.15)(M - 1) - (1.48 \pm 0.09)$) [87], $M = (Na + K + 2Ca)/(Al \times Si)(mol)$) [88] and crustal thickness ($H = [Sr/Y + (42.03 \pm 6.28)]/(1.49 \pm 0.15)$) [89,90]. During the early Paleozoic, a significant increase in crustal growth occurred due to the melting of subducting oceanic crust, resulting in the emplacement of large amounts of granitoids in the SBOB from 452 to 424 Ma [6]. This process was accompanied by a gradual decrease in zircon saturation temperatures and an increase in crustal thickness (Figure 11). The A-type granites, which were formed between 415 and 397 Ma [4,42], exhibit a higher Zr saturation temperature range of 755–831 °C and a thinner crust thickness ranging from 32 to 28 km (Figure 12). These findings suggest that the SBOB had already transitioned into an extensional setting during the early Devonian.

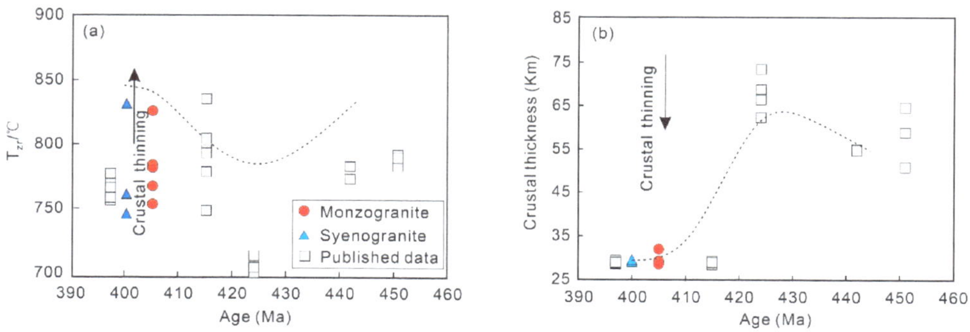

Figure 12. T_{zr} vs. Age (**a**) and Crustal thickness vs. Age (**b**) plots for Paleozoic granitoids in the SBOB. Published data from [4,6,42].

In summary, the Luotuoquan A-type granites yield U-Pb ages of approximately 404–399 Ma, indicating post-collision extensional setting during the early Devonian. This evidence indicates a post-collision extension setting in the early Devonian and implies the closure of the Beishan Ocean prior to this time.

6. Conclusions

(1) The zircon U–Pb ages of the Luotuoquan monzogranite and syenogranite are 399.4 ± 5.1 Ma and 404.9 ± 1.4 Ma, respectively.

(2) The petrographic and geochemical signatures of the Luotuoquan monzogranite and syenogranite indicate they are A-type granites and were emplaced in a post-collision extensional setting. Furthermore, these granites are the result of partial melting primarily from Mesoproterozoic crusts composed mainly of metagreywackes.

(3) The occurrence of early Devonian granitoids suggests that SBOB had already undergone extensional tectonics following the closure of the Beishan Ocean during this period.

Author Contributions: Conceptualization: W.C. and M.Z.; Data curation: Q.Y., X.F. and J.Z.; Investigation: W.C., G.T. and X.L.; Methodology: M.Z. and G.T.; Resources: W.C. and M.Z.; Writing–original draft preparation: W.C., M.Z. and G.T.; Writing–review and editing: W.C. and M.Z.; Visualization: Q.Y. and X.F.; Supervision: X.L.; Project administration: M.Z. and Q.Y. All authors have read and agreed to the published version of the manuscript.

Funding: This research was supported by the Natural Science Foundation of Hebei Province (D2020403044) and the Opening Foundation of Hebei Key Laboratory of Strategic Critical Mineral Resources (HGU-SCMR2304).

Data Availability Statement: All the data are presented in the paper.

Acknowledgments: We sincerely appreciate the editors and reviewers for their constructive comments, which helped to significantly improve the manuscript.

Conflicts of Interest: The authors declare no conflict of interest.

References

1. Windley, B.F.; Alexeiev, D.; Xiao, W.; Kröner, A.; Badarch, G. Tectonic models for accretion of the Central Asian Orogenic Belt. *J. Geol. Soc.* **2007**, *164*, 31–47. [CrossRef]
2. Xiao, W.J.; Windley, B.F.; Yuan, C.; Sun, M.; Han, C.M.; Lin, S.F.; Chen, H.L.; Yan, Q.R.; Liu, D.Y.; Qin, K.Z.; et al. Paleozoic multiple subduction-accretion processes of the southern Altaids. *Am. J. Sci.* **2009**, *309*, 221–270. [CrossRef]
3. Xiao, W.J.; Windley, B.F.; Sun, S.; Li, J.L.; Huang, B.C.; Han, C.M.; Yuan, C.; Sun, M.; Chen, H.L. A Tale of Amalgamation of Three Permo-Triassic Collage Systems in Central Asia, Oroclines, Sutures, and Terminal Accretion. *Annu. Rev. Earth Planet. Sci.* **2015**, *43*, 477–507. [CrossRef]

4. Li, S.; Wang, T.; Tong, Y.; Hong, D.W.; Ouyang, Z.X. Identification of the Early Devonian Shuangfengshan A-type granites in Liuyuan area of Beishan and its implications to tectonic evolution. *Acta Petrol. Mineral.* **2009**, *28*, 407–422. (In Chinese with English Abstract)
5. Xiao, W.J.; Mao, Q.G.; Windley, B.F.; Han, C.M.; Qu, J.F.; Zhang, J.E.; Ao, S.J.; Guo, Q.Q.; Cleven, N.R.; Lin, S.F. Paleozoic multiple accretionary and collisional processes of the Beishan orogenic collage. *Am. J. Sci.* **2010**, *310*, 1553–1594. [CrossRef]
6. Mao, Q.G.; Xiao, W.J.; Fang, T.H.; Wang, J.B.; Han, C.M.; Sun, M.; Yuan, C. Late Ordovician to early Devonian adakites and Nb-enriched basalts in the Liuyuan area, Beishan, NW China, implications for early Paleozoic slab-melting and crustal growth in the southern Altaids. *Gondwana Res.* **2012**, *22*, 534–553. [CrossRef]
7. Ao, S.J.; Xiao, W.J.; Han, C.M.; Li, X.H.; Qu, J.F.; Zhang, J.E.; Guo, Q.Q.; Tian, Z.H. Cambrian to early Silurian ophiolite and accretionary processes in the Beishan collage, NW China, implications for the architecture of the Southern Altaids. *Geol. Mag.* **2012**, *149*, 606–625. [CrossRef]
8. Ao, S.J.; Xiao, W.J.; Windley, B.F.; Mao, Q.G.; Han, C.M.; Zhang, J.E.; Yang, L.K.; Geng, J.Z. Paleozoic accretionary orogenesis in the eastern Beishan orogen, constraints from zircon U-Pb and 40Ar/39Ar geochronology. *Gondwana Res.* **2015**, *30*, 224–235. [CrossRef]
9. Cleven, N.; Lin, S.; Guilmette, C.; Xiao, W.J.; Davis, B. Petrogenesis and implications for tectonic setting of Cambrian suprasubduction-zone ophiolitic rocks in the central Beishan orogenic collage, Northwest China. *J. Asian Earth Sci.* **2015**, *113*, 369–390. [CrossRef]
10. He, Z.Y.; Zong, K.Q.; Jiang, H.Y.; Xiang, H.; Zhang, Z.M. Early Paleozoic tectonic evolution of the southern Beishan orogenic collage, Insights from the granitoids. *Acta Petrol. Sin.* **2014**, *30*, 2324–2338. (In Chinese with English Abstract)
11. Li, S.; Wilde, S.A.; Wang, T. Early Permian post-collisional high-K granitoids from Liuyuan area in southern Beishan orogen, NW China, Petrogenesis and tectonic implications. *Lithos* **2013**, *179*, 99–119. [CrossRef]
12. Song, D.F.; Xiao, W.J.; Han, C.M.; Li, J.L.; Qu, J.F.; Guo, Q.Q.; Lin, L.N.; Wang, Z.M. Progressive accretionary tectonics of the Beishan orogenic collage, southern Altaids, Insights from zircon U–Pb and Hf isotopic data of high-grade complexes. *Precambrian Res.* **2013**, *227*, 368–388. [CrossRef]
13. Tian, Z.; Xiao, W.; Windley, B.F.; Lin, L.N.; Han, C.; Zhang, J.E.; Wan, B.; Ao, S.J.; Song, D.F.; Feng, J.Y. Structure, age, and tectonic development of the Huoshishan–Niujuanzi ophiolitic mélange, Beishan, southernmost Altaids. *Gondwana Res.* **2014**, *25*, 820–841. [CrossRef]
14. Wang, Q.S.; Zhang, J.; Cao, S.S.; Yu, L.D.; Sun, H.W. The element geochemistry, zircon U-Pb chronology and Nd-Hf isotope of the granitoids in Mazhuangshan area, Gansu Province. *Acta Petrol. Sin.* **2020**, *36*, 1445–1460. (In Chinese with English Abstract)
15. Zheng, R.; Wu, T.R.; Zhang, W.; Cao, X.U. Early Devonian Tectono-Magmatic Events in the Middle Beishan, Gansu Province, Evidence from Chronology and Geochemistry of Gongpoquan Granite. *Acta Sci. Nat. Univ. Pekin.* **2012**, *48*, 603–605.
16. Zheng, R.G.; Li, J.Y.; Xiao, W.J. Mid-Paleozoic ridge subduction in the Central Beishan of the southern Altaids, evidence from geochemical, Sr-Nd and zircon U-Pb-Hf-O isotopic data of Gongpoquan volcanic rocks. *J. Geol. Soc.* **2019**, *176*, 755–770. [CrossRef]
17. He, Z.Y.; Klemd, R.; Zhang, Z.M.; Zong, K.Q.; Sun, L.X.; Tian, Z.L.; Huang, B.T. Mesoproterozoic continental arc magmatism and crustal growth in the eastern Central Tianshan Arc Terrane of the southern Central Asian Orogenic Belt, Geochronological and geochemical evidence. *Lithos* **2015**, *236*, 74–89. [CrossRef]
18. Jiang, H.Y.; He, Z.Y.; Zong, K.Q.; Zhang, Z.M.; Zhao, Z.D. Zircon U–Pb dating and Hf isotopic studies on the Beishan complex in the southern Beishan orogenic belt. *Acta Petrol. Sin.* **2013**, *29*, 3949–3967. (In Chinese with English Abstract)
19. Liu, Q.; Zhao, G.; Sun, M.; Eizenhöfer, P.R.; Han, Y.; Hou, W.; Zhang, X.; Wang, B.; Liu, D.; Xu, B. Ages and tectonic implications of Neoproterozoic ortho- and paragneisses in the Beishan Orogenic Belt, China. *Precambrian Res.* **2015**, *266*, 551–578. [CrossRef]
20. Yuan, Y.; Zong, K.Q.; He, Z.Y.; Klemd, R.; Liu, Y.S.; Hu, Z.C.; Guo, J.L.; Zhang, Z.M. Geochemical and geochronological evidence for a former early Neoproterozoic microcontinent in the South Beishan Orogenic Belt, southernmost Central Asian Orogenic Belt. *Precambrian Res.* **2015**, *266*, 409–424. [CrossRef]
21. Zong, K.Q.; Klemd, R.; Yuan, Y.; He, Z.Y.; Guo, J.L.; Shi, X.L.; Liu, Y.L.; Hu, Z.C.; Zhang, Z.M. The assembly of Rodinia: The correlation of early Neoproterozoic (ca. 900 Ma) high-grade metamorphism and continental arc formation in the southern Beishan Orogen, southern Central Asian Orogenic Belt (CAOB). *Precambrian Res.* **2017**, *290*, 32–48. [CrossRef]
22. He, S.P.; Ren, B.S.; Yao, W.G.; Fu, L.P. The division of tectonic units of Beishan area, Gansu-Inner Mongolia. *Northwestern Geol.* **2002**, *35*, 30–40. (In Chinese with English Abstract)
23. Zhang, Y.Y.; Guo, Z.J. Accurate constraint on formation and emplacement age of Hongliuhe ophiolite, boundary region between Xinjiang and Gansu Provinces and its tectonic implications. *Acta Petrol. Sin.* **2008**, *24*, 803–809. (In Chinese with English Abstract)
24. Yang, H.Q.; Zhao, G.B.; Li, Y.; Jiang, H.B.; Tan, W.J.; Ren, H.N.; Wang, X.A.; Xiong, Z.Y.; Zhang, X.P. The relationship between Paleozoic tectonic setting and mineralization in Xinjiang-Gansu-Inner Mongolia juncture area. *Geol. Bull. China* **2012**, *31*, 413–421. (In Chinese with English Abstract)
25. Nie, F.J.; Jiang, S.H.; Zhao, S.M.; Bai, D.M.; Liu, Y.; Zhao, Y.M.; Wang, X.Y.; Su, X.X. *Metallogenic Studies and Ore Prospecting in the Conjunction Area of Inner Mongolia Autonomous Region, Gansu Province and Xinjiang Uygur Autonomous Region (Beishan Mt), Northwest China*; Beijing Geological Publishing House: Beijing, China, 2002; pp. 1–408. (In Chinese with English Abstract)
26. Zheng, R.G.; Wu, T.R.; Zhang, W.; Xu, C.; Meng, Q.P. Late Paleozoic subduction system in the southern Central Asian Orogenic Belt, evidences from geochronology and geochemistry of the Xiaohuangshan ophiolite in the Beishan orogenic belt. *J. Asian Earth Sci.* **2013**, *62*, 463–475. [CrossRef]

27. He, Z.Y.; Klemd, R.; Yan, L.L.; Zhang, Z.M. The origin and crustal evolution of microcontinents in the Beishan orogen of the southern Central Asian Orogenic Belt. *Earth-Sci. Rev.* **2018**, *185*, 1–14. [CrossRef]
28. Song, D.F.; Xiao, W.J.; Han, C.M.; Tian, Z.H.; Li, Y.C. Accretionary processes of the central segment of Beishan, Constraints from structural deformation and 40Ar-39Ar geochronology. *Acta Petrol. Sin.* **2018**, *34*, 2087–2098. (In Chinese with English Abstract)
29. Sun, H.R.; Lv, Z.C.; Yu, X.F.; Li, Y.S.; Du, Z.Z.; Lv, X.; Gong, F.Y. Early Mesozoic tectonic evolution of Beishan orogenic belt, Constraints from chronology and geochemistry of the Late Triassic diabase dyke in Liuyuan area, Gansu province. *Acta Petrol. Sin.* **2020**, *36*, 1755–1768. (In Chinese with English Abstract)
30. Bai, R.L.; Liu, X.F.; Zhou, H.L. Genesis and tectonic setting of granite in the central region of the Beishan orogenic belt (Gansu section, China). *Acta Petrol. Sin.* **2020**, *36*, 1731–1754. (In Chinese with English Abstract)
31. Chen, C.; Teng, X.J.; Pan, Z.L.; Xiu, D.; He, J.Y.; Zhao, H.P.; Zhang, H.; Zhang, J.L.; Yang, J. LA-ICP-MS zircon U-Pb age of A-type granite from the Shibanjing area of middle Beishan orogenic belt, Inner Mongolia, and its constraint on closure time of Beishan Ocean. *Geol. Bull. China* **2020**, *39*, 1448–1460. (In Chinese with English Abstract)
32. Ding, J.; Han, C.; Xiao, W.; Wang, Z.; Song, D. Geochronology, geochemistry and Sr-Nd isotopes of the granitic rocks associated with tungsten deposits in Beishan district, NW China, Central Asian Orogenic Belt, Petrogenesis, metallogenic and tectonic implications. *Ore Geol. Rev.* **2017**, *89*, 441–462. [CrossRef]
33. Fu, P.F.; Xia, D.; Gong, X.P.; Sun, Q.L.; Yang, Y.; Wei, J.; Zhao, Q. LA-ICP-MS Zircon U-Pb Age, Geochemical Characteristics and Geological Significance of Qingbaishan Granite in Beishan Area, Xinjiang. *Xinjiang Geol.* **2020**, *38*, 25–31. (In Chinese with English Abstract)
34. Lei, R.X.; Wu, C.Z.; Gu, L.X.; Zhang, Z.Z.; Chi, G.X.; Jiang, Y.H. Zircon U–Pb chronology and Hf isotope of the Xingxingxia granodiorite from the Central Tianshan zone (NW China), Implications for the tectonic evolution of the southern Altaids. *Gondwana Res.* **2011**, *20*, 582–593. [CrossRef]
35. Pan, Z.; Zhang, H.; Chen, C.; Wang, S.; Li, Q. Zircon U-Pb geochronology and Lu-Hf isotope compositions of porphyritic syenite granite from Aobaohuturen in Beishan, Inner Mongolia and its tectonic implication. *China J. Geol.* **2017**, *52*, 301–316. (In Chinese with English Abstract)
36. Guo, Q.; Xiao, W.; Hou, Q.; Windley, B.F.; Han, C.; Tian, Z.; Song, D. Construction of Late Devonian Dundunshan arc in the Beishan orogen and its implication for tectonics of southern Central Asian Orogenic Belt. *Lithos* **2014**, *184*, 361–378. [CrossRef]
37. Yuan, Y.; Zong, K.Q.; He, Z.Y.; Klemd, R.; Jiang, H.Y.; Zhang, W.; Liu, Y.S.; Hu, Z.C.; Zhang, Z.M. Geochemical evidence for Paleozoic crustal growth and tectonic conversion in the Northern Beishan Orogenic Belt, southern Central Asian Orogenic Belt. *Lithos* **2018**, *302*, 189–202. [CrossRef]
38. Yu, J.Y.; Li, X.M.; Wang, G.Q.; Wu, P.; Yan, Q.J. Zircon U-Pb ages of Huitongshan and Zhangfangshan ophiolite in Beishan of Gansu-Inner Mongolia border area and their significance. *Geol. Bull. China* **2012**, *31*, 2038–2045. (In Chinese with English Abstract)
39. Yu, F.S.; Li, J.B.; Wang, T. The U-Pb Isotopic Age of Zircon from Hongliuhe Ophiolites in East Tianshan Mountains, Northwest China. *Acta Geosci. Sin.* **2006**, *27*, 213–216. (In Chinese with English Abstract)
40. Zhao, Z.H.; Guo, Z.J.; Wang, Y. Geochronology, geochemical characteristics and tectonic implications of the granitoids from Liuyuan area, Beishan, Gansu province, northwest China. *Acta Petrol. Sin.* **2007**, *23*, 1847–1860. (In Chinese with English Abstract)
41. Guo, Z.J.; Shi, H.Y.; Zhang, Z.C.; Zhang, J.J. The tectonic evolution of the south Tianshan paleo-oceanic crust inferred from the spreading structures and Ar-Ar dating of the Hongliuhe ophiolite, NW China. *Acta Petrol. Sin.* **2006**, *22*, 95–102. (In Chinese with English Abstract)
42. Li, S.; Wang, T.; Tong, Y.; Wang, Y.B.; Hong, D.W.; Ouyang, Z.X. Zircon U-Pb age, origin and its tectonic significances of Huitongshan Devonian K-feldspar granites from Beishan orogen, NW China. *Acta Petrol. Sin.* **2011**, *27*, 3055–3070. (In Chinese with English Abstract)
43. Liu, Y.S.; Hu, Z.C.; Gao, S.; Günther, D.; Xu, J.; Gao, C.G.; Chen, H.H. In situ analysis of major and trace elements of anhydrous minerals by LA-ICP-MS without applying an internal standard. *Chem. Geol.* **2008**, *257*, 34–43. [CrossRef]
44. Liu, Y.S.; Hu, Z.C.; Zong, K.Q.; Gao, C.G.; Gao, S.; Xu, J.; Chen, H.H. Reappraisement and refinement of zircon U-Pb isotope and trace element analyses by LA-ICP-MS. *Chin. Sci. Bull.* **2010**, *55*, 1535–1546. (In Chinese with English Abstract) [CrossRef]
45. Ludwig, K.R. *ISOPLOT 3.00: A Geochronological Toolkit for Microsoft Excel*; Berkeley Geochronology Center: Berkeley, CA, USA, 2003.
46. Hu, Z.C.; Liu, Y.S.; Gao, S.; Liu, W.; Yang, L.; Zhang, W.; Tong, X.; Lin, L.; Zong, K.Q.; Li, M.; et al. Improved in situ Hf isotope ratio analysis of zircon using newly designed X skimmer cone and Jet sample cone in combination with the addition of nitrogen by laser ablation multiple collector ICP-MS. *J. Anal. At. Spectrom.* **2012**, *27*, 1391–1399. [CrossRef]
47. Zhang, W.; Hu, Z.C.; Spectroscopy, A. Estimation of Isotopic Reference Values for Pure Materials and Geological Reference Materials. *At. Spectrosc.* **2020**, *41*, 93–102. [CrossRef]
48. Whison, M. *Igneous Petrogenesis*; Wnwin Hyman: London, UK, 1989.
49. Peccerillo, A.R.; Taylor, S.R. Geochemistry of eocene calc-alkaline volcanic rocks from the Kastamonu area, Northern Turkey. *Contrib. Mineral. Petrol.* **1976**, *58*, 63–81. [CrossRef]
50. Maniar, P.D.; Piccoli, P.M. Tectonic discrimination of granitoids. *Geol. Soc. Am. Bull.* **1989**, *101*, 635–643. [CrossRef]
51. Wang, L.H.; Yang, J.G.; Xie, C.L.; Lei, Y.X.; Jin, S. Geochronology and geochemistry of haergentoukoubu granites in the beishan Area, Gansu, China and their geological significanc. *Acta Geol. Sin.* **2009**, *83*, 377–387. (In Chinese with English Abstract)

52. Zhu, J.; Lv, X.; Peng, S. U-Pb zircon geochronology, geochemistry and tectonic implications of the early Devonian granitoids in the Liuyuan area, Beishan, NW China. *Geosci. J.* **2016**, *20*, 609–625. (In Chinese with English Abstract) [CrossRef]
53. Sun, S.S.; McDonough, W.F. Chemical and isotopic system atics of oceanic basalts, implications for mantle compositi on and processes. *Lond. Geol. Soc. Spec. Publ.* **1989**, *42*, 313–345. [CrossRef]
54. Zhang, J.J.; Sun, S.J.; Zhang, L.P.; Deng, J.H.; Li, S.; Sun, W.D. Origin of transitional I-A-type syenite and its relationship to A-type intrusions in the Luzong Basin, the Lower Yangtze River Belt: Insights from geochemistry. *Chem. Geol.* **2023**, *626*, 121458. [CrossRef]
55. King, P.L.; White, A.J.R.; Chappell, B.W.; Allen, C.M. Characterization and origin of aluminous A-type granites from the Lachlan Fold Belt, southeastern Australia. *J. Petrol.* **1997**, *38*, 371–391. [CrossRef]
56. Whalen, J.B.; Currie, K.L.; Chappell, B.W. A-type granites, geochemical characteristics, discrimination and petrogenesis. *Contrib. Mineral. Petrol.* **1987**, *95*, 407–419. [CrossRef]
57. Frost, B.; Barnes, C.; Collins, W.; Arculus, R.; Ellis, D.; Frost, C.A. Geochemical Classification for Granitic Rocks. *J. Petrol.* **2001**, *42*, 2033–2048. [CrossRef]
58. Collins, W.J.; Beams, S.D.; White, A.J.R.; Chappell, B.W. Nature and origin of A-type granites with particular reference to southeastern Australia. *Contrib. Mineral. Petrol.* **1982**, *80*, 189–200. [CrossRef]
59. Turner, S.P.; Foden, J.D.; Morrison, R.S. Derivation of some A-type magmas by fractionation of basaltic magma, an example from the Padthaway Ridge, South Australia. *Lithos* **1992**, *28*, 151–179. [CrossRef]
60. Beyth, M.; Stern, R.J.; Altherr, R.; Kröner, A. The late Precambrian Timna igneous complex, southern Israel, evidence for comagmatic-type sanukitoid monzodiorite and alkali granite magma. *Lithos* **1994**, *31*, 103–124. [CrossRef]
61. Han, B.F.; Wang, S.G.; Jahn, B.M.; Hong, D.W.; Kagami, H.; Sun, Y.L. Depleted-mantle source for the Ulungur River A-type granites from North Xinjiang, China, geochemistry and Nd–Sr isotopic evidence, and implications for Phanerozoic crustal growth. *Chem. Geol.* **1997**, *138*, 135–159. (In Chinese with English Abstract) [CrossRef]
62. Mushkin, A.; Navon, O.; Halicz, L.; Hartmann, G.; Stein, M. The petrogenesis of A-type magmas from the Amram Massif, southern Israel. *J. Petrol.* **2003**, *44*, 815–832. [CrossRef]
63. Konopelko, D.; Biske, G.; Seltmann, R.; Eklund, O.; Belyatsky, B. Hercynian post-collisional A-type granites of the Kokshaal Range, southern Tien Shan, Kyrgyzstan. *Lithos* **2007**, *97*, 140–160. [CrossRef]
64. Mingram, B.; Trumbull, R.B.; Littman, S.; Gerstenberger, H. A petrogenetic study of anorogenic felsic magmatism in the Cretaceous Paresis ring complex, Namibia, evidence for mixing of crust and mantle-derived components. *Lithos* **2000**, *54*, 1–22. [CrossRef]
65. Qiu, J.S.; Wang, D.Z.; Brent, I.A.M. Geochemistry and petrogenesis of the I- and A-type composite granitemasses in the coastal area of Zhejiang and Fujian province. *Acta Petrol. Sin.* **1999**, *15*, 237–246. (In Chinese with English Abstract)
66. Yang, J.H.; Wu, F.Y.; Chung, S.L.; Wilde, S.A.; Chu, M.F. A hybrid origin for the Qianshan A-type granite, northeast China, Geochemical and Sr–Nd–Hf isotopic evidence. *Lithos* **2006**, *89*, 89–106. [CrossRef]
67. Clemens, J.D.; Holloway, J.R.; White, A.J.R. Origin of an A-type granite: Experimental constraints. *Am. Mineral.* **1986**, *71*, 317–324.
68. Creaser, R.A.; Price, R.C.; Wormald, R.J. A-type granites revisited, assessment of a residual-source model. *Geology* **1991**, *19*, 163–166. [CrossRef]
69. Anderson, J.L.; Bender, E.E. Nature and origin of Proterozoic A-type granitic magmatism in the southwestern United States of America. *Lithos* **1989**, *23*, 19–52. [CrossRef]
70. Wu, F.Y.; Sun, D.Y.; Li, H.; Jahn, B.M.; Wilde, S.A. A-type granites in northeastern China, age and geochemical constraints on their petrogenesis. *Chem. Geol.* **2002**, *187*, 143–173. [CrossRef]
71. Jung, S.; Pfander, J.A. Source composition and melting temperatures of orogenic granitoids, constraints from CaO/Na_2O, Al_2O_3/TiO_2 and accessory mineral saturation thermometry. *Northwest Med. Educ.* **2007**, *19*, 19–52. [CrossRef]
72. Sylvester, P.J. Post-collisional strongly peraluminous granites. *Lithos* **1998**, *45*, 29–44. [CrossRef]
73. Belousova, E.; Griffin, W.; Oreilly, S.Y.; Fisher, N. Igneous zircon, Trace element composition as an indicator of source rock type. *Contrib. Mineral. Petrol.* **2002**, *143*, 602–622. [CrossRef]
74. Wu, F.Y.; Li, X.H.; Zheng, Y.F.; Gao, S. Lu-Hf isotopic systematics and thier applications in petrology. *Acta Petrol. Sin.* **2007**, *23*, 185–220. (In Chinese with English Abstract)
75. Altherr, R.; Holl, A.; Hegner, E.; Langer, C.; Kreuzer, H. High-potassium, calc-alkaline I-type plutonism in the European Variscides, northern Vosges (France) and northern Schwarzwald (Germany). *Lithos* **2000**, *50*, 51–73. [CrossRef]
76. Altherr, R.; Siebel, W. I-type plutonism in a continental back-arc setting, Miocene granitoids and monzonites from the central Aegean Sea, Greece. *Contrib. Mineral. Petrol.* **2002**, *143*, 397–415. [CrossRef]
77. Long, X.; Yuan, C.; Sun, M.; Zhao, G.; Xiao, W.; Wang, Y.; Yang, Y.; Hu, A. Archean crustal evolution of the northern Tarim craton, NW China, Zircon U–Pb and Hf isotopic constraints. *Precambrian Res.* **2010**, *180*, 272–284. [CrossRef]
78. Hou, Q.Y.; Wang, Z.; Liu, J.B.; Wang, J.; Li, D.P. Geochemistry Characteristics and SHRIMP Dating of Yueyashan Ophiolite in Beishan Orogen. *Geoscience* **2012**, *265*, 1008–1018.
79. Shi, Y.; Zhang, W.; Kroener, A.; Li, L.; Jian, P. Cambrian ophiolite complexes in the Beishan area, China, southern margin of the Central Asian Orogenic Belt. *J. Asian Earth Sci.* **2017**, *153*, 193–205. [CrossRef]
80. Tian, J.; Xin, H.T.; Teng, X.J.; Duan, X.L.; Cheng, X.Y.; Zhang, Y.; Ren, B.F. The determination of volcanic rocks in Upper Devonian Dundunshan Formation in the Baiyunshan area of Beishan orogenic belt, Inner Mongolia. *Acta Petrol. Sin.* **2020**, *36*, 509–525. (In Chinese with English Abstract)

81. Thorkelson, D.J.; Breitsprecher, K. Partial melting of slab window margins: Genesis of adakitic and non-adakitic magmas. *Lithos* **2005**, *79*, 25–41. [CrossRef]
82. Xiu, D.; Chen, C.; Zhuan, S.D.; Shen, Z.Y.; Wang, J.G.; Cheng, Z.; Zhang, L.G.; Wang, S.; Yang, X.P.; Hou, D.H.; et al. Zircon U-Pb age and petrogenesis of tonalite-quartz diorite in the Shibanjing area, central Beishan orogenic belt, and its constraint on subduction of the ancient oceanic basin. *Geol. Bull. China* **2018**, *37*, 975–986. (In Chinese with English Abstract)
83. Yu, J.Y.; Guo, L.; Li, J.X.; Li, Y.G.; Smithies, R.H.; Wingate, M.T.D.; Meng, Y.; Chen, S.F. The petrogenesis of sodic granites in the Niujuanzi area and constraints on the Paleozoic tectonic evolution of the Beishan region, NW China. *Lithos* **2016**, *256*, 250–268. [CrossRef]
84. Wu, P.; Wang, G.Q.; Li, X.M.; Yu, J.Y.; Kang, L. The age of Niujuanzi ophiolite in Beishan area of Gansu Province and its geological significance. *Geol. Bull. China* **2012**, *31*, 2032–2037. (In Chinese with English Abstract)
85. Pearce, J.A. Sources and settings of granitic rocks. *Episodes* **1996**, *19*, 120–125. [CrossRef]
86. Eby, G.N. The A-type granitoids: A review of their occurrence and chemical characteristics and speculations on their petrogenesis. *Lithos* **1990**, *26*, 115–134. [CrossRef]
87. Boehnke, P.; Watson, E.B.; Trail, D.; Harrison, T.M.; Schmitt, A.K. Zircon saturation re-revisited. *Chem. Geol.* **2013**, *351*, 324–334. [CrossRef]
88. Watson, E.B.; Harrison, T.M. Zircon Saturation Revisited, Temperature and Composition Effect in Avariety of Crustal Magmas Types. *Earth Planet. Sci. Lett.* **1983**, *64*, 295–304. [CrossRef]
89. Zhan, Q.Y.; Zhu, D.C.; Wang, Q.; Cawood, P.A.; Xie, J.C.; Li, S.M.; Wang, R.; Zhang, L.L.; Zhao, Z.D.; Dong, J. Constructing the Eastern Margin of the Tibetan Plateau During the Late Triassic. *J. Geophys. Res. Solid Earth* **2018**, *123*, 10449–104459. [CrossRef]
90. Yu, J.W.; Wang, Z.W.; Zhu, T.C.; Xing, K.; Meng, W.; Zhou, J.L.; Li, Y.G. Carboniferous Crustal Evolution in Northern Erenhot, Constraintfrom Zircon U-Pb Hf Isotopes of Volcano clastic Rocks in Halatumiao Group. *J. Jilin Univ. (Earth Sci. Ed.)* **2022**, *54*, 1153–1173. (In Chinese with English Abstract)

minerals

MDPI

Article

Petrogenesis and Tectonic Evolution of Huashigou Granitoids in the South Qilian Orogen, NW China: Constraints from Geochronology, Geochemistry, and Sr–Nd–Hf–O Isotopes

Yuxi Wang [1,2], Wanfeng Chen [3,*], Jinrong Wang [4], Zhilei Jia [5], Qingyan Tang [4] and Pengfei Di [6]

1 Gold Mine Resource Exploration and Utilization Technology Innovation Center of Gansu Province, Lanzhou 730050, China; wangyx15@lzu.edu.cn
2 The Third Institute of Geology and Minerals Exploration, Gansu Provincial Bureau of Geology and Minerals Exploration and Development, Lanzhou 730050, China
3 College of Geography and Tourism, Hunan University of Arts and Science, Changde 415000, China
4 Key Laboratory of Mineral Resources in Western China, School of Earth Sciences, Lanzhou University, Lanzhou 730000, China; jrwang@lzu.edu.cn (J.W.); tangqy@lzu.edu.cn (Q.T.)
5 Gansu Provincial Bureau of Geology and Mineral Exploration and Development, Lanzhou 730000, China; kaka20100502@outlook.com
6 Gansu Assessment Center of Mineral Resources Reserves, Lanzhou 730000, China; dipf15@lzu.edu.cn
* Correspondence: chenwanfeng@huas.edu.cn

Abstract: The origin of granitic rocks from the South Qilian orogenic belt is of great significance for understanding the continental tectonic framework of Western China. Currently, scholars have different opinions on the tectonic evolution of the South Qilian. Huashigou granite, which exhibits multiple intrusive episodes, is a suitable example for studying the tectonic evolution of the South Qilian. New zircon U–Pb ages and the whole-rock elemental and Sr–Nd–Hf–O isotopic compositions of Huashigou granitic rocks are presented here to investigate their petrogenesis and discuss the tectonic implications for the evolution of the South Qilian orogenic belt. LA-ICP-MS zircon U–Pb dating yielded crystallization ages of 368.7 ± 3.5 Ma, 261.5 ± 0.63 Ma, and 262.2 ± 1.4 Ma for granodiorites from the Hua1 pluton, quartz diorites from the Hua2 pluton, and porphyritic tonalites from the Hua4 pluton, respectively. Late Devonian granodiorites from the Hua1 pluton belonged to the metaluminous calc-alkaline series and were characterized by an enrichment in LREEs, a depletion in HREEs, negative Eu anomalies, and Sr/Y ratios of 9.17 to 11.67. They showed $(^{87}Sr/^{86}Sr)_i$ ratios of 0.712356 to 0.71195, $\varepsilon_{Nd}(t)$ values of −6.56 to −6.14, and an $\varepsilon_{Hf}(t)$ value of −2.06. Middle Permian granitic rocks mainly consisted of quartz diorites and porphyritic tonalites, which are part of the metaluminous tholeiitic series and weakly peraluminous tholeiitic series, respectively. Quartz diorites from the Hua2 pluton were characterized by an enrichment in LREEs, depletions in HREEs and HESEs, weak negative Eu anomalies, and Sr/Y ratios of 13.25 to 14.79. They showed $(^{87}Sr/^{86}Sr)_i$ ratios of 0.705905 to 0.705971, $\varepsilon_{Nd}(t)$ values of +0.78 to +0.82, and a $\delta^{18}O_{V\text{-}SMOW}$ value of 12.4‰. Porphyritic tonalites of the Hua4 pluton were characterized by an enrichment in LREEs, depletions in HREEs and HESEs, weak negative Eu anomalies, and Sr/Y ratios of 9.22 to 12.74. They showed $(^{87}Sr/^{86}Sr)_i$ ratios of 0.719528, $\varepsilon_{Nd}(t)$ values of −8.57, and a $\delta^{18}O_{V\text{-}SMOW}$ value of 11.8‰. We can conclude that Late Devonian granodiorites were derived from the partial melting of enriched and shallow-depth crustal materials, whereas Middle Permian granitic rocks were formed by the delamination of a thickened lower crust after the closure of the Paleo-Tethys Ocean, which caused the underplating of mantle-derived basaltic magma, inducing the partial melting of the lower crust at different depths. Our results show that there were at least two important stages of compressional and extensional tectonic switches in the South Qilian orogenic belt during the Late Paleozoic Era, and the evolution of Altyn Tagh–Qilian–North Qaidam had evident stages.

Keywords: granitoids; geochemistry; Sr–Nd–Hf–O isotopes; tectonic evolution; Qilian orogenic belt

Citation: Wang, Y.; Chen, W.; Wang, J.; Jia, Z.; Tang, Q.; Di, P. Petrogenesis and Tectonic Evolution of Huashigou Granitoids in the South Qilian Orogen, NW China: Constraints from Geochronology, Geochemistry, and Sr–Nd–Hf–O Isotopes. *Minerals* **2024**, *14*, 71. https://doi.org/10.3390/min14010071

Academic Editors: Jaroslav Dostal, Jiafu Chen, Nan Ju and Zhonghai Zhao

Received: 20 October 2023
Revised: 15 December 2023
Accepted: 18 December 2023
Published: 6 January 2024

1. Introduction

The Qilian orogenic belt, located at the northeastern margin of the Tibetan Plateau, is located in the western part of the central orogenic belt that runs across China from east to west [1–5]. The Qaidam Block is located to the south of the belt. The Tarim Craton is located northwest of the belt, delineated by the Altyn Tagh Fault (Figure 1a). A majority of studies have been conducted to investigate the evolution of the Altyn Tagh–Qilian–North Qaidam tectonic system [6–27]. These studies focus on the discovery of an ultrahigh-pressure metamorphic belt (UHPB) in the North Qaidam and Altyn Tagh regions and its comparison to the Altyn Tagh–Qilian–North Qaidam tectonic system [3,28–33]. As a ubiquitous intrusive mass in the orogenic belt, granitoids have recorded much information about plate tectonics and crustal evolution [12,34–40]. Additionally, there are similarities between the tectonic units on both sides of Altyn Tagh.

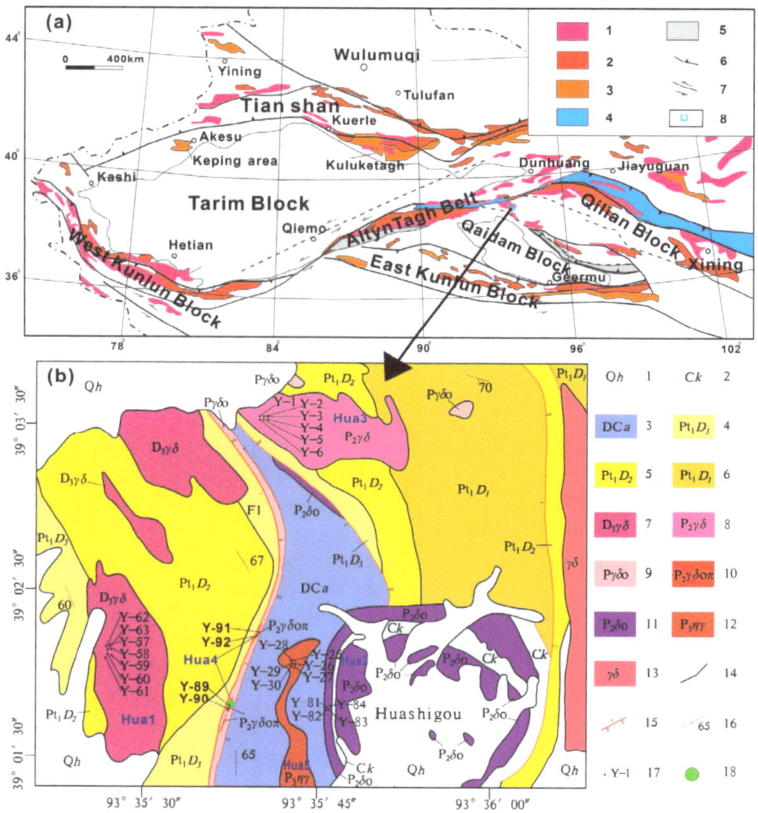

Figure 1. Tectonic location (**a**) and geological sketch map (**b**) of the South Qilian Orogen. (**a**): 1. Archean–Paleoproterozoic basement; 2. Mesoproterozoic rock; 3. Neoproterozoic rock; 4. Early Paleozoic subduction collision complex belt; 5. Early Paleozoic subduction hyperplasia complex belt; 6. thrust fault; 7. strike-slip fault; and 8. study area. (**b**): 1. Quaternary; 2. Carboniferous Yanghugou *Group*; 3. Devonian–Carboniferous Amunike *Group*; 4. Paleoproterozoic Dakendaban *Group* (the Third Rock *Group*); 5. Paleoproterozoic Dakendaban *Group* (the Second Rock *Group*); 6. Paleoproterozoic Dakendaban *Group* (the First Rock *Group*); 7. Late Devonian granodiorites (Hua1); 8. Middle Permian granitoids (Hua3); 9. quartz mica diorites; 10. Middle Permian quartz mica porphyritic diorites (Hua4); 11. Middle Permian quartz diorites (Hua2); 12. Late Permian monzogranites (Hua5); 13. granodiorites; 14. geological boundary; 15. faults; 16. attitude; 17. sample locations; and 18. copper deposits.

Previous studies have mainly focused on Early Paleozoic granitic rocks in the North Qilian orogenic belt, while studies on Late Paleozoic granitoids are relatively scarce. Since the closure of the Qilian orogenic belt in the Mid–Late Paleozoic, there has been a paucity of pertinent research findings. Consequently, Late Paleozoic granitic rocks serve as a pivotal means to unveil not only their geodynamic implications but also the geological evolution of the Qilian orogenic belt.

In this study, we present results from petrographic, geochronologic, and whole-rock elemental and Sr–Nd–Hf–O isotopic composition analyses in an attempt to investigate the petrogenesis of Paleozoic Huashigou granitoids in the South Qilian along the northeastern margin of the Tibetan Plateau. These data could be used to infer and determine the age, genesis, and tectonic environment of the rocks formed during that period in the South Qilian. The data were used to evaluate the crustal evolution of the Qilian Orogen in the Late Paleozoic.

2. Regional Geology and Petrography

The Huashigou area is located at the southwest margin of the Qilian Orogen, which is at the junction of the Altyn Tagh–Qilian–North Qaidam tectonic belt. Paleoproterozoic (Dakendaban *Group*), Devonian (Amunike *Formation*), and Carboniferous (Keluke *Formation*) strata are widely distributed in the study area. The Paleoproterozoic Dakendaban *Group* represents a major stratigraphic formation developed in this area. The Devonian and Carboniferous strata are mainly distributed at the center of the study area. Faults extend in the north–south and north–west directions. The Huashigou area consists of five types of granitic rocks, including granodiorite, biotite granodiorite, porphyritic tonalite, quartz diorite, and monzonitic granite (Figure 1b).

Three granodiorite intrusions exist, with sizes ranging from 2 to 0.1 km^2, which intruded into Paleoproterozoic metamorphic rocks (Dakendaban *Group*). The Hua1 pluton is composed of plagioclase (47%), quartz (20%), albite (18%), biotite (10%), hornblende (5%), and accessory minerals, including apatite, zircon, sphene, and magnetite. Plagioclases are wide and tabular and are intensively altered, with sericitization, decarbonation, and clayzation alterations. Some plagioclase grains wrapped in microcline have an obvious albite reaction rim. Quartzes are in a xenomorphic–granular form, and some of them show wavy extinction in rare situations.

The Hua2 pluton is mainly located south of the study area, with extensive exposure intruding in the Paleoproterozoic Dakendaban *Group* and Carboniferous Keluke *Group*, with a long strip layout and small areas in general. The pluton exhibits a subhedral–granular texture and a blocky structure and consists of plagioclase (70%), normal hornblende (15%), quartz (10%), and biotite (5%), with minor accessory minerals, including apatite, zircon, sphene, and magnetite. Plagioclases are mainly tabular and dotted with small amounts of secondary minerals, such as clay, microcryptocrystalline zoisite, sericite, and carbonates. Hornblende is intensively altered, with actinolitization along fractures.

The Hua3 pluton is located north of the study area, covering about 2 km^2 in a nearly elliptical shape. With a fine-grained granite texture, the pluton mainly has a blocky structure and is composed of plagioclase (49%), orthoclase (15%), quartz (18%), biotite (15%), microcline (3%), and accessory minerals, including apatite, zircon, and magnetite. The tabular biotite is slightly wrapped in orthoclase, and the plagioclase has a visible polysynthetic twin and a mostly rapakivi structure. LA-ICP-MS zircon U–Pb dating indicated that the pluton was emplaced in the Middle Permian (261.1 $\pm$ 3.8 Ma [41]).

The irregular zonal Hua4 pluton is mainly distributed along the F1 Fault. The pluton shows typical porphyritic textures, and its phenocrysts include plagioclase and quartz. Its groundmass is composed of plagioclase, quartz, sericite, and biotite, with microgranular, cataclastic, and blocky structures. The plagioclase phenocrysts are in lenticular and elliptical shapes, and some parts have a wide tabular shape, with particle sizes ranging from 0.6 to 1.0 mm. Quartz phenocrysts have an elliptical shape, aggregating along the fractures with a sericite particle size of $\pm$1.0 mm.

The Hua5 pluton intruded into the Devonian Amunike *Group*, with an exposed surface area of about 0.4 km^2. The pluton displays a fine-grained granitic and blocky structure and consists of plagioclase (45%), quartz (23%), biotite (12%), perthite (12%), microcline (7%), and minor accessory minerals, such as apatite, sphene, zircon, and magnetite. LA-ICP-MS zircon U–Pb dating indicated that the pluton was formed in the Late Permian (252.0 ± 2.1 Ma [42]).

3. Analytical Methods

Sixteen samples were collected from the granitic plutons in the Huashigou area, including seven from the Hua1 pluton (YQ57–YQ63), four from the Hua2 pluton (YQ81–84), and four from the Hua4 pluton (YQ89–92). The locations of the analyzed samples are shown in Figure 1b.

3.1. Zircon U–Pb Age Analytical Method

Firstly, zircon grains were separated using conventional magnetic techniques and handpicked using a binocular microscope (BX51, Olympus, Tokyo, Japan). The zircon grains were mounted in an epoxy mount and polished down to a half-section. Prior to the isotope analysis, all grains were photographed under transmitted and reflected light and subsequently examined using cathode luminescence (CL) images (JSM-6510, JEOL, Tokyo, Japan). The zircon LA-ICP-MS U–Pb analyses were performed on a 7500 ICP-MS (Agilent, Santa Clara, CA, USA) with a UP193SS LA (New Wave, Seattle, WA, USA) at the Key Laboratory of Lithospheric Tectonics and Exploration of the Ministry of Education, China University of Geosciences (Beijing), China. The samples were analyzed in runs, with four GJ-1 zircon standards measured every ten to twelve samples. The analytical data were then reduced, calculated, and plotted using the Squid 1.0 and Isoplot programs (version 4.00.08.09.16, by KR Ludwig, Berkeley Geochr. Ctr., Berkeley, CA, USA) [43]. The compositions of common Pbs were corrected following the Andersen method [44]. Errors during the individual analyses are cited at 1σ, and the weighted age uncertainties are quoted at a confidence level of 95%.

3.2. Whole-Rock, Sr–Nd–O, and Hf Isotopic Analytical Method

The whole-rock major elements, trace elements, and Sr–Nd–Hf isotopes were determined at the State Key Laboratory of Continental Dynamics, Northwest University, China. The samples for the whole-rock chemical analysis were crushed to 200 mesh. The major-element contents of the samples were analyzed using the RIX-2100 (RIGAKU, Tokyo, Japan) X-ray fluorescence spectrometer (XRF). The analytic uncertainties were generally less than 2% for most of the major elements and 2% to 5% for the major elements with <1 wt.% contents. Trace elements and rare-earth elements (REEs) were analyzed using the Elan-6100DRC model ICP-MS (Perkin Elmer, Waltham, MA, USA). About 50 mg of the powder samples was dissolved in high-pressure Teflon bombs using a HF + HNO_3 mixture. The analytical precision was better than 5%. The Nd, Sr, and Hf isotopic analyses were performed using MC-ICP-MS (Nu Instruments, North Wales, UK), and the isotopic ratios were, respectively, normalized to $^{143}Nd/^{144}Nd$ = 0.7219, $^{87}Sr/^{86}Sr$ = 8.3721, and $^{176}Hf/^{177}Hf$ = 0.7325. The oxygen-stable isotopes of the whole-rock samples were determined at the Institute of Mineral Resources, Chinese Academy of Geological Science (Beijing, China). The oxygen isotopic compositions ($\delta^{18}O$) were analyzed using the BrF_5 method and measured using the MAT-253 mass spectrometer (MAT-253, Thermo Fisher Scientific, Waltham, MA, USA), according to international standards (SMOW), with a precision of ±0.2 ‰.

4. Results

Fifteen samples were collected for the major- and trace-element analyses, of which three samples were used for zircon U–Pb dating, six samples were used for Sr–Nd isotopic analyses, one sample was used for Hf isotopic analyses, and two samples were used for O isotopic analyses. The detailed data are presented in Tables 1–3.

Table 1. Major- and trace-element compositions of Huashigou granitoids.

Lithology				Hua1					Hua2					Hua4	
Sample	YQ-57	YQ-58	YQ-59	YQ-60	YQ-61	YQ-62	YQ-63	YQ-81	YQ-82	YQ-83	YQ-84	YQ-89	YQ-90	YQ-91	YQ-92
Major elements (%)															
SiO_2	67.78	68.48	67.81	68.87	68.16	68.35	68.44	52.41	53.32	52.43	52.54	67.28	66.43	63.64	67.01
TiO_2	0.41	0.40	0.48	0.42	0.44	0.42	0.41	1.58	1.40	1.57	1.57	0.50	0.49	0.55	0.48
Al_2O_3	15.77	15.29	15.43	15.26	15.41	15.52	15.45	19.69	19.66	19.67	19.68	16.07	16.34	16.62	16.18
TFe_2O_3	3.69	3.62	4.01	3.67	3.83	3.68	3.60	9.23	8.71	9.01	8.91	3.95	3.83	4.34	3.72
MnO	0.06	0.06	0.06	0.06	0.06	0.06	0.06	0.09	0.11	0.09	0.09	0.03	0.04	0.06	0.04
MgO	1.22	1.19	1.38	1.20	1.31	1.25	1.21	3.18	2.99	3.25	3.14	1.81	1.73	2.50	2.09
CaO	3.80	3.48	3.72	3.39	3.65	3.66	3.56	8.07	7.97	8.11	7.91	1.65	2.08	2.93	1.82
Na_2O	3.79	3.60	3.53	3.57	3.65	3.68	3.65	4.81	4.75	4.91	4.77	4.13	4.39	3.50	4.41
K_2O	2.71	3.06	2.83	3.06	3.02	2.99	3.01	0.64	0.63	0.65	0.68	3.77	3.63	3.80	3.61
P_2O_5	0.11	0.13	0.15	0.13	0.14	0.13	0.13	0.23	0.24	0.24	0.22	0.14	0.14	0.12	0.14
LOI	0.95	0.90	0.94	0.82	0.61	0.67	0.77	0.31	0.54	0.34	0.29	0.95	1.06	1.47	0.86
TOTAL	100.29	100.21	100.34	100.45	100.28	100.41	100.29	100.24	100.32	100.27	99.8	100.28	100.16	99.53	100.36
$Mg^{\#}$	37.24	37.10	38.18	36.98	38.03	37.87	37.62	38.20	38.12	39.29	38.74	45.12	44.77	50.83	50.20
A/CNK	1.07	1.05	1.07	1.06	1.04	1.05	1.05	1.06	1.07	1.04	1.07	1.15	1.11	1.12	1.13
Trace elements ($\times 10^{-6}$)															
Rb	90.18	105.55	95.42	107.30	106.60	105.01	105.51	16.78	17.93	16.61	18.61	146.57	140.96	190.47	164.27
Cs	2.44	2.92	2.50	3.15	3.44	3.31	3.36	0.88	0.89	0.87	0.91	4.52	4.31	6.09	5.11
Sr	226.50	211.00	218.50	205.30	210.30	213.10	210.50	585.70	573.90	584.60	593.10	172.40	197.90	217.30	194.90
Ba	690.00	702.30	777.70	685.20	700.70	679.20	666.00	191.40	177.60	194.00	203.50	1066.10	993.00	1015.30	1103.40
Sc	7.81	8.08	9.25	8.28	8.77	8.01	8.23	27.04	24.30	27.61	25.72	10.40	10.62	9.33	10.30
Nb	14.02	14.82	16.40	15.90	15.31	14.68	14.69	11.41	11.29	11.34	11.80	13.81	13.30	13.59	13.18
Ta	0.98	1.11	1.14	1.28	1.10	1.14	1.08	0.53	0.60	0.55	0.57	1.00	1.00	0.93	0.95
Zr	196.99	200.86	195.42	194.65	223.69	192.78	230.61	213.65	227.58	215.93	194.50	171.99	184.44	143.45	155.80
Hf	4.82	5.08	4.91	5.04	5.66	4.89	5.72	4.86	5.23	4.91	4.42	4.42	4.70	3.64	4.07
U	1.59	1.62	1.75	1.61	1.50	1.46	1.89	0.54	0.77	0.58	0.44	3.48	3.02	2.50	1.85
Th	13.76	19.08	18.32	20.48	18.26	17.15	19.12	2.59	2.59	2.41	2.99	14.53	16.05	9.89	13.48
Cr	7.12	7.69	8.68	6.81	7.53	7.16	6.70	5.25	3.67	4.14	5.07	13.32	10.51	9.50	10.30
Ni	6.15	6.34	7.21	5.88	6.47	6.20	6.15	8.80	6.65	7.46	9.45	11.71	10.71	11.99	9.59
La	39.88	47.78	49.15	55.70	51.45	41.47	48.44	22.45	21.54	22.81	24.04	35.27	43.04	39.44	28.11
Ce	71.71	83.73	90.00	99.07	89.04	73.19	84.75	56.02	55.19	56.33	56.94	61.80	72.08	64.31	49.20
Pr	7.01	8.53	8.96	9.72	8.97	7.48	8.42	7.93	8.01	8.00	7.88	6.44	7.44	6.40	5.17
Nd	23.53	28.52	30.33	32.69	30.10	25.57	28.36	36.66	38.06	36.75	35.55	22.45	25.07	22.19	18.42
Sm	3.92	4.70	5.06	5.20	4.90	4.39	4.60	8.63	9.04	8.74	8.27	3.89	4.19	3.65	3.37
Eu	1.09	1.04	1.11	1.03	1.07	1.05	1.04	2.40	2.36	2.42	2.37	0.82	0.95	1.58	0.82
Gd	3.58	4.23	4.55	4.59	4.36	4.04	4.12	8.43	8.67	8.60	8.10	3.45	3.83	3.27	3.20
Tb	0.53	0.62	0.67	0.67	0.64	0.60	0.60	1.30	1.35	1.33	1.25	0.52	0.57	0.46	0.50
Dy	3.25	3.74	4.04	3.99	3.78	3.65	3.57	7.79	7.96	7.94	7.44	3.13	3.46	2.66	3.11
Ho	0.65	0.75	0.80	0.80	0.76	0.74	0.73	1.54	1.59	1.57	1.48	0.64	0.70	0.55	0.65
Er	1.91	2.24	2.36	2.41	2.27	2.18	2.13	4.32	4.45	4.37	4.13	1.92	2.10	1.72	1.91
Tm	0.30	0.35	0.37	0.38	0.35	0.35	0.34	0.62	0.64	0.62	0.59	0.30	0.33	0.28	0.30
Yb	1.99	2.32	2.42	2.49	2.33	2.29	2.23	3.75	3.97	3.77	3.62	2.03	2.15	2.01	1.95
Lu	0.31	0.36	0.37	0.38	0.36	0.35	0.35	0.55	0.58	0.55	0.52	0.32	0.34	0.33	0.30
Y	19.41	22.74	23.82	24.23	22.90	22.25	21.88	42.36	43.32	42.57	40.10	18.71	20.98	17.05	19.87
∑REE	569.67	672.88	708.91	765.86	708.83	606.44	670.30	701.28	710.84	708.67	692.53	519.60	600.28	543.21	441.35
$(La/Yb)_N$	14.39	14.77	14.59	16.05	15.82	12.97	15.56	4.29	3.89	4.33	4.77	12.46	14.39	14.09	10.33
δEu	0.89	0.72	0.71	0.64	0.71	0.76	0.73	0.86	0.82	0.85	0.89	0.68	0.72	1.40	0.76

Note: The contents of major elements, REEs, and trace elements were analyzed at the State Key Laboratory of Continental Dynamics, Northwest University. An XRF analysis was adopted to analyze major elements, and the experimental instrument used (RIX2100) was manufactured by the Japanese company RIGAKU. Trace elements were analyzed and graded using the Elan6100DRC model ICP-MS, supplied by Perkin Elmer, an American-based company. The relative error and relative standard deviation of the majority of elements were less than 5%.

Table 2. Sr–Nd–Hf–O isotopic compositions of Huashigou granites.

Sample	$^{87}Sr/^{86}Sr$	2s	Sr ($\times 10^{-6}$)	Rb ($\times 10^{-6}$)	^{86}Sr ($\times 10^{-6}$)	$^{87}Rb/^{86}Sr$	$(^{87}Sr/^{86}Sr)_i$		U–Pb Age/Ga		
YQ-57	0.718408	0.000121	227	90.2	0.0985	1.153	0.712356		0.3687		
YQ-63	0.719577	0.000114	210	106	0.0985	1.452	0.711936		0.3687		
YQ-83	0.706211	0.000018	585	16.6	0.0986	0.082	0.705905		0.2618		
YQ-84	0.706309	0.000023	593	18.6	0.0986	0.091	0.705971		0.2618		
YQ-91	0.729009	0.000021	217	190	0.0984	2.542	0.719528		0.2622		
YX-1	0.732882	0.000124	144	154	0.0984	3.094	0.721339		0.2622		
Sample	$^{143}Nd/^{144}Nd$	2s	Nd ($\times 10^{-6}$)	Sm ($\times 10^{-6}$)	^{144}Nd ($\times 10^{-6}$)	$^{147}Sm/^{144}Nd$	$(^{143}Nd/^{144}Nd)_i$	$\varepsilon_{Nd}(t)$	T_{2DM} (Ga)	t_{CHUR} (Ga)	
YQ-57	0.512070	0.000013	23.5	3.92	0.2380	0.1006	0.511827	−6.555202	1.6	0.900661	
YQ-63	0.512068	0.000007	28.4	4.60	0.2380	0.9080	0.511815	−6.790724	1.6	0.905239	
YQ-83	0.512587	0.000008	36.7	8.74	0.2380	0.1437	0.512341	0.778167	1.0	0.146910	
YQ-84	0.512584	0.000007	35.6	8.27	0.2380	0.1406	0.512343	0.824325	1.0	0.146764	
YQ-91	0.512032	0.000006	22.2	3.65	0.2380	0.0993	0.511861	−8.569046	1.7	0.948493	
YX-1	0.512072	0.000006	29.2	4.83	0.2380	0.1001	0.511899	−7.817007	1.7	0.893627	
Sample	$^{176}Hf/^{177}Hf$	2s	Lu ($\times 10^{-6}$)	Hf ($\times 10^{-6}$)	^{177}Hf ($\times 10^{-6}$)	^{176}Hf (10^{-6})	^{176}Lu	$^{176}Lu/^{177}Hf$	$(^{176}Hf/^{177}Hf)_i$	$\varepsilon_{Hf}(t)$	T_{DM} (Ga)
YQ-83	0.282543	0.000009	0.35	5.72	0.185953	0.052604	0.02584	0.0084777	0.28248	−2.07584	1.4

Sample	Lithology
YQ-83	Quartz diorite
YQ-91	Quartz mica porphyritic diorite

Note: The Sr–Nd–Hf isotopic compositions were analyzed at the State Key Laboratory of Continental Dynamics, Northwest University, while the composition of oxygen isotopes ($\delta^{18}O$) was determined at the Institute of Mineral Resources, Chinese Academy of Geological Science.

Table 3. LA-ICP-MS U–Pb zircon data of Huashigou granites.

Spots	^{238}U ($\times 10^{-6}$)	^{232}Th ($\times 10^{-6}$)	Th/U	Isotope Ratio						Isotope Age			
				$^{206}Pb/^{238}U$	1σ	$^{207}Pb/^{235}U$	1σ	$^{207}Pb/^{206}Pb$	1σ	$^{206}Pb/^{238}U$	1σ	$^{207}Pb/^{235}U$	1σ
YQ57-01	315	190	0.60	0.05889	0.00087	0.48024	0.01794	0.05914	0.00222	369	5	398	12
YQ57-02	350	116	0.33	0.06042	0.00088	0.43998	0.01578	0.05280	0.0019	378	5	370	11
YQ57-03	150	97	0.65	0.06003	0.00108	0.46993	0.03886	0.05678	0.00481	376	7	391	27
YQ57-05	316	274	0.87	0.05920	0.00087	0.46580	0.01765	0.05706	0.00217	371	5	388	12
YQ57-06	339	224	0.66	0.05871	0.00087	0.42920	0.01595	0.05301	0.00198	368	5	363	11
YQ57-07	252	232	0.92	0.06073	0.00090	0.47908	0.01839	0.05721	0.00221	380	5	397	13
YQ57-08	355	326	0.92	0.05921	0.00087	0.43751	0.01630	0.05359	0.00201	371	5	368	12
YQ57-09	293	223	0.76	0.05810	0.00086	0.45036	0.01849	0.05621	0.00232	364	5	378	13
YQ57-12	361	204	0.56	0.06005	0.00089	0.48010	0.01811	0.05798	0.00220	376	5	398	12
YQ57-13	439	288	0.66	0.05781	0.00085	0.42800	0.01604	0.05369	0.00202	362	5	362	11
YQ57-15	284	195	0.69	0.06071	0.00091	0.45580	0.01880	0.05445	0.00226	380	6	381	13
YQ57-16	348	297	0.85	0.05931	0.00089	0.47418	0.01820	0.05798	0.00224	371	5	394	13
YQ57-17	290	173	0.60	0.05908	0.00091	0.43250	0.01788	0.05308	0.00222	370	6	365	13
YQ57-18	427	293	0.69	0.05886	0.00088	0.44790	0.01736	0.05518	0.00215	369	5	376	12
YQ57-19	273	157	0.57	0.05829	0.00088	0.46156	0.01955	0.05742	0.00245	365	5	385	14
YQ57-20	440	415	0.94	0.05592	0.00084	0.45384	0.01804	0.05886	0.00236	351	5	380	13
YQ57-21	404	259	0.64	0.05727	0.00086	0.43694	0.01781	0.05533	0.00227	359	5	368	13
YQ57-22	413	271	0.66	0.05881	0.00088	0.43850	0.01720	0.05407	0.00214	368	5	369	12
YQ57-23	162	108	0.67	0.05978	0.00094	0.47134	0.02617	0.05717	0.00320	374	6	392	18
YQ57-25	330	230	0.70	0.05761	0.00090	0.42060	0.01812	0.05295	0.00231	361	5	356	13
YQ81-01	177	153	0.86	0.04155	0.00062	0.29453	0.01115	0.05141	0.00196	262	4	262	9
YQ81-02	239	228	0.95	0.04167	0.00060	0.29783	0.01107	0.05183	0.00193	263	4	265	9
YQ81-03	236	281	1.19	0.04182	0.00061	0.29759	0.01127	0.05161	0.00196	264	4	265	9
YQ81-04	99	72	0.73	0.04197	0.00066	0.29891	0.01434	0.05165	0.00250	265	4	266	11
YQ81-05	151	123	0.81	0.04105	0.00061	0.29311	0.01254	0.05178	0.00223	259	4	261	10
YQ81-06	217	204	0.94	0.04174	0.00061	0.30703	0.01496	0.05335	0.00260	264	4	272	12
YQ81-07	86	62	0.72	0.04313	0.00069	0.30454	0.02352	0.05120	0.00397	272	4	270	18
YQ81-08	149	121	0.81	0.04087	0.00062	0.28327	0.01129	0.05026	0.00202	258	4	253	9
YQ81-09	49	17	0.35	0.04204	0.00075	0.29870	0.02003	0.05152	0.00349	265	5	265	16
YQ81-10	150	129	0.86	0.04095	0.00062	0.29902	0.01291	0.05295	0.00230	259	4	266	10
YQ81-11	136	82	0.60	0.04043	0.00062	0.28644	0.01301	0.05138	0.00235	255	4	256	10
YQ81-12	120	103	0.85	0.04057	0.00065	0.28818	0.01395	0.05151	0.00252	256	4	257	11
YQ81-13	135	113	0.84	0.04076	0.00063	0.29006	0.01336	0.05161	0.00240	258	4	259	11
YQ81-14	125	108	0.87	0.04255	0.00066	0.30327	0.01373	0.05169	0.00236	269	4	269	11
YQ81-15	105	87	0.82	0.04228	0.00068	0.31220	0.01887	0.05355	0.00326	267	4	276	15
YQ81-16	159	136	0.85	0.04097	0.00064	0.29132	0.01328	0.05156	0.00237	259	4	260	10
YQ81-17	130	126	0.97	0.04318	0.00067	0.29919	0.01248	0.05024	0.00212	273	4	266	10
YQ81-18	175	148	0.85	0.04153	0.00062	0.29472	0.01278	0.05146	0.00224	262	4	262	10
YQ81-19	154	131	0.85	0.03991	0.00061	0.28186	0.01182	0.05122	0.00216	252	4	252	9
YQ81-20	161	140	0.87	0.04046	0.00062	0.29061	0.01533	0.05209	0.00276	256	4	259	12
YQ81-21	125	97	0.77	0.04172	0.00066	0.29840	0.01384	0.05187	0.00243	263	4	265	11
YQ81-22	117	102	0.87	0.04233	0.00066	0.30153	0.01521	0.05166	0.00263	267	4	268	12
YQ81-23	55	18	0.32	0.03984	0.00075	0.28010	0.03220	0.05099	0.00589	252	5	251	26
YQ81-24	114	89	0.78	0.04082	0.00066	0.28811	0.01330	0.05119	0.00239	258	4	257	10

Table 3. *Cont.*

Spots	^{238}U ($\times10^{-6}$)	^{232}Th ($\times10^{-6}$)	Th/U	Isotope Ratio						Isotope Age			
				$^{206}Pb/^{238}U$	1σ	$^{207}Pb/^{235}U$	1σ	$^{207}Pb/^{206}Pb$	1σ	$^{206}Pb/^{238}U$	1σ	$^{207}Pb/^{235}U$	1σ
YQ81-25	40	16	0.40	0.03988	0.00083	0.28361	0.03353	0.05157	0.00614	252	5	254	27
YQ81-26	83	32	0.38	0.04179	0.0007	0.29816	0.01652	0.05175	0.00290	264	4	265	13
YQ81-27	100	83	0.83	0.04162	0.00067	0.29657	0.01594	0.05168	0.00280	263	4	264	12
YQ81-28	135	115	0.86	0.04160	0.00065	0.29434	0.01348	0.05131	0.00237	263	4	262	11
YQ81-29	76	53	0.70	0.04171	0.00073	0.29603	0.01754	0.05147	0.00309	263	5	263	14
YQ81-30	141	129	0.91	0.04253	0.00067	0.30127	0.01447	0.05138	0.00249	268	4	267	11
YQ81-31	195	179	0.92	0.04061	0.00062	0.28495	0.01207	0.05088	0.00217	257	4	255	10
YQ81-32	234	210	0.90	0.04081	0.00061	0.28668	0.01135	0.05095	0.00203	258	4	256	9
YQ81-33	49	20	0.40	0.04214	0.00079	0.29406	0.02197	0.05061	0.00383	266	5	262	17
YQ81-34	124	103	0.83	0.04197	0.00067	0.29729	0.01325	0.05137	0.00232	265	4	264	10
YQ81-35	223	206	0.92	0.04100	0.00062	0.29325	0.01179	0.05187	0.00210	259	4	261	9
YQ81-36	55	17	0.31	0.04195	0.00077	0.31005	0.02283	0.05360	0.00399	265	5	274	18
YQ81-37	77	58	0.76	0.04211	0.00073	0.30065	0.02180	0.05178	0.00379	266	5	267	17
YQ81-39	129	105	0.82	0.04163	0.00066	0.29548	0.01465	0.05147	0.00258	263	4	263	11
YQ81-40	133	106	0.80	0.04243	0.00067	0.31463	0.01978	0.05378	0.00340	268	4	278	15
YQ81-41	167	158	0.95	0.04050	0.00063	0.29041	0.01346	0.05200	0.00243	256	4	259	11
YQ81-42	45	20	0.45	0.04111	0.00081	0.28552	0.05593	0.05037	0.00989	260	5	255	44
YQ81-43	128	108	0.85	0.04167	0.00066	0.29561	0.01395	0.05144	0.00245	263	4	263	11
YQ81-44	290	334	1.15	0.04080	0.00061	0.29106	0.01125	0.05173	0.00201	258	4	259	9
YQ89-01	96	68	0.71	0.04075	0.00065	0.28996	0.01582	0.05161	0.00284	33	4	259	12
YQ89-02	153	135	0.88	0.04133	0.00060	0.29282	0.01196	0.05138	0.00211	261	4	261	9
YQ89-03	158	198	1.25	0.04029	0.00060	0.28692	0.01443	0.05164	0.00261	255	4	256	11
YQ89-04	70	43	0.62	0.04014	0.00067	0.28873	0.02078	0.05217	0.00378	254	4	258	16
YQ89-05	75	56	0.75	0.04080	0.00077	0.28745	0.02053	0.05110	0.00370	258	5	257	16
YQ89-06	92	75	0.81	0.04123	0.00066	0.29329	0.01421	0.05159	0.00253	260	4	261	11
YQ89-07	154	134	0.87	0.04044	0.00062	0.29230	0.01215	0.05242	0.00220	256	4	260	10
YQ89-08	79	74	0.94	0.04265	0.00071	0.30072	0.01687	0.05113	0.00290	269	4	267	13
YQ89-09	101	124	1.24	0.04169	0.00065	0.28222	0.01242	0.04909	0.00218	263	4	252	10
YQ89-10	86	72	0.83	0.04121	0.00070	0.29268	0.01854	0.05151	0.00329	260	4	261	15
YQ89-11	112	109	0.97	0.03968	0.00061	0.28116	0.01389	0.05139	0.00256	251	4	252	11
YQ89-12	133	157	1.19	0.04169	0.00063	0.30397	0.01298	0.05288	0.00227	263	4	269	10
YQ89-13	107	88	0.82	0.04228	0.00066	0.29920	0.01478	0.05133	0.00256	267	4	266	12
YQ89-14	120	112	0.94	0.04158	0.00064	0.29572	0.01355	0.05157	0.00238	263	4	263	11
YQ89-15	71	69	0.97	0.04040	0.00075	0.28671	0.01988	0.05147	0.00362	255	5	256	16
YQ89-16	50	25	0.49	0.04174	0.00077	0.29940	0.02315	0.05202	0.00406	264	5	266	18
YQ89-18	277	332	1.20	0.04179	0.00060	0.29723	0.01035	0.05158	0.00180	264	4	264	8
YQ89-19	110	98	0.89	0.04258	0.00068	0.29946	0.01645	0.05100	0.00282	269	4	266	13
YQ89-20	130	102	0.78	0.04169	0.00064	0.29648	0.01480	0.05158	0.00259	263	4	264	12
YQ89-21	156	94	0.60	0.04102	0.00061	0.29012	0.01148	0.05129	0.00204	259	4	259	9
YQ89-22	150	132	0.88	0.04170	0.00064	0.29589	0.01280	0.05145	0.00224	263	4	263	10
YQ89-23	137	120	0.88	0.04067	0.00063	0.28716	0.01239	0.05120	0.00223	257	4	256	10
YQ89-24	159	151	0.95	0.04117	0.00063	0.29021	0.01297	0.05112	0.00230	260	4	259	10
YQ89-25	220	238	1.08	0.04086	0.00060	0.29023	0.01096	0.05151	0.00195	258	4	259	9
YQ89-27	148	134	0.90	0.04272	0.00063	0.30245	0.01298	0.05134	0.00221	270	4	268	10
YQ89-28	122	111	0.91	0.04204	0.00064	0.29958	0.01387	0.05168	0.00241	265	4	266	11
YQ89-30	94	90	0.96	0.04198	0.00067	0.29843	0.01473	0.05155	0.00257	265	4	265	12
YQ89-31	97	77	0.80	0.04149	0.00065	0.29636	0.01567	0.05180	0.00276	262	4	264	12
YQ89-32	88	66	0.75	0.04180	0.00068	0.29623	0.01546	0.05139	0.00271	264	4	263	12
YQ89-34	58	18	0.31	0.04209	0.00074	0.29816	0.01956	0.05137	0.00341	266	5	265	15
YQ89-35	83	81	0.98	0.04170	0.00068	0.29736	0.01812	0.05172	0.00318	263	4	264	14
YQ89-36	58	29	0.50	0.04095	0.00078	0.28665	0.02375	0.05076	0.00425	259	5	256	19
YQ89-37	47	30	0.64	0.04207	0.00086	0.31336	0.04643	0.05402	0.00808	266	5	277	36
YQ89-38	253	264	1.04	0.04198	0.00060	0.30187	0.01175	0.05215	0.00203	265	4	268	9
YQ89-39	107	88	0.82	0.04285	0.00067	0.30212	0.01287	0.05113	0.00220	270	4	268	10
YQ89-40	183	139	0.76	0.04186	0.00062	0.29311	0.01096	0.05078	0.00191	264	4	261	9
YQ89-41	167	210	1.26	0.04140	0.00061	0.29503	0.01168	0.05168	0.00206	262	4	263	9
YQ89-42	210	211	1.01	0.04211	0.00061	0.28727	0.01011	0.04947	0.00175	266	4	256	8
YQ89-43	218	219	1.00	0.04169	0.00061	0.29750	0.01151	0.05175	0.00201	263	4	264	9
YQ89-44	106	114	1.08	0.04212	0.00068	0.31096	0.03124	0.05354	0.00539	266	4	275	24

Notes: The analytical results are reported with a 1σ error. Zircon U–Pb dating with a beam size of 30–40 μm was conducted using LA-ICP-MS at the Key Laboratory of Lithospheric Tectonics and Exploration of the Ministry of Education from the China University of Geosciences (Beijing).

4.1. Geochronology

In this study, three granitic plutons from the Huashigou area were chosen for zircon U–Pb dating. The zircons used were primarily colorless to transparent and euhedral to subeuhedral, with elongated to stubby grains (50–250 μm long with length-to-width ratios of 2:1–4:1). Typical magmatic oscillatory zonation was common, and several inherited cores were observed (see insets in Figure 2). The contents of Th (16–415 $\times$ 10^{-6}) and U (40–440 $\times$ 10^{-6}) were different among these zircons, and the Th/U ratios were consistent with a magmatic origin [45–48]. The detailed analytical results are listed in Table 3. The U–Pb concordia diagrams for Huashigou granitic rocks are shown in Figure 3.

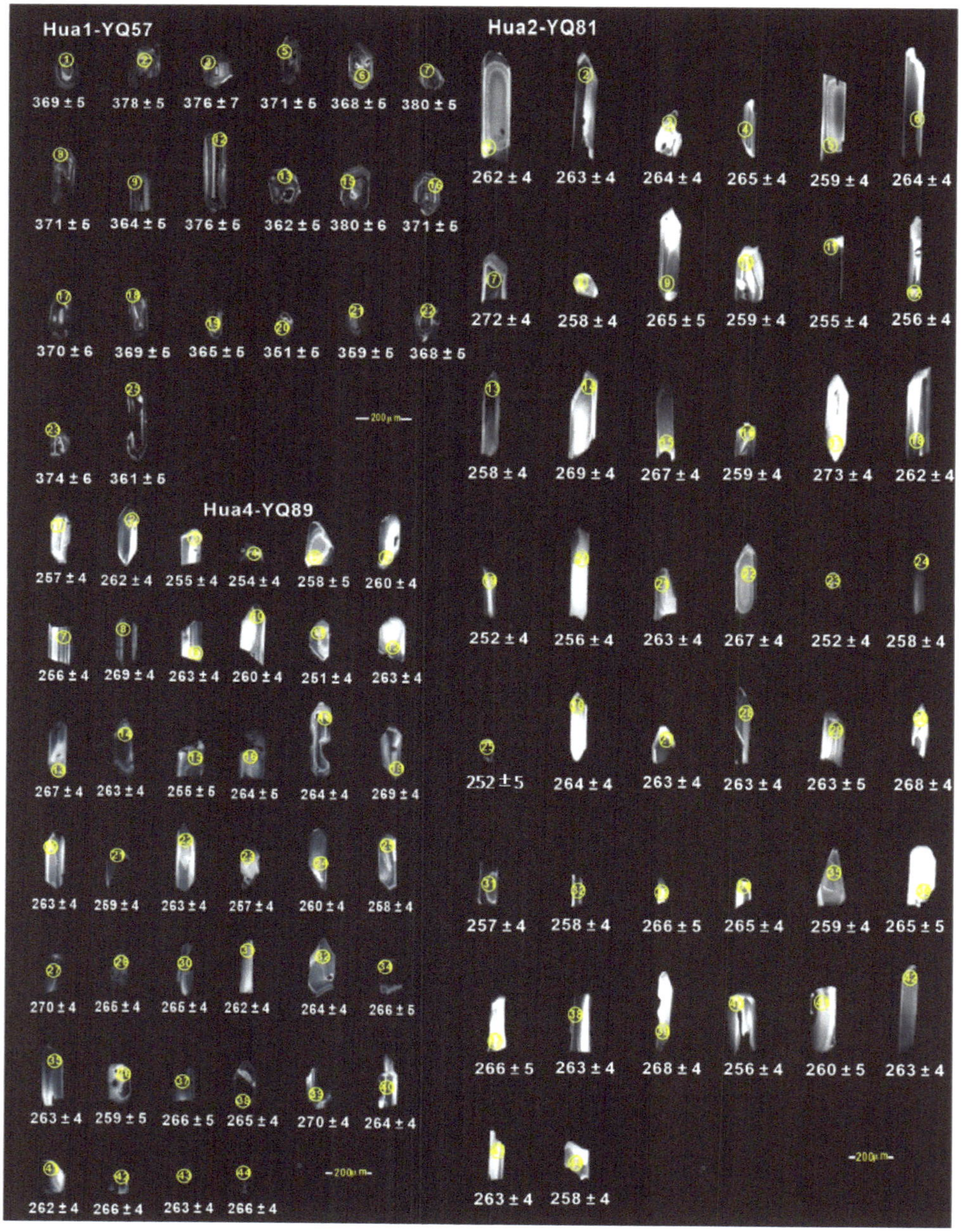

Figure 2. Cathodoluminescence images of zircon grains of granitic rocks from the Huashigou area (ages in Ma).

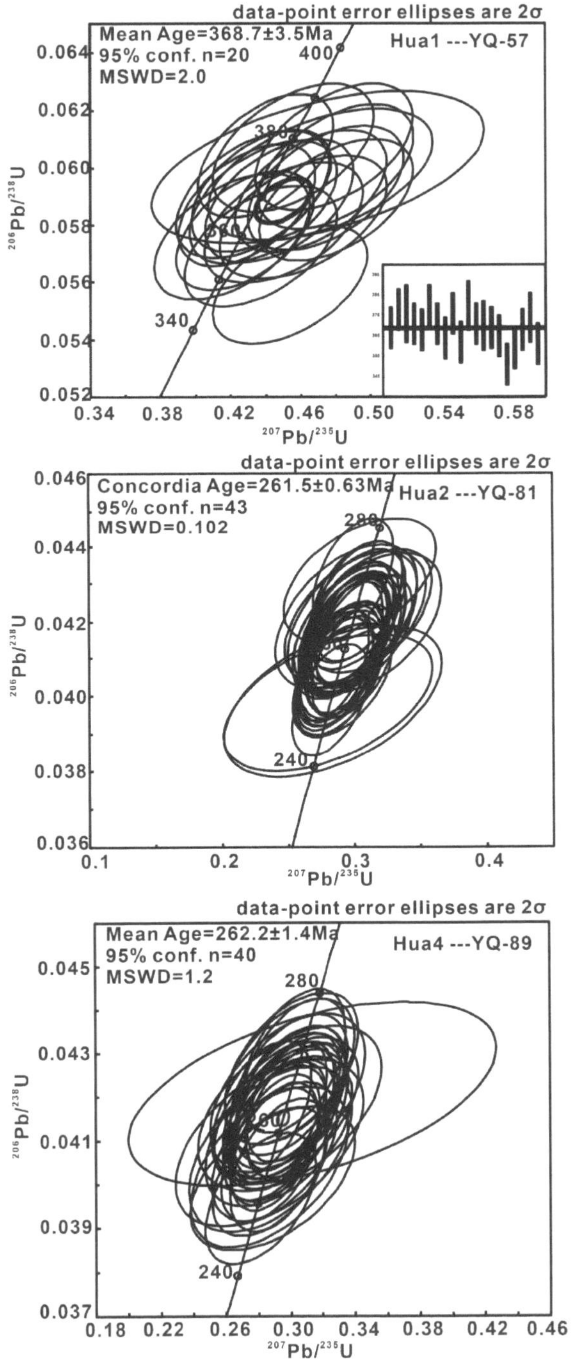

Figure 3. Zircon concordia diagram of granitic rocks. The weighted mean $^{206}Pb/^{238}U$ ages for samples YQ-57, YQ-81, and YQ-89 are 368.7 ± 3.5 Ma (MSWD = 2.0; 2σ), 261.5 ± 0.63 Ma (MSWD = 0.102; 2σ), and 262.2 ± 1.4 Ma (MSWD = 1.2; 2σ), respectively.

4.2. Whole-Rock Geochemistry

4.2.1. Hua1 Pluton

Seven samples of granitic rocks (YQ57–YQ63) from the Hua1 pluton were analyzed for major- and trace-element compositions. Loss on ignition (LOI) for all samples was less than 1 wt.%. The granitic samples showed high contents of SiO_2 (67.78–68.48 wt.%) and Al_2O_3 (15.26–15.77 wt.%) but low contents of MgO (1.19–1.38 wt.%) with low $Mg^{\#}$ (about 38). Their contents of CaO and TiO_2 were 3.39–3.80 wt.% and 0.40–0.48 wt.%, respectively. The Hua1 pluton samples showed total alkali contents ranging from 6.36 to 6.67 wt.% and a slight sodium enrichment (Na_2O/K_2O = 1.16–1.39). They belong to the calc-alkaline series in the SiO_2-K_2O diagram (Figure 4a). Their A/CNK ratios ranged from 0.97 to 0.99, indicating that the Hua1 granites are metaluminous (Figure 4b).

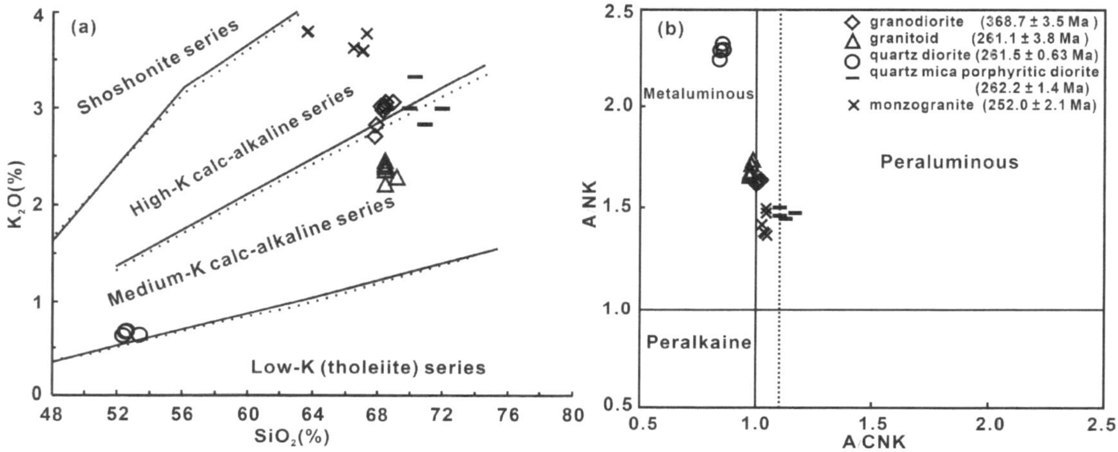

Figure 4. (**a**) Diagram of SiO_2-K_2O contents; (**b**) ANK-ACNK diagram of granitic rocks from the Huashigou area. A/NK = $Al_2O_3/Na_2O + K_2O$; A/CNK = $Al_2O_3/CaO + Na_2O + K_2O$.

The Late Devonian granodiorites showed low Sr contents (205.34–226.54 $\times 10^{-6}$) and high Y contents (19.41–23.82 $\times 10^{-6}$), with Sr/Y ratios of 9.17–11.67. The Nb/U and Nb/Ta ratios of the Late Devonian granodiorites were 7.77–10.21 and 12.42–14.39, respectively, which are comparable to those of the Earth's crust, i.e., 9–12 [49–52] and 12–13 [53]. The Late Devonian granodiorite samples contained relatively high REE contents (ΣREE = 570–766 $\times 10^{-6}$) and were relatively enriched in light rare-earth elements (LREEs) with moderate LREE/HREE fractionation ($(La/Yb)_N$ = 12.99–16.05 and $(La/Sm)_N$ = 6.1–6.91). Additionally, they showed a flat distribution pattern of heavy rare-earth elements (HREEs) and negative Eu anomalies (δEu = 0.64–0.89). In a primitive-mantle-normalized trace element spider diagram (Figure 5a), the samples showed a notable enrichment in Rb, K, La, Th, and Ce and a relative depletion in Ba, Sr, and Cs. In terms of high-field-strength elements (HFSEs), a notable depletion in Nb, Ta, and Ti and a relative enrichment in Zr and Hf can be observed for these samples.

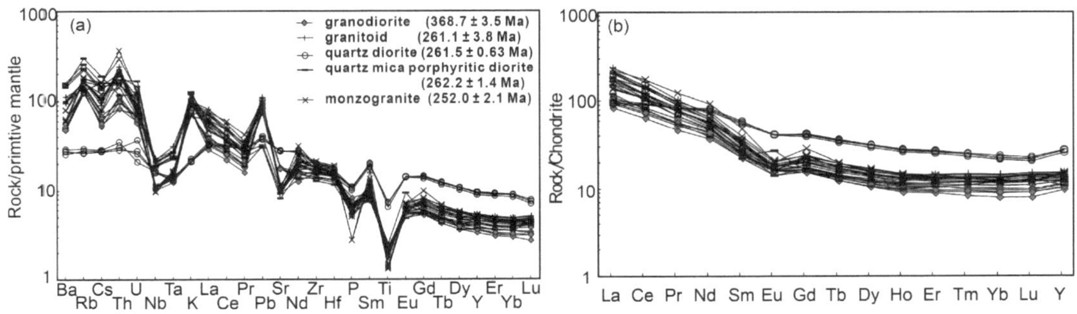

Figure 5. Primitive-mantle-normalized trace element spider diagrams for granitoids (**a**); chondrite-normalized REE patterns of granitic rocks from the Huashigou area (**b**) (PM values and chondrite values are derived from [54]).

4.2.2. Hua2 Pluton

The quartz diorites (YQ81–YQ84) from the Hua2 pluton showed relatively low contents of SiO_2 (52.41–53.32 wt.%) and high contents of Al_2O_3 (19.67–19.69 wt.%), TiO_2 (1.4–1.58 wt.%), TFe_2O_3 (8.71–9.23 wt.%), MgO (2.99–3.25 wt.%), and CaO (7.91–8.11 wt.%). Their total alkali contents ranged from 5.38 wt.% to 5.56 wt.%, and the quartz diorites were evidently sodium-enriched and potassium-depleted (Na_2O/K_2O = 7.01–7.55); thus, they were plotted in the calc-alkaline field in a SiO_2-K_2O diagram (Figure 4a). Based on the plot of the A/NK vs. A/CNK diagram, all of the samples are peraluminous (Figure 4b).

As can be seen from the primitive-mantle-normalized diagram (Figure 5a), the quartz diorites were enriched in LILEs, Ba, Rb, Th, and U and depleted in HFSEs, Nb, Ta, and Ti, with high Sr (573.85–593.13 $\times 10^{-6}$) and Y (40.1–43.32 $\times 10^{-6}$) contents and a low Sr/Y ratio (13.25–14.79). As shown in the chondrite-normalized distribution pattern (Figure 5b), the quartz diorites were slightly enriched in LREEs ($(La/Yb)_N$ = 3.89–4.76, $(La/Sm)_N$ = 1.54–1.88, and $(Gd/Yb)_N$ = 1.81–1.89), with slightly negative Eu anomalies (δEu = 0.81–0.89).

4.2.3. Hua4 Pluton

The Middle Permian porphyritic tonalites (YQ89–YQ92) from the Hua4 pluton showed high contents of SiO_2 (63.64–68.51 wt.%), Al_2O_3 (14.89–16.62 wt.%), Mg (MgO = 0.71%–2.5 wt.%, $Mg^{\#}$ = 45–51), and total alkali ($Na_2O + K_2O$ = 7.3–8.02 wt.%) and low contents of CaO (1.47–2.93 wt.%). In the K_2O vs. SiO_2 diagram (Figure 4a), all of the samples are plotted as part of the calc-alkaline series. Based on the plot of A/NK vs. A/CNK, all of the samples are peraluminous (Figure 4b).

As can be seen in the primitive-mantle-normalized diagram (Figure 5a), the porphyritic tonalites were enriched in Rb, Ba, and Th and relatively depleted in Nb, Ta, and Ti, with low contents of Sr (172.43–217.28 $\times 10^{-6}$) and Y (17.05–20.98 $\times 10^{-6}$) and a low Sr/Y ratio (9.22–12.74). As shown in the chondrite-normalized distribution pattern (Figure 5b), the porphyritic tonalites were enriched in LREEs ($(La/Yb)_N$ = 10.34–14.36, $(La/Sm)_N$ = 5.38–6.98, and $(Gd/Yb)_N$ = 1.35–1.47), with negative Eu anomalies (δEu = 0.68–0.76).

4.3. Whole-Rock Sr–Nd–O and Hf Isotopes

Whole-rock Sr–Nd–O isotope data are presented in Table 2. The Late Devonian granodiorites (YQ57 and YQ63) from the Hua1 pluton showed high values for the initial $^{87}Sr/^{86}Sr$ ratio (0.712356 and 0.71195) and low values for the initial $^{143}Nd/^{144}Nd$ ratio (0.511827 and 0.511815), with negative $\varepsilon_{Nd}(t)$ values of −6.56 and −6.14 and a model age (T_{DM}) of 1.6 Ga. In addition, the sample YQ63 showed an initial $^{176}Hf/^{177}Hf$ ratio value of 0.282484, with an $\varepsilon_{Hf}(t)$ value of −2.06 and a crustal model age of 1.4 Ga.

The Middle Permian granites from the Hua2 pluton showed high I_{Sr} values of 0.705906 and 0.705970 for quartz diorites (YQ83 and YQ84). Porphyritic tonalite (YQ91) showed the highest $(^{87}Sr/^{86}Sr)_{ir}$ value of 0.719527. The quartz diorites (YQ83 and YQ84) showed $\varepsilon_{Nd}(t)$ values of +0.78 and +0.82. The porphyritic tonalite (YQ91) from the Hua4 pluton showed the lowest $\varepsilon_{Nd}(t)$ value of −8.57. In addition, the values of $\delta^{18}O_{V\text{-}SMOW}$ for quartz diorite (YQ83) from the Hua2 pluton and porphyritic tonalite (YQ91) from the Hua4 pluton were all relatively higher (12.4‰ and 11.8‰, respectively).

5. Discussion

5.1. Petrogenesis

5.1.1. Petrogenesis of the Hua1 Pluton

The Late Devonian granodiorites (YQ57–YQ63) from the Hua1 pluton were rich in Al but poor in Mg, and they were slightly sodium-enriched and potassium-depleted. Additionally, they belong to the metaluminous calc-alkaline series (Figure 6). They showed high contents of Cs, Ba, Rb, Th, and Pb, low contents of Nb, Ta, Sr, and Ti (Figure 5a), high total REEs and LREEs, and a flat HREE pattern (Figure 5b) [34,55,56]. The flat pattern of HREEs implies that there is no garnet residual in the source region. The low Sr and negative Eu anomalies suggest fractional crystallization of plagioclase during magma evolution, indicating that the depth of partial melting is shallow [57]. The negative anomalies of P and Ti, high I_{Sr} values, and negative $\varepsilon_{Nd}(t)$ and $\varepsilon_{Hf}(t)$ values are potentially the result of the fractional crystallization of apatite and Fe-Ti oxide in the plagioclase-stable region. The $\varepsilon_{Nd}(t)$ vs. $\varepsilon_{Hf}(t)$ diagram is similar to that of the lower crust (Figure 7). The Sr–Nd–Hf isotopic compositions of granites suggest that they may have been derived from the partial melting of the underplated lower crust.

5.1.2. Petrogenesis of the Hua4 Pluton and Hua2 Pluton

Middle Permian granodiorites resemble the features of the continental crust because of their higher contents of SiO_2, Al_2O_3, and K_2O, as well as because they are rich in Ba, Rb, and Th but poor in Nb, Ta, Ti, Sr, etc., and have a higher positive Pb anomaly relative to REEs (Figure 5a), with high I_{Sr}, negative $\varepsilon_{Nd}(t)$, and high $\delta^{18}O_{V\text{-}SMOW}$ values. The fractional crystallization of plagioclase occurred during the process of rock formation, as evidenced by a slightly negative Eu anomaly (Figure 5b). The negative P and Ti anomalies imply the fractional distillation of apatite and Fe-Ti oxide. In addition, the low Nb/U ratio and Nb content are related to the contamination of materials from the upper crust [53,58]. The high content of Sr and the low contents of Yb and Y in the granodiorites were similar to the contents of those elements [59,60]. However, the isotopic characteristics of low $Mg^{\#}$ content (≈40), high I_{Sr}, low I_{Nd}, and positive $\varepsilon_{Nd}(t)$ (+5.51, Sample YQ3) values are clearly different from the isotopic characteristics of typical adakites formed by the partial melting of juvenile subduction slabs [59–64]; however, they are close to the isotopic characteristics of adakites generated by the partial melting of the underplated basaltic lower crust (Figure 5a) [65–67]. Thus, these adakites were formed in crustal environments with thickened slabs at high temperatures after magmas were contaminated by upper crustal materials during magma ascent and certain fractional crystallization of plagioclases [41].

As for Middle Permian granodiorites, the Nb/U and Nb/Ta ratios were slightly higher than those of the continental crust, and the positive $\varepsilon_{Nd}(t)$ values indicate that granodiorites originated from the partial melting of the underplated basaltic lower crust with the involvement of mantle materials. The low Nb/U ratio of quartz mica porphyritic diorite also implies magma contamination with upper crust materials during its ascent. The $\delta^{18}O_{V\text{-}SMOW}$ values of quartz diorite and quartz mica porphyritic diorite were significantly higher than the typical mantle values (5.4 ± 0.3 ‰ [68]), indicating that these granitic rocks are derived from the partial melting of the lower crust. With regard to this, we speculate that Middle Permian granitic rocks are the products of post-collisional slab break-off [69].

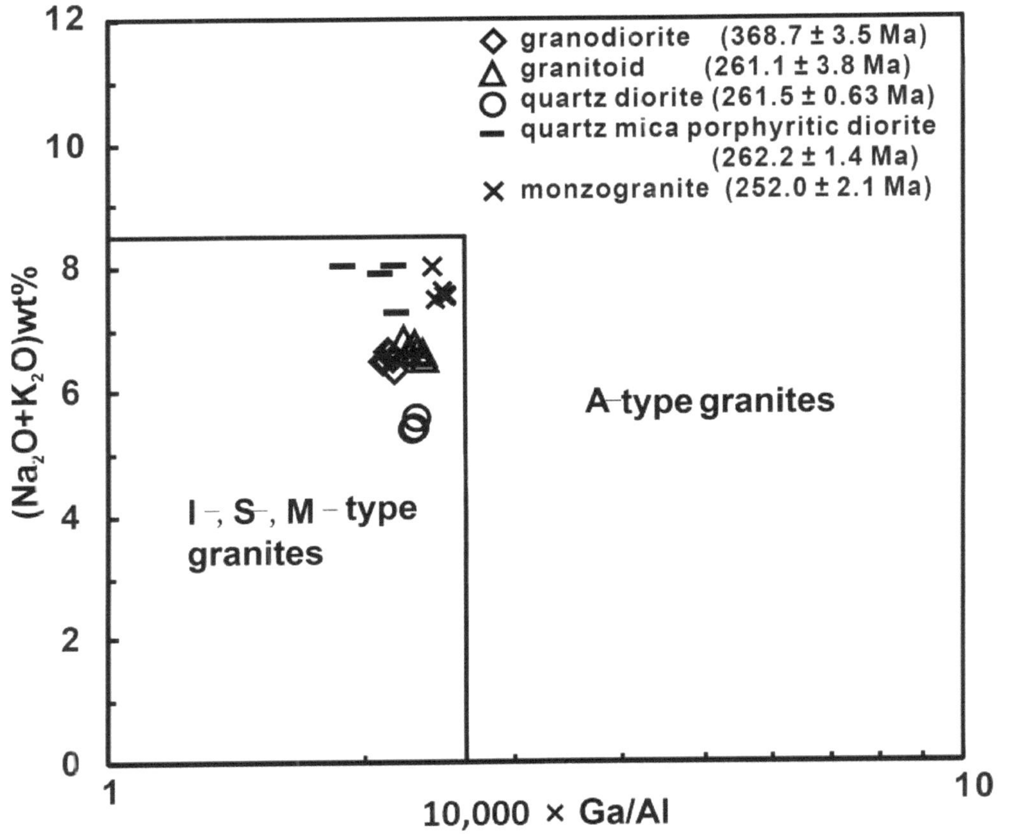

Figure 6. Plot of 10,000 × Ga/Al vs. (Na_2O + K_2O) in Huashigou granites.

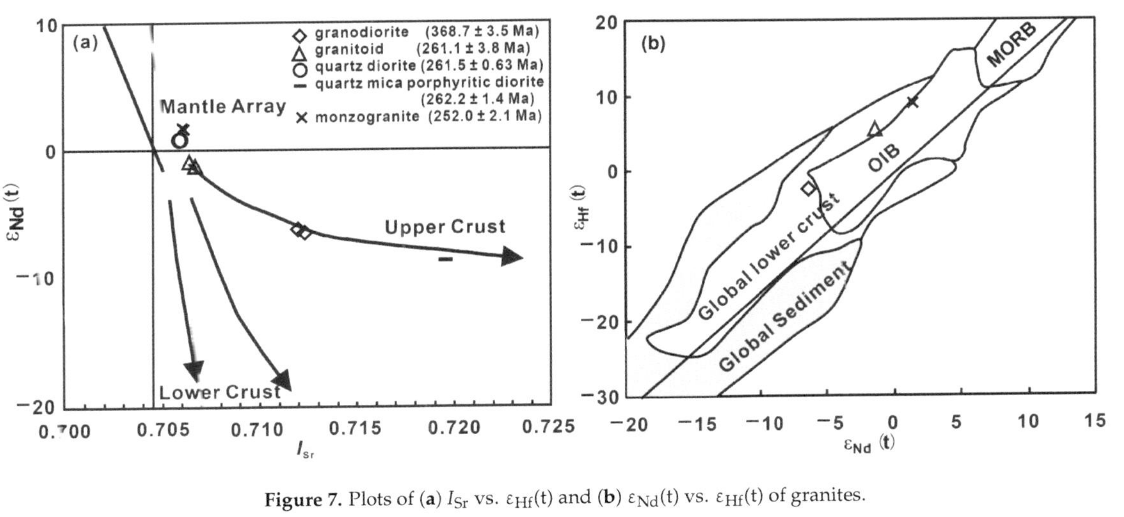

Figure 7. Plots of (**a**) I_{Sr} vs. ε_{Hf}(t) and (**b**) ε_{Nd}(t) vs. ε_{Hf}(t) of granites.

The geochemistry characteristics of Middle Permian granitoids are similar to those of Late Permian monzogranites from the Hua5 pluton [42] because of their high contents of Si and Al and low contents of Ti and Mg, as well as Cs, Ba, Rb, Th, and Pb enrichment and Nb, Ta, Sr, and Ti depletion. The Sr–Nd–Hf isotopic compositions showed that monzogranites derived from the partial melting of the underplated Mid-Neoproterozoic basaltic lower crust from the underplating heating of Late Paleozoic basaltic magma [42].

The $\varepsilon_{Nd}(t)$ values of Middle Permian quartz diorite and Late Permian monzogranite were greater than 0 [42], which is similar to some granites formed in environments with extensional tectonics. Granitic rocks with these characteristics have been found only in several areas in the world, such as the New England Batholith [70], northeastern US, southern Quebec of Canada [71,72], the Central Asia–Mongolia fold belt [73–75], etc. These two types of granitic rocks in this area have even lower $\varepsilon_{Nd}(t)$ values. Generally, the characteristic of $\varepsilon_{Nd}(t) < 0$ indicates that a depleted mantle component played an important role in magma formation. In addition, the geochemical characteristics of Middle Permian quartz diorite and Late Permian monzogranite did not show any evidence of crust–mantle mixing in the source region [42]. Comparing this to other granites with $\varepsilon_{Nd}(t) > 0$, granites in this area had lower $\varepsilon_{Nd}(t)$ (closer to 0) values. Therefore, the characteristic of the $\varepsilon_{Nd}(t)$ value, which is slightly greater than 0, may indicate that Middle Permian quartz diorite and Late Permian monzogranite were derived from the partial melting of basic materials in the lower crust formed by underplating.

5.2. Tectonic Setting of Huashigou Granitoids

Wu et al. [76] considered that granite in Phase 4 is related to isostatic adjustments between and among different rock bodies after orogenesis. Leake et al. [77] identified that, during the continental crash with the orogenic belt, the granitic emplacements caused by local temperatures and stresses while intrusions and tensility slumps happened in many ancient orogenic belts in the world were synchronal, e.g., the Sevier orogenic belt in southeast California, USA [78]; Guitiriz granite in the Variscan belt in Northern Spain [79]; and the Proterozoic orogenic belt in South Greenland. Foster et al. [80] showed that rapid exhumation could cause the decompression and partial melting of the lower or upper mantle, leading to the formation of Eocene intrusive rocks. For Late Devonian Huashigou granitoids, the geochemical characteristics indicate that they were formed from the partial melting of the lower crust, and their formation age is the same as that of Phase 4 granitic magma in the North Qaidam, as classified by [76]. In addition, the granites from Tula in the South Altyn Tagh Fault and Yusupuleke Tagh belong to the same period [39,76]. All of them were formed in the background of extensional tectonics after orogenesis. In Haker's diagram (Figure 8), granitic rocks from different eras show favorable correlations, and their contents of Al_2O_3, MgO, TiO_2, and TFeO decrease gradually as the SiO_2 content increases, indicating a distinct negative correlation. K_2O and Na_2O show a positive correlation, and Rb, Sr, and Zr increase gradually as SiO_2 contents increase. Additionally, Ba shows a negative correlation (Figure 9). These characteristics showed that the sources of different types of granitic rocks in the Huashigou area are similar. As shown in Figure 10, all samples are plotted in a post-collisional tectonic setting. Combined with the regional tectonic setting, we believe that the rocks were formed during a post-collisional tectonic setting where the previously subducted slab underwent detachment, leading to slab break-off magmatism and the formation of a significant number of granitic rocks.

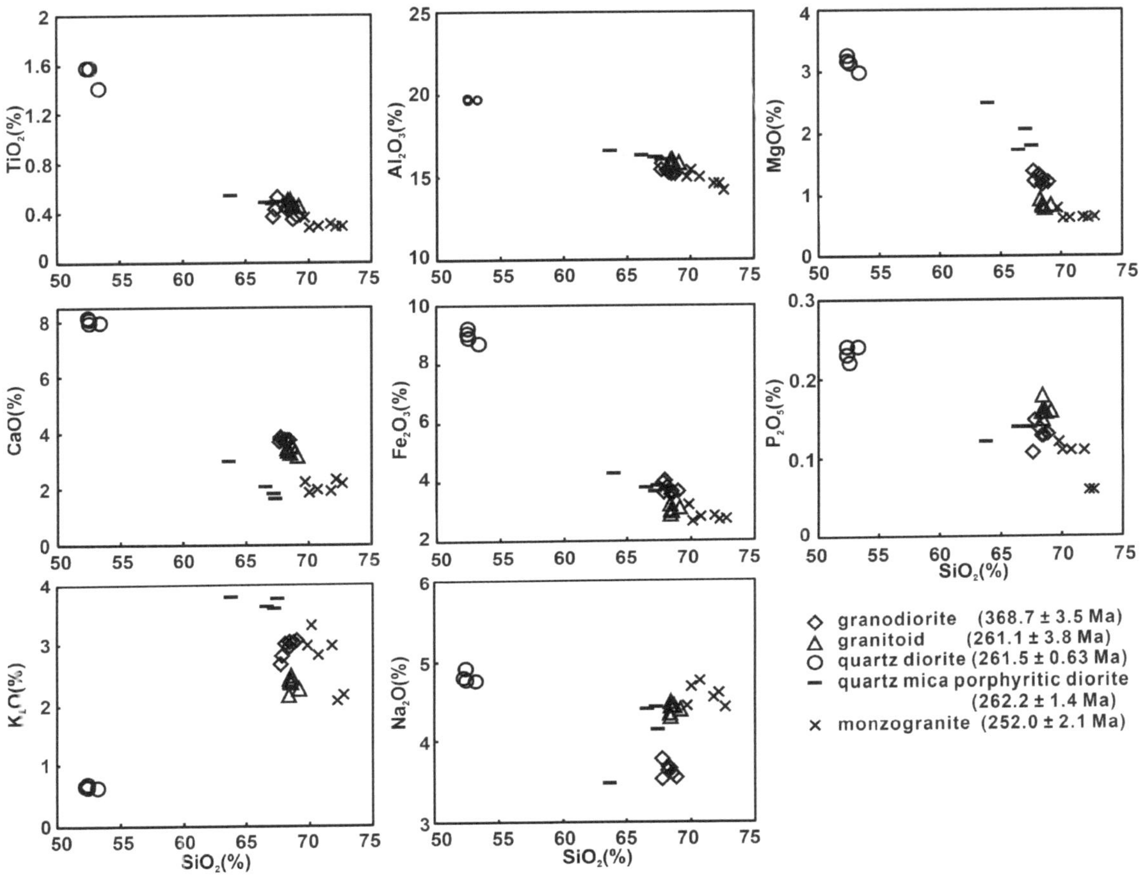

Figure 8. Harker diagrams of Huashiou granitic rocks.

Mid–Late Permian granitoids are comparable to Phase 5 Paleozoic granites in the North Qaidam, as determined by Wu et al. [76], due to the closing of the Paleo-Tethys Ocean, causing the regional tectonic setting to change from tensility to shrinkage. The continental subduction happened before the crust thickened rapidly. In addition, the delamination of the lower crust induced the underplating of basaltic magma, which further elicited the partial melting of the lower crust. Eventually, Mid–Late Permian granitoids were formed. Possessing characteristics of adakites, Mid–Late Permian granodiorites were formed under pressures greater than 1.5 GPa [64,72–87], indicating that the crust thickness could be over 50 km [87,88]. Quartz diorites with high Sr and Yb contents were formed under medium or relatively high pressures, while quartz mica porphyritic diorites with high Sr and Yb contents were formed under medium pressures. The formation depth is estimated to be around 30 km. The depth of the source regions of partial melting decreases gradually from Middle Permian granodiorite to quartz diorite to quartz mica porphyritic diorite. Furthermore, Late Permian monzogranite is similar to granite, with low Sr but high Yb contents. Granite formed under low pressure and potentially at normal crustal thicknesses (≈30 km). Therefore, the thickness of the continental crust in the study area might have changed from thick to normal during the Late Permian.

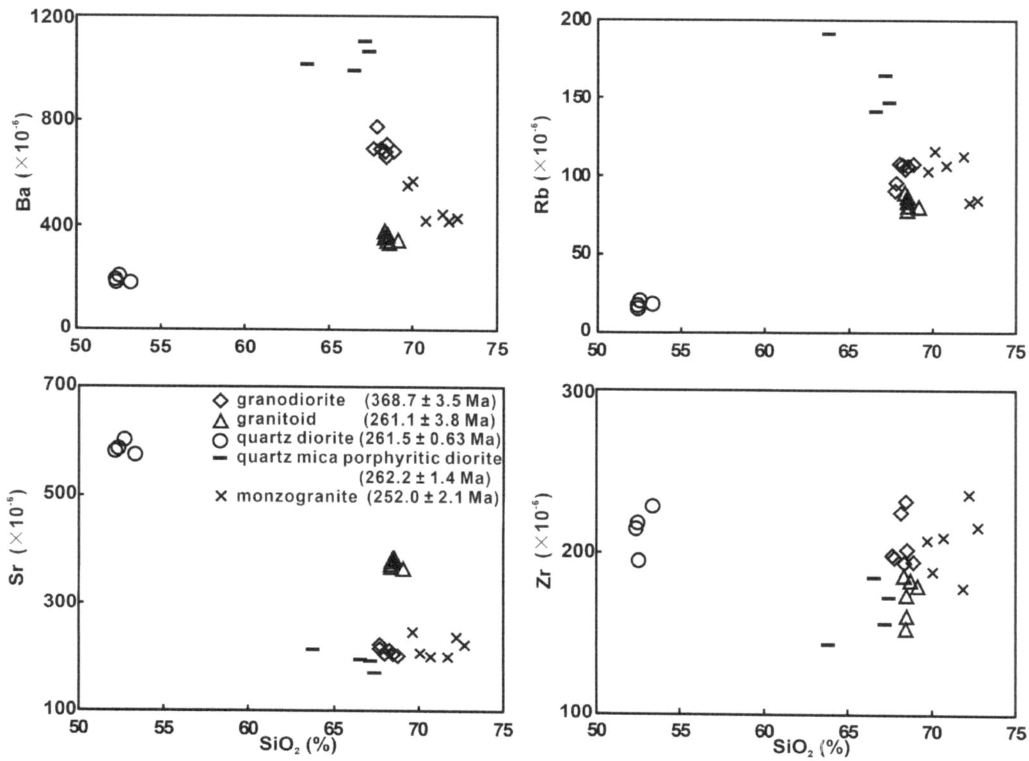

Figure 9. Trace element variation diagrams of granitic rocks from the Huashigou area.

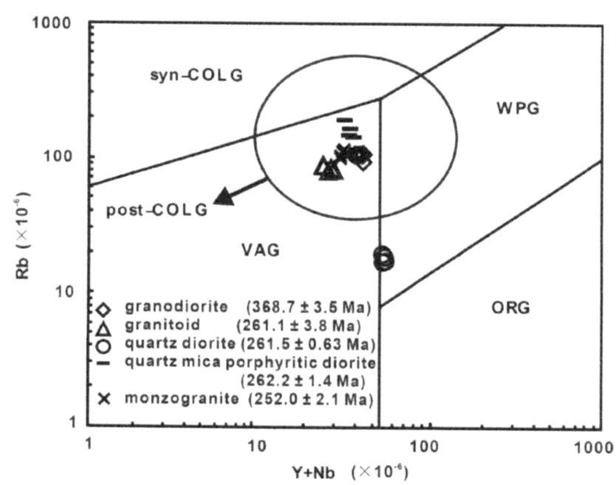

Figure 10. Rb − (Y + Nb) diagram for discriminating the tectonic setting of granitic rocks from the Huashigou area. VAG: volcanic arc granite; ORG: ocean ridge granite; WPG: within-plate granite; syn-COLG: syn-collisional granite; post-COLG: post-collisional granite (after [81]).

It can be concluded that there were at least two important compressional–extensional tectonic transformation phases (c. 369 Ma and c. 262–252 Ma, respectively), with intrusions of granitic magma, which happened in the study area during the period of the Late Paleozoic. Additionally, this demonstrated that the evolution of the Altyn Tagh–Qilian–

North Qaidam tectonic mechanism was evidently divided into phases. The magmatic activities and geochemical characteristics of granites indicate that the crust thickness before c. 369 Ma may have been less than 35 km, with low contents of Sr and high contents of Y. This corresponds to Late Devonian continental lava and volcanic pyroclastic deposits (products of tectonic evolution in the post-orogenic period) developed in the North Qaidam and East Kunlun regions, which are related to extensions of the continental crust [89]. After a short period of extension in the Devonian, the North Margin of the Qinghai–Tibet Plateau was influenced by tectonic movements in the Paleo-Tethys tectonic domain, especially subductions [89]. The ancient Tethys Ocean's collision transpired between the Carboniferous–Permian continents of Asia and the Gondwana continent as the crust was renewed and intensified the compression. The crust underwent rapid thickening, reaching a thickness of up to 50 km in the Mid–Late Permian (262~252 Ma). Simultaneously, there was a transition phase in the tectonic system from compression to extension, and the basaltic lower crust was partially melted due to the heating of underplated basaltic magma, causing the formation of Mid–Late Permian granites. In conclusion, magmatic activities in the study area in the Early Paleozoic–Devonian were mainly affected by the Altyn Tagh–Qilian–North Qaidam tectonic mechanism, and those of the Carboniferous–Mesozoic–Cenozoic were restricted by the tectonic mechanism of the Paleo-Tethys Plate.

5.3. Tectonic Implications

5.3.1. Tectonic System of Altyn Tagh

Altyn Tagh is an important tectonic belt in the Northern Qinghai–Tibet Plateau, which includes a calc-alkaline granitic body dated at c. 575 Ma [90]. These granodiorites were formed at the active continental margin, meaning that plate subduction happened between the end of the Neoproterozoic and the beginning of the Paleozoic. In the Early Paleozoic, the Altyn Tagh region had started the ocean–continent evolution, as evidenced by two major ophiolite belts of such age. The c. 508–449 Ma Hongliugou–Lapeiquan ophiolite [91,92] is comparable to the Mangnai ophiolite, dated at c. 481 Ma [91]. In recent years, the discovery of high-pressure and ultrahigh-pressure metamorphic rocks in southern Altyn Tagh revealed the existence of a UHP metamorphic belt [7,11,12,28,29,93], formed ca. 504–486 Ma [94], which is 40–60 Ma older than that of eclogite in the UHP metamorphic belt in the North Qaidam Margin. Post-collisional extension was the major tectonic evolution process in the Altyn Tagh region, with weak magmatic activity and small rock masses [89].

5.3.2. North Qaidam–Qilian Tectonic System

North Qaidam–Qilian is a Cambrian–Ordovician orogenic belt. The Southern Qilian region is a small, active ocean basin. The oceanic crust subducted toward the Middle Qilian continental crust, and the ocean basin soon closed and formed the tectonic pattern of today. In recent years, the discovery of eclogites in the Xitie Mountain, Dachaidan, Yuka River, Dulan, and Shaliu River [9,12,95] proved that, during ca. 495–466 Ma, the subducted oceanic crust towed the Qaidam continental crust, causing it to subduct continuously. As a result, different types of rocks in Northern Qaidam experienced Early Paleozoic UHP metamorphism [12,17,18]. Furthermore, coesite and pseudomorph in eclogite and paragneiss [9,12], garnet peridotite, and diamond inclusions, as well as the phenomenon that clinopyroxene and rutile were exsolved in garnet [7], are consistent with these results. Several rocks were subducted to a mantle depth of 100–200 km. In the Early Silurian–Devonian, the continental crust began an exhumation process because of the subducted oceanic crust's earlier delamination and violent crustal uplift during orogenesis, with magmatic activities in the crust. Molasse formations were broadly developed in the Early Devonian [19,21,96]. Furthermore, quartz diorite of the Aolao River in North Qaidam, small rock bodies southeast of the Bagaqaidam Lake, Datouyangou rock bodies in the Dachaidan region, and Late Devonian granodiorites were formed by the partial melting that occurred during extensions between different blocks in the orogenic zone after the UHP metamorphic belt was subducted and exhumed, indicating the uplifting stage of the North

Qaidam–Qilian region after its orogenesis. This orogenesis stopped in the Carboniferous when the marine-paralic continental deposition belt covered the whole Qilian Caledonides tectonic belt [21]. In the Late Carboniferous, intracontinental subduction happened in North Qaidam as the Paleo-Tethys Ocean closed. Large amounts of granodiorite–monzogranite magmas were created in the middle–lower crust [97], such as the Sanchagou rock body in the west section of North Qaidam [21] and Late Permian monzogranite. These granites were formed under local tectonic stress changes from extension to compression, leading to intracontinental subduction and the upward intrusion of deep-mantle-derived materials. The middle–lower crust was eventually remelted.

5.3.3. Tectonic Setting of Altyn Tagh and North Qaidam–Qilian

Ge et al. [98] considered the tectonic units on both sides of the Altyn Tagh Fault before the Cenozoic Era to be comparable to the Alashan–Beishan–Tarim Block, the North Qilian–Hexi Corridor–Manjiaer Rift Trough, the Middle Qilian Uplift–Middle Uplift of Tarim, the South Qilian–North Qaidam–South Tarim, and the East Kunlun–West Kunlun. Based on a study of eclogites in the North Qaidam and South Altyn Tagh UHP metamorphic belts, Yang et al. [9] concluded that the South Altyn Tagh UHP metamorphic belt was a westward extension of the Northern Qaidam UHP metamorphic belt, with a stagger of 400 km by the Altyn Tagh Fault. Chen et al. [99] considered the east section of Altyn Tagh and the west section of the Qilian Mountain to be similar and comparable, estimating that the total displacement of the sinistral strike-slip of the Altyn Tagh Fault is about 400 km. Research on the metamorphic ages of eclogites has been conducted in Altyn Tagh and North Qaidam. Liu et al. [93] found that the peak metamorphic age of rocks in the Altyn Tagh high-pressure–ultrahigh-pressure metamorphic belt (c. 504–480 Ma) was c. 40–60 Ma earlier than that of Northern Qaidam (c. 461–420 Ma), which is c. 30–55 Ma earlier than the retrograde metamorphic age. Thus, it is considered that eclogites in Altyn Tagh and Northern Qaidam do not belong to the same high-pressure–ultrahigh-pressure metamorphic belts.

Late Paleozoic granites in Northern Qaidam–Qilian can be distinguished in the Late Devonian (c. 385–367 Ma) and Permian (c. 275–252 Ma). However, Late Paleozoic granites are rare in Altyn Tagh. Chen et al. [99] considered that the Early Paleozoic Era (c. 490–385) was the active stage due to magmatic activities in the north margin of Altyn Tagh. Hence, the granite in Altyn Tagh was not found in the study area, which may be caused by the different active ages of the Altyn Tagh and North Qaidam UHP metamorphic belts. Accordingly, the South Altyn Tagh and North Qaidam UHP metamorphic belts may not belong to one UHP metamorphic belt.

5.3.4. Nd Model Age and Attribution of Precambrian Basement in the Qilian Block

The Proterozoic crystalline basements within the Qilian orogenic belt predominantly emerge in the Yemanan Mountains in the western segment, Huangyuan in the central portion, and Maxianshan in the eastern section. These formations comprise high-grade metamorphic and granitic rocks. Generally, there are two models of the tectonic affinity of the Precambrian basement in the Qilian orogenic belt. Researchers consider that the Qilian Block belongs to the North China Craton [100–102]. However, geochemical and isotopic geochronological results led some researchers to consider that the Qilian Block was genetically related to the Yangtze Block. By analyzing the age spectrum of detrital and magmatic zircons in the Qilian Block, we concluded that the Qilian Block and Yangtze Block may have been in a uniform block in the Neoproterozoic Era and both belonged to Gondwana [33,103,104]. Through the study of Precambrian basement rocks in the Qilian orogenic belt and Pb–Nd isotopic compositions of granites, Zhang et al. [88] concluded that the Qilian orogenic belt is similar to the Yangtze Block and expanded the northwest boundary of the Yangtze Block to the north of the Qilian orogenic belt.

Based on the Nd isotope model ages of granites, we concluded that the basement of the North China Plate was formed in the Archean. The most important crustal growth period was between 2.8 and 2.6 Ga, and the less important crustal growth period was from 1.8 to 2.2 Ga [105]. Simultaneously, through comprehensive data processing and analysis, Wu et al. [106] determined that the Nd isotope model ages of the North China Craton are mainly between 2.4 Ga and 3.6 Ga, and the Nd isotope model ages of the whole South China Craton are mainly between 0.9 Ga and 2.3 Ga, while the intermediate acid igneous rocks in the Yangtze Block are between 0.7 and 2.2 Ga [72]. Furthermore, the Nd model ages of Huashigou granitoids are evidently divided into two phases: 0.9–1.1 Ga, represented by Middle Permian granodiorite, Middle Permian quartz diorite, and Late Permian monzogranite, and 1.6~1.7 Ga, represented by Late Devonian granodiorite and Middle Permian porphyritic tonalite [41,42]. This confirmed that the Mesoproterozoic Era was a crustal growth period in the Qilian region. The results indicate that the crustal growth history of the Qilian orogenic belt is similar to that of the South China Craton but different from that of the North China Craton. Furthermore, the Nd isotope model ages indicated that the Qilian Block and Yangtze Block might have been in a uniform block in the Mesoproterozoic Era.

The results highlighted two stages of granitic production in the Huashigou area, which included Late Devonian granodiorite (c. 369 Ma), Middle Permian biotite granodiorite, quartz diorite, porphyritic tonalite (c. 262–261 Ma), and Late Permian monzonitic granite (252 Ma).

6. Conclusions

(1) Zircon U–Pb dating yielded crystallization ages of 368.7 ± 3.5 Ma, 261.5 ± 0.63 Ma, and 262.2 ± 1.4 Ma for the Hua1, Hua2, and Hua4 plutons, respectively.

(2) Trace elements and Sr–Nd–Hf–O isotopes indicated that Huashigou granitoids were derived from the partial melting of the lower crust.

(3) Huashigou granitoids were formed in the orogenic–post-orogenic extensional tectonic setting, and they were derived from different tectonic levels or crustal depths. Late Devonian granitoids were formed from the decompression melting of enriched and shallow-depth crustal materials. Middle–Late Permian granitoids resulted from the partial melting of the basaltic lower crust through heating by underplating basaltic magma. Additionally, they may have been formed by slab break-off magmatism.

Author Contributions: Y.W.: conceptualization, investigation, visualization, data analysis, writing—original draft, writing—review and editing. W.C.: writing—original draft, writing—review and editing. J.W.: investigation, writing—review and editing. Z.J.: investigation, writing—original draft. Q.T.: writing—original draft, writing—review and editing. P.D.: investigation, writing—review and editing. All authors have read and agreed to the published version of the manuscript.

Funding: This study was funded by the projects of the Science and Technology Program of Gansu Province (22YF7GA050), the Gansu Province Youth Talent Project (2020RCXM152), the Central Government Guides Local Funds for Scientific and Technological Development (YDZX20216200001297), the Second Comprehensive Scientific Expedition to the Qinghai–Tibet Plateau (2019QZKK0704), PhD Research Startup Foundation of Hunan University of Arts and Science (No.23BSQD05, No.23BSQD07, and No.21BSQD32), the National Key Research and Development Program of China (2023YFC2908700), the National Natural Science Foundation of China (U2244204), Key Talent Projects in the Gansu Province (2023), and Gansu Province Key Research and Development Program–Industrial Projects (23YFGA0001).

Data Availability Statement: The original contributions presented in the study are included in the article.

Acknowledgments: We appreciate three anonymous reviewers for their constructive suggestions, which greatly improved the manuscript.

Conflicts of Interest: This manuscript has not been published elsewhere and is not under consideration by another journal. The authors have approved the manuscript and agree with its submission to *Minerals*. There are no conflicts of interest to declare.

References

1. Yin, H.F.; Zhang, K.X. Evolution and characteristics of the Central Orogenic Belt. *Earth Sci.-J. China Univ. Geosci.* **1998**, *23*, 437–442. (In Chinese with English abstract)
2. Lu, S.N.; Yu, H.F.; Li, H.K.; Chen, Z.H.; Wang, H.C.; Zhang, C.L.; Xiang, Z.Q. Early Paleozoic suture zones and tectonic divisions in the "Central China Orogen". *Geol. Bull. China* **2006**, *25*, 1368–1380. (In Chinese with English abstract)
3. Xu, Z.Q.; Yang, J.S.; Li, H.B.; Yao, J.X. The early Paleozoic terrene framework and the formation of the high-pressure (HP) and ultra-high pressure (UHP) metamorphic belts at the Central Orogenic Blet (COB). *Acta Geol. Sin.* **2006**, *80*, 1793–1806. (In Chinese with English abstract)
4. Qiu, H.N.; Wijbrans, J.R. The Paleozoic metamorphic history of the Central orogenic belt of China from 40Ar/39Ar geochronology of eclogite garnet fluid inclusions. *Earth Planet. Sci. Lett.* **2008**, *268*, 501–514. [CrossRef]
5. Yang, J.S.; Xu, Z.Q.; Ma, C.Q.; Wu, C.L.; Zhang, J.X.; Wang, Z.Q.; Wang, G.C.; Zhang, H.F.; Dong, Y.P.; Lai, S.C. Compound orogeny and scientific problems concerning the Central Orogenic Blet of China. *Geol. China* **2010**, *37*, 1–11. (In Chinese with English abstract)
6. Liu, L.; Che, Z.C.; Luo, J.H.; Wang, Y.; Gao, Z.J. Recognition and implication of eclogite in the western Altun Mountains, Xinjiang. *Chin. Sci. Bull.* **1997**, *42*, 931–934. (In Chinese with English abstract) [CrossRef]
7. Liu, L.; Chen, D.L.; Zhang, A.D.; Sun, Y.; Wang, Y.; Yang, J.Q.; Luo, J.H. Ultrahigh pressure gneissic K-feldspar garnet clinopyroxenite in the Altyn Tagh, NW China: Evidence from clinopyroxene exsolution in garnet. *Sci. China (D)* **2005**, *48*, 1000–1010. (In Chinese with English abstract) [CrossRef]
8. Sobel, E.R.; Arnaud, N.; Jolivet, M.; Ritts, B.D.; Brunei, M. Jurassic to Cenozoic exhumation history of the Altyn Tagh range, northwest China, constrained by $^{40}Ar/^{39}Ar$ and apatite fission track thermochronology. *Paleoz. Mesoz. Tecton. Evol. Cent. East. Asia* **2001**, *194*, 247–267.
9. Yang, J.S.; Xu, Z.Q.; Zhang, J.X.; Chu, C.Y.; Zhang, R.Y.; Liou, J.G. Tectonic significance of early Paleozoic high-pressure rocks in Altun-Qaidam-Qilian mountains, northwest China. *Mem.-Geol. Soc. Am.* **2001**, *194*, 151–170.
10. Yang, J.S.; Wu, C.L.; Zhang, J.X.; Shi, R.D.; Meng, F.C.; Wooden, J.; Yang, H.Y. Protolith of eclogites in the north Qaidam and Altun UHP terrane, NW China: Earlier oceanic crust? *J. Asian Earth Sci.* **2006**, *28*, 185–204. [CrossRef]
11. Zhang, J.X.; Zhang, Z.M.; Xu, Z.Q.; Yang, J.S.; Cui, J.W. Petrology and geochronology of eclogites from the western segment of the Altyn Tagh, northwestern China. *Lithos* **2001**, *56*, 187–206. [CrossRef]
12. Zhang, J.X.; Yang, J.S.; Xu, Z.Q.; Meng, F.C.; Li, H.B.; Shi, R.D. Evidence for UHP metamorphism of eclogites from the Altun Mountains. *Chin. Sci. Bull.* **2002**, *47*, 751–755. (In Chinese with English abstract) [CrossRef]
13. Zhang, J.X.; Mattinson, C.G.; Meng, F.C.; Wan, Y.S. An early palaeozoic HP/HT granulite-garnet peridotite association in the south altyn tagh, nw china: P-T history and U-Pb geochronology. *J. Metamorph. Geol.* **2005**, *23*, 491–510. [CrossRef]
14. Zhang, J.X.; Yang, J.S.; Mattinson, C.G.; Xu, Z.Q.; Meng, F.C.; Shi, R.D. Two contrasting eclogite cooling histories, North Qaidam HP/UHP terrane, western China: Petrological and isotopic constraints. *Lithos* **2005**, *84*, 51–76. [CrossRef]
15. Zhang, J.X.; Meng, F.C.; Li, J.P.; Mattinson, C.G. Coesite in eclogite from the North Qaidam Mountains and its implications. *Chin. Sci. Bull.* **2009**, *54*, 1105–1110. (In Chinese with English abstract) [CrossRef]
16. Wan, Y.S.; Xu, Z.Q.; Yang, J.S.; Zhang, J.X. The Precambrian High-grade basement of the Qilian Terrane and Neighboring areas: Its ages and compositions. *Acta Geosci. Sin.* **2003**, *24*, 319–324. (In Chinese with English abstract)
17. Song, S.G.; Niu, Y.L. Ultra-deep origin of garnet peridotite from the North Qaidam ultrahigh-pressure belt, Northern Tibetan Plateau, NW China. *Am. Mineral.* **2004**, *89*, 1330–1336. [CrossRef]
18. Song, S.G.; Zhang, L.; Niu, Y.L.; Su, L.; Jian, P.; Liu, D. Geochronology of diamond-bearing zircons from garnet peridotite in the North Qaidam UHPM belt, Northern Tibetan Plateau: A record of complex histories from oceanic lithosphere subduction to continental collision. *Earth Planet. Sci. Lett.* **2005**, *234*, 99–118. [CrossRef]
19. Song, S.G.; Zhang, L.; Niu, Y.L.; Su, L.; Song, B.; Liu, D. Evolution from oceanic subduction to continental collision: A case study from the Northern Tibetan Plateau based on geochemical and geochronological data. *J. Petrol.* **2006**, *47*, 435–455. [CrossRef]
20. Song, S.G.; Su, L.; Li, X.H.; Niu, Y.L.; Zhang, L. Grenville-age orogenesis in the Qaidam-Qilian block: The link between South China and Tarim. *Precambrian Res.* **2012**, *220*, 9–22. [CrossRef]
21. Song, S.G.; Niu, Y.L.; Su, L. Tectonics of the North Qilian orogen, NW China. *Gondwana Res.* **2013**, *23*, 1378–1401. [CrossRef]
22. Yang, S.; Li, Z.; Chen, H.; Santosh, M.; Dong, C.; Yu, X. Permian bimodal dyke of Tarim Basin, NW China: Geochemical characteristics and tectonic implications. *Gondwana Res.* **2007**, *12*, 113–120. [CrossRef]
23. Xiao, W.J.; Windley, B.F.; Yong, Y.; Yan, Z.; Yuan, C.; Liu, C.Z.; Li, J.L. Early Paleozoic to Devonian multiple-accretionary model for the Qilian Shan, NW China. *J. Asian Earth Sci.* **2009**, *35*, 323–333. [CrossRef]
24. Dong, S.L.; Li, Z.; Gao, J.; Zhu, L. Progress of studies on early Paleozoic tectonic framework and crystalline rock geochronology in Altun-Qilian-Kunlun Orogen. *Geol. Rev.* **2013**, *59*, 731–746.
25. Wang, C.; Liu, L.; Yang, W.Q.; Zhu, X.H.; Cao, Y.T.; Kang, L.; Chen, S.F.; Li, R.S.; He, S.P. Provenance and ages of the altyn complex in altyn tagh: Implications for the early neoproterozoic evolution of northwestern china. *Precambrian Res.* **2013**, *230*, 193–208. [CrossRef]
26. Yu, S.Y.; Zhang, J.X.; del Real, P.G.; Zhao, X.L.; Hou, K.J.; Gong, J.H.; Li, Y.S. The Grenvillian orogeny in the Altun–Qilian–North Qaidam mountain belts of northern Tibet Plateau: Constraints from geochemical and zircon U–Pb age and Hf isotopic study of magmatic rocks. *J. Asian Earth Sci.* **2013**, *73*, 372–395. [CrossRef]

27. Zhang, G.; Zhang, L.; Christy, A.G. From oceanic subduction to continental collision: An overview of HP–UHP metamorphic rocks in the North Qaidam UHP belt, NW China. *J. Asian Earth Sci.* **2013**, *63*, 98–111. [CrossRef]
28. Liu, L.; Sun, Y.; Xiao, P.X.; Che, Z.C.; Luo, J.H.; Chen, D.L.; Wang, Y. Discovery of ultrahighpressure magnesite-bearing garnet lherzolite (>3.8 GPa) in the Altyn Tagh, Northwest China. *Chin. Sci. Bull.* **2002**, *47*, 881–886. (In Chinese with English abstract) [CrossRef]
29. Liu, L.; Sun, Y.; Luo, J.H.; Wang, Y.; Chen, D.L.; Zhang, A.D. Ultra-high pressure metamorphism of granitic gneiss in the Yinggelisayi area, Altun Mountains, NW China. *Sci. China (D)* **2004**, *47*, 338–346. (In Chinese with English abstract) [CrossRef]
30. Zhang, A.D.; Liu, L.; Sun, Y.; Chen, D.L.; Wang, Y.; Luo, J.H. SHRIMP U-Pb zircon ages for the UHP metamorphosed granitoid gneiss in Altyn Tagh and their geological significance. *Chin. Sci. Bull.* **2004**, *49*, 2527–2532. (In Chinese with English abstract) [CrossRef]
31. Yu, S.Y.; Zhang, J.X.; Gong, J.H. Zr-in-rutile thermometry in HP/HP granulite in the Bashiwake area of the South Altun and its geological implications. *Earth Sci. Front.* **2011**, *18*, 140–150. (In Chinese with English abstract)
32. Li, Y.S.; Zhang, J.X.; Li, S.R.; Yu, S.Y.; Gong, J.H.; Lin, Y.H. Metamorphic evolution of the Bashiwake garnet peridotite from the South Altyn Tagh. *Acta Petrol. Sin.* **2013**, *29*, 2073–2092. (In Chinese with English abstract)
33. Xu, W.C.; Zhang, H.F.; Liu, X.M. U-Pb zircon dating constraints on formation time of Qilian high-grade metamorphic rock and its tectonic implication. *Sci. Bull.* **2007**, *52*, 531–538. (In Chinese with English abstract) [CrossRef]
34. Barbarin, B. A review of the relationships between granitoid types, their origins and their geodynamic environments. *Lithos* **1999**, *46*, 605–626. [CrossRef]
35. Qi, X.X.; Li, H.B.; Wu, C.L.; Yang, J.S.; Zhang, J.X.; Meng, F.C.; Shi, R.D.; Chen, S.Y. SHRIMP U-Pb zircon dating for Qiashikansayi granodio-rite, the Northern Altyn Tagh Mountains and its geological implications. *Sci. Bull.* **2005**, *50*, 440–445. (In Chinese with English abstract) [CrossRef]
36. Qi, X.X.; Wu, C.L.; Li, H.B. SHRIMP U-Pb age of zircons from Kazisayi granite in the northern Alty Tagh mountains and its significations. *Acta Petrol. Sin.* **2005**, *21*, 859–866. (In Chinese with English abstract)
37. Wang, J.R.; Guo, Y.S.; Fu, S.M., Chen, J.L.; Qin, X.F.; Zhang, H.P.; Yang, Y.J. Early Paleozoic adakitic rocks in Heishishan, Gansu and their significance for tectonodynamics. *Acta Petrol. Sin.* **2005**, *21*, 977–985. (In Chinese with English abstract)
38. Kusky, T.; Li, J.H.; Santosh, M. The Paleoproterozoic North Hebei Orogen: North China's collisional suture with the Columbia supercontinent. *Gondwana Res.* **2007**, *12*, 4–28. [CrossRef]
39. Wang, C.; Liu, L.; Zhang, A.D.; Yang, W.Q.; Cao, Y.T. Geochemistry and petrography of early Paleozoic Yusupuleke Tagh rapakivi-textured granite complex, South Altyn: An example for magma mixing. *Acta Petrol. Sin.* **2008**, *24*, 2809–2819. (In Chinese with English abstract)
40. Xiao, W.J.; Windley, B.F.; Allen, M.B. Paleozoic multiple accretionary and collisional tectonics of the Chinese Tianshan orogenic collage. *Gondwana Res.* **2013**, *23*, 1316–1341. [CrossRef]
41. Jia, Z.L.; Chen, W.F.; Sha, X.; Wang, J.R. Discovery od middle Permian adakitic rocks in south Qilian area, Gansu and impliccations foe tectonics and Cu(Au) mineralizaton. *Geotecton. Metallog.* **2017**, *156*, 222–234. (In Chinese with English abstract)
42. Hu, W.L.; Jia, Z.L.; Wang, J.R.; Hou, K.X.; Wang, S.H. Geochronology and geochemistry characteristics of the granites from the Huashigou area, south Qilian and their tectonic significance. *Geol. J. China Univ.* **2016**, *22*, 242–253. (In Chinese with English abstract)
43. Ralf, L. How Does Water Bind to Metal Surfaces: Hydrogen Atoms Up or Hydrogen Atoms Down? *Angew. Chem. Int. Ed.* **2003**, *42*, 3458–3460.
44. Andersen, T. Correction of common lead in U-Pb analyses that do not report ^{204}Pb. *Chem. Geol.* **2002**, *193*, 59–79. [CrossRef]
45. Vavra, G. On The Kinematics of Zircon Growth and Its Petrogenetic Significance:A Cathodoluminescence Study. *Contrib. Mineral. Petrol.* **1990**, *106*, 90–99. [CrossRef]
46. Paterson, B.A.; Stephens, W.E.; Rogers, G.; Williams, I.S.; Hinton, R.W.; Herd, D.A. The Nature of Zircon Inheritance In Two Granite Plutons. *Earth Sci.* **1992**, *83*, 459–471.
47. Rubatto, D.; Gebauer, D. *Use of Cathodoluminescence for U-Pb Zircon Dating by Ion Microprobe: Some Examples From The Western Alps. Cathodoluminescence In Geosciences*; Springer: Berlin/Heidelberg, Germany, 2000; pp. 373–400.
48. Möller, A.; O'Brien, P.J.; Kennedy, A.; Kroner, A. Linking Growth Episodes of Zircon and Metamorphic Textures To Zircon Chemistry: An Example From The Ultrahigh-Temperature Granulites of Rogaland (Sw Norway). *Geol. Soc. Lond. Spec. Publ.* **2003**, *220*, 65–81. [CrossRef]
49. Taylor, S.R.; McLennan, S.M.; Armstrong, R.L.; Tarney, J. The composition and evolution of the continental crust: Rare earth element evidence from sedimentary rocks [and discussion]. Philosophical Transactions of the Royal Society of London. *Ser. A Math. Phys. Sci.* **1981**, *301*, 381–399.
50. Taylor, S.R.; McLennan, S.M. *The Continental Crust: Its Composition and Evolution*; Blackwell: London, UK, 1985; pp. 57–72.
51. Wedepohl, K.H. Der primäre Erdmantel und die durch Krustenbildung verarmte Mantelzusammensetzung. *Fortschr Min.* **1981**, *59*, 203–205.
52. Weaver, B.L.; Tarney, J. Empirical approach to estimating the composition of the continental crust. *Nature* **1984**, *310*, 575–577. [CrossRef]
53. Barth, M.G.; McDonough, W.F.; Rudnick, R.L. Tracking the budget of Nb and Ta in the continental crust. *Chem. Geol.* **2000**, *165*, 197–213. [CrossRef]

54. Sun, S.S.; Mcdonough, W.F. Chemical and Isotopic Systematics of Oceanic Basalits: Implications for Mantle Composition and Processes. In *Mamgmatis In The Ocean Basins*; Saunders, A.D., Norry, M.J., Eds.; Geological Society Special Publications: London, UK, 1989; Volume 42, pp. 313–345.
55. Condie, K.C. Mafic crustal xenoliths and the origin of the lower continental crust. *Lithos* **1999**, *46*, 95–101. [CrossRef]
56. Davies, G.R.; Macdonald, R. Crustal influences in the petrogenesis of the Naivasha basalt-comendite complex: Combined trace element and Sr-Nd-Pb isotope constraints. *J. Pet.* **1987**, *28*, 1009–1031. [CrossRef]
57. Whalen, J.B.; Currie, K.L.; Chappell, B.W. A-type granites: Geochemical characteristics, discrimination and petrogenesis. *Contrib. Mineral. Petrol.* **1987**, *95*, 407–419. [CrossRef]
58. Hofmann, A.W.; Jochum, K.P.; Seufert, M.; White, W.M. Nb and Pb in oceanic basalts: New constraints on mantle evolution. *Earth Planet. Sci. Lett.* **1986**, *79*, 33–45. [CrossRef]
59. Defant, M.J.; Drummond, M.S. Derivation of some modern arc magmas by melting of young subducted lithosphere. *Nature* **1990**, *347*, 662–665. [CrossRef]
60. Defant, M.J. Reply for comment by R. Conner on the "Evidence suggests slab melting in arc magmas" by M. Defant and P. Kepezhinskas (EOS, 2001, 82:65, 68-69). *Eos Trans. Am. Geophys. Union* **2002**, *66*, 256–257. [CrossRef]
61. Kay, S.M.; Ramos, V.A.; Marquez, M. Evidence in Cerro Pampa volcanic rocks of slab melting prior to ridge trench collision in southern South America. *J. Geol.* **1993**, *101*, 703–714. [CrossRef]
62. Yogodzinski, G.M.; Kay, R.W.; Volynets, O.N.; Koloskov, A.V.; Kay, S.M. Magnesian andesite in the western Aleutian Komandorsky region: Implications for slab melting and processes in the mantle wedge. *Geol. Soc. Am.* **1995**, *107*, 505–519. [CrossRef]
63. Stern, C.R.; Kilian, R. Role of the subducted slab, mantle wedge and continental crust in the generation of adakites from the Austral volcanic zone. *Contrib Min. Pet.* **1996**, *123*, 263–281. [CrossRef]
64. Rapp, R.P.; Shimizu, N.; Norman, M.D.; Applegate, G.S. Reaction between slab-derived melts and peridotite in the mantle wedge: Experimental constraints at 3.8 GPa. *Chem. Geol.* **1999**, *160*, 335–356. [CrossRef]
65. Peacock, S.M.; Rushmer, T.; Thompson, A.B. Partial melting of subducting oceanic crust. *Earth Planet. Sci. Lett.* **1994**, *121*, 227–244. [CrossRef]
66. Muir, R.J.; Weaver, S.D.; Bradshaw, J.D.; Eby, G.N.; Evans, J.A. The Cretaceous Separation Point batholith, New Zealand: Granitoid magmas formed by melting of mafic lithosphere. *J. Geol. Soc.* **1995**, *152*, 689–701. [CrossRef]
67. Petford, N.; Atherton, M. Na-rich partial melts from newly underplated basaltic crust: The Cordillera Blanca Batholith, Peru. *J. Petrol.* **1996**, *37*, 1491–1521. [CrossRef]
68. Kemp, A.I.S.; Hawkesworth, C.J.; Foster, G.L.; Paterson, B.A.; Woodhead, J.D.; Hergt, J.M.; Whitehouse, M.J. Magmatic and crustal differentiation history of granitic rocks from Hf-O isotopes in zircon. *Science* **2007**, *315*, 980–983. [CrossRef] [PubMed]
69. Whalen, J.B.; Hildebrand, R.S. Trace element discrimination of arc, slab failure, and A-type granitic rocks. *Lithos* **2019**, *348–349*, 105179. [CrossRef]
70. DePaolo, D.J. Age dependence of the composition of continental crust: Evidence from Nd isotopic variations in granitic rocks. *Earth Planet. Sci. Lett.* **1988**, *90*, 263–271. [CrossRef]
71. Landoll, J.D.; Foland, K.A. The formation of quartz syenite by crustal contamination at Mont Shefford and other Montere-gian complexes, Quebec. *Can. Mineral.* **1996**, *34*, 301–324.
72. Chen, J.F.; Jahn, B.M. Crustal evolution of southeastern China: Nd and Sr isotopic evidence. *Tectonophysics* **1998**, *284*, 101–133. [CrossRef]
73. Zhao, Z.H.; Wang, Z.G.; Zou, T.R.; Masuda, A. The REE, isotopic composition of O, Pb, Sr and Nd and diagenetic model of granitoids in Altay region. In *New Improve-Ment of Solid Geosciences in Northern Xinjiang*; Tu, G.Z., Ed.; Science Press: Beijing, China, 1993; pp. 239–266. (In Chinese)
74. Zhao, Z.H.; Wang, Z.G.; Zou, T.R.; Masuda, A. Study on petrogenesis of alkali-rich intrusive rocks of Ulungur, Xinjiang. *Geochimica* **1996**, *25*, 205–220. (In Chinese with English abstract)
75. Chen, J.F.; Zhou, T.X.; Xie, Z.; Zhang, X.; Guo, X.S. Formation of positive εNd(T) granitoids from the Alataw Mountains, Xinjiang, China, by mixing and fractional crystallization: Implication for Phanernozoic crustal growth. *Tectonophysics* **2000**, *328*, 53–67. [CrossRef]
76. Wu, S.P.; Wu, C.L.; Wang, M.Y.; Chen, Q.L.; Wooden, J.L. SHRIMP U-Pb zircon dating of the Tula granite pluton on the south side of the Altun Fault and its geological implications. *Acta Geol. Sin.-Engl. Ed.* **2008**, *82*, 409–414.
77. Leake, B.E. Granite magmas: Their sources, initiation and consequences of emplacement. *J. Geol. Soc.* **1990**, *147*, 579–589. [CrossRef]
78. Kula, J.L.; Spell, T.L.; Wells, M.L. Syntectonic Intrusion and Exhumation of a Mesozoic Plutonic Complex in the Late Cretaceous, Granite Mountains, Southeastern California. 2002. Available online: https://www.researchgate.net/publication/313524963_Syntectonic_intrusion_and_exhumation_of_a_Mesozoic_plutonic_complex_in_the_Late_Cretaceous_Granite_Mountains_southeastern_California (accessed on 31 January 2002).
79. Aranguren, A.; Tubia, J.; Bouchez, J.L.; Vigneresse, J.L. The Guitiriz granite, Variscan belt of northern Spain: Extension-controlled emplacement of magma during tectonic escape. *Earth Planet. Sci. Lett.* **1996**, *139*, 165–176. [CrossRef]
80. Foster, D.A.; Schafer, C.; Fanning, C.M.; Hyndman, D.W. Relationships between crustal partial melting, plutonism, orogeny, and exhumation: Idaho–Bitterroot batholith. *Tectonophysics* **2001**, *342*, 313–350. [CrossRef]

81. Pearce, J.A.; Harris, N.B.W.; Tindle, A.G. Trace element discrimination diagrams for the tectonic interpretation of granitic rocks. *J. Petrol.* **1984**, *25*, 956–983. [CrossRef]
82. Hellman, P.L.; Green, T.H. The role of sphene as an accessory phase in the high-pressure partial melting of hydrous mafic compositions. *Earth Planet. Sci. Lett.* **1979**, *42*, 191–201. [CrossRef]
83. Sen, C.; Dunn, T. Dehydration melting of a basaltic composition amphibolite at 1.5 and 2.0 GPa: Implications for the origin of adakites. *Contrib. Mineral. Petrol.* **1994**, *117*, 394–409. [CrossRef]
84. Liu, J.; Bohlen, S.R.; Ernst, W.G. Stability of hydrous phases in subducting oceanic crust. *Earth Planet. Sci. Lett.* **1996**, *143*, 161–171. [CrossRef]
85. Klemme, S.; Blundy, J.D.; Wood, B.J. Experimental constraints on major and trace element partitioning during partial melting of eclogite. *Geochim. Et Cosmochim. Acta* **2002**, *66*, 3109–3123. [CrossRef]
86. Xiao, L.; Clemens, J.D. Origin of potassic (C-type) adakite magmas: Experimental and field constraints. *Lithos* **2007**, *95*, 399–414. [CrossRef]
87. Xiong, X.L.; Adam, J.; Green, T.H. Rutile stability and rutile/melt HFSE partitioning during partial melting of hydrous basalt: Implications for TTG genesis. *Chem. Geol.* **2005**, *218*, 339–359. [CrossRef]
88. Zhang, Q.; Wang, Y.; Li, C.D.; Wang, Y.L.; Jin, W.J.; Jia, X.Q. Granite classification on the basis of Sr and Yb contents and its implications. *Acta Petrol. Sin.* **2006**, *22*, 2249–2269. (In Chinese with English abstract)
89. Cui, J.W.; Tang, Z.M.; Deng, J.F.; Yue, Y.J.; Meng, L.S.; Yu, Q.F.; Li, J.X.; Lai, S.C.; Qi, L.; Guo, G.C.; et al. *Altyn Tagh Fault System*; Geological Publishing House: Beijing, China, 1999; pp. 39–49. (In Chinese)
90. Ma, T.Q.; Wang, X.H. Characteristic and geology meaning of alkaline calcium invade-rock zone at the SW Fringe of Alltun Massif. *Hunan Geol.* **2002**, *21*, 12–16. (In Chinese with English abstract)
91. Liu, L.; Che, Z.C.; Wang, Y.; Luo, J.H.; Wang, J.Q.; Gao, Z.J. The evidence of Sm-Nd isochron age for the early Paleozoic ophiolite in Mangya area, Altun Mountains. *Chin. Sci. Bull.* **1998**, *43*, 754–756. (In Chinese with English abstract) [CrossRef]
92. Yang, J.S.; Shi, R.D.; Wu, C.L.; Su, D.C.; Chen, S.Y.; Wang, X.B.; Woonden, J. Petrology and SHRIMP age of the Hongliugou ophiolite at Milan, north Altun, at the northern margin of the Tibetan plateau. *Acta Petrol. Sin.* **2008**, *24*, 1567–1584. (In Chinese with English abstract)
93. Liu, L.; Wang, C.; Cao, Y.T.; Chen, D.L.; Kang, L.; Yang, W.Q.; Zhu, X.H. Geochronology of multi-stage metamorphic events: Constraints on episodic zircon growth from the UHP eclogite in the South Altyn, NW China. *Lithos* **2012**, *136*, 10–26. [CrossRef]
94. Liu, L.; Wang, C.; Chen, D.L.; Zhang, A.D.; Luo, J.H. Petrology and geochronology of HP–UHP rocks from the South Altyn Tagh, northwestern China. *J. Asian Earth Sci.* **2009**, *35*, 232–244. [CrossRef]
95. Yang, J.S.; Xu, Z.Q.; Zhang, J.X.; Sun, S.G.; Wu, C.L.; Shi, R.D.; Li, H.B.; Mauruce, B. Early Paleozoic North Qaidam UHP metamorphic belt on the north-eastern Tibetan plateau and a paired subduction model. *Terra Nova* **2002**, *14*, 397–404.
96. Wu, C.L.; Wooden, J.L.; Robinson, P.T.; Gao, Y.H.; Wu, S.P. Geocchemistry and zircon SHRIMP U-Pb dating of granitoids from the west segment of the North Qaidam. *Sci. China Earth Sci.* **2009**, *52*, 1771–1790. (In Chinese with English abstract) [CrossRef]
97. Hao, G.J.; Lu, S.N.; Wang, H.C.; Xin, H.T.; Li, H.K. The pre-Devonian tectonic framework in the northern margin of Qaidam basin and geological evolution of Olongbuluck palaeo block. *Earth Sci. Front.* **2004**, *11*, 115–122. (In Chinese with English abstract)
98. Ge, X.H.; Ye, H.W.; Liu, Y.J.; Liu, J.L.; Pan, H.X.; Ren, S.M. Research Progress of Altyn Fault In Western China. *Earth Sci. Front.* **2000**, *7*, 243–244. (In Chinese with English abstract)
99. Chen, X.H.; Gehrels, G.; Wang, X.F.; Yang, F.; Chen, Z.L. Granite from north Altyn Tagh, NW China: U-Pb geochronology and tectonic setting. *Bull. Mineral. Petrol. Geochem.* **2003**, *22*, 294–298. (In Chinese with English abstract)
100. Zuo, G.C.; Liu, J.C. The evolution of tectonic of early Paleozoic in north Qilian range, China. *Sci. Geol. Sin.* **1987**, *1*, 14–24. (In Chinese with English abstract)
101. Feng, Y.M.; He, S.P. *Geotectonics and Orogeny of the Qilian Mountains, China*; Geological Publishing House: Beijing, China, 1996; pp. 84–101. (In Chinese)
102. Wilde, S.A.; Zhao, G.; Sun, M. Development of the North China Craton during the late Archaean and its final amalgamation at 1.8 Ga: Some speculations on its position within a global Palaeoproterozoic supercontinent. *Gondwana Res.* **2002**, *5*, 85–94. [CrossRef]
103. Wang, Z.Q.; Jiang, C.F.; Yan, Q.R.; Yan, Z. Accretion and collision orogenesis in the West Kunlun Mountains. China. *Gondwana Res.* **2001**, *4*, 843–844.
104. Tung, K.; Yang, H.J.; Yang, H.Y.; Liu, D.; Zhang, J.; Wan, Y.; Tseng, C.Y. SHRIMP U-Pb geochronology of the zircons from the Precambrian basement of the Qilian Block and its geological significances. *Chin. Sci. Bull.* **2007**, *52*, 2687–2701. (In Chinese with English abstract) [CrossRef]
105. Zhu, B.Q. Geochemical evidence for the southern China block being a part of Gondwana. *J. Southeast Asia Sci.* **1994**, *9*, 319–329.
106. Wu, F.Y.; Zhao, G.C.; Wilde, S.A.; Sun, D. Nd isotopic constraints on crustal formation in the north China craton. *J. Asian Earth Sci.* **2005**, *24*, 523–545. [CrossRef]

minerals

MDPI

Article

Geochronology, Geochemical Characterization and Tectonic Background of Volcanic Rocks of the Longjiang Formation in the Lengjimanda Plate Area, Middle Da Hinggan Mountains

Shi-Chang Wang [1], Yu-Jie Hao [1,2,*], Lu Shi [3,*], Zhen Tang [3] and Shuang Zhu [4]

1 College of Earth Sciences, Jilin University, Changchun 130061, China; wsc1873818566@126.com
2 Key Laboratory of Mineral Resources Evaluation of Northeast Asia, Ministry of Natural Resources, Changchun 130061, China
3 Shenyang Center, China Geological Survey, Shenyang 110034, China; tangzhen@mail.cgs.gov.cn
4 The First Geological Survey of Jilin Province, Changchun 130061, China; zhushuang0523@163.com
* Correspondence: haoyujie@jlu.edu.cn (Y.-J.H.); shilu@mail.cgs.gov.cn (L.S.)

Abstract: The Lengjimanda plate is situated in the middle section of the Da Hinggan mountains, in the eastern section of the Tianshan Xingmeng orogenic belt. To determine the formation age of the volcanic rocks in the Longjiang formation in this area, to explore their origin and tectonic background, and to reconstruct the geodynamic evolution of the region, this study conducted petrological, zircon U–Pb geochronological, geochemical, and isotopic analyses of the volcanic rocks in the Longjiang formation. The Longjiang formation's volcanic rocks are primarily composed of trachyandesite, trachyte trachydacite, and andesite, which are intermediate basic volcanic rocks. They are enriched in large-ion lithophile elements, are depleted in high-field-strength elements, are significantly fractionated between light and heavy rare earth elements, and exhibit a moderate negative Eu anomaly in most samples. The results of the LA–ICP–MS zircon U–Pb dating indicate that the volcanic rocks in this group were formed in the Early Cretaceous period at 129.1 ± 0.82 Ma. The zircon $\varepsilon_{Hf}(t)$ ranges from +1.13 to +43.77, the t_{DM2} ranges from +655 to +1427 Ma, the initial Sr ratio $(^{87}Sr/^{86}Sr)_i$ ranges from 0.7030 to 0.7036, and the $\varepsilon_{Nd}(t)$ ranges from +2.1 to +6.6. Based on the geochemical compositions and isotopic characteristics of the rocks, the initial magma of the volcanic rocks in the Longjiang formation originated from the partial melting of basaltic crustal materials, with a source material inferred to be depleted mantle-derived young crustal. These rocks were formed in a superimposed post-collisional and continental arc environment, possibly associated with the Mongol-Okhotsk Ocean closure and the oblique subduction of the Pacific plate. This study addresses a research gap regarding the volcanic rocks of the Longjiang formation in this area. Its findings can be applied to exploration and prospecting in the region.

Keywords: Longjiang volcanic rock; Da Hinggan mountains; zircon U–Pb dating; petro-geochemistry; Sr–Nd–Pb isotope; Lu–Hf isotope

Citation: Wang, S.-C.; Hao, Y.-J.; Shi, L.; Tang, Z.; Zhu, S. Geochronology, Geochemical Characterization and Tectonic Background of Volcanic Rocks of the Longjiang Formation in the Lengjimanda Plate Area, Middle Da Hinggan Mountains. *Minerals* **2024**, *14*, 719. https://doi.org/10.3390/min14070719

Academic Editors: Aleksei V. Travin and Anastassia Yurievna Borisova

Received: 15 May 2024
Revised: 10 July 2024
Accepted: 11 July 2024
Published: 16 July 2024

1. Introduction

The Central Asian orogeny is a large Phanerozoic orogenic belt located between the North China, Tarim, and Siberian plates [1,2]. The Da Hinggan mountain region is situated in the eastern part of the Central Asian orogenic belt, between the Mongolia-Okhotsk suture line and the Xilamulun Yanji suture zone. It is a significant region, where the Paleozoic Paleo-Asian Oceanic metallogenetic tectonic domain is strongly superimposed, compounded, and transformed by the Mesozoic and Neo-Mesozoic Pacific metallogenetic tectonic domains [3–5]. The Paleozoic was characterized by the collision of small land masses and the closure of the Paleo-Asian Ocean [6,7], whereas the Mesozoic was characterized by the superposition and modification of the Paleo-Pacific plate and the Mongolian-Okhotsk plate [8,9]. Multiple phases of volcanic eruptions have resulted in favorable metallogenic geological conditions in the area, with a variety of rock types [10] and extensive development of Mesozoic volcanic and granitic rocks [1,11]. The volcanic rocks are mainly Early Cretaceous rhyolite and basaltic andesite, distributed in facies-like formations [12]. The rock assemblage of the Mesozoic granites is primarily composed of granodiorite, monzogranite, and syenogranite, with a presence of A-type granites [13]. Several researchers have investigated the chronology and tectonic background of Mesozoic magmatism in the Da Hinggan mountains, and a basic understanding of the chronological framework delineation of different contemporaneous volcanic rocks has been achieved [14–16]; however, there are still some controversies regarding the tectonic background of these mountains, such as the subduction orogeny of the Ancient Pacific plate [17,18], and the post-orogenic extension in the Mongol-Okhotsk ocean [19,20]. Studies on the Da Hinggan mountains' Mesozoic stratigraphy have focused on the Tamulangou formation [21], Manketou Ebo formation [22], Manitou formation [23,24], and Baiyingaolao formation [25,26]. There are fewer studies on the Longjiang formation, and these studies do not provide a unified understanding of the era in which the Longjiang formation's volcanic rocks were formed [27,28]. Furthermore, these works mainly focused on the chronology and whole-rock geochemistry and lacked comprehensive studies of the Sr–Nd–Pb and Lu–Hf isotopes [29,30]. The strong stability of zircon minimizes the influence of later geological events on its Hf isotopic composition, while its extremely low Lu content allows us to accurately determine its Hf isotopic composition at the time of its formation. These properties make zircon an important tool for studying crustal evolution and tracing rock source regions [31–33]. Sr-Nd-Pb isotopic analyses are important in determining the petrogenesis of a formation, as they indicate the magma source and reveal the evolutionary pathways of rocks [34–37]. This study aimed to address the research gap concerning the volcanic rocks of the Longjiang formation. The objectives were to determine the formation age of the Longjiang formation's volcanic rocks, investigate their genesis and tectonic background, and reconstruct the geodynamic evolution of the region. This paper presents detailed petrological, zircon U–Pb geochronological, geochemical, and isotopic chemical analyses of the volcanic rocks of the Longjiang formation conducted through extensive field geological surveys using a geological map of the Lengjimanda plate at a scale of 1:25,000. The objective of this research is to provide reliable data for understanding the dynamic history of the Da Hinggan mountains.

2. Geological Background

The Lengjimanda plate area in the middle of the Da Hinggan mountains is located at the junction of the Xing'an block and the Songnen block in the eastern part of the Central Asian orogenic belt, which belongs to the Zhalantun Duobaoshan island arc (Figure 1a). During the Paleozoic, it was in the accretionary stage of the ancient Asian oceanic continental margin, forming an inland sedimentary basin. During the Mesozoic, it experienced the superposition, complexation, and transformation of the Paleo-Pacific tectonic domain and the Mongolian-Okhotsk oceanic tectonic domain [38], forming an inland volcanic basin [39]. In the Da Hinggan mountain region, the area was affected by the subduction of the Mongolian-Okhotsk oceanic plate during the Triassic. The study area experienced uplift and orogenic movements, resulting in the development of fragmented alteration or thrust faults in both the near-EW and near-NE directions [40]. During the Cretaceous, the area was influenced by the eastward subduction of the Paleo-Pacific plate towards the East Asian continent, forming volcanic rift basins and uplifts that were mainly distributed in the NNE direction. Due to the influence of multiple tectonic environments and superpositions, three main stages of structures developed in the study area. The first stage is characterized by NE reverse faults, the second stage by NNE shear zones, and the third stage by near-EW normal faults.

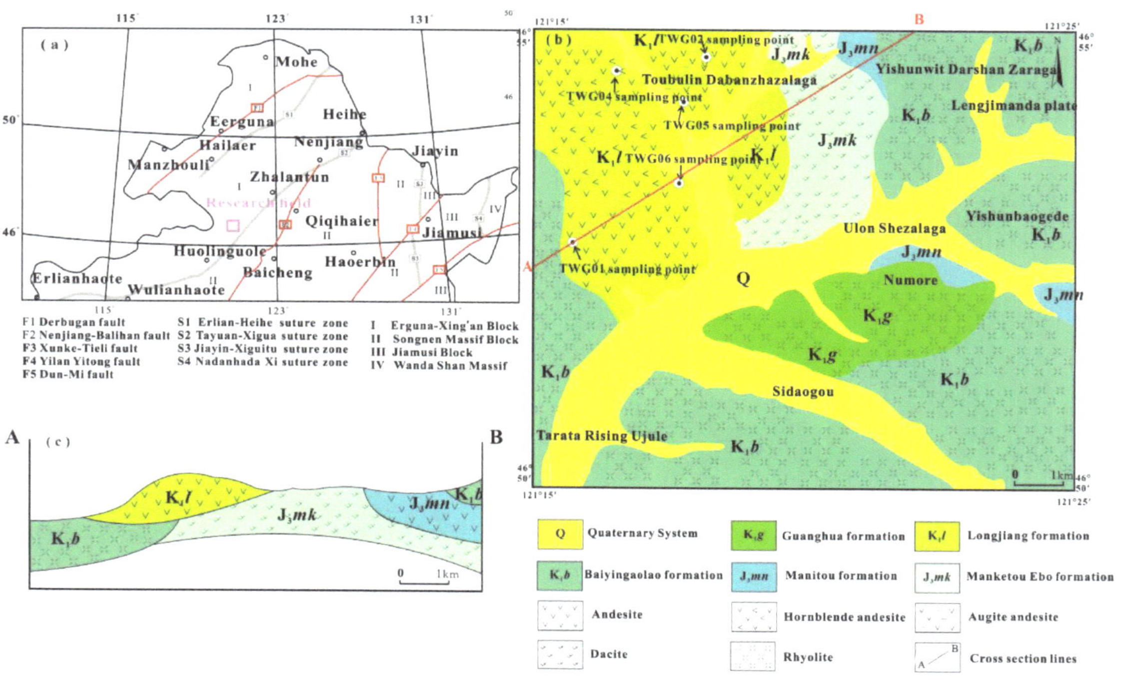

Figure 1. (**a**) The sample structural map in the Northeast region [40]; (**b**) plate geological map of LingjiManda plate, Da Hinggan Mountains [41]; and (**c**) geological cross-section view of Lengjimanda plate.

The region is primarily characterized by Early Cretaceous intrusive rocks derived from the upper crust, with intrusion depths ranging from moderate to shallow. Several lithologies can be recognized, including granite quartz-bearing diorite, quartz-bearing monzonite, and granodiorite belonging to the high-potassium calc-alkaline series. Three types of vein rocks developed in the study area. They are mainly gray-black andesite, light pinkish-red syenite, and gray-white granite aplites. These three varieties occur together in the same location and have a NE distribution pattern, suggesting that they may have been formed by intrusion along a tectonic line.

The study area mainly consists of middle Jurassic Lower Cretaceous formations, which are the Wanbao formation (J_2wb), Manketou Ebo formation (J_3mk), Manitou formation (J_3mn), Baiyingaolao formation (K_1b), Longjiang formation (K_1l), and Guanghua formation (K_1g). The Longjiang formation is a suite of intermediate volcanic rocks characterized by different lithological types, including effusive basalt, explosive basaltic breccia tuff (comprising crystal and lithic fragments), volcaniclastic flow breccia tuff of a basaltic composition, and subvolcanic andesite. In the western part of the study area, particularly in the Tubulin Daban Zhalaga section, the Longjiang formation mainly consists of a suite of neutral to slightly alkaline volcanic rocks, with lithological assemblages including porphyritic rock, porphyritic andesite, andesite, andesitic brecciated tuff with angular lithic fragments, and locally developed andesitic volcaniclastic breccia (Figure 1b).

Sample TWG01 of the Longjiang formation was collected from the Toubulingda Zalaga area in the western part of Lengjimanda Banxi. The sampling coordinates are 46° 52′48″ N, 121°15′36″ E (Figure 2b). The weathered surface appears yellowish-brown, while the fresh surface is light gray. It exhibits a volcanic breccia structure with 30% angular clasts and 10% rock fragments composed of andesite. Crystals include 10% plagioclase and 20%

alkali feldspar. The volcanic glassy matrix is reddish-brown, flame-like, with 50% content, showing trachyandesite under indoor microscopic identification.

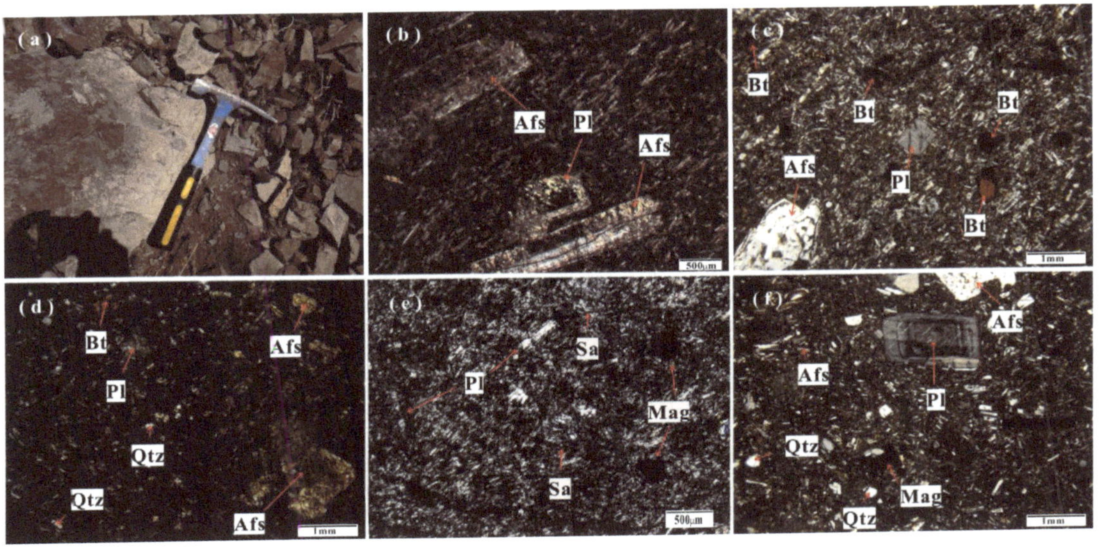

Figure 2. Hand specimen photo (**a**) and microscope photos (**b**–**f**) of the Lengjimanda plate. (**a**) TWG05; (**b**) TWG01; (**c**) TWG02; (**d**) TWG04; (**e**) TWG05; (**f**) TWG06, Sa: sanidine; Pl: plagioclase; Mag: magnetite; Afs: alkalifeldspar; Bt: biotite; Qtz: quartz.

Sample TWG02 of the Longjiang formation was collected from the same area, with coordinates 46° 54′36″ N, 121°17′24″ E (Figure 2c). The weathered surface is yellowish-brown, and the fresh surface is light yellow with patchy structures and an interwoven matrix. The phenocrysts include 20% alkali feldspar, 10% plagioclase, and 5% biotite. The matrix has a microcrystalline interwoven structure of plagioclase, identified indoors as biotite trachyte rock.

Sample TWG04 of the Longjiang formation was collected nearby, with coordinates 46° 53′59″ N, 121°50′11″ E (Figure 2d). The weathered surface is light gray, and the fresh surface is dark gray with patchy structures and an interwoven matrix. Phenocrysts include 20% alkali feldspar, 10% plagioclase, and 5% biotite. The matrix has a microcrystalline interwoven structure of plagioclase, identified indoors as biotite trachyandesite.

Sample TWG05 of the Longjiang formation was collected with coordinates 46°53′30″ N, 121°52′15″ E (Figure 2e). The fresh surface is gray-brown with patchy and vesicular structures. Under polarized light microscopy, phenocrysts constitute approximately 5% to 10% of the sample, ranging from 2 to 4 mm in size, mainly composed of translucent feldspar and plagioclase. The feldspar phenocrysts exhibit tabular self-shaped structures, while the plagioclase phenocrysts are elongated tabular with semi- to self-shaped structures, some showing zoning. Phenocrysts and matrix display weak orientation, with the matrix predominantly glassy or cryptocrystalline, about 80%, occasionally showing fine needle-columnar plagioclase in microcrystalline interwoven structures.

Sample TWG06 of the Longjiang formation was collected with coordinates 46° 53′21″ N, 121°52′18″ E (Figure 2f). The weathered surface is yellowish-brown, and the fresh surface is dark gray with patchy structures and an interwoven matrix. Phenocrysts include 15% alkali feldspar, 20% plagioclase, and 2% magnetite. Alkali feldspar fills gaps between plagioclase microcrystals, identified indoors as trachyandesite.

3. Materials and Methods

The trachyandesite (TWG01, TWG04, and TWG06) and trachyte trachydacite (TWG02, TWG05) samples were picked up from the Toubulin-Dabanzhazalaga area in the western part of the study area (Figure 1b). They were subjected to zircon LA–ICP–MS dating (except sample TWG02), whole rock major and trace element analyses, and Sr–Nb–Pb isotopic analyses (except sample TWG02). Zircon Lu–Hf isotope analyses were performed on a coarse-grained andesite sample (TWG01).

3.1. Major and Trace Element Analyses

Rock geochemical analysis was conducted at the Shenyang geological survey center, China geological survey. Fresh and unweathered samples were selected and then ground to a very fine power of about 80–200 μm for major and trace element analysis. A Rigaku Primus II X-ray fluorescence spectrometer (XRF-1500) was used for the determination of major element contents and analysis standards, according to Zhuo [42]. Chemical titration was used for the determination of FeO, and chemical mass analysis was used for the determination of loss on ignition, with analytical accuracy better than 1% to 5%. The trace and rare earth elements were determined using an Agilent 7500a inductively coupled plasma mass spectrometer (X-Series 2) with analytical accuracy better than 2% to 5%.

3.2. LA–ICP–MS Zircon U–Pb Dating

The separation, mounting, and microscopic image acquisition of zircon grains were conducted at the State Key Laboratory of Geological Processes and Mineral Resources at the Geosciences China University (Wuhan). First, samples were crushed to a particle size of 80–100 mesh using conventional methods. After magnetic and organic heavy liquid separations, zircon grains with consistent colors, crystal shapes, sizes, and low degrees of abrasion were selected. The selected zircon grains were then mounted on an epoxy resin, polished to half of their thickness, and coated with a gold film. The zircons selected from various samples exhibited morphological differences, including short prismatic and long prismatic shapes. Based on the CL images (Figure 3), zircons exhibiting the greatest degree of euhedry and oscillatory zoning, as well as the fewest inclusions, alterations, and cracks, were selected for experimentation. Zircon U–Pb dating was performed at the Qingdao National Laboratory for Marine Science and Technology, Chinese Academy of Sciences, using an Agilent 7500a LA–ICP–MS system coupled to a resolution M50 laser ablation system. The experiment employs helium as the carrier gas for the ablation material, and NIST 610 glass was used as the instrument monitoring standard. The age determination was calibrated using standard zircon 91500 and monitoring standard zircon Plešovice, and the laser spot diameter was 32 μm at a frequency of 7 Hz. The experimental procedures were like those described in Yuan [43]). Glitter software was used for signal selection, correction, and quantitative standardization of the experimental data, and ordinary Pb correction was performed using the method proposed by Andersen [44]. Age calculations and concordance plots were performed using Isoplot.

The measured age of the Plešovice zircon $^{206}Pb/^{238}U$ in the experiment was 342.21 ± 5.53 Ma, which is consistent with the standard value (337.13 ± 0.37 Ma; Sláma [45]) within the error range. The 91500 zircon $^{206}Pb/^{238}U$ age was 1062.44 ± 15.44 Ma, which is consistent with the standard value (1062.4 ± 0.8 Ma; Wiedenbeck [34]) within the error margin, indicating high accuracy and reliability.

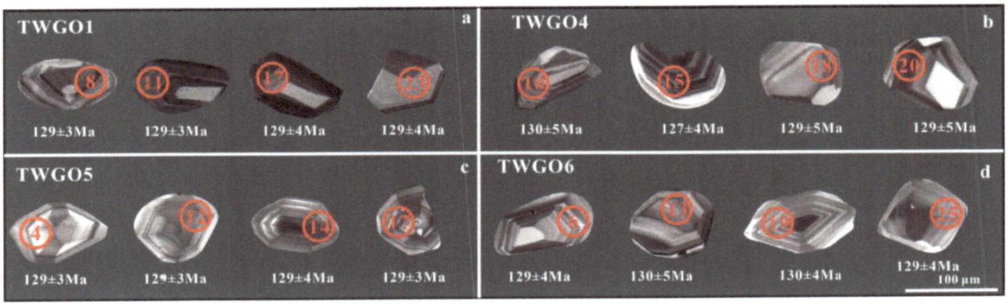

Figure 3. U–Pb zircons cathodoluminescence images from the volcanic rocks of the Longjiang formation at Lengjimanda plate. Circles indicate the locations of U–Pb analysis on zircons, along with their respective zircon numbers, and age information listed below. (**a**) TWG01, (**b**) TWG04, (**c**) TWG05, (**d**) TWG06.

3.3. The Sr–Nd–Pb Isotope Analysis

The pre-treatment and mass spectrometric determination of Sr-Nd-Pb isotopic analysis of volcanic rocks from the Longjiang formation were conducted at the State Key Laboratory for Mineral Deposits Research, Nanjing University, focusing on the ore-forming mechanisms of endogenous metal deposits. After dissolving the whole rock sample with a mixture of HF and HNO_3, AG50W×8 cation exchange resin was used for the separation and purification of Sr and Nd using different eluents. For specific details of the procedure, please refer to Wei [46].

Sr-Nd isotopic analyses were performed using thermal ionization mass spectrometry (TIMS) on a solid isotope mass spectrometer. For the Sr isotope ratio determination, instrumental mass fractionation correction was performed using a reference value of $^{86}Sr/^{88}Sr$ = 0.1194 with NIST SRM 987 as the external standard. For the Nd isotope analysis, the instrument mass fractionation correction was performed using a reference value of $^{146}Nd/^{144}Nd$ = 0.7219 with the Nd isotope international standard material JNdi-1 as the external standard. The measurement accuracy was within 5 ppm for $^{143}Nd/^{144}Nd$ and within 10 ppm for $^{87}Sr/^{86}Sr$.

For Pb isotope analysis, the sample was dissolved in a mixture of HNO_3 and HF solutions in a volume ratio of 1:4. The Pb isotopes were then separated and purified using HBr and cation exchange resin. Pb isotopic testing and analysis were performed using thermal ionization mass spectrometry (TIMS), with instrumental mass fractionation correction using NIST Pb-981 and isotopic fractionation correction using NBS 981. The precision of the measurements was $^{206}Pb/^{204}Pb < 0.2\%$.

3.4. Zircon Lu–Hf Isotope Analysis

A micro-area Hf isotopic analysis was conducted on the samples based on the LA-ICP-MS zircon U–Pb isotopic analysis. Zircon Hf isotope analyses were performed at the State Key Laboratory for Mineral Deposits Research, Nanjing University. The experimental analysis used a Neptune MC–ICP–MS coupled with a COMPEx GeoLas Pro 193 nm ArF excimer laser ablation system. A laser spot diameter of 32 μm was used based on the size of the zircon grains, with a laser pulse frequency of 7 Hz. Helium was used as the ablation carrier gas. Zircon 91500 and MT were used as external reference materials. The obtained $^{176}Hf/^{177}Hf$ ratio was determined to be 0.282948 ± 0.000038, which is within the error margin of the recommended value (0.282302 ± 8; Goolaerts [47]). Further details of the analytic procedure can be found in Wu [48].

4. Results

4.1. Major and Trace Element Compositions

The results of the major and trace element analyses of the volcanic rocks from the Longjiang formation are presented in Table 1.

Table 1. Major and trace element compositions of Lengjimanda plate of the Longjiang formation volcanic rocks.

Sample	TWG01YQ1	TWG01YQ2	TWG01YQ3	TWG01YQ4	TWG01YQ5	TWG02YQ1	TWG02YQ2	TWG02YQ3	TWG02YQ4	TWG02YQ5	TWG02YQ6	TWG04YQ1	TWG04YQ2
Lithology			Trachyandesite					Trachyte Trachydacite				Trachyandesite	
Major elements (wt%)													
SiO_2	60.44	60.59	60.13	61.11	60.42	65.76	66.28	65.38	65.68	65.66	66.23	61.24	61.70
TiO_2	1.09	1.09	1.11	1.09	1.07	0.63	0.60	0.62	0.59	0.62	0.61	0.84	0.83
Al_2o_3	16.70	16.60	16.95	16.57	16.75	16.94	17.08	16.95	17.10	16.97	16.93	17.13	17.11
$Fe_2O_3{}^T$	7.37	7.59	7.35	7.37	7.40	4.12	3.90	4.22	4.00	4.12	4.05	6.49	6.42
MnO	0.10	0.10	0.11	0.12	0.10	0.08	0.06	0.07	0.07	0.10	0.06	0.11	0.10
MgO	2.17	2.02	2.07	1.92	2.08	1.11	1.18	1.25	1.04	0.85	0.99	2.22	2.22
CaO	4.12	4.31	4.48	4.31	4.32	3.26	2.54	3.42	3.24	3.20	3.11	5.01	4.53
Na_2O	4.66	4.51	4.80	4.76	4.67	4.39	4.15	4.40	4.59	4.66	4.53	3.97	4.23
K_2O	2.99	2.85	2.63	2.42	2.86	3.55	4.04	3.52	3.51	3.62	3.37	2.92	2.77
P_2O_5	2.99	2.85	2.63	2.42	2.86	3.55	4.04	3.52	3.51	3.62	3.37	2.92	2.77
LOI	1.63	1.39	1.63	1.54	1.42	2.08	2.46	1.48	1.43	1.73	1.66	1.58	1.39
Total	98.31	98.76	98.44	98.24	98.61	97.26	97.40	98.42	98.24	98.28	98.41	98.30	98.03
Na_2O/K_2O	1.56	1.58	1.82	1.96	1.63	1.24	1.03	1.25	1.31	1.29	1.34	1.36	1.53
A/NK	1.53	1.58	1.58	1.59	1.55	1.53	1.52	1.53	1.50	1.46	1.52	1.77	1.72
A/CNK	0.91	0.90	0.90	0.91	0.90	1.00	1.08	0.98	0.99	0.97	1.01	0.91	0.94
Trace elements (ppm)													
Sc	19.20	21.93	26.67	24.39	22.56	10.49	14.97	13.90	14.16	15.76	16.25	11.74	11.13
V	96.45	91.59	99.92	107.02	97.75	34.14	33.49	33.52	31.59	33.71	34.96	77.86	76.55
Cr	22.14	16.36	15.62	21.94	29.67	17.17	9.28	13.93	16.44	12.73	11.58	11.75	12.43
Co	27.50	27.90	30.65	27.42	30.30	11.91	12.92	12.19	12.99	10.98	12.03	17.59	16.50
Ni	19.70	19.47	22.90	20.40	23.87	12.20	12.03	13.22	13.78	11.87	18.08	5.98	3.79
Ga	23.28	23.44	22.20	23.31	22.84	20.05	20.58	21.64	20.34	20.47	20.75	18.47	18.59
Rb	61.53	62.45	61.01	54.50	61.37	90.81	107.72	93.00	97.73	95.13	88.03	72.54	69.80
Y	55.62	61.80	63.96	55.58	67.30	45.66	48.02	46.42	46.84	42.65	47.50	18.88	17.13
Zr	251.11	250.60	259.72	247.95	252.33	235.48	238.68	235.88	238.24	238.13	239.84	203.48	204.37
Nb	10.11	10.49	10.09	9.72	10.50	10.10	10.33	9.86	10.16	10.21	9.56	7.63	6.89
Ba	788.13	823.75	827.68	610.15	777.25	907.84	945.25	893.44	865.27	908.96	894.87	774.71	736.84
La	56.61	59.28	60.62	55.31	58.54	51.26	53.97	52.85	54.13	51.15	54.58	21.47	20.40
Ce	121.92	122.72	133.76	123.23	122.00	109.67	115.47	112.28	115.53	103.93	114.63	44.60	41.55
Nd	67.34	71.49	73.50	67.18	74.14	54.61	58.06	56.34	57.07	53.08	57.78	23.94	22.75
Sm	13.83	14.75	15.13	14.01	15.40	10.80	11.12	10.92	10.63	10.51	10.92	4.93	4.81
Eu	2.43	2.51	2.52	2.17	2.51	2.32	2.46	2.33	2.39	2.24	2.36	1.24	1.24

Table 1. *Cont.*

Sample	TWG01YQ1	TWG01YQ2	TWG01YQ3	TWG01YQ4	TWG01YQ5	TWG02YQ1	TWG02YQ2	TWG02YQ3	TWG02YQ4	TWG02YQ5		TWG02YQ6	TWG04YQ1	TWG04YQ2
Lithology			Trachyandesite						Trachyte Trachydacite				Trachyandesite	
Gd	9.52	10.46	10.50	9.61	10.37	8.18	8.46	8.25	8.54	7.41		8.43	3.96	3.61
Tb	0.93	0.98	1.02	0.90	1.04	0.73	0.75	0.73	0.76	0.75		0.75	0.53	0.53
Dy	7.36	7.89	8.00	7.14	8.23	5.77	5.94	5.79	5.83	5.84		6.00	3.21	3.16
Ho	1.08	1.20	1.17	1.08	1.13	0.87	0.89	0.88	0.87	0.81		0.89	0.63	0.57
Er	3.35	3.61	3.62	3.38	3.59	2.76	2.95	2.89	2.97	2.58		2.93	2.03	2.00
Tm	0.42	0.45	0.45	0.43	0.46	0.37	0.39	0.39	0.39	0.36		0.36	0.28	0.26
Yb	2.84	3.02	3.26	2.92	3.02	2.51	2.59	2.67	2.55	2.56		2.60	1.83	1.83
Lu	0.40	0.46	0.46	0.42	0.45	0.37	0.39	0.40	0.40	0.34		0.39	0.29	0.27
Ta	0.98	0.99	1.15	1.02	1.15	1.20	1.07	1.07	0.95	1.08		1.15	0.54	0.38
Th	5.06	4.82	5.01	4.56	4.85	4.56	5.16	5.40	5.60	6.82		4.74	7.15	6.72
U	2.32	2.36	2.52	2.41	2.30	2.16	2.29	2.41	2.46	2.03		2.47	1.60	1.48
ΣREE	304.68	316.19	331.91	304.35	319.59	264.42	278.26	271.12	276.64	255.04		277.41	114.82	108.51
LREE	278.78	288.13	303.43	278.48	291.30	242.86	255.90	249.12	254.33	234.39		255.06	102.07	96.27
HREE	25.90	28.06	28.48	25.87	28.29	21.56	22.36	22.00	22.31	20.65		22.35	12.75	12.24
LREE/HREE	10.77	10.27	10.65	10.76	10.30	11.27	11.44	11.32	11.40	11.35		11.41	8.01	7.86
La_N/Yb_N	14.31	14.08	13.35	13.59	13.89	14.68	14.92	14.18	15.23	14.31		15.05	8.42	8.00
δEu	0.61	0.59	0.58	0.54	0.57	0.73	0.74	0.72	0.74	0.74		0.73	0.83	0.87
δCe	0.96	0.93	0.98	0.99	0.90	0.98	0.98	0.98	0.99	0.95		0.97	0.96	0.94
Major elements (wt%)														
SiO_2	61.01	61.28	62.42	65.12	65.59	66.19	65.92	65.97	58.83	60.76	62.74	61.64	62.87	64.60
TiO_2	0.85	0.83	0.82	0.61	0.58	0.55	0.56	0.58	0.88	0.88	0.83	0.84	0.81	0.79
Al_2O_3	17.07	17.09	16.49	16.54	16.26	16.23	16.23	16.29	18.07	17.34	16.42	16.66	16.23	15.56
$Fe_2O_3{}^T$	6.53	6.39	6.50	4.91	4.77	4.43	4.71	4.69	6.56	6.73	6.40	6.57	6.30	6.12
MnO	0.11	0.11	0.09	0.08	0.09	0.08	0.09	0.09	0.10	0.10	0.10	0.09	0.09	0.09
MgO	2.46	2.33	1.99	1.58	1.31	1.33	1.32	1.47	3.27	2.82	2.32	2.50	2.40	2.20
CaO	5.22	5.08	3.78	3.06	2.48	2.74	2.43	2.93	4.68	3.94	4.08	4.11	4.21	3.92
Na_2O	3.79	4.00	4.88	4.38	5.07	4.87	4.90	4.33	3.38	4.16	4.85	4.34	3.00	3.66
K_2O	2.89	2.87	2.93	3.70	3.86	3.55	3.80	3.63	3.95	3.04	2.05	3.02	3.82	2.87
P_2O_5	2.89	2.87	2.93	3.70	3.86	3.55	3.80	3.63	3.95	3.04	2.05	3.02	3.82	2.87
LOI	1.48	1.32	1.57	1.51	1.45	1.55	1.49	1.68	3.34	2.47	1.94	2.37	3.04	2.58
Total	98.14	98.27	98.04	98.57	98.52	97.89	98.10	98.48	96.38	96.93	97.97	97.51	96.76	97.26
Na_2O/K_2O	1.31	1.39	1.66	1.18	1.32	1.37	1.29	1.19	0.86	1.37	2.37	1.44	0.79	1.28

Table 1. *Cont.*

Sample	TWG01YQ1	TWG01YQ2	TWG01YQ3	TWG01YQ4	TWG01YQ5	TWG02YQ1	TWG02YQ2	TWG02YQ3	TWG02YQ4	TWG02YQ5		TWG02YQ6	TWG04YQ1	TWG04YQ2
Lithology			**Trachyandesite**						**Trachyte**	**Trachydacite**			**Trachyandesite**	
A/NK	1.82	1.76	1.47	1.47	1.30	1.37	1.33	1.47	1.83	1.71	1.61	1.60	1.79	1.70
A/CNK	0.90	0.90	0.91	0.98	0.95	0.96	0.98	0.99	0.98	1.00	0.93	0.93	0.97	0.96
Trace elements (ppm)														
Sc	10.93	11.70	11.34	8.61	7.68	7.82	7.92	8.45	12.82	13.00	12.32	13.00	12.33	12.16
V	83.90	78.07	72.74	50.22	42.42	44.10	47.06	44.66	70.84	75.19	79.00	73.99	67.90	72.73
Cr	10.58	9.72	8.18	9.33	6.68	7.55	4.60	6.97	8.97	8.98	9.61	8.25	5.87	10.68
Co	15.25	16.36	15.20	10.76	9.22	9.84	9.69	9.92	17.42	18.16	16.67	17.38	16.94	16.25
Ni	4.87	4.10	3.04	0.85	0.12	1.61	0.66	0.22	4.58	4.52	3.82	4.00	3.91	4.00
Ga	18.61	19.24	17.81	17.63	17.54	17.75	19.43	17.34	18.49	18.50	16.45	16.71	15.09	16.75
Rb	70.28	70.83	73.37	98.83	97.18	91.25	99.49	95.51	88.08	71.83	40.45	70.13	88.08	67.41
Y	16.73	18.27	17.86	18.62	17.52	18.33	17.79	18.19	15.67	14.79	14.90	15.78	15.66	14.28
Zr	201.17	200.88	191.20	258.32	272.94	272.23	273.57	269.53	192.03	193.80	184.57	182.28	175.72	168.28
Nb	6.84	6.50	7.07	8.31	8.75	8.92	8.20	7.92	6.78	6.77	6.90	7.37	6.53	6.90
Ba	676.71	698.50	795.31	932.24	1000.00	1000.00	959.58	961.40	1100.00	866.52	741.55	808.76	110.00	732.53
La	19.32	20.50	20.55	23.03	22.45	23.82	23.41	23.46	20.52	20.16	19.88	20.24	19.21	19.49
Ce	40.38	43.01	42.20	46.33	47.94	48.45	47.40	48.49	43.54	42.19	41.22	42.74	40.82	40.42
Nd	21.79	23.46	23.06	23.48	23.08	23.61	23.25	23.61	23.28	23.36	22.18	23.21	22.28	21.79
Sm	4.37	4.86	4.64	4.48	4.52	4.54	4.52	4.49	4.74	4.57	4.55	4.65	4.60	4.24
Eu	1.16	1.08	1.20	1.15	1.13	1.17	1.08	1.10	1.19	1.24	1.17	1.11	1.02	1.11
Gd	3.62	3.83	3.75	3.63	3.55	3.56	3.46	3.53	3.77	3.55	3.64	3.72	3.62	3.45
Tb	0.49	0.54	0.53	0.50	0.51	0.50	0.50	0.50	0.52	0.49	0.50	0.52	0.50	0.46
Dy	2.88	3.10	3.10	3.04	3.02	2.97	2.92	2.90	2.86	2.98	2.71	2.85	2.76	2.58
Ho	0.56	0.60	0.61	0.60	0.55	0.60	0.56	0.59	0.53	0.49	0.50	0.52	0.52	0.47
Er	1.87	2.00	1.98	2.02	1.97	2.00	2.00	2.01	1.67	1.57	1.56	1.65	1.63	1.51
Tm	0.26	0.28	0.27	0.29	0.30	0.29	0.28	0.30	0.22	0.19	0.21	0.21	0.22	0.19
Yb	1.72	1.84	1.83	1.98	1.98	2.03	1.89	1.98	1.39	1.25	1.29	1.34	1.35	1.23
Lu	0.26	0.28	0.28	0.31	0.29	0.32	0.30	0.31	0.20	0.18	0.18	0.19	0.19	0.17
Ta	0.35	0.51	0.46	0.47	0.48	0.70	0.38	0.42	0.40	0.41	0.34	0.38	0.33	0.27
Th	6.21	6.35	6.44	7.62	8.53	8.17	8.22	8.17	4.62	4.11	3.77	4.30	4.19	4.07
U	1.51	1.58	1.47	1.85	1.90	1.99	1.91	1.99	1.38	1.18	1.48	1.43	1.46	1.17
ΣREE	103.99	111.17	109.72	116.90	117.05	119.99	117.61	119.36	110.18	108.21	105.04	108.57	104.15	102.47
LREE	92.34	98.68	97.36	104.52	104.89	107.71	105.69	107.25	99.03	97.51	94.46	97.58	93.37	92.41
HREE	11.66	12.49	12.36	12.38	12.16	12.28	11.92	12.11	11.16	10.70	10.58	10.99	10.78	10.06

Table 1. *Cont.*

Sample	TWG01YQ1	TWG01YQ2	TWG01YQ3	TWG01YQ4	TWG01YQ5	TWG02YQ1	TWG02YQ2	TWG02YQ3	TWG02YQ4	TWG02YQ5		TWG02YQ6	TWG04YQ1	TWG04YQ2
Lithology			**Trachyandesite**						**Trachyte**	**Trachydacite**			**Trachyandesite**	
LREE/HREE	7.92	7.90	7.88	8.44	8.62	8.77	8.87	8.86	8.88	9.11	8.93	8.87	8.66	9.19
La_N/Yb_N	8.05	7.98	8.06	8.33	8.15	8.42	8.89	8.52	10.58	11.53	11.04	10.86	10.22	11.35
δEu	0.86	0.74	0.85	0.84	0.83	0.86	0.80	0.82	0.83	0.91	0.85	0.79	0.74	0.86
δCe	0.96	0.95	0.94	0.94	1.01	0.96	0.95	0.97	0.97	0.93	0.95	0.97	0.96	0.95

The SiO_2 content of the volcanic rock samples from the Longjiang formation ranges from 58.83% to 66.28%, with an average of 63.17%. The total alkali content (Na_2O+K_2O) ranges from 6.50% to 8.80%, with an average of 7.29%. In the $N_2O+K_2O-SiO_2$ diagram (Figure 4), the samples' chemical compositions fall into the fields of trachyandesite, trachyte trachydacite, andesite, and dacite; in the SiO_2 vs. K_2O diagram (Figure 5a), the samples mainly fall into the calc-alkaline to high-K calc-alkaline series. The aluminum-rich Al_2O_3 content ranges from 15.26% to 18.07%, with an average of 16.75%. The aluminum saturation index A/CNK ranges from 0.90 to 1.08, with an average of 0.95. In the A/NK-A/CNK diagram (Figure 5b), the sample's chemical composition mainly falls into the sub-aluminous rock category; the TiO_2 content ranges from 0.55% to 1.11%, with an average of 0.79%. The $Fe_2O_3{}^T$ content ranges from 4.00% to 7.37%, with an average of 5.78%. The MgO content ranges from 0.85% to 3.27%, with an average of 1.87%. The $Mg^{\#}$ content ranges from 28.75 to 50.32, with an average of 37.85. The CaO content ranges from 2.43% to 5.08%, with an average of 3.80%. The Thorpe index σ ranges from 1.56 to 3.65, with an average of 2.96. These values indicate that the volcanic rocks of the Longjiang formation in the Lengjimanda plate are a series of intermediate basic volcanic rocks belonging to the calc-alkaline series.

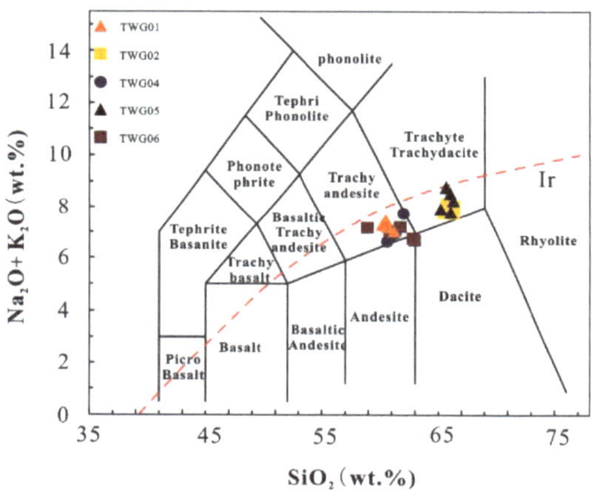

Figure 4. Diagram of (N_2O+K_2O) vs. SiO_2 of volcanic rocks of Longjiang formation [49].

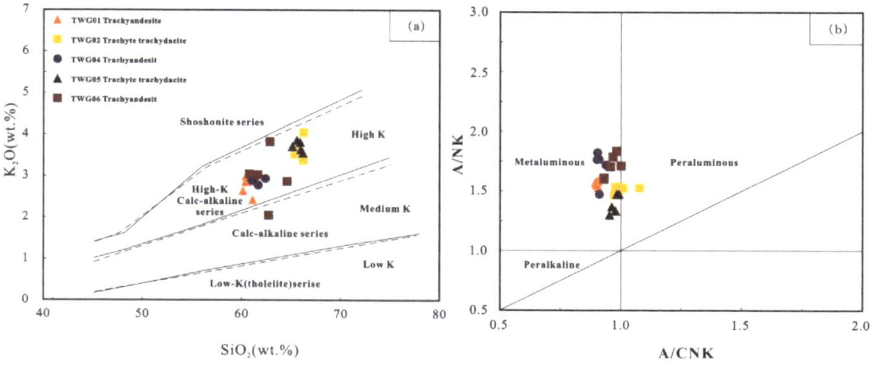

Figure 5. Major elements classification diagrams of Longjiang formation volcanic rocks, Lengjimanda plate. (**a**) K_2O vs. SiO_2 diagram [50]; (**b**) A/NK vs. A/CNK diagram [50].

These volcanic rocks are relatively enriched in large-ion lithophile elements (such as Rb, K, Ba, U, and LREE) and depleted in high-field-strength elements (such as Nb, Ta, Zr,

Hf, and HREE). The Rb/Sr values range from 0.09 to 0.19, with an average of 0.13 (Figure 6), which is higher than the Rb/Sr values of the primitive mantle (0.03), E-MORB (0.033), and OIB (0.047), but lower than those of the crustal source magma (>0.5). The Th/Pb values range from 1.9 to 5.3, with an average of 3.17, which is between those of the crustal source rocks and the primitive mantle. The Nb/Ta values range from 8.33 to 26.03, with an average of 14.51, which is also between those of the crustal source rocks and the primitive mantle. The rocks are characterized by high Sr (average 634.22 ppm), low Y (average 31.55 ppm), low Yb (average 2.12 ppm), and high Sr/Y (average 27.59) values.

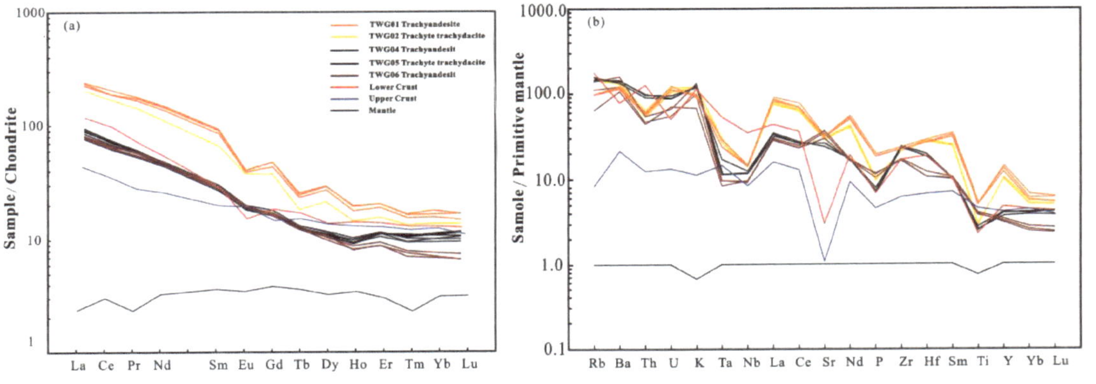

Figure 6. (**a**) Chondrite-normalized REE distribution patterns [51] and (**b**) primitive mantle-normalized spider diagrams [52] of the Longjiang formation volcanic rocks, Lengjimanda plate.

The total rare earth element content of these volcanic rocks from the Longjiang formation ranges from 56.23 ppm to 331.91 ppm, with an average of 184.35 ppm. The light rare earth element content ranges from 92.41 ppm to 333.91 ppm, with an average of 167.52 ppm, while the heavy rare earth element content ranges from 10.06 ppm to 28.48 ppm, with an average of 16.83 ppm. The LREE/HREE ratio ranges from 7.86 to 11.44, with an average of 9.55. The enrichment of light REEs and depletion of heavy REEs are indicated by the right-skewed distribution of REE patterns when they are normalized to chondrite (Figure 6). The fractionation of light and heavy REE is evident, with the fractionation coefficients $(La/Yb)_N$ ranging from 7.98 to 15.23, with an average of 11.33. The δEu values range from 0.54 to 1.19, and except for a few samples with positive Eu anomalies, most samples show moderately negative Eu anomalies.

4.2. Zircon U–Pb Isotopic Dating

The results of the zircon U–Pb isotopic analyses of the volcanic rock samples from the Longjiang formation are shown in Table 2.

The ^{232}Th content ranges from 44 ppm to 1126 ppm, while the ^{238}U content ranges from 70 ppm to 1157 ppm. The ^{232}Th/^{238}U ratio ranges from 0.199 to 1.711. Based on the characteristics observed in the CL images (Figure 3), it is suggested that these zircons are magmatic (Supplementary Materials).

Of the 30 analyzed points on the trachyandesite (TGW01), 27 measurement spots show good age consistency, except for 5 zircon grains with significant age anomalies. In the concordance plot of the zircon U–Pb ages (Figure 7a), all the obtained zircon U–Pb ages lie along or near the concordance line after discordant points are excluded. The ^{206}Pb/^{238}U weighted mean age is 129.1 $\pm$ 0.65 Ma (MSWD = 0.93, n = 27), indicating an Early Cretaceous age.

Table 2. Zircon LA–ICP–MS U–Pb data of the volcanic rocks from Lengjimanda plate of the Longjiang formation.

Sample	Lithology	Concentration (ppm)			Ratio						Age/Ma			
		^{232}Th	^{238}U	Th/U	^{207}Pb/^{206}Pb	±%	^{207}Pb/^{235}U	±%	^{206}Pb/^{238}U	±%	^{207}Pb/^{235}U	2σ	^{206}Pb/^{238}U	2σ
TWG01-01	Trachyandesite	46	70	0.657	0.04751	0.00484	0.14564	0.01571	0.0203	0.00064	138	14	130	4
TWG01-02		208	591	0.352	0.05698	0.00229	0.15885	0.00687	0.02023	0.0005	150	6	129	3
TWG01-03		225	593	0.379	0.05458	0.00237	0.14969	0.00697	0.02021	0.0005	142	6	129	3
TWG01-04		185	351	0.529	0.04789	0.0021	0.13013	0.00608	0.0202	0.0005	124	5	129	3
TWG01-05		260	296	0.878	0.05145	0.00257	0.1447	0.00771	0.02035	0.00052	137	7	130	3
TWG01-06		222	301	0.738	0.04909	0.00281	0.129	0.00781	0.02021	0.00053	123	7	129	3
TWG01-07		775	453	1.711	0.05501	0.00228	0.15136	0.00675	0.02024	0.0005	143	6	129	3
TWG01-08		123	147	0.834	0.05014	0.00297	0.1356	0.00852	0.02018	0.00053	129	8	129	3
TWG01-09		57	80	0.71	0.07026	0.01095	0.19627	0.03278	0.02025	0.00093	182	28	129	6
TWG01-10		202	232	0.87	0.04973	0.00259	0.13746	0.00764	0.02034	0.00052	131	7	130	3
TWG01-11		98	215	0.459	0.04682	0.00326	0.13114	0.00965	0.02024	0.00055	125	9	129	3
TWG01-12		110	155	0.712	0.05788	0.00384	0.15738	0.01111	0.02021	0.00056	148	10	129	4
TWG01-13		111	107	1.034	0.04724	0.00334	0.13109	0.00979	0.0202	0.00055	125	9	129	4
TWG01-14		57	105	0.545	0.04748	0.00319	0.13425	0.00956	0.02021	0.00055	128	9	129	3
TWG01-15		251	626	0.401	0.04927	0.00169	0.13916	0.00521	0.02026	0.00049	132	5	129	3
TWG01-16		196	256	0.766	0.05052	0.00249	0.14164	0.00747	0.02022	0.00051	135	7	129	3
TWG01-17		46	99	0.466	0.04704	0.00382	0.13702	0.01175	0.02025	0.00058	130	10	129	4
TWG01-18		422	789	0.535	0.04594	0.00153	0.13004	0.00474	0.02023	0.00049	124	4	129	3
TWG01-19		43	64	0.679	0.04992	0.00448	0.14338	0.01359	0.02024	0.00061	136	12	129	4
TWG01-20		159	213	0.744	0.05181	0.00261	0.14772	0.00797	0.02019	0.00051	140	7	129	3
TWG01-21		265	373	0.71	0.04994	0.00209	0.14159	0.00643	0.02023	0.0005	135	6	129	3
TWG01-22		618	1157	0.534	0.05277	0.00154	0.15015	0.00491	0.02023	0.00048	142	4	129	3
TWG01-23		59	99	0.6	0.04949	0.00361	0.13722	0.01061	0.02025	0.00056	131	9	129	4
TWG01-24		203	259	0.783	0.04776	0.00631	0.14546	0.02029	0.02011	0.00073	138	18	128	5
TWG01-25		125	157	0.798	0.05601	0.0043	0.17888	0.01475	0.02028	0.00059	167	13	130	4
TWG01-26		50	77	0.647	0.04413	0.00482	0.12089	0.01383	0.02018	0.00065	116	13	129	4
TWG01-27		181	253	0.716	0.05329	0.00277	0.14754	0.00825	0.02023	0.00052	140	7	129	3
TWG01-28		181	192	0.943	0.04598	0.00519	0.13447	0.01598	0.02016	0.00067	128	14	129	4
TWG01-29		105	116	0.912	0.05187	0.00365	0.1547	0.0116	0.0202	0.00056	146	10	129	4
TWG01-30		242	205	1.185	0.06303	0.00427	0.18018	0.01313	0.02016	0.00057	168	11	129	4

Table 2. *Cont.*

Sample	Lithology	Concentration (ppm)			Ratio						Age/Ma			
		^{232}Th	^{238}U	Th/U	^{207}Pb/^{206}Pb	±%	^{207}Pb/^{235}U	±%	^{206}Pb/^{238}U	±%	^{207}Pb/^{235}U	2σ	^{206}Pb/^{238}U	2σ
TWG04-1	Trachyandesite	44	48	0.922	0.15379	0.00562	8.12828	0.32538	0.38309	0.01152	2245	36	2091	54
TWG04-2		706	949	0.744	0.05827	0.00192	0.48677	0.01815	0.06055	0.00169	403	12	379	10
TWG04-3		882	1118	0.789	0.04837	0.00227	0.13442	0.00669	0.02014	0.00058	128	6	129	4
TWG04-4		124	222	0.56	0.05296	0.00429	0.14817	0.01207	0.02028	0.00065	140	11	129	4
TWG04-5		545	829	0.657	0.04985	0.00237	0.13901	0.00699	0.02021	0.00058	132	6	129	4
TWG04-6		180	264	0.683	0.05155	0.0039	0.14345	0.01095	0.02018	0.00064	136	10	129	4
TWG04-7		506	856	0.591	0.05584	0.00246	0.1561	0.00735	0.02027	0.00058	147	6	129	4
TWG04-8		413	479	0.862	0.05517	0.00266	0.15355	0.00781	0.02018	0.00059	145	7	129	4
TWG04-9		132	192	0.684	0.05611	0.00546	0.15638	0.01514	0.02021	0.00071	148	13	129	4
TWG04-10		335	343	0.976	0.0505	0.00575	0.14168	0.01599	0.02034	0.00074	135	14	130	5
TWG04-11		150	348	0.431	0.22667	0.00849	1.1992	0.04775	0.03836	0.00118	800	22	243	7
TWG04-12		304	390	0.777	0.0489	0.00519	0.13624	0.01438	0.0202	0.00071	130	13	129	5
TWG04-13		507	850	0.597	0.05193	0.0028	0.14567	0.00819	0.02035	0.00061	138	7	130	4
TWG04-14		478	761	0.629	0.04914	0.00317	0.13669	0.00903	0.02018	0.00062	130	8	129	4
TWG04-15		784	1083	0.724	0.05212	0.00283	0.14269	0.00807	0.01987	0.0006	135	7	127	4
TWG04-16		112	123	0.911	0.05116	0.00744	0.14163	0.02033	0.02009	0.00083	135	18	128	5
TWG04-17		134	201	0.667	0.05667	0.00509	0.15692	0.01409	0.0201	0.00069	148	12	128	4
TWG04-18		340	417	0.815	0.06105	0.00598	0.17061	0.01661	0.02028	0.00073	160	14	129	5
TWG04-19		390	686	0.568	0.04551	0.00291	0.12634	0.00831	0.02015	0.00062	121	7	129	4
TWG04-20		95	154	0.621	0.0586	0.00676	0.16277	0.01855	0.02016	0.00078	16	129	5	146
TWG05-1	Trachyte Trachydacite	80	121	0.661	0.05421	0.00451	0.15118	0.01256	0.02022	0.00063	143	11	129	4
TWG05-2		103	178	0.578	0.05556	0.00324	0.15484	0.00913	0.02021	0.00056	146	8	129	4
TWG05-3		316	371	0.852	0.04549	0.00222	0.12682	0.00636	0.02022	0.00054	121	6	129	3
TWG05-4		232	343	0.678	0.04889	0.00233	0.13639	0.00669	0.02023	0.00054	130	6	129	3
TWG05-5		213	316	0.673	0.05233	0.00253	0.1459	0.00724	0.02022	0.00054	138	6	129	3
TWG05-6		128	215	0.595	0.05284	0.00312	0.14831	0.00883	0.02035	0.00057	140	8	130	4
TWG05-7		188	269	0.702	0.04665	0.00245	0.12989	0.00697	0.02019	0.00054	124	6	129	3
TWG05-8		127	168	0.756	0.06221	0.00303	0.1735	0.00862	0.02023	0.00054	163	7	129	3
TWG05-9		112	155	0.723	0.04756	0.00439	0.13178	0.01207	0.0201	0.00062	126	11	128	4
TWG05-10		77	168	0.459	0.0515	0.00393	0.14361	0.01092	0.02022	0.00059	136	10	129	4

Table 2. *Cont.*

Sample	Lithology	Concentration (ppm)			Ratio						Age/Ma			
		^{232}Th	^{238}U	Th/U	^{207}Pb/^{206}Pb	±%	^{207}Pb/^{235}U	±%	^{206}Pb/^{238}U	±%	^{207}Pb/^{235}U	2σ	^{206}Pb/^{238}U	2σ
TWG05-11	Trachyte Trachydacite	113	157	0.72	0.06375	0.00394	0.17798	0.01101	0.02025	0.00057	166	9	129	4
TWG05-12		267	354	0.757	0.04806	0.0021	0.13436	0.00607	0.02028	0.00052	128	5	129	3
TWG05-13		180	299	0.602	0.0542	0.00275	0.15163	0.00783	0.02029	0.00054	143	7	130	3
TWG05-14		101	145	0.696	0.05973	0.00405	0.16683	0.0113	0.02026	0.00057	157	10	129	4
TWG05-15		330	406	0.812	0.04791	0.00215	0.13301	0.00612	0.02014	0.00051	127	5	129	3
TWG05-16		80	172	0.465	0.06122	0.0034	0.17024	0.00949	0.02017	0.00054	160	8	129	3
TWG05-17		179	272	0.658	0.07161	0.00372	0.20022	0.01046	0.02028	0.00055	185	9	129	3
TWG05-18		70	132	0.534	0.05234	0.00473	0.14607	0.01306	0.02025	0.00062	138	12	129	4
TWG05-19		122	241	0.505	0.05364	0.00339	0.14898	0.00942	0.02015	0.00055	141	8	129	4
TWG05-20		182	303	0.602	0.05304	0.00261	0.14746	0.00737	0.02017	0.00052	140	7	129	3
TWG06-1	Trachyandesite	73	145	0.504	0.05341	0.00588	0.14797	0.0161	0.02009	0.0007	140	14	128	4
TWG06-2		62	131	0.474	0.05027	0.00566	0.14041	0.01565	0.02026	0.0007	133	14	129	4
TWG06-3		182	277	0.658	0.0509	0.00364	0.14292	0.01028	0.02036	0.00061	136	9	130	4
TWG06-4		80	141	0.569	0.0582	0.00368	0.1628	0.0104	0.02028	0.00059	153	9	130	4
TWG06-5		71	133	0.531	0.04539	0.00491	0.12732	0.01368	0.02034	0.00067	122	12	130	4
TWG06-6		97	156	0.622	0.04747	0.00462	0.13202	0.01278	0.02017	0.00065	126	11	129	4
TWG06-7		164	227	0.723	0.06413	0.00441	0.17815	0.01228	0.02015	0.00061	167	11	129	4
TWG06-8		85	118	0.726	0.04662	0.00479	0.12975	0.01323	0.02018	0.00066	124	12	129	4
TWG06-9		451	503	0.896	0.0542	0.003	0.15143	0.00856	0.02026	0.00057	143	8	129	4
TWG06-10		416	432	0.962	0.06778	0.0026	0.18953	0.00772	0.02028	0.00055	176	7	129	3
TWG06-11		311	320	0.972	0.05628	0.00432	0.1572	0.01205	0.02026	0.00062	148	11	129	4
TWG06-12		967	677	1.43	0.05893	0.00216	0.16419	0.00645	0.02021	0.00054	154	6	129	3
TWG06-13		85	149	0.57	0.03933	0.00583	0.11011	0.01614	0.02031	0.00076	106	15	130	5
TWG06-14		253	291	0.868	0.09994	0.00652	0.27891	0.01801	0.02024	0.00064	250	14	129	4
TWG06-15		307	373	0.823	0.04796	0.00401	0.13424	0.0112	0.0203	0.00063	128	10	130	4
TWG06-16		189	351	0.539	0.04445	0.00469	0.12308	0.01287	0.02008	0.00067	118	12	128	4
TWG06-17		289	423	0.684	0.05073	0.00281	0.14148	0.00798	0.02022	0.00057	134	7	129	4
TWG06-18		80	135	0.593	0.0475	0.00508	0.13168	0.01397	0.0201	0.00068	126	13	128	4
TWG06-19		64	123	0.519	0.05107	0.00496	0.14296	0.01379	0.0203	0.00066	136	12	130	4
TWG06-20		104	156	0.666	0.05726	0.00496	0.15965	0.01373	0.02022	0.00065	150	12	129	4
TWG06-21		179	220	0.815	0.05182	0.00355	0.14473	0.00997	0.02025	0.00059	137	9	129	4
TWG06-22		75	149	0.504	0.06617	0.00556	0.18432	0.01533	0.0202	0.00066	172	13	129	4
TWG06-23		122	189	0.648	0.05238	0.00394	0.14676	0.01104	0.02032	0.00061	139	10	130	4
TWG06-24		141	162	0.87	0.05283	0.00505	0.14759	0.01396	0.02026	0.00067	140	12	129	4
TWG06-25		265	360	0.735	0.04851	0.00283	0.13479	0.008	0.02015	0.00056	128	7	129	4

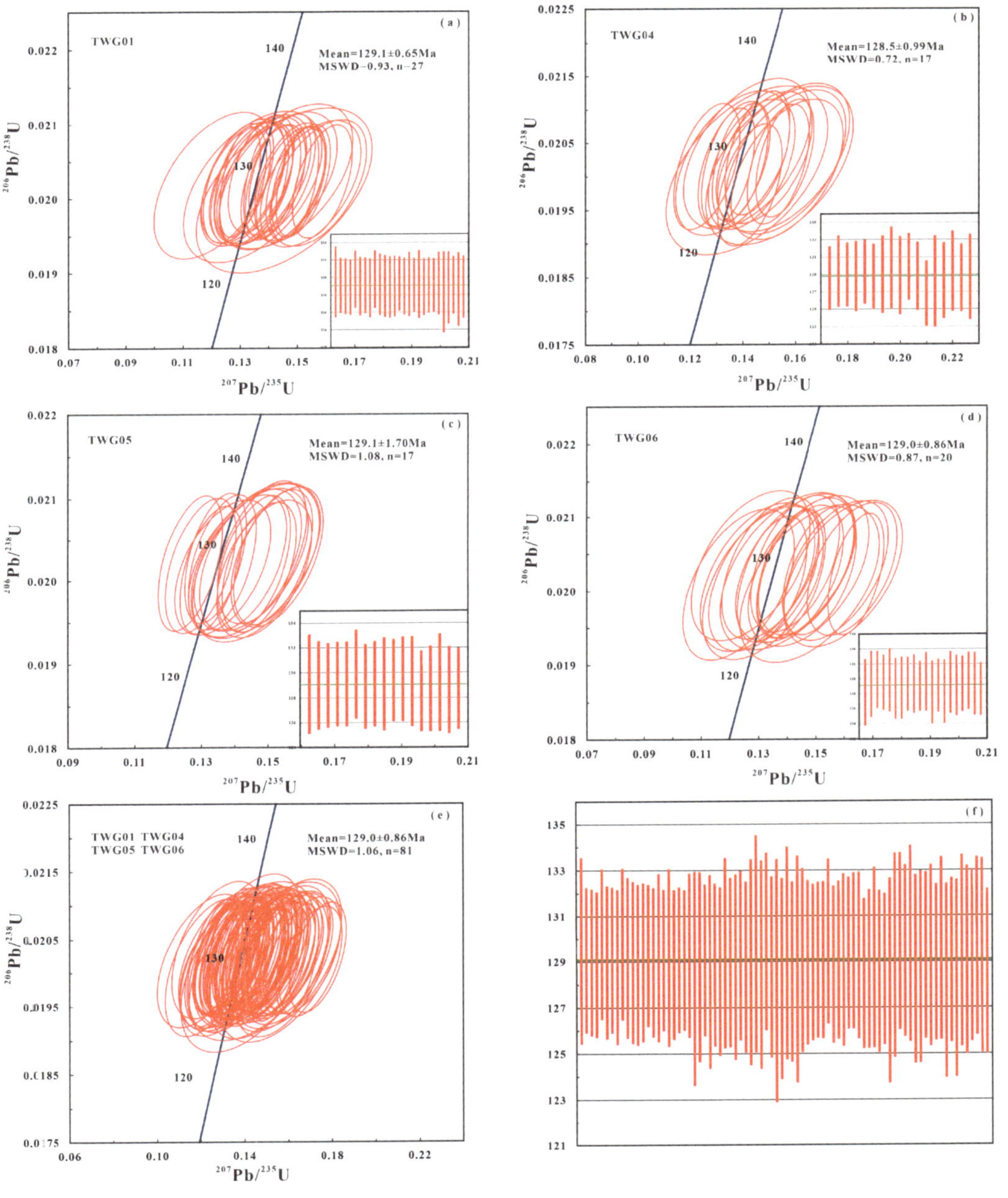

Figure 7. Concordant and weighted diagrams of the zircon U–Pb age of the Longjiang formation volcanic rocks, Lengjimanda plate. Error ellipses are shown for 1-sigma level of uncertainty. (**a**) TWG01, (**b**) TWG04, (**c**) TWG05, (**d**) TWG06, (**e**) Longjiang formation volcanic rocks, (**f**) weighted diagrams of the Longjiang formation volcanic rocks.

Of the 20 analyzed points in the trachyandesite sample (TWG04), 17 measurement spots show good age consistency. In the concordance plot of the zircon U–Pb ages (Figure 7b), all the obtained zircon U–Pb ages lie on or near the concordance line after discordant points have been excluded. The weighted mean age of $^{206}Pb/^{238}U$ is 128.5 ± 0.99 Ma (MSWD = 0.72, *n* = 17), indicating an Early Cretaceous age.

Of the 20 analyzed points in the trachyte trachydacite sample (TWG05), 17 zircon measurement points show good age consistency. In the concordance plot of the zircon U–Pb ages (Figure 7c), all the obtained zircon U–Pb ages lie on or near the concordance line after discordant points are excluded. The weighted mean age of $^{206}Pb/^{238}U$ is determined to be 129.1 ± 1.70 Ma (MSWD = 1.08, *n* = 17), indicating an Early Cretaceous age.

Of the 25 analyzed points in the trachyandesite sample (TWG06), 20 zircon measurement points show good age consistency. In the concordance plot of the zircon U–Pb ages (Figure 7d), all the obtained zircon U–Pb ages lie on or near the concordance line after discordant points are excluded. The weighted mean age of $^{206}Pb/^{238}U$ is determined to be 129.0 ± 0.86 Ma (MSWD = 0.87, *n* = 20), indicating an Early Cretaceous age.

The volcanic rock samples from various sampling locations in the Longjiang formation have a total of 95 analytical points. Among them, 14 spots show significant deviations in their zircon surface ages due to poor separation and a complex composition, while the remaining 81 spots show good age consistency. In the concordance plot of the zircon U–Pb ages (Figure 7f), all the obtained zircon U–Pb ages lie along or near the concordance line after discordant points are excluded. The weighted mean age of $^{206}Pb/^{238}U$ is determined to be 129.0 ± 0.86 Ma (MSWD = 1.06, *n* = 81), indicating an Early Cretaceous age.

4.3. Sr-Nd-Pb Isotopic Dating

The Sr–Nd–Pb isotopic compositions of the volcanic rocks of Longjiang formation volcanic rocks are shown in Table 3. The initial Sr and Nd isotopic compositions of the samples in the study area are corrected based on their corresponding zircon U–Pb ages of 129 Ma.

Table 3. Sr–Nd–Pb isotopic data of the Longjiang formation volcanic rocks.

Sample	Age	Rb	Sr	$^{87}Rb/^{86}Sr$	$(^{87}Sr/^{86}Sr)_i$	Sm	Nd	$^{147}Sm/^{144}Nd$	$^{143}Nd/^{144}Nd$	$(^{143}Nd/^{144}Nd)_i$	εNd(*t*)	t_{DM2} (Ma)	$^{206}Pb/^{204}Pb$	$^{207}Pb/^{204}Pb$	$^{208}Pb/^{204}Pb$	$^{238}U/^{204}Pb$	Δβ	Δγ
	Ma	ppm				ppm												
TWG01	129	61.5	658	0.26384	0.70357	13.83	67.34	0.12903	0.5126	0.51221	2.1	893	18.429	15.537	38.204	9.34	13.75	24.92
TWG04	129	72.5	652	0.31428	0.70352	4.93	23.94	0.12943	0.51277	0.51221	5.5	621	18.386	15.523	38.142	9.32	12.87	23.64
TWG05	129	98.8	557	0.50094	0.70299	4.48	23.48	0.11987	0.51281	0.51221	6.6	544	18.389	15.519	38.133	9.31	12.59	23.13
TWG06	129	88.1	776	0.32055	0.70352	4.74	23.28	0.12786	0.51274	0.51221	4.9	671	18.387	15.516	38.143	9.3	12.41	23.59

These samples have relatively low initial $(^{87}Sr/^{86}Sr)_i$ ratios, ranging from 0.70299 to 0.70357, and initial $(^{143}Nd/^{144}Nd)_i$ values ranging from 0.51260 to 0.51281, with an average of 0.512363. The $\varepsilon_{Nd}(t)$ values are all positive, ranging from +2.1 to +6.6, and the t_{DM1} (Ma) ranges from 551 Ma to 992 Ma, while the t_{DM2} ranges from 544 Ma to 839 Ma. The Pb isotope data in the study area indicate that the $^{206}Pb/^{204}Pb$, $^{207}Pb/^{204}Pb$, and $^{208}Pb/^{204}Pb$ ratios of the volcanic rocks of the Longjiang formation range from 18.397 to 18.429, 15.516 to 15.54, and 38.189 to 38.2, respectively, and are like the Pb isotopic compositions of the Mesozoic volcanic rocks in the southern Da Hinggan mountains [53].

4.4. Zircon Lu–Hf Isotopes Dating

The results of the zircon Lu–Hf isotope analyses of the volcanic rocks of the Longjiang formation in the Manda Banke area in the central section of the Da Hinggan mountains are listed in Table 4. The $^{176}Lu/^{177}Hf$ ratios of the samples in the study area range from 0.0009 to 0.0020, all of which are positive, with an average of 0.0013, which is less than 0.002. This indicates extremely low radiogenic Hf accumulation after zircon formation. Except for one sample (TWG01-5) with a negative $\varepsilon_{Hf}(t)$, the volcanic rocks of the Longjiang formation in the study area have positive $\varepsilon_{Hf}(t)$ values ranging from +1.31 to +43.77, with an average of +8.47. The one-stage Hf model ages (t_{DM1}) range from 538 Ma to 1059 Ma, with an average of 733 Ma, while the two-stage model ages (t_{DM2}) range from 655 Ma to 1427 Ma, with an average of 893 Ma.

Table 4. Zircon Lu–Hf isotope data of Lengjimanda plate of the Longjiang formation volcanic rocks.

Sample	$^{176}Hf/^{177}Hf$	2σ	$^{176}Lu/^{177}Hf$	1σ	$^{176}Yb/^{177}Hf$	1σ	Age (Ma)	$\varepsilon_{Hf}(0)$	$\varepsilon_{Hf}(t)$	t_{DM1}(Ma)	t_{DM2}(Ma)	$f_{Lu/Hf}$
TWG01-2	0.283935	0.000043	0.003236	0.000004	0.076078	0.000195	126	41.3	43.77	1059	1305	−0.9
TWG01-3	0.282729	0.000039	0.00176	0.000018	0.041094	0.000427	130	−1.4	1.31	751	911	−0.95
TWG01-4	0.282749	0.000049	0.001616	0.000012	0.037553	0.000303	129	−0.7	2.01	719	874	−0.95
TWG01-5	0.282444	0.000034	0.002004	0.000044	0.046304	0.001074	132	−11.5	−8.76	1168	1427	−0.94
TWG01-6	0.282819	0.000045	0.001756	0.000009	0.040134	0.000245	135	1.8	4.6	621	746	−0.95
TWG01-8	0.282862	0.000038	0.001333	0.000012	0.029231	0.000249	130	3.3	6.03	553	668	−0.96
TWG01-10	0.282806	0.000038	0.000989	0.000026	0.021776	0.000614	129	1.3	4.08	627	768	−0.97
TWG01-11	0.282852	0.000041	0.001193	0.000008	0.027001	0.000203	127	2.9	5.63	565	687	−0.96
TWG01-16	0.282868	0.000033	0.000959	0.000016	0.021678	0.000368	130	3.5	6.3	538	655	−0.97

5. Discussion

5.1. The Age of the Longjiang Formation's Volcanic Rocks

Several previous studies have determined to resolve the age of the volcanic rocks in the Longjiang formation. However, as mentioned above, these studies have mainly focused on the northern part of the Great Xing'an mountains. Li [15] used laser fusion 40Ar/39Ar dating and zircon U–Pb dating to determine that the Longjiang formation in the Longjiang basin, located between the Dayangshu basin in the north and the Lujiaobao depression in the south, is of an Early Cretaceous age of 122 to 126 Ma; Ye [30] obtained an age of 124.63 Ma for the Longjiang formation's volcanic rocks in the Dongpo Xinliuzhan area of the northern section of the Da Hinggan mountains using LA-ICP-MS zircon U–Pb dating; Chen [54] obtained an age of 120.43 Ma for the volcanic rocks of the Longjiang formation in the Shiwuliqiao gold mining area, Heilongjiang province, by performing zircon U–Pb dating on andesite samples. This age belongs to the Early Cretaceous. We obtained relatively accurate ages for the volcanic rocks of the Longjiang formation.

The samples used for dating in the present study are trachyandesite (TWG01, TWG04, and TWG06) and trachyte trachydacite (TWG05). The CL images (Figure 3) show that these zircons have typical zoning structures, and the Th/U ratios range from 0.1989 to 1.7105, indicating that these zircons are of a magmatic origin. The LA–ICP-MS zircon U–Pb weighted mean age of 81 sample points is determined to be 129.0 ± 0.86 Ma (MSWD of 1.06, Figure 7f). This age is consistent with the previous dating results of 120–126 Ma [40] and is within the error range. Based on a thorough analysis of the above studies, it is concluded that the volcanic rocks of the Longjiang formation were formed during the Early Cretaceous.

5.2. The Genesis of the Longjiang Formation's Volcanic Rocks

There are currently several theories of the genesis of calc-alkaline andesites: the mixing of felsic magmas from the crust and mafic magmas from the mantle [55]; the separation and crystallization of mafic magmas from the mantle [56,57]; partial melting of the mantle wedge triggered by subduction fluids or melts [58,59]; and partial melting of lower crustal material triggered by the underplating of mafic magma of a mantle origin [60,61]. Our analysis of the major and trace elements in the samples (Table 1) indicates that the dominant volcanic rocks in the study area are trachyandesite, trachyte trachydacite, and andesite. These volcanic rocks collectively form a suite of intermediate calc-alkaline volcanic rock combinations. The Mg# values (averaging 37.85), Cr content (averaging 12.11 ppm), and Ni contents (averaging 8.80 ppm) of these volcanic rocks are relatively low, much lower than the Mg# values (65~75) of mantle-derived magmas and the Cr and Nd contents (185 ppm and 16 ppm, respectively) of the crust [52,62]. Furthermore, there are no dark mafic enclaves in the volcanic rocks within the study area, indicating that their formation cannot be attributed to the mixing of crustal and mantle-derived magmas [63]. The volcanic rocks in the study area have relatively high K/Rb ratios, with an average of 342 (>110), Zr/Hf ratios with an average of 39 (>30), and Nb/Ta ratios with an average of 15 (>5) [52,64], which indicates that the studied volcanic rocks are not highly fractionated plutonic rocks (Wu et al., 2004; Li et al., 2018). In the La vs. La/Yb plot and La vs. La/Sm plot (Figure 8), the andesite samples show a partial melting trend rather than fractional crystallization, indicating a weak fractional crystallization process. This suggests that the volcanic rocks in the study area cannot be attributed to mantle-derived magma by means of fractional crystallization [63].

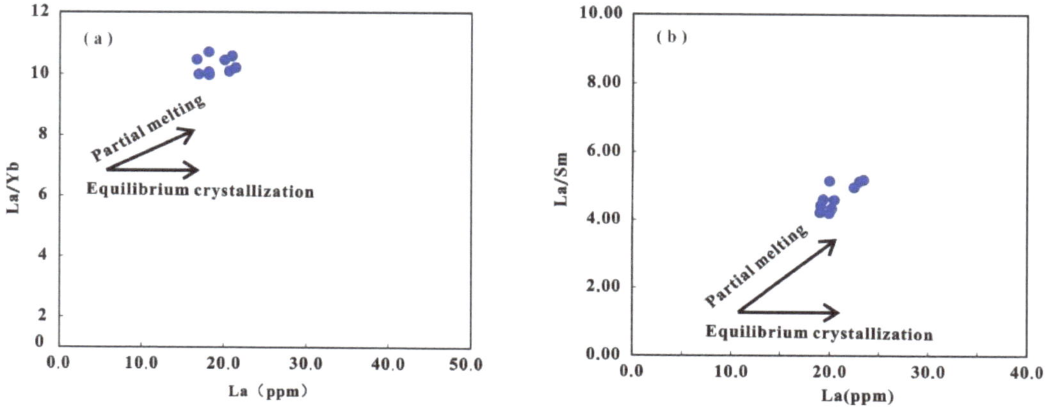

Figure 8. Illustration of magmatic evolution of volcanic rocks in the Longjiang formation of Lengjimanda plate [65]. (**a**) La/Yb vs. La and (**b**) La/Sm vs. La.

The volcanic rocks of the Longjiang formation show relatively high contents of SiO_2 (average 63.17%), Al_2O_3 (average 16.75%), and K_2O (average 3.22%), and low contents of MgO (average 1.87%) and $Mg^{\#}$ (average 37.85). These characteristics are significantly different from magmas produced by the partial melting of mantle material and are closer to magmas produced by the partial melting of lower crustal material ($Mg^{\#}$ < 40) [66]. The volcanic rocks of the Longjiang formation in the study area exhibit a moderately negative Eu anomaly, with average Sr/Y and La/Yb ratios of 27.6 and 15.8, respectively. These ratios are not consistent with the characteristics of adakite rocks (>40, >20) [67–69]. In the Sr/Y-Y diagram and the $(La/Yb)_N$ vs. $(Yb)_N$ diagram (Figure 9), all sample compositions fall within the typical island arc rock field. This indicates that the volcanic rocks of the Longjiang formation are predominantly island arc volcanic rocks rather than adakite rocks, and the volcanic rocks in question exhibit characteristics of large ion lithophile element enrichment and high-field-strength element depletion, as seen in the rare earth element distribution pattern and trace element spider diagram (Figure 6). This pattern is similar to that of the lower crust, indicating that the rocks may have originated from the partial melting of basaltic crustal materials. Additionally, the average ratios of Rb/Sr, Ti/Zr, and Ti/Y in the volcanic rock samples from the Longjiang formation in the study area are 0.13, 20.91, and 199.00, respectively. These values are significantly higher than the range of ratios found in mantle-derived magmas (0.03, <20, <100) [52,64,70]. We infer that the primary magma of the volcanic rocks in the Lengjimanda plate of the Longjiang formation originated from the partial melting of basaltic crustal materials rather than the mantle.

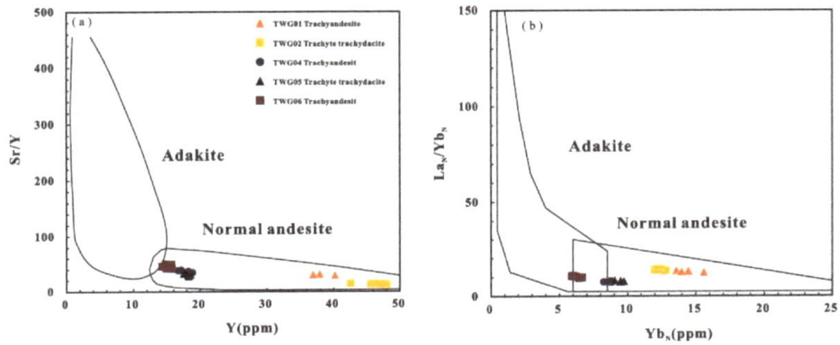

Figure 9. Adakite discrimination diagram of Longjiang formation volcanic rocks, Lengjimanda plate. (**a**) Sr/Y vs. Y diagram; (**b**) $(La/Yb)_N$ vs. (Yb) n diagram [71].

Zircon Hf isotope studies conducted in the Lengjimanda plate indicate positive $\varepsilon_{Hf}(t)$ values ranging from +1.31 to +43.77, with an average of +8.47, except for one sample with negative values; the two-stage Hf model ages (t_{DM2}) range from 655 to 1427 Ma. The $\varepsilon_{Hf}(t)$ values are relatively uniform and high, indicating that the rocks are derived directly from the mantle or from basaltic crust formed by the melting of depleted mantle. The presence of a negative $\varepsilon_{Hf}(t)$ value of −8.76 suggests the possible involvement of ancient crust in the rock formation process. In the $\varepsilon_{Hf}(t)$ vs. age diagram (Figure 10), all the samples' compositions fall within the range of granulite meteorites and depleted mantle, which suggests that the magma may have originated from the partial melting of basaltic crustal materials that were newly generated from the depleted mantle during the Neoproterozoic to Phanerozoic period [35,72,73]. This hypothesis is consistent with the large-scale crustal accretion event that occurred in the Da Hinggan mountain region during the Neoproterozoic to Phanerozoic period, as reported by Sui [74] and Wang [75]. The Longjiang formation's volcanic rocks in the study area exhibit higher average ratios of Nb/Ta and Zr/Hf, 14.51 and 39.39, respectively, compared to the average values in the crust (13.00, 33.00) [64,76]; these values are similar to the average values in the mantle (17.50, 37.00) [52]. According to Shi [77], the magma source area is composed of recently generated crustal materials from the depleted mantle.

The study area's volcanic rock samples display $^{87}Sr/^{86}Sr$ ratios ranging from 0.703 to 0.7036, all of which are below 0.705. The $\varepsilon_{Nd}(t)$ values range from +2.1 to +6.6, and the $\varepsilon_{Sr}(t)$ values range from −10.7 to −19.2. In Figure 11's $\varepsilon_{Nd}(t)$ vs. $\varepsilon_{Sr}(t)$ diagram, all the volcanic rock sample compositions from the Longjiang formation in the study area fall within quadrant II, which indicates that the magma source of the volcanic rocks in the area is related to the depleted mantle, as noted by Hou [78] and Shao [79]. The Pb isotope data from the study area indicate that the volcanic rocks of the Longjiang formation exhibit uranium lead enrichment and thorium lead depletion. The $^{207}Pb/^{204}Pb$ ratios range from 15,516 to 15,540, which are greater than 15,530; the $^{208}Pb/^{204}Pb$ ratios range from 38,189 to 38,200, which are lower than 39,000; and the $^{206}Pb/^{204}Pb$ ratios range from 18,397 to 18,429, which are greater than 18,000. In the genetic classification diagram of $\Delta\gamma$ vs. $\Delta\beta$ (Figure 12a) by Zhu [80] and in the tectonic model diagrams of $^{206}Pb/^{204}Pb$ vs. $^{207}Pb/^{204}Pb$ and $^{206}Pb/^{204}Pb$ vs. $^{208}Pb/^{204}Pb$ (Figure 12b–d), the samples are distributed in the lead region of the subduction zone, where the crust and mantle mix; this suggests that the magma may have originated from the partial melting of basaltic crustal materials that were newly generated from the depleted mantle. This is consistent with the findings of Tan [70], Gao [81], and Jia [82] regarding the petrogenesis of Early Cretaceous volcanic rocks in the Soren and Handagai areas of the central Da Hinggan mountains.

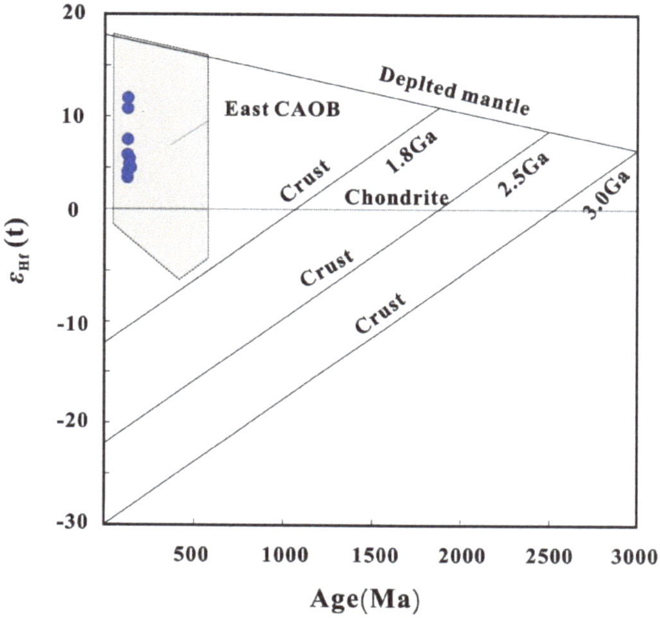

Figure 10. ε_{Hf} (t) vs. age diagram of Longjiang formation volcanic rocks, Lengjimanda plate. East CAOB: East Asian Orogenic Belt Eastern part [21].

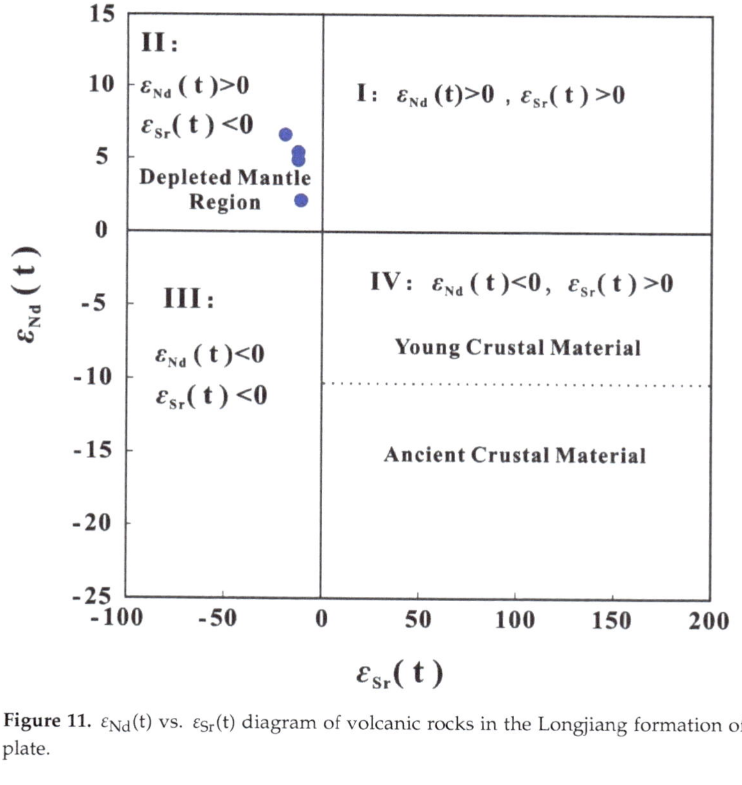

Figure 11. ε_{Nd}(t) vs. ε_{Sr}(t) diagram of volcanic rocks in the Longjiang formation of Lengjimanda plate.

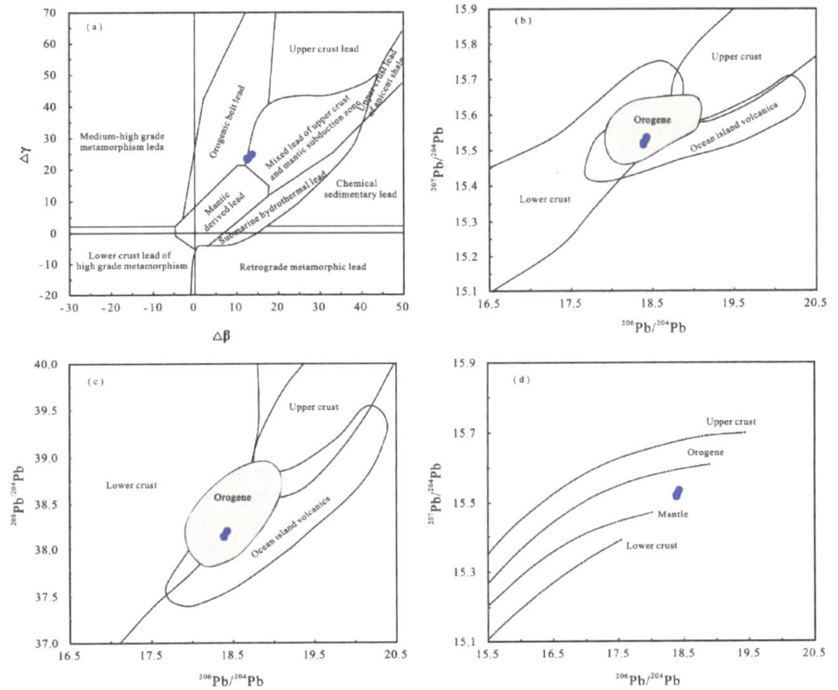

Figure 12. Pb isotope composition diagram of Longjiang formation volcanic rocks, Lengjimanda plate. Orogene: Orogenic belt. ((**a**), [83]; (**b**,**c**), [84]; (**d**), [80]).

Based on the geochemical characteristics and zircon Hf analyses, as well as the whole-rock Sr–Nd–Pb isotopes, we conclude that the volcanic rocks of the Longjiang formation in the Lengjimanda plate region were generated through the partial melting of the lower crust, and the magma's source material is believed to bebasaltic crustal material that was newly generated from the depleted mantle.

5.3. Tectonic Background

There are different theories regarding the tectonic environment that was responsible for the formation of the Mesozoic volcanic rocks in the Da Hinggan mountain region. They include the subduction of the ancient Pacific plate [20,85], a mantle plume origin [12,86], and the subduction of the Mongol-Okhotsk Ocean [11,87]. However, closure of the Mongol-Okhotsk Ocean and subduction of the Pacific plate have been identified as the combined cause in various studies [18,88–92].

Previous data suggest that during the Mesozoic era, the Da Hinggan mountain region underwent the superposition and reformation of the ancient Pacific plate and the Mongolian-Okhotsk plate [88,92]. Early studies indicate that the Mongolian-Okhotsk Ocean closed during the Middle Late Jurassic period [93] and had little influence on the late Mesozoic magmatic activity in the Da Hinggan mountain region. Recently, researchers provided evidence for the existence of a southward subduction of the Mongolian-Okhotsk plate during the Mesozoic era; for instance, Badarch [94] divided the tectonic blocks and established a new stratigraphic column. Discovering the presence of Permo-Triassic accretionary terranes in the Mongolian-Okhotsk suture zone, Chen [95] identified Late Triassic Inner Mongolia Taipingchuan porphyry copper-molybdenum deposits. Zeng [59] obtained geochemical data and concluded that the Badaguan metamorphic complex in the Erguna area was formed due to the subduction of the Mongolian-Okhotsk Ocean towards the Erguna block. These studies provide evidence for the southward subduction of the Mongolian-Okhotsk plate during the Mesozoic era, and many researchers believe that the Late Mesozoic magmatic activity in the Da Hinggan mountain area is related to the extensional environment after the closure of the Mongol-Okhotsk Ocean. This view is supported by Meng [87], Ying [96], Sun [97], Li [98], and Si [99]. Additionally, numerous scholars posit that the Mesozoic volcanic rocks in the Da Hinggan mountains region are closely linked to the subduction of the ancient Pacific plate. Kong conducted a study on the tectonic background of the volcanic rocks in the Khorchin area and concluded that they were primarily influenced by the subduction of the ancient Pacific plate. Tang [41] summarized the petrological composition and geochemical characteristics of Mesozoic Cenozoic volcanic rocks in the Northeast Asian margin and suggested that the ancient Pacific plate underwent low-angle subduction towards Northeast Asia in the Early Cretaceous. Yang [100] conducted research on the paleontology, regional comparisons, and chronology in the Da Hinggan mountain area and concluded that the Early Cretaceous volcanic rocks (145–110 Ma) in the Da Hinggan mountain region are generally distributed in an NNE direction, which is consistent with the mechanism of subduction of the ancient Pacific plate.

During the late Early Cretaceous period, volcanic rocks were widely distributed across the Da Hinggan mountain area, which exhibited a typical bimodal volcanic rock assemblage [101]. The region also showed an extensional environment at the regional scale [92,100,102]. The volcanic activity in the Da Hinggan mountain area during this period was influenced by the closure of the Mongol-Okhotsk suture zone [7,103] and the subduction of the ancient Pacific plate beneath the Eurasian continent [9,104,105]. Our study area's volcanic rocks consist mainly of a suite of Early Cretaceous (129.00 $\pm$ 0.78 Ma) intermediate volcanic rocks. These rocks have a high SiO_2 content and are enriched in large-ion lithophile elements while being depleted in high-field-strength elements. Furthermore, there is an obvious fractionation of light and heavy rare earth elements. In the microelement tectonic environment discrimination diagram of the Longjiang formation's volcanic rocks in the study area (Figure 13), they mainly exhibit characteristics of a post-collisional environment, while some samples also show features of a continental arc environment. This indicates that the Early Cretaceous volcanic rocks in the Longjiang formation were formed in a superimposed environment of post-collision and continental arc. Combined with a regional background analysis, it is believed that the tectonic setting in the study area may be related to the closure of the Mongol-Okhotsk Ocean and the oblique subduction of the ancient Pacific plate. This view is supported by Zhang [17], Zhang [36], Zhang [106], and Xu [7], who suggested that the Da Hinggan mountain area was affected by the closure of the Mongol-Okhotsk Ocean and the oblique subduction of the ancient Pacific plate during the Early Cretaceous.

The tectonic background of the volcanic rocks formed in the Longjimanda plate of the Heilongjiang Group reflects the alternating influence of the Mongol-Okhotsk oceanic plate and the ancient Pacific plate on the Mesozoic volcanic rocks of the Da Hinggan mountains. A synthesis of previous research indicates that the Mesozoic volcanic rocks in the Greater Khingan Range were primarily influenced by the closure of the Mongol-Okhotsk Ocean during the Late Jurassic to Early Cretaceous period. During the Early Cretaceous period, the region may have been affected by the closure of the Mongol-Okhotsk Ocean and the subduction of the ancient Pacific plate. Subsequently, during the Early Cretaceous, they were primarily influenced by the subduction of the ancient Pacific plate.

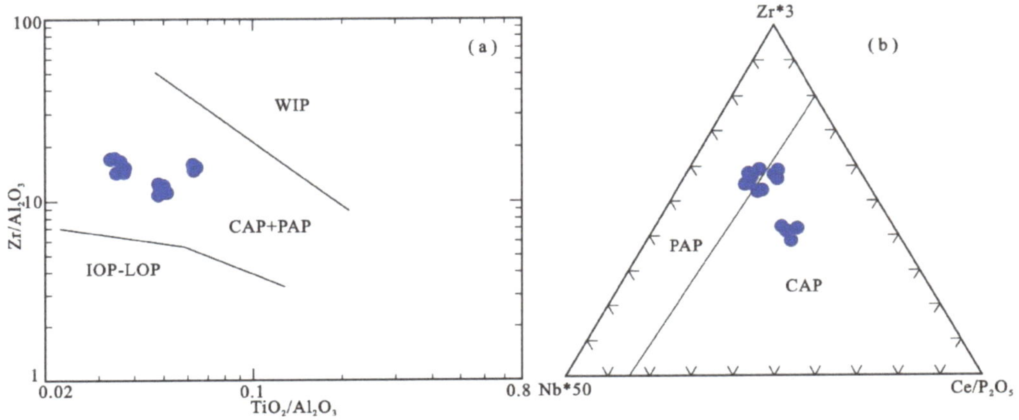

Figure 13. Structural identification diagram of volcanic rocks in Longjiang formation, Lengjimanda plate [107]. (**a**) Intraplate environment (WIP), (**b**) continental arc environment (CAP), post-collision environment (PAP), initial ocean arc environment (IOP), late ocean arc environment (LOP).

6. Conclusions

The volcanic rocks of the Longjiang formation in the Lengjimanda plate in the middle section of the Da Hinggan mountains are predominantly intermediate basic volcanic rocks, comprising porphyritic rock, porphyritic andesite, and andesite. The LA-ICP-MS zircon U–Pb analysis suggested an age of 129.0 ± 0.86 Ma (MSWD = 1.06, n = 81), which suggests that these rocks were formed during the Early Cretaceous period. Petrological, geochemical, and isotopic studies of the Longjiang formation's volcanic rocks indicate that they are the result of the partial melting of the lower crust. The source material for these rocks is newly generated basaltic crustal material from a depleted mantle. These rocks are believed to have formed in a setting of post-collision and continental arc superposition. Their tectonic background may be related to the closure of the Mongolian-Okhotsk Ocean and the oblique subduction of the ancient Pacific plate, as indicated by the regional tectonic evolution history.

This study posits that the regional volcanic eruption event that occurred approximately 130 million years ago in the Lengjimanda plate coincided with the transitional period of the Da Hinggan mountains from rapid uplift to slow uplift, as well as the period of oblique subduction of the ancient Pacific plate beneath East Asia. This finding has significant implications for the analysis of the Early Cretaceous basin's evolution and stratigraphic subdivision in this region. The research on the tectonic background of the Lengjimanda plate presented in this paper demonstrates the alternating influence of the Mesozoic Mongol-Okhotsk oceanic plate and the ancient Pacific plate on the timing of the Da Hinggan mountain region. The Mesozoic volcanic rocks in the Da Hinggan mountains were significantly influenced by the closure of the Mongol-Okhotsk Ocean during the Late Jurassic to Early Cretaceous period. During the Early Cretaceous period, these rocks may have been affected by two distinct geological processes: the closure of the Mongol-Okhotsk Ocean and the subduction of the ancient Pacific plate. Subsequently, during the early Early Cretaceous, they were primarily influenced by the subduction of the ancient Pacific Ocean. This information is useful in delineating the periods in which the Mongol-Okhotsk oceanic plate and the ancient Pacific plate influenced the Da Hinggan mountains.

This article presents a petrological, zircon U–Pb geochronological, geochemical, and isotope chemical analysis of the volcanic rocks in the Longjiang formation, thereby determining the formation age of these rocks in the region. The article discusses the origin and tectonic background of the volcanic rocks, reconstructs the regional geodynamic evolution, fills the research gap concerning the Longjiang volcanic rocks, and provides reliable data on the dynamic history of the Da Hinggan mountain region. Moreover, the study provides avenues for further geological exploration and geophysical detection in the region.

Supplementary Materials: The following supporting information can be downloaded at: https://www.mdpi.com/article/10.3390/min14070719/s1, Table S1: Zircon LA–ICP–MS U–Pb data of the volcanic rocks from Lengjimanda plate of the Longjiang formation. Table S2: Major and trace element compositions of Lengjimanda plate of the Longjiang formation volcanic rocks. Table S3: Sr-Nd-Pb isotopic data of the Longjiang Formation volcanic rocks. Table S4: Zircon Lu–Hf isotope data of Lengjimanda plate of the Longjiang Formation volcanic rocks.

Author Contributions: Methodology, S.Z.; investigation, Z.T.; data curation, L.S.; writing—original draft preparation, S.-C.W.; writing—review and editing, Y.-J.H. All authors have read and agreed to the published version of the manuscript.

Funding: This research received no external funding.

Data Availability Statement: The authors confirm that the data supporting the findings of this study are available within the article and its Supplementary Materials.

Acknowledgments: Thank you very much to all the teachers for their contributions during the paper writing process. Teacher Yu Jie Hao provided review and annotation guidance throughout the writing of the paper. Teacher Lu Shi provided significant assistance with data processing for the article. Teacher Zhen Tang provided much help during field geological surveys. Teacher Shuang Zhu provided valuable assistance with paper revisions and writing techniques. This work was supported by the geological survey project of China "Regional Geological Survey of Ergun Area, Inner Mongolia 1:50,000 geological map including Lengjimanda Plate, Jirigen Linchang, Ximalatu, Wulanzhao, Bayantao Sea" (DD20160048-02).

Conflicts of Interest: The authors declare no conflict of interest.

References

1. Wu, F.Y.; Sun, D.Y.; Lin, Q. Genesis and crustal accretion of the Epigenetic granites in Northeast China. *J. Petrol.* **1999**, 22–30.
2. Hong, D.W.; Wang, S.K.; Xie, X.L.; Zhang, J.S.; Wang, T. Relationship between supercontinental evolution and continental crustal growth from positive ε(Nd)-valued granites in Central Asia. *J. Geol.* **2003**, 203–209.
3. Kuzmin, M.L.; Abramovich, G.Y.A.; Dril, S.L.; Kravchinsky, V.Y.A. The Mongolian-Okhotsk suture as the evidence of Late Paleozoic-Mesozoic collisional processes in Central Asia. In Proceedings of the 30th IGC, Beijing, China, 4–14 August 1996; pp. 1–261.
4. Zhao, X.X.; Coe, R.S. Paleomagnetic constraints on the paleogeography of China: Implications for Gondwanaland. In Proceedings of the 30th IGC, Beijing, China, 4–14 August 1996; pp. 1–231.
5. Zhao, Z. Late Paleozoic Magmatism and Its Tectonic Significance in Northern Daxinganling. Ph.D. Thesis, Jilin University, Changchun, China, 2011.
6. Sengor, A.; Natal'in, B. Paleotectonics of Asia: Fragments of a synthesis. *Tecton. Evol. Asia* **1996**, 1–129.
7. Xu, W.L.; Wang, F.; Pei, F.P.; Meng, E.; Tang, J.; Xu, M.J.; Wang, W. Mesozoic tectonic regime and regional mineralization context in northeast China: Constraints from temporal and spatial variations of Mesozoic volcanic assemblages. *J. Petrol.* **2013**, *30*, 339–353.
8. Xu, W.L.; Ji, W.Q.; Pei, F.P.; Meng, E.; Yu, Y.; Yang, D.B.; Zhang, X.Z. Triassic volcanism in eastern Heilongjiang and Jilin provinces, NE China: Chronology, geochemistry, and tectonic implications. *J. Asian Earth Sci.* **2009**, *34*, 392–402. [CrossRef]
9. Wu, F.Y.; Sun, D.Y.; Ge, W.C.; Zhang, Y.B.; Grant, M.L.; Wilde, S.A.; Jahn, B.M. Geochronology of the Phanerozoic granitoids in northeastern China. *J. Asian Earth Sci.* **2011**, *41*, 1–30. [CrossRef]
10. Li, S.T.; Yang, S.K.; Wu, C.L.; Huang, J.F.; Cheng, S.T.; Xia, W.C.; Zhao, G.Y. Late Mesozoic rifting in northeastern China and the Northeast Asian fault basin system. *Chin. Sci. (Ser. B) Chem. Biol. Agron. Med. Geol.* **1987**, 185–195.
11. Lin, Q.; Ge, W.C.; Cao, L.; Sun, D.Y.; Lin, J.K. Geochemical characterization of Mesozoic bimodal volcanic rocks in the Daxinganling. *Geochemistry* **2003**, 208–222.
12. Ge, W.C.; Lin, Q.; Sun, D.Y.; Wu, F.W.; Won, J.-K.; Lee, W.Y.; Chen, M.S.; Yoon, S.-H. Geochemical characterization of Mesozoic basalts in Daxing'anling: Evidence for crust-mantle interaction. *J. Petrol.* **1999**, 396–406.
13. Ge, W.C.; Wu, F.Y.; Zhou, C.Y.; Zhang, J.H. Zircon U-Pb ages and geologic significance of Mesozoic granites from the Wulanhot area, central Daxinganling. *J. Petrol.* **2005**, *21*, 14.
14. Zhang, L.C.; Chen, Z.G.; Zhou, X.H.; Ying, K.F.; Wang, F.; Zhang, Y.T. Deep source zones and tectono-magmatic evolution of Early Cretaceous volcanic rocks in the Genhe area, Daxing'anling: Sr-Nd-Pb-Hf isotope geochemical constraints. *J. Petrol.* **2007**, *61*, 1594–1612.
15. Li, Y.F.; Bian, X.F.; Gao, X.Y.; Chen, S.W.; Ding, Q.H. Laser total-melt $^{40}Ar/^{39}Ar$ dating of Mesozoic volcanic rocks in the Longjiang Basin, northern Daxinganling. *Geol. Bull.* **2013**, 32, 1212–1223.

16. Zhang, P.C.; Peng, B.; Zhao, J.Z.; Zhang, Z.F.; Zhao, J.; Guo, G.F. Zircon U-Pb chronology, geochemistry and metallogenic significance of the Hai Liyu extraordinarily high-divergence granitoids from the southern Daxinganling section. *J. Rock Mineral.* **2022**, *41*, 1029–1046.
17. Sui, Z.M.; Ge, W.C.; Xu, X.C.; Zhang, J.H. Characteristics of the Late Paleozoic post orogenic granitoids from Station XII, Daxinganling and their geological significance. *J. Petrol.* **2006**, 2679–2686.
18. Chen, Z.G.; Zhang, L.C.; Zhou, X.C.; Wan, B.; Ying, K.F.; Wang, F. Chronological and geochemical characterization of volcanic rock sections from Xinyuqi, Manzhouli. *J. Petrol.* **2006**, *22*, 2971–2986.
19. Zhang, Y.B.; Sun, S.H.; Honma, H.J.; Mao, Q. Mixing of Mesozoic acidic magmas in the Linxi area, southern Daxinganling. *J. Petrol.* **2003**, *19*, 16.
20. Zhang, J.H. Chronology and Geochemistry of Mesozoic Volcanic Rocks in Daxing'anling. Ph.D. Thesis, China University of Geosciences, Wuhan, China, 2009.
21. Yang, B.H.; Wang, W.D.; Yan, Y.S.; Wei, X.Y.; Ge, C.B. Genesis and mantle enrichment of volcanic rocks in the Tamulangou formation, Xinlin District, northern Daxinganling. *Geol. Rev.* **2016**, *62*, 16.
22. Ji, Z.; Ge, W.C.; Yang, H.; Zhao, G.C.; Wang, Q.H.; Bi, J.H. Chronology and geochemistry of volcanic rocks from the Manketou Ebo formation, middle Daxinganling. In *Abstracts from the 15th Annual Academic Conference of the Geological Association of China, Mineral and Rock Geochemistry Division (Volume 1)*; The Geological Association of Canada: St. John's, NL, Canada, 2015.
23. Shang, Y.H.; Lu, S. Chronology, geochemical characteristics and petrogenesis of the Manitu formation, Daxing'anling. *Heilongjiang Univ. Sci. Technol.* **2017**, *27*, 6.
24. Zhang, L.Q.; Chen, B.; Cao, Y.B.; Wang, J.H.; Huang, Z.J.; Chen, G.Y. Chronology and petrogeochemical characterization of the volcanic rocks of the Manitu formation, Horqin Youyiqian Banner, Inner Mongolia. *Mod. Min.* **2017**, *33*, 5.
25. Zhang, L.T.; Li, S.C.; Zhao, Q.Y.; Li, X.F.; Wang, L.; Li, Z.H. Age of formation and geochemical characteristics of volcanic rocks of the Baiyin Gao Lao formation, middle Daxinganling. *World Geol.* **2015**, 44–54.
26. Nie, L.J.; Jia, H.M.; Wang, C.; Lu, X.B. Chronology, geochemistry and its geological significance of rhyolite from the Baiyin Gao Lao formation, middle Daxinganling. *World Geol.* **2015**, *34*, 296–304.
27. Ding, Q.H.; Chen, S.W.; Shang, L.; Li, Y.F.; Wang, J. New understanding of the Lower Cretaceous Longjiang formation in eastern Daxinganling. *Geol. Resour.* **2014**, *23*, 8.
28. Zhang, C.; Wu, X.W.; Zhang, Y.J.; Guo, W.; Quan, J.Y. LA-ICP-MS zircon U-Pb ages of alkali flowstone from Guanghua formation, Longjiang Basin, northern Daxinganling and their geological significance. *Geol. Bull.* **2017**, *36*, 11.
29. Li, Y.F.; Gao, X.Y.; Bian, X.F.; Chen, S.W.; Ding, Q.H. LA-ICP-MS zircon U-Pb ages, geochemical characteristics and their geological significance of Mesozoic volcanic rocks in the Longjiang Basin, northern Daxinganling. *Geol. Bull.* **2013**, 1195–1211.
30. Ye, S.P.; Wei, J.; Sha, X.B. Characteristics and tectonic environment of volcanic rocks of the Longjiang formation in the Xinliu Station area, Daxing'anling. *World Nonferrous Met.* **2017**, 200+202.
31. Amelin, Y.; Lee, D.C.; Halliday, A.N.; Pidgeon, R.T. Nature of the earth's earliest crust from hafnium isotopes in single detrital zircons. *Nature* **1999**, *399*, 1497–1503. [CrossRef]
32. Amelin, Y.; Lee, D.C.; Halliday, A.N. Early-middle archaean crustal evolution deduced from lu-hf and u-pb isotopic studies of single zircon grains. *Geochim. Cosmochim. Acta* **2000**, *64*, 4205–4225. [CrossRef]
33. Griffin, W.L.; Belousova, E.; Se, J.; Sy VA, E.O.; Sr, S.; Nj, P. The hf isotope composition of cratonic mantle: LAM-MC-ICPMS analysis of zircon megacrysts in kimberlites. *Geochim. Cosmochim. Acta J. Geochem. Soc. Meteorit. Soc.* **2000**, *64*, 133–147. [CrossRef]
34. Wiedenbeck, M. Three natural zircon standards for U-Th-Pb, Lu-Hf, trace element and ree analyses. *Geostand. Newsl.* **1995**, *19*, 1–23. [CrossRef]
35. Wu, F.Y.; Li, X.H.; Zheng, Y.F.; Gao, G. Lu-Hf isotope system and its petrological applications. *J. Petrol.* **2007**, *23*, 36.
36. Zhang, Y.T.; Zhang, L.C.; Ying, K.F.; Zhou, X.H.; Wang, F.; Hou, Q.L. Geochemical and source-area characterization of Early Cretaceous volcanic rocks in the Tahe area, northern Daxinganling. *J. Petrol.* **2007**, *23*, 2811–2822.
37. Zhang, J.H.; Ge, W.C.; Wu, F.Y.; Wilde, S.A.; Yang, J.H.; Liu, X.M. Large-scale Early Cretaceous volcanic events in the northern Great Xing'an Range, Northeastern China. *Lithos* **2008**, *102*, 138–157. [CrossRef]
38. Li, J.Y. New understanding of some geotectonic problems in northeast China and neighboring areas. *Geol. Rev.* **1998**, 339–347.
39. Li, J.Y.; Qu, J.F.; Zhang, J. New progress in the study of orogenic process and metallogenic geological background of Tianshan-Xingmeng orogenic mountains. *Geol. Bull.* **2013**, 207–219.
40. Wu, T. Geochemical Characterization and Tectonic Context of Early Cretaceous Neutral Volcanic Rocks from the Zalantun Area, Inner Mongolia. Master's Thesis, Jilin University, Changchun, China, 2018.
41. Tang, J.; Xu, W.L.; Wang, F.; Ge, W.C. Subduction history of the Paleo-Pacific plate under Eurasia: Mesozoic-Paleocene magma record from the Northeast Asian land margin. *Sci. China Earth Sci.* **2018**, *48*, 35.
42. Zhuo, S.; Ji, A. X-ray fluorescence spectroscopy analysis. *Anal. Lab.* **2003**, 102–108.
43. Yuan, H.L.; Gao, S.; Liu, X.M.; Li, H.M.; Günther, D.; Wu, F.Y. Accurate U-Pb age and trace element determinations of zircon by laser ablation-inductively coupled plasma-mass spectrometry. *Geostand. Geoanalytical Res.* **2004**, *28*, 353–370. [CrossRef]
44. Andersen, T. Correction of common lead in U-Pb analyses that do not report ^{204}Pb. *Chem. Geol.* **2002**, *192*, 59–79. [CrossRef]
45. Sláma, J.; Košler, J.; Condon, D.J.; Crowley, J.L.; Gerdes, A.; Hanchar, J.M.; Horstwood, M.S.; Morris, G.A.; Nasdala, L.; Norberg, N. Plešovice zircon—A new natural reference material for U–Pb and Hf isotopic microanalysis. *Chem. Geol.* **2008**, *249*, 1–35. [CrossRef]

46. Wei, P.; Gao, J.F.; Zhao, K.D.; Ling, H.F.; Jiang, S.C. Rapid and effective separation of Rb-Sr and Sm-Nd using DCTA and HIBA. *J. Nanjing Univ.* **2005**, 445–450.
47. Goolaerts, A.; Mettielli, N.; Dejong, J. Hf and Lu isotopic reference values for the zircon standard 91500 by MC-ICP-MS. *Chem. Geol.* **2004**, *206*, 1–9. [CrossRef]
48. Wu, F.Y.; Lin, J.Q.; Wilde, S.A.; Zhang, X.O.; Yang, J.H. Nature and significance of the Early Cretaceous giant igneous event in eastern China. *Earth Planet. Sci. Lett.* **2005**, *233*, 103–119. [CrossRef]
49. Irvine, T.N.; Baragar, W.R.A. A Guide to the Chemical Classification of the Common Volcanic Rocks. *Can. J. Earth Sci.* **1971**, *8*, 523–548. [CrossRef]
50. Maniar, P.D.; Piccoli, P.M. Tectonic discrimination of granitoids. *Geol. Soc. Am. Bull.* **1989**, *101*, 635–643. [CrossRef]
51. Sun, S.S.; McDonough, W.F. Chemical and isotopic systematics of oceanic basalts: Implications for mantle composition and processes. In *Magmatism in the Ocean Basins. Geological Society*; Sanders, A.D., Norry, M.J., Eds.; Special Publications: London, UK, 1989; Volume 42, pp. 313–345.
52. McDonough, W.F.; Sun, S.S. The composition of the Earth. *Chem. Geol.* **1995**, *120*, 223–253. [CrossRef]
53. Wu, H.Y.; Zhang, L.C.; Zhou, X.H.; Chen, C.G. Chronological and geochemical characterization and genesis of Late Mesozoic andesites from the middle Daxinganling. *J. Petrol.* **2008**, *24*, 1339–1352.
54. Chen, Z.; Li, X.W.; Zhang, S.J.; Zhou, C.F.; Li, Y.F.; Guo, L. Geochemical characterization and tectonic background analysis of volcanic rocks in the Longjiang formation, Shiwuliqiao Gold Mine, Heilongjiang. *Geol. Resour.* **2019**, 413–422.
55. Kawabata, H.; Shuto, K. Magma mixing recorded in intermediate rocks associated with high-Mg andesites from the setouchi volcanic belt, Japan: Implications for archean ttg formation. *Volcanol. Geotherm. Res.* **2005**, *140*, 241–271. [CrossRef]
56. Streck, M.J.; Leeman, W.P.; Chesley, J.T. High-magnesian andesite from mount shasta: A product of magma mixing and contamination, not a primitive melt: Comment and replyreply. *Geology* **2007**, *35*, e147. [CrossRef]
57. Qian, Q.; Hermann, R. formation of high-mg diorites through assimilation of peridotite by monzodiorite magma at crustal depths. *J. Petrol.* **2011**, *51*. [CrossRef]
58. Hanyu, T.; Tatsumi, Y.; Nakai, S.; Chang, Q.; Miyazaki, T.; Sato, K.; Tani, K.; Shibata, T.; Yoshida, T. Contribution of slab melting and slab dehydration to magmatism in the Japanese arc. *Geochim. Cosmochim. Acta* **2006**, *70*, A229. [CrossRef]
59. Zeng, W.S.; Zhou, J.B.; Dong, C.; Cao, J.L.; Wang, B. A record of Mongolia-Okhotsk ocean subduction: Evidence from the Badaiguan metamorphic mafic rocks in the Erguna region. *J. Petrol.* **2014**, *30*, 1948–1960.
60. Jung, S.; Hoernes, S.; Mezger, K. Synorogenic melting of mafic lower crust: Constraints from geochronology, petrology and Sr, Nd, Pb and O isotope geochemistry of quartz diorites (Damara orogen, Namibia). *Contrib. Mineral. Petrol.* **2002**, *143*, 551–566. [CrossRef]
61. Ji, Z.; Ge, W.C.; Yang, H.; Bi, J.H.; Yu, Q.; Dong, Y. Late Triassic Andean-type andesites in the middle Daxinganling: A product of southward subduction of the Mongolia-Okhotsk Oceanic Plate. *J. Petrol.* **2018**, *34*, 14.
62. Jung, S.; Masberg, P. Major- and trace-element systematics and isotope geochemistry of Cenozoic mafic volcanic rocks from the Vogelsberg (central Germany): Constraints on the origin of continental alkaline and tholeiitic basalts and their mantle sources. *J. Volcanol. Geotherm. Res.* **1998**, *86*, 151–177. [CrossRef]
63. Wei, B.Y.; Wu, M.H.; Liu, B.; Jiang, J.B.; Liang, W. Chronology, geochemical characterization and tectonic significance of andesites in the Yamansu area, Eastern Tien Shan. *South. Met.* **2020**, *50*, 14–32.
64. Taylor, S.R.; McLennan, S.M. *The Continental Crust: Its Composition and Evolution*; Blackwell Scientific Publication: Oxford, UK, 1985; pp. 1–132.
65. Keppler, H. Constraints from partitioning experiments on the composition of subduction-zone fluids. *Nature* **1996**, *380*, 237–240. [CrossRef]
66. Li, C.D.; Zhang, F.Q.; Miao, L.C.; Jie, H.Q.; Xu, Y.W. Late Permian high-magnesium andesite shrimp zircon chronology and geochemical characterization of the Seruohe Late Permian in Jilin. *J. Petrol.* **2007**, *23*, 767–776.
67. Defant, M.J.; Drummond, M.S. Derivation of Some Modern Arc Magmas by Melting of Young Subducted Lithosphere. *Nature* **1990**, *347*, 662–665. [CrossRef]
68. Zhu, D.C.; Duan, L.P.; Liao, Z.L.; Pan, G.T. Differentiation of two types of adakites. *Mineral. Rocks* **2002**, *22*, 5.
69. Zhang, Q.; Wang, Y.; Wang, Y.L. Edakite and the tectonic environment. *Geotecton. Mineral.* **2003**, *27*, 101–108.
70. Li, C.N. *Trace Element Petrology of Igneous Rocks*; China University of Geosciences Press: Qinhuangdao, China, 1992.
71. Drummond, M.S.; Bordelon, M.; de Boer, J.Z.; Defant, M.J.; Bellon, H.; Feigenson, M.D. Igneous petrogenesis and tectonic setting of plutonic and volcanic rocks of the Cordillera de Talamanca, Costa Rica-Panama, Central American arc. *Am. J. Sci.* **1995**, *295*, 875–919. [CrossRef]
72. Tan, H.Y.; He, Z.H.; Chen, F.; Du, Y.D.; Ren, Z.H. Zircon U-Pb ages, geochemical characteristics and tectonic significance of volcanic rocks from the Baiyin Gao Lao formation, Solun area, middle Daxinganling. *Geol. Bull.* **2017**, *36*, 893–908.
73. Hao, Y.J.; Ren, Y.S.; Shi, Y.F.; Shang, Q.Q.; Sun, Z.M.; Gao, Y.; Yang, Q. Age of formation, petrogenesis and tectonic background of the granite porphyry of the porphyry-type tin polymetallic deposits in Hekou Forest, Wundashan District, Heilongjiang Province. *J. Petrol.* **2020**, 837–855.
74. Sui, Z.M.; Ge, W.C.; Wu, F.Y.; Zhang, J.H.; Xu, X.C.; Cheng, R.Y. Zircon U-Pb ages, geochemical characteristics and genesis of Jurassic granitic rocks from northeastern Daxinganling. *J. Petrol.* **2007**, *23*, 20.

75. Wang, Y.L.; Ao, G.; Wang, H.P.; Wang, Z.J.; Wang, Q.; Zhong, M.S. Zircon U-Pb chronology, geochemistry and tectonic significance of Early Cretaceous granitoids from the Solun area, middle Daxinganling. *Geosci. Technol. News* **2019**, *13*.
76. Barth, M.G.; Mcdonough, W.F.; Rudnick, R.L. Tracking the budget of Nb and ta in the continental crust. *Chem. Geol.* **2000**, *165*, 197–213. [CrossRef]
77. Shi, L.; Tang, Z.; Zheng, C.Q.; Qin, T.; Zhang, L.D.; Wang, Y. Petrogenesis and tectonic context of Early Cretaceous neutral volcanic rocks in the Chaihe-Mushroom gas region, central Daxinganling. *Geol. Bull.* **2018**, *37*, 13.
78. Hou, M.L.; Jiang, S.C.; Jiang, Y.H.; Ling, H.F. S-Pb isotope geochemistry and Rb-Sr isotope chronology of the Penglai gold mineralization in Jiaodong. *J. Petrol.* **2006**, 2525–2533.
79. Shao, J.A.; Mou, B.L.; Zhu, H.Z.; Zhang, C.Q. Deep origin and background of Mesozoic mineralization in the south-central section of the Daxinganling. *J. Petrol.* **2010**, 649–656.
80. Zhu, B.Q. Shell-mantle chemical inhomogeneities and the study of block geochemical boundaries. *Geol. Front.* **1998**, *5*, 72–82.
81. Gao, L.F.; He, Z.H.; Sui, Z.M.; Lu, K.J.; Liu, F.Y. Age, geochemical characteristics of volcanic rocks of the Manketou Ebo formation and their tectonic setting in the Solun area, middle Daxinganling. *Geol. Bull.* **2018**, *37*, 14.
82. Jia, H.M.; Jing, Y.; Wang, Q.H.; Lan, L.X.; Jing, J.H. Genesis of Early Cretaceous granites in the Handaigai area, middle Daxinganling and their geological significance. *World Geol.* **2019**, *38*, 10.
83. Zartman, R.E.; Haines, S.M. The plumbotectonic model for Pb isotopic systematics among major terrestrial reservoirs-A case for bi-directional transport. *Geochim. Cosmochim. Acta* **1988**, *52*, 1327–1339. [CrossRef]
84. Chu, S.X.; Liu, J.M.; Xu, J.H.; Wei, H.; Chai, H.; Tong, K.G. Zircon U-Pb dating, petrogenesis and tectonic significance of granodiorite diorite from the Sanminggou iron and copper deposit, Heilongjiang. *J. Petrol.* **2012**, *28*, 433–450.
85. Ma, J.J.; Dae-hyuk. Preliminary study of Mesozoic volcanic rocks in Heilongjiang Province. *Heilongjiang Geol.* **1991**, *2*, 16.
86. Lin, W.G. Late mesozoic calc-alkaline volcanism of post-orogenic extension in the northern Da Hinggan Mountains, northeastern China. *J. Volcanol. Geotherm. Res.* **2003**, *121*, 115–135.
87. Meng, Q.R. What drove late Mesozoic extension of the northern China–Mongolia tract? *Tectonophysics* **2003**, *369*, 155–174. [CrossRef]
88. Zorin, Y.A. Geodynamics of the western part of the mongolia–okhotsk collisional belt, trans-baikal region (russia) and mongolia. *Tectonophysics* **1999**, *306*, 33–56. [CrossRef]
89. Guo, F.; Fan, W.M.; Wang, Y.J.; Lin, G. Late Mesozoic bimodal volcanism in the southern section of the Daxinganling. *J. Petrol.* **2001**, 161–168.
90. Wang, F.; Zhou, X.H.; Zhang, L.C.; Ying, J.F.; Zhang, Y.T.; Wu, F.Y.; Zhu, R.X. Late mesozoic volcanism in the great Xing 'an range (ne China): Timing and implications for the dynamic setting of ne Asia. *Earth Planet. Sci. Lett.* **2006**, *251*, 179–198. [CrossRef]
91. Zhang, L.C.; Zhou, X.H.; Ying, J.F.; Wang, F.; Guo, F.; Wan, B. Geochemistry and Sr-Nd-Pb-Hf isotopes of early cretaceous basalts from the great xinggan range, ne china: Implications for their origin and mantle source characteristics. *Chem. Geol.* **2008**, *256*, 12–23. [CrossRef]
92. Xu, M.J.; Xu, W.L.; Wang, F.; Gao, F.H.; Yu, J.J. Chronology and geochemistry of Early Jurassic granitic rocks in central Xiaoxinganling and their tectonic significance. *J. Petrol.* **2013**, *15*.
93. Shao, J.A.; Zhang, L.Q.; Mu, B. Tectono-thermal evolution of middle-south section of the Da Hinggan Mountains. *Sci. China Ser. D: Earth Sci.* **1998**, *41*, 570–579. [CrossRef]
94. Badarch, G.; Cunningham, W.D.; Windley, B.F. A new terrane subdivision for Mongolia: Implications for the Phanerozoic crustal growth of Central Asia. *J. Asian Earth Sci.* **2002**, *21*, 87–110. [CrossRef]
95. Chen, C.G.; Zhang, L.C.; Lu, B.C.; Li, Z.L.; Wu, H.Y.; Xiang, P. Age, geochemistry and geological significance of the mineralized porphyry of the Taipingchuan copper-molybdenum mine, Inner Mongolia. *J. Petrol.* **2010**, *26*, 1437–1449.
96. Ying, J.F.; Zhou, R.H.; Zhang, R.C.; Wang, R. Geochronological framework of mesozoic volcanic rocks in the great xing'an range, ne China, and their geodynamic implications. *J. Asian Earth Sci.* **2010**, *39*, 786–793. [CrossRef]
97. Sun, D.Y.; Gou, J.; Ren, Y.S.; Fu, C.L.; Wang, L.; Liu, S. U-Pb ages and geochemistry of zircon from volcanic rocks of the Manitou formation, southern Manchuria. *J. Petrol.* **2011**.
98. Li, S.C.; Xu, Z.Y.; Liu, Z.H.; Li, Y.F.; Wang, X.A.; Zhang, C. Age and geochemical characterization of zircon U-Pb from volcanic rocks of the Manitou formation, middle Daxing'anling, LA-ICP-MS. *Geol. Bull.* **2013**, *32*, 9.
99. Si, Q.L. Chronology and Geochemistry of Late Mesozoic Volcanic Rocks in the Zalantun Area, Middle Daxinganling. Ph.D. Doctoral Dissertation, Northeastern University, Boston, MA, USA, 2018.
100. Yang, Y.J.; Yang, X.P.; Jiang, B.; Wang, Y.; Pang, X.J. Spatial and temporal distribution of Mesozoic volcanic stratigraphy in Daxing'anling and response to subduction of the Mongolian-Okhotsk and Paleo-Pacific plates. *Geol. Front.* **2022**, *29*, 115–131.
101. Yu, J.J.; Wang, F.; Xu, W.L.; Gao, F.H.; Pei, F.P. Early jurassic mafic magmatism in the lesser xing'an–zhangguangcai range, Ne China, and its tectonic implications: Constraints from zircon U-Pb chronology and geochemistry. *Lithos* **2012**, *142–143*, 256–266. [CrossRef]
102. Wu, F.Y.; Sun, D.Y.; Li, H.; Jahn, B.M.; Wilde, S.A. A-type granites in northeastern China: Age and geochemical constraints on their petrogenesis. *Chem. Geol.* **2002**, *187*, 143–173. [CrossRef]
103. Kravchinsky, V.A.; Cogn, J.P.; Harbert, W.P.; Kuzmin, M.I. Evolution of the mongol-okhotsk ocean as constrained by new palaeomagnetic data from the mongol-okhotsk suture zone, Siberia. *Geophys. R. Astron. Soc.* **2010**, *148*, 34–57. [CrossRef]

104. Metelkin, D.V.; Gordienko, I.V.; Klimuk, V.S. Paleomagnetism of Upper Jurassic basalts from Transbaikalia: New data on the time of closure of the Mongol-Okhotsk Ocean and Mesozoic intraplate tectonics of Central Asia. *Russ. Geol. Geophys.* **2007**, *48*, 825–834. [CrossRef]
105. Peng, Y.W.; Chen, Y.J. Location of the tectonic boundary between the Jihei orogenic belt and the Kaiyuan-Shanzhen Town section of the North China Platform. *World Geol.* **2007**, 1–6+74.
106. Zhang, X.Z.; Ma, Y.X.; Chi, X.G.; Zhang, F.X.; Sun, Y.W.; Guo, Y.E. Issues related to the tectonic evolution of the Phanerozoic in Northeast China and East Inner Mongolia. *J. Jilin Univ. Earth Sci. Ed.* **2012**, *42*.
107. Müller, D.; Rock, N.M.S.; Groves, D.I. Geochemical discrimination between shoshonitic and potassic volcanic rocks in different tectonic settings: A pilot study. *Mineral. Petrol.* **1992**, *46*, 259–289. [CrossRef]

minerals

MDPI

Article

Origin and Geological Implications of Monzogranites and Rhyolitic Porphyries in the Wunugetu Porphyry Copper–Molybdenum Deposit, Northeast China: Evidence from Zircon U-Pb-Hf Isotopes and Whole-Rock Geochemistry

Qingshuang Wang [1,2], Yanchen Yang [1,*], Qiulin Fu [3], Zhongyue Zhang [4], Xiaodan Guo [5], Taotao Wu [2], Lu Chai [2], Yongheng Zhou [2] and Yonghai An [5]

1 College of Earth Sciences, Jilin University, Changchun 130061, China; wangqingshuang1988@163.com
2 Shenyang Geological Survey Center of China Geological Survey, Shenyang 110034, China; wutaotao@mail.cgs.gov.cn (T.W.); chailu@mail.cgs.gov.cn (L.C.); zhouyongheng@mail.cgs.gov.cn (Y.Z.)
3 Liaoning Province Institute of Geology and Mineral Resources Co., Ltd., Shenyang 110034, China; woshifuqiulin@163.com
4 Liaoning Earthquake Agency, Shenyang 110031, China; 13390583679@163.com
5 Liaoning Province Geology and Mineral Group Energy Geology Co., Ltd., Shenyang 110032, China; guoxiaodan63@163.com (X.G.); liwaca@163.com (Y.A.)
* Correspondence: yyc@jlu.edu.cn

Abstract: The Wunugetu deposit, a large-scale porphyry copper–molybdenum deposit, is located in the southern Erguna block. Its ore bodies are primarily found within monzogranites, granite porphyries, and biotite monzogranites. Additionally, the deposit contains late-stage intrusive dykes of rhyolitic porphyries. This study examined the deposit's monzogranites and rhyolitic porphyries using lithogeochemistry, zircon U-Pb dating, and Hf isotopic analysis. The main findings include: (1) Zircon U-Pb dating showed that the monzogranites formed around 209.0 ± 1.0 Ma, whereas the rhyolitic porphyries in the northern portion formed around 170.49 ± 0.81 Ma, suggesting magmatic activity in the deposit spanned from the Late Triassic to the Middle Jurassic. (2) The monzogranites exhibited high silicon content (73.16–80.47 wt.%) and relatively low aluminum content (10.98–14.37 wt.%). They are enriched in alkalis (content: 3.42–10.10 wt.%) and deficient in magnesium and sodium, with aluminum saturation indices (A/CNK) ranging from 1.1 to 2.9. In addition, the monzogranites are enriched in large-ion lithophile elements (LILEs) such as Rb, K, and Ba and deficient in high-field-strength elements (HFSEs) like Nb, P, and Ti. (3) The monzogranites have low Zr + Nb + Ce + Y contents of (151.3–298.6 ppm) $\times 10^{-6}$ and 10,000 × Ga/Al ratios varying between 1.20 and 2.33, suggesting that they are characteristic of I-type granites. (4) Positive zircon εHf(t) values ranging from +0.3 to +7.6 in both rhyolitic porphyry and monzogranite samples, increasing with younger emplacement ages, imply that the deposit's rocks originated from magmatic mixing between mantle-derived mafic magmas and remelts of the juvenile crust. Considering these results and the regional geological evolution, this study proposes that the Wunugetu deposit was formed in an active continental margin setting and was influenced by the Late Triassic–Middle Jurassic southeastward subduction of the Mongol-Okhotsk Ocean.

Keywords: Cu-Mo deposit; lithogeochemistry; I-type granite; Mongolia-Okhotsk Ocean; Wunugetu

Citation: Wang, Q.; Yang, Y.; Fu, Q.; Zhang, Z.; Guo, X.; Wu, T.; Chai, L.; Zhou, Y.; An, Y. Origin and Geological Implications of Monzogranites and Rhyolitic Porphyries in the Wunugetu Porphyry Copper–Molybdenum Deposit, Northeast China: Evidence from Zircon U-Pb-Hf Isotopes and Whole-Rock Geochemistry. *Minerals* **2024**, *14*, 310. https://doi.org/10.3390/min14030310

Academic Editor: Massimo D'Antonio

Received: 22 January 2024
Revised: 13 March 2024
Accepted: 13 March 2024
Published: 15 March 2024

1. Introduction

The Central Asian Orogenic Belt (CAOB), located between the Siberian Craton and the North China Craton, is recognized as one of the largest Phanerozoic orogenic belts globally. Its extensive geological history features the accretion of continental crust induced by oceanic subduction, the convergence of multiple plates, intracontinental orogeny, and associated multistage tectono-magmatic activity. Hence, the deep magmatic activity and mineralization related to the CAOB's formation and evolution have garnered significant interest from geologists worldwide [1–8]. Within the CAOB, multiple world-class porphyry copper deposits, such as Oyu Tolgoi, Erdenet, and Aktogai–Aiderly, have been identified, positioning this region as a premier porphyry copper belt on a global scale, also referred to as the Central Asian Metallogenic Belt [9]. This belt, along with the Circum-Pacific and Tethyan metallogenic belts, constitutes one of the three principal global copper metallogenic provinces [10]. Situated in the eastern segment of the CAOB, Northeast China emerged from the amalgamation of various microblocks, including the Erguna, Xing'an, Songnen-Zhangguangcai Range, Jiamusi, and Xingkai blocks [11], highlighting its importance as a notable copper-bearing region in China.

The Wunugetu deposit, situated in the southern Erguna block of Northeast China (Figure 1a), represents a significant porphyry copper–molybdenum deposit within the eastern CAOB. Since its identification, the deposit has garnered considerable attention from the geological community [12–14]. Extensive studies of this deposit focus on its geological characteristics, the isotopic geochronology and lithogeochemistry of the ore-hosting surrounding rocks, and ore-forming fluids [12,15–17]. However, debates persist regarding the metallogenic periods and geodynamic setting of the Wunugetu deposit. Wang Zhitian's (1988) initial K-Ar dating of monzogranite porphyry stocks within the deposit yielded an age of 138 Ma [12], leading to the hypothesis of a Late Yanshanian origin. Qin Kezhang (1999), through Rb-Sr dating of the ore-hosting monzogranite porphyries and K-Ar dating of altered sericite, proposed that the deposit formed between 180 and 190 Ma, indicating an Early Jurassic age of this copper–molybdenum deposit [15]. Advancements in analytical methods allowed Tan Gang (2010) to determine a formation age of 177 Ma using Re-Os isotopic dating of the molybdenite [17]. These studies demonstrate that the Wunugetu deposit experienced intense magmatic activity and mineralization events during the Early Jurassic. However, there is limited information on Late Triassic and Middle Jurassic intrusions within the deposit. This study delves into the geological, ore-body, and ore characteristics of the Wunugetu deposit. Employing zircon U-Pb dating, Hf isotope analysis, and whole-rock geochemistry of the ore-hosting porphyritic monzogranites and the late-stage intrusive rhyolitic porphyries, this study aims to constrain the formation periods of ore-forming rock masses and identify the genetic types, source traits, and tectonic setting of the rock masses and late-stage dykes. Additionally, the new data can be used to effectively elucidate the geodynamic context of copper–molybdenum mineralization in the Wunugetu deposit.

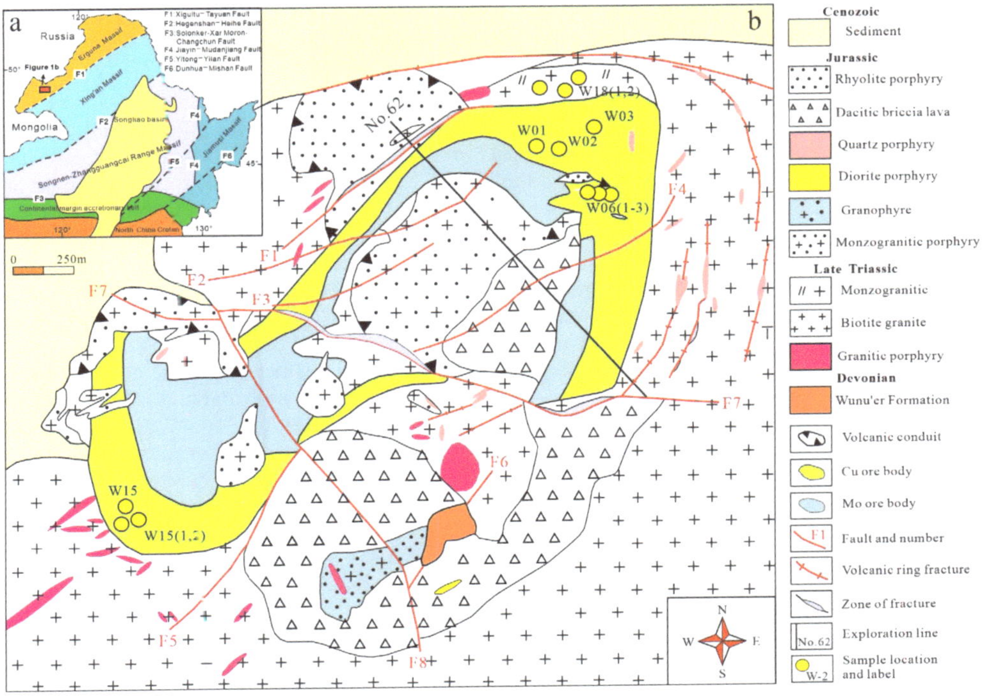

Figure 1. (**a**) Tectonic location map of the Wunugetu porphyry copper–molybdenum deposit; (**b**) Geological map of the Wunugetu deposit showing the main lithologies and faults with a volcanic edifice [18].

2. Geological Setting

2.1. Regional Geological Setting

Located about 24 km southwest of Manzhouli City, the Wunugetu deposit is positioned in the central part of the Erguna block, nestled between the Mongol-Okhotsk suture zone (MOSZ) and the Xiguitu-Tayuan fault (Figure 1a). In the Paleozoic era, the Erguna block, influenced by the evolution of the Paleo-Asian Ocean tectonic system, experienced the complex amalgamation of several microblocks within the CAOB [19,20]. From the Late Paleozoic to the Mesozoic, the block fell under the tectonic influence of the Mongol-Okhotsk Ocean tectonic domain [19–23]. The superposition of multiple tectonic domains has endowed this block with highly developed regional fault structures, primarily including a series of E–W- and NE–NNE-trending faults, as well as secondary faults. The oldest exposed strata within the Erguna block consist mainly of Precambrian strata, overlain by Paleozoic and Mesozoic layers. The Precambrian strata are chiefly made up of a crystalline basement composed mainly of schists, leptites, and quartzites from the Neoproterozoic Jiageda Formation (Ptj). The Paleozoic sequence mainly comprises volcanic and terrigenous clastic rocks from the Middle Devonian Wunu'er Formation, while the Mesozoic strata encompass intermediate-mafic volcanic and pyroclastic rocks from the Middle Jurassic Tamulangou Formation, intermediate-acid volcanic rocks from the Shangkuli Formation, and acidic volcanic and pyroclastic rocks from the Upper Jurassic Manketouebo and Manitu formations. The complex process of tectonic evolution has resulted in frequent regional magmatic activity, with Mesozoic granites being the most extensively developed. The region hosts several significant deposits, including the Badaguan copper–molybdenum deposit, the Wunugetu deposit, the Jiawula and Halasheng lead–zinc–silver deposits, and the Chaganbulagen silver–lead–zinc deposit, all of which are intimately linked to the Mesozoic magmatic activity [15,18,24,25].

2.2. Geological Setting of the Wunugetu Deposit

The Wunugetu deposit was discovered by the No. 706 Team of the Heilongjiang Nonferrous Metal Geological Survey in 1978. By 2006, the deposit had 1.106 million tons of copper and 0.362 million tons of molybdenum identified, establishing it as the most significant large-scale porphyry copper–molybdenum deposit within the Erguna block [13]. In 2008, the China National Gold Group Co., Ltd. (Beijing, China) acquired the Wunugetu deposit to further its exploration and production endeavors.

The exposed strata in the Wunugetu deposit encompass the Devonian Wunu'er Formation's andesites and crystalline limestones (southwest), Upper Jurassic volcanic rocks (north), and extensive Quaternary unconsolidated deposits (Figure 1b).

The Wunugetu deposit underwent two distinct magmatic phases. The first phase, during the Late Triassic, produced biotite granites, monzogranites, and granite porphyries, which together cover about 60% of the deposit's total area. The biotite granites and monzogranites, forming batholiths, extend across the deposit and its vicinity, covering an exposed area of approximately 110 km^2. These rocks are crucial as the ore-hosting rock masses and surrounding rocks of the deposit.

The second magmatic phase, in the Early Jurassic, led to the formation of a volcanic edifice. The volcanic rocks mainly consist of andesitic agglomerates, while the lithologic assemblages of intrusions primarily comprise monzogranite porphyries, quartz porphyries, diorite porphyrites, and rhyolitic porphyries. The rhyolitic porphyries, intruding into the mineral-bearing monzogranite porphyries, show no signs of mineralization or alteration zones. The monzogranite porphyries are distributed in the central portion of the deposit. They are disrupted by the late-stage F7 fault, making them a key ore-hosting rock mass. Additionally, the outcrops in the deposit include minor quantities of dacitic breccias, quartz porphyries, and diorite porphyrites.

Fault structures are prominent in the Wunugetu deposit, including NE-, NNW-, and nearly E–W-trending faults, along with ring-shaped faults in volcanic areas. The NE-trending and ring-shaped faults developed during the mineralization period, whereas the NW-oriented F8 fault emerged in the later stages of mineralization. These faults cut through ore bodies, disrupting their continuity. The E–W-trending F7 fault divides the ore bodies into southern and northern segments.

3. Ore Body Characteristics

The ore bodies within the Wunugetu deposit are arranged in a ring-like pattern orienting NWE, with monzogranite porphyries at their core. The area of mineralization spans roughly 2600 m in length and about 1400 m in width. These ore bodies have a strike of 50° and a dip direction of WN. The F7 fault causes a horizontal displacement of approximately 700 m, thereby splitting the ore bodies into two distinct segments: the southern segment spans about 800 m in width, while the northern segment extends approximately 900 m across. Both segments are subjected to hydrothermal alteration within the deposit. The Wunugetu deposit exhibits typical alteration zones of porphyry-type mineralization, covering an area of approximately 5 km^2. These zones roughly comprise potassic (Or), quartz-sericite (Qs), and illite-hydromica (IH) alteration zones [15]. The primary alterations observed include silicification, potassic feldspathization, sericitization, illitization, muscovitization, carbonatization, and kaolinization, with chloritization and epidotization being relatively less common (Figure 2).

The potassic zone (Or), centrally located within the deposit, is found predominantly in the monzogranite porphyry rock masses. The alteration minerals characteristic of this zone are mainly K-feldspar and biotite, with sericite present in smaller quantities (Figure 2a). Molybdenite is the principal type of mineralization in this zone, along with notable occurrences of chalcopyrite and pyrite mineralizations.

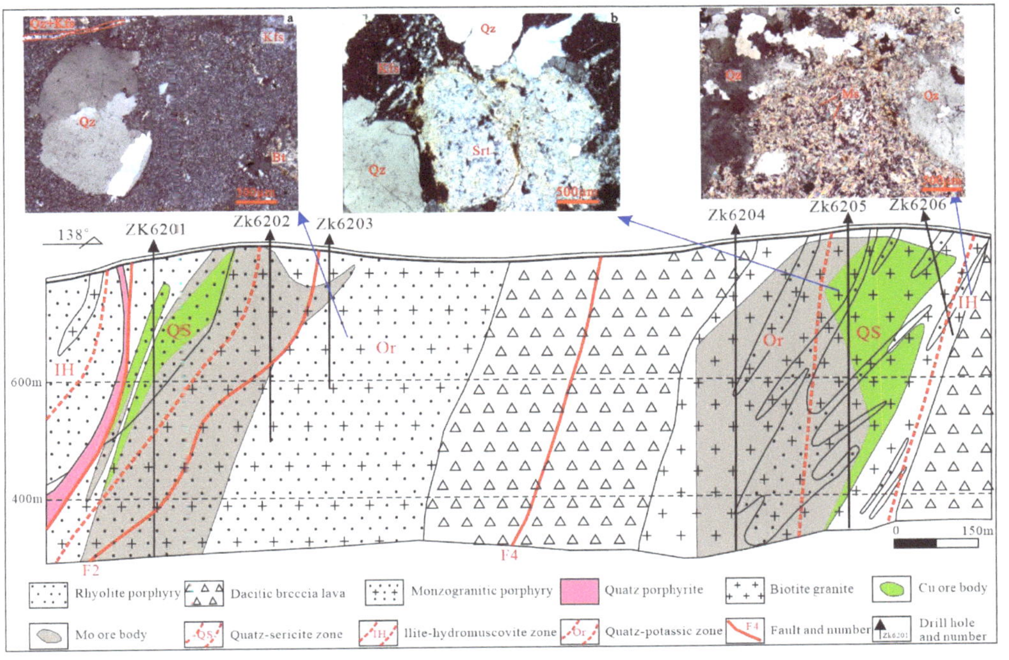

Figure 2. Northwest–southeast section (620 line) through the Wunugetu deposit showing the relationship of host monzogranitic porphyry, alteration, and mineralization [24]. Micrographs of potassium feldspathization (**a**), quartz-sericitization (**b**), and illite-hydromica (**c**). Bt—biotite, Qz—quartz, Pl—plagioclase, Kfs—feldspar, Srt—sericite, Ms—Muscovite.

The quartz-sericite zone (Qs) extends through the biotite granites and their contact zone with the monzogranite porphyries. This zone is rich in alteration minerals such as quartz, sericite (Figure 2b), and carbonate. The ore minerals in this zone include molybdenite, chalcopyrite, malachite, pyrite, and azurite.

The illite-hydromica zone (IH), located within biotite granites on the periphery of the deposit and rhyolites in the northern part, features alteration minerals like illite, hydromuscovite (Figure 2c), carbonate, and kaolinite. This zone also exhibits lead and zinc mineralizations.

4. Sampling and Methodology

4.1. Sampling and Lithofacies Characteristics of Samples

A total of 14 samples were gathered, comprising three monzogranite samples (WNG01, WNG02, and WNG03), four rhyolitic porphyry samples (WNG06, WNG06-1, WNG06-2, and WNG06-3; Figure 3a) from the deposit's northern area, four monzogranite samples (WNG15-1, WNG15-2, WNG15-3, and WNG15; Figure 3b) from the southern portion, and three additional monzogranite samples (WNG18-1, WNG18-2, and WNG18-3) from the deposit's periphery. The specific sampling locations are depicted in Figure 1b.

Figure 3. Field outcrops, hand specimens, and microscopic photos of the monzogranites and rhyolitic porphyries in the Wunugetu deposit. Qz—quartz, Pl—plagioclase, Kfs—feldspar, Srt—sericite. (**a**) Field outcrops and sampling locations of rhyolitic porphyries; (**b**) Sampling location of monzogranite, with fine veins of pyrite and chalcopyrite; (**c**) Monzogranite hand specimens; (**d**) Rhyolitic porphyry hand specimen; (**e**) Microscopic photos of rhyolitic porphyry; (**f**) Microscopic photos of monzogranites.

4.2. Zircon U-Pb Dating

Zircon monomineralic separation, target preparation, cathodoluminescence (CL) imaging, and zircon U-Pb dating were performed at the laboratory of Yandu Zhongshi Testing Technology Co., Ltd. (Beijing, China) The procedures utilized a New Wave UP213 laser ablation system and an M90 inductively coupled plasma mass spectrometer (ICP-MS) manufactured by Analytik Jena AG. Zircon reference materials 91500 and Plesovice employed exhibited the values recommended [26,27], as calculated using ZSkits software ver. 1.1.0. The quantification of trace elements in zircon utilized SRM610 and Si as the external and internal standards, respectively [28]. For U-Pb isotopic dating, zircon reference material 91500 was used as an external standard for isotopic fractionation correction. The U-Pb age concordia diagrams and the weighted average ages of the zircon samples were generated using Isoplot/Exver3 [29]. Additionally, the correction of common lead in U-Pb analyses was performed following Andersen's (2002) method [30].

4.3. In-Situ Lu-Hf Isotopes of Zircon

In-situ zircon Lu-Hf isotopic analyses were performed utilizing a Neptune Plus MC-ICP-MS manufactured by Thermo-Fisher and a New Wave UP213 laser ablation system, with the procedures and calibration methods akin to those proposed by Wu Fuyuan et al. (2007) [31]. Laser ablation was executed for 31 s at a frequency of 8 Hz and an energy density of 16 J/cm^2. Considering the exceptionally low ^{176}Lu/^{177}Hf ratios in zircon, typically below 0.002, the isotopic interference of ^{176}Lu on ^{176}Hf was deemed negligible. The isotopic fractionation factor of Yb was calculated using the average ^{173}Yb/^{172}Yb ratios of various test points, followed by removal of the isobaric interference of ^{176}Yb on ^{176}Hf. The ^{173}Yb/^{172}Yb ratio was established at 1.35274.

4.4. Whole-Rock Major- and Trace-Element Analyses

Complete silicate analysis was performed at the laboratory of Yandu Zhongshi Testing Technology Co., Ltd. Major element analysis utilized the acid dissolution method in conjunction with a Zetium X-ray fluorescence spectrometer (XRF) from PANalytical, with analytical precision errors under 0.1%. Trace element analysis was carried out with an M90 ICP-MS from Analytik Jena AG, with analytical errors maintained below 10%.

5. Results

5.1. Petrography

The rhyolitic porphyries, as stocks, intrude into granite porphyries, characterized by porphyritic textures and blocky structures. Their phenocrysts, making up 10% of the composition, consist of quartz (5%) and feldspar (5%), with particle sizes between 0.4 and 1.4 mm. Their matrix, constituting 90%, is mainly feldspar and quartz. It manifests felsitic textures, with radial or fan-shaped grains scattering locally (Figure 3c,e).

The monzogranites comprise plagioclase (35%), alkali feldspar (30%), and quartz (35%), with consistent mineral grain sizes ranging from 1.6 to 3.0 mm. The plagioclase is hypidiomorphic, tabular, and granular, featuring sericite alterations on their surfaces. In contrast, the alkali feldspar appears as xenomorphic grains (Figure 3d,f).

The monzogranite porphyries are composed of plagioclase (30%), alkali feldspar (30%), quartz (32%) (Figure 3d), sericite (5%), and pyrite (3%). The grains are uniform in size, ranging from 1.6 to 3.0 mm. The plagioclase is subhedral, appearing in tabular and granular shapes, with sericite pseudomorphing its surfaces. The alkali feldspar occurs as xenomorphic grains, comprising perthitic feldspar with irregularly vein-like, spotted, and patchy patterns. It exhibits weak sericitization on surfaces, with sericite replacing along the perthite (Figure 3f). The pyrite occurs mainly as euhedral to anhedral grains distributed among the minerals' intergranular spaces.

5.2. Zircon U-Pb Ages

The LA-ICP-MS zircon U-Pb dating data are shown in Table 1, partial zircon CL images are illustrated in Figure 4a,b, and the concordia diagrams with weighted average age curves are presented in Figure 5a,b. Zircon grains from rhyolitic porphyry sample WNG06 ranged from idiomorphic to hypidiomorphic columnar shapes, with sizes ranging from 50 to 150 μm. These grains were primarily short columnar, followed by long columnar, with length/width ratios from 1:1 to 2.5:1 and minor fissures occasionally visible. Oscillatory zoning was observed in the CL images of these zircon grains. Their Th/U ratios spanned from 0.74 to 1.34, all above 0.1, indicating a magmatic origin for these grains. Analysis of 22 test points on this sample yielded ^{206}Pb/^{238}U ages between 169 and 173 Ma, corresponding to weighted average ages of 170.49 ± 0.81 Ma (MSWD = 0.68).

Table 1. LA-ICP-MS zircon U-Pb dating data of Wunugetu rhyolite porphyry and monzogranites.

Test Point Number	Th ($\times10^{-6}$)	U ($\times10^{-6}$)	Th/U	Isotope Atomic Ratio								Isotopic Age (Ma)							
				$^{207}Pb/^{206}Pb$	1σ	$^{207}Pb/^{235}U$	1σ	$^{206}Pb/^{238}U$	1σ	$^{208}Pb/^{232}Th$	1σ	$^{207}Pb/^{206}Pb$	1σ	$^{207}Pb/^{235}U$	1σ	$^{206}Pb/^{238}U$	1σ	$^{208}Pb/^{232}Th$	1σ
Sample WNG06																			
WNG06-08	801	724	1.11	0.05012	0.00097	0.18207	0.00359	0.02636	0.00023	0.00812	0.00013	200	45	170	3	168	1	163	3
WNG06-26	1007	932	1.08	0.04959	0.00107	0.18035	0.00413	0.02638	0.00029	0.00808	0.00022	176	51	168	4	168	2	163	4
WNG06-21	516	618	0.83	0.05038	0.00126	0.18452	0.00511	0.02653	0.00036	0.00852	0.00038	213	58	172	4	169	2	171	8
WNG06-23	614	689	0.89	0.04959	0.00120	0.18107	0.00453	0.02654	0.00032	0.00795	0.00031	176	57	169	4	169	2	160	6
WNG06-13	722	783	0.92	0.04972	0.00148	0.18326	0.00615	0.02663	0.00035	0.00826	0.00024	182	69	171	5	169	2	166	5
WNG06-20	997	821	1.22	0.05074	0.00137	0.18636	0.00469	0.02664	0.00030	0.00824	0.00038	229	62	174	4	169	2	166	8
WNG06-30	630	692	0.91	0.04925	0.00095	0.18037	0.00333	0.02665	0.00032	0.00867	0.00028	160	45	168	3	170	2	175	6
WNG06-01	607	624	0.97	0.05027	0.00117	0.18590	0.00398	0.02678	0.00026	0.00855	0.00018	208	54	173	3	170	2	172	4
WNG06-22	475	602	0.79	0.04981	0.00140	0.18458	0.00591	0.02679	0.00040	0.00874	0.00042	186	65	172	5	170	3	176	8
WNG06-25	715	727	0.98	0.04887	0.00110	0.18015	0.00403	0.02680	0.00032	0.00860	0.00031	142	53	168	3	170	2	173	6
WNG06-27	826	823	1.00	0.04911	0.00100	0.18165	0.00409	0.02682	0.00031	0.00813	0.00020	153	48	169	4	171	2	164	4
WNG06-16	561	598	0.94	0.05070	0.00120	0.18792	0.00463	0.02682	0.00026	0.00861	0.00026	227	55	175	4	171	2	173	5
WNG06-14	457	570	0.80	0.04917	0.00155	0.18074	0.00574	0.02682	0.00033	0.00859	0.00026	156	74	169	5	171	2	173	5
WNG06-15	502	570	0.88	0.05059	0.00132	0.18667	0.00475	0.02683	0.00033	0.00805	0.00025	222	61	174	4	171	2	162	5
WNG06-09	341	458	0.74	0.05001	0.00172	0.18565	0.00656	0.02686	0.00034	0.00831	0.00021	195	80	173	6	171	2	167	4
WNG06-29	664	701	0.95	0.04942	0.00124	0.18396	0.00511	0.02702	0.00045	0.00880	0.00026	168	59	171	4	172	3	177	5
WNG06-03	589	651	0.90	0.05033	0.00105	0.18768	0.00423	0.02706	0.00029	0.00889	0.00015	210	48	175	4	172	2	179	3
WNG06-05	865	748	1.16	0.05059	0.00094	0.18864	0.00365	0.02708	0.00029	0.00868	0.00013	222	43	175	3	172	2	175	3
WNG06-28	304	411	0.74	0.04884	0.00155	0.18165	0.00558	0.02709	0.00035	0.00857	0.00026	140	74	169	5	172	2	172	5
WNG06-10	789	765	1.03	0.05071	0.00114	0.18990	0.00419	0.02712	0.00025	0.00902	0.00027	228	52	177	4	172	2	181	5
WNG06-02	2099	1566	1.34	0.04953	0.00118	0.18585	0.00464	0.02717	0.00031	0.00864	0.00013	173	56	173	4	173	2	174	3
WNG06-12	804	740	1.09	0.04900	0.00158	0.18321	0.00601	0.02720	0.00040	0.00917	0.00033	148	76	171	5	173	3	184	7
Sample WNG15																			
WNG15-10	667	959	0.63	0.05252	0.00294	0.23892	0.01295	0.03299	0.00046	0.01035	0.00012	308	131	218	11	209	3	208	2
WNG15-27	1137	2309	0.47	0.05220	0.00287	0.23680	0.01276	0.03290	0.00035	0.01033	0.00010	294	128	216	10	209	2	208	2
WNG15-04	569	1121	0.48	0.05198	0.00200	0.23667	0.00878	0.03302	0.00033	0.01037	0.00009	285	90	216	7	209	2	209	2
WNG15-11	91	236	0.36	0.05154	0.00135	0.23341	0.00653	0.03280	0.00043	0.01136	0.00035	265	40	213	5	208	3	228	7
WNG15-18	201	423	0.45	0.05138	0.00148	0.22978	0.00675	0.03239	0.00040	0.01002	0.00035	258	45	210	6	205	2	202	7
WNG15-19	313	586	0.49	0.05129	0.00157	0.23114	0.00799	0.03256	0.00052	0.00972	0.00041	254	50	211	7	207	3	196	8
WNG15-30	209	462	0.42	0.05112	0.00090	0.23118	0.00456	0.03284	0.00034	0.01027	0.00023	246	27	211	4	208	2	207	5
WNG15-29	565	1261	0.43	0.05090	0.00075	0.23156	0.00421	0.03290	0.00030	0.01024	0.00026	236	25	211	3	209	2	206	5
WNG15-08	1453	2266	0.60	0.05084	0.00073	0.23251	0.00365	0.03308	0.00041	0.01070	0.00029	233	17	212	3	210	3	215	6
WNG15-14	233	259	0.85	0.05074	0.00144	0.22922	0.00690	0.03277	0.00040	0.01044	0.00021	229	47	210	6	208	2	210	4
WNG15-17	274	544	0.46	0.05076	0.00113	0.23267	0.00560	0.03316	0.00039	0.01080	0.00038	230	34	212	5	210	2	217	8
WNG15-28	1611	2523	0.60	0.05057	0.00087	0.23200	0.00487	0.03311	0.00043	0.01046	0.00046	221	26	212	4	210	3	210	9
WNG15-05	250	641	0.37	0.05054	0.00092	0.23153	0.00511	0.03309	0.00032	0.01065	0.00041	220	33	211	4	210	2	214	8
WNG15-12	443	1048	0.39	0.05050	0.00137	0.23097	0.00583	0.03313	0.00038	0.01092	0.00037	218	37	211	5	210	2	220	7
WNG15-24	425	1234	0.32	0.05048	0.00114	0.23252	0.00509	0.03332	0.00039	0.01015	0.00047	217	30	212	4	211	2	204	9
WNG15-16	1041	1784	0.56	0.05028	0.00068	0.22845	0.00376	0.03288	0.00032	0.01002	0.00031	208	21	209	3	209	2	202	6
WNG15-09	2274	2216	0.97	0.04959	0.00399	0.22341	0.01758	0.03267	0.00055	0.01032	0.00015	176	183	205	15	207	3	208	3
WNG15-06	339	481	0.67	0.04950	0.00086	0.22857	0.00460	0.03340	0.00027	0.01013	0.00027	171	32	209	4	212	2	204	5

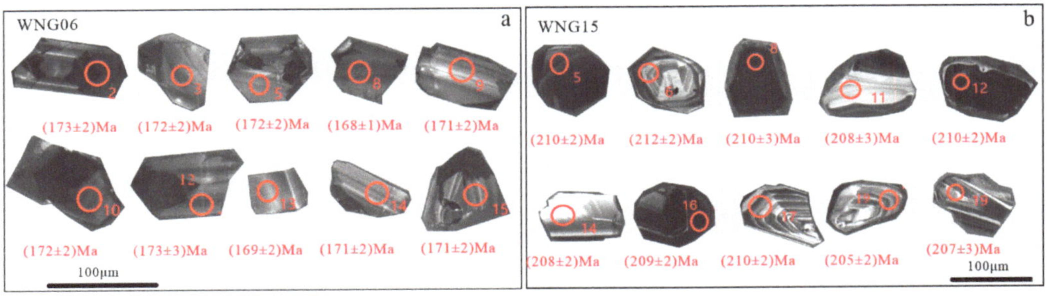

Figure 4. Cathodoluminescence images of zircons in the Wunugetu deposit. (**a**) Rhyolite Porphyry; (**b**) Monzogranite.

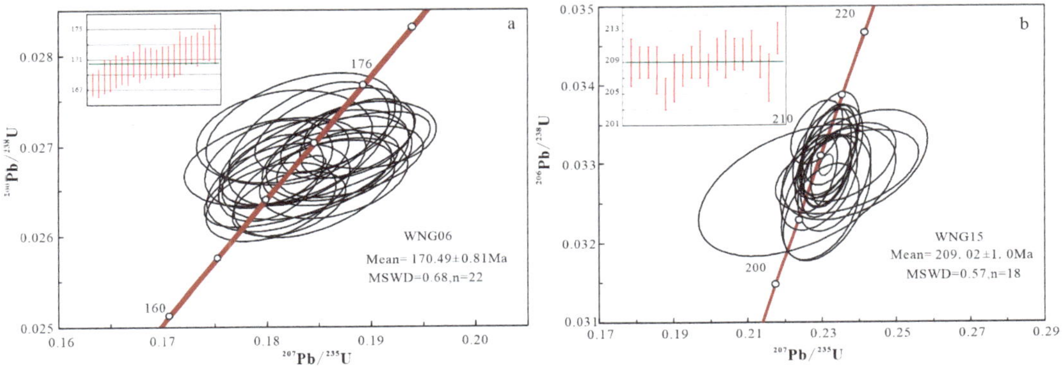

Figure 5. Zircon LA-ICP-MS U-Pb concordia diagrams from Wunugetu copper–gold deposit. (**a**) Rhyolitic porphyries; (**b**) Monzogranite.

Zircon grains from monzogranite sample WNG15 displayed shapes from idiomorphic to hypidiomorphic columnar, with sizes varying between 50 to 200 μm. These grains were predominantly short columnar and secondarily long columnar, with length/width ratios ranging between 1:1 and 3:1. Oscillatory zoning was apparent in the CL images of the zircon grains. The Th/U ratios of the zircons spanned from 0.32 to 0.97, all above the 0.1 threshold, indicating a magmatic origin. Analysis of 18 test points on this sample revealed $^{206}Pb/^{238}U$ ages between 205 to 218 Ma, with weighted average ages of 209.02 ± 1.0 Ma (MSWD = 0.57).

5.3. In-Situ Lu-Hf Isotopes of Zircon

The in-situ Lu-Hf isotopic analysis results of zircon are summarized in Table 2. The Hf isotopic data were calculated using U-Pb ages obtained from the same zircon test points. Zircon grains from samples WNG06 and WNG15 exhibited $^{176}Hf/^{177}Hf$ ratios within the ranges of 0.282742–0.282874 and 0.282658–0.282784, respectively, corresponding to εHf(t) values ranging from +2.4 to +10.8 for WNG06 and from +0.3 to +4.6 for WNG15. The derived two-stage model ages (T_{DM2}) for these samples were between 727 to 1058 Ma for WNG06 and 949 to 1227 Ma for WNG15.

Table 2. Zircon Hf isotope analysis results of Wunugetu rhyolite porphyry and monzogranites.

Sample	Age (Ma)	$^{176}Yb/^{177}Hf$	2σ	$^{176}Lu/^{177}Hf$	2σ	$^{176}Hf/^{177}Hf$	2σ	ε_{Hf} (0)	ε_{Hf} (t)	T_{DM1} (Ma)	T_{DM2} (Ma)	$f_{Lu/Hf}$
WNG06-002	173	0.092751	0.000975	0.002781	0.000029	0.282836	0.000023	2.3	5.7	620	850	−0.92
WNG06-003	172	0.097305	0.001427	0.002702	0.000047	0.282847	0.000025	2.7	6.1	601	823	−0.92
WNG06-005	172	0.084428	0.000322	0.002441	0.000032	0.282742	0.000026	−1.1	2.4	751	1058	−0.93
WNG06-008	168	0.087212	0.001147	0.002416	0.000046	0.282782	0.000021	0.4	3.8	691	970	−0.93
WNG06-009	171	0.081093	0.001396	0.002398	0.000048	0.282865	0.000025	3.3	6.8	569	781	−0.93
WNG06-010	172	0.073125	0.000371	0.002032	0.000028	0.282855	0.000020	3.0	6.5	578	800	−0.94
WNG06-012	173	0.099545	0.000903	0.002832	0.000046	0.282858	0.000027	3.0	6.5	587	800	−0.91
WNG06-013	169	0.093224	0.000430	0.002751	0.000027	0.282788	0.000027	0.6	4.0	690	959	−0.92
WNG06-014	171	0.077629	0.000628	0.002241	0.000018	0.282889	0.000027	4.1	7.6	532	727	−0.93
WNG06-015	171	0.083756	0.000953	0.002540	0.000047	0.282874	0.000028	3.6	7.1	558	762	−0.92
WNG15-005	210	0.062887	0.001353	0.002038	0.000049	0.282684	0.000023	−3.1	1.2	828	1166	−0.94
WNG15-006	212	0.081818	0.002085	0.002546	0.000050	0.282658	0.000020	−4.0	0.3	877	1227	−0.92
WNG15-008	210	0.110316	0.000582	0.003513	0.000064	0.282673	0.000028	−3.5	0.6	878	1202	−0.89
WNG15-011	208	0.047510	0.003141	0.001509	0.000117	0.282732	0.000023	−1.4	3.0	746	1053	−0.95
WNG15-012	210	0.097143	0.001717	0.003026	0.000081	0.282774	0.000022	0.1	4.3	716	972	−0.91
WNG15-014	208	0.092412	0.002386	0.002985	0.000046	0.282784	0.000020	0.4	4.6	700	949	−0.91
WNG15-016	209	0.064028	0.001103	0.001971	0.000037	0.282751	0.000016	−0.7	3.6	729	1015	−0.94
WNG15-017	210	0.085413	0.000815	0.002533	0.000018	0.282736	0.000019	−1.3	3.0	762	1052	−0.92
WNG15-018	205	0.076844	0.000813	0.002362	0.000023	0.282747	0.000015	−0.9	3.3	742	1028	−0.93
WNG15-019	207	0.070343	0.000818	0.002212	0.000025	0.282740	0.000019	−1.1	3.1	750	1043	−0.93

5.4. Whole-Rock Major and Trace Element Analyses

The analytical outcomes for major, trace, and rare earth elements (REEs) across 12 rock samples are compiled in Table 3. Regarding major elements, these samples contained 70.16–80.47 wt.% SiO_2, 10.98–14.37 wt.% Al_2O_3 (rhyolite), 0.17–3.77 wt.% Na_2O, 2.95–9.36 wt.% K_2O, 0.06–0.99 wt.% CaO, 0.14–0.37 wt.% MgO, and 3.42–10.10 wt.% Na_2O + K_2O, indicating high silicon content and depletion of sodium generally. The total alkali–silica (TAS) diagram shows that most samples fell within the granite field (Figure 6), in line with petrographic results.

Table 3. Analysis results of major elements (wt.%), rare earth elements, and trace elements (ppm) in Wunugetu rhyolite porphyry and monzogranites.

No.3	W06-1	W06-2	W06-3	W01	W02	W03	W15-1	W15-2	W15-3	W18-1	W18-2	W18-3
Rock Type	Porphyry Rhyolite			Monzogranites								
					Major element (wt.%)							
SiO_2	76.35	75.77	77.45	73.44	73.16	80.47	75.36	76.69	77.71	74.40	75.13	74.92
Al_2O_3	12.86	12.87	12.72	14.25	14.37	11.38	13.01	12.89	10.98	13.52	13.24	12.81
Na_2O	0.23	0.23	0.23	0.74	0.74	0.17	0.22	0.30	1.59	3.77	3.55	3.73
K_2O	5.91	5.33	5.00	9.36	8.24	3.25	5.05	7.15	5.09	3.94	4.14	2.95
CaO	0.09	0.12	0.09	0.07	0.11	0.06	0.06	0.07	0.15	0.99	0.93	0.57
P_2O_5	0.04	0.04	0.04	0.09	0.08	0.05	0.05	0.06	0.05	0.09	0.08	0.07
TiO_2	0.05	0.05	0.05	0.17	0.17	0.07	0.10	0.09	0.16	0.21	0.18	0.22
MgO	0.17	0.18	0.17	0.17	0.26	0.34	0.34	0.14	0.37	0.27	0.25	0.23
MnO	0.19	0.28	0.11	0.03	0.02	0.03	0.04	0.03	0.04	0.04	0.04	0.03
TFe_2O_3	1.29	1.93	1.03	0.54	0.84	1.44	2.29	0.51	1.34	1.43	1.32	2.89
LOI	2.34	2.82	2.49	0.77	1.38	2.11	2.89	1.46	1.91	0.73	0.74	1.39
TOTAL	99.51	99.63	99.38	99.63	99.36	99.37	99.40	99.38	99.38	99.39	99.58	99.81
					Trace element (ppm)							
Rb	62.98	121.78	113.08	117.95	113.48	77.48	146.83	138.72	171.18	158.50	166.50	122.31
Ba	203.79	225.28	839.21	757.42	606.85	303.08	696.06	1041.38	1112.03	604.78	551.03	916.73
Th	2.77	6.09	5.93	15.38	20.62	13.02	19.03	23.97	11.00	13.17	10.23	20.39
U	0.25	0.80	0.58	2.49	2.38	14.72	4.12	7.38	1.62	2.70	2.84	3.21
Ta	0.27	0.51	0.51	1.93	1.88	2.31	1.65	1.00	0.76	1.29	1.08	0.82
Nb	4.05	8.34	7.92	19.01	17.78	13.71	14.18	13.20	12.22	18.78	14.98	12.12
Sr	23.30	50.70	68.98	130.77	109.44	7.36	28.35	80.05	126.96	167.97	157.60	214.75
Zr	24.93	56.69	53.11	109.40	92.35	81.56	95.83	148.81	162.38	89.83	114.23	201.24

Table 3. *Cont.*

No.3	W06-1	W06-2	W06-3	W01	W02	W03	W15-1	W15-2	W15-3	W18-1	W18-2	W18-3
Rock Type	**Porphyry Rhyolite**			**Monzogranites**								
Hf	1.33	2.56	2.48	4.19	3.50	3.63	3.88	4.74	4.59	3.38	3.93	5.30
La	7.21	15.25	16.25	23.06	32.07	29.87	36.42	46.27	37.12	10.74	10.88	54.28
Ce	15.21	32.38	34.44	46.74	64.09	60.42	69.61	82.46	70.73	20.56	20.21	74.95
Pr	1.77	4.10	4.44	5.47	7.29	6.92	8.00	9.15	8.63	2.93	3.28	7.93
Nd	5.76	13.86	15.31	17.99	23.33	22.66	26.14	32.66	32.29	10.51	11.63	23.94
Sm	0.99	2.39	2.63	3.47	4.18	3.91	4.13	5.19	5.83	2.65	2.87	3.26
Eu	0.18	0.41	0.56	0.65	0.67	0.74	0.74	0.94	1.12	0.57	0.57	0.78
Gd	0.78	1.95	1.99	3.56	4.66	4.10	4.04	5.07	5.53	2.98	2.81	3.09
Tb	0.10	0.24	0.21	0.59	0.79	0.70	0.56	0.66	0.83	0.53	0.48	0.35
Dy	0.46	1.16	0.92	3.68	5.03	4.54	3.01	3.62	4.79	3.52	2.95	1.74
Ho	0.09	0.23	0.19	0.80	1.15	1.03	0.65	0.77	0.97	0.80	0.64	0.39
Er	0.28	0.69	0.58	2.54	3.59	3.30	1.99	2.46	2.98	2.52	2.02	1.30
Tm	0.04	0.10	0.08	0.43	0.60	0.59	0.35	0.41	0.50	0.45	0.36	0.22
Yb	0.17	0.51	0.40	2.77	4.30	4.33	2.41	2.75	3.30	2.89	2.46	1.56
Lu	0.04	0.10	0.09	0.49	0.63	0.67	0.42	0.49	0.53	0.49	0.41	0.28
Y	2.50	6.25	5.12	21.35	30.48	29.67	18.14	21.48	26.15	22.18	17.81	10.26
ΣREE	33.08	73.38	78.07	112.24	152.37	143.79	158.48	192.91	175.14	62.13	61.57	174.06
LREE	31.13	68.39	73.63	97.39	131.62	124.53	145.03	176.67	155.71	47.96	49.44	165.14
HREE	1.95	4.99	4.44	14.85	20.76	19.27	13.45	16.24	19.43	14.16	12.13	8.92
LREE/HREE	15.96	13.70	16.59	6.56	6.34	6.46	10.78	10.88	8.01	3.39	4.08	18.51
La_N/Yb_N	31.23	21.48	29.37	5.97	5.35	4.95	10.82	12.07	8.07	2.67	3.17	25.01
δEu	0.64	0.58	0.75	0.57	0.46	0.57	0.55	0.56	0.60	0.62	0.62	0.75
δCe	1.05	1.00	0.99	1.02	1.03	1.03	1.00	0.98	0.97	0.90	0.83	0.89

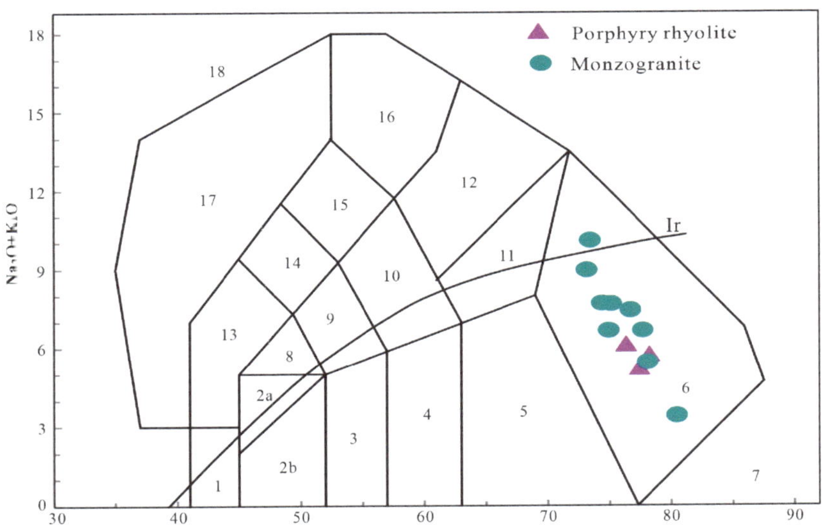

Figure 6. TAS discrimination diagram of igneous rocks in the Wunugetu deposit. Ir—Irvine boundary [32]; above is alkaline, below is subalkaline. 1—Peridotgabbro; 2a—Alkali Gabbro; 2b—Subalkaline Gabbro; 3—Gabbroic Diorite; 4—Diorite; 5—Granodiorite; 6—Granite; 7—Quartzolite; 8—Monzogabbro; 9—Monzodiorite; 10—Monzosyenite; 11—Quartz Monzonite; 12—Syenite; 13—Foid Gabbro; 14—Foid Monzodiorite; 15—Foid Monzosyenite; 16—Foidolite Syenite; 17—Foidolite; 18—Tawite/Urtite/Italite.

Regarding major elements, the monzogranite samples contained 73.16–80.47 wt.% SiO_2, 10.98–14.37 wt.% Al_2O_3, 0.17–3.77 wt.% Na_2O, and 2.95–9.36 wt.% K_2O. Notably, samples WNG01, WNG02, and WNG15-2 exhibited elevated K_2O levels, suggesting the potential influence of potassic feldsparization within the deposit. The CaO varied from 0.06 to 0.99 wt.%, MgO from 0.14 to 0.37 wt.%, and total alkalis (Na_2O + K_2O) from 3.42 to 10.10 wt.%, indicating high silica and low sodium. All the samples fell within the granite field on the TAS diagram (Figure 6).

The rhyolitic porphyry samples revealed SiO_2 between 75.77 wt.% and 77.45 wt.%, Al_2O_3 from 10.98 to 14.37 wt.%, Na_2O at 0.23 wt.%, K_2O from 5.00 to 5.91 wt.%, CaO between 0.09 and 0.12 wt.%, MgO from 0.17 to 0.18 wt.%, and Na_2O + K_2O from 5.23 to 6.14 wt.%, also demonstrating high silica and low sodium traits. These samples fell within the granite field on the TAS diagram (Figure 6).

In the K_2O–SiO_2 diagram (Figure 7a), the samples fell within the high-K calc-alkaline to shoshonitic series zones, with one point in the calc-alkaline series zone, indicating the rhyolite and monzogranite samples' classification belongs to these series. All rock samples, with aluminum saturation index (A/CNK) values from 1.1 to 2.9, fell within the peraluminous zone on the A/NK vs. A/CNK diagram (Figure 7b).

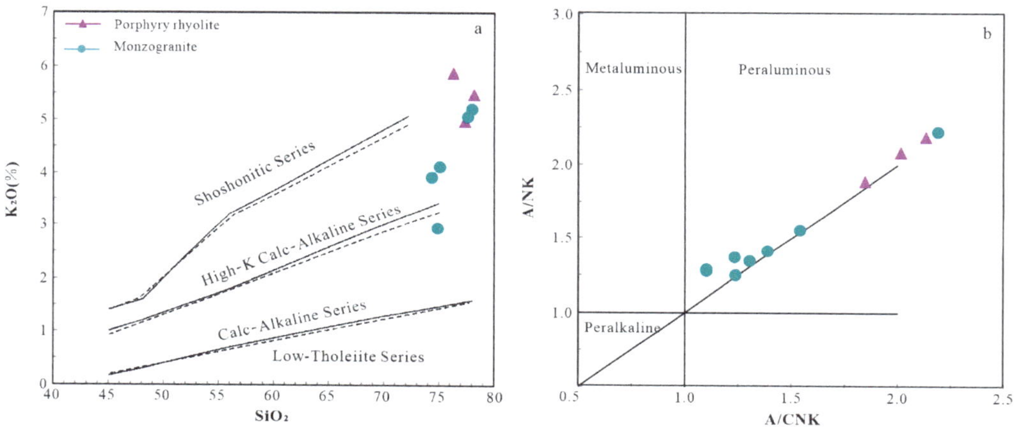

Figure 7. SiO_2–K_2O (**a**) [33] and A/CNK–A/NK (**b**) [34] diagrams of monzogranites and rhyolitic porphyries in the Wunugetu deposit.

5.5. Rare Earth and Trace Elements

The rhyolite samples from the Wunugetu deposit showed total REE (ΣREE) contents ranging from 33.08 ppm to 78.07 ppm, LREE/HREE ratios between 13.70 and 15.96, and $(La/Yb)_N$ ratios from 21.48 to 31.23, indicating a general enrichment in light rare earth elements (LREEs) and a relative depletion in heavy rare earth elements (HREEs). Their chondrite-normalized REE patterns were rightward curves, with negative Eu anomalies (Eu/Eu*= 0.58–0.75; Figure 8a; Table 3).

The monzogranite samples from the Wunugetu deposit yielded ΣREE contents from 61.57 ppm to 192.91 ppm, LREE/HREE ratios from 3.39 to 18.51, and $(La/Yb)_N$ ratios from 2.67 to 25.01, showing an enrichment in LREEs. Their chondrite-normalized REE patterns were also rightward curves, indicative of negative Eu anomalies (Eu/Eu* = 0.46–0.75; Figure 8a; Table 3).

As revealed by the primitive mantle-normalized profiles, both rhyolitic porphyry and monzogranite samples exhibited similar patterns, both showing significant enrichment in LILEs such as Rb, K, Ba, and elements like Th, U, La, but depletion in HFSEs like Nb, P, and Ti (Figure 8b; Table 3). Both rock types had relatively low Sm/Nd ratios (rhyolitic porphyry: 0.17; monzogranite: 0.14 to 0.19, averaging 0.19), Rb/Sr ratios (rhyolitic porphyry: 1.64 to 2.70, averaging 2.24; monzogranite: 0.90 to 10.53, averaging 2.59), and Rb/Ba ratios (rhyolitic porphyry: 0.13 to 0.54, averaging 0.33; monzogranite: 0.13 to

0.30, averaging 0.20), along with higher K/Rb ratios (rhyolitic porphyry: 363.32 to 778.55, averaging 503.03; monzogranite: 200.48 to 658.60, averaging 353.81).

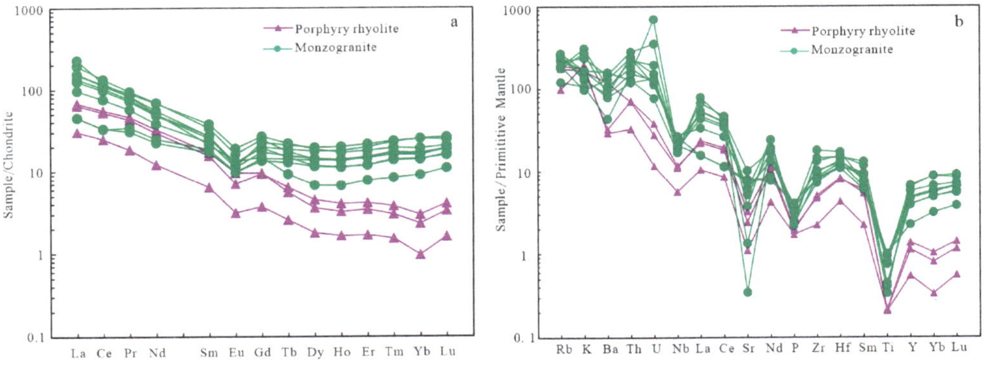

Figure 8. Chondrite-normalized REE patterns (**a**) [35] and spidergram of primitive mantle-normalized REEs (**b**) [36] of monzogranite and rhyolitic porphyries in the Wunugetu deposit.

6. Discussion

6.1. Diagenetic and Metallogenic Epochs

Mao Jingwen et al. (2005) proposed that metallization in eastern China predominantly occurred during the periods of 180–188 Ma, 135–144 Ma, 115–127 Ma, and 94–105 Ma [37]. Drawing from extensive isotopic dating data, Xu Wenliang et al. (2013) identified six distinct phases of mineralization-related volcanic activity in Northeast China: 220–280 Ma, 173–190 Ma, 158–166 Ma, 138–145 Ma, 106–133 Ma, and 88–97 Ma [38]. In the eastern Xing'an-Mongolia orogenic belt, principal metallization events are dated to 165–195 Ma and 110–115 Ma [39]. The Erguna block, a critical segment of the eastern CAOB, underwent notably intense Mesozoic magmatic episodes, categorized into seven stages: ~246 Ma (Late Permian to early Middle Triassic), ~225 Ma (late Middle Triassic to early Late Triassic), ~205 Ma (late Late Triassic to early Early Jurassic), ~185 Ma (Early Jurassic), ~155 Ma (Late Jurassic), ~137 Ma (early Early Cretaceous), and ~125 Ma (late Early Cretaceous) [40]. This block has revealed considerable porphyry copper mineralization, with several copper–molybdenum deposits like Badaguan, Taipingchuan, Wunugetu, and Naomingte in Mongolia being uncovered. Specifically, the Badaguan deposit's quartz porphyries and granodiorite porphyries exhibit weighted average zircon U-Pb ages of 217.6 ± 2.6 Ma and 206.5 ± 1.6 Ma, respectively, while its molybdenite shows Re-Os ages of 222.4 ± 3.3 Ma [41,42]. The Taipingchuan deposit's ore-forming granodiorite porphyries have average zircon U-Pb ages of 202.0 ± 5.7 Ma, with molybdenite revealing an average Re-Os age of 203 Ma [43]. In Mongolia's Naomingte deposit, the ore-hosting granodiorites demonstrate a diagenetic age of 166 Ma. These findings suggest that the porphyry copper mineralization in the Erguna block primarily occurred from the Late Triassic to the Middle Jurassic, consistent with the Mesozoic magmatic activity timelines proposed by the aforementioned researchers.

The ore bodies in the Wunugetu deposit are primarily found within granite porphyries, monzogranites, quartz monzogranite porphyries, and biotite monzogranites, indicating a strong linkage with these rock types. Prior studies, leveraging isotopic geochronology, have suggested that the ore bodies and their encompassing rock masses share similar formation periods, highlighting Early Jurassic diagenesis and mineralization processes [13,15,44]. Contrarily, this study reveals that an ore-hosting monzogranite sample, WNG15, produced LA-ICP-MS zircon U-Pb ages of 209.0 ± 1.0 Ma, pointing to a late Late Triassic formation. Additionally, rhyolitic porphyries located in the northern part of the deposit, which intrude into the ore bodies without causing mineralization or alteration, implying that they are dykes intruding in the late mineralization stage. These rhyolitic porphyries have shown LA-ICP-MS zircon U-Pb ages of 170.49 ± 0.81 Ma, indicating their formation during the early Middle Jurassic. The diverse ore features within the Wunugetu deposit imply a multi-stage mineralization process. Molybdenum mineralization predominantly occurs in joints and fissures, with the Re content in molybdenite being influenced by various factors such as lithology, paragenetic minerals, and temperature conditions [44,45]. Consequently, the molybdenite age may not precisely reflect the copper mineralization epoch. Considering the petrogenetic ages of monzogranite and rhyolitic porphyry in the deposit, along with chronological testing data on these rock masses obtained by previous researchers, the authors propose that the magmatic activity at Wunugetu Mountain spanned from 209.0 Ma to 170.5 Ma. When combined with the petrogenetic age range from 180 to 188 Ma for the ore-hosting monzogranitic porphyry in the deposit [15,17] and isochron ages of 177.4 to 179.8 Ma for the deposit's molybdenite [17,44], it can be inferred that the ore-bearing rock-bodies in the Wunugetu region were formed from the Late Triassic to the Early Jurassic, with the mineralization process predominantly occurring during the Early Jurassic.

6.2. Origins of Rocks and Source Characteristics

The genetic types of magmatic rocks are pivotal for interpreting the tectonic setting related to their origin and the characteristics of their magma sources. By examining mineral compositions and elemental geochemical traits, alongside considerations such as geotectonic context, variances in magma sources, and magma evolution, it is well-recognized that granitic rocks can be categorized into M, I, S, and A types [31,46,47]. In the case of the Wunugetu deposit under study, a microscopic analysis of monzogranite samples reveals the absence of key aluminous indicator minerals like cordierite, garnet, and muscovite, ruling out their classification as S-type granites [48]. The granitic rocks within the deposit and its vicinity are characterized by high silicon levels, potassium enrichment, and relatively high Th content. M-type granites, known to result from the remelting of the oceanic crust under island arcs, typically have mantle-derived magma sources, thus featuring low K_2O and Th contents [49,50], a stark contrast to the Late Triassic to Middle Jurassic rocks found in the Wunugetu deposit. Additionally, no contemporaneous M-type granites have been identified in the study area, indicating that the Mesozoic volcanic and magmatic rocks of the Wunugetu deposit are not M-type magmatic rocks. On the discrimination diagrams for magma origin (Figure 9a–c), the samples predominantly cluster around the I- and S-type granite zones, with only a couple aligning with the A-type granite category. Nonetheless, highly differentiated I- and S-type granitic samples with SiO_2 content exceeding 72% share similar geochemical characteristics to A-type magmatic rocks [47,51].

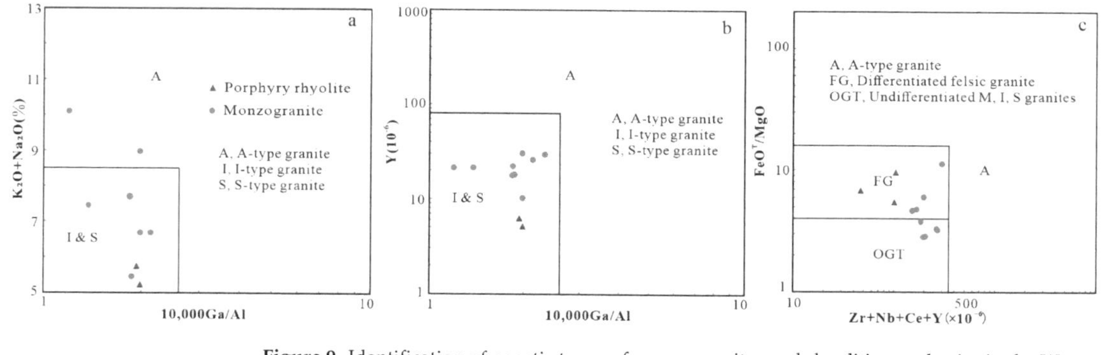

Figure 9. Identification of genetic types of monzogranites and rhyolitic porphyries in the Wunugetu deposit [47]. (**a**) $K_2O + N_2O$–Ga/Al diagram; (**b**) Y–Ga/Al diagram; (**c**) FeO^T/MgO–Zr + Nb + Ce + Y diagram.

The monzogranites within the Wunugetu deposit are characterized by high silicon content (73.16–80.47 wt.%), relatively low aluminum content (10.98–14.37 wt.%), enrichment in alkalis (3.42–10.10 wt.%), and depletion in magnesium and sodium, with A/CNK ratios (Al_2O_3/(CaO + Na_2O + K_2O) ranging from 1.1 to 2.9. On the A/NK–A/CNK diagram, the major element composition classifies these monzogranites as peraluminous to quasi-aluminous rocks. Additionally, these rocks, with low FeO^T, MgO, and CaO contents, are enriched in Rb and Th and depleted in Ba, P, and Ti (Figure 8b), indicative of significant crystallization and differentiation from primitive magmas [52,53]. The monzogranites also display low Zr + Nb + Ce + Y content (151.3 ppm–298.6 ppm, average: 198.6 ppm) and 10,000 × Ga/Al ratios (1.20–2.33, average: 2.07), sharply contrasting with the typical characteristics of A-type granitic magmatic rocks (Zr + Nb + Ce + Y content: >400 ppm; 10,000 × Ga/Al ratios: >2.6). As illustrated in the Zr + Nb + Ce + Y–FeO^T/MgO diagram (Figure 9c), most samples fell into the highly differentiated magma zone, indirectly confirming that the Wunugetu deposit's granitic rocks are highly differentiated magmatic rocks. Despite A/CNK ratios from 1.1 to 2.9 suggesting peraluminous to quasi-aluminous nature in the A/NK–A/CNK diagram, previous studies have substantiated that I-type magmatic rocks can exhibit peraluminous and even strongly peraluminous features [54,55]. As revealed by experiments, apatite exhibits low solubility and is prone to saturate undergoing fractional crystallization during magma evolution in I-type felsic magmas, with the P_2O_5 content diminishing with an increase in the SiO_2 content in rocks. Conversely, S-type granites have high apatite solubility, with P_2O_5 content increasing with the SiO_2 content [56]. Thus, the correlation between SiO_2 and P_2O_5 contents serves as a critical indicator for determining the type of granitic magma. As shown in the SiO_2–P_2O_5 diagram (Figure 10a), the magmatic rock samples from the Wunugetu deposit exhibited low P_2O_5 content (0.04%–0.09%) and a negative P_2O_5–SiO_2 correlation, classifying them as I-type granites. For S-type felsic magmas, Th and Y contents inversely correlate with Rb content, but in I-type magmas, they increase with Rb content [57]. Therefore, the evolutionary trends of Th and Y in magmas can be used to distinguish between S- and I-type granites. The positive correlation of the Th and Y contents with Rb content in the Rb-Y (Figure 10b) and Rb-Th (Figure 10c) diagrams indicates that the deposit's magmatic rocks are characteristic of I-type granites. The aforementioned microscopic observations reveal the absence of aluminum-rich silicate minerals like muscovite, cordierite, and garnet in monzogranites. All these lead to the conclusion that the Late Triassic magmatic rocks of the Wunugetu deposit are highly differentiated I-type rocks.

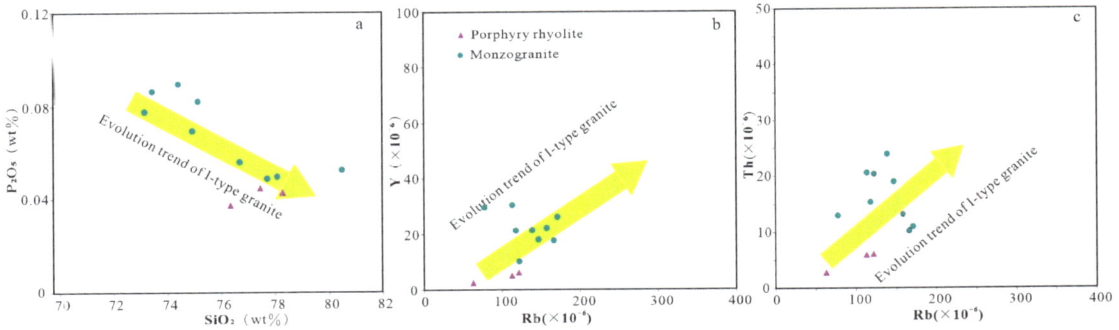

Figure 10. Diagrams of (**a**) SiO_2-P_2O_3, (**b**) of Rb-Y, (**c**) and Rb-Th of monzogranites and rhyolitic porphyries in the Wunugetu deposit [57].

The Middle Jurassic rhyolitic porphyry samples from the deposit all fell into the I- and S-type granite zones in the 10,000 × Ga/Al vs. Na_2O + K_2O, 10,000 × Ga/Al vs. Y, and Zr + Nb + Ce + Y vs. FeO^T/MgO diagrams. The Rb-Y (Figure 10b) and Rb-Th (Figure 10c) diagrams, where Th and Y contents show a positive correlation with the Rb content, suggest that the rhyolite porphyries also have the geochemical characteristics of I-type granites.

These Middle Jurassic rhyolitic porphyries are characterized by high silica (75.77–77.45 wt.%), elevated potassium (5.005.91 wt.%), and an A/CNK ratio of 1.85 to 2.14, all above the 1.1 threshold. Their chondrite-normalized REE patterns exhibit pronounced HREE depletion, hinting at the potential residual effects of garnet. Positioned within the I- and S-type granite zones in geochemical diagrams like (10,000 × Ga/Al) vs. (Na_2O + K_2O), (10,000 × Ga/Al) vs. Y, and (Zr + Nb + Ce + Y) vs. (FeO^T/MgO) (Figure 10b,c), the samples also displayed an increase in P_2O_5 with SiO_2 in the SiO_2 vs. P_2O_5 diagram (Figure 10a), indicating that the rhyolite porphyries in the study area exhibit geochemical traits of S-type granites.

In-situ Lu-Hf isotopic analysis of zircon from these samples revealed εHf(t) values between +2.4 and +7.6. Moreover, all these samples fell within the zone of the εHf(t) range for CAOB's felsic rocks as shown in the t vs. εHf(t) diagram (Figure 11a). Their T_{DM2} ages ranged from 727 to 1058 Ma, suggesting the rhyolitic porphyries' primitive magmas originated from the partial melting of the Meso- to Neoproterozoic accreted lower crust.

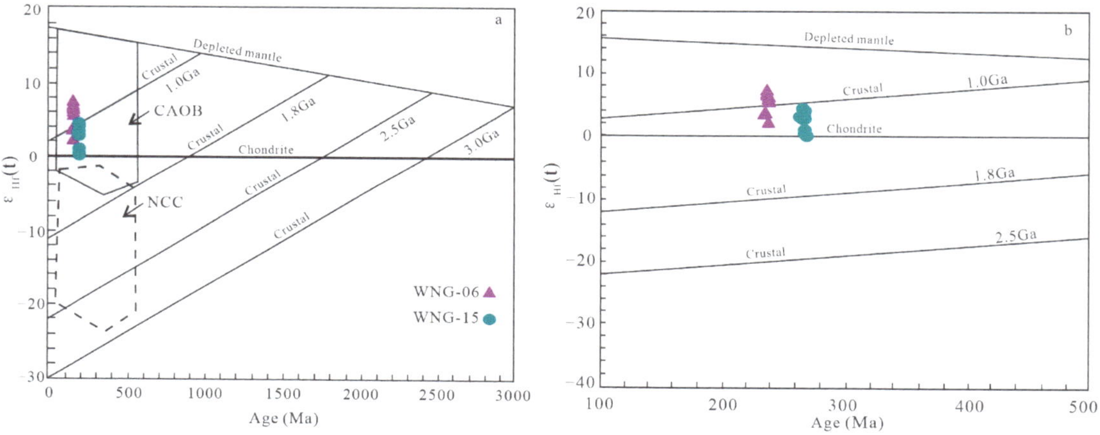

Figure 11. Hf isotopic characteristics (**a**), after Yang et al., 2006 [58] and εHf(t) vs. U-Pb ages for zircons (**b**) of monzonites and rhyolitic porphyries from the Wunugetu deposit.

The Late Triassic monzogranite samples from the Wunugetu deposit yielded zircon εHf(t) values ranging from +0.3 to +4.6. On the t-εHf(t) diagram (Figure 11a), these samples clustered around the chondrite zone, displaying εHf(t) values aligned with those of CAOB's felsic rocks. This suggests that the study area is a part of the CAOB. The T_{DM2} ages of these samples spanned 949 to 1227 Ma, indicating that the original magma of the rhyolite porphyries in the Wunugetu deposit is derived from partial melting of the Meso- to Neoproterozoic accreted lower crust.

In summary, both the Late Triassic I-type granites and the Early Jurassic rhyolitic porphyries from the Wunugetu deposit demonstrate positive zircon εHf(t) values. These values show an increasing trend with a decrease in the emplacement age, suggesting that the magmatic source for the study area's monzogranites is derived from the partial melting of the Mesoproterozoic accreted lower crust.

6.3. Diagenetic and Metallogenic Tectonic Settings

The Wunugetu deposit, nestled within the eastern CAOB and the central MOSZ), is distinguished as the largest porphyry deposit in the Kherlen (Mongolia)-Erguna (China) block. Delving into the tectonic setting of rocks within this deposit is crucial for deepening our understanding of the MOSZ's tectonic evolution and contributes to the exploration of the regional geological settings of porphyry copper mineralization and the tectonic evolution of the eastern CAOB. Previous studies on the evolution of the Mongol-Okhotsk Ocean generally indicate that this ocean evolved from the Eopaleozoic to the Mesozoic [59], undergoing bidirectional subduction toward the north and the south [22,60].

All samples of the Late Triassic magmatic rocks from the Wunugetu deposit fell within the volcanic arc-syn-collisional granite zone on the Y vs. Nb tectonic discrimination diagram (Figure 12a), with a majority situated within the volcanic arc granite zone on the Y + Nb − Rb diagram (Figure 12b). These Late Triassic magmatic rocks display the characteristics of I-type granites. It is believed that the highly differentiated I-type granites were formed in a subduction environment [61] or during the post-orogenic phase [62]. Considering their geochemical similarities with coeval magmatic rocks exposed in the northern Erguna block [40], this study posits that the study area was in an active continental margin environment subjected to subduction during the Late Triassic. This is associated with the southward subduction of the Mongol-Okhotsk Ocean sandwiched between the Neopaleozoic Siberian and Sino-Korean-Mongolian plates.

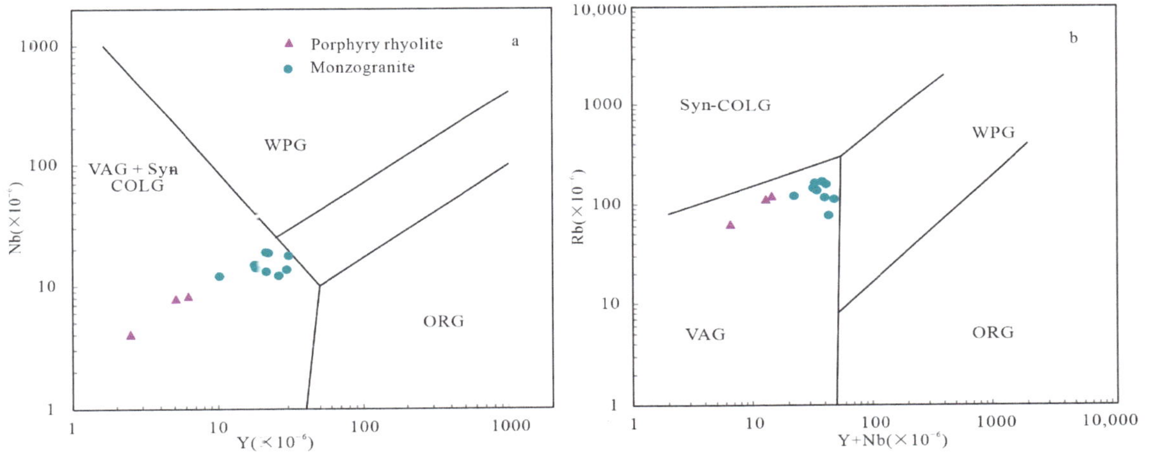

Figure 12. Diagrams of monzonites and rhyolitic porphyries from the Wunugetu deposit. (**a**) Y vs. Nb; (**b**) Y + Nb vs. Rb [63].

As illustrated by the Y-Nb tectonic discrimination diagram (Figure 12a), all samples of the Middle Jurassic rhyolitic porphyries from the Wunugetu deposit fall within the volcanic arc-syn-collisional granite zone. As shown in the Y + Nb vs. Rb (Figure 12b) and tectonic discrimination diagrams, these samples occupied the volcanic arc granite zone. Recent studies have unveiled a suite of Middle Jurassic (168 Ma) muscovite granites in the Sunwu area, adjacent to the study area. These granites originated from the partial melting of the thickened continental crust, resulting from the continent-continent collision during the closure of the Mongol-Okhotsk Ocean [64]. Additionally, the geochemical traits of the S-type granitic rhyolitic porphyry within the deposit suggest their formation in a continent-continent collisional context.

In summary, the geochemical characteristics of the monzogranites and rhyolitic porphyries in the Wunugetu deposit suggest that the Mongol-Okhotsk Ocean was still subjected to subduction during the Late Triassic. By the Middle Jurassic, following the ocean's closure, the ocean transitioned to a syn-collisional tectonic setting (Figure 13).

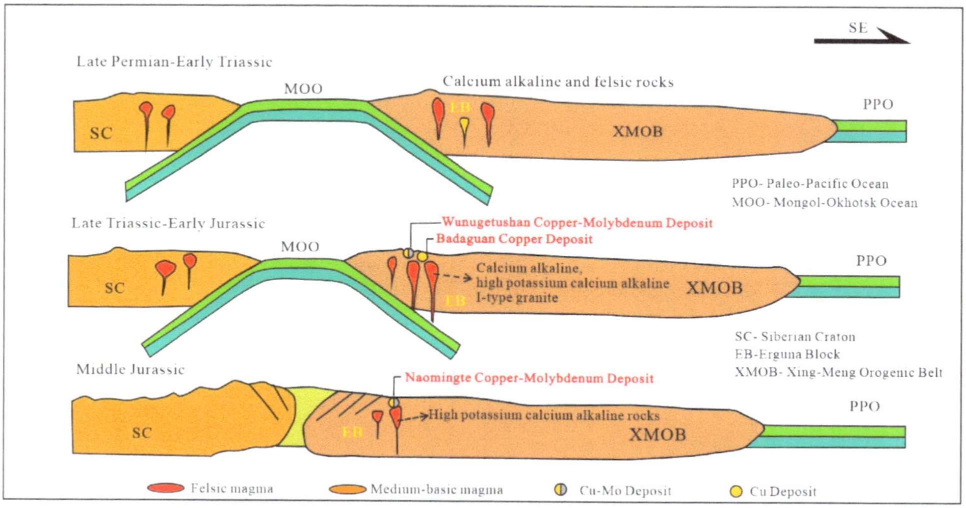

Figure 13. Diagram showing the evolutionary model of mineralization in the Mongolian-Okhotsk Ocean [65].

7. Conclusions

(1) The formation age of the monzogranites in the Wunugetu deposit is established at 209.02 ± 1.0 Ma, while the rhyolitic porphyries are dated at 170.49 ± 0.81 Ma. This indicates that the deposit's magmatic activity primarily spanned from the Late Triassic to the Middle Jurassic, with the major ore-bearing intrusions dating from the Late Triassic to the Early Jurassic.

(2) The Triassic monzogranites from the Wunugetu deposit exhibit the geochemical characteristics of I-type granites, including high silica and alkali content, enrichment in Rb and Th, and depletion in Ba, P, and Ti. The rhyolitic porphyries are characterized by high silica and potassium content, with trace elements exhibiting enrichment in LREEs and depletion in HREEs, aligning with the geochemical traits of S-type granites. Both the monzogranites and rhyolitic porphyries within the deposit exhibit positive zircon εHf(t) values, indicating their source regions underwent the partial melting of juvenile crustal materials.

(3) The Wunugetu porphyry copper–molybdenum deposit was formed within an active continental margin context, influenced by the southeastward subduction of the Mongol-Okhotsk Ocean from the Late Triassic to the Early Jurassic. Following the Mongol-Okhotsk Ocean's closure during the Middle Jurassic, the Wunugetu region transitioned to a syn-collisional tectonic setting.

Author Contributions: Conceptualization, Q.W. and Y.Y.; methodology, Q.W.; validation, Y.Y., Q.F. and Z.Z.; formal analysis, Q.W., Q.F., Z.Z. and X.G.; investigation, Z.Z., X.G., T.W., L.C. and Y.Z.; data curation, Q.W.; writing—original draft preparation, Q.W.; writing—review and editing, Y.A.; supervision, Y.Y.; project administration, Q.F.; funding acquisition, Y.Y. All authors have read and agreed to the published version of the manuscript.

Funding: This work was supported by the Investigation and Evaluation of Copper and Lithium Resources in Mongolia (No. DD20230586) from the China Geological Survey, the Northeast Geological Science and Technology Innovation Center Regional Innovation Fund (No. QCJJ2022-38).

Data Availability Statement: The data presented in this study are available in Tables 1–3.

Acknowledgments: The authors express their deep gratitude towards the two anonymous reviewers and the academic editor for their insightful and invaluable suggestions.

Conflicts of Interest: Qiulin Fu is an employee of Liaoning Province Institute of Geology and Mineral Resources Co., Ltd. and Xiaodan Guo is an employee of Liaoning Province Geology and Mineral Group Energy Geology Co., Ltd. We declare that we have no financial and personal relationships with other people or organizations that can inappropriately influence our work, there is no professional or other personal interest of any nature or kind in any product, service and/or company that could be construed as influencing the position presented in, or the review of, the manuscript entitled.

References

1. Sengör, A.M.C.; Natal'in, B.A.; Burtman, V.S. Evolution of the Altaid tectonic collage and Palaeozoic crustal growth in Eurasia. *Nature* **1993**, *364*, 299–307. [CrossRef]
2. Jahn, B.M.; Wu, F.Y.; Chen, B. Massive granitoid generation in Central Asia:Nd isotope evidence and implication forcontinental growth in the Phanerozoic. *Episodes* **2000**, *23*, 82–92. [CrossRef]
3. Jahn, B.M.; Wu, F.Y.; Chen, B. Granitoids of the Central Asian Orogenic Belt and continental growth in the Phanerozoic. *Earthand Environ. Sci. Trans. R. Soc. Edinb.* **2000**, *91*, 181–193.
4. Badarch, G.; Cunningham, W.D.; Windley, B.F. A new terrane subdivision for Mongolia: Implications for the Phanerozoic crustal growth of central Aisa. *J. Asian Earth Sci.* **2002**, *21*, 87–110. [CrossRef]
5. Buslov, M.M.; Saphonova, I.Y.; Watanabe, T.; Obut, O.T.; Fujiwara, Y.; Iwata, K.; Sugai, Y.; Smirnova, L.V.; Kazansky, A.Y. Evolution of the Paleo-Asian Ocean (Altai–Sayan region, Central Asia) and collision of possible Gondwana-derived terranes with the southern marginal part of the Siberian continent. *Geosci. J.* **2001**, *5*, 203–224. [CrossRef]
6. Windley, B.F.; Alexeiev, D.; Xiao, W.J.; Kröner, A.; Badarch, G. Tectonic models for accretion of the Central Asian Orogenic Belt. *J. Geol. Soc.* **2007**, *164*, 31–47. [CrossRef]
7. Wu, F.Y.; Sun, D.Y.; Ge, W.C.; Zhang, Y.B.; Grant, M.L.; Wilde, S.A.; Jahn, B.M. Geochronology of the Phanerozoic granitoids in northeastern China. *J. Asian Earth Sci.* **2011**, *41*, 1–30. [CrossRef]
8. Cai, K.D.; Sun, M.; Yuan, C.; Zhao, G.C.; Xiao, W.J.; Long, X.P.; Wu, F.Y. Carboniferous mantle-derived felsic intrusion in the Chinese Altai, NW China: Implications for geodynamic change of the accretionary orogenic Belt. *Gondwana Res.* **2012**, *22*, 681–698. [CrossRef]
9. Tu, G.Z. Preliminary discussion on the Central Asian metallogenic domain. *Geol. Sci.* **1999**, *34*, 397–404, (In Chinese with English abstract).
10. Xiao, W.; Song, D.; Windley, B.F.; Li, J.; Han, C.; Wan, B.; Zhang, J.; Ao, S.; Zhang, Z. Research progresses of the accretionary processes and metallogenesis of the Central Asian Orogenic Belt. *Sci. China Earth Sci.* **2019**, *11*, 14–19, (In Chinese with English abstract).
11. Wang, Q.S.; Wei, Y.L.; Gao, J.; Xiong, X.L.; Wang, J.B.; Guo, Z.J.; Li, J.T.; Sun, M. Zircon U-Pb ages and geochemistry of granitoid in the Yuejinshan copper-gold deposit, NE China: Constraints on the petrogenesis and metallogenesis. *Minerals* **2021**, *11*, 1206. [CrossRef]
12. Wang, Z.T.; Qin, K.Z. Geological and geochemical feature and the source of metallogenetie matter of lower crust porphyry Cu–Mu deposit in Wunugetu Mountain. *Min. Depos.* **1988**, *7*, 3–15, (In Chinese with English abstract).
13. Tan, G. *Research on the Ore-Forming Process of the Wunugetu Porphyry Copper Molybdenum Deposit in Inner Mongolia*; Chinese Academy of Geological Sciences: Beijing, China, 2011; (In Chinese with English abstract).
14. Li, N. *The Characteristics and Mineralization Age of Intrusive Rocks in the Wunugetu Copper Molybdenum Deposit, Inner Mongolia*; China University of Geosciences: Beijing, China, 2013; (In Chinese with English abstract).
15. Qin, K.Z.; Li, H.M.; Li, W.S.; Ishihara, S. Diagenesis and Metallogenic Era of Wunugetu Porphyry Copper Molybdenum Deposit in Inner Mongolia. *Geol. Rev.* **1999**, *2*, 180–185, (In Chinese with English abstract).
16. Chen, Z.G.; Zhang, L.C.; Wan, B.; Zhang, Y.T.; Wu, H.Y. Geochemical characteristics and geological significance of low Sr Yb type ore-forming porphyry in the Wunugetu porphyry copper molybdenum deposit, Inner Mongolia. *Acta Petrol. Sin.* **2008**, *24*, 115–128, (In Chinese with English abstract).
17. Tan, G.; Chang, G.X.; She, H.Q.; Li, J.W.; Zhang, D.Q.; Yang, X.C.; Zhang, B.; Xiang, A.P.; Dong, Y.J. Rhenium osmium isotope dating of molybdenite in Wunugetu porphyry copper molybdenum deposit, Inner Mongolia and its geological significance. *Miner. Depos.* **2010**, *29*, 506–508, (In Chinese with English abstract).
18. Chen, Z.G.; Zhang, L.C.; Wan, B.; Wu, H.Y.; Cleven, N. Geochronology and geochemistry of the Wunugetu porphyry Cu–Mo deposit in NE China, and their geological significance. *Ore Geol. Rev.* **2011**, *43*, 92–105. [CrossRef]
19. Sengör, A.M.C.; Natal'in, B.A. Paleotectonics of Asia: Fragments of a synthesis. In *The Tectonic Evolution of Asia*; Yin, A., Harrison, T.M., Eds.; Cambrige University Press: Cambridge, UK, 1996; pp. 486–641.
20. Li, J.Y. Permian geodynamic setting of Northeast China and adjacent regions: Closure of the Paleo-Asian ocean and subduction of the Paleo-Pacific plate. *J. Asian Earth Sci.* **2006**, *26*, 207–224. [CrossRef]
21. Tang, J.; Xu, W.L.; Wang, F.; Zhao, S.; Wang, W. Early Mesozoic southward subduction history of the Mongol–Okhotsk oceanic plate: Evidence from geochronology and geochemistry of Early Mesozoic intrusive rocks in the Erguna Massif, NE China. *Gondwana Res.* **2016**, *31*, 218–240. [CrossRef]

22. Tang, J.; Xu, W.L.; Wang, F.; Wang, W.; Xu, M.J.; Zhang, Y.H. Geochronology and geochemistry of Early–Middle Triassic magmatism in the Erguna Massif, NE China: Constraints on the tectonic evolution of the Mongol–Okhotsk Ocean. *Lithos* **2014**, *184–187*, 1–16. [CrossRef]
23. Li, Y.; Xu, W.L.; Wang, F.; Tang, J.; Zhao, S.; Guo, P. Geochronology and geochemistry of late Paleozoic–early Mesozoic igneous rocks of the Erguna Massif, NE China: Implications for the early evolution of the Mongol–Okhotsk tectonic regime. *J. Asian Earth Sci.* **2017**, *144*, 205–224. [CrossRef]
24. Hou, Z.S. *The Genesis and Tectonic Background of the Badaguan Copper Molybdenum Deposit in the Erguna Area of Inner Mongolia*; Jilin University: Changchun, China, 2014; (In Chinese with English abstract).
25. Niu, S.D. *Magmatic Intrusion Sequence and Mineralization of the Jiawula Lead-Zinc Silver Deposit in the Daxing'anling Mountains*; China University of Geosciences: Beijing, China, 2017; (In Chinese with English abstract).
26. Wiedenbeck, M.; Allé, P.; Corfu, F.; Griffin, W.L.; Meier, M.; Oberli, F.; von Quadt, A.; Roddick, J.C.; Spiegel, W. Three natural zircon standards for U–Th–Pb, Lu–Hf, trace element and REE analyses. *Geostand. Newsl.* **1995**, *19*, 1–23. [CrossRef]
27. Slama, J.; Košler, J.; Condon, D.J.; Crowley, J.L.; Gerdes, A.; Hanchar, J.M.; Horstwood, M.S.A.; Morris, G.A.; Nasdala, L.; Norberg, N.; et al. Plešovice zircon—A new natural reference material for U–Pb and Hf isotopic microanalysis. *Chem. Geol.* **2008**, *249*, 1–35. [CrossRef]
28. Liu, Y.S.; Hu, Z.C.; Zong, K.Q.; Gao, C.G.; Gao, S.; Xu, J.; Chen, H.H. Reappraisement and refinement of zircon U-Pb isotope and trace element analyses by LA-ICP-MS. *Chin. Sci. Bull.* **2010**, *55*, 1535–1546. [CrossRef]
29. Ludwig, K.R. *Isoplot/Ex Version 3.00: A Geochronological Toolkit for Microsoft Excel*; Berkeley Geochronology Center: Berkeley, CA, USA, 2003.
30. Andersen, T. Correction of Common Lead in U-Pb Analyses that Do Not Report ^{204}Pb. *Chem. Geol.* **2002**, *192*, 59–79. [CrossRef]
31. Wu, F.Y.; Li, X.H.; Zheng, Y.F.; Gao, S. Lu-Hf isotopic systematics and their applications in petrology. *Acta Petrol. Sin.* **2007**, *23*, 185–220, (In Chinese with English abstract).
32. Middlemost, E.A.K. Naming materials in the magma/igneous rock system. *Earth Sci. Rev.* **1994**, *37*, 215–224. [CrossRef]
33. Peccerillo, R.; Taylor, S.R. Geochemistry of eocene calc-alkaline volcanic rocks from the Kastamonu area, Northern Turkey. Contrib. *Miner. Petrol.* **1976**, *58*, 63–81. [CrossRef]
34. Maniar, P.D.; Piccoli, P.M. Tectonic discrimination of granitoids. *Geol. Soc. Am. Bull.* **1989**, *101*, 635–643. [CrossRef]
35. Boynton, W.V. Geochemistry of the rare earth elements: Meteorite studies. In *Rare earth Element Geochemistry*; Henderson, P., Ed.; Elsevier: Amsterdam, The Netherlands, 1984; pp. 63–114.
36. Sun, S.S.; McDonough, W.F. Chemical and Isotopic Systematics of Oceanic Basalts: Implications for Mantle Composition and Processes. *Geol. Soc. Lond. Spec. Publ.* **1989**, *42*, 313–345. [CrossRef]
37. Mao, J.W.; Xie, G.Q.; Zhang, Z.H.; Li, X.F.; Wang, Y.T.; Zhang, C.Q.; Li, Y.F. Mesozoic large-scale metallogenic pulses in North China and corresponding geodnamic settings. *Acta Prtrologica Sin.* **2005**, *21*, 169–188.
38. Xu, W.L.; Wang, F.; Pei, F.P.; Meng, E.; Tiang, J.; Xu, M.J.; Wang, W. Mesozoic tectonic regimes and regional ore-forming background in NE China: Constraints from spatial and temporal variations of Mesozoic volcanic rock associations. *Acta Petrol. Sin.* **2013**, *29*, 339–353.
39. Sun, J.G.; Zhang, Y.; Xing, S.W.; Zhao, K.Q.; Zhang, Z.J.; Bai, L.A.; Ma, Y.B.; Liu, Y.S. Genetic types ore-forming age and geodynamic setting of endogenic molybdenum deposits in the eastern edge of Xing-Meng orogenic belt. *Acta Prtrologica Sin.* **2012**, *28*, 1317–1332.
40. Tang, J. *Chronology and Geochemistry of Mesozoic Igneous Rocks in the Erguna Block: Constraints on the Tectonic Evolution of the Mongolian Okhotsk Suture Zone*; Jilin University: Changchun, China, 2016; (In Chinese with English abstract).
41. Zhang, J.; Shao, J.; Zhou, Y.H.; Bao, Q.Z.; Wang, H.B. Geochemical characteristics and U-Pb, Re Os ages of the Badaguan copper molybdenum deposit in Chenbalhu Banner, Inner Mongolia. *Geol. Bull. China* **2016**, *35*, 1388–1399, (In Chinese with English abstract).
42. Wang, X.; Duan, M.X.; Ren, Y.S.; Hou, Z.S.; Sun, D.Y.; Hao, Y.J. Fluid Inclusion Characteristics and Metallogenic Era of the Badaguan Copper Molybdenum Deposit in the Erguna Area of Inner Mongolia. *J. Jilin Univ. (Earth Sci. Ed.)* **2016**, *46*, 1354–1367, (In Chinese with English abstract).
43. Chen, Z.G.; Zhang, L.C.; Lu, B.Z.; Li, Z.L.; Wu, H.Y.; Xiang, P.; Huang, S.W. Geochronology and Geochemistry of the Taipingchuan Copper-Molybdenite Deposit in Inner Mongolia, and Its Geological Significances. *Acta Petrol. Sinca* **2010**, *26*, 1437–1449, (In Chinese with English abstract).
44. Wang, Y.H.; Zhao, C.B.; Zhang, F.F.; Liu, J.J.; Wang, J.P.; Peng, R.M.; Liu, B. SIMS zircon U–Pb and molybdenite Re–Os geochronology, Hf isotope, and whole-rock geochemistry of the Wunugetu porphyry Cu–Mo deposit and granitoids in NE China and their geological significance. *Gondwana Res.* **2015**, *28*, 1228–1245. [CrossRef]
45. Yang, Z.F.; Luo, Z.H.; Lu, X.X.; Cheng, L.L.; Huang, F. Discussion on the Tracer Sources of Re Content in Molybdenum Ores. *Miner. Depos.* **2011**, *30*, 654–674, (In Chinese with English abstract).
46. Chappell, B.W.; White, A.J.R. Two contrasting granite types. *Pac. Geol.* **1974**, *8*, 173–174.
47. Whalen, J.B.; Currie, K.L.; Chappell, B.W. A-type granites: Geochemical characteristics, discrimination and petrogenesis. *Contrib. Mineral. Petrol.* **1987**, *95*, 407–419. [CrossRef]
48. Pitcher, W.S. *The Nature and Origin of Granite*, 2nd ed.; Springer: Dordrecht, The Netherlands, 1997.

49. Chen, J.L.; Guo, Y.S.; Fu, S.M. Progress in Granite Research—Overview of ISMA Granite Classification. *Gansu Geol.* **2004**, *13*, 67–73, (In Chinese with English abstract).
50. Freund, S.; Haase, K.M.; Keith, M.; Beier, C.; Garbe-Schionberg, D. Constraints on the formation of geochemically variable plagiogranite intrusions in the Troodos Ophiolite, Cyprus. *Contrib. Mineral. Petrol.* **2014**, *167*, 978. [CrossRef]
51. Wu, F.Y.; Jahn, B.M.; Wilde, S.A.; Lo, C.H.; Yui, T.F.; Lin, Q.; Ge, W.C.; Sun, D.Y. Highly fractionated I-type granites in NE China (I): Geochronology and petrogenesis. *Lithos* **2003**, *66*, 241–273. [CrossRef]
52. Sylvester, P.J. Post-collisional strongly peraluminous granites. *Lithos* **1998**, *45*, 29–44. [CrossRef]
53. Clemens, J.D. S-type granitic magmas-petrogenetic issues, models and evidence. *Earth-Sci. Rev.* **2003**, *61*, 1–18. [CrossRef]
54. Sisson, T.W.; Ratajeski, K.; Hankins, W.B.; Glazner, A.F. Voluminous granitic magmas from common basaltic sources. *Contrib. Mineral. Petrol.* **2005**, *148*, 635–661. [CrossRef]
55. Chappell, B.W.; Bryant, C.J.; Wyborn, D. Peraluminous I-type granites. *Lithos* **2012**, *153*, 142–153. [CrossRef]
56. Watson, E.B.; Capobianco, C.J. Phosphorus and the rare earth elements in felsic magmas: An assessment of the role of apatite. *Geochim. Cosmochim. Acta* **1981**, *45*, 2349–2358. [CrossRef]
57. Chappell, B.W. Aluminium saturation in I- and S-type granites and the characterization of fractionated haplogranites. *Lithos* **1999**, *46*, 535–551. [CrossRef]
58. Yang, J.H.; Wu, F.Y.; Shao, J.A.; Wilde, S.A.; Xie, L.W.; Liu, X.M. Constraints on the timing of uplift of the Yanshan fold and thrust belt, North China. *Earth Planet. Sci. Lett.* **2006**, *246*, 336–352. [CrossRef]
59. Xu, W.L.; Sun, C.Y.; Tang, J.; Luan, J.P.; Wang, F. Basement nature and tectonic evolution of the Xing'an-Mongolian orogenic belt. *Earth Sci.* **2019**, *44*, 1620–1646, (In Chinese with English abstract).
60. Xu, W.L.; Pei, F.P.; Wang, F.; Meng, E.; Ji, W.Q.; Yang, D.B.; Wang, W. Spatial–temporal relationships of Mesozoic volcanic rocks in NE China: Constraints on tectonicoverprinting and transformations between multiple tectonic regimes. *J. Asian Earth Sci.* **2013**, *74*, 167–193. [CrossRef]
61. Zhu, D.C.; Mo, X.X.; Wang, L.Q.; Zhao, Z.D.; Niu, Y.L.; Zhou, C.Y.; Yang, Y.H. Genesis of the Chayu highly differentiated I-type granite in eastern Gangdise, Xizang: Zircon U-Pb geochronology, geochemistry and Sr Nd Hf isotope constraints. *Sci. Sin. (Terrae)* **2009**, *39*, 833–848, (In Chinese with English abstract).
62. Wu, F.Y.; Liu, X.C.; Ji, W.Q.; Wang, J.M.; Yang, L. Highly fractionated granites: Recognition and research. *Sci. China (Earth Sci.)* **2017**, *60*, 1201–1219. [CrossRef]
63. Pearce, J.A.; Harris, N.B.W.; Tindle, A.G. Trace element, discrimination diagrams for the tectonic interpretation of granitic rocks. *J. Petrol.* **1984**, *25*, 956–983. [CrossRef]
64. Li, Y.; Ding, L.L.; Xu, W.L.; Wang, F.; Tang, J.; Zhao, S.; Wang, Z.J. Chronology and geochemistry of the Middle Jurassic muscovite granite in the Sunwu area: Constraints on the closure time of the Mongol-Okhotsk Ocean. *Acta Petrol. Sin.* **2015**, *31*, 56–66.
65. Han, R.; Qin, K.Z.; Su, S.Q.; Groves, D.I.; Zhao, C.; Hui, K.X.; Meng, Z.J. An Early Cretaceous Ag-Pb-Zn mineralization at Halasheng in the South Erguna Block, NE China: Constraints from U-Pb and Rb-Sr geochronology, geochemistry and Sr-Nd-Hf isotopes. *Ore Geol. Rev.* **2020**, *122*, 103526. [CrossRef]

minerals

Article

Genesis and Prospecting Potential of the Da'anhe Skarn Au Deposit in the Central of the Lesser Xing'an Range, NE China: Evidence from Skarn Mineralogy, Fluid Inclusions and H-O Isotopes

Chuntao Zhao [1,2,3], Fanting Sun [3,*], Jinggui Sun [3], Jianping Wang [1,2,*], Jilong Han [3,4], Xiaolei Chu [3], Chenglin Bai [3,5], Dongmei Yu [1,2], Zhikai Xu [3], Lei Yi [1,2] and Shan Hua [6]

1 Key Laboratory of Green and High-End Utilization of Salt Lake Resources, Qinghai Institute of Salt Lakes, Chinese Academy of Sciences, Xining 810008, China; zhaoctao22@isl.ac.cn (C.Z.); ydm2011@isl.ac.cn (D.Y.); yilei@isl.ac.cn (L.Y.)
2 Qinghai Provincial Key Laboratory of Geology and Environment of Salt Lakes, Xining 810008, China
3 College of Earth Science, Jilin University, Changchun 130061, China; sunjinggui@jlu.edu.cn (J.S.); hanjilong@cugb.edu.cn (J.H.); chuxiaolei19@163.com (X.C.); 3057220016@email.cugb.edu.cn (C.B.); xuzk@qdio.ac.cn (Z.X.)
4 Development and Research Center of China Geological Survey, Beijing 100037, China
5 School of Earth Sciences and Resources, China University of Geosciences, Beijing 100083, China
6 The Ninth Geological Brigade, Hebei Bureau of Geology and Mineral Resources Exploration, Xingtai 054000, China; m15903218986@163.com
* Correspondence: sunft21@mails.jlu.edu.cn (F.S.); wang_ktz2022@126.com (J.W.)

Citation: Zhao, C.; Sun, F.; Sun, J.; Wang, J.; Han, J.; Chu, X.; Bai, C.; Yu, D.; Xu, Z.; Yi, L.; et al. Genesis and Prospecting Potential of the Da'anhe Skarn Au Deposit in the Central of the Lesser Xing'an Range, NE China: Evidence from Skarn Mineralogy, Fluid Inclusions and H-O Isotopes. *Minerals* **2024**, *14*, 214. https://doi.org/10.3390/min14030214

Academic Editor: Alfons M. Van den Kerkhof

Received: 7 December 2023
Revised: 11 February 2024
Accepted: 15 February 2024
Published: 20 February 2024

Abstract: Skarn Au deposits exist in the circum-pacific metallogenic belt. Interestingly, the Da'anhe Au deposit is the only independent skarn gold deposit in the Lesser Xing'an Range. To determine the metallogenic mechanism and prospecting potential of the Da'anhe deposit, we performed skarn mineralogy, fluid inclusion (FI) and H-O isotope analyses. The results show the following: (1) The Da'anhe deposit is a calcareous reduced skarn Au deposit that formed between an Early Jurassic gabbroic diorite and the Permian Tumenling Formation marble. Its metallogenic process includes five stages: the early skarn stage (Stage I_1), late skarn stage (Stage I_2), early quartz-sulfide stage (Stage II_1), late quartz-sulfide stage (Stage II_2) and quartz-carbonate stage (Stage II_3). Gold precipitated in Stage II_1 and Stage II_2. (2) The initial ore-forming fluid was derived from magmatic water and featured a high temperature and intermediate to high salinity. After boiling and mixing, the fluid eventually changed to a low-temperature and low-salinity reducing fluid dominated by meteoric water. (3) The formation depth of the Au orebodies was 2.27–3.11 km, and the orebodies were later lifted to the surface (<500 m). The potential for finding skarn Au deposits in the study area is limited. (4) The distinctive nature of the ore-related magma (i.e., source, reducing conditions and high water content) was key to the formation of the Da'anhe skarn gold deposit.

Keywords: mineralization; skarn mineralogy; H-O isotopes; fluid inclusion; Da'anhe Au deposit; lesser Xing'an range

1. Introduction

Skarn deposits are important sources of iron, copper, lead, zinc, molybdenum, gold, tungsten and other metal minerals worldwide. In the circum-pacific metallogenic belt, gold production from skarn deposits has exceeded 1000 tons [1,2]. Skarn gold deposits in China are distributed in various tectonic provinces and are hosted in rocks ranging in age from the Palaeoproterozoic to the Cretaceous. At least seventy deposits have been identified as gold or gold-bearing skarn deposits in China [3]. Northeastern China, located in the western circum-pacific metallogenic belt, was affected by the subduction of the

Paleo-Pacific plate and extension after the collision and orogeny between the Laurasian and Gondwanan continents. The Lesser Xing'an Range in northeastern China experienced extensive magmatism and mineralization during the Palaeozoic and Mesozoic, particularly the latter which was an important epoch for numerous skarn polymetallic deposits [4–6]. Strong tectonic-magmatic activity provided favourable conditions for the formation of skarn Au deposits in this region [7,8].

However, most of these deposits are calcareous skarn deposits that are composed mainly of iron, copper lead and zinc and are genetically related to the metasomatism of intermediate-acidic plutonic rocks and marble. Examples of such deposits include the Ergu Fe-Zn polymetallic deposit [9–12], Cuihongshan Fe-Cu-W-Mo polymetallic deposit [13–15], and Xulaojiugou Pb-Zn polymetallic deposit [16,17]. The medium-sized Da'anhe gold deposit is the only skarn Au deposit. At present, several geologists have performed detailed studies on the metallogenic mechanisms and processes of the skarn Fe-Cu-Pb-Zn deposits in this region [13,17,18], but research on the Da'anhe Au deposit has focused mainly on geological investigations and the chronologic and geochemical characteristics of plutonic rocks [19–21]. Identifying the genetic and mineralization processes of the Da'anhe deposit, and comparing them with those of the skarn Fe-Cu-Pb-Zn deposit in this region would be beneficial and is necessary for researching the metallogenic specificity of skarn deposits. Therefore, we recently carried out detailed field geological investigations, skarn mineralogy analysis, fluid inclusion (FI) analysis, coupled with H-O isotope research on the deposits, aiming to further determine the metallogenic mechanism and metallogenic process and provide a reference for prospecting for skarn Au deposits in the study area.

2. Regional Geological Characteristics

The Da'anhe deposit is located in the eastern section of the Xingmeng orogenic belt and is sandwiched between the Siberian, North China and Paleo-Pacific plates. This superimposed tectonic regime has experienced the evolution of the Palaeo-Asian Ocean tectonic system and the Xingmeng orogeny in the Palaeozoic and the subduction of the Paleo-Pacific plate and Okhotsk Ocean plate in the Mesozoic [22–25] (Figure 1A,B) [26,27]. The multi-stage tectonic-magmatic activities resulted in the Lesser Xing'an Range being a typical basin-mountain region with endogenetic metal development related to tectonic-magmatic activity [28–30]. The stratigraphy in this region is dominated by the silicoferrite of the Dongfengshan Formation (821–801 Ma [31]) and overlying Cambrian–Ordovician neritic facies sedimentary rocks (limestone, dolomite marble, carbonaceous slate, etc.; ~500 Ma [32]), Permian terrigenous clastic-carbonate rocks (amphibolite, marble, tuff, felsic schist, slate, etc.; 262–286 Ma [10]), Mesozoic continental volcanic rocks, and Palaeogene fluvial and lacustrine clastic sedimentary rocks. The study area experienced at least five stages of magmatism: middle Cambrian-Late Ordovician (508~450 Ma [30]), middle-late Permian (272~260 Ma [29]), Early-Middle Triassic (252~234 Ma [29]), Late Triassic-Early Jurassic (225~197 Ma [28,29,33]) and Late Jurassic (191~175 Ma [34,35]). Voluminous Phanerozoic granitoids, including abundant I-type granites and a few A- and S-type granites are present in this region [36]. In addition to these granitoids, rare mafic-ultramafic intrusions are also present in the region. Among them, Late Triassic and Early Jurassic granites are the most developed and widely distributed. The large-scale mineralization of porphyry molybdenum and skarn deposits mainly occurred at 186~176 Ma [24], and epithermal gold mineralization mainly occurred at 110~95 Ma [24]. Many porphyry molybdenum deposits in different scales (Luming, Huojihe, Cuiling, and Gaogangshan) formed in the Mesozoic granitic complex (Figure 1C) [28]. Many skarn deposits (such as the Cuihongshan, Ergu and Xulaojiugou deposits) formed in the contact belts between Mesozoic granite and Palaeozoic carbonate strata (Figure 1C) [28]. Most epithermal deposits are located in the Mesozoic to Cenozoic basins that are composed of continental volcanic and sedimentary rocks, such as the Dong'an, Gaosongshan, Yongxin, Tuanjiegou, and Yidonglinchang deposits. In this study, the Da'anhe gold deposit is essentially consistent with the mineralization of the widely developed skarn polymetallic deposits and porphyry molybdenum deposits, which formed during the late Triassic to Early-Middle Jurassic (203.4~179.9 Ma) [10,16,28] (Figure 1C) [28].

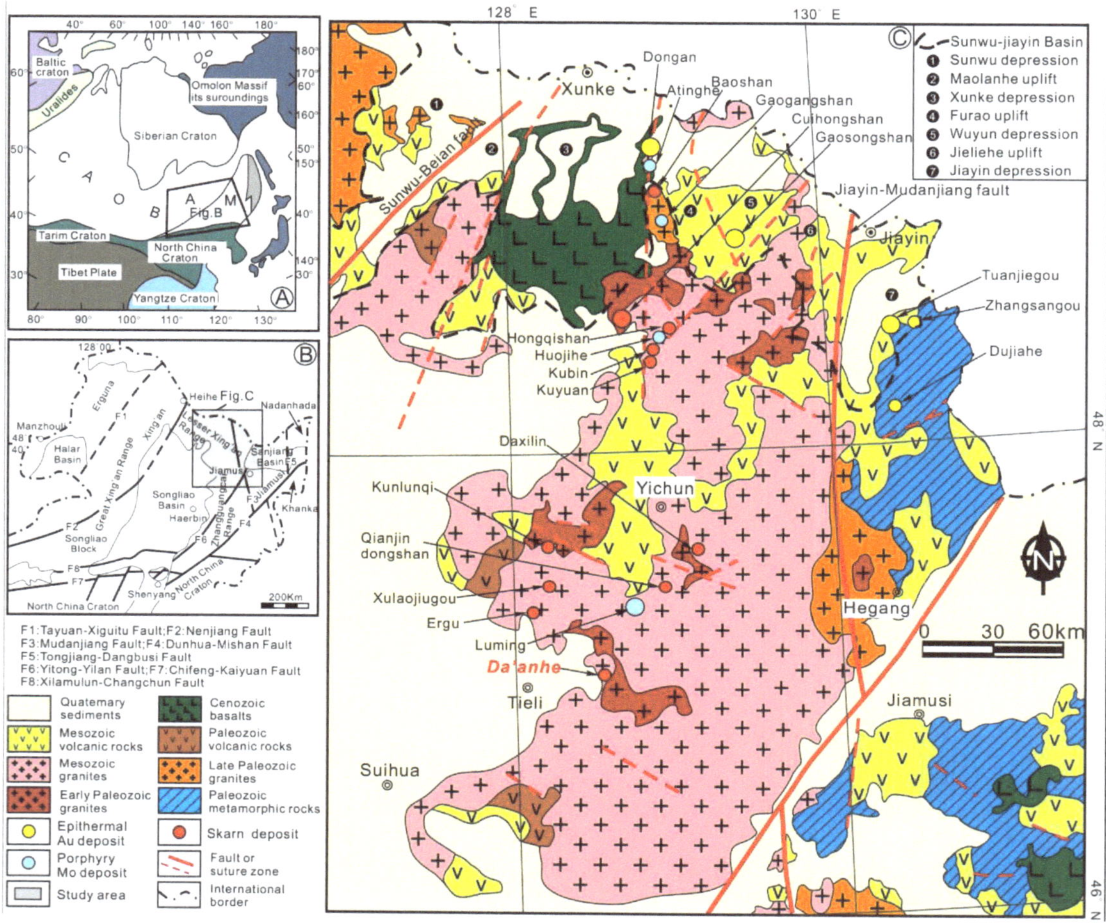

Figure 1. (**A**) Location of Northeast China with respect to the main tectonic units of China and Russia, modified after Wu et al. [26]. 'A' and 'M' represent the Altaids and Manchurids, respectively. (**B**) Tectonic sketch maps of Northeast China, modified after Wu et al. [27]. (**C**) Simplified geological map of the Lesser Xing'an range, modified after Wu et al. [28].

3. Deposit Geology

3.1. Stratigraphy, Magmatism and Structure

The Da'anhe deposit is located approximately 30 km northeast of the Tieli city, Heilongjiang Province, and is tectonically part of the southern Yichun-Yanshou magmatically and tectonically active zone [19–21]. The strata in the mining area include the Permian Tumenling Formation and the Jurassic Tai'antun Formation (Figure 2) [21]. The Tumenling Formation is distributed in the Sanjiaoshan area of the mining area in a discontinuous belt trending in the NE direction. The rock types include slate, metamorphic quartz sandstone, marble and andesitic crystal tuff. Marble is distributed in metamorphic quartz sandstone strata with lenticular or irregular shapes and surrounds Au ore bodies. The Tai'antun Formation is less exposed in the mining area and is distributed in the northeastern part of the mining area in the form of residual bodies. The rock types are mainly andesitic tuff and andesite, which unconformably overlie the Tumenling Formation.

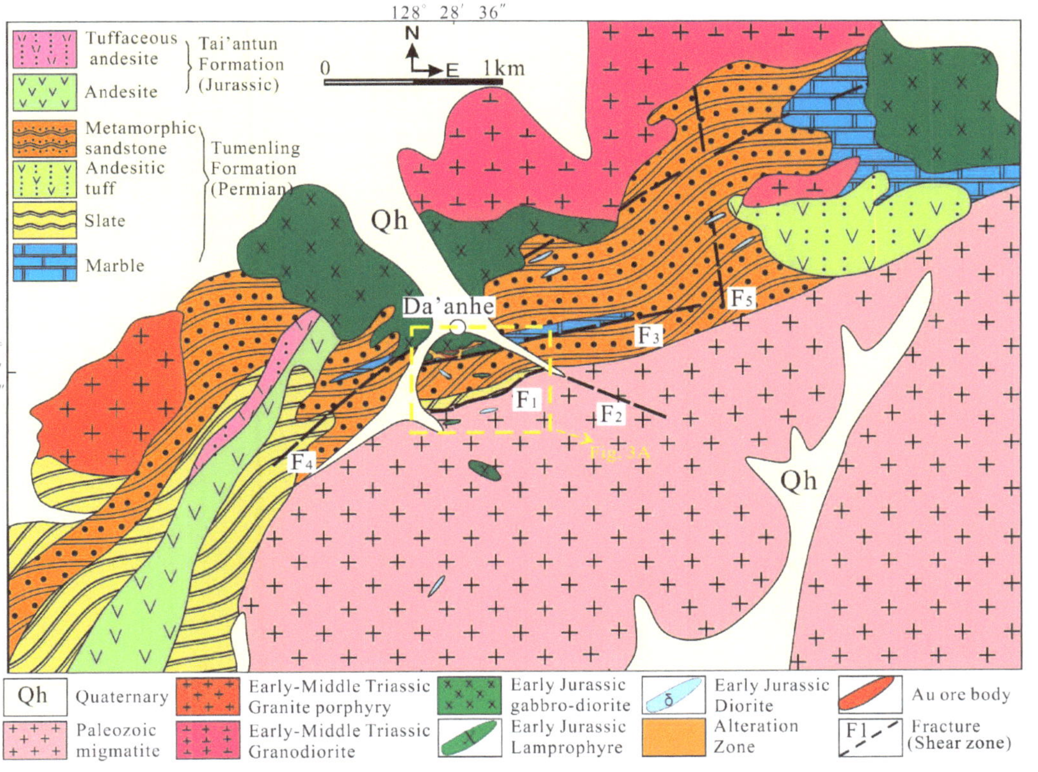

Figure 2. Geological map of the Da'anhe Au deposit (Figure 2 modified according to Yang [21]. Figure 3A detailed in Figure 3).

The magmatic rocks in the mining area are mainly Palaeozoic migmatite granite and Mesozoic granodiorite, granodiorite porphyry and gabbroic diorite (Figure 2) [21]. Palaeozoic migmatite is exposed in the southern mining area in the form of a large-scale batholith. The granodiorite and granodiorite porphyry formed in the Early-Middle Triassic and were emplaced in the form of stocks in the Tumenling Formation along the NE-trending Jinniugou-Da'anhedongshan fault in the northern part of the mining area. The gabbroic diorite on both sides of the ore body has undergone significant alteration (see Section 3.2.1) and formed endoskarn, which occurs in the inner belt of the skarn, indicating a close relationship with mineralization. In addition, late lamprophyre and fine-grained diorite dikes are present in the mining area.

The structures in the mining area are mainly faults and folds. The faults in the mining area mainly include the NE-trending Jinniugou-Da'anhedongshan fault, the NW-trending Da'anhe fault and the nearly NW-trending Da'anhenanshan fault. These faults are mainly manifested as breccia zones, shear zones and schistosity zones in the mining area. The Jinniugou-Da'anhedongshan fault and Da'anhe fault intersect and become the main ore controlling structures in the mining area. The large fold in the region is the Shenshu complex anticline, and the mining area and the deposit are located on the southeastern limb of the NE-striking Sanjiaoshan syncline within the Shenshu anticlinorium (Figure 2) [21].

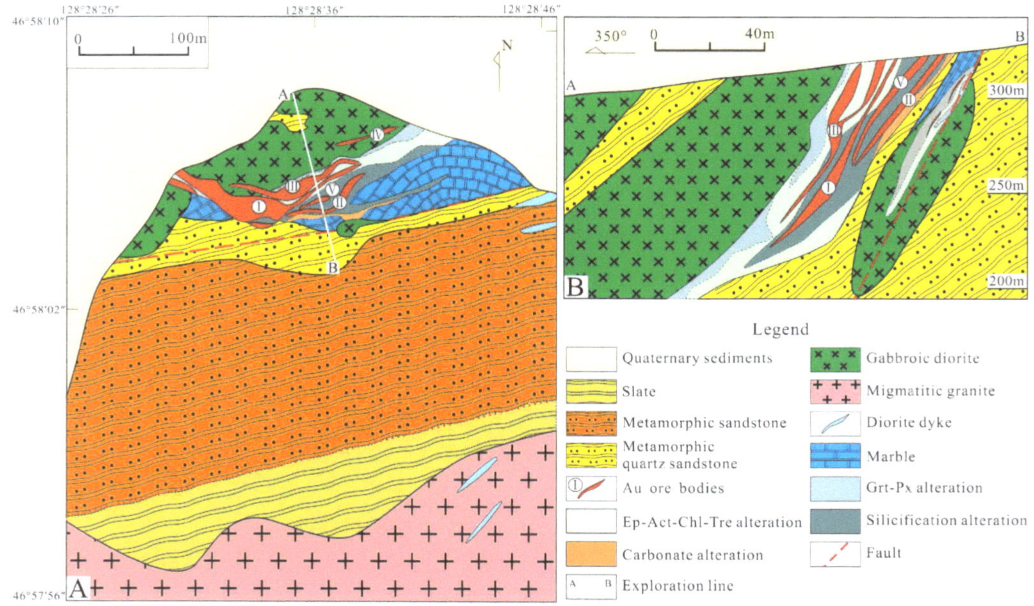

Figure 3. (**A**) Geological map of the Da'anhe Au deposit; (**B**) Profile of exploration line 4 of the Da'anhe Au deposit. ((**A**,**B**) modified according to Xue [20] and Yang [21]). Note: F1 and F3: Da'anhenanshan fault; F2: Da'anhe fault; F4: Jinniugou-Da'anhedongshan fault. Grt-garnet; Px-diopside; Ep-epidote; Act-actinolite; Tr-tremolite; Chl-chlorite.

3.2. *Mineralization*

The gold orebodies in the Da'anhe deposit mainly developed in the skarn belt between the gabbroic diorite (Figures 3, 4A and 5A) and the marble of the Tumenling Formation in the central and western mining areas, and small amounts of altered gabbroic diorite and marble developed on both sides of the skarn belt. The orebodies are controlled by shear zones and occur in NE- or NW-striking lenses or veins dipping steeply between 42 and 87 degrees. The ore bodies are generally approximately 25 m long, and some ore bodies can reach lengths of up to 168 m. At present, 12 industrial orebodies have been identified in the Da'anhe deposit, and the main orebodies are the No. I, II and III orebodies. The three main orebodies occur as veins in the inner and outer belts of the skarn zone, and the average grades of gold are 11.93 g/t, 46.78 g/t and 5.93 g/t, respectively.

3.2.1. Alteration of the Surrounding Rock

The wallrock alteration in the mining area is relatively well developed. The main alteration types are Grt-Px alteration (Figures 3 and 4B–D), Ep-Act-Chl-Tr alteration (Figures 3 and 4E,F), sericitization (Figure 4D), silicification and carbonatization (Figures 3 and 4L). Among them, Grt-Px alteration and silicification are most closely related to gold mineralization. Grt-Px alteration occurs mainly between the gabbroic diorite and the marble of the Tumenling Formation. Silicification occurs mainly in the outer belt of the skarn, and silicified veins of different colours and sizes cut through the early skarn minerals and sulfide veins. Carbonatization occurs in the late stage of mineralization and occurs in the tectonically active part in the form of veins. In addition, potassic alteration and tremolite alteration occur sporadically in the mining area.

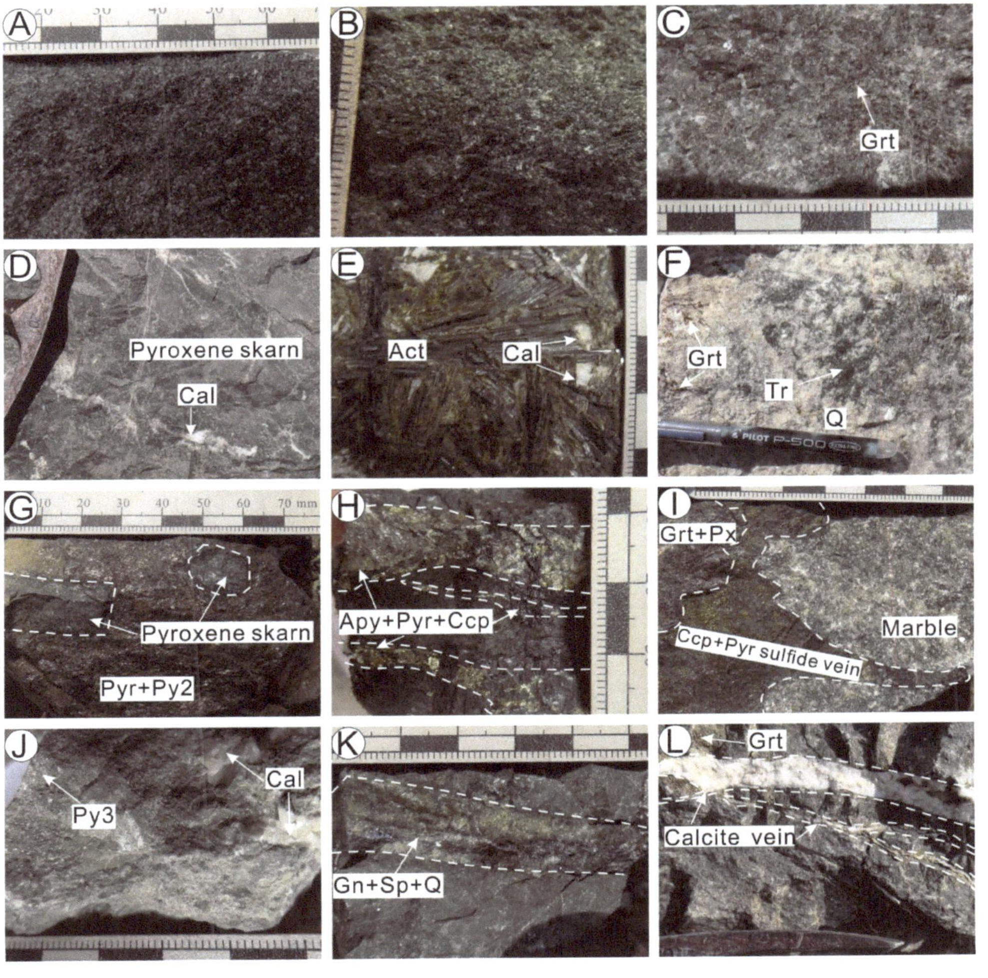

Figure 4. Field geological characteristics of the Da'anhe gold deposit. (**A**). Unaltered gabbroic diorite; (**B**). Skarnized gabbroic diorite hand specimen photographs; (**C**). Photographs of garnet skarn hand specimens; (**D**). Diopside skarn was cut through by late carbonate veins; (**E**). Photographs of skarn hand specimens from actinolite skarn; (**F**). The garnet in skarn metasomatized by late tremolite and quartz began to appear; (**G**). The pyroxene skarn formed in the dry skarn stage was broken into a breccia shape under the influence of tectonism and cemented by late sulfide; (**H**). Sulfide vein penetrating skarn containing arsenopyrite pyrrhotite chalcopyrite; (**I**). Sulfide veins containing chalcopyrite pyrrhotite cut through garnet pyroxene skarn and marble; (**J**). Granular pyrite ore with good crystallinity cut through by late carbonate vein; (**K**). Quartz vein containing galena sphalerite; (**L**). The carbonate veins of the late mineralization cut through the early garnet skarn. Note: Grt-garnet; Px-diopside;Act-actinolite; Tr-tremolite; Q-quartz; Cal-calcite; Ccp-chalcopyrite; Apy-arsenopyrite; Pyr-pyrite; Py2-pyrite coexisting with quartz; Py3-pyrite coexisting with calcite; Gn-galena; Sp-sphalerite.

Figure 5. Microscopic characteristics of rocks and ores in the Da'anhe deposit (cross-polarized light). (**A**). Microphotograph of gabbroic diorite; (**B**). Microscopic characteristics of gabbroic diorite in skarn alteration; (**C**). Garnet and pyroxene were metasomatized by quartz and tremolite and were cut through by the late carbonate vein; (**D**). Sericitization alteration; (**E**). Skarn alteration of marble; (**F**). Garnet was replaced by pyrite bearing quartz veinlets and calcite veinlets in later stage; (**G**). Magnetite shows its granular structure after replacement by pyrite and pyrrhotite; (**H**). The early pyrite and pyrrhotite were replaced by the later pyrite and brass bearing quartz veins; (**I**). Chalcopyrite, pyrrhotite and epidote coexist; (**J**). Sphalerite and galena coexist, and metasomatic arsenopyrite; (**K**). Semiautomorphic pyrite particles in late mineralization; (**L**). Encapsulated gold developed in sphalerite. Note: Hbl-hornblende; Pl-plagioclase; Px-pyroxene; Bt-biotite; Ep-epidote; Cal-calcite; Grt- garnet; Tr- tremolite; Qz-quartz; Ser-sericite; Pyr-pyrite; Py1-early pyrite; Py2-pyrite coexisting with quartz; Py3-pyrite coexisting with calcite; Mag-magnetite; Ccp-chalcopyrite; Apy2-late arsenopyrite; Au-natural gold; Gn-galena; Sp-sphalerite.

3.2.2. Characteristics and Stages of Mineralization

The ore minerals in the ores are widely distributed and mainly include arsenopyrite (Figure 4H), pyrrhotite (Figure 4G–I), chalcopyrite (Figure 4H,I), pyrite (Figure 4G,J), sphalerite (Figure 4K), native gold, electrum, and small amounts of magnetite, galena (Figure 4K), malachite and bismuth telluride. Among them, gold occurs mainly in the form of native gold or electrum in skarn minerals, quartz or sulfide grains. The ore occurs in banded (Figure 4H), vein (Figure 4I,K) and sparse disseminated (Figure 4J) forms, followed by dense massive and brecciated (Figure 4G) forms. According to the cross-cutting relationships of the veins and the mineral assemblage characteristics under the microscope, the metallogenic process of the Da'anhe deposit can be divided into askarn period and a quartz-sulfide period, with a total of five stages. (1) In the early skarn stage (Stage I_1), high-temperature anhydrous skarn minerals, such as garnet and diopside minerals, formed (Figure 5B,C). (2) The late skarn stage (Stage I_2) did not have a major impact on the mining area and mainly formed water-bearing skarn minerals such as actinolite, tremolite and epidote (Figure 5C,E), and a small amount of magnetite (Figure 5G). (3) In the early quartz-sulfide stage (Stage II_1), skarn minerals were largely metasomatized, and gangue minerals, such as quartz, epidote, chlorite and sericite, were formed (Figure 5C,D). In addition to arsenopyrite, the ore minerals were mainly iron-copper sulfides, including pyrrhotite, pyrite and chalcopyrite (Figure 5G–K), and a small amount of native gold precipitation began in the late part of this stage. (4) The late quartz-sulfide stage (Stage II_2) was the main metallogenic stage of gold. Gold precipitated as native gold or electrum between skarn minerals and quartz or metal sulfide grains (Figure 5L). In this stage, the gangue minerals are mainly quartz and calcite, and the ore minerals are mainly pyrite, galena and sphalerite (Figure 5J,K). (5) In the quartz carbonate stage (Stage II_3), only small amounts of pyrite, sphalerite and other sulfides formed, and many quartz-calcite veins formed, cutting through the early skarn minerals or quartz sulfide veins (Figure 5F), signifying the end of mineralization [20,21].

4. Samples and Analytical Methods

4.1. Electron Probe Analysis

Initially, garnet-pyroxene skarn samples collected from the surface of the Da'anhe deposit were cut into pieces for microprobe analysis. On the basis of detailed microscopic identification, garnet and pyroxene were selected and analysed by an electron probe (EPMA-1600, Japan Shimadzu Company, Kyoto, Japan) at the Jilin University Testing and Analysis Center. Standard samples of 53 kinds of minerals were obtained from the SPI Company, Lakewood, WA, USA. The element analysis range was 5B-92U, the electron beam current was stable at 1.5×10^{-3}/h, the accelerating voltage was 15 kV, the current was 4.5 nA, the beam spot was less than 1 μm, and the correction method was ZAF. Geokit software (GeokitPro version) was used to calculate the occupation and endmember composition of each element in the mineral [37].

4.2. Temperature Measurements and Laser Raman Analysis of Fluid Inclusions (FIs)

The analysed samples consisted of garnet and pyroxene from the early skarn stage, epidote from the late skarn stage, quartz (Q_1) from the early quartz-sulfide stage, quartz (Q_2) from the late quartz-sulfide stage and calcite from the quartz-carbonate stage. Samples for the FI study were collected from the surface of the Da'anhe deposit. For the selected samples, first, FI thin sections (thickness ~0.3 mm) were prepared, and then the type, distribution, gas-liquid ratio and daughter minerals of Fis were carefully identified under a high-magnification microscope. From the above observations, the primary inclusions were selected for the temperature measurement experiment, and the analysis was performed in the FI laboratory at the Institute of Geology and Geophysics of the Chinese Academy of Sciences. The experiment was carried out on a THMSG 600 (Linkam Company, Salfords, UK) heating and cooling stage. The measurable temperature ranged from −196 to 600 °C, and the temperature increase and decrease rate was approximately 3 °C/min. Near the phase transition temperature, the rate was maintained at 0.2~0.4 °C/min. The measurement errors of the homogenization temperature and freezing point temperature were ±2 °C and ±0.1 °C, respectively. The analysis procedure is described in detail in Fan et al. [38]. The salinity of the salt solution was calculated using the freezing point temperature [39].

To systematically determine the fluid composition during the mineralization process, representative single FIs captured during each stage were selected for gas phase composition analysis under nondestructive conditions. The experimental instrument used was a LahKam laser confocal Raman spectrometer (produced by Horiba Jobin Yvon, Montpellier, France), and the excitation wavelength was 532 nm. The test was performed by the authors in the FI laboratory at the Institute of Geology and Geophysics, Chinese Academy of Sciences.

4.3. H-O Isotope Analysis

The samples used for H-O isotope testing were garnet (2 samples, Stage I_1), epidote (1 sample, Stage I_2), quartz (1 sample, Stage II_2) and calcite (1 sample, Stage II_3). Samples for H-O isotope analysis were collected from the surface of the Da'anhe deposit. Initially, each selected sample was crushed and ground to 40–80 mesh, and a pure sample with a purity greater than 99% was selected for analysis under a binocular microscope. The samples were tested after washing and desorbing water and secondary FIs. The test was completed at the Beijing Experimental Center of Nuclear Industry. The experimental analysis instrument was a Finnigan-MAT253 mass spectrometer. Oxygen isotope analysis was carried out using the BrF5 method. Mineral oxygen was extracted by the reaction of BrF5 with oxygen-containing minerals under vacuum and high-temperature conditions and converted into CO_2 gas by burning with a thermally resistant graphite rod. The analytical accuracy was $\pm 0.2‰$. The H isotopes of the FIs were analysed by the zinc method with an accuracy of $\pm 2‰$. The international standard for H and O isotopes was SMOW.

To calculate the $\delta^{18}O_{V\text{-}SMOW}$ values of fluids in equilibrium with silicate minerals, the following fractionation equations were used: the garnet-water equation of Taylor [40], the epidote-water equation of Zheng [41], the quartz-water equation of Clayton et al. [42], and the calcite-water equation of Zheng [41].

5. Results

5.1. Electron Probe Analytical Results for the Skarn Minerals

5.1.1. Garnet

Only one type of garnet has been found in the Da'anhe mining area. In hand specimens, the garnet is light red to reddish-brown with particle sizes of 0.3~1.5 mm, and it is often associated with pyroxene and replaced by later minerals (Figure 6A). The garnet is homogeneous under orthogonal polarization, shows total extinction and a metasomatic residual structure, and exhibits a heteromorphic anhedral morphology. The results of the electron probe analysis of the garnet are listed in Table 1. The results show that the main components of the garnet are SiO_2 (average = 37.84%), CaO (average = 35.56%), FeO (average = 12.13%) and Al_2O_3 (average = 12.20%), with small amounts of TiO_2, MnO, MgO and trace Cr_2O_3. The garnet in the Da'anhe deposit is grossular ($Gr_{44.93\text{–}74.76}Ad_{21.89\text{–}52.89}$), with small amounts of andradite, uvarovite, pyrope and spessartite. On the garnet endmember discrimination diagram, the garnet plots within the range of typical skarn Au deposits worldwide (Figure 6C) [1].

5.1.2. Pyroxene

Pyroxene is another important skarn mineral in the Da'anhe deposit that is associated with garnet and is mostly replaced by silicate minerals, quartz, calcite and other late water-bearing minerals. There is a single type of pyroxene, and the pyroxene is usually dark green in hand specimens. Bright interference colours are observed under orthogonally polarized light. The minerals exhibit high protrusions, surface cracks, and euhedral and subhedral granular textures (Figure 6B) and the particle sizes are generally 0.5~2 mm under a polarization microscope. The electron probe analytical results for pyroxene are listed in Table 2. The analytical results show that pyroxene is mainly composed of SiO_2 (average = 54.34%), CaO (average = 25.20%), MgO (average = 11.60%) and FeO (average = 9.05%); small amounts of Al_2O_3, MnO and Na_2O; and trace amounts of TiO_2, Cr_2O_3 and K_2O (Table 2). The pyroxene in the Da'anhe deposit is diopside rich ($Di_{65.31\text{–}70.25}Hd_{28.67\text{–}33.65}$), with small amounts of hedenbergite and johannsenite, belonging to the diopside-hedenbergite solid solution series. In addition, on the discrimination diagram of pyroxene endmembers, the pyroxene plots in the range of typical skarn Au deposits worldwide (Figure 6D).

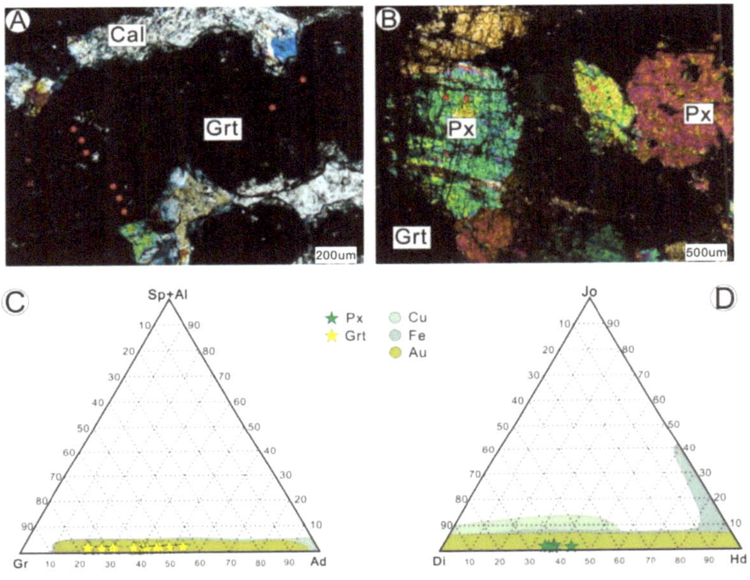

Figure 6. (**A**,**B**): Skarn mineral microscopic characteristics of the Da'anhe gold deposit (cross-polarized light); (**C**,**D**): Composition of garnet (data taken from Table 1) and pyroxene (data taken from Table 2) from gold skarn deposits in the study area and comparison with garnet and pyroxene from other types of skarn deposits in the world [1]. The red dot represents the testing location.

Table 1. Electron microprobe analyses of garnet from the Da'anhe deposits.

Sample	Unit	1-1	1-2	1-3	1-4	2-1	2-2	2-3	2-4	1-4-1	1-4-2	1-4-3	1-4-4	2-1-1	2-1-2	2-1-3	2-1-4
SiO_2	wt.%	37.60	38.45	38.10	37.29	36.85	36.84	37.82	37.68	38.09	37.89	38.05	39.06	38.14	38.48	38.36	38.14
TiO_2	wt.%	0.79	0.56	0.64	0.76	1.01	0.69	0.94	0.70	0.59	0.35	0.96	0.12	1.01	0.80	0.87	0.91
Al_2O_3	wt.%	11.25	13.24	12.79	11.24	9.10	9.72	11.99	12.48	13.21	11.76	12.55	14.98	14.35	16.02	15.79	15.87
Cr_2O_3	wt.%	0.07	0.00	0.09	0.04	0.17	0.01	0.06	0.00	0.00	0.00	0.00	0.00	0.18	0.12	0.18	0.21
FeO	wt.%	13.51	10.50	11.50	12.88	15.75	14.98	12.35	11.57	11.44	13.60	12.08	9.41	8.15	6.94	6.97	6.83
MnO	wt.%	0.19	0.16	0.22	0.19	0.33	0.28	0.16	0.29	0.12	0.26	0.11	0.25	0.17	0.19	0.17	0.15
MgO	wt.%	0.18	0.14	0.11	0.13	0.23	0.16	0.21	0.13	0.07	0.09	0.11	0.03	0.59	0.60	0.56	0.68
CaO	wt.%	35.47	36.04	35.82	35.95	34.30	35.41	35.64	35.89	35.78	35.11	36.00	35.84	35.07	35.48	35.80	35.45
Na_2O	wt.%	0.10	0.05	0.02	0.04	0.02	0.07	0.05	0.03	0.00	0.02	0.02	0.04	0.10	0.08	0.06	0.05
K_2O	wt.%	0.02	0.03	0.01	0.00	0.00	0.03	0.00	0.01	0.01	0.00	0.01	0.00	0.03	0.12	0.00	0.04
Total	wt.%	99.17	99.15	99.28	98.52	97.76	98.20	99.21	98.78	99.31	99.08	99.88	99.74	97.77	98.82	98.75	98.33
Based on 12 oxygen atoms																	
Si	apfu	2.97	3.01	2.99	2.97	2.98	2.96	2.98	2.98	2.98	2.99	2.97	3.02	3.01	2.99	2.98	2.98
Ti	apfu	0.05	0.03	0.04	0.05	0.06	0.04	0.06	0.04	0.03	0.02	0.06	0.01	0.06	0.05	0.05	0.05
Al	apfu	1.05	1.22	1.18	1.06	0.87	0.92	1.11	1.16	1.22	1.09	1.15	1.36	1.33	1.47	1.45	1.46
Cr	apfu	0.00	0.00	0.01	0.00	0.01	0.00	0.00	0.00	0.00	0.00	0.00	0.00	0.01	0.01	0.01	0.01
Fe3+	apfu	0.89	0.69	0.75	0.86	1.06	1.01	0.81	0.76	0.75	0.90	0.79	0.61	0.54	0.45	0.45	0.45
Fe2+	apfu	0.00	0.00	0.00	0.00	0.00	0.00	0.00	0.00	0.00	0.00	0.00	0.00	0.00	0.00	0.00	0.00
Mn	apfu	0.01	0.01	0.01	0.01	0.02	0.02	0.01	0.02	0.01	0.02	0.01	0.02	0.01	0.01	0.01	0.01
Mg	apfu	0.02	0.02	0.01	0.02	0.03	0.02	0.02	0.01	0.01	0.01	0.01	0.00	0.07	0.07	0.06	0.08
Ca	apfu	3.01	3.02	3.01	3.07	2.97	3.05	3.01	3.04	3.00	2.97	3.01	2.97	2.96	2.95	2.98	2.97
Ura	wt.%	0.22	0.00	0.28	0.13	0.53	0.03	0.20	0.00	0.00	0.00	0.00	0.00	0.56	0.38	0.55	0.63
Ad	wt.%	44.12	33.82	37.23	41.57	52.89	48.92	40.09	37.34	37.24	44.90	39.03	30.54	26.49	22.28	22.22	21.89
Pyr	wt.%	0.69	0.52	0.43	0.51	0.90	0.63	0.81	0.49	0.26	0.37	0.44	0.13	2.27	2.27	2.10	2.59
Spe	wt.%	0.41	0.34	0.48	0.41	0.75	0.61	0.35	0.62	0.25	0.58	0.23	0.55	0.37	0.42	0.37	0.33
Gro	wt.%	54.55	65.32	61.57	57.37	44.93	49.81	58.56	61.55	62.24	54.14	60.30	68.79	70.32	74.66	74.76	74.56
Alm	wt.%	0.00	0.00	0.00	0.00	0.00	0.00	0.00	0.00	0.00	0.00	0.00	0.00	0.00	0.00	0.00	0.00

Note: Ura-Uvarovite; Ad-Andradite; Pyr-Pyrope; Spe-Spessartite; Gro-Grossular; Alm-Almandine.

Table 2. Electron microprobe analyses of pyroxene from the Da'anhe deposits.

Sample	Unit	DAH-1-1	DAH-1-2	DAH-1-3	DAH-1-4	DAH-1-5
SiO_2	wt.%	54.11	54.22	53.95	54.59	54.84
TiO_2	wt.%	0.00	0.00	0.04	0.03	0.01
Al_2O_3	wt.%	0.15	0.34	0.34	0.14	0.16
Cr_2O_3	wt.%	0.03	0.05	0.08	0.03	0.02
FeO	wt.%	8.62	8.72	10.10	8.62	9.21
MnO	wt.%	0.33	0.13	0.31	0.33	0.35
MgO	wt.%	11.88	11.79	11.05	11.90	11.40
CaO	wt.%	25.51	25.24	24.99	25.28	24.98
Na_2O	wt.%	0.11	0.18	0.16	0.08	0.10
K_2O	wt.%	0.05	0.00	0.01	0.00	0.04
Total	wt.%	100.78	100.68	101.02	100.99	101.11
			Based on 6 oxygen atoms.			
Al(iv)	apfu	0.00	0.00	0.00	0.00	0.00
Al(vi)	apfu	0.01	0.01	0.01	0.01	0.01
Ti	apfu	0.00	0.00	0.00	0.00	0.00
Cr	apfu	0.00	0.00	0.00	0.00	0.00
Fe^{3+}	apfu	0.00	0.00	0.00	0.00	0.00
Fe^{2+}	apfu	0.27	0.27	0.32	0.27	0.29
Mn	apfu	0.01	0.00	0.01	0.01	0.01
Mg	apfu	0.66	0.65	0.61	0.66	0.63
Ca	apfu	1.02	1.01	1.00	1.00	0.99
Na	apfu	0.01	0.01	0.01	0.01	0.01
K	apfu	0.00	0.00	0.00	0.00	0.00
Di	wt.%	70.21	70.25	65.31	70.21	67.83
Hd	wt.%	28.67	29.30	33.65	28.70	30.99
Jo	wt.%	1.12	0.45	1.04	1.09	1.18
Mn/Fe	——	0.02	0.01	0.02	0.02	0.02

Note: Di-Diopside; Hd-Hedenbergite; Jo-Johannsenite.

5.2. Results of FI Analysis

5.2.1. Petrographic and Microthermometric Results for FIs

In this study, a petrographic microscope was used to observe the FIs in garnet, diopside, epidote, quartz and calcite in the Da'anhe mining area. Among them, the FIs in quartz and calcite were the most common, followed by those in garnet and diopside, and the FIs in epidote were the least common. Most of the FIs were primary FIs, with a small number of secondary FIs developed along mineral fissures. The sizes of the FIs vary and are concentrated in the range of 3~15 μm, and the maximum size is more than 20 μm. The morphologies are mainly elliptical and irregular, with minor long strip and spindle morphologies. According to the phase characteristics of FIs at room temperature, the FIs in the Da'anhe gold deposit can be divided into five types: daughter mineral-bearing three-phase FIs (S type), liquid-rich FIs (W1 type), gas-rich FIs (W2 type), pure gas FIs (PV) and pure liquid FIs (PL).

(1) S-type FIs: S-type FIs developed in the early skarn stage and late skarn stage and those that exhibit oval, spindle or irregular shapes. These FIs in garnet are larger (Figure 7A), and the FIs in epidote are smaller. The overall sizes range from 3~20 μm, and the gas-liquid ratios vary from 15% to 35%. Combined with the morphology of transparent daughter minerals in FIs and the findings of previous research [20,21], we believe that the crystalline minerals are mainly halite and sylvinite. These FIs are generally less common, accounting for approximately 10% of the total number of inclusions, and coexist with the W1-, W2- and PV-type inclusions (Figure 7A,C).

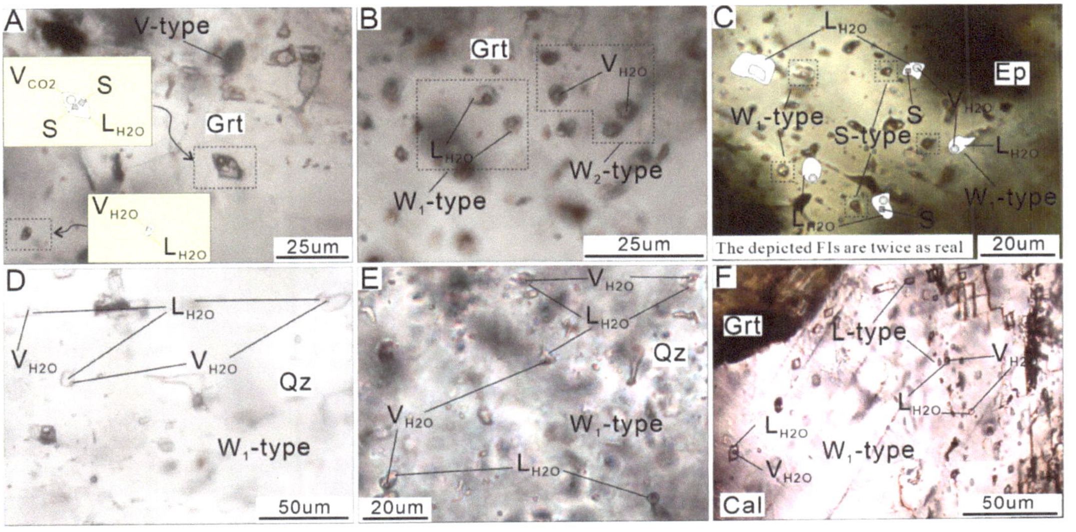

Figure 7. Microscopic photographs of fluid inclusions in the Da'anhe deposit. (**A**): W1-, S- and PV-type FIs coexist in garnet; (**B**): W1- and W2-type FIs coexist in garnet; (**C**): W1- and S-type FIs coexist in epidote; (**D**): W1-type FIs in early quartz-sulfide stage; (**E**): W1-type FIs in quartz of late quartz-sulfide stage; (**F**): W1- and PL-type FIs coexist in calcite of quartz-carbonate stage.

(2) W1-type FIs: These FIs formed throughout the metallogenic stage and are observed in garnet, diopside, epidote, quartz and calcite, especially in quartz. Most of the FIs are oval and irregularly shaped, range from 3~18 μm, and account for approximately 60% of the total number of FIs. The gas-liquid ratios of W1-type inclusions vary greatly from 5%~40%, and the gas-liquid ratios in the early quartz-sulfide stage inclusions are the lowest (Figure 7A–F).

(3) W2-type FIs: These FIs developed during all stages, except the quartz-carbonate stage, and appear in large quantities in quartz. The FIs are mostly uniform in size, with oval and irregular shapes and a small number of long strips, range in size from 5 to 15 μm, and account for approximately 25% of the total number of inclusions. The gas-liquid ratios of the W2 inclusions range between 65% and 80% (Figure 7B).

(4) PV-type and PL-type FIs: These two types of inclusions are less common in each mineralization stage and account for approximately 5% of the total number of inclusions. Among them, the PV-type FIs mainly developed during the skarn stage, while the PL-type FIs mainly developed during the quartz carbonate stage of late mineralization. The FIs are mostly elliptical and cuboid shaped, and a few are irregular or long strips, with sizes ranging from 2 to 15 μm (Figure 7A,F).

On the basis of petrographic observations, temperature measurements during the heating and cooling stages were carried out for the FIs in each mineralization. Most of the FIs used for these temperature measurement stages appear in groups, indicating that they are primary FIs. A total of 164 uniform temperature data points were obtained. The test results are listed in Table 3.

Table 3. Microscopic thermometry results of fluid inclusions in the Da'anhe gold deposit.

Stage	Sample	Minerals	Type	Gas-Liquid Ratio (V_G/V_T%)	T_d (°C)	$T_{m\text{-}ice}$ (°C)	T_h (°C)	Salinity (wt.% NaCl Equiv.)	Density (g/cm^3)	Number	Pressure (MPa)	Depth (Km)
I$_1$	**DAH-1**	**Grt**	**W1**	10~35		−12.7~−7.8	447~515	11.5~16.7	0.56~0.61	11	49.5~64.1	5.60~6.46
			W2			−10.8~−8.8	468~511	12.7~14.8	0.55~0.59	8	52.5~60.8	5.79~6.27
			S	15~30	393~431		477~503	46.7~51.0	1.56~1.70	5		
	DAH-1	Di	W1	15~40		−13.1~−8.5	420~520	12.3~17.1	0.57~0.66	11	46.7~65.3	5.43~6.52
			W2			−11.7~−7.6	453~511	11.2~15.8	0.57~0.58	6	48.7~62.2	5.56~6.35
			S	15~30	409~431		481~505	38.0~42.0	1.62~1.70	4		
I$_2$	DAH-2	Ep	W1	10~35		−8.5~−5.8	358~431	8.94~12.3	0.64~0.72	17	35.6~47.9	3.56~5.51
			W2			−7.3~−5.7	356~418	8.81~10.9	0.63~0.72	12	35.2~43.6	3.52~5.24
			S	10~30	336~366		385~428	41.1~43.9	1.45~1.50	5		
II$_1$	DAH-3	Q$_1$	W1	15~35		−6.3~−3.7	279~366	5.99~9.60	0.71~0.80	19	24.5~37.3	2.45~3.73
			W2	55~60		−6.4~−4.4	305~357	7.01~9.73	0.73~0.78	10	28.2~36.5	2.82~3.65
II$_2$	DAH-4	Q$_2$	W1	10~30		−5.7~−2.5	226~305	4.17~8.81	0.80~0.87	15	18.0~30.2	1.8~3.02
			W2	60~70		−4.2~−2.5	247~300	4.17~5.62	0.78~0.84	11	20.0~27.2	2.0~2.72
II$_3$	DAH-5	Cal	W1	15~25		−2.0~−0.5	137~247	0.87~3.37	0.82~0.94	30	9.25~18.8	0.93~1.88

Note: Grt-Garnet; Di-Diopside; Ep-Epidote; Q$_1$-Quartz from early quartz sulfide stage; Q$_2$-Quartz from late quartz sulfide stage; Cal-Calcite. T_d-disappearance temperature of daughter minerals; $T_{m\text{-}ice}$-freezing point temperatures; T_h-homogenization temperature; V_G-Volume of gas; V_T-Total Volume.

Early skarn stage: The FIs used for temperature measurements during the heating and cooling stages include garnet and diopside. The FIs are mainly W1- and S-type FIs, with a small number of W2- and PL-type FIs. In the same field of view, W1-, S- and PV-type FIs coexist (Figure 7A), indicating that the fluid underwent boiling during the evolution process. When the W1-type FIs are heated to 420~520 °C, they homogenize to the liquid phase (homogenization temperature, T_h). The freezing point temperatures ($T_{m\text{-}ice}$) of this type of inclusion vary from −13.1 °C to −7.8 °C and are obtained by cooling. After calculation, the corresponding salinities range from 11.5~17.1 wt.% NaCl equiv., and the densities range from 0.56~0.66 g/cm^3. When the W2-type FIs are heated to 453~511 °C, they uniformly transform to the gas phase. The $T_{m\text{-}ice}$ values of the W2-type FIs vary from −11.7 to −7.6 °C and are obtained by cooling. The salinities of the corresponding fluids range from 11.2~15.8 wt.% NaCl equiv. and the densities range from 0.55~0.59 g/cm^3. The daughter minerals in the S-type FIs disappear when heated to 393~431 °C (disappearance temperature of daughter minerals, T_d) and homogenize to the liquid phase when heated to 477~503 °C. Based on these results, the salinities and densities of the fluids are 38.0~51.0 wt.% NaCl equiv. and 1.56~1.70 g/cm^3, respectively (Figures 7 and 8A–D).

Late skarn stage: The FIs used for temperature measurements using the heating and cooling stage are from epidote. The FIs that developed in this stage were mainly W1, W2 and S-type FIs. The coexistence of different types of FIs in the same field of view is observed in epidote (Figure 7C), indicating that fluid boiling also occurred during this stage. When the W1-type FIs homogenize into the liquid phase, the temperature is 358~431 °C, the freezing point temperatures vary from −8.5 to −5.8 °C, the corresponding salinities are 8.94~12.3 wt.% NaCl equiv., and the densities are 0.64~0.72 g/cm^3 (Figures 7 and 8E,F). W2-type FIs homogenize into the gas phase at temperatures of 356~418 °C, the freezing point temperature varies from −7.3 to −5.7 °C, and their corresponding salinities are 8.81~10.9 wt.% NaCl equiv. and the density is 0.63~0.72 g/cm^3. The daughter minerals in the S-type inclusions disappear at 336~366 °C, and they homogenize into the liquid phase at 385 to 428 °C. After calculation, their corresponding salinities range from 41.1~43.9 wt.% NaCl equiv., and their densities are 1.45~1.50 g/cm^3 (Figures 7 and 8E,F).

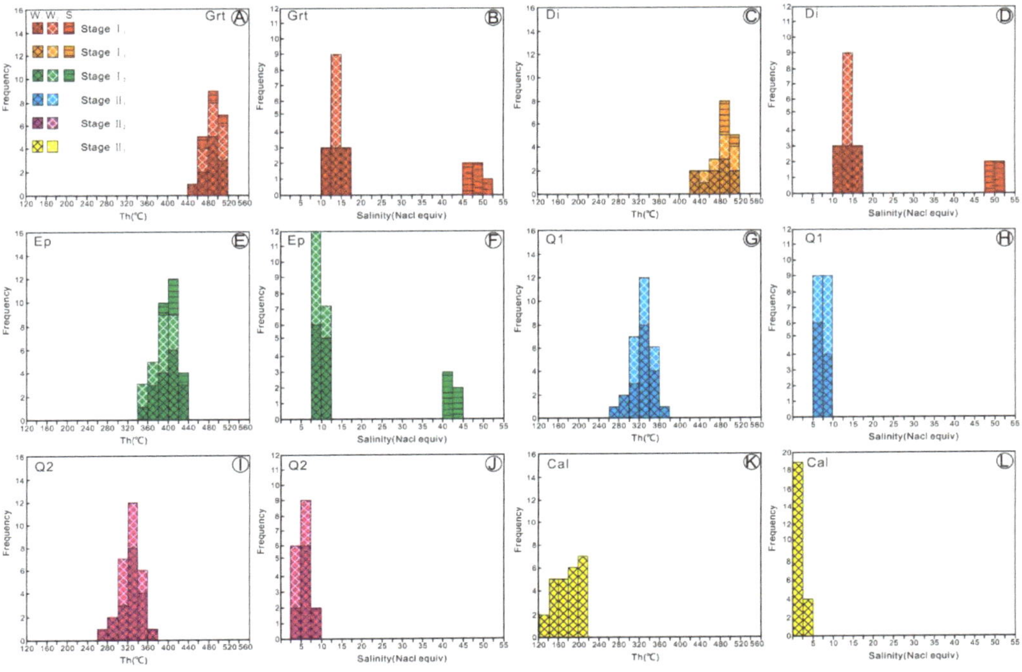

Figure 8. Histograms of total homogenization temperatures (T_h) and salinities of FIs in different stages. (**A–D**): Stage I_1; (**E,F**): Stage I_2; (**G,H**): Stage II_1; (**I,J**): Stage II_2; (**K,L**): Stage II_3.

Early quartz-sulfide stage: The FIs used for temperature measurements using the heating and cooling stage are from quartz. The FIs are mainly W1 and W2 type FIs, and no S-type FIs are found (Figure 7). The W1-type FIs were heated and homogenized to the liquid phase. The homogenization temperatures range from 279 to 366 °C, the freezing point temperatures vary from −6.3 to −3.7 °C, the calculated corresponding salinities are 5.99~9.60 wt.% NaCl equiv., and the densities are 0.71~0.80 g/cm^3. For the W2-type FIs, the homogenization temperatures, freezing point temperatures, and corresponding salinities and densities are 305~357 °C, −6.4~−4.4 °C, 7.01~9.73 wt.% NaCl equiv., and 0.73~0.78 g/cm^3, respectively (Figure 8G,H).

Late quartz-sulfide stage: The FIs used for temperature measurements using the heating and cooling stage are from quartz. Petrography revealed that the types of FIs in this stage are similar to those in the early quartz-sulfide stage, i.e., mainly W1- and W2-type FIs (Figure 7E). In this stage, the W1-type FIs were heated and homogenized to the liquid phase. The homogenization temperatures range from 226 to 305 °C, and the freezing point temperatures vary from −5.7 to −2.5 °C. After calculation, their corresponding salinities range from 4.17~8.81 wt.% NaCl equiv., and their densities range from 0.80~0.87 g/cm^3. For the W2-type FIs, the homogenization temperatures range from 247~300 °C, the freezing point temperatures vary from −4.2 to −2.5 °C, their corresponding salinities are 4.17~5.62 wt.% NaCl equiv., and their densities are 0.78~0.84 g/cm^3 (Figure 8I,J).

Quartz-carbonate stage: The FIs developed in calcite veins were selected for microscopic analysis. The petrography shows that only W1- and PL-type FIs developed during this stage. The homogenization temperatures of the W1-type FIs range from 137 to 247 °C, and the freezing point temperatures vary from −2.0~−0.5 °C. After calculation, their corresponding salinities are 0.87~3.37 wt.% NaCl equiv., and their fluid densities are 0.82~0.94 g/cm^3 (Figure 8K,L).

5.2.2. Laser Raman Spectral Analysis Results

To further understand the compositions of the FIs, the authors of this paper selected representative FIs from different mineralization stages of the Da'anhe gold deposit for Laser Raman spectral analysis. Because W2-type FIs are not well developed, W1-type FIs were used.

The Laser Raman spectral results show that the gas phase components of the W1-type FIs in the early skarn stage are CH_4 and H_2O (Figure 9A,B). The CH_4 and H_2O peaks of the W1-type FIs in the late skarn stage are weaker than those in the early skarn stage, and CO_2 is present (Figure 9C). The gas phase composition of the W1-type FIs in the early quartz-sulfide stage is mainly H_2O with a small amount of CH_4 (Figure 9D). In the late quartz-sulfide stage and quartz-carbonate stage, the gas phase components in the W1-type FIs are CH_4 and H_2O, and the peak intensity of CH_4 gradually increases (Figure 9E,F).

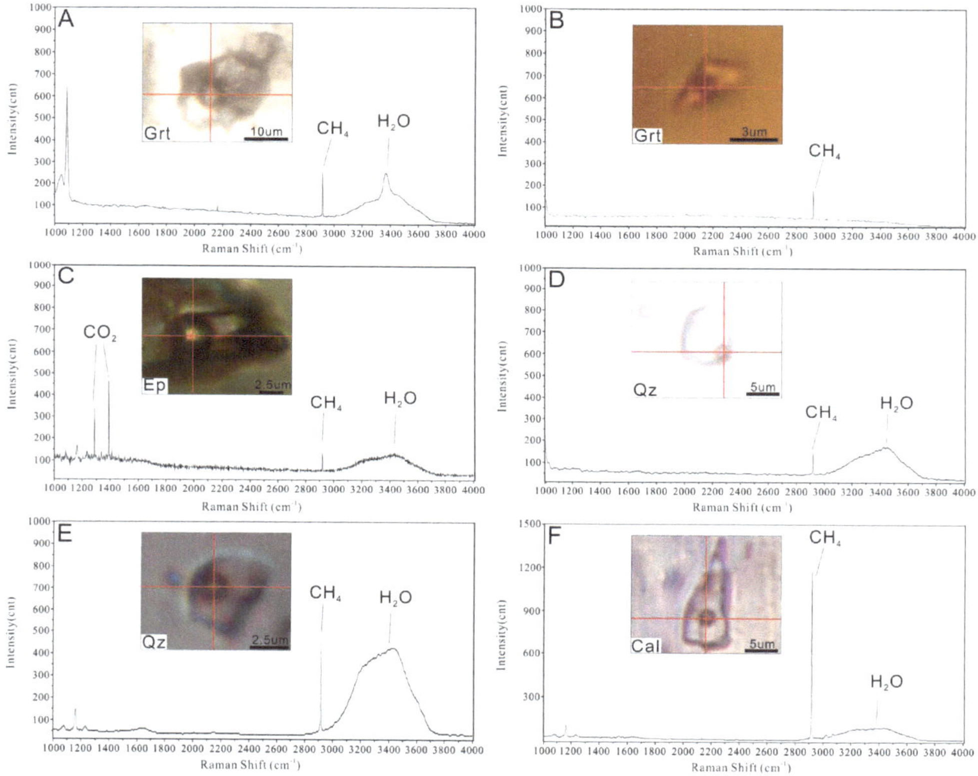

Figure 9. Representative Laser Raman spectra of FIs in garnet, epidote, quartz and calcite grains of different stages from the Da'anhe deposit. (**A**,**B**): Spectrum for vapor phase of W1−type FIs hosted in garnet grain of the dry skarn stage; (**C**): Spectrum for vapor phase of W1−type FIs hosted in epidote grain of the wet skarn stage; (**D**): Spectrum for vapor phase of W1−type FIs hosted in quartz grain of the early quartz sulfide stage; (**E**): Spectrum for vapor phase of W1−type FIs hosted in quartz grain of the late quartz sulfide stage; (**F**): Spectrum for vapor phase of W1−type FIs hosted in calcite grain of the quartz-carbonate stage.

In general, the gas components in the ore-forming fluid of the Da'anhe deposit are mainly CH_4, H_2O and a small amount of CO_2. The process of mineralization is commonly accompanied by a transition from an oxidizing environment to a reducing environment. Therefore, we suggest that the large amount of CH_4 present during the skarn stage was provided by carbonate rocks, and that the CO_2 in the fluid may have escaped in large quantities. As the mineralization entered the early quartz sulfide stage, the abundance of CH_4 gradually decreased, and that of CO_2 gradually increased. During the late quartz sulfide stage and quartz carbonate stage, a large amount of CO_2 in the fluid was consumed, the ore-forming system transformed into a reducing environment, and large amounts of CH_4 accumulated. Therefore, the initial ore-forming fluid was a H_2O-CO_2-NaCl system (Figure 9).

5.3. H-O Isotope Analytical Results

The results of mineral H-O isotope tests for different metallogenic stages in the Da'anhe gold deposit (Figure 10, Table 4) [2,43] reveal that the δD_{H2O} values of garnet samples in the early skarn stage vary from −89.1‰ to −85.9‰, and the $\delta^{18}O_{V\text{-}SMOW}$ values vary from 6.1‰ to 7.0‰. Combined with the results of FI temperature measurements (the average value is 490 °C), the $\delta^{18}O_{H2O}$ values of the fluid are calculated and vary from 7.57‰ to 8.47‰. In the late skarn stage, the δD_{H2O} value of epidote is −73.3‰, and the $\delta^{18}O_{V\text{-}SMOW}$ value is 5.4‰. After calculation, the $\delta^{18}O_{H2O}$ value of the fluid is 5.71‰ (T_h is 393 °C). The δD_{H2O} values of the quartz samples in the late quartz-sulfide stage vary from −79.3‰ to −102.5‰, and the $\delta^{18}O_{V\text{-}SMOW}$ values range from 8.7‰ to 14.1‰. After recalculation, the $\delta^{18}O_{H2O}$ values of the fluid vary from 0.25‰ to 5.65‰ (T_h is 261 °C). The δD_{H2O} values of calcite samples in the quartz-carbonate stage are concentrated at −116‰, and the $\delta^{18}O_{V\text{-}SMOW}$ value is 11.6‰. The calculated $\delta^{18}O_{H2O}$ value of the fluid is −0.77‰ (T_h is 190 °C).

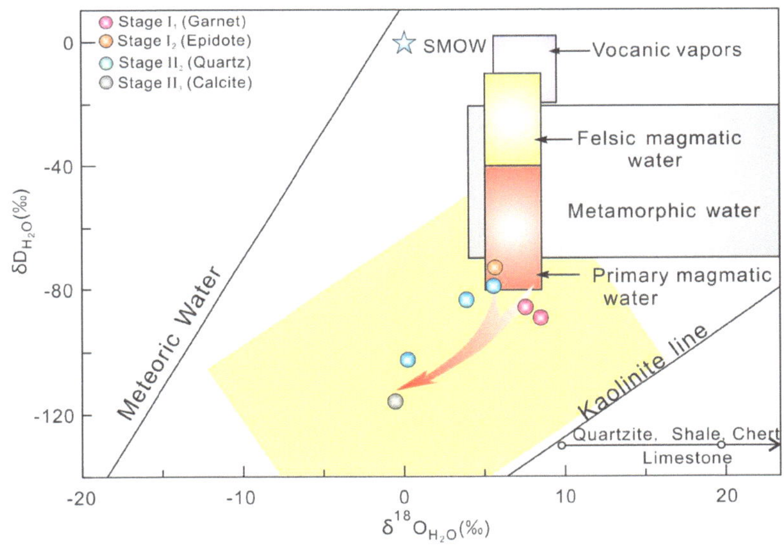

Figure 10. Hydrogen and oxygen isotopic compositions of the ore−forming fluid in the Da'anhe deposit (data taken from Table 4, modified after Hedenquist and Lowenstern [2]).

Table 4. Hydrogen and oxygen isotopic compositions of fluids from the Da'anhe Au deposit.

Sample	Mineral	*Th* (°C)	$\delta^{18}O_{V\text{-}SMOW}$ (‰)	$\delta^{18}O_{H2O}$ (‰)	δD_{H2O} (‰)	Data Sources
19DAH-2	Garnet	490	6.1	7.57	−85.9	This study
19DAH-3			7.0	8.47	−89.1	
19DAH-3	Epidote	393	5.4	5.71	−73.3	
19DAH-7	Quartz	261	8.7	0.25	−102.5	
DKDA3			12.4	3.95	−83.4	[43]
DKDA4			14.1	5.65	−79.3	
19DAH-7	Calcite	190	11.6	−0.77	−116	This study

6. Discussion

6.1. Implications of Skarn Minerals for Mineralization

The chemical composition of skarn minerals can be characterized by physical and chemical conditions, such as temperature, pressure, depth, pH and oxygen fugacity during their formation, which can provide important information on the formation environment of deposits [44–46]. It is generally believed that andradite and magnesium-rich diopside are formed in high-temperature environments with strong oxidation reactions and weak acidity, while grossular and hedenbergite are often formed in environments with high temperature, weak oxidation, reduction and alkalinity [45,47,48]. The garnet in the Da'anhe gold deposit is mainly grossular (Figure 6C), and pyroxene belongs to the diopside-hedenbergite solid solution series (Figure 6D). Both coexist in the outer belt of the skarn, and the pyroxene content is greater than the garnet content. The above results show that the Da'anhe deposit formed in a reducing environment related to an intrusive body and is a typical calcareous reduced skarn Au deposit. In addition, our team conducted electron probe analysis on the hornblende and biotite in the gabbroic diorite, and the results showed that the gabbro-diorite magma related to mineralization had reducing characteristics (ΔFMQ = −0.95~0.60, with an average value of 0.17; log(fO_2) = −15.65~−14.68, with an average value of −15.11 [49]), which indicates that the Da'anhe deposit is a reduced skarn gold deposit.

6.2. Source of the Ore-Forming Materials

Previous studies have shown that skarn deposits in the Lesser Xing'an Range generally lack sulfate minerals [49]. Therefore, the $\delta^{34}S$ values of the metal sulfide minerals in the ore can be used to represent the $\delta^{34}S$ values of the ore-forming fluid of the deposit [50]. To determine the source of ore-forming materials in the Da'anhe gold deposit, S-Pb isotope data from metal sulfides in the Da'anhe gold deposit were collected and are listed in Tables 5 and 6. The S-Pb isotope data show that the $\delta^{34}S$ values of pyrrhotite in the Da'anhe deposit vary from −0.6‰ to 0.5‰ (with an average of 0.10‰), the $\delta^{34}S$ values of pyrite vary from −2.9‰ to −0.6‰ (with an average of −3.07‰), and the arithmetic average of the sulfides is −0.83‰. The $\delta^{34}S$ values of sulfides in the Da'anhe deposit are within the range of typical porphyry-skarn deposits (−5.0‰ to 5.0‰, [51]) (Figure 11) [50,52], indicating that the sulfur in the Da'anhe deposit may have mainly come from the magmatic system. The Pb isotope values of the metal sulfides in the Da'anhe deposit vary greatly, and the $^{206}Pb/^{204}Pb$, $^{207}Pb/^{204}Pb$ and $^{208}Pb/^{204}Pb$ ratios are 18.099~18.568, 15.532~15.580 and 37.909~38.176, respectively. In the $^{207}Pb/^{204}Pb$-$^{206}Pb/^{204}Pb$ projection (Figure 12) [53], the sample points are mainly located in the lower crust area. According to the $^{208}Pb/^{204}Pb$-$^{206}Pb/^{204}Pb$ diagram, the sample points are located mainly on both sides of the orogenic belt evolution line, and some of them are located on the upper crust evolution line, which indicates that the Pb in the Da'anhe sulfide has the characteristics of mixed lead. These results imply that the ore-forming materials may have been derived from crust-mantle mixing. The Sr-Nd isotopes of the medium-grained gabbroic diorite related to the mineralization of the Da'anhe gold deposit have been studied by Yang et al. [54] (εNd(t) = 0.8~1.9, εSr(t) = 75.9~77.2, and (^{87}Sr/^{86}Sr)$_i$ = 0.70962~0.70972, T_{DM2} = 815–1076 Ma); their results indicate that the mineralization originated from Neoproterozoic lower crust metasomatized by fluids or weakly enriched mantle and that the ascending material was contaminated by young crustal materials during the process of ascent and emplacement [54]. This result further shows that the ore-forming material of the Da'anhe deposit was derived from gabbro-dioritic magma.

Table 5. Sulfur isotopic composition of sulfides from the Da'anhe Au deposit.

Sample	Mineral	δ^{34}S‰	Data Sources
Dah4-Pyr	Pyr	0.5	
Dah5-Pyr	Pyr	0.5	
Dah7	Pyr	−0.6	
Dah4-Py	Py	−0.6	[21]
Dah5-Py	Py	−2.9	
Dah6-Py	Py	−2.7	
Dah6-Pyr	Pyr	0	

Note: Pyr-Pyrrhotite; Py-Pyrite.

Table 6. Lead isotopic composition of sulfides from the Da'anhe Au deposit.

Sample	Mineral	^{206}Pb/^{204}Pb	^{207}Pb/^{204}Pb	^{208}Pb/^{204}Pb	Data Sources
Dah4-Pyr	Pyr	18.308	15.543	38.139	
Dah4-Py	Py	18.508	15.578	38.176	
Dah5-Pyr	Pyr	18.303	15.536	38.125	
Dah5-Py	Py	18.099	15.556	38.071	[21]
Dah6Py	Py	18.317	15.532	38.098	
Dah6Pyr	Pyr	18.568	15.580	37.909	
Dah7	Pyr	18.455	15.577	38.103	

Note: Pyr-Pyrrhotite; Py-Pyrite.

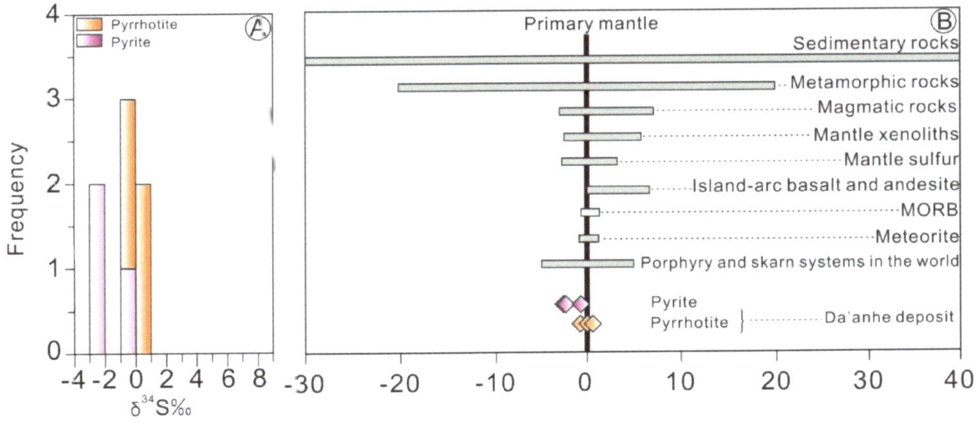

Figure 11. (**A**) Histogram of the sulfur isotopic compositions of sulfides from the Da'anhe deposit; (**B**) Ranges of δ^{34}S values of selected geologically important sulfur reservoirs [50,52] and the Da'anhe deposit (data taken from Table 5). The orange and purple squares represent magnetite and pyrite, respectively. The gray squares represent different geologically important sulfur reservoirs.

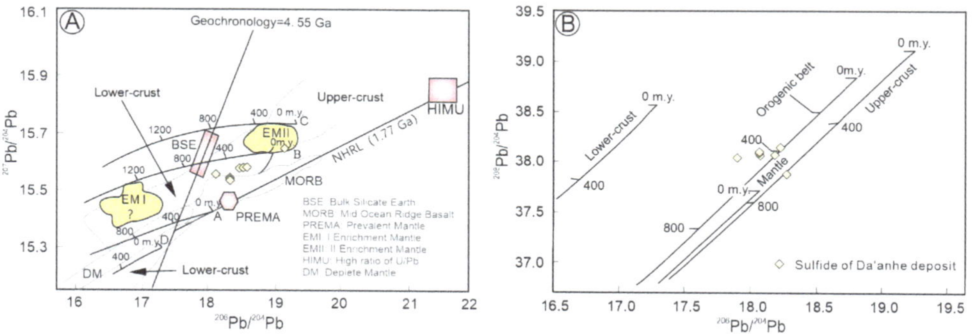

Figure 12. Diagram of lead isotopic compositions of the Da'anhe deposit. (**A**)—$^{207}Pb/^{204}Pb$ v.s. $^{206}Pb/^{204}Pb$; (**B**)—$^{208}Pb/^{204}Pb$ v.s. $^{206}Pb/^{204}Pb$ (data taken from Table 6, modified after Zartman and Doe [53]).

6.3. *Source and Evolution of Ore-Forming Fluids*

For the source of ore-forming fluids in skarn deposits, magmatic water is generally the dominant source, followed by meteoric water [55,56], and the duration and amount of meteoric water can vary according to the geological characteristics of different deposits. Several geologists [55,57] have suggested that the different alteration characteristics of the early skarn stage and late skarn stage reflect the dominant roles of magmatic water and meteoric water, respectively. Meinert et al. [58,59] suggested that these two processes may be the result of the evolution of a single magmatic hydrothermal system. The FIs in the skarn minerals of the Da'anhe gold deposit are mainly W1- and W2-type FIs and a small number of S-type FIs. According to the crystal form, the daughter crystals are mainly halite and sylvinite. Laser Raman spectroscopy shows that the gas phases involved in the early skarn stage were H_2O and CH_4 and that CO_2-containing FIs began to appear in the late skarn stage. The homogenization temperatures of the FIs that developed in the early skarn and late skarn stages are 393~515 °C and 336~431 °C, respectively, and their corresponding salinities are 11.2~51.0 wt.% NaCl equiv. and 8.81~43.9 wt.% NaCl equiv., respectively. The above characteristics indicate that the initial ore-forming fluid was an H_2O-NaCl-CO_2 system with a high temperature and high salinity. The main types of FIs in the quartz-sulfide period are W1 and W2. Laser Raman analysis revealed that the gas phases in the FIs are mainly H_2O and CH_4. The homogenization temperatures of the FIs from the early to late stages are 279~366 °C, 226~305 °C and 137~247 °C. The corresponding salinities are 5.99~9.73 wt.% NaCl equiv., 4.17~8.81 wt.% NaCl equiv. and 0.87~3.37 wt.% NaCl equiv., respectively. The above characteristics indicate that the ore-forming fluids in this period were part of a medium- to low temperature and low salinity H_2O-CH_4 system. The H-O isotope data show that the δD_{H2O} and $\delta^{18}O_{H2O}$ values ($\delta D_{H2O} = -89.1‰ \sim -73.3‰$; $\delta^{18}O_{H2O} = 5.71‰ \sim 8.47‰$) of the skarn minerals are close to the range of magmatic water during the skarn subperiod (Figure 10) [2], and the δD_{H2O} and $\delta^{18}O_{H2O}$ values ($\delta D_{H2O} = -79.3‰ \sim -116‰$; $\delta^{18}O_{H2O} = -0.77‰ \sim 5.65‰$) of quartz in the quartz-sulfide period show a trend of gradual migration towards those of meteoric water (Figure 10) [2]. These results indicate that the initial ore-forming fluid of the Da'anhe gold deposit was derived from magmatic water that dissolved into the gabbro-dioritic magma. With continuous mineralization, the fluid mixed with meteoric water during the early stage of the quartz-sulfide period, and the mixing proportion gradually increased during the subsequent mineralization process until the end of mineralization at which point the system was dominated by meteoric water.

6.4. Metallogenic Pressure and Depth Estimation

To better understand the physical and chemical conditions involved in the mineralization process of the Da'anhe deposit, we calculated the pressure and depth at each stage.

Previous studies [60,61] have suggested that the trapping pressure can be estimated only when the exact trapping temperature is known or when FIs are trapped during phase separation. The coexistence of S-type, W1-type and W2-type FIs with similar temperature ranges occurred during the skarn period of the Da'anhe gold deposit, indicating that fluid immiscibility occurred during the early stage of mineralization. Previous studies have shown that cooling to a temperature below 400 °C leads to a transition from ductile to brittle rock behaviour, and therefore causes a change from lithostatic to hydrostatic pressure conditions [62]. The FIs in the quartz-sulfide period formed below 400 °C (137–366 °C), which indicates that the metallogenic system changed from lithostatic to hydrostatic pressure conditions. Therefore, the trapping pressures and depth were estimated using Flincor software [60]. The calculated results show that the metallogenic pressures during each mineralization stage (I_1, I_2, II_1, II_2 and II_3) in the Da'anhe deposit were 46.7~65.3 MPa, 35.2~47.9 MPa, 24.5~37.3 MPa, 18.0~30.2 MPa and 9.25~18.8 MPa, corresponding to mineralization depths of 5.43~6.52 km (average of 6.03 km), 3.52~5.51 km (average of 4.44 km), 2.45~3.73 km (average of 3.11 km), 1.80~3.02 km (average of 2.27 km) and 0.93~1.88 km (average of 1.36 km), respectively. As mentioned above, gold precipitated mainly during the quartz sulfide stage, so the depth of gold deposit formation ranged from 2.27~3.11 km. At present, the orebodies are close to the surface (<500 m), indicating strong uplift and denudation after the mineralization period. This result also corresponds to the denudation depth of approximately 3.0 km in the Lesser Xing'an Range since the Jurassic [6]. Based on the above characteristics, we suggest that the Da'anhe deposit has undergone severe erosion, making it difficult to identify industrial ore bodies at depth (Figure 13).

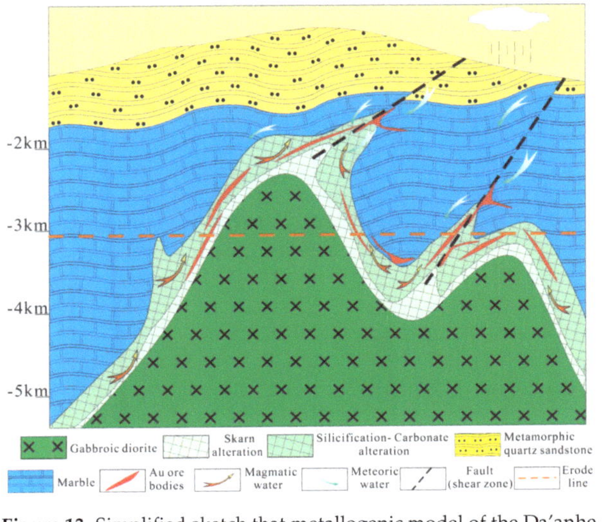

Figure 13. Simplified sketch that metallogenic model of the Da'anhe Au deposit.

6.5. Metallogenic Mechanism and Process

In magmatic-hydrothermal metallogenic systems, fluid boiling and fluid mixing are considered the most effective ways to restrict metal enrichment and precipitation [63]. The S-Pb isotopic characteristics of the metal sulfides and the H-O isotopic characteristics of the gangue minerals indicate that the initial ore-forming materials and ore-forming fluids of the Da'anhe gold deposit originated from gabbro-dioritic magma related to mineralization (Figures 10–12). The temperature of the magma chamber in the deep part of the Da'anhe deposit was very high, and the surrounding rock exhibited ductile characteristics, thereby preventing interactions between the magma solution fluid and the external fluid. [62]. When the initial ore-bearing fluid rose to the shallow part of the upper crust along with the magma and met the carbonate (approximately 6.03 km), ore-bearing fluids with high temperatures (T = 393~520 °C) and intermediate to high salinities (S = 11.2%~51.0% NaCl equiv.) coexisted, and the reducing fluid underwent double metasomatism with the carbonate to form grossular and diopside (Stage I_1) (Figure 4C,D). The chemical reactions are as follows:

$$3CaCO_3 + Al_2O_3 + 3SiO_2 = Ca_3Al_2(SiO_4)_3(\text{grossular}) + 3CO_2\uparrow$$

$$CaCO_3 + MgCO_3 + 2SiO_2 = CaMgSi_2O_6(\text{diopside}) + 2CO_2\uparrow$$

Garnet and pyroxene contain three types of FIs (Figure 7A,B): S-type, W1-type and W2-type FIs, indicating that immiscibility caused by fluid boiling occurred in the early stage of mineralization. This process resulted in phase separation and the coexistence of different types of FIs in similar temperature ranges (Figure 8A–D) [64–66]. In the late skarn stage (Stage I_2), the ore-bearing hydrothermal solution continued to rise (reaching approximately 4.44 km). Although the overall temperature of the hydrothermal solution decreased during this time, it was still part of the hydrothermal system with high temperatures (T = 336~431 °C) and high salinities (S = 2.06%~43.9% NaCl equiv.) (Figure 8E,F). The FI evidence indicates that the ore-forming fluid began to shift to a subcritical state during this stage, and anhydrous skarn minerals, such as garnet and pyroxene, which formed during the early stage, underwent degenerative alteration and were replaced by water-bearing skarn minerals, such as epidote (Figure 5B,E), tremolite (Figures 4F and 5C), actinolite (Figure 4E) and chlorite. This process was accompanied by the precipitation of a small amount of magnetite (Figure 5G). The equations for the reactions in the late skarn stage is as follows [67]:

$$Ca_3Fe_2(SiO_4)_3\ (\text{andradite}) + 4CO_2 + 3/2O_2 = 4CaCO_3 + 3SiO_2 + Ca_2FeSi_3O_{13}\ (\text{epidote}) + Fe_3O_4\ (\text{magnetite})$$

$$5CaFe_2(SiO_4)_3\ (\text{hedenbergite}) + 3CO_2 + H_2O = Ca_2(Mg, Fe)_5Si_8O_{22}(OH)_2\ (\text{actinolite}) + 3CaCO_3 + 2SiO_2$$

Studies have shown that the process of skarn formation produces a large amount of free space, resulting in the release of pressure in the metallogenic system as well as boiling [10]. We found that the skarn minerals that formed early in the mining area were broken into breccia and cemented by quartz, chalcopyrite and pyrite (Figure 4G), indicating a sudden release of pressure during the skarn period. The petrographic study of FIs in the Da'anhe deposit shows that different types of FIs in skarn minerals (garnet, pyroxene and epidote) developed during the skarn stage (Figure 7A–C). According to the Th-salinity correlation diagram of FIs, high- and low-salinity FIs coexist within similar temperature intervals (Figure 8A–F), indicating that the ore-forming fluids experienced boiling [61].

The quartz-sulfide stage (Stage II_1 and Stage II_2) was the gold mineralization stage (Figures 4I and 5L). When the gold started to precipitate, the ore-forming fluid was part of a medium- to high-temperature and medium- to low-salinity reducing system (T = 226~366 °C; S = 4.17%~9.73% NaCl equiv., with CH_4 and H_2O as the main gases) (Figures 8G–J and 9D,E). The ranges of temperature and pressure were relatively large. Studies have shown that gold migrates mainly in the form of chloride complexes in low-to high-temperature hydrothermal systems [68,69], and contributes to the control of the stability of metal complexes in ore-forming fluids because a decrease in temperature leads to the decomposition of metal chlorides and the precipitation of minerals [61,70]. Therefore, the solubility of Au in an ore-bearing fluid gradually decreases as the ore-forming fluid temperature decreases, and Au gradually precipitates in the ore-forming system.

In addition, fluid boiling during the skarn period caused the H_2O, H_2S, HCl and CO_2 components in the gas phase from escape to the original uniform fluid phase; this process resulted in an increase in the pH of the fluid, an increase in the concentrations of reduced sulfur and metal elements, and a decrease in the chloride ion concentration [71–73]. In addition, the fluid that separated from the magma underwent a water-rock reaction with the surrounding rock (Figure 10), causing CH_4 to enter the mineralization system, leading to a continued decrease in the oxygen fugacity of the fluid [74–76]. Changes in the physical and chemical conditions of ore-forming fluids can promote the decomposition of metal complexes and eventually lead to the reaction of Fe^{3+}, Cu^{2+}, Pb^{2+}, Zn^{2+} and other metal ions with HS^- ions in solution, forming small amounts of metal sulfides (pyrite, pyrrhotite, chalcopyrite, sphalerite, etc.) [69,77].

In the quartz-carbonate stage (Stage II_3), after boiling, mixing and water-rock reactions in the previous stages, the initial ore-bearing hydrothermal fluid changed to a low-temperature and low-salinity system (T = 137~247 °C; S = 0.87%~3.37% NaCl equiv.) (Figure 8K,L). At this time, meteoric water entered and dominated the metallogenic system (Figure 10), and large amounts of quartz and calcite began to precipitate from the fluid, forming quartz-carbonate veins and cutting through skarns and orebodies (Figure 4L) and marking the end of mineralization.

Overall, the mineralization in the Da'anhe deposit experienced an evolutionary process from high temperatures and intermediate to high salinities to low temperatures and low salinities. Fluid boiling and mixing were the mechanisms driving gold precipitation.

6.6. *Significant Indicators for Identifying Different Types of Skarn Deposits*

Numerous skarn deposits have developed in NE China, and skarn gold deposits are mainly associated with copper deposits (such as the Laozuoshan Cu-Au deposit and Yuejinshan Cu-Au deposit). The Da'anhe deposit is the only independent skarn gold deposit in this region. By analysing the genesis and mineralization characteristics of the skarn deposits in the region, we suggest that the reasons for the formation of the Da'anhe independent skarn gold deposit are as follows.

Skarn deposits in northeastern China mainly formed in the late Permian-Early Triassic period (Baoshan Cu deposit [10]; Laozuoshan Cu-Au deposit [78]), Jurassic (Cuihongshan Fe-polymetallic deposit [14]; Ergu Fe-Zn polymetallic deposit [12]), and Early Cretaceous (Huanggangliang Fe-Sn deposit [79]; Haobugao Pb-Zn deposit [48]; Xieertala Fe-Zn deposit [80]; Tongshan Cu-Mo deposit [81]), while independent skarn gold deposits formed only in the Early Jurassic period (184–203 Ma, [19,20,43]. From the perspective of magmatic properties related to mineralization, the gabbro-dioritic magma related to the mineralization of the Da'anhe deposit originated from lower crust or mantle wedge metasomatized by subduction fluid, and such magmas are characterized by reducing conditions and a high water content (ΔFMQ = 0.17, log(fO_2) = -15.11, water content = 6.80 wt.%, [49]). The magmas related to the mineralization of the skarn Fe-Cu-Pb-Zn-Ag-W-Sn deposit were mostly felsic magmas (Chaobuleng Fe-polymetallic deposit [82]; Xulaojiugou Pb-Zn deposit [16]) or highly differentiated alkaline granitic magmas (Cuihongshan Fe-polymetallic deposit [14]), which were mainly derived from partial melting of the crust. These magmas were typically weakly oxidized to oxidized and had a high water content (Xieertala Fe-Zn deposit [80]; Ergu Fe-Cu polymetallic deposit [49]). In terms of the temperature during metal precipitation, the temperature during the precipitation of skarn Fe-Cu deposits is relatively high, corresponding to medium- to high-temperature mineralization (235–402 °C, Ergu Fe-Cu polymetallic deposit [11]; 200–350 °C, Handagai Fe-Cu deposit [83]). The temperature during the precipitation of skarn Pb-Zn-Ag deposits varies significantly (Xiaoxilin Pb-Zn deposit, 188.6–398.3 °C [43]; Baiyinnuoer, ~350 °C [84]), The temperature at which gold precipitated in the Da'anhe deposit was mainly in the medium temperature range (226–305 °C). Therefore, we propose that the properties of the magma and ore-forming fluids may contribute to the specific mineralization of skarn deposits, which has guiding significance for identifying different types of skarn deposits.

7. Conclusions

(1) The Da'anhe deposit is a calcareous reduced skarn Au deposit related to the metasomatism of gabbro-dioritic magma and marble in the middle-deep part of the crust. The metallogenic process involved skarn and quartz-sulfide periods. Gold precipitated mainly in the late quartz sulfide stage and to a lesser extent in.

(2) The initial ore-forming fluid of the Da'anhe deposit was derived from the gabbro-dioritic magma. Boiling and mixing were the main mechanisms driving gold and sulfide precipitation. The ore-forming material was mainly derived from the gabbro-dioritic magma.

(3) The Da'anhe deposit was formed at depths ranging from 2.27 to 3.11 km, and is a medium to deep skarn gold deposit. Combined with regional observations, the rocks in the study area have been uplifted to the shallow surface (<500 m), and its prospecting potential is limited.

(4) The distinctive nature of the ore-forming magma (source, reducing conditions and high water content, ΔFMQ = 0.17, log(fO_2) = −15.11, water content = 6.80 wt.%) was key to the formation of the Da'anhe skarn gold deposit.

Author Contributions: Conceptualization, C.Z., F.S., J.W. and J.H.; Data curation, J.H., X.C., C.B., Z.X., L.Y. and S.H.; Formal analysis, J.S., J.W., D.Y., Z.X. and L.Y.; Funding acquisition, J.S. and J.W.; Investigation, C.Z., F.S., J.S., X.C. and C.B.; Methodology, C.Z., J.S., J.W. and X.C.; Project administration, C.Z., F.S., J.H. and S.H.; Resources, C.Z., F.S., J.S., J.W. and X.C.; Software, J.H., D.Y. and L.Y.; Supervision, C.Z., F.S., J.W., C.B. and Z.X.; Validation, C.B.; Visualization, J.S., J.H. and D.Y.; Writing—original draft, C.Z., F.S., J.S. and J.W. All authors have read and agreed to the published version of the manuscript.

Funding: This work was supported by the Second Tibetan Plateau Scientific Expedition and Research Program (STEP) (2022QZKK0201); the National Natural Science Foundation of China (Grant No. U20A2088; 42202068); the Kunlun Talented People of Qinghai Province, High-end Innovation and Entrepreneurship talents (Grant to Zhao Chuntao; E140DZ3901); the Special Research Assistant of Chinese Academy of Sciences (Grant to Zhao Chuntao); the Natural Science Foundation of Qinghai Province (No. 2020-ZJ-939Q); the Chinese Academy of Sciences Youth Innovation Promotion Association (No. 2023453).

Data Availability Statement: Data are contained within the article.

Acknowledgments: We sincerely thank guest editors Zhonghai Zhao, Jiafu Chen and Ju Nan for their efforts on this issue. Sincere thanks are extended to the three anonymous reviewers for their constructive comments.

Conflicts of Interest: No conflicts of interest exist in the submission of this manuscript, and the manuscript is approved by all authors for publication. I would like to declare on behalf of my coauthors that the work described was original research that has not been published previously, and is not under consideration for publication elsewhere, in whole or in part. All the listed authors have approved the manuscript that is enclosed.

References

1. Meinert, L.D. Skarns and skarn deposits. *Geosci Can.* **1992**, *19*, 145–162.
2. Hedenquist, J.W.; Lowenstern, J.B. The role of magmas in the formation of hydrothermal ore deposits. *Nature* **1994**, *370*, 519–527. [CrossRef]
3. Chen, Y.J.; Chen, H.Y.; Zaw, K.; Pirajno, F.; Zhang, Z.J. Geodynamic settings and tectonic model of skarn gold deposits in China: An overview. *Ore Geol. Rev.* **2007**, *31*, 139–169. [CrossRef]
4. Sun, J.G.; Zhang, Y.; Han, S.J.; Men, L.J.; Li, Y.X.; Chai, P.; Yang, F. Timing of formation and geological setting of low-sulphidation epithermal gold deposits in the continental margin of NE China. *Int. Geol. Rev.* **2013**, *55*, 608–632. [CrossRef]
5. Zhang, Y.; Sun, J.G.; Xing, S.W.; Zhang, Z.J. Geochronology and geochemistry of the Cuihongshan Fe-polymetallic deposit, northeastern China: Implications for ore genesis and tectonic setting. *Can. J. Earth Sci.* **2018**, *55*, 475–489. [CrossRef]
6. Li, Y.Q.; Zhang, D.H.; Sun, X.D.; Lv, C.L. Application of apatite fission-track analysis and zircon U-Pb geochronology to study the hydrothermal ore deposits in the Lesser Hinggan Range: Exhumation history and implications for mineral exploration. *J. Geochem. Explor.* **2019**, *199*, 141–164. [CrossRef]

7. Zhao, Z.H.; Chen, J.; Cheng, B.B.; Liu, Y.W.; Qiao, K.; Cui, X.M.; Yin, Y.C.; Li, C.L. Spatial analysis of structure and metal mineralization based on fractal theory and fry analysis: A case study in Nenjiang−Heihe metallogenic belt. *Minerals* **2023**, *13*, 313. [CrossRef]
8. Sun, F.T.; Zhao, C.T.; Sun, J.G.; Bai, C.L.; Liu, Y.; Chu, X.L.; Xu, Z.K.; Han, J.L. Superimposed mineralization and emplacement paleodepth of the Laozuoshan Cu-Au deposit in the Jiamusi block: Evidence from garnet U-Pb geochronology and mineral geochemistry. *J. Geochem. Explor.* **2023**, *250*, 107237. [CrossRef]
9. Ouyang, H.G.; Che, X.G.; Zhou, Z.H. ^{40}Ar-^{39}Ar dating of Ergu Fe-polymetalic skarn deposit in Yichun igneous belt and its geological implications. *Miner. Depos.* **2016**, *35*, 1035–1046. (In Chinese with English abstract)
10. Ren, L. *Study on the Diagenesis Mechanisms and Metallogenic Model of Skarn-Type Fe-Cu(Mo)Polymetallic Deposit in the Lesser Xing'an Range, NE China*; Jilin University: Changchun, China, 2017.
11. Zhao, C.T.; Sun, J.G.; Chu, X.L.; Qin, K.Z.; Ren, L.; Xu, Z.K.; Liu, Y.; Han, J.L.; Bai, C.L.; Shu, W. Metallogeny of the Ergu Fe-Zn polymetallic deposit, central Lesser Xing'an Range, NE China: Evidence from skarn mineralogy, fluid inclusions and H-O-S-Pb isotopes. *Ore Geol. Rev.* **2021**, *135*, 104227. [CrossRef]
12. Zhao, C.T.; Sun, J.G.; Liu, Y.; Chu, X.L.; Xu, Z.K.; Han, J.L.; Li, W.Q.; Ren, L.; Bai, C.L. Constraints of magmatism on the Ergu Fe-Zn polymetallic metallogenic system in the central Lesser Xing'an Range, NE China: Evidence from geochronology, geochemistry and Sr-Nd-Pb-Hf isotopes. *Geol. Mag.* **2021**, *158*, 1862–1890. [CrossRef]
13. Fei, X.H.; Zhang, Z.C.; Cheng, Z.G.; Wilde, S.A. Highly differentiated magmas linked with polymetallic mineralization: A case study from the Cuihongshan granitic intrusions, Lesser Xing'an Range, NE China. *Lithos* **2018**, *302–303*, 158–177. [CrossRef]
14. Chen, X.; Liu, J.J.; Carranza, E.J.M.; Zhang, D.H.; Collin, A.S.; Yang, B.; Xu, B.W.; Zhai, D.G.; Wang, Y.H.; Wang, J.P.; et al. Geology, geochemistry, and geochronology of the Cuihongshan Fe-polymetallic deposit, Heilongjiang Province, NE China. *Geol. J.* **2019**, *54*, 1254–1278. [CrossRef]
15. Zeng, G.P.; Hu, X.L.; Yao, S.Z.; Dong, Z.L.; Zhou, M.; Yang, J. Two Epochs of Mo Mineralization in Cuihongshan Fe-Mo-Polymetallic Ore field, Heilongjiang Province. *Earth Sci.* **2021**, *47*, 309–324. (In Chinese with English abstract)
16. Hu, X.L.; Ding, Z.J.; He, M.C.; Yao, S.Z.; Zhu, B.P.; Shen, J.; Chen, B. Two epochs of magmatism and metallogeny in the Cuihongshan Fe-polymetallic deposit, Heilongjiang Province, NE China: Constrains from U-Pb and Re-Os geochronology and Lu-Hf isotopes. *J. Geochem. Explor.* **2014**, *143*, 116–126. [CrossRef]
17. Ji, R. *Geological Characterisitics and Genesis Mechanism of the Xulaojiugou Pb-Zn Deposit, Heilongjiang Province*; Jilin University: Changchun, China, 2015.
18. Fei, X.H.; Zhang, Z.C.; Cheng, Z.G.; Santosh, M. Factors controlling the crystal morphology and chemistry of garnet in skarn deposits: A case study from the Cuihongshan polymetallic deposit, Lesser Xing'an Range, NE China. *Am. Miner.* **2019**, *104*, 1455–1468. [CrossRef]
19. Tang, C.; Chai, P.; Sun, J.G.; Wang, Q.H.; Chen, X.S.; Li, Y.X.; Yang, F.; Liu, Y.S. SHRIMP U-Pb zircon age of gabbro in Da'anhe gold deposit and its geological implications of Yichun, Heilongjiang Province. *Glob. Geol.* **2011**, *30*, 173–179. (In Chinese with English abstract)
20. Xue, M.X. *Metallgenesis of Endogenic Gold Deposits in Heilongjiang Province*; Jilin University: Changchun, China, 2012.
21. Yang, F. *The Study on Mineralization and Metallogenic Background of Da'anhe Gold Deposit in Yichun, Heilongjiang Province*; Jilin University: Changchun, China, 2013.
22. Jahn, B.M.; Wu, F.Y.; Chen, B. Granitoids of the Central Asian Orogenic Belt and continental growth in the Phanerozoic. *Earth Environ. Sci. Trans. R. Soc. Edinb.-Earth Sci.* **2000**, *91*, 181–193.
23. Windley, B.F.; Alexeiev, D.; Xiao, W.J.; Kroner, A.; Badarch, G. Tectonic models for accretion of the Central Asian Orogenic Belt. *J. Geol. Soc.* **2007**, *164*, 31–47. [CrossRef]
24. Sun, J.G.; Zhang, Y.; Xing, S.W.; Zhao, K.Q.; Zhang, Z.J.; Bai, L.A.; Ma, Y.B.; Liu, Y.S. Genetic types, ore-forming age and geodynamic setting of endogenic molybdenum deposits in the eastern edge of Xing-Meng orogenic belt. *Acta Petrol. Sin.* **2012**, *28*, 1317–1332. (In Chinese with English abstract)
25. Zhao, Z.H.; Qiao, K.; Liu, Y.W.; Chen, J.; Li, C.L. Geochemical data mining by integrated multivariate component data analysis: The Heilongjiang Duobaoshan area (China) case study. *Minerals* **2022**, *12*, 1035. [CrossRef]
26. Wu, F.Y.; Yang, J.H.; Lo, C.H.; Wilde, S.A.; Sun, D.Y.; Jahn, B.M. The Heilongjiang Group: A Jurassic accretionary complex in the Jiamusi Massif at the western Pacific margin of northeastern China. *Isl. Arc* **2007**, *16*, 156–172. [CrossRef]
27. Wu, F.Y.; Jahn, B.M.; Wilde, S.; Sun, D.Y. Phanerozoic crustal growth. U-Pb and Sr-Nd isotopic evidence from the granites in northeastern China. *Tectonophysics* **2000**, *328*, 89–113. [CrossRef]
28. Wu, F.Y.; Sun, D.Y.; Ge, W.C.; Zhang, Y.B.; Grant, M.L.; Wilde, S.A.; Jahn, B.M. Geochronology of the Phanerozoic granitoids in northeastern China. *J. Asian Earth Sci.* **2011**, *41*, 1–30. [CrossRef]
29. Wei, H.Y. *The Geochronology and Petrogenesis of Granite in Yichun-Hegang Area, Heilongjiang Province*; Jilin University: Changchun, China, 2012.
30. Wang, Z.W.; Xu, W.L.; Pei, F.P.; Wang, F.; Guo, P. Geochronology and geochemistry of early Paleozoic igneous rocks of the Lesser Xing'an Range, NE China: Implications for the tectonic evolution of the eastern Central Asian Orogenic Belt. *Lithos* **2016**, *261*, 144–163. [CrossRef]

31. Gao, F.H.; Wang, F.; Xu, W.L.; Yang, Y. Age of the "Paleoproterozoic" Dongfengshan group in the lesser Xing'an range, NE China, and its tectonic implications: Constraints from zircon U-Pb geochronology. *J. Jilin Univ. (Earth Sci. Ed.)* **2013**, *43*, 440–456. (In Chinese with English abstract)
32. Li, J.Y. *Formation and Evolution of the Crust in the Northern Changbaishan*; Geological Publishing House: Beijing, China, 1999. (In Chinese)
33. Sun, D.Y.; Wu, F.Y.; Gao, S. LA-ICP-MS Zircon U-Pb age of the Qingshui pluton in the East Xiao Hinggan Mountains. *Acta Geosci. Sin.* **2004**, *25*, 213–218. (In Chinese with English abstract)
34. Yu, J.J.; Wang, F.; Xu, W.L.; Gao, F.H.; Pei, F.P. Early Jurassic mafic magmatism in the Lesser Xing'an-Zhangguangcai Range, NE China, and its tectonic implications: Constrains from zircon U-Pb chronology and geochemistry. *Lithos* **2012**, *142–143*, 256–266. [CrossRef]
35. Xu, M.J.; Xu, W.L.; Wang, F.; Gao, F.H.; Yu, J.J. Geochronology and geochemistry of the Early Jurassic granitoids in the central Lesser Xing'an Range, NE China and its tectonic implications. *Acta Petrol. Sin.* **2013**, *29*, 354–368, (In Chinese with English abstract).
36. Han, J.L.; Sun, J.G.; Liu, Y.; Ren, L.; Wang, C.S.; Zhang, X.T.; He, Y.P.; Yu, R.D.; Lu, Q. Jurassic granitic magmatism in the lesser Xing'an-Zhangguangcai ranges of NE China: The Dong'an example. *Int. Geol. Rev.* **2019**, *61*, 2143–2163. [CrossRef]
37. Lu, Y.F. Geokit—A geochemical toolkit for Microsoft Excel. *Geochimica* **2004**, *33*, 459–464. (In Chinese with English abstract)
38. Fan, H.R.; Xie, Y.H.; Zhai, M.G.; Jin, C.W. A three stage fluid flow model for Xiaoqinling lode gold metallogenesis in the He'nan and Shanxi Provinces, central China. *Acta Petrol. Sin.* **2003**, *19*, 260–266. (In Chinese with English abstract)
39. Liu, B.; Duan, G.X. The density and isochoric formulae for NaCl-H_2O fluid inclusion (salinity $\leq$ 25wt%) and their applications. *Acta Miner. Sin.* **1987**, *7*, 345–352. (In Chinese with English abstract)
40. Taylor, B.E. *Origin and Significance of C-O-H Fluids in the Formation of Ca-Fe-Si Skarn, Osgood Mountains, Humboldt County, Nevada*; Stanford University: Palo Alto, CA, USA, 1976.
41. Zheng, Y.F. Calculation of oxygen isotope fractionation in hydroxyl-bearing silicates. *Earth Planet. Sci. Lett.* **1993**, *120*, 247–263. [CrossRef]
42. Clayton, R.N.; O'Neil, J.R.; Mayeda, T.K. Oxygen isotope exchange between quartz and water. *J. Geophys. Res.* **1972**, *77*, 3057–3067. [CrossRef]
43. Tan, H.Y. *Metallogenetic Series and Prospecting Assessment in Lesser Xing'an Range-Zhangguangcai Range Metallogenic Belt of Heilongjiang Province*; China University of Geosciences (Beijing): Beijing, China, 2013.
44. Titley, S.R. "Pyrometasomatic": An alteration type. *Econ. Geol.* **1973**, *68*, 1326–1329. [CrossRef]
45. Kwak, T.A. Hydrothermal alteration in carbonate-replacement deposits. *Geol. Assoc. Can. Short Course Notes* **1994**, *11*, 381–402.
46. Meinert, L.D.; Dipple, G.M.; Nicolescu, S. World skarn deposits. *Econ. Geol.* **2005**, *1905–2005*, 299–336.
47. Lu, H.Z.; Liu, Y.M.; Wang, C.L.; Xu, Y.Z.; Li, H.Q. Mineralization and fluid inclusion study of the Shizhuyuan W-Sn-Bi-Mo-F skarn deposit, Hunan Province, China. *Econ. Geol.* **2003**, *98*, 955–974. [CrossRef]
48. Hong, J.X.; Zhang, H.Y.; Zhai, D.G.; Li, D.F.; Zhang, Y.L.; Liu, J.J. The geochronology of the Haobugao skarn Zn-Pb deposit (NE China) using garnet LA-ICP-MS U-Pb dating. *Ore Geol. Rev.* **2021**, *139*, 104437. [CrossRef]
49. Zhao, C.T. *Mineralization, Metallogenic Model and Geodynamic Setting of Skarn Type Au-Fe-Cu Polymetallic Deposits in the Central of Heilongjiang Province*; Jilin University: Changchun, China, 2022.
50. Ohmoto, H.; Rye, R.O. Isotopes of Sulfurand Carbon. In *Geochemistry of Hydrothermal Ore Deposits*, 2nd ed.; Barnes, H.L., Ed.; Wiley-Interscience: New York, NY, USA, 1979.
51. Bowman, J.R. Stable-isotope systematics of skarns. *Miner. Assoc. Can. Short Course* **1998**, *26*, 99–145.
52. Ohmoto, H. Systematics of sulfur and carbon isotopes in hydrothermal ore deposits. *Econ. Geol.* **1972**, *67*, 551–578. [CrossRef]
53. Zartman, R.E.; Doe, B.R. Plumbotectonics-the model. *Tectonophysics* **1981**, *75*, 135–162. [CrossRef]
54. Yang, F.; Sun, J.G.; Fu, J. Geochemistry and its geological implication of the Da'anhe deposit and ore-forming rocks in Heilongjiang Province. *Geol. Resour.* **2014**, *23*, 419–423. (In Chinese with English abstract)
55. Einaudi, M.T.; Meinert, L.D.; Newberry, R.J. Skarn deposits. *Econ. Geol.* **1981**, *75*, 317–391.
56. Soloviev, S.G.; Kryazhev, S.G.; Dvurechenskaya, S.S. Geology, mineralization, stable isotope geochemistry, and fluid inclusion characteristics of the Novogodnee-Monto oxidized Au-(Cu) skarn and porphyry deposit, Polar Ural, Russia. *Miner. Depos.* **2013**, *48*, 603–627. [CrossRef]
57. Kodĕra, P.; Lexa, J.; Fallick, A.E. Formation of the Vysoka-Zlatno Cu-Au skarn-porphyry deposit, Slovakia. *Miner. Depos.* **2010**, *45*, 817–843. [CrossRef]
58. Meinert, L.D.; Hefton, K.K.; Mayes, D.; Tasiran, I. Geology, zonation, and fluid evolution of the Big Gossan Cu-Au skarn deposit, Ertsberg district, Irian Jaya. *Econ. Geol.* **1997**, *92*, 509–534. [CrossRef]
59. Meinert, L.D.; Hedenquist, J.W.; Satoh, H.; Matsuhisa, Y. Formation of anhydrous and hydrous skarn in Cu-Au ore deposits by magmatic fluids. *Econ. Geol.* **2003**, *98*, 147–156. [CrossRef]
60. Brown, P.E.; Hagemann, S.G. MacFlincor and its application to fluids in archean lode-gold deposit. *Geochim. Cosmochim. Acta* **1995**, *59*, 3943–3952. [CrossRef]
61. Roedder, E.; Bodnar, R.J. Geologic pressure determinations from fluid inclusion studies. *Earth Planet. Sci. Lett.* **1980**, *8*, 263–301. [CrossRef]
62. Fournier, R.O. Hydrothermal processes related to movement of fluid from plastic into brittle rock in the magmatic-epithermal environment. *Econ. Geol.* **1999**, *94*, 1193–1211. [CrossRef]

63. Simmons, S.F.; White, N.C.; John, D. Geological characteristics of epithermal precious and base metal deposits. *Econ. Geol.* **2005**, *1905–2005*, 485–522.
64. Klemm, L.M.; Pettke, T.; Heinrich, C.A. Hydrothermal evolution of the El Teniente deposit, Chile: Porphyry Cu-Mo ore deposition from low-salinity magmatic fluids. *Econ. Geol.* **2007**, *102*, 1021–1045. [CrossRef]
65. Chen, H.; Chen, Y.; Baker, M.J. Evolution of ore-forming fluids in the Sawayaerdun gold deposit in the Southwestern Chinese Tianshan metallogenic belt, Northwest China. *J. Asian Earth Sci.* **2012**, *49*, 131–144. [CrossRef]
66. Pirajno, F.; Mernagh, T.P.; Huston, D.; Creaser, R.A.; Seltmann, R. The Mesoproterozoic Abra polymetallic sedimentary rock-hosted mineral deposit, Edmund Basin, Western Australia. *Ore Geol. Rev.* **2015**, *76*, 442–462. [CrossRef]
67. Berman, R.G.; Brown, T.H. Heat capacity of minerals in the system Na_2O- K_2O-CaO-MgO-FeO-Fe_2O_3-Al_2O_3-SiO_2-TiO_2-H_2O-CO_2: Representation, estimation, and high temperature extrapolation. *Contrib. Miner. Petrol.* **1985**, *89*, 168–183. [CrossRef]
68. Xie, G.D. Research progresses on the transport forms and depositional mechanisms of gold. *Geoscience* **1994**, *8*, 357–363. (In Chinese with English abstract)
69. Zhang, D.H. Overview of research on the ore depositional mechanisms in ore-forming fluid. *Geol. Sci. Technol. Inf.* **1997**, *16*, 53–58. (In Chinese with English abstract)
70. Barnes, H.L. *Geochemistry of Hydrothermal Ore Deposits*, 3rd ed.; Pan Books: New York, NY, USA, 1997.
71. Han, J.L.; Sun, J.G.; Liu, Y.; Zhang, X.T.; He, Y.P.; Yang, F.; Chu, X.L.; Wang, L.L.; Wang, S.; Zhang, X.W.; et al. Genesis and age of the Toudaoliuhe breccia-type gold deposit in the Jiapigou mining district of Jilin Province, China: Constraints from fluid inclusions, H-O-S-Pb isotopes, and sulfide Rb-Sr dating. *Ore Geol. Rev.* **2020**, *118*, 103356. [CrossRef]
72. Han, J.L.; Sun, J.G.; Zhang, X.T.; Liu, Y.; Xu, Z.K.; Wang, S.; Xu, Z.T.; Li, X. Comparison of vein- and breccia-type Au-mineralization in the giant Jiapigou mining district of Northeast China. *Ore Geol. Rev.* **2022**, *150*, 105173. [CrossRef]
73. Han, J.L.; Deng, J.; Zhang, Y.; Sun, J.G.; Wang, Q.F.; Zhang, Y.M.; Zhang, X.T.; Liu, Y.; Zhao, C.T.; Yang, F.; et al. Au mineralization-related magmatism in the giant Jiapigou mining district of Northeast China. *Ore Geol. Rev.* **2022**, *141*, 104638. [CrossRef]
74. Wood, S.A.; Samson, I.M. Solubility of ore minerals and complexation of ore metals in hydrothermal solutions. *Rev. Econ. Geol.* **1998**, *10*, 33–80.
75. Chang, Z.S.; White, N.C.; Crowe, R.W.A.; Woodhouse, W.; Whalan, G.; Wilson, N. Geological characteristics and genesis of the Caijiaying Zn-Au deposit, China. Mineral Deposit Research for a High-Tech World: Society for Geology Applied to Mineral Deposits (SGA). In Proceedings of the 12th Biennial Meeting, Uppsala, Sweden, 12–15 August 2013; Volume 4, pp. 1535–1538.
76. Chang, Z.S.; Shu, Q.H.; Meinert, L.D. Skarn deposits of China. Soc. *Econ. Geol. Spec.* **2019**, *22*, 189–234.
77. Du, Y.S.; Tian, S.H.; Li, X.J. Contrast in fluid mineralization between Tianmashan and datuanshan ore deposits, Tongling, Anhui. *Earth Sci. J. China Univ. Geosci.* **2000**, *25*, 433–437. (In Chinese with English abstract)
78. Bai, C.L.; Sun, J.G.; Zhao, C.T.; Qin, K.Z.; Liu, Y.; Chu, X.L.; Xu, Z.K.; Li, Y.X. The multiple mineralizations and geodynamic settings of the Laozuoshan Cu-Au deposit in the Jiamusi Massif, NE China: Zircon U-Pb geochronological, elemental and Hf isotopic geochemical evidence. *Ore Geol. Rev.* **2021**, *137*, 104291. [CrossRef]
79. Zhai, D.G.; Liu, J.J.; Zhang, H.Y.; Yao, M.J.; Wang, J.P.; Yang, Y.Q. S-Pb isotopic geochemistry, U-Pb and Re-Os geochronology of the Huanggangliang Fe-Sn deposit, Inner Mongolia, NE China. *Ore Geol. Rev.* **2014**, *59*, 109–122. [CrossRef]
80. Zhai, D.G.; Liu, J.J.; Zhang, H.Y.; Wang, J.P.; Su, L.; Yang, X.A.; Wu, S.H. Origin of oscillatory zoned garnets from the Xieertala Fe-Zn skarn deposit, northern China: In situ LA-ICP-MS evidence. *Lithos* **2014**, *190–191*, 279–291. [CrossRef]
81. Zhang, Y.; Xing, S.W.; Song, Q.H.; Wang, Y.; Yu, Z.T.; Du, X.H.; Ma, Y.B.; Zhang, Z.J. Re-Os and U-Pb geochronology of porphyry and skarn types copper deposits in Jilin Province, NE China. *Resour. Geol.* **2015**, *65*, 394–404. [CrossRef]
82. Wu, C.; Wang, B.R.; Zhou, Z.G.; Wang, G.S.; Zuza, A.V.; Liu, C.F.; Jiang, T.; Liu, W.C.; Ma, S.W. The relationship between magma and mineralization in Chaobuleng iron polymetallic deposit, Inner Mongolia. *Gondwana Res.* **2017**, *45*, 228–253. [CrossRef]
83. Zhou, Z.H.; Mao, J.W.; Che, H.W.; Ouyang, H.G.; Ma, X.H. Metallogeny of the Handagai skarn Fe-Cu deposit, northern Great Xing'an Range, NE China: Constraints on fluid inclusions and skarn genesis. *Ore Geol. Rev.* **2017**, *80*, 623–644. [CrossRef]
84. Shu, Q.H.; Lai, Y.; Wei, L.M.; Sun, Y.; Wang, C. Fluid inclusion study of the Baiyinnuo'er Zn-Pb deposit, south segment of the Great Xing'an Mountain, norheastern China. *Acta Petrol. Sin.* **2011**, *27*, 1467–1482. (In Chinese with English abstract)

minerals

MDPI

Article

Geochemistry, Chronology and Tectonic Implications of the Hadayang Schists in the Northern Great Xing'an Range, Northeast China

Fuchao Na, Weimin Song *, Yingcai Liu, Junyu Fu, Yan Wang and Wei Sun

Shenyang Center of China Geological Survey, Shenyang 110034, China; nafuchao@mail.cgs.gov.cn (F.N.); sylyingcai@126.com (Y.L.); fjyzxy@163.com (J.F.); wy68413@163.com (Y.W.); soohiboy@126.com (W.S.)
* Correspondence: swmws@126.com

Abstract: The Late Paleozoic tectonic evolution of the Xing'an block in the eastern Central Asian orogenic belt has long been the subject of debate. In this paper, a comprehensive study of U-Pb zircon ages, Lu-Hf isotopes and whole-rock elemental analyses was carried out on Hadayang schists. Representative samples of the epidote-biotite-albite schist and biotite-albite schist yielded the weighted mean $^{206}Pb/^{238}U$ ages of 360 ± 2 Ma and 355 ± 3 Ma, respectively. This indicated the presence of Late Devonian–Early Carboniferous intermediate-basic rocks in the eastern Xing'an block. The Hadayang schists exhibited a Na-rich, tholeiitic and calc-alkaline affinity in composition with low Mg# (35.2–53.0), Cr (23.7–86.5 ppm), Ni (21.1–40.0 ppm) and Co (12.1–30.6 ppm). They were characterized by enrichment of LILEs, depletion of HFSEs and highly positive zircon $\varepsilon_{Hf}(t)$ values (the average values were +8.93 and +9.29, respectively). The magma source of the Hadayang schists was a mantle that consisted of both spinel and garnet lherzolite, with a partial melting degree of 1%–5%, and it had undergone fractional crystallization of olivine, orthopyroxene and plagioclase. The Hadayang schists, together with other Late Devonian–Early Carboniferous intermediate-basic magmatic rocks in the eastern Xing'an block, were formed in an intracontinental extension tectonic setting similar to that of the North American Basin and Range basalt. Moreover, Late Devonian–Early Carboniferous ophiolite under a similar tectonic background in the western Xing'an block has been reported. We believe that the Xing'an block would have been in the stage of intracontinental extension during the Late Devonian–Early Carboniferous.

Keywords: eastern Xing'an block; late Devonian–early Carboniferous; Hadayang schists; petrogenesis; tectonic implications

Citation: Na, F.; Song, W.; Liu, Y.; Fu, J.; Wang, Y.; Sun, W. Geochemistry, Chronology and Tectonic Implications of the Hadayang Schists in the Northern Great Xing'an Range, Northeast China. *Minerals* **2023**, *13*, 1379. https://doi.org/10.3390/min13111379

Academic Editor: Manuel Francisco Pereira

Received: 14 August 2023
Revised: 23 October 2023
Accepted: 25 October 2023
Published: 28 October 2023

1. Introduction

The Central Asian orogenic belt (CAOB) is located between the Siberian craton and the Sino-Korean craton, and is composed of a series of micro-continental blocks and suture zones [1–5] (Figure 1a). The Xing'an–Mongolia orogenic belt (XMOB) is located in the eastern section of the CAOB, and is made up of several microcontinents, including the Erguna, Xing'an, Songnen and Jiamusi–Khanka blocks [6–11] (Figure 1b). Although much progress has been made in the study area of the tectonic pattern and evolutionary history of the XMOB in recent years, there are still many controversies, among which the dating of the integration of the Xing'an (XB) and Songnen blocks (SB) is the most prominent. For example, Xu et al. [6] argued that the XB and SB combined along the Angin Sum–Xilinhot–Heihe line before the Devonian [7,12]; Liu et al. [8] proposed that these two blocks closed along the Hegenshan–Moguqi–Heihe line during the late Early Carboniferous and early Late Carboniferous [9,13–15]; Xu et al. [11] believed that the XB and SB combined along the Xilinhot–Ulanhot–Nenjiang–Heihe line, and the time limit was the late Early Carboniferous [16]. In addition, the time limit of convergence is also dated to the Late Devonian to Early Carboniferous [17,18], to before the Permian [19,20] and to the Triassic [21,22].

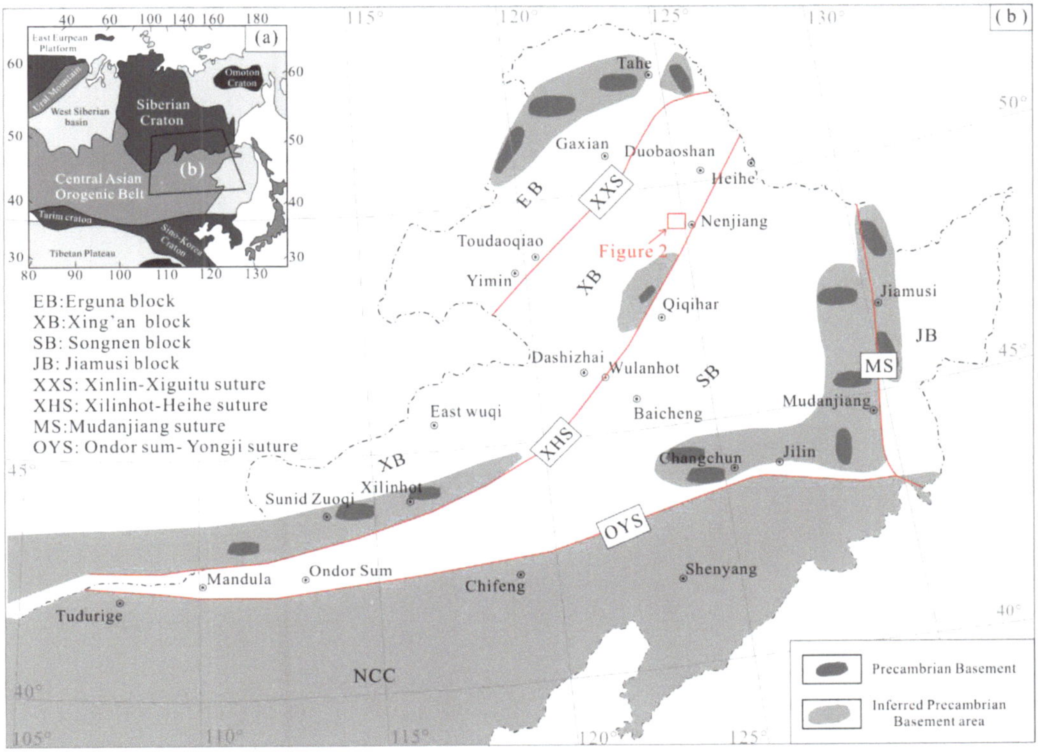

Figure 1. Pre-Devonian blocks in XMOB, northeast China: (**a**) location of the CAOB; (**b**) blocks in the XMOB (modified after [7]).

Through the analysis of the above different views, it can be found that one of the focuses of the debate is whether there was subduction in the ocean basin between the XB and the SB in the late Paleozoic, and the study of the tectonic background of the late Paleozoic igneous rock in the XB can provide important evidence for solving the above problems. Compared with granitic rocks, mafic magmatic rocks originate directly from partial melting of mantle and the geochemistry is less affected by late transformation, so they become ideal objects for the derivation of mantle source characteristics and regional tectonic background [23]. Therefore, we report zircon U-Pb geochronology, Hf isotopes and whole-rock geochemistry studies on late Devonian–early Carboniferous metamorphic intermediate-basic volcanic rocks in the southern XB, to determine their origins and tectonic setting. This study provides a more detailed understanding of the evolution history of the XB and SB in the late Paleozoic. Furthermore, it provides important constraints for the study of the tectonic evolution of the XMOB.

2. Geological Settings and Sampling

The study area was located in the Hadayang area of Molidawa Banner in the eastern part of the Inner Mongolia Autonomous Region and it tectonically belongs to the southern part of the XB (Figure 2). As the oldest block in the XMOB, the Precambrian record of the EB is well developed, and the oldest rock exposed is a set of Neoarchean–Paleoproterozoic granitic gneiss [24–27]. The Xinghuadukou Group, which was previously considered to be of Paleo–Mesoproterozoic age, may have formed in the Neoproterozoic [27–31].

In addition, Neoproterozoic magmatism is also well developed in the EB [31–34]. The XB is traditionally based on Precambrian strata such as the Xinghuadukou Group, the Jiageda Formation and the Wolegen Group, but recent research shows that these strata were formed in the Paleozoic [35–37]. In recent years, a series of Neoarchean–Paleoproterozoic magmatic rocks [38–42] and metamorphic rock series [43] have been identified in the Ulanhot–Longjiang area, suggesting the existence of Precambrian basement in the XB. After the convergence of the EB and XB along the XXS in the early Paleozoic [44–49] (Figure 1b), the Erguna-Xing'an block (EXB) was covered by the Early Paleozoic–Lower Permian marine and continental alternative deposition, the Upper Permian continental sedimentary strata and the Mesozoic volcaniclastic rocks, and developed multistage Paleozoic–Mesozoic magmatic intrusion. The SB is located in the southeast of the EXB, which is bounded by the XHS (Figure 1b). The Dongfengshan Group and Tadong Group in the eastern SB, which are traditionally considered as "Paleoproterozoic", were formed in the Neoproterozoic [50–52], accompanied by contemporaneous magmatic intrusions [53–56]. The western SB is mostly covered by late Mesozoic–Cenozoic volcaniclastic sediments.

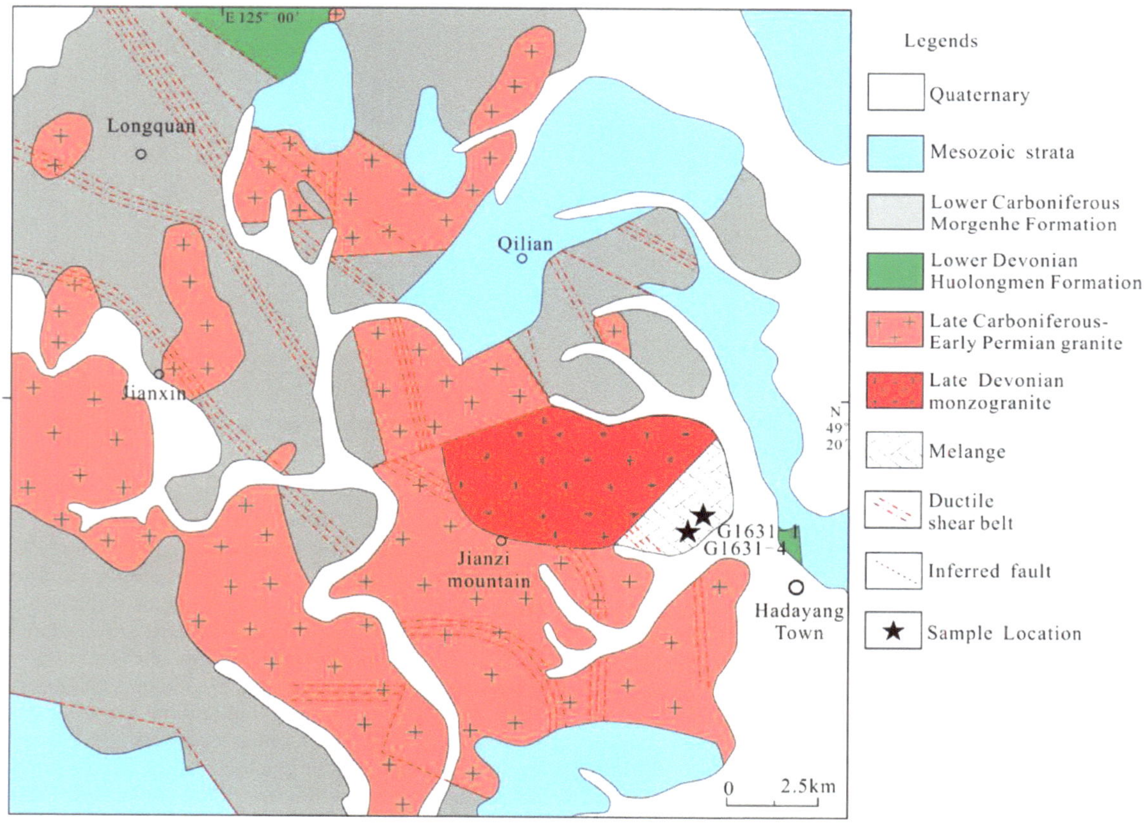

Figure 2. Simplified geological map of Hadayang area and locations of samples used in this study.

The late Paleozoic strata in the study area include the Lower Devonian Huolongmen Formation and the Lower Carboniferous Morgenhe Formation (Figure 2). The Lower Devonian Huolongmen Formation is mainly composed of clastic sedimentary rocks with a small amount of intermediate volcanic rocks, and is affected by a poorly developed ductile deformation. The formation age of chlorite muscovite schist (protolith is volcanic rock) is 414 Ma [57]. The Lower Carboniferous Morgenhe Formation is mainly composed of weakly metamorphic intermediate-acid volcanic rocks, which are generally mylonitic. The zircon U-Pb age of the volcanic rocks is 353 to 352 Ma [58,59]. Late Paleozoic intrusive rocks include Late Devonian mylonitic syenite granite [60], Late Carboniferous–Early Permian granodiorite, syenogranite and alkali feldspar granite. The above are mostly covered by Mesozoic volcanic rocks and Cenozoic detrital sedimentary rocks. In addition to the above features, there is also a Hadayang melange developed in the study area [61,62]. The matrix in the melange is mainly composed of schist, while blocks consist of schist, mafic–ultramafic rocks and meta-acidic volcanic rocks [62].

Two samples were collected from the blocks of Hadayang melange for age dating, and the sample locations are shown in Figure 2. At each of these two sites, six additional samples were collected for geochemical and petrological analysis. Sample G1631-1, an epidote-biotite-albite schist (49°18′28″ N, 125°06′33″ E, Figures 2 and 3a-b), showed porphyroblastic, lepidoblastic texture and schistose structure (Figure 3c). The mineral assemblage consisted of albite (~27%), biotite (~30%), chlorite (~8%), epidote (~20%), quartz (~8%) and opaque mineral (~7%) (Figure 3c). The content of the remaining phenocryst was about 5%, and the composition included opaque mineral aggregates and albite. The matrix included biotite, chlorite, epidote, quartz and albite. Sample G1631-4, a biotite-albite schist (49°18′25″ N, 125°06′31″ E), was located near Sample G1631-1 (Figures 2 and 3d-e). Petrographically, the biotite-albite schist had porphyroblastic, lepidoblastic texture and schistose structure, consisting of biotite (~15%), chlorite (~8%), epidote (~12%), albite (~40%) and quartz (~25%) (Figure 3f). The crystalloblasts were composed of oligoclase and albite, and the oligoclase was semi-euhedral wide plate with matrix components interspersed, and content of less than 1%. The albite was granular and contained matrix minerals biotite, chlorite and epidote inclusions, accounting for about 5%. The matrix included biotite, chlorite, epidote, quartz and albite (Figure 3f).

Figure 3. Field occurrence and microscope views of the Hadayang schists. (**a–c**) Epidote-biotite-albite schist. (**d–f**) Biotite-albite schist.

3. Analytical Methods

3.1. Major and Trace Element Analysis

The major and trace element analyses were carried out at the Shenyang Mineral Resources Supervision and Inspection Center, Shenyang, China. Whole-rock major element concentrations were determined by X-ray fluorescence spectrometry, yielding an analytical precision better than 2%. Trace element and rare earth element (REE) concentrations were determined by ICP–MS, yielding an analytical precision better than 5% for elements with concentrations of >10 ppm, <8% for elements with concentrations of <10 ppm and <10% for the transition metals.

3.2. Zircon U–Pb Dating

U-Pb dating analyses were conducted by LA-ICP-MS at Beijing Createch Testing Technology Co., Ltd. (Beijing, China). The detailed operating conditions for the laser ablation system and the ICP-MS instrument and data reduction were the same as described by Hou et al. [63]. Laser sampling was performed using an ESI NWR 193 nm laser ablation system. An AnalytikJena PQMS Elite ICP-MS (Analytik Jena AG, Jena, Germany) instrument was used to acquire ion-signal intensities. Helium was applied as a carrier gas. Argon was used as the make-up gas and mixed with the carrier gas via a T-connector before entering the ICP. Each analysis incorporated a background acquisition of approximately 15–20 s (gas blank) followed by 45 s of data acquisition from the sample. Off-line raw data selection and integration of background and analyte signals, and time-drift correction and quantitative calibration for U-Pb dating, were performed by ICPMSDataCal [64]. Zircon GJ-1 was used as the external standard for U-Pb dating, and was analyzed twice every 5–10 analyses. Time-dependent drifts of U-Th-Pb isotopic ratios were corrected using a linear interpolation (with time) for every 5–10 analyses according to the variations of GJ-1 (i.e., 2 zircon GJ-1 + 5–10 samples + 2 zircon GJ-1) [64]. Uncertainty of preferred values for the external standard GJ-1 was propagated to the ultimate results of the samples. In all analyzed zircon grains, the common Pb correction was not necessary due to the low signal of common ^{204}Pb and high ^{206}Pb/^{204}Pb. U, Th and Pb concentration was calibrated by NIST 610. Concordia diagrams and weighted mean calculations were made using Isoplot/Ex_ver3. The zircon Plesovice was dated as unknown samples and yielded a weighted mean ^{206}Pb/^{238}U age of 337.5 ± 2.4 Ma (2SD, n = 5), which was in good agreement with the recommended ^{206}Pb/^{238}U age of 337.13 ± 0.37 Ma (2SD) [65].

3.3. Lu–Hf Isotope Analysis

Zircon in situ Hf isotope analysis was carried out using a RESOlution SE 193 laser-ablation system attached to a Thermo Fisher Scientific Neptune Plus MC-ICP-MS at Beijing Createch Testing Technology Co., Ltd (Beijing, China). Instrumental conditions and data acquisition protocols were as described by Hou et al. [66]. A stationary spot used a beam diameter of ~55 μm. Helium was used as the carrier gas to transport ablated material from the laser-ablation cell after which it was mixed with Ar prior to entering the ICP-MS torch. ^{176}Lu/^{175}Lu = 0.02658 and ^{176}Yb/^{173}Yb = 0.796218 ratios were determined to correct for the isobaric interferences of ^{176}Lu and ^{176}Yb on ^{176}Hf. For instrumental mass bias correction, Yb isotope ratios were normalized to ^{172}Yb/^{173}Yb = 1.35274 and Hf isotope ratios to ^{179}Hf/^{177}Hf = 0.7325 using an exponential law. The mass bias behavior of Lu was assumed to follow that of Yb; the mass bias correction protocol was as described by Hou et al. [66].

4. Analytical Results

4.1. Whole-Rock Geochemistry

Results of the major element, trace element and REE analyses of the samples from the Hadayang schists are listed in Table 1.

Table 1. Analytical results for the major (w%) and trace elements ($\times 10^{-6}$) of Hadayang schists.

Lithology	Epidote-Biotite-Albite Schist						Biotite-Albite Schist					
Sample No.	G1631-1.1	G1631-1.2	G1631-1.3	G1631-1.4	G1631-1.5	G1631-1.6	G1631-4.1	G1631-4.2	G1631-4.3	G1631-4.4	G1631-4.5	G1631-4.6
SiO_2 (%)	49.28	50.55	50.47	50.16	49.46	51.15	58.07	58.36	58.82	55.44	57.72	56.23
Al_2O_3 (%)	16.75	16.56	16.76	16.51	16.99	16.73	15.98	15.84	16.17	16.46	16.27	16.30
Fe_2O_3 (%)	8.08	7.80	7.18	7.97	7.95	7.61	2.97	2.73	2.65	2.89	5.57	5.18
FeO (%)	3.05	3.28	3.32	3.14	3.41	3.32	3.73	4.00	4.09	4.40	2.43	2.97
CaO (%)	12.68	11.54	9.51	11.27	12.09	11.42	4.43	4.93	4.24	5.30	5.93	6.97
MgO (%)	3.31	3.59	3.99	3.45	3.19	3.28	3.96	3.83	4.01	4.40	3.04	3.17
K_2O (%)	0.17	0.18	0.73	0.26	0.09	0.02	0.76	0.86	1.40	0.90	1.38	1.97
Na_2O (%)	0.32	0.42	1.28	0.47	0.26	0.03	4.82	3.86	4.33	3.97	2.77	2.68
TiO_2 (%)	1.80	1.70	1.79	1.68	1.77	1.66	1.08	1.02	1.02	1.06	1.58	1.34
P_2O_5 (%)	1.07	1.02	1.12	1.02	1.05	1.03	0.61	0.57	0.58	0.61	0.75	0.69
MnO (%)	0.14	0.15	0.13	0.15	0.16	0.15	0.18	0.17	0.13	0.19	0.13	0.15
LOI (%)	3.48	3.45	3.45	3.45	3.36	3.47	3.24	3.59	2.67	4.05	2.23	2.26
SUM	100.14	100.26	99.75	99.53	99.78	99.86	99.83	99.75	100.11	99.67	99.81	99.90
FeO^T	10.33	10.29	9.79	10.32	10.57	10.18	6.40	6.45	6.47	7.01	7.44	7.63
Mg#	36.51	38.50	42.25	37.53	35.14	36.62	52.61	51.58	52.63	52.97	42.30	42.74
Na_2O/K_2O	1.96	2.31	1.76	1.84	2.98	1.56	6.35	4.47	3.08	4.43	2.01	1.36
Na_2O+K_2O	0.51	0.62	2.09	0.76	0.36	0.04	5.77	4.91	5.88	5.09	4.26	4.76
σ	0.03	0.04	0.46	0.06	0.02	0.00	1.95	1.37	1.99	1.73	1.12	1.55
Y	27.63	27.19	28.93	27.84	26.50	27.18	28.98	26.43	26.03	27.01	43.88	36.10
La	34.09	33.83	30.10	31.19	32.41	32.91	44.17	36.95	38.22	43.43	54.62	49.92
Ce	75.52	75.02	68.90	71.33	72.55	72.89	93.02	80.91	82.54	93.07	119.90	109.83
Pr	9.97	9.50	9.39	9.51	9.56	9.69	10.92	10.07	10.20	11.56	14.78	12.36
Nd	47.38	48.12	46.47	46.41	45.61	46.99	55.88	47.35	47.19	53.38	66.40	58.93
Sm	11.51	11.46	11.63	11.25	11.13	11.25	12.39	11.05	11.07	11.84	15.27	12.42
Eu	2.54	2.39	2.76	2.58	2.41	2.51	3.17	2.90	2.97	3.22	2.93	2.78
Gd	7.49	7.54	7.80	7.65	7.48	7.68	7.91	7.21	7.41	7.88	10.35	8.31
Tb	1.18	1.20	1.25	1.20	1.15	1.21	1.25	1.15	1.14	1.21	1.68	1.41
Dy	6.10	6.05	6.47	6.21	5.91	6.05	6.47	5.93	5.95	6.31	9.14	7.19
Ho	1.13	1.07	1.20	1.19	1.11	1.11	1.20	1.15	1.11	1.16	1.71	1.34
Er	3.05	2.86	3.18	2.98	2.95	3.07	3.59	3.15	3.05	3.17	4.92	4.09
Tm	0.46	0.42	0.46	0.46	0.42	0.46	0.52	0.48	0.47	0.50	0.76	0.66
Yb	2.28	2.25	2.41	2.26	2.15	2.25	2.93	2.63	2.59	2.58	3.91	3.36

Table 1. *Cont.*

Lithology	Epidote-Biotite-Albite Schist						Biotite-Albite Schist					
Sample No.	G1631-1.1	G1631-1.2	G1631-1.3	G1631-1.4	G1631-1.5	G1631-1.6	G1631-4.1	G1631-4.2	G1631-4.3	G1631-4.4	G1631-4.5	G1631-4.6
Lu	0.31	0.31	0.34	0.33	0.31	0.30	0.38	0.37	0.36	0.37	0.55	0.47
ΣREE	202.99	202.01	192.34	194.53	195.14	198.36	243.82	211.29	214.24	239.68	306.92	273.07
LREE	181.00	180.32	169.25	172.27	173.66	176.24	219.56	189.22	192.18	216.50	273.90	246.24
HREE	21.98	21.69	23.10	22.27	21.48	22.12	24.25	22.07	22.06	23.18	33.02	26.83
L/H	8.23	8.31	7.33	7.74	8.09	7.97	9.05	8.57	8.71	9.34	8.29	9.18
La_N/Yb_N	10.73	10.78	8.97	9.89	10.80	10.49	10.82	10.09	10.58	12.06	10.02	10.65
δCe	0.99	1.01	1.00	1.01	1.00	0.99	1.01	1.01	1.00	1.00	1.01	1.05
δEu	0.78	0.74	0.83	0.80	0.76	0.78	0.92	0.93	0.94	0.96	0.67	0.79
Li	18.58	20.34	27.07	22.73	20.42	21.77	25.75	25.34	26.55	29.34	36.92	40.63
Sc	14.27	14.17	15.34	15.52	15.04	15.69	11.35	12.81	12.47	12.44	13.23	12.99
V	401.29	379.56	347.99	361.07	417.77	400.93	107.69	126.32	111.54	131.04	202.25	166.49
Cr	39.74	43.51	29.62	42.25	41.89	39.04	86.51	82.63	75.99	82.93	23.71	36.53
Co	24.30	24.63	30.59	27.44	25.92	27.25	12.06	12.83	13.02	13.17	16.46	15.25
Ni	21.15	22.40	21.70	21.98	22.28	23.29	34.82	39.55	36.45	40.00	23.84	24.71
Ga	23.20	23.44	21.69	22.00	24.98	23.42	18.47	19.86	18.39	19.02	20.92	20.71
Rb	8.83	8.46	31.45	13.85	5.57	3.97	26.70	30.04	48.56	30.16	45.00	59.66
Sr	1100	1200	893	995	1400	1300	566	586	554	5408	944	942
Zr	87.75	61.90	71.92	73.22	88.55	80.19	185.02	176.05	166.34	168.27	27.95	51.58
Nb	10.84	10.27	11.11	10.74	9.10	8.95	15.52	14.40	14.34	15.77	30.61	23.04
Ba	78.40	106.57	146.80	113.68	82.57	71.06	308.28	357.57	419.52	400.07	474.22	591.36
Hf	0.84	0.75	0.80	0.81	0.76	0.83	0.78	0.66	0.80	0.74	0.83	0.64
Ta	0.85	0.80	0.85	0.76	0.87	0.77	0.67	0.64	0.68	0.63	1.60	1.07
Th	17.66	7.07	6.77	5.97	5.97	5.51	7.27	6.82	7.21	7.56	7.74	6.39
U	2.11	2.20	2.24	2.13	2.14	2.08	2.44	2.33	2.16	2.40	3.19	2.29

4.1.1. The Epidote-Biotite-Albite Schists

The epidote-biotite-albite schists had 49.28-51.15 wt% SiO_2, 3.19–3.99 wt% MgO, 15.84–16.99 wt% Al_2O_3, 1.66–1.80 wt% TiO_2, 0.03–1.28 wt% Na_2O, 0.02–0.73 wt% K_2O, 9.51–12.68 wt% CaO and 9.79–10.57 wt% total FeO. They had Mg# of [Mg# = 100 × $MgO_{molar}/(MgO_{molar} + FeO_{molar})$] 35.14–42.25. The Na_2O/K_2O ratios ranged from 1.56~2.98, indicating that the samples belonged to the sodium series. As can be seen from the plot of SiO_2 vs. $K_2O + Na_2O$ and Nb/Y vs. Zr/TiO_2*0.0001 diagrams (Figure 4a-b), all of these metavolcanic rocks plot in the basalt and basaltic andesite field. On a plot of SiO_2 vs. $^{T}FeO/MgO$ (Figure 4c), the samples plot in the tholeiitic field.

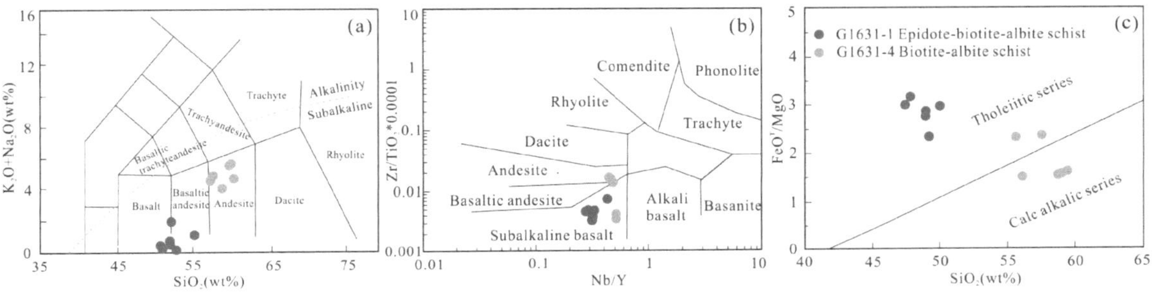

Figure 4. Classification and series diagrams of the Hadayang schists. (**a**) Plot of SiO_2 vs. $K_2O + Na_2O$ (after [67]), with the alkalic/subalkalic boundary of [68] shown; (**b**) plot of Zr/TiO_2*0.0001 vs. Nb/Y (after [69]); (**c**) plot of $^{T}FeO/MgO$ vs. SiO_2 (after [70]).

The epidote-biotite-albite schists had total rare earth element (REE) concentrations from 194.53 to 202.99 ppm, with an average value of 197.56 ppm. On the chondrite-normalized REE diagram (Figure 5a), the light rare earth elements (LREEs) are moderately enriched relative to the heavy rare earth elements (HREEs) with $(La/Yb)_N$ ratios ranging from 8.97 to10.78, while the HREEs are relatively flat. These metavolcanic rocks had slightly negative Eu anomalies (δEu = 0.74–0.83). The primitive-mantle-normalized incompatible element diagrams (Figure 5b) show that the epidote-biotite-albite schists were relatively enriched in large ion lithophile elements (LILEs) (such as Rb, Ba, Th and U) relative to high-field-strength elements (HFSEs) (such as Nb, P and Ti).

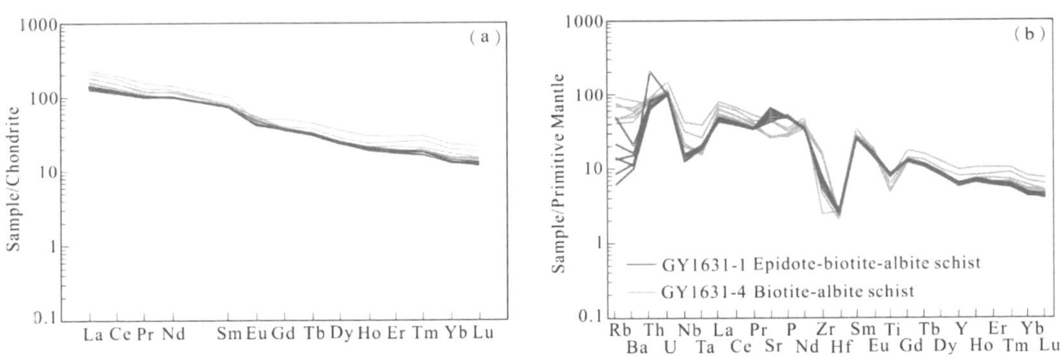

Figure 5. Chondrite-normalized REE patterns (**a**) and primitive-mantle-normalized trace element spidergrams (**b**) for the Hadayang schists. Normalization values are taken from [71,72].

4.1.2. The Biotite-Albite Schists

The biotite-albite schists had SiO_2 of 57.59–60.69 wt% and high TiO_2 (1.05–1.62 wt%), ∑FeO (6.63–7.81 wt%) and Al_2O_3 (16.47–17.21 wt%). Their alkalis ($K_2O + Na_2O$) ranged from 4.26 to 5.77 wt%, and the low K_2O/Na_2O ratio (0.16 to 0.74) indicated that the samples belonged to the sodium series. All of the samples plot in the andesite field on a SiO_2 vs. $Na_2O + K_2O$ diagram (Figure 4a). As can be seen from the plot of $Zr/TiO_2 \times 0.0001$ vs. Nb/Y (Figure 4b), most of the samples plot in the subalkaline basalt and andesite fields. On the other hand, on the plot of FeO^T/MgO vs. SiO_2, except for two samples that fall into the tholeiitic field, the biotite-albite schist samples plot in the calc-alkaline field.

The biotite-albite schists had variable total rare earth element (REE) concentrations from 211.29 to 306.92 ppm, with an average value of 248.17 ppm. These metavolcanic rocks exhibited LREEs enrichment ($(La/Yb)_N$ = 10.02–12.06) and slight negative Eu anomalies (δEu = 0.67–0.96) (Figure 5a). On a primitive-mantle-normalized trace element plot (Figure 5b), the biotite-albite schist samples also appear as enriched in large ion lithophile elements (LILEs) relative to the high-field-strength elements (HFSEs), which were similar to those in the epidote-biotite-albite schists.

4.2. Zircon U-Pb Dating

Cathodoluminescence (CL) images and concordia plots for the analyzed zircons are shown in Figure 6. Zircons from the epidote-biotite-albite schist sample (G1631-1) and biotite-albite schist sample (G1631-4) were mainly euhedral and elongate, with aspect ratios of 1:1–2:1. They showed straight rhythmic stripes, typical oscillatory zoning or sector zoning in CL images (Figure 6), with high Th/U ratios (0.49–1.38, Table 2), which is characteristic of magmatic zircons [73]. In total, 60 concordant $^{206}Pb/^{238}U$ ages, obtained from 70 zircon grains, ranged from 338 to 370 Ma. Two samples yielded similar zircon U-Pb ages around 360 Ma (Figure 7), with weighted mean $^{206}Pb/^{238}U$ ages of 360 $\pm$ 2 Ma (G1631-1, n = 35, MSWD = 1.4) and 355 $\pm$ 3 Ma (G1631-4, n = 35, MSWD = 2.1). The two ages were interpreted as the crystallization ages of the magmatic protolith of the two schists.

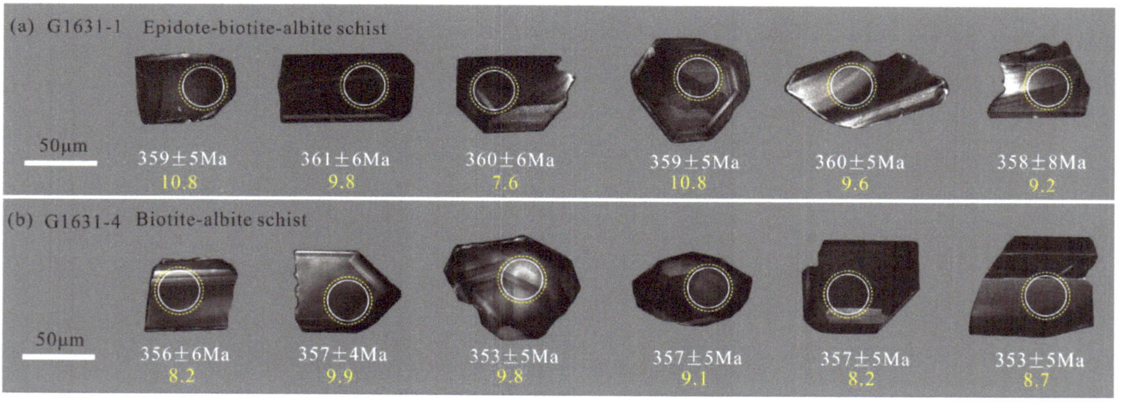

Figure 6. Cathodoluminescence images of zircon grains selected for analysis from the epidote-biotite-albite schist (**a**) and biotite-albite schist (**b**). The smaller white circles represent locations of LA-ICP-MS U-Pb dating, whereas the bigger yellow circles (dotted line) indicate locations of Hf isotopic analyses.

Table 2. LA-ICP-MS zircon U-Pb dating results for the Hadayang schists

Analytical No.	Th (10^{-6})	U (10^{-6})	Th/U	Isotopic Ratio						Age (Ma)						Concordance
				$^{206}Pb/^{238}U$		$^{207}Pb/^{235}U$		$^{207}Pb/^{206}Pb$		$^{206}Pb/^{238}U$		$^{207}Pb/^{235}U$		$^{207}Pb/^{206}Pb$		
				Ratio	1σ	Ratio	1σ	Ratio	1σ	Age	1σ	Age	1σ	Age	1σ	
G1631-1 Epidote biotite albite schist																
1	334.7	354.2	0.94	0.0582	0.0011	0.4358	0.0097	0.0543	0.0008	365	7	367	7	387	32	99.5%
2	551.7	501.4	1.10	0.0601	0.0009	0.4499	0.0080	0.0543	0.0006	376	5	377	6	389	19	99.7%
3	484.3	436.0	1.11	0.0573	0.0008	0.4235	0.0067	0.0537	0.0007	359	5	359	5	367	28	100.0%
4	1201.1	1055.1	1.14	0.0589	0.0010	0.4434	0.0073	0.0547	0.0005	369	6	373	5	398	16	98.9%
5	640.3	554.4	1.15	0.0576	0.0011	0.4249	0.0077	0.0536	0.0007	361	7	360	6	354	30	99.7%
6	1369.8	927.8	1.48	0.0582	0.0010	0.4373	0.0075	0.0545	0.0006	365	6	368	5	391	24	99.2%
7	610.1	586.2	1.04	0.0591	0.0009	0.4304	0.0073	0.0529	0.0007	370	5	364	5	324	30	98.4%
8	550.6	744.5	0.74	0.0585	0.0009	0.4291	0.0062	0.0533	0.0006	366	5	363	4	339	24	99.2%
9	691.4	608.9	1.14	0.0578	0.0008	0.4215	0.0066	0.0528	0.0006	362	5	357	5	320	24	98.6%
10	1105.5	1001.7	1.10	0.0575	0.0008	0.4288	0.0065	0.0541	0.0005	360	5	362	5	372	20	99.4%
11	475.6	522.9	0.91	0.0570	0.0009	0.4269	0.0072	0.0543	0.0006	357	5	361	5	383	21	98.9%
12	1172.7	909.3	1.29	0.0583	0.0008	0.4349	0.0063	0.0541	0.0005	365	5	367	5	376	20	99.5%
13	168.8	288.6	0.59	0.0565	0.0008	0.4316	0.0069	0.0554	0.0007	354	5	364	5	432	5	97.2%
14	230.1	272.5	0.84	0.0563	0.0009	0.4239	0.0084	0.0546	0.0008	353	5	359	6	395	36	98.3%
15	473.3	483.8	0.98	0.0568	0.0008	0.4233	0.0070	0.0540	0.0006	356	5	358	5	372	24	99.4%
16	1575.6	1139.1	1.38	0.0574	0.0009	0.4213	0.0071	0.0532	0.0005	360	6	357	5	345	22	99.2%
17	480.5	534.4	0.90	0.0569	0.0014	0.4268	0.0189	0.0539	0.0013	357	8	361	14	365	49	98.9%
18	656.6	584.9	1.12	0.0571	0.0008	0.4252	0.0068	0.0544	0.0007	358	5	360	5	387	25	99.4%
19	1081.4	877.1	1.23	0.0574	0.0008	0.4238	0.0076	0.0535	0.0006	360	5	359	5	350	32	99.7%
20	740.6	669.2	1.11	0.0558	0.0009	0.4160	0.0083	0.0539	0.0006	350	5	353	6	365	24	99.1%
21	538.6	513.2	1.05	0.0579	0.0010	0.4330	0.0073	0.0543	0.0007	363	6	365	5	383	21	99.5%
22	254.0	298.4	0.85	0.0567	0.0009	0.4244	0.0073	0.0544	0.0008	355	6	359	5	387	33	98.9%
23	1794.4	1046.2	1.72	0.0590	0.0010	0.4334	0.0067	0.0533	0.0005	369	6	366	5	343	22	99.2%
24	683.2	584.2	1.17	0.0572	0.0008	0.4321	0.0075	0.0547	0.0006	359	5	365	5	398	22	98.3%
25	292.0	374.7	0.78	0.0574	0.0008	0.4210	0.0072	0.0532	0.0007	360	5	357	5	345	28	99.2%
26	410.1	476.3	0.86	0.0572	0.0009	0.4236	0.0074	0.0538	0.0007	358	6	359	5	365	23	99.7%
27	188.0	248.3	0.76	0.0579	0.0011	0.4308	0.0082	0.0541	0.0007	363	6	364	7	376	30	99.7%
28	1332.8	971.5	1.37	0.0556	0.0007	0.4076	0.0059	0.0532	0.0006	349	5	347	4	345	19	99.4%
29	334.7	354.2	0.94	0.0582	0.0011	0.4358	0.0097	0.0543	0.0008	365	7	367	7	387	32	99.5%
30	551.7	501.4	1.10	0.0601	0.0009	0.4499	0.0080	0.0543	0.0006	376	5	377	6	389	19	99.7%
31	484.3	436.0	1.11	0.0573	0.0008	0.4235	0.0067	0.0537	0.0007	359	5	359	5	367	28	100.0%
32	1201.1	1055.1	1.14	0.0589	0.0010	0.4434	0.0073	0.0547	0.0005	369	6	373	5	398	16	98.9%
33	640.3	554.4	1.15	0.0576	0.0011	0.4249	0.0077	0.0536	0.0007	361	7	360	6	354	30	99.7%
34	1369.8	927.8	1.48	0.0582	0.0010	0.4373	0.0075	0.0545	0.0006	365	6	368	5	391	24	99.2%
35	610.1	586.2	1.04	0.0591	0.0009	0.4304	0.0073	0.0529	0.0007	370	5	364	5	324	30	98.4%

Table 2. *Cont.*

Analytical No.	Th (10^{-6})	U (10^{-6})	Th/U	Isotopic Ratio						Age (Ma)						Concordance
				$^{206}Pb/^{238}U$		$^{207}Pb/^{235}U$		$^{207}Pb/^{206}Pb$		$^{206}Pb/^{238}U$		$^{207}Pb/^{235}U$		$^{207}Pb/^{206}Pb$		
				Ratio	1σ	Ratio	1σ	Ratio	1σ	Age	1σ	Age	1σ	Age	1σ	
G1631-4 Chlorite epidote biotite albite schist																
1	414.8	339.8	1.22	0.0568	0.0010	0.4355	0.0077	0.0557	0.0008	356	6	367	5	443	32	97.0%
2	268.7	273.3	0.98	0.0571	0.0007	0.4433	0.0102	0.0561	0.0010	358	5	373	7	458	41	95.9%
3	211.8	198.9	1.06	0.0571	0.0008	0.4344	0.0087	0.0552	0.0009	358	5	366	6	420	35	97.8%
4	285.3	300.2	0.95	0.0570	0.0007	0.4223	0.0078	0.0537	0.0008	357	4	358	6	367	33	99.7%
5	161.0	296.5	0.54	0.0581	0.0009	0.4377	0.0092	0.0546	0.0008	364	6	369	7	395	33	98.6%
6	52.5	100.1	0.52	0.0577	0.0009	0.4436	0.0124	0.0558	0.0014	362	6	373	9	456	54	97.0%
7	91.9	110.7	0.83	0.0562	0.0009	0.4195	0.0109	0.0541	0.0012	353	5	356	8	376	53	99.2%
8	321.8	326.1	0.99	0.0570	0.0011	0.4262	0.0100	0.0542	0.0008	358	7	361	7	389	33	99.2%
9	133.0	179.9	0.74	0.0555	0.0008	0.4124	0.0084	0.0540	0.0010	348	5	351	6	372	39	99.1%
10	138.7	284.9	0.49	0.0561	0.0010	0.4062	0.0084	0.0526	0.0009	352	6	346	6	322	37	98.3%
11	209.1	210.6	0.99	0.0551	0.0008	0.4075	0.0072	0.0537	0.0008	346	5	347	5	367	35	99.7%
12	459.3	411.0	1.12	0.0580	0.0016	0.4238	0.0116	0.0531	0.0008	363	10	359	8	345	32	98.9%
13	189.4	209.4	0.90	0.0582	0.0010	0.4482	0.0129	0.0557	0.0013	365	6	376	9	439	50	97.0%
14	348.6	343.1	1.02	0.0587	0.0011	0.4316	0.0087	0.0534	0.0008	368	7	364	6	346	27	98.9%
15	159.3	192.3	0.83	0.0573	0.0010	0.4323	0.0103	0.0547	0.0010	359	6	365	7	398	36	98.3%
16	102.2	129.5	0.79	0.0581	0.0014	0.4235	0.0146	0.0532	0.0016	364	8	359	10	345	67	98.6%
17	151.1	170.2	0.89	0.0570	0.0009	0.4050	0.0103	0.0515	0.0011	357	5	345	8	261	50	96.6%
18	128.2	179.6	0.71	0.0572	0.0009	0.4192	0.0087	0.0532	0.0010	359	5	356	6	339	47	99.2%
19	212.1	226.9	0.93	0.0543	0.0009	0.4159	0.0104	0.0553	0.0010	341	6	353	7.5	433	39	96.5%
20	163.6	203.1	0.81	0.0555	0.0009	0.4156	0.0084	0.0544	0.0009	348	6	353	6	391	39	98.6%
21	380.3	405.1	0.94	0.0553	0.0009	0.4168	0.0085	0.0546	0.0007	347	6	354	6	395	34	98.0%
22	145.1	179.4	0.81	0.0538	0.0007	0.4009	0.0087	0.0539	0.0010	338	4	342	6	367	43	98.8%
23	420.8	585.3	0.72	0.0546	0.0008	0.3950	0.0067	0.0523	0.0006	343	5	338	5	302	24	98.5%
24	110.9	133.7	0.83	0.0570	0.0008	0.4283	0.0118	0.0546	0.0015	357	5	362	8	395	61	98.6%
25	70.2	127.1	0.55	0.0575	0.0007	0.4186	0.0105	0.0526	0.0011	360	5	355	8	322	51	98.6%
26	141.9	153.0	0.93	0.0579	0.0009	0.4224	0.0100	0.0529	0.0011	363	5	358	7	324	48	98.6%
27	122.8	179.0	0.69	0.0572	0.0008	0.4239	0.0085	0.0537	0.0010	359	5	359	6	361	39	100.0%
28	245.0	234.4	1.05	0.0584	0.0012	0.4403	0.0145	0.0547	0.0015	366	7	371	10	467	58	98.6%
29	249.3	250.8	0.99	0.0569	0.0010	0.4200	0.0093	0.0536	0.0010	357	6	356	7	354	47	99.7%
30	369.7	389.9	0.95	0.0563	0.0009	0.4113	0.0065	0.0530	0.0006	353	6	350	5	328	32	99.1%
31	337.6	274.2	1.23	0.0552	0.0008	0.3943	0.0079	0.0516	0.0007	346	5	338	6	265	32	97.7%
32	414.8	339.8	1.22	0.0568	0.0010	0.4355	0.0077	0.0557	0.0008	356	6	367	5	443	32	97.0%
33	268.7	273.3	0.98	0.0571	0.0007	0.4433	0.0102	0.0561	0.0010	358	5	373	7	458	41	95.9%
34	211.8	198.9	1.06	0.0571	0.0008	0.4344	0.0087	0.0552	0.0009	358	5	366	6	420	35	97.8%
35	285.3	300.2	0.95	0.0570	0.0007	0.4223	0.0078	0.0537	0.0008	357	4	358	7	367	33	99.7%

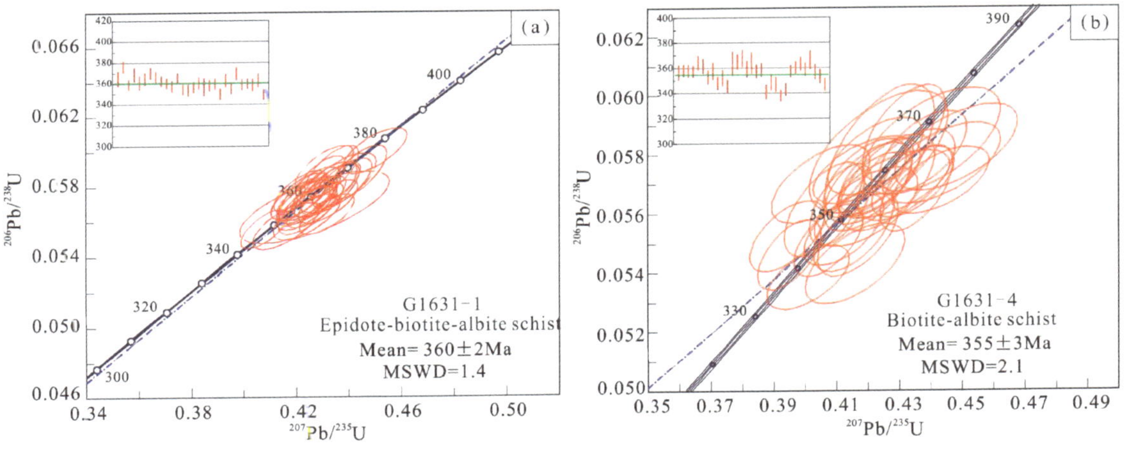

Figure 7. U-Pb concordia and weighted mean ages for the epidote-biotite-albite schist (**a**) and biotite-albite schist (**b**).

4.3. Zircon Hf Isotopies

The zircons from the epidote-biotite-albite schist (sample G1631-1) had homogeneous $^{176}Hf/^{177}Hf$ ratios of 0.282749 to 0.282872, with εHf(t) values of +6.5 to +11.0 (Figure 8), one-stage Hf model (T_{DM1}) ages from 552 to 746 Ma and two-stage Hf model (T_{DM2}) ages from 661 to 902 Ma. Similarly, the zircons from the biotite-albite schist sample (sample G1631-4) also possessed relatively homogeneous Hf isotopic compositions of 0.282783 to 0.282870, εHf(t) values of +7.9 to +10.9 (Figure 8), T_{DM1} ages from 547 to 649 Ma and T_{DM2} ages from 660 to 859 Ma (Table 3). All of these characteristics indicate that the parent magmas of the Hadayang schists were produced from a relatively depleted mantle source.

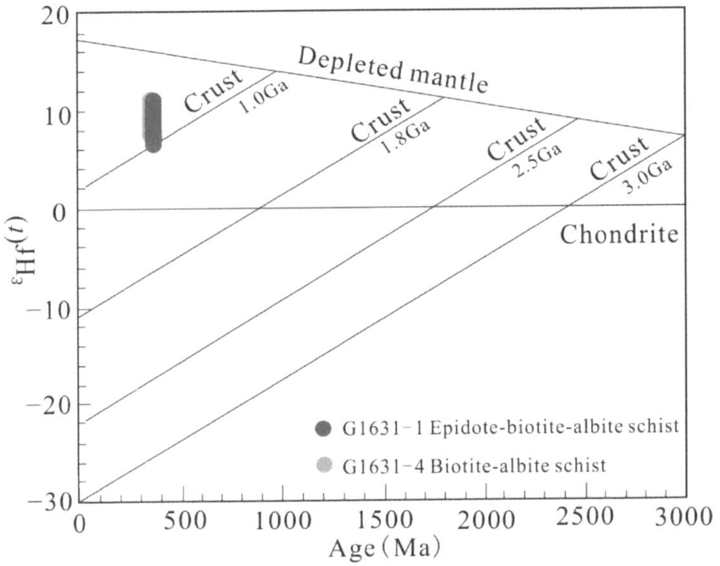

Figure 8. Diagrams of zircon Hf isotopic of the Hadayang schists.

Table 3. In situ zircon Lu-Hf isotopic results for the Hadayang schists.

Analytical Spot	t (Ma)	$^{176}Yb/^{177}Hf$	$^{176}Lu/^{177}Hf$	$^{176}Hf/^{177}Hf$	I_{Hf}	$\varepsilon_{Hf}(0)$	$\varepsilon_{Hf}(t)$	T_{DM1}(Hf) (Ma)	T_{DM2}(Hf) (Ma)	$f_{Lu/Hf}$
G1631-1 Epidote-biotite-albite schist										
1	365	0.065324	0.001772	0.282827	0.282815	2.0	9.6	615	756	−0.95
2	359	0.069656	0.001920	0.282866	0.282853	3.3	10.8	560	672	−0.94
3	361	0.082890	0.002175	0.282798	0.282783	0.9	8.3	665	832	−0.93
4	362	0.053714	0.001471	0.282834	0.282824	2.2	9.8	600	737	−0.96
5	360	0.112169	0.002980	0.282772	0.282752	0.0	7.2	718	902	−0.91
6	357	0.056043	0.001614	0.282853	0.282842	2.9	10.3	576	701	−0.95
7	360	0.059902	0.001785	0.282776	0.282764	0.1	7.6	690	876	−0.95
8	357	0.092929	0.002522	0.282821	0.282804	1.7	9.0	637	787	−0.92
9	358	0.056231	0.001558	0.282795	0.282785	0.8	8.3	657	829	−0.95
10	360	0.088918	0.002387	0.282836	0.28282	2.3	9.6	612	748	−0.93
11	363	0.093148	0.002687	0.282749	0.282731	−0.8	6.5	746	948	−0.92
12	359	0.069427	0.001865	0.282872	0.282859	3.5	11.0	552	661	−0.94
13	360	0.023950	0.000701	0.282780	0.282775	0.3	8.0	664	850	−0.98
14	358	0.064122	0.002020	0.282823	0.282809	1.8	9.2	625	773	−0.94
15	363	0.047436	0.001443	0.282801	0.282791	1.0	8.7	647	812	−0.96
G1631-4 Biotite-albite schist										
1	356	0.050095	0.001393	0.282794	0.282785	0.8	8.3	655	829	−0.96
2	358	0.034306	0.000991	0.282851	0.282844	2.8	10.4	569	695	−0.97
3	358	0.033785	0.000965	0.282789	0.282782	0.6	8.2	656	835	−0.97
4	357	0.071549	0.002065	0.282846	0.282832	2.6	10.0	593	723	−0.94
5	364	0.031583	0.000947	0.282801	0.282794	1.0	8.8	639	804	−0.97
6	353	0.050729	0.001448	0.282840	0.28283	2.4	9.8	592	731	−0.96
7	358	0.105460	0.003007	0.282851	0.282831	2.8	10.0	600	724	−0.91
8	352	0.050404	0.001402	0.282870	0.282861	3.5	10.9	547	660	−0.96
9	359	0.030825	0.000972	0.282847	0.28284	2.7	10.3	574	703	−0.97
10	357	0.059140	0.001785	0.282820	0.282808	1.7	9.1	625	776	−0.95
11	359	0.059194	0.001850	0.282783	0.282771	0.4	7.9	679	859	−0.94
12	357	0.074471	0.002239	0.282795	0.28278	0.8	8.2	669	839	−0.93
13	359	0.062644	0.001924	0.282823	0.28281	1.8	9.2	624	772	−0.94
14	357	0.056902	0.001824	0.282832	0.28282	2.1	9.6	609	751	−0.95
15	353	0.081466	0.002392	0.282814	0.282798	1.5	8.7	645	802	−0.93

5. Discussion

5.1. Protolith Restoration

The trace elements of the Hadayang schists were generally enriched in LILEs and depleted of HFSEs, Sr/Ba > 1, Cr/Ni > 1. Furthermore, the samples retained the crystalloblasts, which indicates that Hadayang schists should be a metaigneous rock. On a plot of A-C-FM diagram (Figure 9a), the epidote-biotite-albite schist samples plot in the calc-silicate rock and quartzite and basic volcanic rocks and ferrous-dolomitic tuff fields, while the biotite-albite schist samples plot in the intermediate-acid volcanic rock field. As can be seen from the plot of K vs. A (Figure 9b), all of the samples were igneous rock. On the other hand, on the (al + fm) − (c + alk) vs. Si diagram (Figure 9c), all the samples were volcanic rocks.

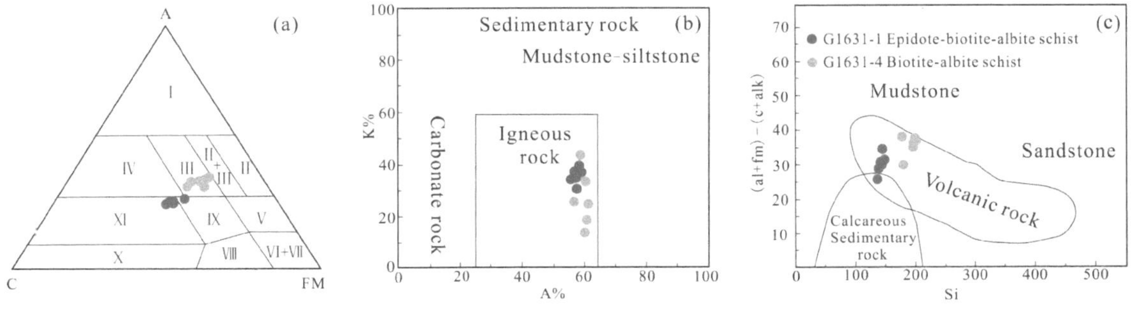

Figure 9. Classification diagrams of the Hadayang schists. (**a**) A-C-FM diagram (after [74]) (I—claystone and acidic volcanic rock; II—claystone and subgrauwacke; II+III—intermediate-felsic volcanic rock and subgrauwacke; III—intermediate volcanic rock; IV—arkose; V—tuffaceous siltstone; IX—basic volcanic rock and ferric dolomitic marlite; XI—calcium silicate rock and quartzite; X—calcium silicate rock; VIII—ultrabasic rock; VI+VII—silico-ferric sedimentary rock and ultrabasic rock); (**b**) A-K diagram (after [75]); (**c**) (al + fm) − (c + alk)-Si diagram (after [75]).

According to the comprehensive analysis of geochemical and petrographic characteristics, the protolith of Hadayang schist was a set of intermediate-basic volcanic rocks, among which the protolith of the epidotite-biotite-albite schist was a basalt, and the protolith of the biotite-albite schist was andesite.

5.2. Geochronological Framework

The Hadayang schists were previously considered to be Precambrian in age [76], but our chronological data indicate that they formed in the Late Devonian to Early Carboniferous period. Although the magmatic activity in this period was relatively weak in the region [77], in recent years, Late Devonian–Early Carboniferous magmatic rocks have been successively identified in the XB, such as in the Yakeshi-Huolongmen, Handagai, Nenjiang and Zhalantun-Zhalaite areas. The Late Devonian–Early Carboniferous magmatic rocks in the Yakeshi-Huolongmen area include rhyolite (ca. 364–379 Ma) [78], basalt (373 ± 5 Ma) [79], diorite (381 ± 2 Ma) [77] and gabbro (ca. 341–344 Ma) [78]. Biotite granite (364 ± 4 Ma) [80] and monzonitic granite (379 ± 4 Ma) [81] are found in the Handagai area. Silicic volcanic rocks (ca. 352–354 Ma) [16,58,59], monzonitic granite (352 ± 1 Ma) [16], syenite granite (381 ± 2 Ma) [60], hornblende gabbro (363 ± 1 Ma) [61] and hornblende rock (362 ± 1 Ma) [61] have been identified in the Nenjiang area. A large number of Late Devonian–Early Carboniferous granites, including alkali feldspar granite, syenite granite, monzonitic granite and granodiorite, with ages ranging from ca. 343 to 379 Ma [13,16,77,78,80–85], were exposed in the Zhalantun-Zhalaite Banner area. In addition, there are some basic–intermediate magmatic rocks scattered in this area, including andesite (ca. 346–362 Ma) [14,86,87], basaltic andesite (373 ± 3 Ma) [88], quartz diorite (ca. 357–359 Ma) [16,78], gabbro diorite (350 ± 1 Ma) [78] and gabbro (353 ± 2 Ma) [59]. In summary, a magmatic activity occurred in the XB during the Late Devonian–Early Carboniferous period, but the tectonic significance of this magmatic event is still controversial.

5.3. Petrogenesis

5.3.1. Crustal Contamination

Before using the elemental and isotopic compositions of the samples to discuss the petrogenesis of the schists, we first assess the effects of post-magmatic alteration. The protolith of the Hadayang schists was an intermediate-basic volcanic rock, which had undergone greenschist facies metamorphism under the influence of regional metamorphism in a later period, which is supported by their LOI content of 2.23–4.05 wt% and the fact that some of the minerals had been partially replaced by chlorite and epidote. After rock metamorphism, high-field-strength elements (HFSEs), such as REEs, Y, Th, U, Zr,

Hf, Ti, Nb, Ta, etc., tend to exhibit inert migration characteristics without significant migration [89], so the characteristics of HFSEs can better reflect the protolith characteristics of metamorphic rocks.

Mantle-derived magma will inevitably be subjected to crustal contamination during the ascent process [90]. The crustal material is characterized by low Nb/La, Sm/Nd, Nb/Ta and Mg#, and high Th/La, La/Sm and SiO_2. Therefore, Mg# is positively correlated with the ratios of Nb/La, Nb/Ta and Sm/Nd, and negatively correlated with Th/La, La/Sm and SiO_2, which can indicate crustal contamination [91]. As can be seen from Figure 10, Mg# was positively correlated with Nb/La, Nb/Ta, Th/La, La/Sm and SiO_2, and negatively correlated with Sm/Nd, which is significantly different from the characteristics of crustal contamination. In addition, the Hadayang schists had significantly depleted Zr and Hf elements (Figure 5b), excluding inherited zircons, and had positive εHf(*t*) values of +5.8 to +9.3, suggesting that crustal contamination was not a significant factor during the magmatic evolution of the Hadayang schists.

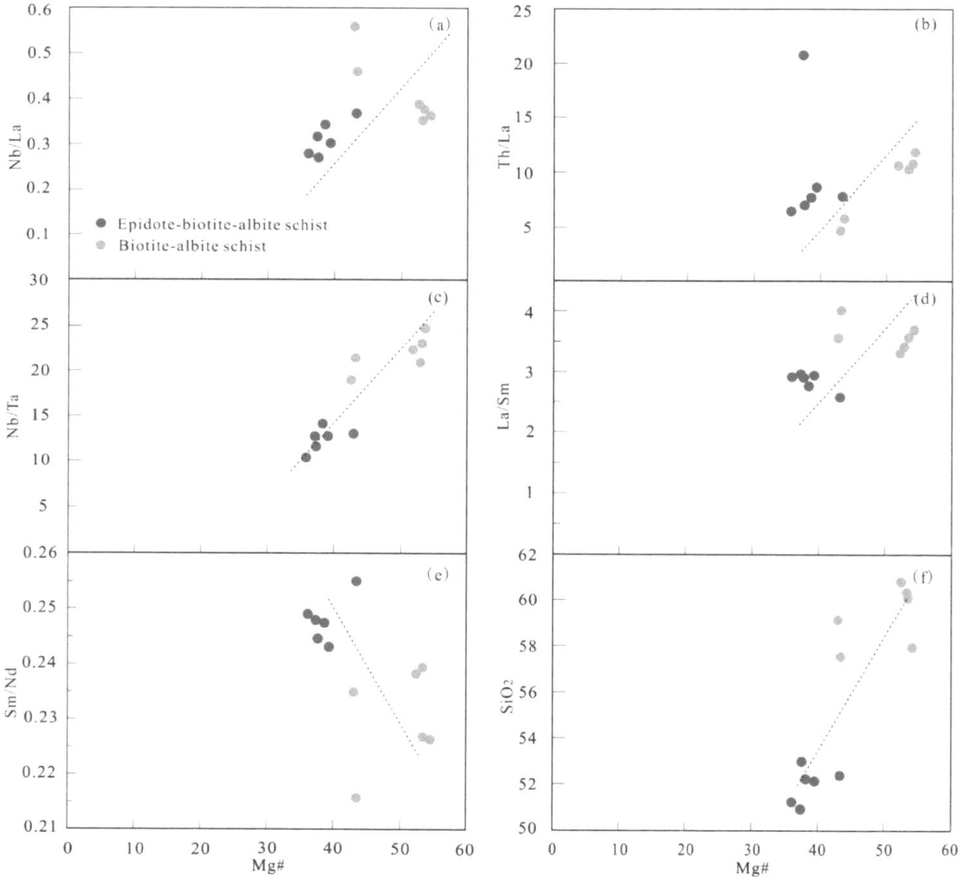

Figure 10. Plots of Nb/La vs. Mg# (**a**), Th/La vs. Mg# (**b**), Nb/Ta vs. Mg# (**c**), La/Sm vs. Mg# (**d**), Sm/Nd vs. Mg# (**e**) and SiO_2 vs. Mg# (**f**) for the Hadayang schists.

5.3.2. Fractional Crystallization

The Hadayang schists had low Mg# (35.2–53.0), Cr (23.7–86.5 ppm), Ni (21.1–40.0 ppm) and Co (12.1–30.6 ppm) contents relative to primitive-mantle-derived magmas, suggesting that our samples were evolved magmas rather than primitive magmas. The negative correlation between Mg# and CaO and FeO^T, and the fact that Mg# was positively correlated with Cr, Ni, SiO_2 and Na_2O, indicated olivine and clinopyroxene fractionation (Figure 11). Based on the negative Eu anomalies, plagioclase fractionation also must have played an important role in the evolution of these magmas.

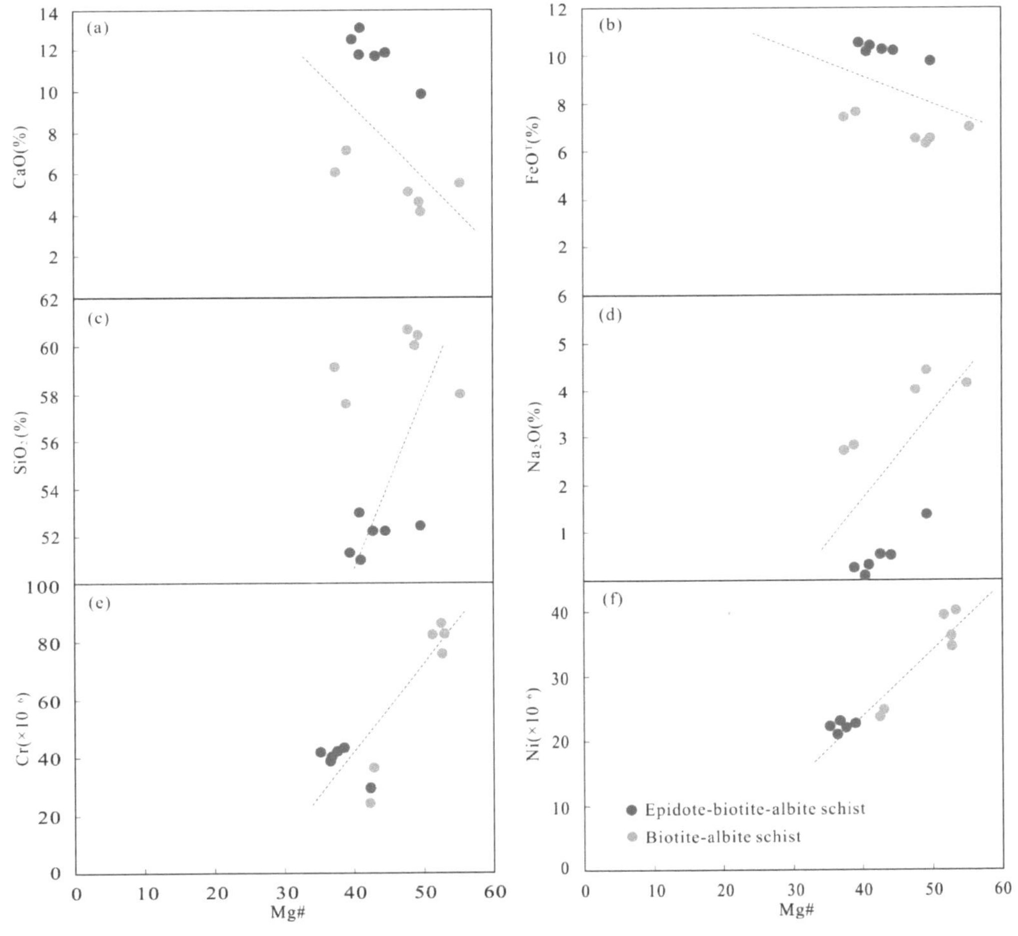

Figure 11. Plots of CaO vs. Mg# (**a**), FeO^T vs. Mg# (**b**), SiO_2 vs. Mg# (**c**), Na_2O vs. Mg# (**d**), Cr vs. Mg# (**e**) and Ni vs. Mg# (**f**) for the Hadayang schists.

5.3.3. Magma Source

The source region of basic volcanic rocks is generally asthenosphere or lithospheric mantle [92]. The low silica and alkali contents of the Hadayang schists, and their high MgO, Ti, Co and Cr contents, all favor a mantle magmatic source. This conclusion is also supported by the positive $\varepsilon_{Hf}(t)$ values (the average values were +8.93 and +9.29, respectively) (Figure 8). The T_{DM1} ages (552 to 781 Ma and 547 to 679 Ma, respectively) indicated that their parental magma fractionated from mantle in the Neoproterozoic period. As mentioned above, the Hadayang schists have LILE enrichment and significant HFSE depletion with moderately fractionated REE patterns and relatively low HREE abundances, suggesting that the components of the magma source may have included aqueous fluids released by dehydration of the subducted oceanic plate and/or partial melts derived from the subducted oceanic crust/sediments. The Ba/Th ratio of the Hadayang schist samples varied significantly, while the $(La/Sm)_N$ ratio was relatively constant (Figure 12b), indicating that fluids played a key role in the enrichment process, and the influence of sediment was small or absent. The wide range of the Sr/Nd (10.1–30.7) ratio values further indicated that fluid action played a major role in the process of mantle-derived magma enrichment. The above characteristics are supported by the Nb/Zr vs. Th/Zr plot (Figure 12a). The abundances and ratios of REEs are widely used to distinguish the source regions of magma and the degree of partial melting [93]. As can be seen in Sm/Yb vs. Sm and Sm/Yb vs. La/Yb plots (Figure 12c, d), the Hadayang schist samples plot mainly near or slightly deviating from the garnet + spinel lherzolite melting curves, and they fall between these two melting curves with starting compositions of primitive mantle, implying a mantle source consisting of spinel and garnet lherzolite, and that it took 1%–5% partial melting of this mantle source to produce the parent magma of the Hadayang schists (Figure 12c,d).

5.4. Tectonic Evolution and Geological Significance

Although a lot of progress has been made in the study of Late Devonian–Early Carboniferous magmatic rocks in the XB, there are still major disputes about their tectonic setting, including the active continental margin or island arc [16,58,59,79,82,83,86,87], back-arc basin [61,88], post-collision extensional [60] and subduction-related extension environment [78,81]. Some authors also believe that there was a transition from passive continental margin to island arc environment during this period [84,94]. In the following discussion, we suggest that the Late Devonian–Early Carboniferous magmatic rocks were formed in an intracontinental extension setting.

As mentioned in an earlier section, the Hadayang schists were characterized by LILE enrichment and significant HFSE depletion, and also exhibited moderately fractionated REE patterns and relatively low HREE abundances, which is also seen in the Late Devonian to Early Carboniferous basic magmatic rocks of the XB [61,78,86,88]. These geochemical characteristics are similar to those arc-type magmatic rocks [98], but there is some uncertainty in judging whether basic igneous rocks are arc magmatic rocks [99–101]. Previous studies have shown that the participation of fluids can change the geochemical characteristics of trace elements, and the precipitation of water-bearing fluid from subducting plates, or the release of water-bearing fluid from deep fluid circulation under the background of subducting plates, may lead to water-rich fluid metasomatism in the mantle source region, forming arc-like trace element characteristics [101–103]. Due to the differences in the contents of Zr, Ti and Sr elements between arc-like basalts and arc basalts, it is very effective to distinguish the two types of basalts by using a discriminant diagram of the above three elements [101]. On the plots of Ti/1000 vs. V, Zr vs. Zr/Y and Zr vs. Ti (Figure 13a,b,d), all of the Late Devonian to Early Carboniferous intermediate-basic magmatic rocks in the XB plot in the intraplate basalt field or deviating from the arc basalt area, similar to the geochemical characteristics of the North American Basin and Range basalt. As can be seen from the plot of Zr/Sm-Sr/Nd-Ti/V (Figure 13c), most of the samples plot in or near the intraplate magma and Basin and Range regions, which is significantly different from the nearly vertical characteristics of arc basalt. Based on the above analysis, we propose that the Late Devonian–Early Carboniferous intermediate-basic magmatic rocks in the eastern XB may have been formed in an intracontinental extension tectonic setting similar to the North American Basin and Range basalt. Coincidentally, the same tectonic background also appears in the western part of the XB. Previous studies have shown that the Erenhot–Hegenshan ophiolite in the XB was formed during the Late Devonian–Early Carboniferous [104–110] and was the product of the upwelling process of the asthenosphere caused by the extension of the lithosphere [111,112]. Based on the identification of basement rocks and other evidence around the ophiolite [113], it was indicated that the ophiolite would have been formed in the continental crust extension zone [6,114]. In summary, we believe that the XB would have been in the stage of intracontinental extension during the Late Devonian–Early Carboniferous.

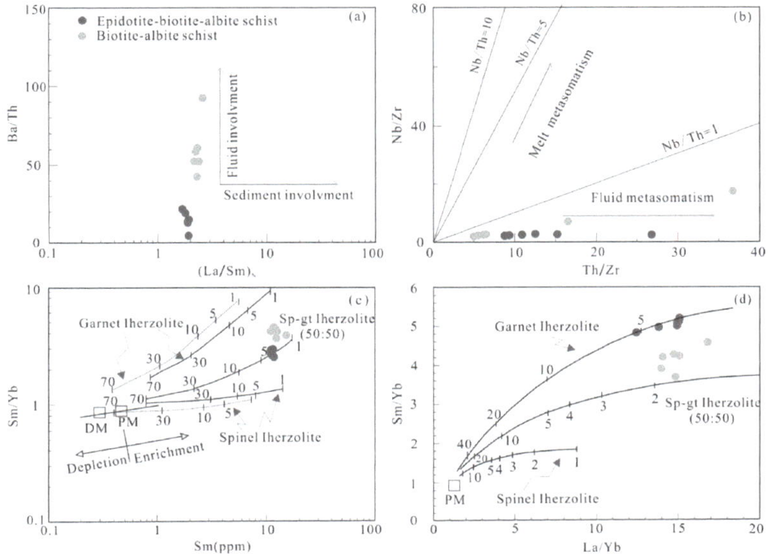

Figure 12. Plots of Nb/Zr vs. Th/Zr ((**a**), after [95]), La/Sm vs. Sm/Yb ((**b**), after [96]), Sm vs. Sm/Yb ((**c**), after [97]) and La/Yb vs. Sm/Yb ((**d**), after [97]) for the Hadayang schists.

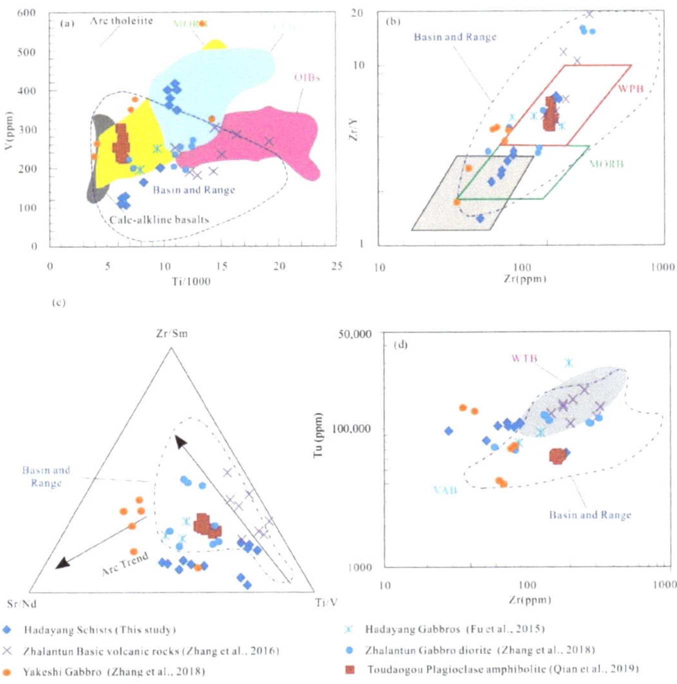

Figure 13. Structural discrimination diagrams of V vs. Ti/1000 ((**a**), after [115]), Zr/Y vs. Zr ((**b**), after [89]), Zr/Sm-Sr/Nd-Ti/V (**c**) and Ti vs. Zr ((**d**), after [116]) for Late Devonian to Early Carboniferous intermediate-basic rocks in the eastern XB. CFB = continental flood basalts; OIBs = ocean-island basalts; MORB = mid-ocean ridge basalt; WPB = within-plate basalts; IAB = island arc basalts; VAB = volcanic arc basalt; WTB = within-plate tholeiitic basalt.

6. Conclusions

1. The protolith of the Hadayang schists was intermediate-basic volcanic rock. LA-ICP-MS U-Pb dating of zircons from the epidote-biotite-albite schist and biotite-albite schist yielded crystallization ages of 360 $\pm$ 2 Ma (MSWD = 1.4) and 355 $\pm$ 3 Ma (MSWD = 2.1). This, combined with the presence of contemporaneous magmatic rocks in the region, indicates an important magmatic event in the eastern XB during the Late Devonian–Early Carboniferous.
2. The magma source of the Hadayang schists was a mantle that consisted of both spinel and garnet lherzolite, with a partial melting degree of 1%–5%, and which underwent fractional crystallization of olivine, orthopyroxene and plagioclase.
3. The Late Devonian–Early Carboniferous intermediate-basic magmatic rocks in the eastern XB were formed in an intracontinental extension setting, which is consistent with the tectonic setting of contemporaneous ophiolites in the western section. The Xing'an block would have been in the stage of intracontinental extension during the Late Devonian–Early Carboniferous.

Author Contributions: Conceptualization, F.N. and W.S. (Weimin Song); methodology, J.F.; software, Y.L.; validation, Y.W. and W.S. (Wei Sun); formal analysis, F.N.; investigation, J.F.; resources, J.F.; data curation, Y.L.; writing—original draft preparation, F.N. and W.S. (Weimin Song); writing—review and editing, Y.L. and J.F.; visualization, Y.W. and W.S. (Wei Sun); supervision, F.N.; project administration, J.F.; funding acquisition, W.S. (Weimin Song) and J.F. All authors have read and agreed to the published version of the manuscript.

Funding: This research was funded by the Geological Survey Project of China (Grant No. DD20230224 and DD20190360).

Data Availability Statement: The authors confirm that the data generated or analyzed during this study are provided in full within the published article.

Conflicts of Interest: The authors declare no conflict of interest.

References

1. Şengör, A.; Natal'In, B.; Burtman, V. Evolution of the Altaid tectonic collage and Palaeozoic crustal growth in Eurasia. *Nature* **1993**, *364*, 299–307. [CrossRef]
2. Jahn, B.-M.; Griffin, W.; Windley, B. Continental growth in the Phanerozoic: Evidence from Central Asia. *Tectonophysics* **2000**, *328*, 7–10. [CrossRef]
3. Jahn, B.-M. The Central Asian Orogenic Belt and growth of the continental crust in the Phanerozoic. *Geol. Soc. Lond. Spec. Publ.* **2004**, *226*, 73–100. [CrossRef]
4. Kröner, A.; Windley, B.; Badarch, G.; Tomurtogoo, O.; Hegner, E.; Jahn, B.M.; Gruschka, S.; Khain, E.V.; Demoux, A.; Wingate, M.T.D.; et al. Accretionary growth and crust formation in the Central Asian Orogenic Belt and comparison with the Arabian-Nubian shield. *Geol. Soc. Am. Mem.* **2007**, *200*, 461. [CrossRef]
5. Windley, B.F.; Alexeiev, D.; Xiao, W.; Kröner, A.; Badarch, G. Tectonic models for accretion of the Central Asian Orogenic Belt. *J. Geol. Soc.* **2007**, *164*, 31–47. [CrossRef]
6. Xu, B.; Zhao, P.; Bao, Q.Z.; Zhou, Y.H.; Wang, Y.Y.; Luo, Z.W. Preliminary study on the pre-Mesozoic tectonic unit division of the Xing-Meng Orogenic Belt (XMOB). *Acta Petrol. Sin.* **2014**, *30*, 1841–1857, (In Chinese with English Abstract).
7. Xu, B.; Zhao, P.; Wang, Y.Y.; Liao, W.; Luo, Z.W.; Bao, Q.Z.; Zhou, Y.H. The pre-Devonian tectonic framework of Xing'an–Mongolia orogenic belt (XMOB) in north China. *J. Asian Earth Sci.* **2015**, *97*, 183–196. [CrossRef]
8. Liu, Y.; Li, W.; Feng, Z.; Wen, Q.; Neubauer, F.; Liang, C. A review of the Paleozoic tectonics in the eastern part of Central Asian Orogenic Belt. *Gondwana Res.* **2017**, *43*, 123–148. [CrossRef]
9. Liu, Y.; Li, W.; Ma, Y.; Feng, Z.; Guan, Q.; Li, S.; Chen, Z.; Liang, C.; Wen, Q. An orocline in the eastern Central Asian Orogenic Belt. *Earth-Sci. Rev.* **2021**, *221*, 103808. [CrossRef]
10. Zhou, J.B.; Wilde, S.A.; Zhao, G.C.; Han, J. Nature and assembly of microcontinental blocks within the Paleo-Asian Ocean. *Earth-Sci. Rev.* **2018**, *186*, 76–93. [CrossRef]
11. Xu, W.L.; Sun, Y.C.; Tang, J.; Luan, J.P.; Wang, F. Basement nature and tectonic evolution of the Xing'an-Mongolian Orogenic belt. *Earth Sci.* **2019**, *44*, 1620–1646, (In Chinese with English Abstract). [CrossRef]
12. Zhang, J.M.; Xu, B.; Yan, L.J.; Wang, Y.Y. Evolution of the Heihe-Nenjiang Ocean in the eastern Paleo-Asian Ocean: Constraints of sedimentological, geochronological and geochemical investigations from Early-Middle Paleozoic Heihe-Dashizhai Orogenic Belt in the northeast China. *Gondwana Res.* **2020**, *81*, 339–361. [CrossRef]

13. Ma, Y.F. The Late Paleozoic Tectonic Evolution of the Central Great Xing'an Range, NE China. Ph.D. Thesis, Jilin University, Changchun, China, 2019. (In Chinese with English Abstract).
14. Ma, Y.-F.; Liu, Y.-J.; Wang, Y.; Qin, T.; Chen, H.J.; Sun, W.; Zang, Y.Q. Late Carboniferous mafic to felsic intrusive rocks in the central Great Xing'an Range, NE China: Petrogenesis and tectonic implications. *Int. J. Earth Sci.* **2020**, *109*, 761–783. [CrossRef]
15. Liu, Y.J.; Feng, Z.Q.; Jiang, L.W.; Jin, W.; Li, W.M.; Guan, Q.B.; Wen, Q.B.; Liang, C.Y. Ophiolite in the eastern Central Asian Orogenic Belt, NE China. *Acta Pet. Sin.* **2019**, *35*, 3017–3047, (In Chinese with English Abstract). [CrossRef]
16. Li, Y.; Xu, W.-L.; Tang, J.; Sun, C.-Y.; Zhang, X.-M.; Xiong, S. Late Paleozoic igneous rocks in the Xing'an Massif and its amalgamation with the Songnen Massif, NE China. *J. Asian Earth Sci.* **2020**, *197*, 104407. [CrossRef]
17. Shao, J.A. *Crust Evolution in the Middle Part of the Northern Margin of Sino-Korean Plate*; Peking University Press: Beijing, China, 1991; pp. 1–136, (In Chinese with English Abstract).
18. Hong, D.; Huang, H.; Xiao, Y.; Xu, H.; Jin, M. The Permian alkaline granites in Central Inner Mongolia and their geodynamic significance. *Acta Geol. Sin. Engl. Ed.* **1995**, *8*, 27–39. [CrossRef]
19. Sun, D.Y.; Wu, F.Y.; Li, H.M.; Lin, Q. Emplacement age of the postorogenic A-type granites in Northwestern Lesser Xing'an Ranges, and its relationship to the eastward extension of Suolushan-Hegenshan-Zhalaite collisional suture zone. *Chin. Sci. Bull.* **2001**, *46*, 427–432. [CrossRef]
20. Tong, Y.; Hong, D.W.; Wang, T.; Shi, X.J.; Zhang, J.J.; Zeng, T. Spatial and temporal distribution of granitoids in the middle segment of the Sino-Mongolian Border and its tectonic and metallogenic implications. *Acta Geosci. Sin.* **2010**, *31*, 395–412, (In Chinese with English Abstract).
21. Chen, B.; Jahn, B.-M.; Wilde, S.; Xu, B. Two contrasting paleozoic magmatic belts in northern Inner Mongolia, China: Petrogenesis and tectonic implications. *Tectonophysics* **2000**, *328*, 157–182. [CrossRef]
22. Miao, L.C.; Fan, W.M.; Zhang, F.Q.; Liu, D.Y.; Jian, P.; Shi, G.H.; Tao, H.; Shi, Y.Q. Zircon SHRIMP geochronology of the Xinkailing-Kele complex in the northwestern Lesser Xing'an Range, and its geological implications. *Chin. Sci. Bull.* **2004**, *49*, 201–209. [CrossRef]
23. Yarmolyuk, V.V.; Kovalenko, V.I.; Sal'nikova, E.B.; Kovach, V.P.; Kozlovsky, A.M.; Kotov, A.B.; Lebedev, V.I. Geochronology of igneous rocks and formation of the Late Paleozoic south Mongolian active margin of the Siberian continent. *Strat. Geol. Correl.* **2008**, *162*, 162–181. [CrossRef]
24. Sun, L.X.; Ren, B.F.; Zhao, F.Q.; Ji, S.P.; Geng, J.Z. Late Paleoproterozoic magmatic records in the Erguna massif: Evidences from the zircon U-Pb dating of granitic gneisses. *Geol. Bull. China* **2013**, *32*, 341–352, (In Chinese with English Abstract). [CrossRef]
25. Shao, J.; Li, Y.F.; Zhou, Y.H.; Wang, H.B.; Zhang, J. Neo-Archaean magmatic event in Erguna Massif of northeast China: Evidence from the zircon LA-ICP-MS dating of the geneissic monogranite from the drill. *J. Jilin Univ. Earth Sci. Ed.* **2015**, *45*, 364–373, (In Chinese with English Abstract). [CrossRef]
26. Li, G.Y.; Zhou, J.B.; Li, L.; Han, J.; Song, M.C. Paleoproterozoic basement of the Xing'an Block in the eastern Central Asian Orogenic Belt: Evidence from the geochemistry and zircon U–Pb geochronology of granitic gneisses. *Precambrian Res.* **2019**, *331*, 105372. [CrossRef]
27. Feng, Z.Q.; Zhang, Q.H.; Liu, Y.J.; Li, L.; Jiang, L.W.; Zhou, J.P.; Li, W.M.; Ma, Y.F. Reconstruction of Rodinia supercontinent: Evidence from the Erguna Block (NE China) and adjacent units in the eastern Central Asian orogenic Belt. *Precambrian Res.* **2022**, *368*, 106467. [CrossRef]
28. Ge, W.C.; Chen, J.S.; Yang, H.; Zhao, G.C.; Zhang, Y.L.; Tian, D.X. Tectonic implications of new zircon U–Pb ages for the Xinghuadukou Complex, Erguna Massif, northern Great Xing'an Range, NE China. *J. Asian Earth Sci.* **2015**, *106*, 169–185. [CrossRef]
29. Zhao, S.; Xu, W.-L.; Tang, J.; Li, Y.; Guo, P. Timing of formation and tectonic nature of the purportedly Neoproterozoic Jiageda Formation of the Erguna Massif, NE China: Constraints from field geology and U–Pb geochronology of detrital and magmatic zircons. *Precambrian Res.* **2016**, *281*, 585–601. [CrossRef]
30. Xu, J.L.; Zheng, C.Q.; Tajčmanová, L.; Zhong, X.; Han, X.M.; Wang, Z.Y.; Yang, Y. Phase Equilibria Modeling and Zircon Dating for Precambrian Metapelites from the Xinghuadukou Complex in the Lulin Forest of the Erguna Massif, Northeast China. *J. Earth Sci.* **2018**, *29*, 1276–1290. [CrossRef]
31. Liu, H.C.; Li, Y.L.; Wan, Z.F.; Lai, C.K. Early Neoproterozoic tectonic evolution of the Erguna Terrane (NE China) and its paleogeographic location in Rodinia supercontinent: Insights from magmatic and sedimentary record. *Gondwana Res.* **2020**, *88*, 185–220. [CrossRef]
32. Tang, J.; Xu, W.L.; Wang, F.; Wang, W.; Xu, M.J.; Zhang, Y.H. Geochronology and geochemistry of Neoproterozoic magmatism in the Erguna Massif, NE China: Petrogenesis and implications for the breakup of the Rodinia supercontinent. *Precambrian Res.* **2013**, *244*, 597–615. [CrossRef]
33. Zhao, S.; Xu, W.L.; Wang, F.; Wang, W.; Tang, J.; Zhang, Y.H. Neoproterozoic Magmatisms in the Erguna Massif, NE China: Evidence from Zircon U-Pb Geochronology. *Geotecton. Metallog.* **2016**, *40*, 559–573, (In Chinese with English Abstract). [CrossRef]
34. Zhao, S.; Xu, W.L.; Tang, J.; Li, Y.; Guo, Y.P. Neoproterozoic magmatic events and tectonic attribution of the Erguna Massif: Constraints from geochronological, geochemical and Hf isotopic data of intrusive rocks. *Earth Sci.* **2016**, *41*, 1803–1829, (In Chinese with English Abstract). [CrossRef]
35. Sun, W.; Chi, X.-G.; Zhao, Z.; Pan, S.-Y.; Liu, J.-F.; Zhang, R.; Quan, J.-Y. Zircon geochronology constraints on the age and nature of 'Precambrian metamorphic rocks' in the Xing'an block of Northeast China. *Int. Geol. Rev.* **2014**, *56*, 672–694. [CrossRef]

36. Na, F.C.; Song, W.M.; Liu, Y.C.; Wang, Y.; Fu, J.Y.; Sun, W.; Li, L.C.; Wu, J.W.; Ge, J.T.; Tan, H.Y. Chronological study and tectonic significance of Precambrian metamorphic rocks in Zhalantun area of Da Hinggan Mountains. *Geol. Bull. China* **2018**, *37*, 1607–1619, (In Chinese with English Abstract).
37. Na, F.C.; Fu, J.Y.; Song, W.M.; Liu, Y.C.; Wang, Y.; Ge, J.T.; Sun, W.; Yang, F.; Zhang, G.Y.; Zhong, H. Petrological and geochronological study of Keluo Complex in northwestern Lesser Xing'an Range. *Earth Sci.* **2019**, *44*, 3265–3278, (In Chinese with English Abstract) [CrossRef]
38. Zhang, C.; Wu, X.; Guo, W.; Zhang, Y.; Quan, J. Discovery of the 1.8 Ga Granite on the Western Margin of the Songnen Masiff, China. *Acta Geol. Sin. Engl. Ed.* **2017**, *91*, 1497–1498. [CrossRef]
39. Zhang, C.; Wu, X.W.; Liu, Z.H.; Zhang, Y.J.; Guo, W.; Quan, J.Y. Precambrian geological events on the western margin of Songnen massif: Evidence from LA-ICP-MS U-Pb geochronologu of zircons from Paleoproterozoic granite in the Longjiang area. *Acta Petrol. Sin.* **2018**, *34*, 3137–3152, (In Chinese with English Abstract) [CrossRef]
40. Qian, C.; Chen, H.J.; Lu, L.; Pang, X.J.; Qin, T.; Wang, Y. The discovery of Neoarchean granite in Longjiang area, Heilongjiang Province. *Acta Geosci. Sin.* **2018**, *39*, 27–36, (In Chinese with English Abstract) [CrossRef]
41. Wu, X.W.; Zhang, C.; Zhang, Y.J.; Guo, W.; Zhang, C.; Cui, T.R.; Yang, Y.J.; Hu, J.F.; Song, W.B. 2.7 Ga Monzogranite on the Songnen Massif and Its Geological Implications. *Acta Geol. Sin. Engl. Ed.* **2018**, *92*, 1265–1266. [CrossRef]
42. Wu, X.W.; Zhang, C.; Shao, J.; Guo, W.; Li, L.C.; Zhang, G.Y.; Liu, B.S.; Song, W.B. The Neoarchean to Paleoproterozoic magmatic events in Longjiang area and Ulanhot area: Constraint on the attribute of Precambrian geological units in western Songnen massif. *Acta Petrol. Sin.* **2022**, *38*, 2811–2828, (In Chinese with English Abstract)
43. Cheng, Z.X.; Wang, Y.; Qian, C.; Yang, X.P.; Li, Z.H.; Liu, H.W.; Xiao, L. The discovery of the Paleoproterozoic metamorphic rocks in Ulanhot, Inner Mongolia, and its geological significance. *Geol. Bull. China* **2018**, *37*, 1599–1606, (In Chinese with English Abstract)
44. Zhao, P.; Fang, J.Q.; Xu, B.; Chen, Y.; Faure, M. Early Paleozoic tectonic evolution of the Xing-Meng Orogenic Belt: Constraints from detrital zircon geochronology of western Erguna–Xing'an Block, North China. *J. Asian Earth Sci.* **2014**, *95*, 136–146. [CrossRef]
45. Zhou, J.B.; Wang, B.; Wilde, S.A.; Zhao, G.C.; Cao, J.L.; Zheng, C.Q. Geochemistry and U–Pb zircon dating of the Toudaoqiao blueschists in the Great Xing'an Range, northeast China, and tectonic implications. *J. Asian Earth Sci.* **2015**, *97*, 197–210. [CrossRef]
46. Miao, L.C.; Zhang, F.C.; Jiao, S.J. Age, protoliths and tectonic implications of the Toudaoqiao blueschist, Inner Mongolia, China. *J. Asian Earth Sci.* **2015**, *105*, 360–373. [CrossRef]
47. Zhao, L.M.; Takasu, A.; Liu, Y.J.; Li, W.M. Blueschist from the Toudaoqiao Area, Inner Mongolia, NE China: Evidence for the Suture between the Ergun and the Xing'an Blocks. *J. Earth Sci.* **2017**, *28*, 241–248. [CrossRef]
48. Feng, Z.Q.; Liu, Y.J.; Li, Y.R.; Li, W.M.; Wen, Q.B.; Liu, B.Q.; Zhou, J.P.; Zhao, Y.L. Ages, geochemistry and tectonic implications of the Cambrian igneous rocks in the northern Great Xing'an Range, NE China. *J. Asian Earth Sci.* **2017**, *144*, 5–21. [CrossRef]
49. Feng, Z.Q.; Liu, Y.J.; Li, L.; She, H.Q.; Jiang, L.W.; Du, B.Y.; Liu, Y.W.; Li, W.M.; Wen, Q.B.; Liang, C.Y. Subduction, accretion, and collision during the Neoproterozoic-Cambrian orogeny in the Great Xing'an Range, NE China: Insights from geochemistry and geochronology of the Ali River ophiolitic mélange and arc-type granodiorites. *Precambrian Res.* **2018**, *311*, 117–135. [CrossRef]
50. Gao, F.H.; Wang, F.; Xu, W.L.; Yang, Y. Age of the "Paleoproterozoic" Dongfengshan Group in the Lesser Xing'an Range, NE China, and its tectonic implications: Constraints from zircon U-Pb geochronology. *J. Jilin Univ. Earth Sci. Ed.* **2013**, *43*, 440–456, (In Chinese with English Abstract). [CrossRef]
51. Quan, J.Y.; Chi, X.G.; Zhang, R.; Fan, L.F.; Hu, Z.C. LA-ICP-MS U-Pb geochronology of detrital zircon from the Neoproterozoic Dongfengshan Group in Songnen masiff and its geological significance. *Geol. Bull. China* **2013**, *32*, 353–364, (In Chinese with English Abstract). [CrossRef]
52. Wang, F.; Xu, W.L.; Gao, F.H.; Zhang, H.H.; Pei, F.P.; Zhao, L.; Yang, Y. Precambrian terrane within the Songnen–Zhangguangcai Range Massif, NE China: Evidence from U–Pb ages of detrital zircons from the Dongfengshan and Tadong groups. *Gondwana Res.* **2014**, *26*, 402–413. [CrossRef]
53. Luan, J.P.; Xu, W.L.; Wang, F.; Wang, Z.W.; Guo, P. Age and geochemistry of Neoproterozoic granitoids in the Songnen–Zhangguangcai Range Massif, NE China: Petrogenesis and tectonic implications. *J. Asian Earth Sci.* **2017**, *148*, 265–276. [CrossRef]
54. Luan, J.P.; Yu, J.J.; Yu, J.L.; Cui, Y.C.; Xu, W.L. Early Neoproterozoic magmatism and the associated metamorphism in the Songnen Massif, NE China: Petrogenesis and tectonic implications. *Precambrian Res.* **2019**, *328*, 250–268. [CrossRef]
55. Luan, J.P.; Tang, J.; Xu, W.L.; Guo, P.; Long, X.Y.; Xiong, S. Petrogenesis of Neoproterozoic magmatic rocks in the Songnen Massif (northeastern China): Implications for basement composition and crustal growth. *Precambrian Res.* **2022**, *376*, 106687. [CrossRef]
56. Zhang, L.Y.; Wang, Z.W.; Xu, B.; Zou, H.B.; Zhao, P.; Zhang, H. Neoproterozoic–Early Cambrian igneous and sedimentary sequences in the Songliao Block, NE China: Records of Rodinia supercontinent evolution in eastern Central Asian orogenic Belt. *Precambrian Res.* **2022**, *381*, 106865. [CrossRef]
57. Na, F.C.; Fu, J.Y.; Wang, Y.; Yang, F.; Zhang, G.Y.; Liu, Y.C.; Kang, Z. LA-ICP-MS zircon U-Pb age of the chlorite-muscovite tectonic schist in Hadayang, Morin Dawa Banner, Inner Mongolia, and its tectonic significance. *Geol. Bull. China* **2014**, *33*, 1326–1332, (In Chinese with English Abstract). [CrossRef]
58. Zhao, Z.; Chi, X.G.; Pan, S.Y.; Liu, J.F.; Sun, W.; Hu, Z.C. Zircon U-Pb LA-ICP-MS dating of Carboniferous volcanics and its geological significance in the northwestern Lesser Xing'an Range. *Acta Petrol. Sin.* **2010**, *26*, 2452–2464, (In Chinese with English Abstract)

59. Feng, Z.Q.; Li, W.M.; Liu, Y.J.; Jin, W.; Wen, Q.B.; Liu, B.Q.; Zhou, J.P.; Zhang, T.A.; Li, X.Y. Early Carboniferous tectonic evolution of the northern Heihe-Nenjiang-Hegenshan suture zone, NE China: Constraints from the mylonitized Nenjiang rhyolites and the Moguqi gabbros. *Geol. J.* **2018**, *53*, 1005–1021. [CrossRef]
60. Yang, F.; Na, F.C.; Zhang, G.Y.; Wang, Y.; Fu, J.Y.; Sun, W.; Chen, J.S.; Li, B.; Liu, M.; Pang, X.J.; et al. New Discovery of the Devonian Orthoclase Granite of Nenjiang-Heihe Structural belt, China and its Zircon U-Pb Data. *Acta Geol. Sin. Engl. Ed.* **2018**, *92*, 874–875. [CrossRef]
61. Fu, J.Y.; Wang, Y.; Na, F.C.; Sun, W.; Yang, F.; Zhong, H.; Zhang, G.Y.; Liu, Y.C. Zircon U-Pb geochronology and geochemistry of the Hadayang mafic-ultramafic rocks in Inner Mongolia: Constraints on the Late Devonian subduction of Nenjiang-Heihe area, Northeast China. *Geol. China* **2015**, *42*, 1740–1753, (In Chinese with English Abstract).
62. Fu, J.Y.; Wang, Y.; Na, F.C.; Zhang, G.Y.; Liu, Y.C.; Kang, Z. Geology of the Hadayang tectonic mélange in Inner Mongolia: Discovery significance. *Geol. Resour.* **2015**, *24*, 408–413, (In Chinese with English Abstract). [CrossRef]
63. Hou, K.J.; Li, Y.H.; Tian, Y.R. In situ U-Pb zircon dating using laser ablation-multiion counting-ICP-MS. *Miner. Depos.* **2009**, *28*, 481–492, (In Chinese with English Abstract). [CrossRef]
64. Liu, Y.S.; Gao, S.; Hu, Z.C.; Gao, C.G.; Zong, K.Q.; Wang, D.B. Continental and oceanic crust recycling-induced melt-peridotite interactions in the Trans-North China Orogen: U-Pb dating, Hf isotopes and trace elements in zircons from mantle xenoliths. *J. Pet.* **2010**, *51*, 537–571. [CrossRef]
65. Sláma, J.; Kosler, J.; Condon, D.J.; Crowley, J.L.; Gerdes, A.; Hanchar, J.M.; Horstwood, M.S.A.; Morris, G.A.; Nasdala, L.; Norberg, N.; et al. Plešovice zircon—A new natural reference material for U–Pb and Hf isotopic microanalysis. *Chem. Geol.* **2008**, *249*, 1–35. [CrossRef]
66. Hou, K.J.; Yanhe, L.I.; Xie, G.Q. Laser ablation-MC-ICP-MS technique for Hf isotope microanalysis of zircon and its geological applications. *Acta Petrol. Sin.* **2007**, *23*, 2595–2604, (In Chinese with English Abstract). [CrossRef]
67. Le Maitre, R.W.; Streckeisen, A.; Zanettin, B.; Bas, M.J.L.; Bonin, B.; Bateman, P. *Igneous Rocks: A Classification and Glossary of Terms*; Cambridge University Press: Cambridge, UK, 2005; p. 256.
68. Irvine, T.N.; Baragar, W.R.A. A Guide to the chemical classification of the common volcanic rocks. *Can. J. Earth Sci.* **1971**, *8*, 523–548. [CrossRef]
69. Winchester, J.A.; Floyd, P.A. Geochemical discrimination of different magma series and their differentiation products using immobile elements. *Chem. Geol.* **1977**, *20*, 325–343. [CrossRef]
70. Miyashiro, A. Volcanic rock series in island arcs and active continental margins. *Am. J. Sci.* **1974**, *274*, 321–355. [CrossRef]
71. Boynton, W.V. Cosmochemistry of the Rare Earth Elements: Meteorite Studies. *Dev. Geochem.* **1984**, *2*, 63–114. [CrossRef]
72. Sun, S.S.; McDonough, W.F. Chemical and isotopic systematics of oceanic basalts: Implications for mantle composition and processes. *Geol. Soc. Lond.* **1989**, *42*, 313–345. [CrossRef]
73. Rubatto, D.; Gebauer, D. Use of cathodoluminescence for U-Pb Zircon dating by Ion Microprobe: Some examples from the western Alps. In *Cathodoluminescence in Geoscience*; Springer: Berlin/Heidelberg, Germany, 2000; pp. 373–400. [CrossRef]
74. Wu, J.S.; Geng, Y.S.; Xu, H.F.; Jin, L.G.; He, S.Y.; Sun, S.W. Metamorphic geology of the Fuping Group. *Bull. Inst. Geol. Chin. Acad. Geol. Sci.* **1989**, *19*, 231. (In Chinese)
75. Simonen, A. Stratigraphy and sedimentation of the Svecofennidie, Early Archean supracrustal rocks in southwestern Finland. *Bull. Geol. Soc. Finl.* **1953**, *160*, 1–64.
76. Heilongjiang Bureau of Geology and Mineral Resources (HBGMR). *Regional Geology of Heilongjiang Province*; Geological Publishing House: Beijing, China, 1993; pp. 57–734, (In Chinese with English Abstract).
77. Wu, F.Y.; Sun, D.Y.; Ge, W.C.; Zhang, Y.B.; Grant, M.L.; Winde, S.A.; Jahn, B.M. Geochronology of the Phanerozoic granitoids in northeastern China. *J. Asian Earth Sci.* **2011**, *41*, 1–30. [CrossRef]
78. Zhang, Y.; Pei, F.P.; Wang, Z.W.; Xu, W.L.; Li, Y.; Wang, F.; Zhou, Z.B. Late Paleozoic tectonic evolution of the central Great Xing'an Range, northeast China: Geochronological and geochemical evidence from igneous rocks. *Geol. J.* **2018**, *53*, 282–303. [CrossRef]
79. Zhao, Z.; Chi, X.G.; Liu, J.F.; Wang, T.F.; Hu, Z.C. Late Paleozoic arc-related magmatism in Yakeshi region, Inner Mongolia: Chronological and geochemical evidence. *Acta Petrol. Sin.* **2010**, *26*, 3245–3258, (In Chinese with English Abstract).
80. She, H.Q.; Li, J.W.; Xiang, A.P.; Guan, J.D.; Yang, X.C.; Zhang, D.Q.; Tan, G.; Zhang, B. U-Pb ages of the zircons from primary rocks in middle-northern Daxinganling and its implications to geotectonic evolution. *Acta Petrol. Sin.* **2012**, *28*, 571–594, (In Chinese with English Abstract).
81. Ji, Z.; Ge, W.C.; Yang, H.; Zhang, Y.L.; Dong, Y.; Bi, J.H.; Liu, X.W. Geochronology and geochemistry of Late Devonian I- and A-Type granites from the Xing'an Block, NE China: Implications for slab break-off during subduction of the Hegenshan-Heihe Ocean. *J. Earth Sci.* **2022**, *33*, 150–160. [CrossRef]
82. Shi, L.; Zheng, Z.Q.; Yao, W.G.; Li, J.; Cui, F.H.; Gao, F.; Gao, Y.; Xu, J.L. Geochronological framework and tectonic setting of the granitic magmatism in the Chaihe–Moguqi region, central Great Xing'an Range, China. *J. Asian Earth Sci.* **2015**, *113*, 443–453. [CrossRef]
83. Qian, C.; Lu, L.; Qin, T.; Li, L.C.; Chen, H.J.; Cui, T.R.; Jiang, B.; Na, F.C.; Sun, W.; Wang, Y.; et al. The Early Late-Paleozoic granitic magmatism in the Zhalantun region, Northern Great Xing'an range, NE China: Constraints on the timing of amalgamation of Erguna-Xing'an and Songnen Blocks. *Acta Geol. Sin.* **2018**, *92*, 2706–2720, (In Chinese with English Abstract). [CrossRef]

84. Ma, Y.F.; Liu, Y.J.; Qin, T.; Sun, W.; Zang, Y.L.; Zhang, Y.J. Late Devonian to early Carboniferous magmatism in the western Songliao–Xinlinhot block, Northeast China: Implications for eastward subduction of the Nenjiang oceanic lithosphere. *Geol. J.* **2020**, *55*, 2208–2231. [CrossRef]
85. Li, D.X.; Zheng, C.Q.; Liang, C.Y.; Zhou, X.; Yang, Y.; Song, Z.W.; Chen, L.; Geng, Z.Z.; Zhao, Y.L. Genesis and Geological Significance of Granitic Mylonites in Southern Zhalantun, Central Xing'an Range. *Earth Sci.* **2022**, *47*, 3354–3370, (In Chinese with English Abstract). [CrossRef]
86. Zhang, Y.J.; Zhang, C.; Wu, X.W.; Cui, T.R.; Yang, Y.J.; Chen, H.J.; Jiang, B.; Guo, W.; Ma, Y.F. Geochronology and geochemistry of Late Paleozoic marine volcanic from the Zhalanun area in Northern DaHinggan Mountains and its geological significance. *Acta Geol. Sin.* **2016**, *90*, 2706–2720, (In Chinese with English Abstract) [CrossRef]
87. Yang, B.; Zhang, B.; Zhang, Q.K.; Ma, W.; Zhao, M.T.; Chen, S.L.; Yuan, X. Characteristics and geological significance of Early Carboniferous high-Mg andesites in Ma'anshan area, east Inner Mongolia. *Geol. Bull. China* **2018**, *37*, 1760–1771, (In Chinese with English Abstract)
88. Qian, C.; Wang, Y.; Lu, L.; Qin, T.; Li, L.C.; Cui, T.R.; Chen, H.J.; Yang, L. Geochronology, geochemistry and Hf isotopic composition of amphibolite from Zhalantun region in northern Great Xing'an Range and its tectonic significance. *Earth Sci.* **2019**, *44*, 3193–3208, (In Chinese with English Abstract) [CrossRef]
89. Pearce, J.; Norry, M. Petrogenetic implications of Ti, Zr, Y, and Nb variations in volcanic rocks. *Contrib. Miner. Pet.* **1979**, *69*, 33–47. [CrossRef]
90. DePaolo, D.C. Trace element and isotopic effects of combined wallrock assimilation and fractional crystallization. *Earth Planet. Sci. Lett.* **1981**, *53*, 189–202. [CrossRef]
91. Wang, X.C.; Li, Z.X.; Li, X.H.; Li, J.; Xu, Y.G.; Li, X.H. Identification of an ancient mantle reservoir and young recycled materials in the source region of a young mantle plume: Implications for potential linkages between plume and plate tectonics. *Earth Planet. Sci. Lett.* **2013**, *377–378*, 248–259. [CrossRef]
92. Zhang, H.-F.; Sun, M.; Zhou, X.-H.; Ying, J.-F. Geochemical constraints on the origin of Mesozoic alkaline intrusive complexes from the North China Craton and tectonic implications. *Lithos* **2005**, *81*, 297–317. [CrossRef]
93. Dilek, Y.; Furnes, H. Ophiolite genesis and global tectonics: Geochemical and tectonic fingerprinting of ancient oceanic lithosphere. *Geol. Soc. Am. Bull.* **2011**, *123*, 387–411. [CrossRef]
94. Ma, Y.F.; Liu, Y.J.; Wang, Y.; Tang, Z.; Qian, C.; Qin, T.; Feng, Z.Q.; Sun, W.; Zang, Y.Q. Geochronology and geochemistry of the Carboniferous felsic rocks in the central Great Xing'an Range, NE China: Implications for the amalgamation history of Xing'an and Songliao–Xilinhot blocks. *Geol. J.* **2019**, *54*, 482–513. [CrossRef]
95. Elliott, T. Tracers of the slab. In *Inside the Subduction Factory*; Eiler, J., Ed.; AGU: Washington, DC, USA, 2003; pp. 23–45. [CrossRef]
96. Kepezhinskas, P.; McDermott, F.; Defant, M.J.; Hochstaedter, A.; Drummond, M.S.; Hawkesworth, C.J.; Koloskov, A.; Maury, R.C.; Bellon, H. Trace element and Sr-Nd-Pb isotopic constraints on a three-component model of Kamchatka Arc petrogenesis. *Geochim. Cosmochim. Acta* **1997**, *61*, 577–600. [CrossRef]
97. Aldanmaz, E.; Pearce, J.A.; Thirlwall, M.F.; Mitchell, J.G. Petrogenetic evolution of late Cenozoic, post-collision volcanism in western Anatolia, Turkey. *J. Volcanol. Geotherm. Res.* **2000**, *102*, 67–95. [CrossRef]
98. Pearce, J.A.; Harris, N.B.W.; Tindle, A.G. Trace element discrimination diagrams for the tectonic interpretation of granitic rocks. *J. Pet.* **1984**, *25*, 956–983. [CrossRef]
99. Xia, L.-Q. The geochemical criteria to distinguish continental basalts from arc related ones. *Earth-Sci. Rev.* **2014**, *139*, 195–212. [CrossRef]
100. Wang, J.R.; Pan, Z.J.; Zhang, Q.; Chen, W.F.; Yang, J.; Jiao, S.T.; Wang, S.H. Intra-continental basalt data mining: The diversity of their constituents and the performance in basalt discrimination diagrams. *Acta Petrol. Sin.* **2016**, *32*, 1919–1933, (In Chinese with English Abstract)
101. Wang, X.C.; Wilde, S.A.; Xu, B.; Pang, C.J. Origin of arc-like continental basalts: Implications for deep-Earth fluid cycling and tectonic discrimination. *Lithos* **2016**, *261*, 5–45. [CrossRef]
102. Sobolev, A.A.; Hofmann, A.W.; Kuzmin, D.V.; Yaxley, G.M.; Arndt, N.; Ai, E. The amount of recycled crust in sources of mantle derived melts. *Science* **2007**, *316*, 412–417. [CrossRef] [PubMed]
103. Ivanov, A.V.; Litasov, K.D. The deep water cycle and flood basalt volcanism. *Int. Geol. Rev.* **2014**, *56*, 1–14. [CrossRef]
104. Jian, P.; Kröner, A.; Windley, B.F.; Shi, Y.R.; Zhang, W.; Zhang, L.Q.; Yang, W.R. Carboniferous and Cretaceous mafic–ultramafic massifs in Inner Mongolia (China): A SHRIMP zircon and geochemical study of the previously presumed integral "Hegenshan ophiolite". *Lithos* **2012**, *142–143*, 48–66. [CrossRef]
105. Song, S.G.; Wang, M.M.; Xu, X.; Wang, C.; Niu, W.L.; Allen, M.B.; Su, L. Ophiolites in the Xing'an-Inner Mongolia accretionary belt of the CAOB: Implications for two cycles of seafloor spreading and accretionary orogenic events. *Tectonics* **2015**, *34*, 2221–2248. [CrossRef]
106. Zhang, Z.C.; Li, K.; Li, J.F.; Tang, W.H.; Chen, Y.; Luo, Z.W. Geochronology and geochemistry of the Eastern Erenhot ophiolitic complex: Implications for the tectonic evolution of the Inner Mongolia–Daxinganling Orogenic Belt. *J. Asian Earth Sci.* **2015**, *97*, 279–293. [CrossRef]
107. Huang, B.; Fu, D.; Li, S.C.; Ge, M.C.; Zhou, W.X. The age and tectonic implications of the Hegenshan ophiolite in Inner Mongolia. *Acta Petrol. Sin.* **2016**, *32*, 158–176, (In Chinese with English Abstract)

108. Yang, J.; Zhang, Z.; Chen, Y.; Yu, H.; Qian, X. Ages and origin of felsic rocks from the Eastern Erenhot ophiolitic complex, southeastern Central Asian Orogenic Belt, Inner Mongolia China. *J. Asian Earth Sci.* **2017**, *144*, 126–140. [CrossRef]
109. Li, Y.J.; Wang, G.H.; Santosh, M.; Wang, J.F.; Dong, P.P.; Li, H.Y. Supra-subduction zone ophiolites from Inner Mongolia, North China: Implications for the tectonic history of the southeastern Central Asian Orogenic Belt. *Gondwana Res.* **2018**, *59*, 126–143. [CrossRef]
110. Liu, J.F.; Li, J.Y.; Zhang, W.; Zhang, J.; Zhao, S.; Yin, D. Newly discovered late Devonian and early Carboniferous ophiolite fragments in the Diyanmiao mélange in southeastern Inner Mongolia: Implications for the late Paleozoic tectonic evolution of the southeastern Central Asian Orogenic Belt. *Lithos* **2022**, *408–409*, 106566. [CrossRef]
111. Jian, P.; Liu, D.Y.; Kröner, A.; Windley, B.F.; Shi, Y.R.; Zhang, W.; Zhang, F.Q.; Miao, L.C.; Zhang, L.Q.; Tomurhuu, D. Evolution of a Permian intraoceanic arc–trench system in the Solonker suture zone, Central Asian Orogenic Belt, China and Mongolia. *Lithos* **2010**, *118*, 169–190. [CrossRef]
112. Shao, J.A.; Zhang, L.L.; Zhou, X.H.; Zhang, L.Q.; Tang, K.D. A further study on the ophiolite in Hegenshan, Inner Mongolia. *Acta Pet. Sin.* **2019**, *35*, 2864–2872, (In Chinese with English Abstract) [CrossRef]
113. Zhang, Y.J.; Xu, B.; Tian, Y.J.; Wang, Z.W. The Late Paleozoic extending processes of Xing'an-Mongolia Orogenic Belt (XMOB): Evidence from Carboniferous-Permian sedimentary basin in northeast of Erenhot, Inner Mongolia. *Acta Petrol. Sin.* **2018**, *34*, 3083–3100, (In Chinese with English Abstract)
114. Xu, B.; Wang, Z.W.; Zhang, L.Y.; Wang, Z.H.; Yang, Z.Y.; He, Y. The Xing-Meng Intracontinent Orogenic Belt. *Acta Petrol. Sin.* **2018**, *34*, 2819–2844, (In Chinese with English Abstract).
115. Shervais, J.W. Ti-V plots and the petrogenesis of modern and ophiolitic lavas. *Earth Planet. Sci. Lett.* **1982**, *59*, 101–118. [CrossRef]
116. Pearce, J.A. A user's guide to basalt discrimination diagrams. In *Trace Element Geochemistry of Volcanic Rocks: Applications for Massive Sulphide Exploration*; Wyman, D.A., Ed.; Short Course Notes; Geological Association of Canada: St. John's, NL, Canada, 1996; pp. 79–113.

minerals

MDPI

Article

Ore Genesis of the Lower Urgen Porphyry Molybdenum Deposit in the Northern Great Xing'an Range, Northeast China: Constraints from Molybdenite Re-Os Dating, Fluid Inclusions, and H-O-S-Pb Isotopes

Guangliang Zhang [1], Wei Xie [2,3,*], Shouqin Wen [1], Qingdong Zeng [3], Lingli Zhou [4], Hui Wang [1], Kailun Zhang [1], Tieqiao Tang [5] and Pengcheng Ma [6]

1 Key Laboratory of Ministry of Education on Safe Mining of Deep Metal Mines, Department of Geology, Northeastern University, Shenyang 110819, China; zhangguangliang@duchengmining.com (G.Z.); wenshouqin@mail.neu.edu.cn (S.W.); 2000993@stu.neu.edu.cn (H.W.); 2000998@stu.neu.edu.cn (K.Z.)
2 College of Earth Science and Engineering, Shandong University of Science and Technology, Qingdao 266590, China
3 Key Laboratory of Mineral Resources, Institute of Geology and Geophysics, Chinese Academy of Sciences, Beijing 100029, China; zengqingdong@mail.iggcas.ac.cn
4 Department of Earth Sciences, VU Amsterdam, De Boelelaan 1105, 1081 HV Amsterdam, The Netherlands; l.zhou@vu.nl
5 China Water Resources Bei Fang Investigation, Design & Research Co., Ltd., Tianjin 300222, China; tang_tq@bidr.com.cn
6 China Railway Design Corporation, Tianjin 300308, China; mapengcheng01@crdc.com
* Correspondence: xiewei20@mails.ucas.ac.cn

Citation: Zhang, G.; Xie, W.; Wen, S.; Zeng, Q.; Zhou, L.; Wang, H.; Zhang, K.; Tang, T.; Ma, P. Ore Genesis of the Lower Urgen Porphyry Molybdenum Deposit in the Northern Great Xing'an Range, Northeast China: Constraints from Molybdenite Re-Os Dating, Fluid Inclusions, and H-O-S-Pb Isotopes. *Minerals* **2023**, *13*, 1189. https://doi.org/10.3390/min13091189

Academic Editors: Jiafu Chen, Nan Ju, Zhonghai Zhao and Alfons M. Van den Kerkhof

Received: 9 July 2023
Revised: 17 August 2023
Accepted: 8 September 2023
Published: 10 September 2023

Abstract: The Lower Urgen molybdenum deposit (44,856 t Mo @ 0.141%), situated in the northern Great Xing'an Range, is a newly discovered porphyry molybdenum deposit. Mineralization is characterized by veinlet-disseminated- and vein-type quartz–sulfide orebodies primarily occurring in the cupola of the Early Cretaceous granite porphyry stock. In this study, we present a detailed description of the ore geology, molybdenite Re-Os dating, H-O-S-Pb isotopic compositions, and fluid inclusion (FI) analyses including petrography, laser Raman, and microthermometry to precisely constrain the timing of ore formation, the origin of ore-forming fluids and materials, as well as the metal precipitation mechanism. Molybdenite Re-Os dating yielded two model ages of 141.2 ± 1.5 and 147.7 ± 1.7 Ma, coeval with the regional Late Jurassic–Early Cretaceous molybdenum metallogenesis. The hydrothermal process can be divided into three stages: the quartz–molybdenite(–pyrite) stage, quartz–polymetallic sulfide stage, and quartz–carbonate stage. Four types of FIs were distinguished for quartz, including two-phase liquid-rich (L-type), saline (S-type), CO_2-rich (C1-type), and CO_2-bearing (C2-type) FIs. Microthermometric data showed that the homogenization temperatures and salinities from the early to late stages were 240–430 °C, 5.0–11.9, and 30.1–50.8 wt% NaCl equiv.; 180–280 °C and 3.0–9.1 wt% NaCl equiv.; and 120–220 °C and 0.2–7.9 wt% NaCl equiv., respectively, suggesting a decreasing trend. H-O isotopic compositions indicate that the ore-forming fluids were initially of magmatic origin with the increasing incorporation of meteoric water. S-Pb isotopic compositions indicate that the ore-forming materials originated from granitic magmas, and the mineralization is genetically related to the ore-bearing granite porphyry stock in the deposit. Fluid immiscibility and fluid–rock interaction are collectively responsible for the massive deposition of molybdenite in stage 1, whereas fluid mixing and immiscibility played a critical role in the deposition of polymetallic sulfide in stage 2.

Keywords: molybdenite Re-Os dating; fluid inclusions; H-O-S-Pb isotopes; lower urgen molybdenum deposit; NE China

1. Introduction

Porphyry deposits are commonly spatially, temporally, and genetically related to intermediate–acid hypabyssal porphyry intrusions [1], and provide more than 75%, 50%, and 20% of the world's molybdenum, copper, and gold resources, respectively [2–5]. Mineralization mainly occurs as veinlets and disseminations concentrated in or near the cupolas of porphyry intrusions, and is related to the hydraulic fracturing resulting from fluid exsolution from magmas [3,6,7]. Taking the economic importance of porphyry molybdenum deposits into account, the study of the genesis and metallogenic models of these deposits has always been an essential subject in ore deposit research [8–16].

China has the largest molybdenum resources worldwide, accounting for about 54% of the world's known economic molybdenum resources [17], which appear in six predominant molybdenum metallogenic provinces, i.e., NE China, Qinling-Dabie, Yanliao, Sanjiang, the middle-lower Yangtze River Valley, and South China [18]. NE China is situated in the eastern Central Asian Orogenic Belt (CAOB) tectonically. A long-term tectonomagmatic evolutionary history has made it a significant polymetallic metallogenic province around the world [19–25] (Figure 1). More than 11.4 Mt of molybdenum has now been discovered in NE China, including 4 super-large (>0.5 Mt Mo), 11 large (0.1–0.5 Mt Mo), 22 medium (0.01–0.1 Mt Mo), and 18 small (<0.01 Mt Mo) molybdenum deposits, marking NE China as the largest molybdenum metallogenic province in China [26] (Figure 1c).

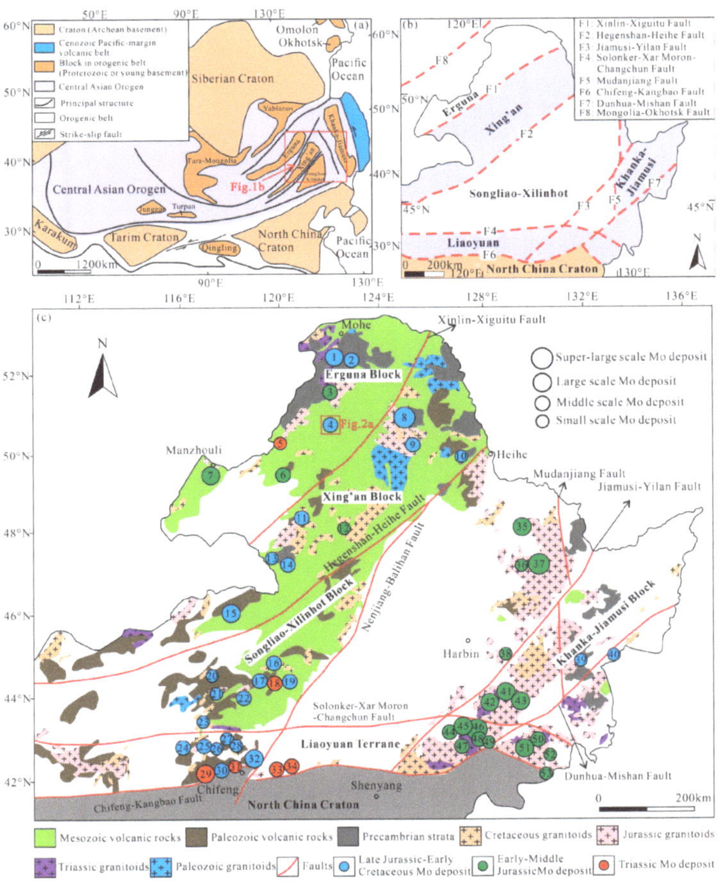

Figure 1. (**a**) Simplified tectonic map of the Central Asian Orogenic Belt (modified from [27,28]). (**b**) Schematic tectonic map of NE China (modified from [29]). (**c**) Distribution of major molybdenum

deposits in NE China (modified from [20]). Deposit locations: 1—Jiliuhe; 2—Fukeshan; 3—Taipingchuan; 4—Lower Urgen; 5—Badaguan; 6—Zishi; 7—Wunugetushan; 8—Chalukou; 9—Daheishan; 10—Yezhugou; 11—Taipinggou; 12—Baojiagou; 13—Nanxingan; 14—Jiazishan; 15–Diyanqinamu; 16—Aolunhua; 17—Banlashan; 18—Laojiagou; 19—Haolibao; 20—Hashitu; 21—Erbadi; 22—Yangchang; 23—Longtoushan; 24—Hongshanzi; 25—Xiaodonggou; 26—Liutiaogou; 27—Gangzi; 28—Tuohe; 29—Chehugou; 30—Nianzigou; 31—Yuanbaoshan; 32—Jiguanshan; 33—Baimashi; 34—Kulitu; 35—Huojihe; 36—Cuiling; 37—Luming; 38—Wudaoling; 39—Kanchuangou; 40—Jinchanggou; 41—Fuanpu; 42—Jidetun; 43—Dashihe; 44—Houdaomu; 45—Xingshan; 46—Daheishan; 47—Sangedingzi; 48—Sifangdianzi; 49—Xinhualong; 50—Liushengdian; 51—Sanchazi; 52—Dongfeng; 53—Shirengou.

The Lower Urgen deposit in Inner Mongolia, NE China, is a newly discovered porphyry molybdenum deposit, with estimated reserves of 44,856 t of molybdenum at an average grade of 0.141% [30]. The geochronology, petrogenesis, and tectonic context of the barren granites in the mining area have been the primary subjects of earlier investigations of the deposit [31]. Nonetheless, the direct timing of molybdenum mineralization, ore-forming fluids and the metal source, and metallogenic processes remains enigmatic. In accordance with a thorough field investigation and petrographic observations of thin sections, we present comprehensive descriptions of the geology, mineralization stages, and hydrothermal alteration in the Lower Urgen molybdenum deposit. The new molybdenite Re-Os geochronology, fluid inclusion (FI) study (petrography, laser Raman, and microthermometry analyses), and isotope geochemistry (H-O-S-Pb) for the deposit are investigated. By combining our results with available geochronology and geodynamic research in the region, we aim to (1) restrict the timing of molybdenum mineralization in the Lower Urgen molybdenum deposit; (2) determine ore-forming fluids and the metal source; and (3) decipher fluid evolution processes and ore precipitation mechanisms. This study not only contributes to the establishment of a plausible metallogenetic model for the Lower Urgen molybdenum deposit, but also has important implications for the comprehension of regional molybdenum metallogenesis and the future molybdenum prospecting and exploration.

2. Regional Geology

The tectonic components of NE China include the Erguna Block in the northwest, the Xing'an and Songliao-Xilinhot blocks in the central part, the Khanka-Jiamusi Block in the southeast, and the Liaoyuan Terrane in the south [28] (Figure 1). Major faults separating these tectonic units include the Hegenshan-Heihe, Solonker-Xar Moron-Changchun, Xinlin-Xiguitu, and Mudanjiang faults [28,29] (Figure 1b). NE China has experienced a multi-stage Phanerozoic tectonic evolutionary history, including the subduction and final closure of the Palaeo-Asian Ocean (PAO) during the Paleozoic to Early Mesozoic [28,32–35], the south-directed subduction and closure of the Mongol-Okhotsk Ocean (MOO), and the west-directed subduction and subsequent rollback of the Palaeo-Pacific Ocean (PPO) plate during the Mesozoic [29,36,37]. Due to the overprinting of multiple tectonic domains, NE China experienced extensive tectonomagmatic events, accompanied by intense metallogenesis [19–25]. Three episodes of Mesozoic molybdenum mineralization are recognized in the region, including the Triassic, Early–Middle Jurassic, and Late Jurassic–Early Cretaceous [18,24]. The major location of the Triassic molybdenum deposits is along the Solonker-Xar Moron-Changchun Fault (SXCF), with minor occurrences in the northern and central Great Xing'an Range (NCGXR). The Triassic molybdenum deposits near the SXCF were formed in a syn-collision compressional to post-collision extensional environment induced through the tectonic evolution of the PAO [18,38–43], whereas the Triassic molybdenum deposits in the NCGXR were formed in a subduction-dominated compressional environment induced through the SE-directed subduction of the MOO plate [44–48]. The Early–Middle Jurassic molybdenum deposits are concentrated in the Lesser Xing'an-

Zhangguangcai Range (LXZR) and sparsely distributed in the NCGXR, which were formed in a compressional environment controlled by the overprinting influence of the SE-directed subduction of the MOO plate in the west and the NW-directed subduction of the PPO plate in the east [49–64]. Molybdenum deposits from the Late Jurassic to Early Cretaceous are mostly concentrated in the GXR, with sporadic occurrences in the LXZR, which were formed in a post-orogenic extensional environment triggered by the PPO and MOO tectonic domains [56,63,65–88].

The NCGXR comprises two main units, i.e., the Erguna and Xing'an blocks (Figure 1c). The amalgamation of the Erguna and Xing'an blocks occurred along the Xinlin-Xiguitu Fault in the Early Palaeozoic induced by the closure of the PAO [28,89–92]. Volcanic-sedimentary rocks of the Xinghuadukou, Ergunahe, Jiageda, and Luomahu groups; Neoproterozoic magmatic rocks; and minor Palaeoproterozoic gneisses compose the basement of the Erguna Block [29,93,94]. Multiple Late Palaeozoic–Late Mesozoic episodes of accretion and collision occurred on the Erguna Block, resulting in the formation of tremendous intrusions, volcanics, and volcaniclastic rocks [95,96]. Widespread Mesozoic igneous rocks; Early Palaeozoic basement rocks composed of the Wolegen, Xinkailing, Zalantun, and Fengshuigouhe groups; together with the Xinhuadukou Complex and dispersed Palaeozoic plutons and sediments constitute the Xing'an Block, the primary component of the GXR [29]. Neoproterozoic granite is locally distributed in the western region [97]. Mesozoic volcanics are categorized by the Tamulangou, Manketouebo, Manitu, Baiyingaolao, and Meiletu Formations (Fms), from oldest to youngest [93]. The Palaeozoic strata were extensively emplaced through substantial Palaeozoic to Mesozoic granitic intrusions. The latest geochronological evidence indicates that the magmatism within the NCGXR took place in two periods: the Palaeozoic granites consisting of granite, monzogranite, granodiorite, quartz diorite, and syenogranite mainly distributed in the northeastern part with zircon U-Pb ages of 517–420 and 359–250 Ma; and the Mesozoic granites, comprising monzogranite, syenogranite, granodiorite, granite porphyry, quartz porphyry, and quartz monzogranite, which are more widespread than the Palaeozoic intrusions and yielded a zircon U-Pb age of 250–118 Ma [29,98–104] (Figure 1c).

3. Ore Deposit Geology

The Lower Urgen molybdenum deposit (120°27′30″ E, 50°22′42″ N), situated in the central NCGXR, discovered in 2017, is still in exploration and not in production (Figure 1c). The strata exposed in the mining area comprise the Tamulangou Fm of the Middle Jurassic, Manketouebo Fm of the Upper Jurassic, Baiyingaolao Fm of the Lower Cretaceous, and Quaternary [30] (Figure 2). The Tamulangou Fm comprises intermediate–basic volcanics including trachyandesite, andesite, and basaltic-trachyandesite. The Manketouebo Fm comprises felsic volcanics including volcanic breccia, rhyolite, and volcanic agglomerate. The Baiyingaolao Fm is locally dispersed in the northwestern part, which is composed of rhyolite, rhyolitic breccia lava, and tuff [30]. Mesozoic granitoids include quartz monzonite porphyry, syenite porphyry, and granite porphyry. The Middle Jurassic quartz monzonite porphyry outcrops are scattered in the western and central regions. The Early Cretaceous syenite porphyry locally occurs in the western region, while the Early Cretaceous granite porphyry outcrops are mostly distributed in the eastern region [30] (Figure 2). Zhang et al. [31] obtained precise zircon U-Pb ages of the barren quartz monzonite porphyry and syenite porphyry of 165.0 ± 2.5 Ma and 125.5 ± 1.9 Ma, respectively. Mineralization is characterized by concealed veinlet-disseminated- and vein-type quartz–sulfide orebodies mostly hosted in the cupola of the Early Cretaceous granite porphyry (Figures 2c and 3). Ore minerals are mainly molybdenite and pyrite, as well as subordinate chalcopyrite, sphalerite, arsenopyrite, and galena (Figure 3). K-feldspar, quartz, and sericite are the predominant gangue minerals, accompanied by epidote, biotite, and calcite (Figures 3g and 4). Molybdenite occurs as medium-coarse grained (1.2–7 mm) crystals with platy and leaf-like forms in ore-bearing quartz veins and granite porphyry (Figure 3a–c). Pyrite is present as euhedral to subhedral octahedral–dodecahedral crystals, and usually coexists with molybdenite, chalcopyrite, sphalerite, and galena in granular aggregate forms (Figure 3). Sphalerite occurs as medium-coarse grained (1.5–7 mm) anhedral crystals, and is often intergrown with galena, chalcopyrite, and pyrite (Figure 3e,f,i). Chalcopyrite appears as fine-medium grained (0.4–1.3 mm) irregular crystals. It often coexists with pyrite and occurs in small amounts with galena and sphalerite (Figure 3d,e,i). Arsenopyrite is observed as rhomboid euhedral crystals (Figure 3d). Galena is observed as fine-medium-grained (0.6–4.5 mm) irregular crystals, and is often coeval with sphalerite, pyrite, and chalcopyrite in quartz veins (Figure 3e,f,i). Superposition mineralization in the cupola of the granite porphyry is ubiquitous, and no obvious metallogenic zonation is observed.

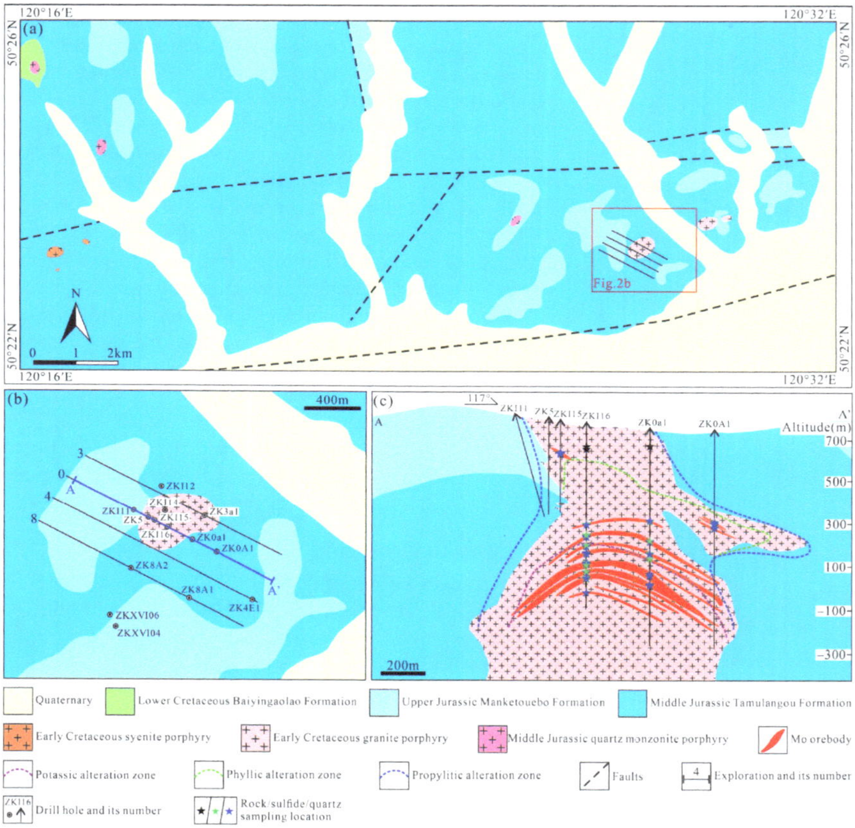

Figure 2. (**a**) Schematic regional geological map of the Lower Urgen molybdenum deposit. (**b**) Geological sketch map of the Lower Urgen molybdenum deposit. (**c**) Schematic geological cross-section of no. 0 exploration line (modified after [30]).

Intensive hydrothermal alteration is a defining characteristic of the Lower Urgen molybdenum deposit. There are three alteration zones recognized according to field observations and thin-section petrography: an intense potassic alteration zone at depth, a phyllic alteration zone at mid-shallow depths, and a propylitic alteration zone near the surface and periphery of the deposit. The potassic and phyllic alteration zones are all distributed in the granite porphyry, while the propylitic alteration zone mainly occurs in the intermediate–mafic volcanics around the granite porphyry. The spatial relationship of the three alteration zones is delineated in Figure 2c. The potassic alteration has a high temperature and is featured by a mineral assemblage of K-feldspar, biotite, and quartz [1,3] (Figure 4a). Quartz–molybdenite veinlets are distributed in the granite porphyry with intense potassic alteration (Figure 4a). The phyllic alteration is distinguished by the replacement of primary K-feldspar and plagioclase in the granite porphyry with quartz and sericite (Figure 4e). Quartz–pyrite–chalcopyrite, quartz–galena–sphalerite, and quartz–molybdenite(–pyrite) veins are usually developed in the phyllic alteration zone (Figure 4b,c). The propylitic alteration generally represents low-temperature hydrothermal alteration, which is featured by plagioclase replaced by epidote [1,3] (Figure 4d,f).

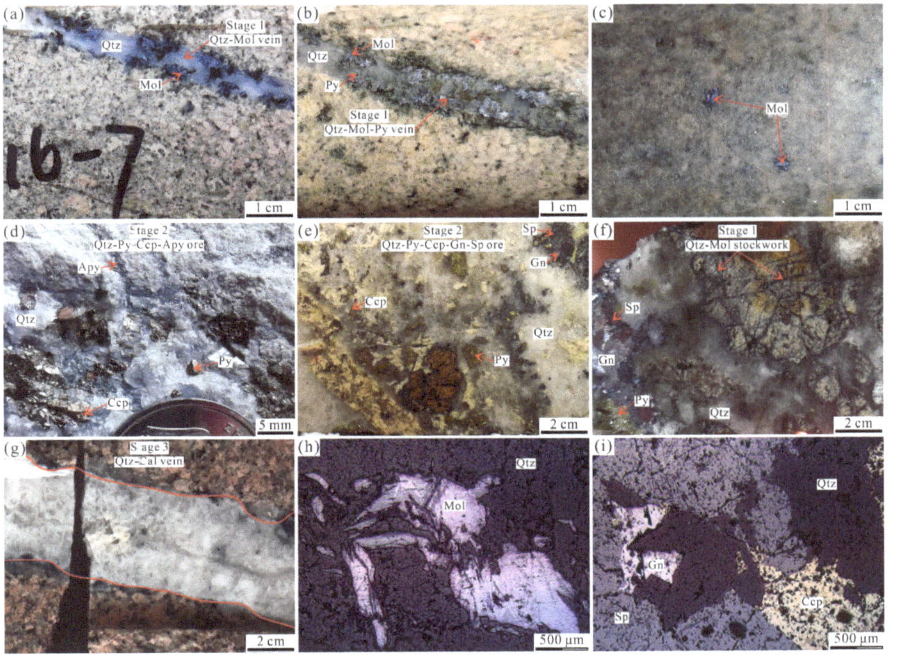

Figure 3. Representative photographs (**a**–**g**) and photomicrographs (**h**,**i**) of the mineralization in the Lower Urgen molybdenum deposit. (**a**) Quartz–molybdenite vein of stage 1 in the granite porphyry. (**b**) Quartz–molybdenite–pyrite vein of stage 1 in the granite porphyry. (**c**) Molybdenite occurs as disseminations. (**d**) Quartz–pyrite–chalcopyrite–arsenopyrite ore of stage 2. (**e**) Quartz–pyrite–chalcopyrite–galena–sphalerite ore of stage 2. (**f**) The brecciated fragments of granite porphyry containing quartz–molybdenite stockworks of stage 1 hosted by a quartz–pyrite–galena–sphalerite vein of stage 2. (**g**) Quartz–calcite vein of stage 3. (**h**) Molybdenite. (**i**) Galena intergrown with sphalerite. Sphalerite replacing chalcopyrite along crystal margins. Abbreviations: Mol, molybdenite; Py, pyrite; Ccp, chalcopyrite; Gn, galena; Sp, sphalerite; Apy, arsenopyrite; Qtz, quartz; Cal, calcite.

There are three main stages of hydrothermal processes in the Lower Urgen molybdenum deposit according to crosscutting relationships and mineral paragenetic sequences, including the quartz–molybdenite(–pyrite) stage (stage 1), quartz–polymetallic sulfide stage (stage 2), and quartz–carbonate stage (stage 3).

Stage 1 is the main molybdenum mineralization stage. Molybdenite, pyrite, quartz, and sericite are developed in this stage. Various types of veins were formed, including quartz–molybdenite veins and veinlets (Figures 3a and 4a), quartz–molybdenite stockworks (Figure 3f), and quartz–molybdenite–pyrite veins (Figures 3b and 4b). Disseminated molybdenite ores can also be found (Figure 3c).

Stage 2 is featured by an assemblage of chalcopyrite, sphalerite, arsenopyrite, pyrite, galena, quartz, sericite, and epidote. The most common sulfide is pyrite, which appears as fine-medium-grained euhedral to subhedral crystals and aggregates, which often form quartz–sulfide veins with chalcopyrite, galena, sphalerite, and quartz, i.e., quartz–pyrite–chalcopyrite–arsenopyrite veins (Figure 3d), quartz–chalcopyrite–pyrite veins (Figure 4c), quartz–pyrite–chalcopyrite–sphalerite–galena veins (Figure 3e), and quartz–pyrite–sphalerite–galena veins (Figure 3f). Galena and sphalerite are often intergrown, and they form quartz–galena–sphalerite veins and veinlets (Figure 4b,d). The brecciated fragments of granite porphyry containing stage 1 quartz–molybdenite stockworks can be found in stage 2 quartz–pyrite–sphalerite–galena veins (Figure 3f). Additionally, stage 2 quartz–galena–sphalerite veins cut stage 1 quartz–molybdenite–pyrite veins (Figure 4b).

Stage 3 represents the termination of the hydrothermal activity in the Lower Urgen molybdenum deposit. During this stage, quartz and calcite make up the majority of the veins, and sulfides are no longer found in this veins' stage (Figure 3g).

Figure 4. Representative hand specimen (**a**–**d**) and photomicrographs (crossed-polarized light) (**e**,**f**) of gangue minerals from the Lower Urgen molybdenum deposit. (**a**) Quartz–molybdenite veinlets of stage 1 occur in the granite porphyry with potassic alteration. (**b**) Quartz–molybdenite–pyrite veins of stage 1 cut by quartz–galena–sphalerite of stage 2. These ore veins occur in the granite porphyry with phyllic alteration. (**c**) Quartz–pyrite–chalcopyrite vein of stage 2 occurs in the granite porphyry with phyllic alteration. (**d**) Quartz–galena–sphalerite veinlets of stage 2 are spatially related to silicification and propylitic alteration. (**e**) Phyllic-altered granite porphyry is characterized by quartz and sericite replacing plagioclase along fractures and crystal margins. (**f**) Propylitic-altered andesite is featured with epidote replacing plagioclase. Abbreviations: Qtz, quartz; Kfs, K-feldspar; Pl, plagioclase; Ser, sericite; Ep, epidote; Mol, molybdenite; Py, pyrite; Gn, galena; Sp, sphalerite; Ccp, chalcopyrite.

4. Sampling and Analytical Methods

4.1. Samples

Samples for this study were all collected from drillholes, containing three granite porphyry samples, twenty-nine sulfide samples, and twenty-eight quartz samples. The sample locations are indicated in Figure 2c. All granite porphyry samples (T0a1-1, T0a1-2, and T16-1) were fresh and used for whole-rock Pb isotope analyses. Twenty-nine sulfide samples comprised sixteen molybdenite samples, seven pyrite samples, four chalcopyrite samples, and two galena samples. Sixteen molybdenite samples were obtained from stage 1 molybdenite(–pyrite)-bearing quartz veins, while the other sulfide samples were all collected from stage 2 sulfide-bearing quartz veins. Two molybdenite samples (G5 and LE-1) were used for Re-Os isotope analyses. Six molybdenite samples (LU16-1, LU16-2, LU16-3, LU16-4, LU16-5, and LU16-6), two pyrite samples (LU16-7 and LU16-8), two chalcopyrite samples (LU16-9 and LU16-10), and one galena sample (LU16-11) were used for S isotope analyses. Eight molybdenite samples (D17-2, D17-3, D17-4, D17-5, D17-7, D0A1-1, D0A1-2, and D0a1-1), five pyrite samples (G10-2, G11-2, G13, G14, and D17-1), two chalcopyrite samples (D17-2 and G8-2), and one galena sample (G8-3) were used for Pb isotope analyses. The quartz samples were obtained from veins of all stages, including seven samples from stage 1 molybdenite(–pyrite)-bearing quartz veins, nine samples from stage 2 sulfide-bearing quartz veins, and five samples from stage 3 quartz–carbonate veins. Seven quartz samples of stage 1 veins (B16-1, B16-2, B16-3, B16-4, B16-5, B16-6, B16-7), seven quartz samples of stage 2 veins (D14-1, D16-1, D16-4, D16-6, D16-7, D16-8, and D16-9), and three quartz samples of stage 3 veins (G8-1, G10-1, and G11-1) were used for H-O isotope analyses. Seven quartz samples of stage 1 veins (BG16-1, BG16-2, BG16-3, BG16-4, BG16-5, BG16-6, and BG16-7), two quartz samples of stage 2 veins (BG8-1 and BG14-1), and two quartz samples of stage 3 veins (BG8-2 and BG8-3) were used for the FI study. For investigations of FIs in quartz cores, thirty-six double-polished thin sections (200 μm thick) were prepared.

4.2. Molybdenite Re-Os Isotope Dating

Re-Os isotope analyses were carried out at the State Key Laboratory of Ore Deposit Geochemistry, Institute of Geochemistry, Chinese Academy of Sciences in Guiyang, China. A TJA PQ ExCell ICP-MS was applied to measure the contents of Re, ^{187}Re, and ^{187}Os. We followed the analytical procedures reported in [105,106]. The Re-Os model ages were calculated using the formula $t = 1/\lambda\ [\ln (1 + {}^{187}Os/{}^{187}Re)]$, where λ is the decay constant for ^{187}Re (1.666×10^{-11} year^{-1}) [107].

4.3. FIs Laser Raman Spectroscopy

A Renishaw inVia laser Raman probe at the Beijing Createch Testing Technology Co., Ltd. in Beijing, China, was applied to examine the vapor and liquid components of quartz-hosted FIs at room temperature. For detection, a laser power of 20 mW and an argon ion laser with a wavelength of 514.5 nm were used. The spectrum was measured from 1000 to 4000 cm^{-1} with a counting time of 20 s for each scan. Detailed analytical procedures are consistent with those in [108].

4.4. FIs Microthermometry

FI microthermometry was conducted using a Linkam THMS 600 heating-freezing stage with a stage of –195 to 600 °C mounted on a Zeiss microscope at the Beijing Createch Testing Technology Co., Ltd., Beijing, China. The accuracy of temperature observations is 0.2 °C through cooling, ±0.2 °C for 0–100 °C, and ±2 °C for 100–450 °C. To properly record the phase transformation process, the heating rate was decreased to 0.1 °C/min close to the freezing point and 0.2 °C/min close to the homogenization temperature. To prevent unintentional deviations, FI results were only collected on FI assemblages (FIAs) [109]. During heating, seven different temperature types were recorded, including CO_2-ice melting temperatures (Tm_{CO_2}), first ice melting temperatures (Tfm_{ice}), final ice melting temperatures

(Tm_{ice}), final clathrate melting temperatures (Tm_{clath}), CO_2 homogenization temperatures (Th_{CO_2}), halite dissolution temperatures (Tm_H), and final homogenization temperatures (Th). Tm_{clath} was used to determine the salinities of the fluid system composed of NaCl, CO_2, and H_2O [110]. Tm_{ice} and Tm_H were used to obtain the salinities of the fluid system composed of NaCl and H_2O [111,112]. Using the Flincor program, the bulk molar volume of FIs, the density of the bulk fluids and carbonic liquid, and mole fractions of compositions were acquired [113–115].

4.5. Quartz H-O Isotopes

H-O isotope analyses were carried out using a MAT253 mass spectrometer at the Beijing Research Institute of Uranium Geology in Beijing, China. The accuracies of the H and O isotopic measurements are better than ±2‰ and ±0.2‰, respectively. The fractionation formula $1000\ \ln\alpha_{quartz\ water} = 3.38 \times 10^6\ T^{-1}–3.40$ was applied to calculate the $\delta^{18}O_{H_2O}$ of fluids, where T is the Kelvin temperature [116]. The average FI homogenization temperature for each stage was used to determine the $\delta^{18}O_{H_2O}$ values.

4.6. S-Pb Isotopes

The sulfide S-Pb isotope measurements were performed at the State Key Laboratory of Isotope Geochemistry, Guangzhou Institute of Geochemistry, Chinese Academy of Sciences. A Delta V Plus mass spectrometer was used to analyze the S isotopic compositions of sulfides with an analytical accuracy of ±2‰. Sulfide $\delta^{34}S$ values were calculated on sulfur dioxide produced via the traditional extraction method described in [117]. The $\delta^{34}S$ results were recorded using the Canyon Diablo Troilite (CDT).

Thermal ionization mass spectrometry using an ISOPROBE–T mass spectrometer was used to analyze the Pb isotopic compositions of sulfides and granite porphyry. The accuracy for 1 μg of $^{208}Pb/^{206}Pb$ and $^{207}Pb/^{206}Pb$ was greater than 0.005% (2σ). The Pb isotopic analyses are reported with respect to Pb standard reference NBS-981 values, as follows: $^{206}Pb/^{204}Pb = 16.934 \pm 0.007$, $^{207}Pb/^{204}Pb = 15.486 \pm 0.012$, and $^{208}Pb/^{204}Pb = 36.673 \pm 0.033$ [118].

5. Results

5.1. Molybdenite Re-Os Ages

Table 1 lists the results of Re and Os isotope compositions of two molybdenite samples from the Lower Urgen deposit. The ^{187}Re and ^{187}Os contents of the samples vary from 19,468 to 27,503 ppb and 47.958 to 64.540 ppb, respectively. The Re contents of the samples are 31,202–43,934 ppb. The two Re-Os model ages are 141.2 ± 1.5 and 147.7 ± 1.7 Ma.

Table 1. Re–Os isotopic data for molybdenite from the Lower Urgen molybdenum deposit.

Sample No.	Re (ppb)		Os (ppb)		^{187}Re (ppb)		^{187}Os (ppb)		Model Age (Ma)	
	Re	1σ	Os	1σ	^{187}Re	1σ	^{187}Os	1σ	Age	1σ
G5	43,934	1558	0.083	0.008	27,503	975	64.540	0.698	141.2	1.5
LE-1	31,202	528	0.128	0.015	19,468	231	47.958	0.523	147.7	1.7

5.2. FI Petrography

Primary and secondary FIs were identified based on the criteria suggested by [109,119]. All descriptions strictly follow the concept of FIAs [109]. An FIA only applies to FI vacuoles that developed along the identical growth bands in an individual crystal, as well as within the same T-P circumstances. Nonetheless, extremely few FIs are dispersed within or along the growth bands, and the bulk is too small to be quantified (<5 μm). Most FIs that are suitable for measurement appear as isolated, clustered, or randomly distributed (Figure 5). As suggested by [109,119–121], FIs that are away from apparent cracks and appear as isolated, clustered, or randomly distributed forms are considered as primary FIs, and these FIs are selected for further investigation. In contrast, FIs along healed fractures that have apparent secondary causes are not considered. Four different types of FIs are recognized based on phase components at room temperature and phase changes due to variations in temperature, including the two-phase (L + V) liquid-rich (L-type), saline (S-type), CO_2-rich (C1-type), and CO_2-bearing (C2-type) FIs. The descriptions of the petrographic features of each FI type are as follows.

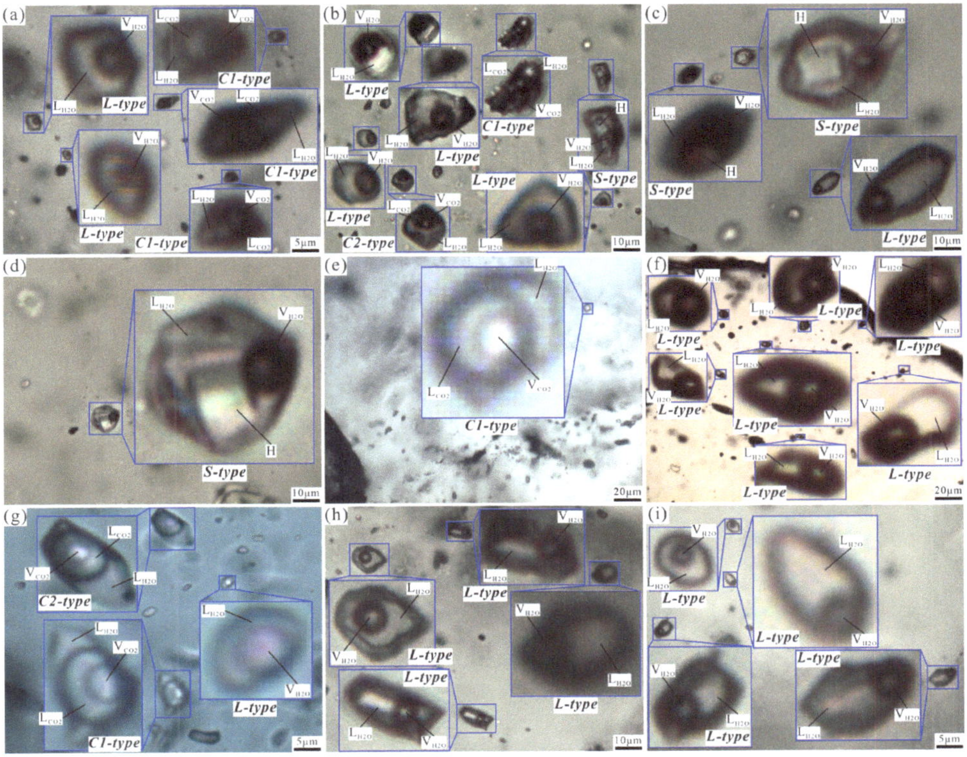

Figure 5. Representative photomicrographs of FIs in quartz veins of different stages at Lower Urgen. (**a**) Phase separation recorded by L- and C1-type FIs at stage 1. (**b**) An immiscibility FIA consisting of L-, S-, C1-, and C2-type FIs at stage 1. (**c**) Coexistence of L- and S-type FIs at stage 1. (**d**) Isolate S-type FI at stage 1. (**e**) Isolate C1-type FI at stage 1. (**f**) An FIA containing many primary L-type FIs at stage 1. (**g**) Phase separation recorded by L-, C1-, and C2-type FIs at stage 2. (**h**) An FIA containing many primary L-type FIs at stage 2. (**i**) An FIA containing four L-type FIs at stage 3. Abbreviations: H, halite; L_{CO_2}, liquid CO_2; V_{CO_2}, vapor CO_2; L_{H_2O}, liquid H_2O; V_{H_2O}, vapor H_2O.

L-type FIs are ubiquitous in all stages of quartz. At room temperature, they have a vapor phase and a liquid phase with $L_{H_2O}/(L_{H_2O} + V_{H_2O})$ ratios of 65–90 vol% (Figure 5). With diameters ranging from 5 to 15 μm, these FIs can be ellipsoidal, rectangular, negative-crystal, or irregular in shape. L-type FIs occur alongside S-, C1-, and C2-type FIs in stage 1 as immiscibility FIAs (Figure 5a–c). They are also intergrown with C1- and C2-type FIs and form immiscibility FIAs in stage 2 quartz crystals (Figure 5g). In quartz veins of stages 1 and 2, L-type FIs are abundant and make up around 50% of all FIs; however, stage 3 veins solely contain L-type FIs.

Only stage 1 quartz veins have S-type FIs, which exhibit high salinities and densities. S-type FIs are composed of a cubic halite daughter mineral, a liquid phase, and a vapor bubble (3–10 vol%). They are elliptical, negative-crystal, or irregular in shape, with diameters of 7 to 15 μm. S-type FIs can exist alone or with L-, C1-, and C2-type FIs to form immiscibility

FIAs in stage 1 (Figure 5b–d). In abundance, S-type FIs account for about 10% of all FIs in stage 1 veins.

C1-type FIs occur as ellipsoidal, circular, or negative-crystal shapes, and have a diameter of 5 to 13 μm. The CO_2 phase content ranges from 60 to 85 vol%. Usually, 60 to 80 vol% of the entire CO_2 phase is made up of vapor CO_2. C1-type FIs commonly appear in the three-phase form (V_{CO_2} + L_{CO_2} + L_{H_2O}), with a small amount of the two-phase form (V_{CO_2} + L_{H_2O}) (Figure 5a,b,e,g). The two-phase FIs are regarded as CO_2-rich FIs as clathrates were observed during freezing runs. In abundance, C1-type FIs frequently coexist as immiscibility FIAs with L-, S-, and C2-type FIs (Figure 5a–c,g), and account for about 25 vol% and 30 vol% of all FIAs in veins of stages 1 and 2, respectively.

At room temperature, C2-type FIs exist in the three-phase (V_{CO_2} + L_{CO_2} + L_{H_2O}) state, and minor FIs display a two-phase (V_{CO_2} + L_{H_2O}) form (Figure 5b,g). During freezing, clathrates are also observed in these two-phase C2-type FIs and can therefore be distinguished from L-type FIs. Typically, vapor CO_2 makes about 60–80 vol% of the total CO_2 phase, and the amount of CO_2 phase varies from 15 to 45 vol%. These FIs are irregular or nearly ellipsoidal in shape with a diameter of 5–15 μm. C2-type FIs are intergrown with L- and C1-type FIs (Figure 5g), together with small quantities of S-type FIs (Figure 5b), and account for approximately 15 vol% and 20 vol% of all FIAs in veins of stages 1 and 2, respectively.

5.3. FI Laser Raman Spectroscopy

Figure 6 illustrates the components of typical FIs from quartz veins of all stages. A substantial volume of H_2O and a negligible amount of CO_2 (1385 cm^{-1}, 1283 and 1388 cm^{-1}, respectively) were detected in both the liquid and vapor phase in L-type FIs of stage 1 (Figure 6a). C1-type FIs of stage 1 contain CO_2 as the only species in both the vapor and liquid CO_2 phases (1284 and 1388 cm^{-1}, 1282 and 1386 cm^{-1}, respectively) (Figure 6b). C2-type FIs of stage 1 have a similar vapor and liquid phase composition to stage 1 C1-type FIs, with CO_2 as the only component of the CO_2 phase (1283 and 1388 cm^{-1}, 1281 and 1386 cm^{-1}, respectively) (Figure 6c). Analogous liquid and vapor phase components, which are predominately made of H_2O and include a little amount of CO_2 (1385 cm^{-1}, 1284 and 1388 cm^{-1}, respectively), are present in S-type FIs of stage 1 (Figure 6d). Consequently, a NaCl–H_2O–CO_2 system is represented by the fluids in the quartz–molybdenite(–pyrite) stage. C1-type FIs of stage 2 show typical CO_2 spectra (1282 and 1387 cm^{-1}) in the vapor CO_2 phase (Figure 6e). The compositions of the vapor and liquid CO_2 phase of C2-type FIs in stage 2 only contain CO_2 (1281 and 1386 cm^{-1}) (Figure 6f). H_2O is the predominant gas species in the vapor phase of L-type FIs in stage 2 quartz, with only trace amounts of CO_2 (1387 cm^{-1}). In contrast, only H_2O was detected in the liquid phase of L-type FIs in stage 2 quartz (Figure 6g). Thus, a NaCl–H_2O–CO_2 system is represented by the fluids in the quartz–polymetallic sulfide stage. L-type FIs of stage 3 lack CO_2 detection and only have H_2O for the compositions of the liquid and vapor phases, demonstrating the characteristic of a NaCl–H_2O system (Figure 6h).

5.4. FI Microthermometry

Only primary FIs were chosen for microthermometry to prevent interpretation errors brought by the post-ore modification. Using the FIA concept, the reliability of microthermometric results has been assessed [109]. Large disparities in homogenization temperatures in spatially close FIs prevent them from being unequivocally classified as an FIA; instead, they are thought to be the result of necking, stretching, or heterogeneous entrapment [120]. The microthermometry data for isolated FIs are considered, barring a significant discrepancy with surrounding FIs [121]. If the homogenization temperature variation within a cluster of a single FI type is below 15 °C, the possibility of post-entrapment modification, generational overlap, or heterogeneous trapping of FIs is ruled out, and it is argued that these FIs represent a uniform fluid entrapment. We thus suggest that those results are trustworthy and that the overall average of a cluster is adequate for additional research [120]. An immiscibility FIA is a group of FIs of various types that possess comparable ranges of homogenization temperature (variation below 15 °C) but distinct homogenization processes. In this case, the complete FIA was represented for statistical analysis by the average of several types of FIs.

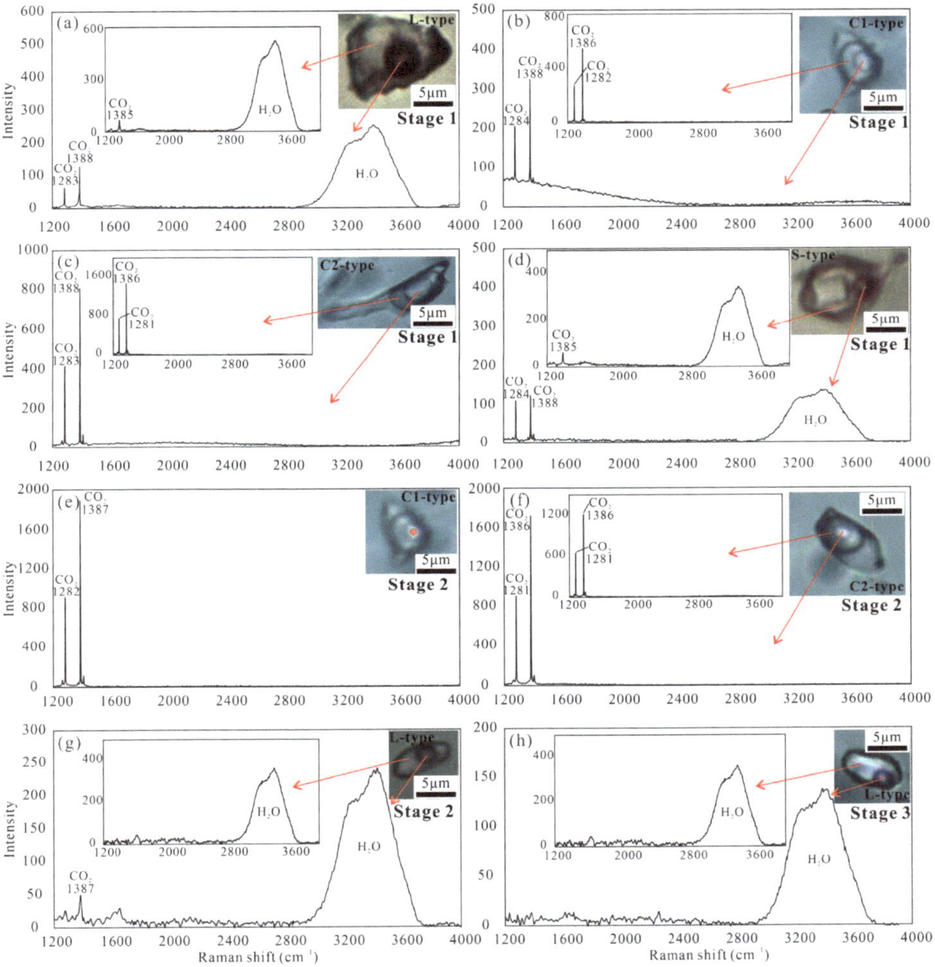

Figure 6. Representative laser Raman spectra for FIs from the Lower Urgen molybdenum deposit. (**a**) Raman spectra of stage 1 L-type FI reflect the presence of H_2O and minor amounts of CO_2. (**b**) Raman spectra of stage 1 C1-type FI indicate the presence of CO_2 in the CO_2 phase (V_{CO_2} + L_{CO_2}). (**c**) Raman spectra of stage 1 C2-type FI reflect the existence of CO_2 in the CO_2 phase (V_{CO_2} + L_{CO_2}). (**d**) Raman spectra of stage 1 S-type FI indicate the presence of H_2O and minor amounts of CO_2. (**e**) Raman spectra of stage 2 C1-type FI reflect the occurrence of CO_2 in vapor CO_2. (**f**) Raman spectra of stage 2 C2-type FI indicate the existence of CO_2 in the CO_2 phase (V_{CO_2} + L_{CO_2}). (**g**) Raman spectra of stage 2 L-type FI reflect the presence of H_2O and minor amounts of CO_2. (**h**) Raman spectra of stage 3 L-type FI indicate the existence of H_2O.

In this investigation, microthermometric results for 320 FIAs of three stages are listed in Table S1 (see the Supplementary Materials) and illustrated in Figure 7. The Tfm_{ice} for L-type FIs of all stages ranges from −21.6 to −20.8 °C, consistent with the eutectic temperature of the H_2O–NaCl system, indicating that most of the dissolved salts are

NaCl [111,122]. As mentioned in Section 5.3, the L- and S-type FIs sometimes display weak CO_2 signals, but the CO_2 characteristics cannot be observed during the petrography observation and microthermometry of these FIs. The detection limit of CO_2 through laser Raman spectroscopy can be as low as 1 bar, while the phase transition of CO_2 can be observed in microthermometry with a CO_2 pressure of at least 10.4 bar (the melting of CO_2 clathrate can be observed) or 45 bar (solid phase of CO_2 melting can be observed) [123]. In summary, the amount of CO_2 in the S-type FIs of stage 1 and L-type FIs of stages 1 and 2 is extremely low and almost negligible, and thus the salinities of these FIs were obtained using the Tm_H and Tm_{ice}, respectively.

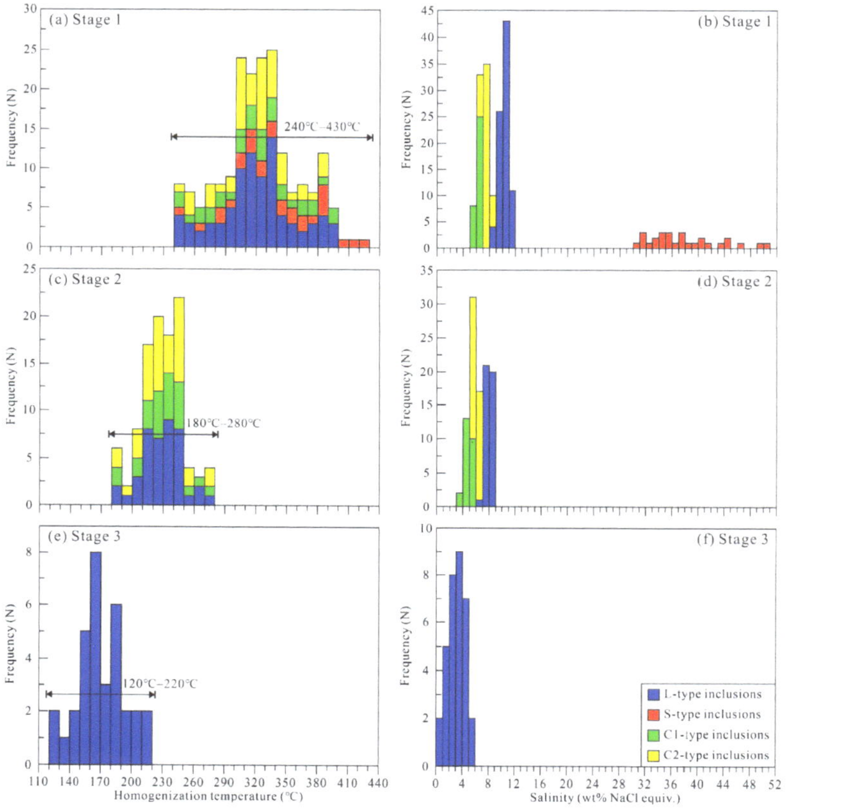

Figure 7. Histograms of homogenization temperatures (**a**,**c**,**e**) and salinities (**b**,**d**,**f**) for FIs in quartz from the Lower Urgen molybdenum deposit according to the concept of FIAs.

The entrapped fluid characteristics are often dramatically changed through post-entrapment modification (dismemberment and/or migration) in quartz-hosted FIs [124,125]. Euhedral quartz crystals that are naturally formed frequently exhibit these modifications in addition to quartz phenocrysts and veins. Although not verified, a satisfactory FI preservation state is recommended if the FIs display constant phase ratios, develop in clear trails or growth bands, and provide densely concentrated microthermometric values [125]. On the contrary, regular negative-crystal forms are neither necessary nor indicative of the preservation of syn-entrapment characteristics of FIs. Unless they adhere to specified growth bands, clusters of randomly scattered FIs often reflect secondary or pseudosecondary FI trails that have dissolved. Audétat [124] demonstrated that the decline in FI volume caused by post-entrapment modification results in a decrease in the homogenization temperature, a steepening of the isochore, and an increase in fluid density. By contrast, in magmatic–hydrothermal systems, water loss from FIs seldom surpasses twenty percent; consequently, the fluid salinities that are determined should, in most instances, be almost accurate. In the current research, FIs inside or along the growth zones of euhedral quartz occur incredibly seldom, and the bulk of them are too small to investigate (<5 μm). Post-entrapment modification may have occurred to varied degrees in clustered, isolated, and randomly dispersed FIs, but it seems inescapable. However, to guarantee that the chosen FIAs only had slight effects of post-entrapment modification, FIAs that displayed obvious post-entrapment modification features, e.g., deformation, movement tracks, stretching, shrinkage, or deteriorating halos, have been excluded [119,124–126]. Nonetheless, the genuine densities and homogenization temperatures of the examined FIs are slightly lower and higher than the calculated data, respectively. Comparatively, FI salinities are frequently less influenced, and most of them tend to have maintained their original salinities. As a result, we primarily concentrate on the salinities of the examined FIs, while the homogenization temperatures are used as a reference.

Stage 1 veins contain L-, S-, C1-, and C2-type FIs. The Th of L-type FIs ranges from 240 to 400 °C, and the Tm_{ice} varies between –8.2 and –5.1 °C (corresponding to 8.0–11.9 wt% NaCl equiv.). The disappearance of the vapor bubbles indicates complete homogenization. The Tm_H of S-type FIs is between 160 and 428 °C (corresponding to 30.1–50.8 wt% NaCl equiv.). Halite dissolution or the disappearance of vapor bubbles is an indicator that S-type FIs homogenize between 245 and 428 °C. The Tm_{CO_2} of C1-type FIs ranges from –59.8 to –57.0 °C. C1-type FIs have a Tm_{clath} of 5.9–7.4 °C and calculated salinities of 5.0–7.5 wt% NaCl equiv. At 14.5–21.5 °C, the CO_2 phase homogenizes to the vapor CO_2 phase. At 240–397 °C, the liquid H_2O phase disappears, signifying the completion of the homogenization. C2-type FIs have a Tm_{CO_2} of –59.8 to –57.0 °C and Tm_{clath} of 5.1–6.8 °C, corresponding to salinities of 6.0–8.8 wt% NaCl equiv. At 20.3–25.9 °C, CO_2 homogenizes to the vapor CO_2 phase, and at 241–395 °C, C2-type FIs finally homogenize to the liquid phase (Figure 7a,b).

L-, C1-, and C2-type FIs can be found in stage 2 quartz veins. Compared to stage 1 FIs, these FIs have a lower Th and lower salinities (Figure 7c,d). L-type FIs have a Tm_{ice} of –5.9 to –3.7 °C, corresponding to 6.0–9.1 wt% NaCl equiv. At 182–278 °C, L-type FIs homogenize and are characterized by the vapor bubbles' disappearance. C1-type FIs yield a Tm_{CO_2} of –59.8 to –57.0 °C, Tm_{clath} of 6.8–8.5 °C, and calculated salinities of 3.0–6.0 wt% NaCl equiv. At 14.6–22.5 °C, the CO_2 phase homogenizes to the vapor CO_2 phase, and at 181–277 °C, the liquid H_2O phase is eliminated, indicating complete homogenization. C2-type FIs contain a Tm_{CO_2} that varies between –59.8 and –56.8 °C and a Tm_{clath} of 6.2–7.9 °C (corresponding to 4.1–7.1 wt% NaCl equiv.). The homogeneity of the CO_2 phase is featured by the elimination of the liquid CO_2 phase, and the Th_{CO2} has a range of 21.9–25.9 °C. Total homogenization occurs when the vapor CO_2 phase is eliminated, and the Th of C2-type FIs has a range of 180–277 °C (Figure 7c,d).

L-type FIs are the only FIs found in quartz–carbonate veins of stage 3, which have Tm_{ice} values between –5.0 and –0.1 °C and NaCl equivalent salinities between 0.2 and 7.9 wt%. At 120–218 °C, these FIs ultimately homogenize to the liquid phase (Figure 7e,f).

5.5. Quartz H-O Isotopic Compositions

The hydrogen and oxygen isotopic compositions of analyzed quartz samples are presented in Table 2 and shown in Figure 8. Seven quartz samples of stage 1 veins have δD and $\delta^{18}O$ values ranging from –141.8 to –136.5‰ and 7.2–8.0‰, respectively. The $\delta^{18}O_{fluid}$ values are determined to be between 1.1 and 1.9‰. The δD values of seven quartz samples collected from stage 2 veins are –144.1 to –129.9‰, the $\delta^{18}O$ values range between 7.3 and 8.1‰, and the calculated $\delta^{18}O_{fluid}$ values are –2.7 to –1.9‰. The δD and $\delta^{18}O$ values in three quartz samples from stage 3 veins are –146.6 to –142.6‰ and 5.8–7.7‰, respectively, and the calculated $\delta^{18}O_{fluid}$ values are between –8.0 and –6.1‰.

Table 2. Hydrogen and oxygen isotopic compositions of quartz from the Lower Urgen molybdenum deposit.

Sample No.	Stage	Mineral	δD (‰)	$\delta^{18}O_{quartz}$ (‰)	T (°C)	$\delta^{18}O_{fluid}$ (‰)
B16-1	Stage 1	Quartz	−140.2	7.6	323	1.5
B16-2	Stage 1	Quartz	−138.3	7.2	323	1.1
B16-3	Stage 1	Quartz	−136.5	7.4	323	1.3
B16-4	Stage 1	Quartz	−140.9	7.9	323	1.8
B16-5	Stage 1	Quartz	−139.0	8.0	323	1.9
B16-6	Stage 1	Quartz	−141.8	8.0	323	1.9
B16-7	Stage 1	Quartz	−139.1	7.6	323	1.5
D14-1	Stage 2	Quartz	−142.7	7.9	229	−2.1
D16-1	Stage 2	Quartz	−135.5	7.6	229	−2.4
D16-4	Stage 2	Quartz	−142.5	7.7	229	−2.3
D16-6	Stage 2	Quartz	−135.4	7.9	229	−2.1
D16-7	Stage 2	Quartz	−129.9	8.1	229	−1.9
D16-8	Stage 2	Quartz	−142.7	7.6	229	−2.4
D16-9	Stage 2	Quartz	−144.1	7.3	229	−2.7
G8-1	Stage 3	Quartz	−146.6	7.7	170	−6.1
G10-1	Stage 3	Quartz	−143.9	5.8	170	−8.0
G11-1	Stage 3	Quartz	−142.6	7.5	170	−6.3

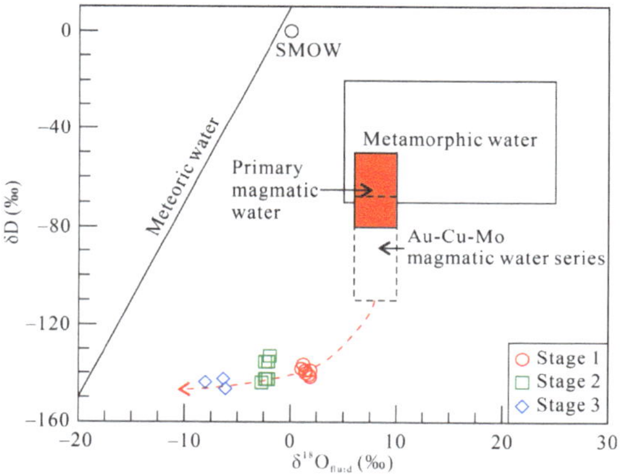

Figure 8. δD versus $\delta^{18}O_{fluid}$ diagram for the Lower Urgen molybdenum deposit (base map after [127]). The magmatic water field of the Au-Cu-Mo series is from [128].

5.6. Sulfide S Isotopic Compositions

In Table 3 and Figure 9, the S isotopic compositions of molybdenite, pyrite, chalcopyrite, and galena collected from the Lower Urgen molybdenum deposit are displayed. Sulfide $\delta^{34}S$ values range from 3.9 to 5.5‰, with an average value of 5.2‰. Six molybdenite samples contain $\delta^{34}S$ values of 5.4–5.5‰. Two pyrite samples and two chalcopyrite samples have $\delta^{34}S$ values of 5.4‰ and 4.8–5.4‰, respectively. One galena sample has a $\delta^{34}S$ value of 3.9‰.

Table 3. Sulfur isotopic compositions of sulfides from the Lower Urgen molybdenum deposit.

Sample No.	Stage	Mineral	$\delta^{34}S$ (‰)
LU16-1	Stage 1	Molybdenite	5.5
LU16-2	Stage 1	Molybdenite	5.4
LU16-3	Stage 1	Molybdenite	5.5

Table 3. *Cont.*

Sample No.	Stage	Mineral	$\delta^{34}S$ (‰)
LU16-4	Stage 1	Molybdenite	5.5
LU16-5	Stage 1	Molybdenite	5.5
LU16-6	Stage 1	Molybdenite	5.5
LU16-7	Stage 2	Pyrite	5.4
LU16-8	Stage 2	Pyrite	5.4
LU16-9	Stage 2	Chalcopyrite	5.4
LU16-10	Stage 2	Chalcopyrite	4.8
LU16-11	Stage 2	Galena	3.9

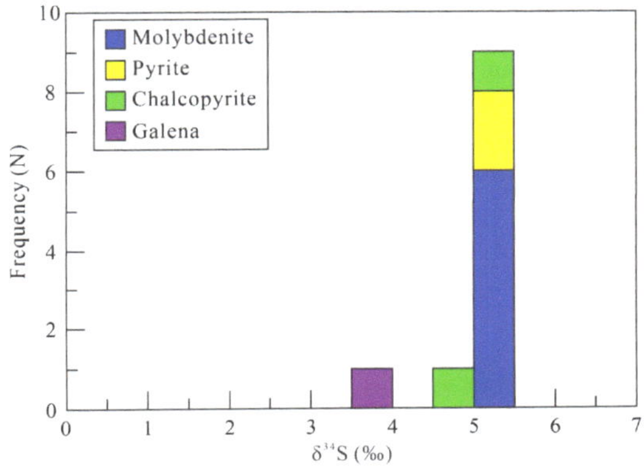

Figure 9. Histogram of sulfur isotopic compositions of metal sulfides from the Lower Urgen molybdenum deposit.

5.7. Sulfides and Whole-Rock Pb Isotopic Compositions

The Pb isotopic compositions of molybdenite, pyrite, chalcopyrite, galena, and granite porphyry are given in Table 4 and Figure 10. The $^{206}Pb/^{204}Pb$, $^{207}Pb/^{204}Pb$, and $^{208}Pb/^{204}Pb$ ratios for eight molybdenite samples are 18.340–18.412, 15.540–15.546, and 38.130–38.212, respectively. Five pyrite samples contain $^{206}Pb/^{204}Pb$, $^{207}Pb/^{204}Pb$, and $^{208}Pb/^{204}Pb$ ratios of 18.336–18.510, 15.542–15.569, and 38.131–38.312, respectively. The $^{206}Pb/^{204}Pb$, $^{207}Pb/^{204}Pb$, and $^{208}Pb/^{204}Pb$ ratios of two chalcopyrite samples are 18.336–18.340, 15.541–15.547, and 38.131–38.143, respectively. One galena sample has $^{206}Pb/^{204}Pb$, $^{207}Pb/^{204}Pb$, and $^{208}Pb/^{204}Pb$ ratios of 18.343, 15.550, and 38.149, respectively. The $^{206}Pb/^{204}Pb$, $^{207}Pb/^{204}Pb$, and $^{208}Pb/^{204}Pb$ ratios of three molybdenum-bearing granite porphyry samples range from 18.535 to 18.599, 15.571 to 15.574, and 38.419 to 38.583, respectively.

Table 4. Pb isotopic composition of sulfide and granite porphyry samples from the Lower Urgen molybdenum deposit.

Sample No.	Stage	Mineral	$^{206}Pb/^{204}Pb$	2σ	$^{207}Pb/^{204}Pb$	2σ	$^{208}Pb/^{204}Pb$	2σ	μ
D17-2	1	Molybdenite	18.340	0.0004	15.542	0.0003	38.137	0.0008	9.36
D17-3	1	Molybdenite	18.340	0.0005	15.543	0.0005	38.153	0.0011	9.36
D17-4	1	Molybdenite	18.355	0.0004	15.544	0.0004	38.164	0.0009	9.36
D17-5	1	Molybdenite	18.343	0.0004	15.540	0.0004	38.135	0.0009	9.35
D17-7	1	Molybdenite	18.340	0.0004	15.540	0.0004	38.130	0.0009	9.35
D0A1-1	1	Molybdenite	18.412	0.0004	15.546	0.0004	38.212	0.0009	9.36
D0A1-2	1	Molybdenite	18.387	0.0004	15.544	0.0004	38.183	0.0010	9.36
D0a1-1	1	Molybdenite	18.363	0.0004	15.543	0.0003	38.170	0.0009	9.36

Table 4. *Cont.*

Sample No.	Stage	Mineral	$^{206}Pb/^{204}Pb$	2σ	$^{207}Pb/^{204}Pb$	2σ	$^{208}Pb/^{204}Pb$	2σ	μ
G10-2	2	Pyrite	18.373	0.0005	15.549	0.0004	38.174	0.0010	9.37
G11-2	2	Pyrite	18.510	0.0004	15.557	0.0004	38.249	0.0010	9.37
G13	2	Pyrite	18.471	0.0004	15.569	0.0003	38.312	0.0008	9.40
G14	2	Pyrite	18.341	0.0003	15.549	0.0003	38.148	0.0008	9.37
D17-1	2	Pyrite	18.336	0.0003	15.542	0.0002	38.131	0.0006	9.36
D17-2	2	Chalcopyrite	18.336	0.0005	15.541	0.0004	38.131	0.0011	9.36
G8-2	2	Chalcopyrite	18.340	0.0004	15.547	0.0003	38.143	0.0008	9.37
G8-3	2	Galena	18.343	0.0004	15.550	0.0004	38.149	0.0009	9.37
T0a1-1	–	Granite porphyry	18.599	0.0005	15.574	0.0005	38.583	0.0013	9.40
T0a1-2	–	Granite porphyry	18.565	0.0005	15.573	0.0004	38.515	0.0012	9.40
T16-1	–	Granite porphyry	18.535	0.0004	15.571	0.0004	38.419	0.0010	9.39

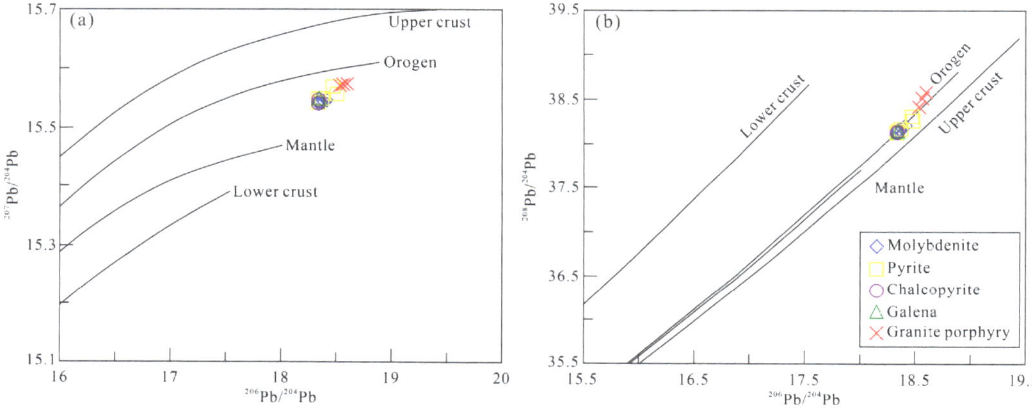

Figure 10. (**a**) $^{207}Pb/^{204}Pb$ versus $^{206}Pb/^{204}Pb$ diagram. (**b**) $^{208}Pb/^{204}Pb$ versus $^{206}Pb/^{204}Pb$ diagram (modified from [129]).

6. Discussion

6.1. Timing of Molybdenum Metallogenesis

From both scientific and economic perspectives, it is crucial to constrain the length and time of magmatic–hydrothermal systems [130]. Precise isotopic dating of hydrothermal minerals is an essential technique for determining the geochronology of hydrothermal processes in relation to intrusive magmatism and for constructing genetic models of hydrothermal deposits. As the primary product of hydrothermal mineralization, molybdenite obtained from ore-bearing quartz veins in molybdenum deposits is suggested as a Re-Os geochronometer for directly dating molybdenum mineralization events. In this contribution, two Re-Os model ages of 141.2 ± 1.5 and 147.7 ± 1.7 Ma were obtained. The wide range of Re-Os model ages could be attributed to Re and Os isotope decoupling, the considerable spatial Re heterogeneity, the post-ore deformation, and the analytical procedure [130–136]. For instance, many explanations based on post-crystallization diffusion of ^{187}Os are proposed to answer the Re and Os decoupling mechanism. ReS_2 inclusions or 3R polytype molybdenite formed during the crystallization of molybdenite would, over time, result in a chemical gradient for ^{187}Os that may homogenize during a thermal event. Decoupling could also result from the exsolution of Os-rich sulfides or diffusion of Os to crystal dislocations within the molybdenite during metamorphism [133,134]. The heterogeneous distribution of Re in molybdenite samples could also lead to the discrepancy of Re-Os model ages, which may be related to the overall abundance of Re in the hydrothermal fluids [136], compositional changes in the hydrothermal fluid with time [132], or partitioning of Re between liquid and vapor [130]. Additionally, a relatively wide range of molybdenite Re-Os model ages has been reported in numerous magmatic–hydrothermal deposits, such as the Haigou Au deposit (311.3–306.7 Ma; [131]), Lala Fe-Cu deposit (1097–1081 Ma; [137]), Kenge Mo deposit (1957–1950 Ma; [138]), Porcupine Mo deposit (1896–1882 Ma; [138]), Kedengjian Cu deposit (99.1–95.7 Ma; [139]), Mada Mo deposit (150.9–136.9 Ma; [140]), Meiziwo W deposit (165.6–156.2 Ma; [141]), Yaoling W deposit (165.9–156.2 Ma; [141]), Banlashan Mo deposit (142.7–131.8 Ma; [26]), Huashandong W deposit (827.8–795.3 Ma; [142]), Jiguanshan Mo deposit (156.2–151.7 Ma; [143]), and Dongbeigou Mo deposit (132.6–127.1 Ma; [144]). Zhang et al. [31] reported that the barren quartz monzonite porphyry and syenite porphyry stocks were emplaced at 165.0 ± 2.5 and 125.5 ± 1.9 Ma, respectively. Our recent geochronology study yielded a zircon U-Pb age of 142.3 ± 0.5 Ma for molybdenum-bearing granite porphyry stock (unpublished data). In recent decades, advances in high-precision dating, numerical simulation, and diffusion modeling mostly argue for short-lived (<1 Ma) timescales for the lifetime of magmatic–hydrothermal systems [7,145–152]. The Re-Os model age of molybdenite sample G5 (141.2 ± 1.5 Ma) is highly consistent with the emplacement time of the molybdenum-bearing granite porphyry stock within the analytical uncertainty. Spatially, the molybdenum mineralization in the deposit is characterized by veinlet-disseminated- and vein-type quartz–sulfide orebodies hosted in the cupola of the ore-bearing granite porphyry stock (Figure 2c). Geochemically, the metal sulfides obtained in this study have similar lead isotopic compositions to those of the ore-bearing granite porphyry stock (discussed in Section 6.3). This evidence collectively demonstrates that the molybdenum mineralization at Lower Urgen was temporally, spatially, and genetically associated with the molybdenum-bearing granite porphyry stock, supporting an intrusion-related origin during the Early Cretaceous. In contrast, the barren quartz monzonite porphyry and syenite porphyry stocks record pre-ore and post-ore magmatic events, respectively. Our findings also indicate that accurate and precise molybdenite Re-Os ages representing the true timing of ore formation need an integrated combination of careful geology, geochemistry, and multiple dating methods.

Considering the Mesozoic intense and widespread molybdenum mineralization in NE China, molybdenum mineralization at Lower Urgen is not a singular instance. By collecting published molybdenite Re-Os ages for Mesozoic molybdenum deposits within NE China, we determine three periods of molybdenum metallogenesis events in the region: Middle–Late Triassic (245–225 Ma), Early–Middle Jurassic (200–165 Ma), and Late Jurassic–Early Cretaceous (160–130 Ma) (Table S2 and Figure 11). The Middle–Late Triassic molybdenum deposits are primarily located in the southern GXR (SGXR) and along the SXCF, with minor occurrences in the NCGXR, including the Laojiagou (234.9 Ma; [41]), Chehugou (245 Ma; [85]), Kulitu (236 Ma; [153]), and Badaguan deposits (228.8–225.1 Ma; [46,48,154]). Molybdenum deposits formed in the Early to Middle Jurassic are abundant within the LXZR, and scattered in the NCGXR, including the Daheishan (169.2–168.0 Ma; [52,155,156]), Xingshan (167.3 Ma; [53]), Jidetun (168.9–168.0 Ma; [54,58,157]), Fu'anpu (168.2–166.9 Ma; [57,158]), Houdaomu (167.5 Ma; [159]), Shuangshan (171.6 Ma; [160]), Jiapigou (196 Ma; [160]), Sancha (183.1 Ma; [160]), Dashihe (186.7 Ma; [161]), Liushengdian (169.3–168.7 Ma; [159,162]), Sifangdianzi (176.6 Ma; [59]), Luming (177.9–176.7 Ma; [61,62]), Huojihe (181.3 Ma; [64]), Xinhualong (171.3 Ma; [159]), Wunugetushan (180.5–177.6 Ma; [49,163]), and Taipingchuan deposits (200.1 Ma; [164]). Most of the molybdenum deposits dated between the Late Jurassic and Early Cretaceous are located in the GXR, including the Chalukou (150–148 Ma; [65,67]), Taipinggou (130.1–129.4 Ma; [72,165]), Jiazishan (147.4 Ma; [166]), Diyanqinamu (156.2 Ma; [167]), Hashitu (150 Ma; [75]), Aolunhua (132–129.4 Ma; [79,168,169]), Shabutai (135.3 Ma; [26]), Haisugou (136.4 Ma; [26]), Banlashan (143–136.1 Ma; [26,169]), Yangchang (138.5 Ma; [169]), Hongshanzi (138.2–137.2 Ma; [83]), Xiaodonggou (138.1–135.5 Ma; [84,170]), Nianzigou (154.3–153 Ma; [85,86,153]), and Jiguanshan deposits (155.4–151.1 Ma; [143,171]). The zircon

U-Pb ages of ore-bearing granodiorites record the granitic magmatism in the Kanchuangou and Jinchanggou deposits at 111.1 [56] and 114 Ma [63], respectively. Unfortunately, no accurate molybdenite Re-Os ages of these two deposits have been reported, and the genetic relationship between molybdenum mineralization and host granodiorites remains to be further investigated. Hence, the molybdenum mineralization in these two deposits is not involved in the present study. The Early Cretaceous molybdenum mineralization in Lower Urgen is in accordance with the Late Jurassic–Early Cretaceous molybdenum metallogenesis in the NCGXR.

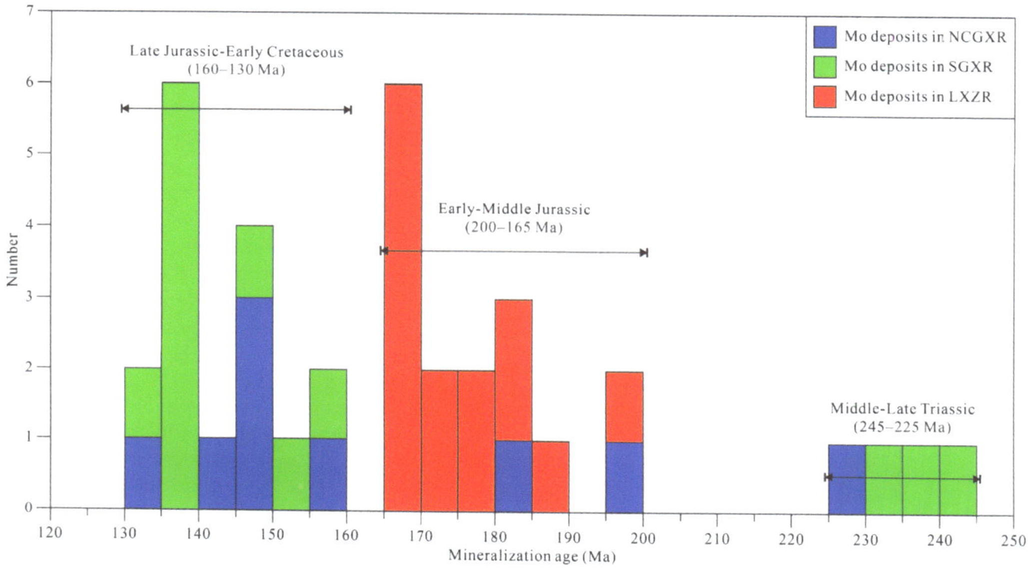

Figure 11. Histogram of molybdenite Re-Os isochron ages for Mesozoic molybdenum deposits in NE China. Abbreviations: NCGXR, northern and central Great Xing'an Range; SGXR, southern Great Xing'an Range; LXZR, lesser Xing'an-Zhangguangcai Range. Data sources are listed in Table S2.

6.2. Source of Mineralizing Fluids

Stage 1 quartz has calculated $\delta^{18}O_{fluid}$ values (1.1 to 1.9‰) that are lower than normal magmatic fluids (7–9‰; [172]) and substantially higher than normal Mesozoic meteoric water (−18‰; [173]). The $\delta^{18}O_{fluid}$ values of the quartz samples from the late quartz–carbonate stage are more depleted (−8.0 to −6.1‰) and comparable to the values of Mesozoic meteoric water [127,173] (Figure 8). This $\delta^{18}O_{fluid}$ variation trend implies the mixing of magmatic fluids with later meteoric water during the mineralization at Lower Urgen. Furthermore, in agreement with the documented δD values of Mesozoic meteoric water in the GXR (−140 to −90‰; [173]), the δD values of all stages (−146.6 to −129.9‰) are much lower than those of magmatic water (−50 to −85‰; [174]). The depletion of δD in the mineralizing fluids could be controlled through (1) fluid boiling or immiscibility [175], (2) magma degassing led by major meteoric water incursion [176], and (3) hydrogen isotope fractionation between the mineralizing fluids and some reduced species (e.g., CH_4). As shown in Figure 5, the coexistence of multiple FI types in stages 1 and 2 indicates intense fluid immiscibility during ore precipitation. The decreasing trend of Th and salinity in the FIs from stage 1 to 3 indicates increasing meteoric water input toward the late fluid system (Figure 7). However, no reduced species such as CH_4 have been detected in the H_2O-dominated ore-forming fluids, which indicates that hydrogen isotope fractionation is likely negligible [127]. Hence, we preliminarily speculate that fluid immiscibility and meteoric water input could explain the δD depletion of ore-forming fluids at Lower Urgen. In the δD versus $\delta^{18}O_{fluid}$ diagram, the calculated $\delta^{18}O_{fluid}$ and δD values also show a clear evolution trend from stage 1 to 3, and the data are plotted between the magmatic and meteoric water fields (Figure 8). Therefore, we conclude that the mineralizing fluids of the Lower Urgen deposit were primarily of magmatic origin with progressive meteoric water input.

6.3. Source of Sulfur and Lead

S isotopes can be used to trace the source of sulfur in sulfide minerals, despite the fact that the entire S isotopic components could be influenced by pH, temperature, ion activity, and f_{O2} [177–179]. The $\delta^{34}S$ data of sulfides can be utilized to determine their origin because no sulfate minerals have been found in the Lower Urgen district. The molybdenite, pyrite, chalcopyrite, and galena have $\delta^{34}S$ values of 5.4–5.5‰, 5.4‰, 4.8–5.4‰, and 3.9‰, respectively, implying that these sulfides have reached sulfide equilibrium fractionation [179]. The sulfides $\delta^{34}S$ values of the Lower Urgen deposit display a restricted range (3.9–5.5‰), which is commonly interpreted as a magma-derived sulfur [180] (Figure 9).

The source of the lead in the mineralizing fluids has been constrained using the lead isotope ratios of sulfides. The Pb isotopic ratios of sulfides can provide information about the source of Pb since sulfide minerals often have minimal radiogenic Pb isotopes and extremely low quantities of Th and U [129,181]. By contrasting the Pb isotope ratios of molybdenite, pyrite, chalcopyrite, galena, and molybdenum-bearing granite porphyry, the Pb source of the Lower Urgen molybdenum deposit can be determined. The sulfide samples have narrow ranges of $^{206}Pb/^{204}Pb$ (18.336–18.510), $^{207}Pb/^{204}Pb$ (15.540–15.569), and $^{208}Pb/^{204}Pb$ (38.130–38.312), implying an individual or successfully mixed lead origin. The Pb isotopic ratios of sulfides in the Lower Urgen deposit are distinct from those of the surrounding Mesozoic strata (e.g., Tamulagou and Manketouebo Fms) ($^{206}Pb/^{204}Pb$ = 18.08–18.32, $^{207}Pb/^{204}Pb$ = 15.46–15.55, and $^{208}Pb/^{204}Pb$ = 37.62–38.10) [93,182], but similar to those of the ore-bearing granite porphyry stock in the deposit ($^{206}Pb/^{204}Pb$ = 18.535–18.599, $^{207}Pb/^{204}Pb$ = 15.571–15.574, and $^{208}Pb/^{204}Pb$ = 38.419–38.583). Lead with high μ values (>9.58) was derived from the upper crust [129], whereas that with values of 8.92 was derived from the mantle [183]. The sulfides and granite porphyry stock from the Lower Urgen deposit yield μ values of 9.35–9.40 and 9.39–9.40, respectively, indicating that lead was basically from the crustal source and an extremely small amount of the mantle source. On the diagram of $^{207}Pb/^{204}Pb$ versus $^{206}Pb/^{204}Pb$ [129], the sulfide and ore-bearing granite porphyry samples are all located between the mantle and orogen evolution curves (Figure 10a). On the diagram of $^{208}Pb/^{204}Pb$ versus $^{206}Pb/^{204}Pb$ [129], most sulfide and ore-bearing granite porphyry samples are distributed on the orogen evolution curve (Figure 10b). These features imply that the metal sulfides and granite porphyry at Lower Urgen have the same lead source, which has been documented in many porphyry deposits [2,3,184], and that the granitic magmatism is probably largely responsible for the molybdenum mineralization in the Lower Urgen deposit.

6.4. Metal Precipitation Mechanisms

Fluid immiscibility, one of the most significant ore-forming mechanisms in magmatic–hydrothermal metallogenic systems, can change the temperature, pH, and salinity of the ore-bearing fluids, thereby reducing the solubility of metals in the fluids [14,185–188]. The occurrence of fluid immiscibility is supported by the following evidence from stage 1 quartz-hosted FIs: (1) L-, S-, C1-, and C2-type FIs have intimate spatial relationships (Figure 5a–c); (2) they exhibit distinct homogenization modes; and (3) these FIs show distinct salinities and comparable homogenization temperature ranges (Figure 7a,b). Similarly, at stage 2, L-, C1-, and C2-type FIs frequently coexist, indicating that they were simultaneously captured and showed different homogenization modes (Figures 5g and 7c,d). Notably, in stages 1 and 2, C1- and C2-type FIs frequently show lower salinities but analogous homogenization temperatures than L-type FIs, which shows that fluid immiscibility occurred before their entrapment [189]. The increased salinity of the residual fluids caused by volatile separation from the ore-bearing fluids as temperature and pressure drops may be the cause of the higher salinities in L-type FIs than those in C1- and C2-type FIs [185,190]. The mineralizing fluids at Lower Urgen underwent fluid immiscibility processes during stages 1 and 2, as reflected by the diagram of salinity versus homogenization temperature (Figure 12). Field evidence, such as the widespread fracture and brecciation of granite porphyry cupola and the overlying intermediate–basic volcanic rocks at Lower Urgen, suggests a dramatic pressure drop that might result in fluid immiscibility, further supporting the existence of fluid immiscibility [119,191–193]. Previous research has demonstrated that a mafic magmatic source or the magmatic–hydrothermal formation procedure of a deep-seated magma chamber often contributes significant amounts of dissolved CO_2 [119,194–198]. In combination with the H-O isotopic compositions of stage 1 quartz, we speculate that the original fluids of the Lower Urgen molybdenum deposit are CO_2-rich magmatic–hydrothermal fluids. As the fluids rose, volatiles (such as CO_2) started to separate from the fluids due to a decrease in pressure and temperature. The sudden supersaturation of the solution was caused by the release of volatiles, which also increased the pH of the solution and caused the decoupling of metal complexes. As a result of the procedure, significant amounts of molybdenite precipitated from stage 1 fluids, while chalcopyrite, pyrite, sphalerite, arsenopyrite, and galena were deposited from stage 2 fluids.

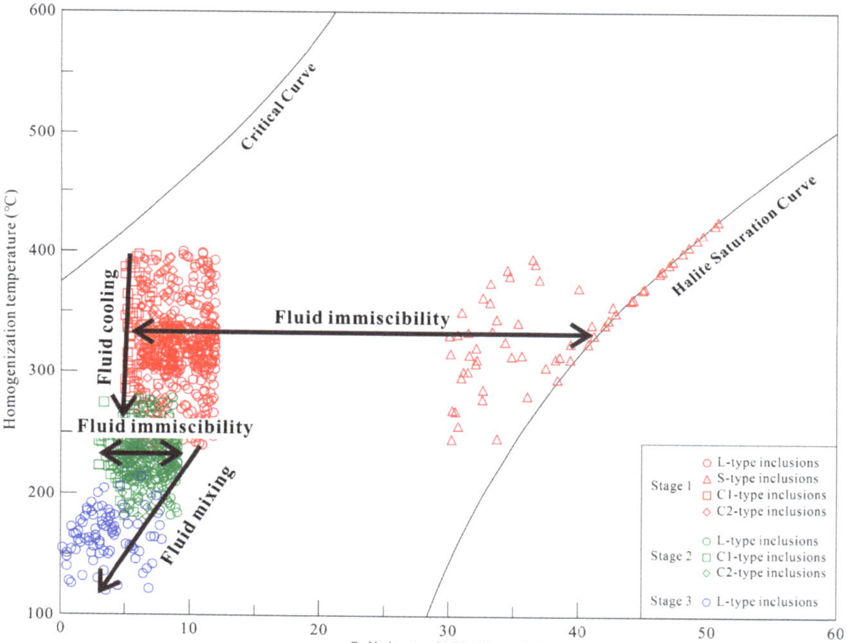

Figure 12. Homogenization temperature versus salinity of FIs from all stages in the Lower Urgen molybdenum deposit.

Hydrothermal mineralization frequently involves fluid–rock interaction, which is essential for the transport and deposition of metal ions, especially in an intrusive-related mineralization system [4,199–201]. Many investigations have demonstrated that changes in the physicochemical properties of fluids as a result of fluid–rock interactions affect metal solubility and mineral precipitation [202–205]. Different hydrothermal alteration types are characteristics of porphyry molybdenum mineralization, and phyllic alteration is frequently seen as a sign of prospecting [201,203,205,206], as indicated by the fact that molybdenum mineralization in the deposit is spatially related to phyllic alteration, and plagioclase and K-feldspar of the ore-bearing granite porphyry tend to be replaced by quartz and sericite in stage 1 (Figure 4b,e). Phyllic alteration lowered the H^+ contents of the hydrothermal fluids and enriched the reduced sulfur (S^{2-}), which mixed with Mo^{2+} to generate molybdenite. The following formulas reflect the pertinent reactions:

$$3KAlSi_3O_8 + 2H^+ \rightarrow KAl_3Si_3O_{10}(OH)_2 \text{ (sericite)} + 6SiO_2 \text{ (quartz)} + 2K^+ \quad (1)$$

$$3NaAlSi_3O_8 + K^+ + 2H^+ \rightarrow KAl_3Si_3O_{10}(OH)_2 \text{ (sericite)} + 6SiO_2 \text{ (quartz)} + 3Na^+ \quad (2)$$

$$H_2S \rightarrow H^+ + HS^- \quad (3)$$

$$HS^- \rightarrow H^+ + S^{2-} \quad (4)$$

Therefore, we provide a preliminary argument that phyllic alteration could benefit the deposition of molybdenite at Lower Urgen. Nevertheless, a more detailed explanation to this conclusion is beyond the scope of our current study and additional alteration mineralogical and geochemical research is necessary before a full understanding can be attained.

Numerous studies have shown that fluid mixing is a crucial factor in the production of base metal sulfide deposits because it can dilute and cool hydrothermal solutions, which causes enormous sulfide mineral depositions [207–211]. Mineral deposition occurs when the solubility of metal ions in the fluids surpasses saturation, and substantial deposition occurs as the solubility is dramatically decreased [208]. In the Lower Urgen deposit, the continuous fluid immiscibility in stages 1 and 2 would have increased the permeability of the host granite porphyry stock, and further accelerated meteoric water circulation. Stage 2 fluids have much lower salinity and homogenization temperatures than stage 1 fluids (Figure 12). Additionally, stage 2 fluids have more depleted $\delta^{18}O_{fluid}$ values than stage 1 fluids (Figure 8). These features collectively demonstrate the existence of fluid mixing processes in stage 2. To summarize, fluid mixing propelled the polymetallic sulfide mineralization in stage 2 at Lower Urgen. The progressive incursion of meteoric water into the magmatic–hydrothermal system might cause considerable cooling, the instability of metal chloride complexes, and sulfide deposition.

In conclusion, stage 1 molybdenite deposition is primarily controlled through fluid immiscibility and fluid–rock interaction. The increasing injection of meteoric water into the hydrothermal system and fluid immiscibility collectively controlled the massive deposition of polymetallic sulfides in stage 2.

6.5. Metallogenic Model for the Lower Urgen Magmatic–Hydrothermal System

Especially for intrusion-related magmatic–hydrothermal deposits, the development of a metallogenic model may successfully direct the investigation of the ore deposit and increase the successful rate of prospecting and prediction [4,212]. We construct the metallogenic model for the Lower Urgen molybdenum deposit upon the above discussion and previous research, as outlined below.

Due to the closure of the MOO, the surveyed region experienced an extensional environment in the Late Jurassic and Early Cretaceous, which promoted the partial melting of the lower crust and produced ore-bearing granitic magmas [25,50,213–218]. As a lithophile element, molybdenum tends to be abundant in the crust but deficient in the mantle [219]. Consequently, a potential source for molybdenum mineralization might be the lower crust under the Lower Urgen region. Metallic elements are partitioned into the melt through partial melting of the protolith. The ore-bearing granite porphyry, which represents the parental rock and served as a constant source of hot, ore-rich fluids for the generation of the hydrothermal mineralization system, was emplaced at about 142.3 Ma. The potassic and phyllic alteration zones were generated by these fluids accumulating at the cupola of the mineralization intrusion and interacting with cooling granite porphyry (Figure 2c). In the meantime, intense and widespread fluid immiscibility was created through the constant ascent and decompression of mineralizing fluids. Stage 1 fluids belong to a NaCl–H_2O–

CO_2 system and contain bimodal salinities (5.0–11.9 and 30.1–50.8 wt% NaCl equiv.) and have medium–high temperatures (240–430 °C). Intense fluid immiscibility and fluid–rock interaction caused an increase in pH and a drop in temperature, which significantly reduced the solubility of molybdenum and aided in the precipitation of molybdenite. The molybdenum mineralization stage was characterized by widespread molybdenum-bearing quartz veinlets and disseminated molybdenite in the granite porphyry. In stage 2, massive meteoric water mixed with the evolved magmatic–hydrothermal fluids through localized regional hydraulic fractured and brecciated zones. Stage 2 fluids have salinities and homogenization temperatures of 3.0–9.1 wt% NaCl equiv. and 180–280 °C, respectively, and belong to a NaCl–H_2O–CO_2 system. The molybdenum mineralization came to an end with the extinction of veinlet-disseminated- and vein-type quartz–molybdenite(–pyrite) ores. The constant infiltration of meteoric water diluted ore-forming fluids and decreased their temperature. This process, combined with extensive fluid immiscibility, collectively led to the large-scale decoupling of base metal complexes and promoted the massive deposition of polymetallic sulfides. Stage 3 fluids belong to a NaCl–H_2O system and have salinities of 0.2 to 7.9 wt% NaCl equiv. and homogenization temperatures of 120 to 220 °C, suggesting a hydrothermal environment with low salinity and temperature. The hydrothermal fluids were governed by meteoric water as the pressure of the system was dropped. The fluids produced quartz and calcite as precipitates. The absence of ore minerals in the veins of this stage suggests that the fluids are no longer useful for mineralization, signaling the termination of the hydrothermal process in the Lower Urgen region.

7. Conclusions

(1) Molybdenite Re-Os dating yields two model ages of 141.2 ± 1.5 and 147.7 ± 1.7 Ma for the Lower Urgen molybdenum deposit. The new geochronological data indicate that the molybdenum mineralization event in the Lower Urgen area is consistent with the intensive and extensive Late Jurassic–Early Cretaceous molybdenum mineralization events in NE China.

(2) Three hydrothermal stages of the Lower Urgen molybdenum deposit can be distinguished: (i) quartz–molybdenite(–pyrite) stage—ore-forming fluids have medium–high temperatures (240–430 °C) and bimodal salinities (5.0–11.9 and 30.1–50.8 wt% NaCl equiv.) and belong to a NaCl–H_2O–CO_2 system; (ii) quartz–polymetallic sulfide stage—ore-forming fluids have medium temperatures (180–280 °C) and low salinities (3.0–9.1 wt% NaCl equiv.) and belong to a NaCl–H_2O–CO_2 system; and (iii) quartz–carbonate stage—fluids have low temperatures (120–220 °C) and low salinities (0.2–7.9 wt% NaCl equiv.) and belong to a NaCl–H_2O system.

(3) Hydrogen–oxygen isotope results indicate that the mineralizing fluids initially had a magmatic source with the progressive incursion of meteoric water. Sulfur–lead isotopic compositions indicate that the mineralizing materials have a genetic relationship with ore-bearing granite porphyry, and sulfur and lead are magmatically derived.

(4) Fluid immiscibility and fluid–rock interaction are the predominant mechanisms for molybdenite deposition in stage 1. Fluid mixing and immiscibility are cooperatively accountable for the massive deposition of polymetallic sulfides in stage 2.

Supplementary Materials: The following supporting information can be downloaded at https://www.mdpi.com/article/10.3390/min13091189/s1, Table S1: Microthermometry data for fluid inclusions in quartz from the Lower Urgen molybdenum deposit; Table S2: Molybdenite Re-Os isochron ages for Mesozoic molybdenum deposits in NE China [26,41,43,46,48,49,52–54,57,59,61,62, 64,65,67,72,73,75,79,81,83–86,143,153–156,158–165,167–171].

Author Contributions: Conceptualization, G.Z. and W.X.; methodology, G.Z., H.W. and K.Z.; formal analysis, W.X.; investigation, G.Z., W.X., S.W., H.W., K.Z., T.T. and P.M.; data curation, G.Z. and W.X.; writing—original draft preparation, G.Z. and W.X.; writing—review and editing, W.X., S.W., Q.Z. and L.Z.; visualization, W.X.; supervision, W.X.; funding acquisition, Q.Z. All authors have read and agreed to the published version of the manuscript.

Funding: This research was funded by the National Natural Science Foundation of China, grant number 91962104.

Data Availability Statement: The data presented in this study are available in the article and Supplementary Materials.

Acknowledgments: We are grateful to three anonymous reviewers for their constructive comments, which greatly help in improving our paper. We would like to thank the managers of the Baojin Mining Group Co., Ltd. for their support of our fieldwork.

Conflicts of Interest: The authors declare no conflict of interest.

References

1. Kirkham, R.V. Porphyry deposits. In *Report of Activities Part B*; Blackadar, R.G., Ed.; Geological Survey of Canada 72-1b: Ottawa, ON, Canada, 1972; pp. 62–64.
2. Bodnar, R.J. Fluid inclusion evidence for a magmatic source for metals in porphyry copper deposits. *Magmas Fluids Ore Depos. Mineral. Assoc. Can. Short. Course* **1995**, *23*, 139–152.
3. Sinclair, W.D. Porphyry deposits. In *Mineral Deposits of Canda: A Synthesis of Major Deposit-Types, District Metallogeny, The Evolution of Geological Provinces, and Exploration Methods*; Special Publication 5; Geological Association of Canada, Mineral Deposits Division: St. John's, NL, Canada, 2007.
4. Sillitoe, R.H. Porphyry copper systems. *Econ. Geol.* **2010**, *105*, 3–41. [CrossRef]
5. Chen, H.Y.; Wu, C. Metallogenesis and major challenges of porphyry copper systems above subduction zones. *Sci. China Earth Sci.* **2020**, *63*, 899–918. [CrossRef]
6. Kooiman, G.J.A.; McLeod, M.J.; Sinclair, W.D. Porphyry tungsten-molybdenum orebodies, polymetallic veins and replacement bodies, and tin-bearing greisen zones in the Fire Tower Zone, Mount Pleasant, New Brunswick. *Econ. Geol.* **1986**, *81*, 1356–1373. [CrossRef]
7. Weis, P.; Driesner, T.; Heinrich, C.A. Porphyry-copper ore shells form at stable pressure-temperature fronts within dynamic fluid plumes. *Science* **2012**, *338*, 1613–1616. [CrossRef]
8. Audétat, A.; Pettke, T.; Heinrich, C.A.; Bodnar, R.J. The composition of magmatic-hydrothermal fluids in barren and mineralized intrusions. *Econ. Geol.* **2008**, *103*, 877–908. [CrossRef]
9. Audétat, A. Source and evolution of molybdenum in the porphyry Mo (–Nb) deposit at Cave Peak, Texas. *J. Petrol.* **2010**, *51*, 1739–1760. [CrossRef]
10. Klemm, L.M.; Pettke, T.; Heinrich, C.A.; Campos, E. Hydrothermal evolution of the El Teniente deposit, Chile: Porphyry Cu-Mo ore deposition from low-salinity magmatic fluids. *Econ. Geol.* **2007**, *102*, 1021–1045. [CrossRef]
11. Klemm, L.M.; Pettke, T.; Heinrich, C.A. Fluid and source magma evolution of the Questa porphyry Mo deposit, New Mexico, USA. *Miner. Depos.* **2008**, *43*, 533. [CrossRef]
12. Landtwing, M.R.; Furrer, C.; Redmond, P.B.; Pettke, T.; Guillong, M.; Heinrich, C.A. The Bingham Canyon porphyry Cu-Mo-Au deposit. III. Zoned copper-gold ore deposition by magmatic vapor expansion. *Econ. Geol.* **2010**, *105*, 91–118. [CrossRef]
13. Rusk, B.G.; Reed, M.H.; Dilles, J.H.; Klemm, L.; Heinrich, C.A. Compositions of magmatic hydrothermal fluids determined by LA-ICP-MS of fluid inclusions from the porphyry copper-molybdenum deposit at Butte. Montana. *Chem. Geol.* **2004**, *210*, 173–199. [CrossRef]
14. Rusk, B.G.; Reed, M.H.; Dilles, J.H. Fluid inclusion evidence for magmatic-hydrothermal fluid evolution in the porphyry copper-molybdenum deposit at Butte, Montana. *Econ. Geol.* **2008**, *103*, 307–334. [CrossRef]
15. Ulrich, T.; Mavrogenes, J.A. An experimental study of the solubility of molybdenum in H_2O and KCl–H_2O solutions from 500 °C to 800 °C, and 150 to 300 MPa. *Geochim. Cosmochimca Acta* **2008**, *72*, 2316–2330. [CrossRef]
16. Zhang, L.; Audétat, A.; Dolejs, D. Solubility of molybdenite (MoS_2) in aqueous fluids at 600–800 °C, 200 MPa: A synthetic fluid inclusion study. *Geochim. Cosmochimca Acta* **2012**, *77*, 175–185. [CrossRef]
17. Mao, J.W.; Pirajno, F.; Xiang, J.F.; Gao, J.J.; Ye, H.S.; Li, Y.F.; Guo, B.J. Mesozoic molybdenum deposits in the east Qinling-Dabie orogenic belt: Characteristics and tectonic settings. *Ore Geol. Rev.* **2011**, *43*, 264–293. [CrossRef]
18. Zeng, Q.D.; Liu, J.M.; Qin, K.Z.; Fan, H.R.; Chu, S.X.; Wang, Y.B.; Zhou, L.L. Types, characteristics, and time–space distribution of molybdenum deposits in China. *Int. Geol. Rev.* **2013**, *55*, 1311–1358. [CrossRef]
19. Zeng, Q.D.; Liu, J.M.; Yu, C.M.; Ye, J.; Liu, H.T. Metal deposits in the Da Hinggan Mountains, NE China: Styles, characteristics, and exploration potential. *Int. Geol. Rev.* **2011**, *53*, 846–878. [CrossRef]
20. Zeng, Q.D.; Liu, J.M.; Chu, S.X.; Wang, Y.B.; Sun, Y.; Duan, X.X.; Zhou, L.L. Mesozoic molybdenum deposits in the East Xingmeng orogenic belt, northeast China: Characteristics and tectonic setting. *Int. Geol. Rev.* **2012**, *54*, 1843–1869. [CrossRef]
21. Zeng, Q.D.; Qin, K.Z.; Liu, J.M.; Li, G.M.; Zhai, M.G.; Chu, S.X.; Guo, Y.P. Porphyry molybdenum deposits in the Tianshan–Xingmeng orogenic belt, northern China. *Int. J. Earth Sci.* **2015**, *104*, 991–1023. [CrossRef]
22. Ouyang, H.G.; Mao, J.W.; Santosh, M.; Zhou, J.; Zhou, Z.H.; Wu, Y.; Hou, L. Geodynamic setting of Mesozoic magmatism in NE China and surrounding regions: Perspectives from spatio-temporal distribution patterns of ore deposits. *J. Asian Earth Sci.* **2013**, *78*, 222–236. [CrossRef]

23. Ouyang, H.G.; Mao, J.W.; Zhou, Z.H.; Su, H.M. Late Mesozoic metallogeny and intracontinental magmatism, southern Great Xing'an Range, northeastern China. *Gondwana Res.* **2015**, *27*, 1153–1172. [CrossRef]
24. Wang, R.L.; Zeng, Q.D.; Zhang, Z.C.; Zhou, L.L.; Qin, K.Z. Extensive mineralization in the eastern segment of the Xingmeng orogenic belt, NE China: A regional view. *Ore Geol. Rev.* **2021**, *135*, 104204. [CrossRef]
25. Xie, W.; Zeng, Q.D.; Wang, R.L.; Wu, J.J.; Zhang, Z.M.; Li, F.C.; Zhang, Z. Spatial-temporal distribution and tectonic setting of Mesozoic W-mineralized granitoids in the Xing-Meng Orogenic Belt, NE China. *Int. Geol. Rev.* **2022**, *64*, 1845–1884. [CrossRef]
26. Shu, Q.H.; Chang, Z.S.; Lai, Y.; Zhou, Y.T.; Sun, Y.; Yan, C. Regional metallogeny of Mo-bearing Deposits in Northeastern China, with new Re-Os Dates of porphyry Mo Deposits in the Northern Xilamulun District. *Econ. Geol.* **2016**, *111*, 1783–1798. [CrossRef]
27. Feng, Z.Q.; Liu, Y.J.; Li, L.; Jin, W.; Jiang, L.W.; Li, W.M.; Wen, Q.B.; Zhao, Y.L. Geochemical and geochronological constraints on the tectonic setting of the Xinlin ophiolite, northern Great Xing'an Range, NE China. *Lithos* **2019**, *326–327*, 213–229. [CrossRef]
28. Liu, Y.J.; Li, W.M.; Feng, Z.Q.; Wen, Q.B.; Neubauer, F.; Liang, C.Y. A review of the Paleozoic tectonics in the eastern part of Central Asian Orogenic Belt. *Gondwana Res.* **2017**, *43*, 123–148. [CrossRef]
29. Wu, F.Y.; Sun, D.Y.; Ge, W.C.; Zhang, Y.B.; Grant, M.L.; Wilde, S.A.; Jahn, B.M. Geochronology of the Phanerozoic granitoids in northeastern China. *J. Asian Earth Sci.* **2011**, *41*, 1–30. [CrossRef]
30. Zhang, G.L. Metallogenic Regularity and Prospect Estimation of Urgen Au-Ag-Pb-Zn-Mo Polymetallic Ore Field in Erguna, Inner Mongolia. Ph.D. Thesis, Northeastern University, Shenyang, China, 2022. (In Chinese with English Abstract).
31. Zhang, G.L.; Wen, S.Q.; Li, J.Y.; Xie, W. Petrogenesis and tectonic implications of alkali-rich porphyry in the Urgen ore field, Erguna Block, Inner Mongolia: Constraints from geochronology and geochemistry. *Geotecton. Metallog.* **2023**, *47*, 428–448. (In Chinese with English Abstract)
32. Sengör, A.M.C.; Natal'in, B.A.; Burtman, V.S. Evolution of the Altaid tectonic collage and Paleozoic crustal growth in Eurasia. *Nature* **1993**, *364*, 299–307. [CrossRef]
33. Eizenhöfer, P.R.; Zhao, G.C.; Zhang, J.; Sun, M. Final closure of the Paleo-Asian Ocean along the Solonker suture zone: Constraints from geochronological and geochemical data of Permian volcanic and sedimentary rocks. *Tectonics* **2014**, *33*, 441–463. [CrossRef]
34. Xiao, W.J.; Windley, B.F.; Hao, J.; Zhai, M.G. Accretion leading to collision and the Permian Solonker suture, Inner Mongolia, China: Termination of the central Asian orogenic belt. *Tectonics* **2003**, *22*, 288–308. [CrossRef]
35. Xiao, W.J.; Santosh, M. The western Central Asian Orogenic Belt: A window to accretionary orogenesis and continental growth. *Gondwana Res.* **2014**, *25*, 1429–1444. [CrossRef]
36. Xu, W.L.; Pei, F.P.; Wang, F.; Meng, E.; Ji, W.Q.; Yang, D.B.; Wang, W. Spatial-temporal relationships of Mesozoic volcanic rocks in NE China: Constraints on tectonic overprinting and transformations between multiple tectonic systems. *J. Asian Earth Sci.* **2013**, *74*, 167–193. [CrossRef]
37. Wilde, S.A. Final amalgamation of the Central Asian Orogenic Belt in NE China: Paleo-Asian Ocean closure versus Paleo-Pacific plate subduction–A review of the evidence. *Tectonophysics* **2015**, *662*, 345–362. [CrossRef]
38. Chen, P.W.; Zeng, Q.D.; Wang, Y.B.; Zhou, T.C.; Yu, B.; Chen, J.Q. Petrogenesis of the Dasuji porphyry Mo deposit at the northern margin of North China Craton: Constrains from geochronology, geochemistry and isotopes characteristics. *Lithos* **2018**, *322*, 87–103. [CrossRef]
39. Chen, P.W.; Zeng, Q.D.; Zhou, T.C. Petrogenesis and Mo prospecting significance of Sadaigoumen granites on the northern margin of the North China Craton. *J. Geochem. Explor.* **2020**, *214*, 106536. [CrossRef]
40. Tang, W.H.; Li, J.; Wang, K.Y.; Li, S.D.; Cai, W.Y.; Liu, H.L.; Wang, Y.C. Ore genesis and tectonic setting of the Laojiagou porphyry molybdenum deposit, Inner Mongolia, China: Evidence from geology, fluid inclusions, H–O isotopes, zircon U–Pb geochronology, and geochemistry. *Arab. J. Geosci.* **2019**, *12*, 731. [CrossRef]
41. Duan, X.X.; Zeng, Q.D.; Yang, Y.H.; Liu, J.M.; Chu, S.X.; Sun, Y.; Zhang, Z.L. Triassic magmatism and Mo mineralization in Northeast China: Geochronological and isotopic constraints from the Laojiagou porphyry Mo deposit. *Int. Geol. Rev.* **2015**, *57*, 55–75. [CrossRef]
42. Sun, Q.F.; Wang, K.Y.; Sun, F.Y.; Lai, C.K.; Zhang, M.; Zhao, C.G.; Sun, L.X. Superimposing porphyry Mo and vein-type Cu-Pb-Zn mineralization in the Panjiaduan deposit, Great Xing'an Range (NE China): Perspective from zircon U-Pb and sphalerite Rb-Sr dating, geochemistry and S-Pb isotopes. *Ore Geol. Rev.* **2021**, *139*, 104538. [CrossRef]
43. Zeng, Q.D.; Yang, J.H.; Liu, J.M.; Chu, S.X.; Duan, X.X.; Zhang, Z.L.; Zhang, W.Q.; Zhang, S. Genesis of the Chehugou Mo-bearing granitic complex on the northern margin of the North China Craton: Geochemistry, zircon U-Pb age and Sr-Nd-Pb isotopes. *Geol. Mag.* **2011**, *149*, 753–767. [CrossRef]
44. Chen, Z.G.; Zhang, L.C.; Lu, B.Z.; Li, Z.L.; Wu, H.Y.; Xiang, P.; Huang, S.W. Geochronology and geochemistry of the Taipingchuan copper-molybdenum deposit in Inner Mongolia, and its geological significances. *Acta Petrol. Sin.* **2010**, *26*, 1437–1449. (In Chinese with English Abstract) [CrossRef]
45. Kang, Y.J.; She, H.Q.; Lai, Y.; Wang, Z.Q.; Li, J.W.; Zhang, Z.H.; Xiang, A.P.; Jiang, Z.S. Evolution of Middle-Late Triassic granitic intrusions from the Badaguan Cu-Mo deposit, Inner Mongolia: Constraints from zircon U-Pb dating, geochemistry and Hf isotopes. *Ore Geol. Rev.* **2018**, *95*, 195–215. [CrossRef]
46. Kang, Y.J.; Wang, Y.J.; Huang, G.J.; She, H.Q.; Xiang, A.P.; Tian, J.; Guo, Z.J.; Dong, X.Z. Study of rock-forming and ore-forming ages of Badaguan porphyry Cu-Mo deposit in Inner Mongolia. *Miner. Depos.* **2014**, *33*, 795–806. (In Chinese with English Abstract)
47. Mi, K.F.; Liu, Z.J.; Li, C.F.; Liu, R.B.; Wang, J.P.; Peng, R.M. Origin of the Badaguan porphyry Cu Mo deposit, Inner Mongolia, northeast China: Constraints from geology, isotope geochemistry and geochronology. *Ore Geol. Rev.* **2017**, *81*, 154–172. [CrossRef]

48. Gao, B.Y.; Zhang, L.C.; Jin, X.D.; Li, W.J.; Chen, Z.G.; Zhu, M.T. Geochronology and geochemistry of the Badaguan porphyry Cu–Mo deposit in Derbugan metallogenic belt of the NE China, and their geological significances. *Int. J. Earth Sci.* **2015**, *105*, 507–519. [CrossRef]
49. Wang, Y.H.; Zhao, C.B.; Zhang, F.F.; Liu, J.J.; Wang, J.P.; Peng, R.M.; Liu, B. SIMS zircon U–Pb and molybdenite Re–Os geochronology, Hf isotope, and whole-rock geochemistry of the Wunugetushan porphyry Cu–Mo deposit and granitoids in NE China and their geological significance. *Gondwana Res.* **2015**, *28*, 1228–1245. [CrossRef]
50. Mi, K.F.; Liu, Z.J.; Liu, R.B.; Li, C.F.; Wang, J.P.; Peng, R.M. U–Pb zircon, geochemical and Sr–Nd–Hf isotopic constraints on age and origin of the intrusions from Wunugetushan porphyry deposit, Northeast China: Implication for Triassic–Jurassic Cu–Mo mineralization in Mongolia–Erguna metallogenic belt. *Int. Geol. Rev.* **2018**, *60*, 496–512. [CrossRef]
51. Zhang, X.J.; Zhang, Y.Y.; Liu, K.; He, X.L.; Wang, S.X.; Jia, W.H.; Zhao, Z.N. Zircon U-Pb and Lu-Hf isotopic dating of magmatic rocks in the Wunugetushan porphyry copper-molybdenum deposit, Inner Mongolia. *Rock. Miner. Anal.* **2022**, *41*, 774–788. (In Chinese with English Abstract)
52. Zheng, W.; Yu, X.F. Geochronological and Geochemical Constraints on the Petrogenesis and Geodynamic Setting of the Daheishan Porphyry Mo Deposit, Northeast China. *Resour. Geol.* **2018**, *68*, 1–21. [CrossRef]
53. Zhou, L.L.; Zeng, Q.D.; Liu, J.M.; Friis, H.; Zhang, Z.L.; Duan, X.X. Geochronology of the Xingshan molybdenum deposit, Jilin Province, NE China, and its Hf isotope significance. *J. Asian Earth Sci.* **2013**, *75*, 58–70. [CrossRef]
54. Wang, Z.G.; Wang, K.Y.; Wan, D.; Konare, Y.; Yang, T.N.; Liang, Y.H. Metallogenic age and hydrothermal evolution of the Jidetun Mo deposit in central Jilin Province, northeast China: Evidence from fluid inclusions, isotope systematics, and geochronology. *Ore Geol. Rev.* **2017**, *89*, 731–751. [CrossRef]
55. Zhang, Y.; Sun, J.G.; Xing, S.W.; Zhao, K.Q.; Ma, Y.B.; Zhang, Z.J.; Wang, Y. Ore-forming granites from Jurassic porphyry Mo deposits, east–central Jilin Province, China: Geochemistry, geochronology, and petrogenesis. *Int. Geol. Rev.* **2016**, *58*, 1158–1174. [CrossRef]
56. Hou, X.G.; Sun, D.Y.; Gou, J.; Yang, D.G. The origin of variable-δ18O zircons in Jurassic and Cretaceous Mo-bearing granitoids in the eastern Xing–Meng Orogenic Belt, Northeast China. *Int. Geol. Rev.* **2019**, *61*, 129–149. [CrossRef]
57. Yu, X.F.; Hou, Z.Q.; Qian, Y.; Li, B.L. Ore–forming fluids, stable isotopes and metallogenic epoch of the Fu'anpu molybdenum deposit in Mid–East Jilin Province. *Geol. Explor.* **2012**, *48*, 1151–1162, (In Chinese with English Abstract).
58. Zhang, Y.; Sun, J.G.; Chen, Y.J.; Zhao, K.Q.; Gu, A.L. Re–Os and U–Pb geochronology of porphyry Mo deposits in central Jilin Province: Mo ore-forming stages in northeast China. *Int. Geol. Rev.* **2013**, *55*, 1763–1785. [CrossRef]
59. Zhang, Y.; Xing, S.W.; Sun, J.G.; Zhao, K.Q.; Wang, Y. Re-Os and U-Pb ages of the Sifangdianzi molybdenite deposit in central Jilin Province, NE China. *Acta Geol. Sin. Engl. Ed.* **2014**, *88*, 1045–1046. [CrossRef] [PubMed]
60. Yang, Y.C.; Han, S.J.; Sun, D.Y.; Guo, J.; Zhang, S.J. Geological and geochemical features and geochronology of porphyry molybdenum deposits in the Lesser Xing'an Range-Zhangguangcai Range metallogenic belt. *Acta Petrol. Sin.* **2012**, *28*, 379–390. (In Chinese with English Abstract)
61. Hu, X.L.; Ding, Z.J.; He, M.C.; Yao, S.Z.; Zhu, B.P.; Shen, J.; Chen, B. A porphyry-skarn metallogenic system in the Lesser Xing'an Range, NE China: Implications from U–Pb and Re–Os geochronology and Sr–Nd–Hf isotopes of the Luming Mo and Xulaojiugou Pb–Zn deposits. *J. Asian Earth Sci.* **2014**, *90*, 88–100. [CrossRef]
62. Cheng, G.H.; Wang, R.L.; Zeng, Q.D.; Guo, Y.P.; Duan, X.X.; Wei, J.J.; Zhang, J.S.; Gao, X.H. Zircon U-Pb ages, Hf isotopes of the granitoids and Re-Os ages of the molybdenites in Luming molybdenum ore area, Heilongjiang Province, and its geological significance. *Acta Petrol. Sin.* **2015**, *31*, 2450–2464, (In Chinese with English Abstract).
63. Sun, J.G.; Zhang, Y.; Xing, S.W.; Zhao, K.Q.; Zhang, Z.J.; Bai, L.A.; Ma, Y.B.; Liu, Y.S. Genetic types, ore-forming age and geodynamic setting of endogenic molybdenum deposits in the eastern edge of Xing-Meng orogenic belt. *Acta Petrol. Sin.* **2012**, *28*, 1317–1332. (In Chinese with English Abstract)
64. Hu, X.L.; Yao, S.Z.; Zeng, G.P.; Liu, W.H.; Zhang, Z.J. Multistage magmatism resulting in large-scale mineralizaion: A case from the Huojihe porphyry Mo deposit in NE China. *Lithos* **2019**, *326-327*, 397–414. [CrossRef]
65. Zhang, C.; Li, N. Geochronology and zircon Hf isotope geochemistry of granites in the giant Chalukou Mo deposit, NE China: Implications for tectonic setting. *Ore Geol. Rev.* **2017**, *81*, 780–793. [CrossRef]
66. Liu, Y.F.; Bagas, L.; Jiang, S.H.; Wang, F.X. The Chalukou deposit in the North Great Xing'an Range of China: A protracted porphyry Mo ore-forming system in a long-lived magmatic evolution cycle. *Ore Geol. Rev.* **2017**, *89*, 171–186. [CrossRef]
67. Liu, J.; Mao, J.W.; Wu, G.; Wang, F.; Luo, D.F.; Hu, Y.Q. Zircon U-Pb and molybdenite Re-Os dating of the Chalukou porphyry Mo deposit in the northern Great Xing'an Range, China and its geological significance. *J. Asian Earth Sci.* **2014**, *79*, 696–709. [CrossRef]
68. Li, Z.Z.; Qin, K.Z.; Li, G.M.; Ishihara, S.; Jin, L.Y.; Song, G.X.; Meng, Z.J. Formation of the giant Chalukou porphyry Mo deposit in northern Great Xing'an Range, NE China: Partial melting of the juvenile lower crust in intra-plate extensional environment. *Lithos* **2014**, *202−203*, 138–156. [CrossRef]
69. Hu, X.L.; Yao, S.Z.; He, M.C.; Ding, Z.J.; Cui, Y.B.; Shen, J.; Chen, B.; Zhu, B.P. Geochemistry, U-Pb Geochronology and Hf Isotope Studies of the Daheishan Porphyry Mo Deposit in Heilongjiang Province, NE China. *Resour. Geol.* **2014**, *64*, 102–116. [CrossRef]
70. Deng, C.Z.; Sun, D.Y.; Han, J.S.; Chen, H.Y.; Li, G.H.; Xiao, B.; Li, R.C.; Feng, Y.Z.; Li, C.L.; Lu, S. Late-stage southwards subduction of the Mongol-Okhotsk oceanic slab and implications for porphyry Cu Mo mineralization: Constraints from igneous rocks associated with the Fukeshan deposit, NE China. *Lithos* **2019**, *326–327*, 341–357. [CrossRef]

71. Deng, C.Z.; Li, G.H. The Cu-Mo mineralization of the Late Jurassic Porphyry in the northern Great Xing'an Range: Constraints from zircon U-Pb ages of the ore-causative granites. *Acta Geol. Sin. (Engl. Ed.)* **2019**, *93*, 236–237. [CrossRef]
72. Wang, S.W.; Wang, J.G.; Zhang, D.; Qi, X.J.; Wu, G.G.; Zhao, P.Z.; Yang, Z.F.; Liu, Y.B. Geochronological study on Taipinggou molybdenum deposit in Da Hinggan Mountain. *Acta Petrol. Sin.* **2009**, *25*, 2913–2923. (In Chinese with English Abstract)
73. Wang, Z.L.; Yang, J.Z.; Zhao, K.Q.; Fu, Y.G.; Duan, Z.; Lin, L.J.; Xi, M.J.; Deng, Y.M.; Lu, G.F. Zircon U-Pb age, geochemical characteristics and geological significance of the granite of Jiazishan Mo(Ag) deposit, Inner Mongolia. *Earth Sci.* **2021**. (In Chinese with English Abstract)
74. Sun, H.R.; Huang, Z.L.; Li, W.B.; Leng, C.B.; Ma, D.Y.; Zhang, X.C. Chronology, geochemistry and Sr–Nd isotope studies of Jurassic intrusions in the Diyanqinamu porphyry Mo mine, central Inner Mongolia, China. *J. Asian Earth Sci.* **2014**, *88*, 85–97. [CrossRef]
75. Zhai, D.G.; Liu, J.J.; Wang, J.P.; Yang, Y.Q.; Zhang, H.Y.; Wang, X.L.; Zhang, Q.B.; Wang, G.W.; Liu, Z.J. Zircon U–Pb and molybdenite Re–Os geochronology, and whole-rock geochemistry of the Hashitu molybdenum deposit and host granitoids, Inner Mongolia, NE China. *J. Asian Earth Sci.* **2014**, *79*, 144–160. [CrossRef]
76. Zeng, Q.D.; Guo, W.K.; Chu, S.X.; Duan, X.X. Late Jurassic granitoids in the Xilamulun Mo belt, Northeastern China: Geochronology, geochemistry, and tectonic implications. *Int. Geol. Rev.* **2016**, *58*, 588–602. [CrossRef]
77. Shang, Z.; Zhou, D.; Chen, Y.Q. Genesis, metallogenetic and tectonic significance of the A-type granites in Hashitu Mo deposit, southern Great Hinggan Range, NE China. *Ore Geol. Rev.* **2021**, *138*, 104388. [CrossRef]
78. Wu, H.Y.; Zhang, L.C.; Wan, B.; Chen, Z.G.; Zhang, X.J.; Xiang, P. Geochronological and geochemical constraints on Aolunhua porphyry Mo–Cu deposit, northeast China, and its tectonic significance. *Ore Geol. Rev.* **2011**, *43*, 78–91. [CrossRef]
79. Ma, X.H.; Chen, B.; Lai, Y.; Lu, Y.H. Petrogenesis and mineralization chronology study on the Aolunhua porphyry Mo deposit, Inner Mongolia, and its geological implications. *Acta Petrol. Sin.* **2009**, *25*, 2939–2950. (In Chinese with English Abstract)
80. Shu, Q.H.; Lai, Y.; Zhou, Y.T.; Xu, J.J.; Wu, H.Y. Zircon U–Pb geochronology and Sr–Nd–Pb–Hf isotopic constraints on the timing and origin of Mesozoic granitoids hosting the Mo deposits in northern Xilamulun district, NE China. *Lithos* **2015**, *238*, 64–75. [CrossRef]
81. Zeng, Q.D.; Liu, J.M. Zircon SHRIMP U-Pb dating and geological significance of the granite porphyry from Banlashan porphyry molybdenum deposit in Xilamulun molybdenum metallogenic belt. *J. Jilin Univ. Earth Sci. Ed.* **2010**, *40*, 828–834. (In Chinese with English Abstract)
82. Zeng, Q.D.; Yang, J.H.; Zhang, Z.L.; Liu, J.M.; Duan, X.X. Petrogenesis of the Yangchang Mo-bearing granite in the Xilamulun metallogenic belt, NE China: Geochemistry, zircon U-Pb ages and Sr-Nd-Pb isotopes. *Geol. J.* **2014**, *49*, 1–14. [CrossRef]
83. Ji, H.W.; Niu, Z.L.; Dong, Q.; Zhang, C. Zircon U-Pb and molybdenite Re-Os dating for Hongshanzi uranium-molybdenum deposit in the Inner Mongolia and its geological significance. *Uranium Geol.* **2021**, *37*, 810–822. (In Chinese with English Abstract)
84. Zeng, Q.D.; Liu, J.M.; Qin, F.; Zhang, Z.L. Geochronology of the Xiaodonggou Porphyry Mo Deposit in Northern Margin of North China Craton. *Resour. Geol.* **2010**, *60*, 192–202. [CrossRef]
85. Zeng, Q.D.; Liu, J.M.; Zhang, Z.L.; Chen, W.J.; Zhang, W.Q. Geology and geochronology of the Xilamulun molybdenum metallogenic belt in eastern Inner Mongolia, China. *Int. J. Earth Sci.* **2011**, *100*, 1791–1809. [CrossRef]
86. Li, Y.; Yang, Y.Q.; Hua, L.; Leng, Y.X.; Xin, J. Geochronology and Tectonic Implications of the Nianzigou Granites and Associated Mo Deposit, Inner Mongolia. *Minerals* **2022**, *12*, 791. [CrossRef]
87. Wu, H.Y.; Zhang, L.C.; Pirajno, F.; Xiang, P.; Wan, B.; Chen, Z.G.; Zhang, X.J. The Jiguanshan porphyry Mo deposit in the Xilamulun metallogenic belt, northern margin of the North China Craton, U–Pb geochronology, isotope systematics, geochemistry and fluid inclusion studies: Implications for a genetic model. *Ore Geol. Rev.* **2014**, *56*, 549–565. [CrossRef]
88. Ge, W.C.; Wu, F.Y.; Zhou, C.Y.; Zhang, J.H. The ore-forming time of porphyry Cu and Mo deposits in eastern segment of the Xing-Meng orogen and its geodynamic significance. *Chinese Sci. Bull.* **2007**, *52*, 2407–2417. (In Chinese) [CrossRef]
89. Ge, W.C.; Wu, F.Y.; Zhou, C.Y.; Abdel Rahman, A.A. Emplacement age of the Tahe granite and its constraints on the tectonic nature of the Ergun block in the northern part of the Da Hinggan Range. *Chinese Sci. Bull.* **2005**, *50*, 2097–2105. [CrossRef]
90. Zhou, J.B.; Wang, B.; Wilde, S.A.; Zhao, G.C.; Cao, J.L.; Zheng, C.Q.; Zeng, W.S. Geochemistry and U–Pb zircon dating of the Toudaoqiao blueschists in the Great Xing' an Range, northeast China, and tectonic implications. *J. Asian Earth Sci.* **2015**, *97*, 197–210. [CrossRef]
91. Feng, Z.Q.; Liu, Y.J.; Jin, W.; Li, W.M.; Wen, Q.B. Is there the Paleoproterozoic komatiite related to mantle plume in the Jifeng area, Northern Great Xing'an Range, NE China? *Int. J. Earth Sci.* **2017**, *106*, 2225–2231. [CrossRef]
92. Miao, L.C.; Zhang, F.; Jiao, S.J. Age, protoliths and tectonic implications of the Toudaoqiao blueschist, Inner Mongolia, China. *J. Asian Earth Sci.* **2015**, *105*, 360–373. [CrossRef]
93. IMBGMR. *Regional Geology of Inner Mongolia*; Geological Publishing House: Beijing, China, 1991. (In Chinese)
94. Zhao, S.; Xu, W.L.; Tang, J.; Li, Y.; Guo, P. Timing of formation and tectonic nature of the purportedly Neoproterozoic Jiageda formation of the Erguna Massif, NE China: Constraints from field geology and U-Pb geochronology of detrital and magmatic zircons. *Precambrian Res.* **2016**, *281*, 585–601. [CrossRef]
95. Li, S.C.; Liu, Z.H.; Xu, Z.Y.; Li, G.; Zhang, C. Age and tectonic setting of volcanic rocks of the Tamulangou Formation in the Great Xing'an Range, NE China. *J. Asian Earth Sci.* **2015**, *113*, 471–480. [CrossRef]

96. Ji, Z.; Ge, W.C.; Wang, Q.H.; Yang, H.; Zhao, G.C.; Bi, J.H.; Dong, Y. Petrogenesis of early Cretaceous volcanic rocks of the Manketouebo formation in the Wuchagou region, central great Xing'an range, NE China, and tectonic implications: Geochronological, geochemical, and Hf isotopic evidence. *Int. Geol. Rev.* **2016**, *58*, 556–573. [CrossRef]
97. Badarch, G.; Cunningham, W.D.; Windley, B.F. A new terrane subdivision for Mongolia: Implications for the Phanerozoic crustal growth of Central Asia. *J. Asian Earth Sci.* **2002**, *21*, 87–110. [CrossRef]
98. Li, S.; Wang, T.; Wilde, S.A.; Tong, Y. Evolution, source and tectonic significance of early Mesozoic granitoid magmatism in the central Asian Orogenic Belt (central segment). *Earth-Sci. Rev.* **2013**, *126*, 206–234. [CrossRef]
99. Gou, J.; Sun, D.Y.; Ren, Y.S.; Liu, Y.J.; Zhang, S.Y.; Fu, C.L.; Wang, T.H.; Wu, P.F.; Liu, X.M. Petrogenesis and geodynamic setting of Neoproterozoic and Late Paleozoic magmatism in the Manzhouli-Erguna area of Inner Mongolia, China: Geochronological, geochemical and Hf isotopic evidence. *J. Asian Earth Sci.* **2013**, *67–68*, 114–137. [CrossRef]
100. Gou, J.; Sun, D.Y.; Yang, D.G.; Tang, Z.Y.; Mao, A.Q. Late Paleozoic igneous rocks of the Great Xing'an Range, NE China: The Tayuan example. *Int. Geol. Rev.* **2017**, *21*, 314–340.
101. Tang, J.; Xu, W.L.; Wang, F.; Wang, W.; Xu, M.J.; Zhang, Y.H. Geochronology and geochemistry of Early-Middle Triassic magmatism in the Erguna Massif, NE China: Constraints on the tectonic evolution of the Mongol-Okhotsk suture belt. *Lithos* **2014**, *184–187*, 1–16. [CrossRef]
102. Sun, D.Y.; Gou, J.; Wang, T.H.; Ren, Y.S.; Liu, Y.J.; Gou, H.Y.; Liu, X.M.; Hu, Z.C. Geochronological and geochemical constraints on the Erguna massif basement, NE China subduction history of the Mongol-Okhotsk oceanic crust. *Int. Geol. Rev.* **2013**, *55*, 1801–1816. [CrossRef]
103. Wang, W.; Tang, J.; Xu, W.L.; Wang, F. Geochronology and geochemistry of Early Jurassic volcanic rocks in the Erguna Massif, northeast China: Petrogenesis and implications for the tectonic evolution of the Mongol–Okhotsk suture belt. *Lithos* **2015**, *218–219*, 73–86. [CrossRef]
104. Tang, J.; Xu, W.L.; Wang, F.; Zhao, S.; Wang, W. Early Mesozoic southward subduction history of the Mongol-Okhotsk oceanic plate: Evidence from geochronology and geochemistry of Early Mesozoic intrusive rocks in the Erguna Massif, NE China. *Gondwana Res.* **2016**, *31*, 218–240. [CrossRef]
105. Du, A.D.; Wu, S.Q.; Sun, D.Z.; Wang, S.X.; Qu, W.J.; Markey, R.; Stein, H.; Morgan, J.; Malinovoskiy, D. Preparation and certification of Re-Os dating reference materials, molybdenite HLP and JDC. *Geostand. Geoanalytical Res.* **2004**, *28*, 41–52. [CrossRef]
106. Shirey, S.B.; Walker, R.J. Carius tube digestion for low-blank Rhenium-Osmium analysis. *Anal. Chem.* **1995**, *67*, 2136–2141. [CrossRef]
107. Smoliar, M.I.; Walker, R.J.; Morgan, J.W. Re–Os ages of group IIA, IIIA, IVA, and IVB iron meteorites. *Science* **1996**, *271*, 1099–1102. [CrossRef]
108. Wopenka, B.; Pasteris, J.D.; Freeman, J.J. Analysis of individual fluid inclusions by Fourier transform infrared and Raman microspectroscopy. *Geochim. Cosmochimca Acta* **1990**, *54*, 519–533. [CrossRef]
109. Goldstein, R.H.; Reynolds, T.J. Systematics of fluid inclusions in diagenetic minerals. *Soc. Sediment. Geol.* **1994**, *31*, 199.
110. Collins, P.L.F. Gas hydrates in CO_2-bearing fluid inclusions and the use of freezing data for estimation of salinity. *Econ. Geol.* **1979**, *74*, 1435–1444. [CrossRef]
111. Bodnar, R.J. Revised equation and table for determining the freezing point depression of H_2O-NaCl solutions. *Geochim. Cosmochimca Acta* **1993**, *57*, 683–684. [CrossRef]
112. Bodnar, R.J.; Vityk, M.O. Interpretation of microthermometric data for H_2O-NaCl fluid inclusions. In *Fluid Inclusion in Minerals: Methods and Applications*; Springer: Berlin/Heidelberg, Germany, 1994; pp. 117–130.
113. Brown, P.E.; Lamb, W.M. P-V-T properties of fluids in the system $H_2O \pm CO_2 \pm$ NaCl: New graphic presentations and implications for fluid inclusion studies. *Geochim. Cosmochimca Acta* **1989**, *53*, 1209–1221. [CrossRef]
114. Duan, Z.; Moller, N.; Weare, J.H. An Equation of State for the CH_4-CO_2-H_2O System: I. Pure Systems from 0 to 1000 °C and 0 to 8000 Bar. *Geochim. Cosmochimca Acta* **1992**, *56*, 2605–2617. [CrossRef]
115. Duan, Z.; Moller, N.; Weare, J.H. An Equation of State for the CH_4-CO_2-H_2O System: II. Mixtures from 50 to 1000 °C and 0 to 1000 bars. *Geochim. Cosmochimca Acta* **1992**, *56*, 2619–2631. [CrossRef]
116. Clayton, R.N.; O'Neil, J.R.; Mayeda, T.K. Oxygen isotope exchange between quartz and water. *J. Geophys. Res.* **1972**, *77*, 3057–3067. [CrossRef]
117. Robinson, B.W.; Kusakabe, M. Quantitative preparation of sulfur dioxide, for $^{34}S/^{32}S$ analyses, from sulfides by combustion with cuprous oxide. *Anal. Chem.* **1975**, *47*, 1179–1181. [CrossRef]
118. Todt, W.; Cliff, R.A.; Hanser, A.; Hofmann, A.W. Re-calibration of NBS lead standards using a ^{202}Pb-^{205}Pb double spike. *Terra Abstr.* **1993**, *5*, 396.
119. Roedder, E. Fluid inclusions. *Geol. Soc. Am. Bull.* **1984**, *12*, 1–644.
120. Chi, G.X.; Lu, H.Z. Validation and representation of fluid inclusion microthermometric data using the fluid inclusion assemblage (FIA) concept. *Acta Petrol. Sin.* **2008**, *24*, 1945–1953. (In Chinese with English Abstract)
121. Chi, G.X.; Diamond, L.W.; Lu, H.Z.; Lai, J.Q.; Chu, H.X. Common Problems and Pitfalls in Fluid Inclusion Study: A Review and Discussion. *Minerals* **2021**, *11*, 7. [CrossRef]
122. Davis, D.W.; Lowenstein, T.K.; Spencer, R.J. Melting behavior of fluid inclusions in laboratory-grown halite crystals in the systems NaCl-H_2O, NaCl-KCl-H_2O, NaCl-$MgCl_2$-H_2O, and NaCl-$CaCl_2$-H_2O. *Geochim. Cosmochimca Acta* **1990**, *54*, 591–601. [CrossRef]

123. Rosso, K.M.; Bodnar, R.J. Microthermometric and Raman spectroscopic detection limits of CO_2 in fluid inclusions and the Raman spectroscopic characterization of CO_2. *Geochim. Cosmochimca Acta* **1995**, *59*, 3961–3975. [CrossRef]
124. Audétat, A. A Plea for More Skepticism toward Fluid Inclusions: Part II. Homogenization via Halite Dissolution in Brine Inclusions from Magmatic-Hydrothermal Systems Is Commonly the Result of Postentrapment Modifications. *Econ. Geol.* **2023**, *118*, 43–55. [CrossRef]
125. Zhang, D.H.; Audétat, A. A Plea for More Skepticism toward Fluid Inclusions: Part I. Postentrapment Changes in Fluid Density and Fluid Salinity Are Very Common. *Econ. Geol.* **2023**, *118*, 15–41. [CrossRef]
126. Audétat, A.; Günther, D. Mobility and H_2O loss from fluid inclusions in natural quartz crystals. *Contrib. Miner. Petr.* **1999**, *137*, 1–14. [CrossRef]
127. Taylor, H.P. The application of oxygen and hydrogen isotope studies to problems of hydrothermal alteration and ore deposition. *Econ. Geol.* **1974**, *69*, 843–883. [CrossRef]
128. Zhang, L.G. *Application of Stable Isotopes in Geological Sciences*; Shanxi Science and Technology Press: Xi'an, China, 1985. (In Chinese)
129. Zartman, R.E.; Doe, B.R. Plumbotectonics: The model. *Tectonophysics* **1981**, *75*, 135–162. [CrossRef]
130. Stein, H.J. Dating and tracing the history of ore formation. In *Treatise in Geochemistry*, 2nd ed.; Turekian, K.K., Holland, H.D., Scott, S.D., Eds.; Elsevier-Pergamon: Oxford, UK, 2014; Volume 12.
131. Zhai, D.G.; Williams-Jones, A.E.; Liu, J.J.; Selby, D.; Li, C.; Huang, X.W.; Qi, L.; Guo, D.H. Evaluating the use of the molybdenite Re-Os chronometer in dating gold mineralization: Evidence from the Haigou deposit, northeastern China. *Econ. Geol.* **2019**, *114*, 897–915. [CrossRef]
132. Stein, H.J.; Markey, R.J.; Morgan, J.W.; Hannah, J.L.; Scherstén, A. The remarkable Re-Os chronometer in molybdenite: How and why it works. *Terra Nova* **2001**, *13*, 479–486. [CrossRef]
133. Stein, H.; Schersten, A.; Hannah, J.; Markey, R. Subgrain-scale decoupling of Re and ^{187}Os and assessment of laser ablation ICP-MS spot dating in molybdenite. *Geochim. Cosmochim. Acta* **2003**, *67*, 3673–3686. [CrossRef]
134. Selby, D.; Creaser, R.A. Macroscale NTIMS and microscale LA-MC-ICP-MS Re-Os isotopic analysis of molybdenite: Testing spatial restrictions for reliable Re-Os age determinations, and implications for the decoupling of Re and Os within molybdenite. *Geochim. Cosmochim. Acta* **2004**, *68*, 3897–3908. [CrossRef]
135. Porter, S.J.; Selby, D. Rhenium-Osmium (Re-Os) molybdenite systematics and geochronology of the Cruachan granite skarn mineralization, Etive Complex: Implications for emplacement chronology. *Scott. J. Geol.* **2010**, *46*, 17–21. [CrossRef]
136. Rathkopf, C.; Mazdab, F.; Barton, I.; Barton, M.D. Grain-scale and deposit-scale heterogeneity of Re distribution in molybdenite at the Bagdad porphyry Cu-Mo deposit, Arizona. *J. Geochem. Explor.* **2017**, *178*, 45–54. [CrossRef]
137. Chen, W.T.; Zhou, M.F. Paragenesis, stable isotopes, and molybdenite Re-Os isotope age of the Lala iron-copper deposit, southwest China. *Econ. Geol.* **2012**, *107*, 459–480. [CrossRef]
138. Lawley, C.J.M.; Selby, D. Re-Os geochronology of quartz-enclosed ultrafine molybdenite: Implications for ore geochronology. *Econ. Geol.* **2012**, *107*, 1499–1505. [CrossRef]
139. Li, J.; Xu, L.L.; Bi, X.W.; Tang, Y.Y.; Sheng, X.Y.; Yu, H.J.; Liu, G.; Ma, R. New titanite U–Pb and molybdenite Re–Os ages for a hydrothermal vein-type Cu deposit in the Lanping Basin, Yunnan, SW China: Constraints on regional metallogeny and implications for exploration. *Miner. Depos.* **2021**, *56*, 441–456. [CrossRef]
140. Girei, M.B.; Li, H.; Vincent, V.I.; Algeo, T.J.; Elatikpo, S.M.; Bute, S.I.; Ahmed, H.A.; Amuda, A.K. Genesis and timing of Mo mineralization in the Mada Ring Complex, north-central Nigeria: Insights from whole-rock geochemistry, Nd-Sr isotopes, zircon U-Pb-Hf isotopes, and molybdenite Re-Os systematics. *Miner. Depos.* **2022**, *57*, 601–620. [CrossRef]
141. Qi, H.W.; Hu, R.Z.; Wang, X.F.; Qu, W.J.; Bi, X.W.; Peng, J.T. Molybdenite Re–Os and muscovite $^{40}Ar/^{39}Ar$ dating of quartz vein-type W–Sn polymetallic deposits in Northern Guangdong, South China. *Miner. Depos.* **2012**, *47*, 607–622. [CrossRef]
142. Huang, X.D.; Huang, D.; Lu, J.J.; Zhang, R.Q.; Ma, D.S.; Jiang, Y.H.; Chen, H.W.; Liu, J.X. Neoproterozoic tungsten mineralization: Geology, chronology, and genesis of the Huashandong W deposit in northwestern Jiangxi, South China. *Miner. Depos.* **2023**, *58*, 771–796. [CrossRef]
143. Wu, H.Y.; Zhang, L.C.; Wan, B.; Chen, Z.G.; Xiang, P.; Pirajno, F.; Du, A.D.; Qu, W.J. Re–Os and $^{40}Ar/^{39}Ar$ ages of the Jiguanshan porphyry Mo deposit, Xilamulun metallogenic belt, NE China, and constraints on mineralization events. *Miner. Depos.* **2010**, *46*, 171–185. [CrossRef]
144. Zhang, P.; Zhao, Y.; Kou, L.L.; Yang, H.Z.; Yang, F.C. Zircon U–Pb and molybdenite Re–Os geochronology of copper–molybdenum deposits in southeast Liaoning Province, China. *Int. Geol. Rev.* **2016**, *58*, 1481–1491. [CrossRef]
145. Buret, Y.; von Quadt, A.; Heinrich, C.; Selby, D.; Walle, M.; Peytcheva, I. From a long-lived upper-crustal magma chamber to rapid porphyry copper-emplacement: Reading the geochemistry of zircon crystals at Bajo de la Alumbrera (NW Argentina). *Earth Planet. Sci. Lett.* **2016**, *450*, 120–131. [CrossRef]
146. Cernuschi, F.; Dilles, J.H.; Grocke, S.B.; Valley, J.W.; Kitajima, K.; Tepley, F.J. Rapid formation of porphyry copper deposits evidenced by diffusion of oxygen and titanium in quartz. *Geology* **2018**, *46*, 611–614. [CrossRef]
147. Li, Y.; Li, X.H.; Selby, D.; Li, J.W. Pulsed magmatic fluid release for the formation of porphyry deposits: Tracing fluid evolution in absolute time from the Tibetan Qulong Cu-Mo deposit. *Geology* **2018**, *46*, 7–10. [CrossRef]
148. Li, Y.; Allen, M.B.; Li, X.H. Millennial pulses of ore formation and an extra-high Tibetan Plateau. *Geology* **2022**, *50*, 665–669. [CrossRef]

149. Li, Y.; Zhang, R.Q.; He, S.; Chiaradia, M.; Li, X.H. Pulsed exsolution of magmatic ore-forming fluids in tin-tungsten systems: A SIMS cassiterite oxygen isotope record. *Miner. Depos.* **2022**, *57*, 343–352. [CrossRef]
150. Von Quadt, A.; Erni, M.; Martinek, K.; Moll, M.; Peytcheva, I.; Heinrich, C.A. Zircon crystallization and the lifetimes of ore-forming magmatic-hydrothermal systems. *Geology* **2011**, *39*, 731–734. [CrossRef]
151. Chiaradia, M.; Schaltegger, U.; Spikings, R.; Wotzlaw, J.F.; Ovtcharova, M. How accurately can we date the duration of magmatic-hydrothermal events in porphyry systems?—An invited paper. *Econ. Geol.* **2013**, *108*, 565–584. [CrossRef]
152. Cathles, L.M.; Erendi, A.H.J.; Barrie, T. How long can a hydrothermal system be sustained by a single intrusive event? *Econ. Geol.* **1997**, *92*, 766–771. [CrossRef]
153. Zhang, L.C.; Wu, H.Y.; Wan, B.; Chen, Z.G. Ages and geodynamic settings of Xilamulun Mo–Cu metallogenic belt in the northern part of the North China Craton. *Gondwana Res.* **2009**, *16*, 243–254. [CrossRef]
154. Li, C.F.; Liu, Z.J.; Mi, K.F.; Wang, J.P.; Liu, R.B.; Zou, J.Y.; Wang, Z.B. Metallogenic age and ore genesis of the Badaguan porphyry copper-molybdenum deposit in Inner Mongolia. *Geol. China* **2014**, *41*, 1253–1269. (In Chinese with English Abstract)
155. Han, C.M.; Xiao, W.J.; Windley, B.F.; Zhao, G.C.; Su, B.X.; Ao, S.J.; Zhang, J.E.; Zhang, Z.Y.; Wan, B.; Cui, B.; et al. Re-Os Age of Molybdenite from the Daheishan Mo Deposit in the Eastern Central Asian Orogenic Belt, NE China. *Resour. Geol.* **2014**, *64*, 379–386. [CrossRef]
156. Wang, C.H.; Song, Q.H.; Wang, D.H.; Li, L.X.; Yu, C.; Wang, Z.G.; Qu, W.J.; Du, A.D.; Ying, L.J. Re-Os isotopic dating of molybdenite from the Daheishan molybdenum deposit of Jilin Province and its geological significance. *Rock. Miner. Anal.* **2009**, *28*, 269–273. (In Chinese with English Abstract)
157. Wang, Z.G.; Wang, K.Y.; Konare, Y.; Yang, T.N.; Liang, Y.H. Metallogenesis and hydrothermal evolution of Middle Jurassic porphyry Mo deposits in the southern Zhangguangcai Range, NE China: Evidence from fluid inclusions, H–O–S isotopes, and Re–Os geochronology. *Int. Geol. Rev.* **2017**, *59*, 1391–1412. [CrossRef]
158. Li, L.X.; Song, Q.H.; Wang, D.H.; Wang, C.H.; Qu, W.J.; Wang, Z.G.; Bi, S.Y.; Yu, C. Re-Os isotopic dating of molybdenite from the Fu'anpu molybdenum deposit of Jilin Province and discussion on its metallogenesis. *Rock. Miner. Anal.* **2009**, *28*, 283–287. (In Chinese with English Abstract)
159. Zhang, Y. Research on Characteristics of Geology, Geochemistry and Metallogenic Mechanism of the Jurassic Molybdenum Deposits in the Mid-East Area of Jilin. Ph.D. Thesis, Jilin University, Changchun, China, 2013. (In Chinese with English Abstract)
160. Zhang, Y.; Sun, J.G.; Xing, S.W.; Zhao, K.Q.; Ma, Y.B. Geochronology and metallogenesis of porphyry Mo deposits in east-central Jilin province, China: Constraints from molybdenite Re–Os isotope systematics. *Ore Geol. Rev.* **2015**, *71*, 363–372. [CrossRef]
161. Ju, N.; Ren, Y.S.; Wang, C.; Wang, H.; Zhao, H.L.; Qu, W.J. Ore genesis and molybdenite Re-Os dating of Dashihe molybdenum deposit in Dunhua, Jilin. *Glob. Geol.* **2012**, *31*, 68–76. (In Chinese with English Abstract)
162. Wang, H.; Ren, Y.S.; Zhao, H.L.; Ju, N.; Qu, W.J. Re-Os dating of molybdenite from the Liushengdian molybdenum deposit in Antun area of Jilin Province and its geological significance. *Acta Geosci. Sin.* **2011**, *32*, 707–715. (In Chinese with English Abstract)
163. Chen, Z.G.; Zhang, L.C.; Wan, B.; Wu, H.Y.; Cleven, N. Geochronology and geochemistry of the Wunugetushan porphyry Cu-Mo deposit in NE China, and their geological significance. *Ore Geol. Rev.* **2011**, *43*, 92–105. [CrossRef]
164. Zhang, L.C.; Gao, B.Y.; Li, W.J.; Chen, Z.G.; Sakyi, P.A.; Jin, X.D. Early Mesozoic tectono-magmatic activity and mineralization in Northeast China: Evidence from Re–Os to U–Pb studies of the Taipingchuan porphyry Cu–Mo deposit in the Derbugan metallogenic belt. *Int. Geol. Rev.* **2014**, *56*, 1837–1851. [CrossRef]
165. Zhai, D.G.; Liu, J.J.; Wang, J.P.; Peng, R.M.; Wang, S.G.; Li, Y.N.; Chang, Z.Y. Re-Os isotopic chronology of molybdenite from the Taipinggou porphyry-type molybdenum deposit in Inner Mongolia and its geological significance. *Geoscience* **2009**, *23*, 262–268. (In Chinese with English Abstract)
166. Wang, Z.L.; Zhao, K.Q.; Yang, J.Z.; Fu, Y.G.; Xi, M.J.; Li, C.; Deng, Y.M.; Lu, G.F. The molybdenite Re-Os dating of the Jiazishan Mo (Ag) deposit, Inner Mongolia and its geological significance. *Acta Geol. Sin.* **2021**, *95*, 3768–3777. (In Chinese with English Abstract)
167. Leng, C.B.; Zhang, X.C.; Huang, Z.L.; Huang, Q.Y.; Wang, S.X.; Ma, D.Y.; Luo, T.Y.; Li, C.; Li, W.B. Geology, Re-Os ages, sulfur and lead isotopes of the Diyanqinamu porphyry Mo deposit, Inner Mongolia, NE China. *Econ. Geol.* **2015**, *110*, 557–574. [CrossRef]
168. Shu, Q.H.; Jiang, L.; Lai, Y.; Lu, Y.H. Geochronology and fluid inclusion study of the Aolunhua porphyry Cu-Mo deposit in Arhorqin area, Inner Mongolia. *Acta Petrol. Sin.* **2009**, *25*, 2601–2614. (In Chinese with English Abstract)
169. Zeng, Q.D.; Liu, J.M.; Zhang, Z.L. Re-Os geochronology of porphyry molybdenum deposit in south segment of Da Hinggan Mountains, Northeast China. *J. Earth Sci.-China* **2010**, *21*, 392–401. [CrossRef]
170. Nie, F.J.; Zhang, W.Y.; Du, A.D.; Jiang, S.H.; Liu, Y. Re-Os isotopic dating on molybdenite separates from the Xiaodonggou porphyry Mo deposit, Hexigten Qi, Inner Mongolia. *Acta Geol. Sin.* **2007**, *81*, 898–905. (In Chinese with English Abstract)
171. Chen, W.J.; Liu, J.M.; Liu, H.T.; Sun, X.G.; Zhang, R.B.; Zhang, Z.L.; Qin, F. Geochronology and fluid inclusion study of the Jiguanshan porphyry Mo deposit, Inner Mongolia. *Acta Petrol. Sin.* **2010**, *26*, 1423–1436. (In Chinese with English Abstract)
172. Hedenquist, J.W.; Lowenstern, J.B. The role of magmas in the formation of hydrothermal ore deposits. *Nature* **1994**, *370*, 519–527. [CrossRef]
173. Zhang, L.G. *Theory of Diagenesis and Metallogenesis and Ore Prospecting*; Beijing University of Technology Press: Beijing, China, 1989. (In Chinese)
174. Barnes, H.L. *Geochemistry of Hydrothermal Ore Deposits*; Wiley: New York, NY, USA, 1979; p. 781.

175. Kodera, P.; Lexa, J.; Rankin, A.H.; Fallick, A.E. Epithermal gold veins in a caldera setting: Banska Hodrusa, Slovakia. *Miner. Depos.* **2005**, *39*, 921–943. [CrossRef]
176. Rye, R.O.; Ohmoto, H. Sufur and carbon isotopes and ore genesis: A review. *Econ. Geol.* **1974**, *69*, 826–842. [CrossRef]
177. Seal, R.R. Sulfur Isotope Geochemistry of Sulfide Minerals. *Rev. Mineral. Geochem.* **2006**, *61*, 633–677. [CrossRef]
178. Zhu, D.C.; Zhao, Z.D.; Niu, Y.L.; Mo, X.X.; Chung, S.L.; Hou, Z.Q.; Wang, L.Q.; Wu, F.Y. The Lhasa Terrane: Record of a microcontinent and its histories of drift and growth. *Earth Planet. Sci. Lett.* **2011**, *301*, 241–255. [CrossRef]
179. Ohmoto, H.; Rye, R.O. Isotope of sulfur and carbon. In *Geochemistry of Hydrothermal Ore Deposits*, 2nd ed.; Barnes, H.L., Ed.; John Wiley and Sons: New York, NY, USA, 1979.
180. Ohmoto, H. Stable isotope geochemistry of ore deposits. *Rev. Mineral. Geochem.* **1986**, *16*, 491–559.
181. Pass, H.E.; Cooke, D.R.; Davidson, G.; Maas, R.; Dipple, G.; Rees, C.; Ferreira, L.; Taylor, C.; Deyell, C.L. Isotope geochemistry of the Northeast Zone, Mount Polley Alkalic Cu-Au-Ag Porphyry Deposit, British Columbia: A case for carbonate assimilation. *Econ. Geol.* **2014**, *109*, 859–890. [CrossRef]
182. Guo, F.; Fan, W.M.; Gao, X.F.; Li, C.W.; Miao, L.C.; Zhao, L.; Li, H.X. Sr-Nd-Pb isotope mapping of Mesozoic igneous rocks in NE China: Constraints on tectonic framework and Phanerozoic crustal growth. *Lithos* **2010**, *120*, 563–578. [CrossRef]
183. Doe, B.R.; Zartman, R.E. Plumbotectonics I, the Phanerozoic. In *Geochemistry of Hydrothermal Ore Deposits*, 2nd ed.; Barnes, H.L., Ed.; John Wiley & Sons Inc.: New York, NY, USA, 1979; pp. 22–70.
184. Cooke, D.R.; Hollings, P.; Walsh, J.L. Giant porphyry deposits: Characteristics, distribution, and tectonic controls. *Econ. Geol.* **2005**, *100*, 801–818. [CrossRef]
185. Xiong, Y.Q.; Shao, Y.J.; Mao, J.W.; Wu, S.C.; Zhou, H.D.; Zheng, M.H. The polymetallic magmatic-hydrothermal Xiangdong and Dalong systems in the W–Sn–Cu–Pb–Zn–Ag Dengfuxian orefield, SE China: Constraints from geology, fluid inclusions, H–O–S–Pb isotopes, and sphalerite Rb–Sr geochronology. *Miner. Depos.* **2019**, *54*, 1101–1124. [CrossRef]
186. Cline, J.S.; Bodnar, R.J. Direct evolution of brine from crystallizing silicic melt at the Questa, New Mexico, Molybdenum deposit. *Econ. Geol.* **1994**, *89*, 1780–1802. [CrossRef]
187. Ulrich, T.; Günther, D.; Heinrich, C.A. The evolution of a porphyry Cu-Au deposit, based on LA-ICP-MS analysis of fluid inclusions: Bajo de la Alumbrera. Argentina. *Econ. Geol.* **2002**, *97*, 1889–1920. [CrossRef]
188. Goldfarb, R.J.; Ayuso, R.A.; Miller, M.L.; Ebert, S.; Marsh, E.E.; Petsel, S.A. The late Cretaceous Donlin Creek gold deposit, southwestern Alaska: Controls on epizonal ore formation. *Econ. Geol.* **2004**, *99*, 643–671. [CrossRef]
189. Shepherd, T.J.; Rankin, A.H.; Alderton, D.H.M. *A Practical Guide to Fluid Inclusion Studies*; Chapman & Hall: London, UK, 1985; p. 239.
190. Wei, W.F.; Hu, R.Z.; Bi, X.W.; Peng, J.T.; Su, W.C.; Song, S.Q.; Shi, S.H. Infrared microthermometric and stable isotopic study of fluid inclusions in wolframite at the Xihuashan tungsten deposit, Jiangxi province, China. *Miner. Depos.* **2012**, *47*, 589–605. [CrossRef]
191. Kwak, T.A.P. Fluid inclusions in skarns (carbonate replacement deposits). *J. Metamorph. Geol.* **1986**, *4*, 363–384. [CrossRef]
192. Lu, H.Z.; Fan, H.R.; Ni, P.; Ou, G.X.; Shen, K.; Zhang, W.H. *Fluid Inclusions*; Science Press: Beijing, China, 2004; p. 444. (In Chinese)
193. Roedder, E.; Bodnar, R.J. Geologic pressure determinations from fluid inclusion studies. *Annu. Rev. Earth Planet. Sci.* **1980**, *8*, 263–301. [CrossRef]
194. Chen, Y.J.; Wang, P.; Li, N.; Yang, Y.F.; Pirajno, F. The collision-type porphyry Mo deposits in Dabie Shan. China. *Ore Geol. Rev.* **2017**, *81*, 405–430. [CrossRef]
195. Hagemann, S.; Lüders, V. P-T-X conditions of hydrothermal fluids and precipitation mechanism of stibnite-gold mineralization at the Wiluna lode-gold deposits, Western Australia: Conventional and infrared microthermometric constraints. *Miner. Depos.* **2003**, *38*, 936–952. [CrossRef]
196. Lowenstern, J.B. Carbon dioxide in magmas and implications for hydrothermal systems. *Miner. Depos.* **2001**, *36*, 490–502. [CrossRef]
197. Soloviev, S.G.; Kryazhev, S.G.; Dvurechenskaya, S.S. Geology, mineralization, and fluid inclusion characteristics of the Meliksu reduced tungsten skarn deposit, Alai Tien Shan, Kyrgyzstan: Insights into conditions of formation and regional links to gold mineralization. *Ore Geol. Rev.* **2019**, *111*, 103003. [CrossRef]
198. Xue, L.W.; Wang, G.W.; Tang, L.; Cao, Y.; Du, J.G.; Du, Y.S.; Luo, L.P.; Cheng, H.T. Genesis and hydrothermal evolution of the Zhazigou skarn W (Mo) deposit, East Qinling, China: Constraints from fluid inclusions and H-O-S-Pb isotopes. *Ore Geol. Rev.* **2021**, *138*, 104374. [CrossRef]
199. Pirajno, F. *Hydrothermal Mineral Deposits*; Springer: Dordrecht, The Netherlands, 2009; p. 1250.
200. Putnis, A. Mineral replacement reactions. *Rev. Mineral. Geochem.* **2009**, *70*, 87–124. [CrossRef]
201. Chen, P.W.; Zeng, Q.D.; Zhou, T.C.; Wang, Y.B.; Yu, B.; Chen, J.Q. Evolution of fluids in the Dasuji porphyry Mo deposit on the northern margin of the North China Craton: Constraints from Microthermometric and LA-ICP-MS analyses of fluid inclusions. *Ore Geol. Rev.* **2019**, *104*, 26–45. [CrossRef]
202. Guillemette, N.; Williams-Jones, A.E. Genesis of the Sb-W-Au deposits at Ixtahuacan, Guatemala: Evidence from fluid inclusions and stable isotopes. *Miner. Depos.* **1993**, *28*, 167–180. [CrossRef]
203. Chen, P.W.; Zeng, Q.D.; Zhou, L.L.; Zhou, T.C. Fluid inclusion evidence for hydrothermal evolution of the Sadaigoumen porphyry Mo deposit on the northern margin of the north China Craton. *Ore Geol. Rev.* **2021**, *134*, 104145. [CrossRef]

204. Zhao, X.Y.; Zhong, H.; Hu, R.Z.; Mao, W.; Bai, Z.J.; Lan, T.G.; Xue, K. Evolution of Multistage Hydrothermal Fluids in the Luoboling Porphyry Cu-Mo Deposit, Zijinshan Ore Field, Fujian Province, China: Insights from LA-ICP-MS Analyses of Fluid Inclusions. *Econ. Geol.* **2021**, *116*, 581–606. [CrossRef]
205. Xie, W.; Zeng, Q.D.; Huang, L.L.; Zhou, L.L.; Fan, H.R.; Wu, J.J.; Wang, R.L.; Zhu, H.P. Composition and evolution of ore-forming fluids in the Sansheng porphyry W-Mo deposit, Inner Mongolia, NE China: Evidence from LA-ICP-MS analysis of fluid inclusions. *Ore Geol. Rev.* **2023**, *158*, 105481. [CrossRef]
206. Hu, D.L.; Jiang, S.Y.; Xiong, S.F.; Dong, J.X.; Wang, K.X. Genesis of the Maogongdong deposit in the Dahutang W-Cu-(Mo) ore field of northern Jiangxi Province, South China: Constraints from mineralogy, fluid inclusions, and H-O-C-S isotopes. *Miner. Depos.* **2022**, *57*, 1449–1468. [CrossRef]
207. Baker, T.; Van Achterberg, E.; Ryan, C.G.; Lang, J.R. Composition and evolution of ore fluids in a magmatic-hydrothermal skarn deposit. *Geology* **2004**, *32*, 117–120. [CrossRef]
208. Seward, T.M.; Barnes, H.L. *Geochemistry of Hydrothermal Ore Deposits*; John Wiley & Sons: New York, NY, USA, 1997.
209. Shu, Q.H.; Chang, Z.S.; Hammerli, J.; Lai, Y.; Huizenga, J.M. Composition and Evolution of Fluids Forming the Baiyinnuo'er Zn-Pb Skarn Deposit, Northeastern China: Insights from Laser Ablation ICP-MS Study of Fluid Inclusions. *Econ. Geol.* **2017**, *112*, 1441–1460. [CrossRef]
210. Shu, Q.H.; Chang, Z.S.; Mavrogenes, J. Fluid compositions reveal fluid nature, metal deposition mechanisms, and mineralization potential: An example at the Haobugao Zn-Pb skarn, China. *Geology* **2021**, *49*, 473–477. [CrossRef]
211. Stoffell, B.; Appold, M.S.; Wilkinson, J.J.; McClean, N.A.; Jeffries, T.E. Geochemistry and evolution of mississippi valley-type mineralizing brines from the tri-state and northern arkansas districts determined by LA-ICP-MS microanalysis of fluid inclusions. *Econ. Geol.* **2008**, *103*, 1411–1435. [CrossRef]
212. Hedenquist, J.W.; Arribas, A.; Reynolds, T.J. Evolution of an intrusion centered hydrothermal system: Far Southest-Lepanto porphyry and epithermal Cu-Au deposit. *Econ. Geol.* **1998**, *93*, 373–404. [CrossRef]
213. Fritzell, E.H.; Bull, A.L.; Shephard, G.E. Closure of the Mongol-Okhotsk Ocean: Insights from seismic tomography and numerical modelling. *Earth Planet. Sci. Lett.* **2016**, *445*, 1–12. [CrossRef]
214. Wang, Z.H.; Ge, W.C.; Yang, H.; Zhang, Y.L.; Bi, J.H.; Tian, D.X.; Xu, W.L. Middle Jurassic oceanic island igneous rocks of the Raohe accretionary complex, northeastern China: Petrogenesis and tectonic implications. *J. Asian Earth Sci.* **2015**, *111*, 120–137. [CrossRef]
215. Gou, J.; Sun, D.Y.; Qin, Z. Late Jurassic–Early Cretaceous tectonic evolution of the Great Xing'an Range: Geochronological and geochemical evidence from granitoids and volcanic rocks in the Erguna Block, NE China. *Int. Geol. Rev.* **2019**, *61*, 1842–1863. [CrossRef]
216. Feng, Z.; Sun, D.Y.; Yue, Y.; Mao, A.Q.; Tian, L.; Sun, C.; Gou, J. Petrogenesis of highly differentiated I-type volcanic rocks: Reinjection of high-temperature magma-An example from Suolun silicic volcanic rocks, central Great Xing'an Range, China. *Geol. J.* **2020**, *55*, 6677–6695. [CrossRef]
217. Zhang, G.L.; Xie, W.; Wen, S.Q.; Gong, E.P.; Guo, R.R.; Tang, T.Q. Petrogenesis and tectonic implications of Late Mesozoic volcanic rocks in the northern and central Great Xing'an Range, NE China: Constraints from geochronology and geochemistry. *Geol. J.* **2020**, *55*, 8282–8308. [CrossRef]
218. Deng, C.Z.; Sun, D.Y.; Han, J.S.; Li, G.H.; Feng, Y.Z.; Xiao, B.; Li, R.C.; Shi, H.L.; Xu, G.Z.; Yang, D.G. Ages and petrogenesis of the Late Mesozoic igneous rocks associated with the Xiaokele porphyry Cu–Mo deposit, NE China and their geodynamic implications. *Ore Geol. Rev.* **2019**, *107*, 417–433. [CrossRef]
219. Rudnick, R.L.; Gao, S. 3.01—Composition of the continental crust. *Treatise Geochem.* **2003**, *3*, 1–64.

minerals

MDPI

Article

Continental Arc Flare-Ups and Crustal Thickening Events in NE China: Insights from Detrital Zircon U-Pb Dating and Trace Elements from the Heilongjiang Complex

Yanchen Pan [1], Mengyu Xu [2], Kai Liu [3,*] and Meng Wang [4]

1 School of Water Resources and Environment, China University of Geosciences, Beijing 100083, China
2 Institute of Geology and Geophysics, Chinese Academy of Sciences, Beijing 100029, China
3 School of Earth Sciences and Resources, China University of Geosciences, Beijing 100083, China
4 Key Laboratory of Western Mineral Resources and Geological Engineering, MOE, School of Earth Science and Resources, Chang'an University, Xi'an 710054, China; wangmeng@chd.edu.cn
* Correspondence: k.liu@cugb.edu.cn

Abstract: Continental arc is characterized by alternant magmatic flare-ups and lulls. From the Permian to the Middle Jurassic period, two flare-ups with a lull developed in NE China, but the tectonic controls that caused the flare-ups remain unclear. Sedimentary rocks of the Heilongjiang Complex were derived from these magmatic rocks; thus, we employed detrital zircon U-Pb dating and trace elements analyses to unravel the regional tectono-magmatic evolution. Eu anomaly, $(Dy/Yb)_N$ and Th/U ratios of the detrital zircons and Sr/Y and $(La/Yb)_N$ of the regional granitoids together indicate the occurrence of two episodes of crustal thickening during the two flare-ups, accompanied with a westward migration of magmatism. We propose that the Permian flare-up was caused by the shallowing subduction from the east, which thickened the upper plate and enhanced the deep crustal melting. During the Middle Triassic period, the mantle wedge was expelled by the flat slab and thickened crust, leading to the magmatic lull. However, the westward subduction of the back-arc oceanic plate occurred before the lull, gradually producing the Jurassic magmatic flare-up and crustal thickening. Closure of the back-arc ocean caused by the outboard Paleo-Pacific oceanic plate subduction was important in the formation of the episodic magmatic flare-ups and crustal thickening in NE China.

Keywords: continental arc; arc magma tempo; detrital zircon; accretionary orogen

Citation: Pan, Y.; Xu, M.; Liu, K.; Wang, M. Continental Arc Flare-Ups and Crustal Thickening Events in NE China: Insights from Detrital Zircon U-Pb Dating and Trace Elements from the Heilongjiang Complex. *Minerals* **2023**, *13*, 1121. https://doi.org/10.3390/min13091121

Academic Editors: Jiafu Chen, Nan Ju, Zhonghai Zhao and Petrus J Le Roux

Received: 17 July 2023
Revised: 19 August 2023
Accepted: 23 August 2023
Published: 25 August 2023

1. Introduction

Continental arc shows periodicity. Although a continental arc have persisted for hundreds of millions of years, its flux significantly varied and showed alternate magmatic flare-ups (70–90 $km^3\ km^{-1}\ Myr^{-1}$) and lulls (<20 $km^3\ km^{-1}\ Myr^{-1}$) [1]. Each flare-up or lull lasted for several tens of millions of years, consisting of the "continental arc tempo" [2]. However, which tectonic process triggered this tempo remains unclear and this limits our understanding of the continental arc magmatism. Because arc magmatism and differentiation were the key to generate continental crust, unravelling the arc tempo will directly help to understand how the continent was formed and evolved [3].

The orogenic cyclicity, or periodicity was defined along the eastern Pacific margin, the American Cordillera [3]. Taking Sierra Nevada as an example, although continental arc magmatism was first initiated in the early Mesozoic era, there were only two main flare-ups, occurring in the Middle to Late Jurassic and the mid-Cretaceous periods, respectively [4]. It was noted that the two flare-ups were closely associated with strong regional compression, crustal thickening, and landward migration of arc magmatism. Based on the data from both the North and South American Cordillera, Decellas et al. (2009) suggested that these aforementioned associations were widespread in the Andes, Sierra Nevada, Canada Coastal

Mountains, and other parts of the American Cordilleran belt; after the landward migration of the arc magmas, they waned and migrated back toward the trench before the next round of flare-up, crustal thickening, and landward migration [3]. Ducea (2001) pointed out a key factor in this orogenic cyclicity, that is, each magmatic flare-up corresponded to more evolved Nd isotopic features (isotope "pull-down") [5]. Therefore, it is indicated that increasing magmatic flux was somehow related to the involvement of enriched, mature continental materials.

Although many tectonic models were proposed, controversies exist in many aspects [1]. First, some studies suggested that the landward arc migration was caused by slab angle shallowing [6], but some others believed that crustal thickening with magmatic addition and regional compression could migrate the arc without slab shallowing [7]. Second, why did the magmas show enriched isotopes during flare-ups? An early explanation was that the isotopes formed due to the "under-thrusting" of the rear-arc continental crust deep into the arc source, which supplied fertile and mature materials to the magmas [3,5]. Recent research proposed that the arc magmas were more enriched simply because they moved to the ancient continental interior and the addition of rear-arc materials was not necessary [8]. Some studies even suggested that it was the mantle input that triggered the magmatic flare-ups based on the juvenile zircon Hf isotopes [9]. Third, which tectonic mechanism drive the cyclicity? Many studies indicate that the crustal thickening transformed the lower continental arc crust to eclogites, and they finally foundered into the mantle, thereby thinning the crust [3,4]. Afterwards, regional compression thickened the crust again and the orogen evolved to the next cycle. According to this viewpoint, cyclicity is self-driven, influenced only by the internal factors of the orogen matter. However, other studies have suggested that external factors dominate the orogeny, such as the change in slab angle, subduction rate, and so on [1,6].

During the Early Permian to Middle Jurassic, N-S trending calc-alkaline magmatic belts developed in NE China (Figure 1), representing long-lived continental arc activities [10–13]. These arc magmas also show typical landward migration trends [11,14], similar to the situations in the American Cordillera, providing us with an opportunity to test the contested views regarding the orogenic cyclicity. However, some important geological parameters, such as crustal thickness, have been largely changed due to the surface processes for >150 Myr; most geological records were also superposed or changed due to the tectonic activities, especially the multiple episodes of regional compression, shearing, and extension.

The Heilongjiang Complex is a Mesozoic accretionary complex in NE China and represents the remnants of the Mudanjiang Ocean [15], a fossil oceanic basin between the Jiamusi and Songliao blocks, with sedimentary materials deposits mainly from the nearby continental arc magmatic rocks [13]. Recently, trace elements with coupled U-Pb ages of detrital zircons become a useful tool to indicate the tectonic evolution of continental arcs [16]. This is because many volcanic rocks and shallow plutonic rocks have been eroded, and many parts of a now-exhumed arc are covered with sediments or vegetations, making them inaccessible for sampling. But zircons could survive both surface erosion and several rounds of recycling low- to middle-grade (hydro-)thermal alteration.

In this study, we employed U-Pb dating and trace elements analyses of detrital zircons from the Heilongjiang Complex, integrated with published data from regional magmatic rocks and detrital zircons, in order to explore the evolution of continental arcs in NE China. We propose a tectonic model to explain the continental arc tempos in NE China and compare the geodynamic mechanism with the American Cordillera.

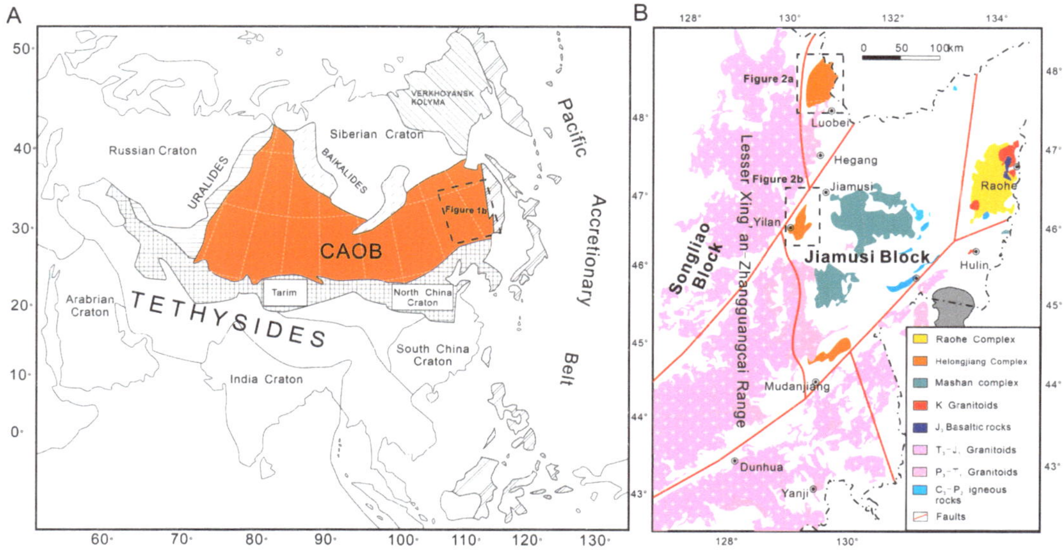

Figure 1. (**A**) Tectonic position of the study area (after [17]); (**B**) simplified geological map of the eastern Heilongjiang Province (after [14]). Our samples were collected from the Heilongjiang Complex, near Luobei and Yilan.

2. Geological Setting and Sampling

The study area is located in NE China, east of the Songliao Basin, tectonically including the eastern Songliao and Jiamusi blocks (Figure 1B).

The Songliao Block include the Lesser Xing'an-Zhangguangcai Range along the eastern margin, and most other parts are covered by the post-Cretaceous strata in the Songliao Basin. Previous studies suggested that the basement of the Songliao Basin is juvenile crust and mainly contains the Paleozoic to Mesozoic meta-volcanic, meta-sedimentary rocks, and granitic gneisses. However, recent studies reported ages of ca. 1.8 Ga from meta-gabbros and meta-granitic breccias from the drill holes [18]. From the Middle Jurassic, the Songliao Basin experienced initial rifting, post-rifting subsidence, and basin inversions [19,20]. The most important volcanism in the basin occurred in the Early Cretaceous (c. 135–100 Ma) [12,21]. The Lesser Xing'an-Zhangguangcai Range is located along the eastern Songliao Block (Figure 1B), and mainly consisted of granitoids with various ages [10,22], including the Early Paleozoic (c. 510–470 Ma), Early Permian (280–290 Ma), Late Permian to Early Triassic (c. 260–240 Ma), Late Triassic to Early Jurassic (220–170 Ma), and middle Cretaceous (c. 100 Ma). The most widespread magmatic episode occurred in the Late Triassic to Early Jurassic [22].

The Bureya-Jiamusi-Khanka Block is cut by the north branches of the Tan-Lu Fault, namely the Dunhua-Mishan and Jiamusi-Yilan faults (Figure 1B); the Jiamusi Block is the central part bounded by the two faults. (Figure 1B). The oldest rocks in the Jiamusi Block have ages ranging from c. 891 to 898 Ma [23]; an almost similar age (c. 933 Ma) is reported in the Bureya Block as well [24]. Thus, the Neoproterozoic could be the basement age of this united continental block. However, the most important basement of the block was formed during the late Pan-African period (c. 500 Ma) [17,25], including granulite-facies metasedimentary and minor meta-mafic rocks, accompanied by the Early Paleozoic granitoids [25]. Along the eastern Jiamusi and Khanka blocks, there are some Paleozoic volcanic and sedimentary strata. The main lithologies are the Early Paleozoic marbles, terrigenous clastic rocks, and the Late Paleozoic volcanic rocks with intervening terrigenous clastic rocks. Calc-alkaline arc-like andesites and dacites firstly occurred in the Early

Permian, indicating the early-stage continental arc in the region [26]. From the Early Permian to the earliest Triassic, N-S trending granitoids with associated volcanic rocks were generated and migrated throughout the whole block from east to west, mainly consisting of calc-alkaline granitoids [11]. However, there are very few magmatic records from the late Early Triassic to Jurassic in the Jiamusi block, although these ages have been reported along the western Bureya and Khanka blocks [27]. Volcanic rocks with minor plutonic rocks increased in the Early Cretaceous and are most abundant from the late Early to early Late Cretaceous in the Jiamusi block, and mainly show intra-plate affinities [12,22].

There are several fragments of Mesozoic accretionary complexes along the western margin of the Jiamusi block (Figure 1B), which are named as the Heilongjiang Complex, representing a lost ocean (the Mudanjiang Ocean) between the Songliao and Jiamusi blocks [15]. The Heilongjiang Complex is composed of mica schist, marble, quartzite, greenschist, blueschist, and amphibolite. Most published data show that the protolith age of the blueschist is Early Permian to Triassic (c. 285 to 213 Ma), whereas the high-pressure metamorphic age ranges from c. 210 to 170 Ma [15]. However, a blueschist sample from Yilan yields a concordant zircon U-Pb age of 142 Ma, which is the only Cretaceous protolith age of rocks from the Heilongjiang Complex so far [28]. There are also some other accretionary complexes exposed in the region, such as the N-S trending Zhangguangcai Complex along the eastern Songliao block [29]. It consists of amphibolite, biotite-plagioclase-gneiss, marble, schist, and meta-volcanic rocks. The metamorphic ages range from the Paleozoic to the Early Mesozoic [29].

Eight meta-sedimentary rock samples were collected from the two areas of the Heilongjiang Complex, near Luobei and Yilan, respectively (Figures 1B and 2). These samples are exposed as strongly deformed matrix (Figures 3 and 4), within which metamorphic basaltic rocks (such as greenschist) were observed and they are commonly intruded by later-stage magmatic rocks (Figure 3). Details of the samples are listed in Table 1, including the sample numbers, GPS locations, and lithologies.

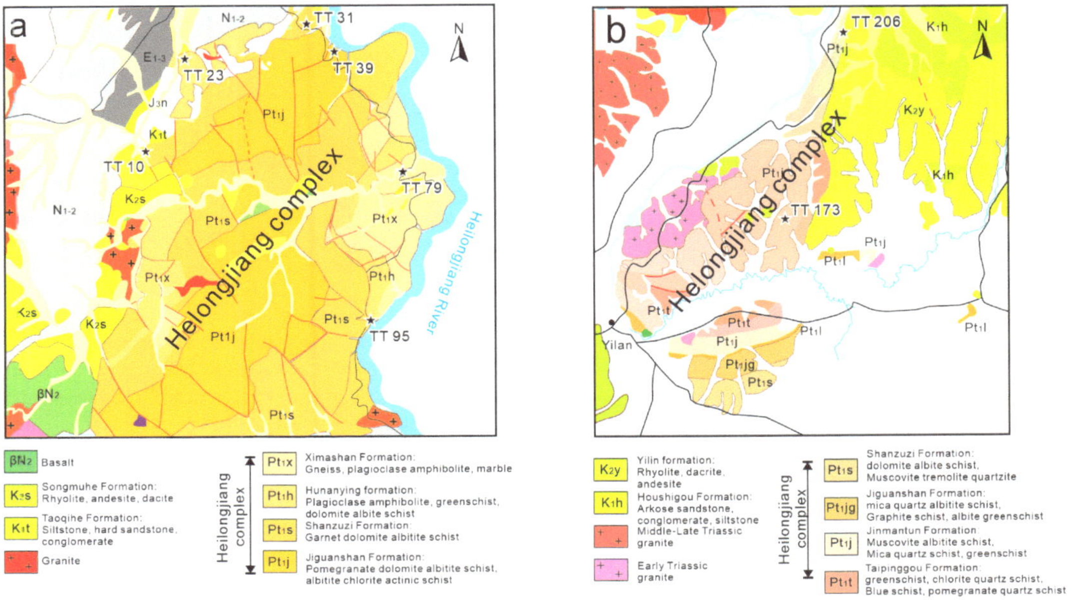

Figure 2. Geological maps of the sampling locations (after local 1:200,000 geological map). See Table 1 for details of the samples. (**a**) Map of the Jiayin-Luobei area and (**b**) the Jiamusi-Yilan area.

Figure 3. Typical outcrops of the sampling locations. (**a**) Gold mine near Wulaga, with the Early Cretaceous granodiorite intruding into the Heilongjiang Complex; (**b**) the Shanzuizi Group (albite mica schist) in Jiayin; (**c**) the Jiguanshan Group (two mica schist) near the mouth of the Jiayin river and there are residual vesicles in the greenschist layer; (**d**,**e**) show mica schist and meta-sandstone to the north of Luobei; (**f**) the Jinmantun Group (two mica quartz schist) near Jiamusi.

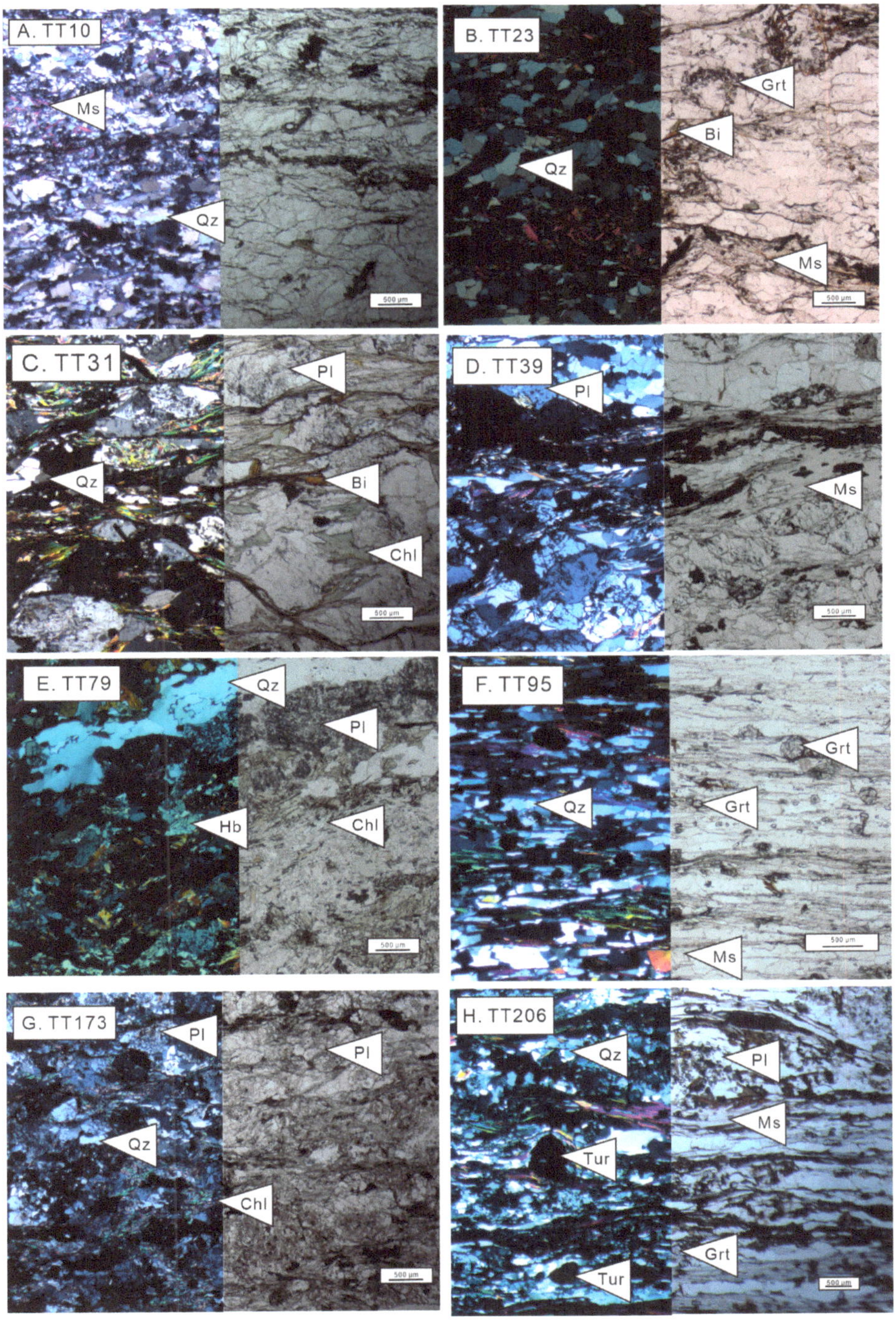

Figure 4. Thin sections of the 8 meta-sedimentary rocks in this study.

Table 1. Sample information of meta-sedimentary rocks from the Heilongjiang Complex.

Sample Number	GPS Location	Area	Descriptions
TT10	E 130°15′2.9″ N 48°22′34.4″	Jiayin-Luobei	Grey, mica schist; Qz (>85%) + muscovite mica (5%–10%) + opaque mineral (3%–5%).
TT23	E 130°19′18.6″ N 48°28′29.9″		Grey, garnet two-mica quartz schist; Qz (65%) + muscovite mica (15%) + biotite (5%) + opaque mineral (3%); accessory minerals: garnet, chlorite.
TT31	E 130°33′14.9″ N 48°30′44.8″		Grey greenish, biotite greenschist; highly deformed; Plagioclase (55%) + Qz (15%) + chlorite (25%) + biotite (5%).
TT39	E 130°36′35″ N 48°28′46.4″		Grey, mica schist; Plagioclase (35%) + Qz (35%) + muscovite mica (25%) + opaque mineral (5%).
TT79	E 130°44′15.6″ N 48°19′58.3″		Dark grey, hornblende greenschist; deformed; Plagioclase (55%) + chlorite (20%) + hornblende (15%) + Qz (10%).
TT95	E 130°40′0.7″ N 48°09′6.5″		Yellowish, garnet mica schist; highly deformed; Qz (75%) + muscovite mica (20%) + garnet (5%).
TT173	E 129°52′23.7″ N 46°27′8″	Yilan-Jiamusi	Dark grey, quartz greenschist; highly deformed; Plagioclase (70%) + Qz (15%) + chlorite (15%).
TT206	E 129°58′4.59″ N 46°40′14.32″		Light grey, garnet muscovite mica schist; highly deformed; plagioclase (40%) + Qz (35%) + muscovite mica (20%) + tourmaline (3%) + garnet (2%).

We first selected detrital zircons from these samples, and then carried out U-Pb dating and trace elements analyses. See the Table S1 for details of the results, Table S2 for the compiled whole-rock geochemical data, and Table S3 for the methods used in this study. According to the temporal variations of zircon geochemical features, combined with the geochemical and geochronological data from the regional magmatic rocks, we discuss the tectonic evolution of the two episodes of continental arc in NE China.

3. Results

Most zircons from our eight samples show inherited cores with oscillatory zonation, indicating their magmatic origins (Figure 5). The internal zonation patterns in the cores differ from each other, e.g., some zoning intervals are wide and some others are narrow (Figure 5). This is because they are detrital zircons which were derived from various source rocks with different magmatic evolution history. However, almost all the grains are characterized by very thin bright rims in the cathode luminescence (CL) images. The rims have irregular boundaries with the magmatic zircon cores (Figure 5), probably caused by fluid-assisted hydrothermal alteration in low temperature metamorphism [30].

Nighty concordant ages yielded in sample TT10 (Figure 6a), ranging from 165 to 527 Ma. Zircons have length/width ratios varying from 1:1 to 4:1, and the lengths range from 80 to 200 μm (Figure 4). The age groups are divided into four categories: 509–527 Ma (2.2%), 436–458 Ma (3.3%), 216–269 Ma (12.1%), and 165–206 Ma (82.4%). The final group constitutes the major age peak (c. 182 Ma) in this sample, whereas the other three groups show very weak peak ages at 253 Ma, 455 Ma, and 509 Ma, respectively. According to the youngest age at c. 165 Ma and the youngest peak age at c. 182 Ma, the maximum depositional age of sample TT10 is the Middle Jurassic.

Zircons from sample TT23 show similar sizes and shapes to those from sample TT10. Nineteen-eight concordant results were obtained (Figure 6b), ranging from 153 Ma to 1874 Ma. More than 90% zircons have ages from 153 to 241 Ma, showing a major peak at c. 176 Ma. The maximum depositional age of sample TT23 should be the Middle Jurassic, based on the youngest zircon grain age at c. 153 Ma.

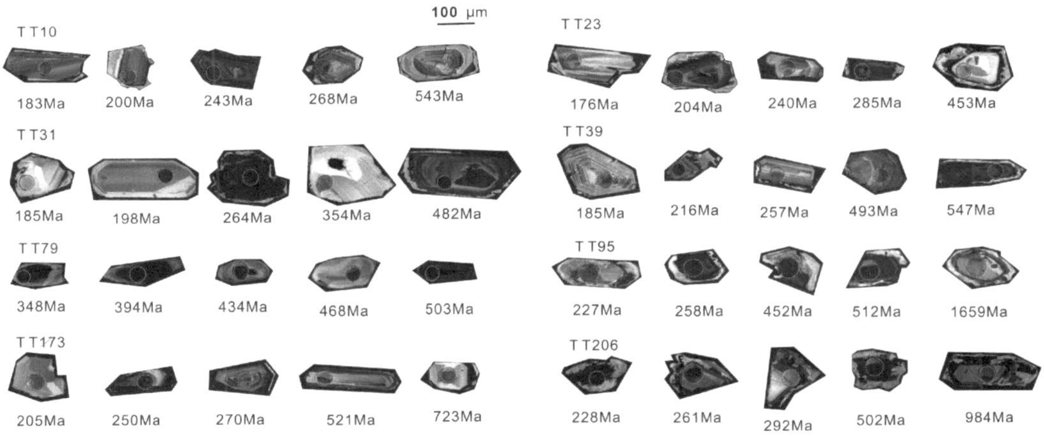

Figure 5. CL images of the analyzed zircons from the 8 meta-sedimentary rocks. Please see the main text for descriptions. Red circles denote the laser spots during the geochronological analyses.

In sample TT31, zircon grains are slightly larger than those from the former two samples, with lengths ranging from 100 to 300 μm (Figure 5). There are 111 concordant ages in sample TT31 (Figure 6c). Three age populations are subdivided from c. 176 to 1474 Ma: 444–486.6 Ma (13.5%), 240.9–272.8 Ma (28.8%), and 176.1–206.1 Ma (27.9%). This sample shows two major peaks at c. 185 and 263 Ma, respectively, with a minor peak at c. 485 Ma. According to the youngest peak age at c. 185 Ma, the maximum depositional age of sample TT10 is the Early Jurassic.

Zircons from sample TT39 show angular shapes and various sizes; their lengths range from 50 to 250 μm (Figure 5). They yield 121 concordant ages, varying from c. 181 to 1785 Ma (Figure 6d), showing three major groups: 443–527 Ma (28.9%), 235.9–284.1 Ma (32.2%), and 181–209 Ma (18.1%). The age spectrum is similar to sample TT31, that is, it has two major peaks at 185 and 257 Ma, respectively, with a weak peak at about 490 Ma. The maximum depositional age of sample TT10 should be the Early Jurassic, according to the youngest peak age at c. 181 Ma.

Zircons from sample TT79 are different from other samples because no low temperature alteration can be recognized. This sample probably experienced higher metamorphic grade because some zircons have metamorphic overgrowth rims (Figure 5). One hundred and seven concordant ages from sample TT79 (Figure 6e) were obtained. They range from 251 to 769 Ma, and thus, sample TT79 is the only one with no Mesozoic age peak in this study. The ages show a quasi-continuous distribution, and the most important age population is 339 to 556 Ma (87%) with a peak age at c. 468 Ma. Although some younger (Permian to Middle Triassic) and older (Proterozoic) ages exist, they are insignificant, indicating that the maximum depositional age of sample TT79 should be in the Early Triassic.

With similar shapes and sizes (lengths from 80 to 200 μm) to other samples (Figure 5), zircons from sample TT95 have wider hydrothermal altered rims and yielded 101 concordant ages, which vary from 221 to 1818 Ma and is subdivided into four groups (Figure 6f): 500–539 Ma (12.7%), 422–461 Ma (8.8%), 236–289.8 Ma (28.4%) and 218–228.6 Ma (10.8%). There are three major peak ages in this sample, 226 Ma, 259 Ma, and 512 Ma. According to the ages of the youngest zircon grain at c. 221 Ma and peak at c. 226 Ma, the maximum depositional age of this sample is the Late Triassic.

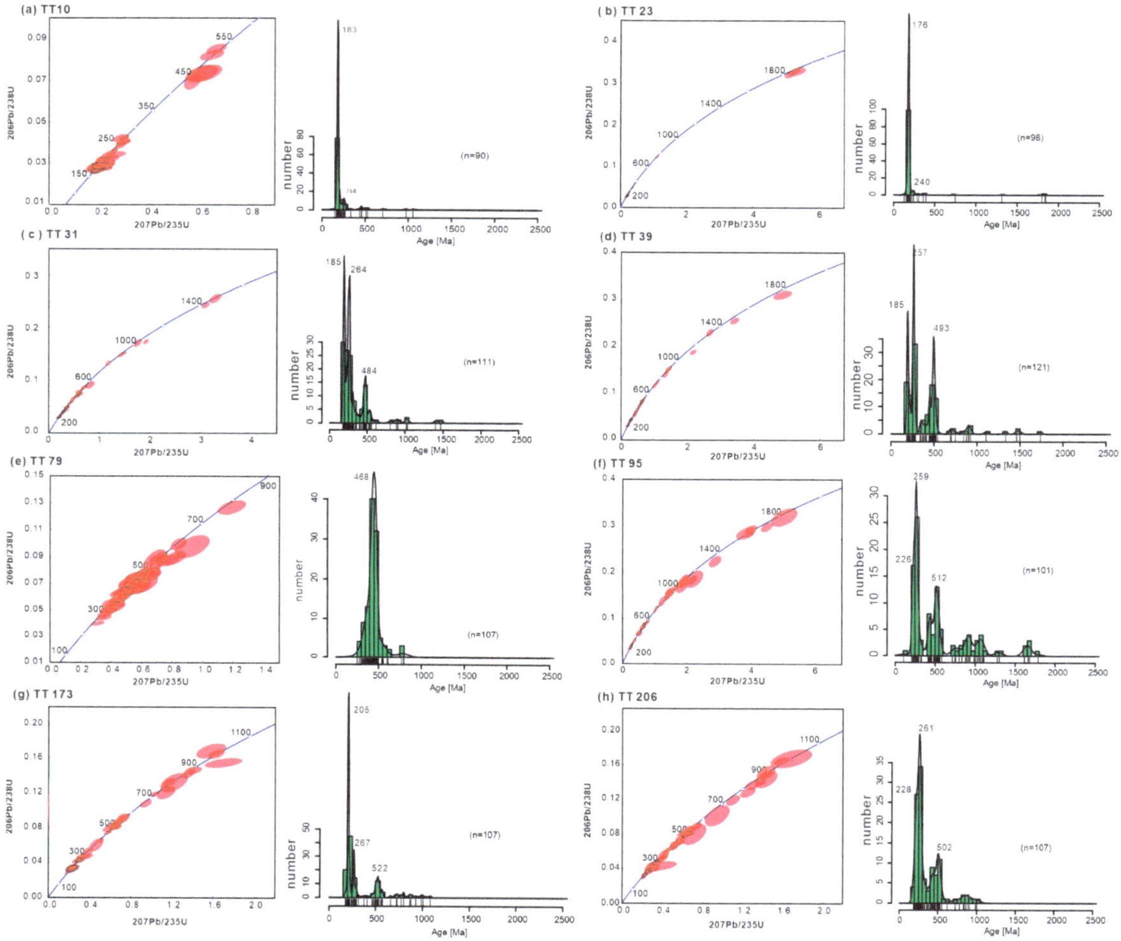

Figure 6. Detrital zircon U-Pb dating results of samples in this study. For each sample, the concordant diagram is shown in the left, and the histogram with a KDE curve is in the right. The figure was illustrated using IsoplotR [31] and refers to [32,33].

The hydrothermal rims of zircons are thin in sample TT173. One hundred and seven grains yielded concordant ages between 177 and 1004 Ma in this sample (Figure 6g), which show three major age populations, 501–537 Ma (10.2%), 250–296 Ma (14.0%) and 177–225 Ma (57.9%) corresponding to three major peak ages at 522, 267, and 205 Ma, respectively. The maximum depositional age of sample TT193 should be the earliest Jurassic, according to the youngest several grains which have ages from 177 to 199 Ma.

The low temperature alteration was remarkable in sample TT206 because of the wide rims with irregular complicated internal structures (Figure 5). We obtained 107 concordant ages from this sample, which range from 185 Ma to 999 Ma (Figure 6h). Three age groups can be recognized in this sample, that is, 829–906 Ma (4.7%), 486–529 Ma (14.0%) and 202–296 Ma (57.0%). Age peak at c. 261 Ma is the most important one relative to the other two at c. 228 Ma and 502 Ma, respectively. Although the youngest age peak is 228 Ma, the maximum depositional age of this sample should be the Early Jurassic, because there are six grains showing ages from 200 to 185 Ma.

4. Discussion

4.1. Types of Detrital Zircon Age Spectra and Provenance Analysis

We recognize three types of age spectra from the 8 meta-sedimentary rock samples in this study. The first one is a unimodal type, including samples TT10 and TT23 (Figure 6). Each of them only contains one major peak having ages of the Early Jurassic, 183 Ma and 176 Ma, respectively. Generally, this kind of age spectrum should be related to the intra-oceanic island arc [34] because oceanic arc-related basins are usually separated from the ancient continents by deep trench, abyssal plain, or middle ocean ridge. These positive/negative topographic anomalies are similar to the communication between oceanic arc-related basins and the old continents. Thus, sediments in these basins are mainly from the juvenile island arc, showing a unimodal peak pattern. However, this is not the case in the Heilongjiang Complex because the zircon grains with ages from Paleozoic to Neoproterozoic cannot be ignored in both samples (TT10 and TT23), indicating that they contain materials not only from the young arc, but also from the old continental basement. The depositional environment is not intra-oceanic arc, but near a continental arc. The second type is trimodal type, showing three major peak ages at the Late Triassic to Early Jurassic, Late Permian to Early Triassic, and Early Paleozoic. The three age peaks of this "trimodal" type can be recognized in the first type (unimodal). However, in the latter type, the Late Triassic to Early Jurassic age group is much more important than any other ages. In the second type, however, the Late Permian to Early Triassic and Early Paleozoic age groups are increasingly significant. The two post-Paleozoic age populations are contrasting. In sample TT173, the Late Triassic to Early Jurassic peak is much higher, but in other cases, the Late Permian to Early Triassic is equally (sample TT31) or even more (samples TT39, TT95, and TT206) important. The third type of age spectrum only contains sample TT79. It is special because the Mesozoic ages are quite fewer in this sample than in all other ones.

The three types of the detrital zircon age spectrum match the magmatic records in the Jiamusi block and the Lesser Xing'an-Zhangguangcai Range, suggesting the source region should be the continental blocks on both sides. For example, the most common age in the Jiamusi block is the Early Paleozoic (c. 530 to 450 Ma), representing the late Pan-African granulite-facies metamorphism and its accompanying or later-stage felsic magmatism [25]. Thus, the third type (sample TT79) probably suggests that it mainly accepted the sediments from the east side, the Jiamusi block, with only very minor materials from the west. Along the western margin of the Jiamusi block, numerous granitic gneisses and weakly deformed granitoids are exposed and show I-type geochemical features. Their ages continuously vary from c. 275 to 245 Ma, with a peak age at 260–250 Ma. This age population is quite common in most samples in this study (samples TT31, TT39, TT95, TT173, and TT206), indicating the important contribution of the western Jiamusi block to the Heilongjiang Complex. However, the Late Triassic to Middle Jurassic (c. 210–170 Ma) magmatism is almost undetectable in the Jiamusi block but occur as the most important ages in the Lesser Xing'an-Zhangguangcai Range [10,22]. Thus, the first and second types of age spectrum should be largely containing materials from this range, namely the eastern Songliao block because age peaks at 170–210 Ma are usually high as or higher than those at 250–260 Ma (Figure 6).

Therefore, our provenance analyses indicate that the Heilongjiang Complex mainly contains deposits of terrigenous materials from both sides, that is, the Jiamusi and Songliao blocks. Although all these samples were collected from the Heilongjiang Complex, the relative contributions of the two blocks differ. Some samples (the first type) mainly consisted of materials from the west side, for example, the samples TT10 and TT23; other samples (the second type) accepted materials from both sides, including samples TT31, TT39, TT95, TT173, and TT206; the sediments of only one sample (TT79, the third type) was mainly supplied by the Jiamusi Block. The provenance complexity of the Heilongjiang Complex is consistent with the nature of the "non-Smith strata" of the accretionary complex. Accretionary complexes are formed after the closure of fossil oceanic basins and thus near the suture zones. They show a complicated "block-in-matrix" structure. Although they are mainly composed of meta-sedimentary rocks, the original sequence has been disturbed.

Rocks that have quite different provenance and metamorphic history are stacked together after the emplacement of the accretionary complex.

4.2. Two Crustal Thickening Events in NE China Revealed by Zircon and Whole-Rock Geochemistry

The provenance analysis result—the detrital zircons in this study are mainly derived from the Songliao and Jiamusi blocks—indicates that the zircon trace elements can reflect the geochemical variations of their sources [16], the regional magmatic rocks. This provides a basis to our following discussions on magmatic and tectonic evolution.

Zircon trace elements are strongly controlled by the magma compositions from which they crystallize, which further record the temperature and pressure conditions influenced by tectonics. For example, zircon Eu anomaly ($Eu_N/\sqrt{Gd_N \times Sm_N}$; the subscript N means normalized to chondrites) [35] is used together with the whole-rock Sr/Y and La/Yb ratios [36] of felsic magmatic rocks to discuss the crustal thickness. The hypothesis is that crustal magmas are usually formed either by differentiation or crustal partial melting [37]. Both processes tend to occur at depths where mantle-derived stall, such as Moho. During the formation of these crustal magmas, the stability fields of some minerals are mainly associated with pressure, such as garnet (rich in HREEs and Y), plagioclase (rich in Sr and Eu^{2+}), and amphibole (rich in middle to heavy REEs). Under higher pressure, amphibole and garnet tend to crystalize and plagioclase will break down. Thus, higher/lower whole-rock Sr/Y and La/Yb ratios indicate higher/lower pressure of felsic magmas, or deeper/shallower Moho depth [16]. Instable plagioclase under high pressure will release more Eu into the magma and increase the magmatic Eu anomaly, which will finally produce zircons with higher Eu anomaly values than those formed under lower pressure [35]. Accordingly, we will discuss the variation of magmatic formation depth by employing these geochemical proxies, based on both our new detrital zircon data and previously published data. Using the compiled geochronological and geochemical data from the regional magmatic rocks, we discuss the tectonic evolution of this region and its implications for the continental flare-ups along the active continental margin.

Detrital zircon trace elements reveal two crustal thickening events (Figure 7A–C), which coeval with the regional magmatic flare-up events: the first one lasted from the Early Permian to Early Triassic and the second mainly occurred in the Early Jurassic (Figure 7F). The median value of Eu/Eu* increased from <0.2 at c. 280 Ma to 0.5 at c. 245 Ma (Figure 7A) and decreased back to 0.2 in the Middle Triassic. Afterwards, the median value of Eu/Eu* again increased slowly to 0.4 at c. 180 Ma. This increase indicate that the crust was probably thickened during the two episodes of magmatic flare-ups, leading to the breakdown of plagioclase in the source and the higher zircon Eu/Eu* values in the magmas. During the magmatic lull from the Middle to Late Triassic, the crust thinned, facilitating the stabilization of plagioclase in the source region.

Similar to the Eu/Eu* values, the $(Dy/Yb)_N$ values in the detrital zircons are also low during the magmatic lull but higher in the flare-ups (Figure 7b). The median of the $(Dy/Yb)_N$ values fluctuated around 0.2 in the Permian and slightly increased from 0.17 to 0.2 in the Late Triassic to Early Jurassic. In the Middle to Late Triassic, it decreased to near 0.15. This means that the middle REEs were depleted relative to the heavy REEs in the magmatic lull, indicating that stabilization of amphiboles weakened and thus the crustal probably thinned.

Zircon Th/U ratio is commonly used to distinguish the magmatic or metamorphic origins of zircons [38], with the former being usually higher than 0.1 and the later being lower. In this study, almost all the zircons (<290 Ma) have Th/U ratios > 0.1, indicating their magmatic origins and thus the temporal change of this value should not be related to any metamorphic processes. In the intermediate to felsic magmas, monazite and allanite are common accessory minerals controlling the Th and U budgets. Specially, Th is more compatible in these minerals than U and thus their fractionation will lead to low Th/U ratio in other minerals. Some studies indicate that the early-stage crystallization of Th-rich minerals relative to zircon occurs in low-temperature magmas, whereas these minerals are instable in high-temperature magmas [39]. Thus, high Th/U ratio probably indicates high magmatic temperature. During the crustal thickening event, the deep crust becomes warmer when the geothermal gradient remains constant; a thicker crust is able to keep the magma reservoir warm, accelerating the Th/U ratios, similar to the examples in the Cretaceous Sikhote-Alin arc magmas [40] and Mesozoic Andes [16,41]. In this study, higher Th/U ratio corresponds to higher zircon Eu/Eu*, whole-rock Sr/Y, and $(La/Yb)_N$, probably representing the crustal thickening events. The low zircon Th/U ratios during the Middle to Late Triassic may have resulted from the crustal magmas produced in shallower crust and thus in lower temperatures.

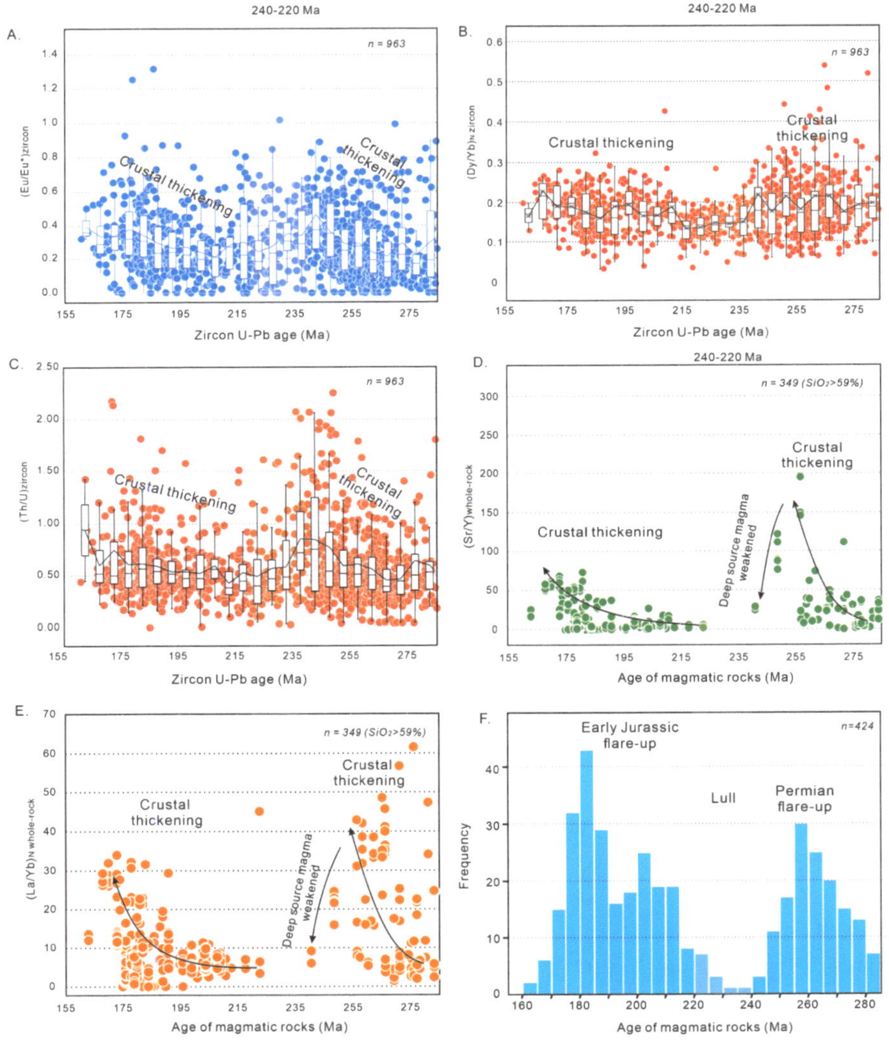

Figure 7. Trace elements geochemical data (shown in different colors) from the regional magmatic rocks and detrital zircons in this study. Part of the detrital zircon data are from [13]. (**A–C**) show Eu anomaly, Th/U and Dy/Yb ratios of zircons varying with zircon ages, respectively; (**D**,**E**) show whole-rock Sr/Y and $(La/Yb)_N$ ratios varying with rock ages, respectively; (**F**) age histogram of magmatic rocks (Permian to Jurassic).

Whole-rock trace elements of the compiled magmatic rocks show consistent trends with the zircon Eu/Eu*, $(Dy/Yb)_N$ and Th/U ratios. Two crustal thickening events are revealed by high Sr/Y and $(La/Yb)_N$ ratios (Figure 7D,E). The first occurred in the Permian, during which the whole-rock Sr/Y ratios increased from 10–40 to >150, and the $(La/Yb)_N$ ratios also increased from <10 to >40. The second lasted from the Late Triassic to Early Jurassic when the Sr/Y ratios increased from <10 to >50, whereas the $(La/Yb)_N$ ratios increased from about 5 to 35. The good coherence between the two ratios indicates that the source of these felsic magmatic rocks changed from low pressure to high, indicating that the crust was possibly thickened at that time [42]. It should be noted that during the Middle to Late Triassic, both ratios decreased remarkably, suggesting that the deep magma source (high pressure) was shut down and only the shallow-level (low pressure) magmas were active.

In a word, trace elements of both the whole rock and detrital zircon reveal two events of crustal thickening in NE China during the Permian and Early Jurassic, respectively. Each thickening event was coeval with a continental flare-up (Figure 7F). Between them, the volume of deep source magmas reduced with the shallow ones remaining active, accompanied by an important magmatic lull (Figure 7F).

4.3. Tectono-Magmatic Evolution in NE China: Not a "Cyclic Orogen"

The tectono-magmatic evolution in NE China shares similarities with the proposed "cyclic orogen", such as in the American Cordillera [3]. The most important fact is that crustal thickening was accompanied by continental arc flare-ups and landward arc migration (Figure 8). During the Permian flare-up, the arc-related andesite and other volcanic rocks in the Erlongshan Formation was distributed along the eastern margin of the Jiamusi block in the Early Permian. At the end of the Permian, most magmatism had migrated to the Yilan-Mudanjiang alignment and the migrating distance was about 180 km (Figure 8A). During the Late Triassic to Middle Jurassic, the magmatic rocks migrated westwards from Hegang-Yilan-Mudanjiang to Tieli-Jilin, and the distance approximated to 150 km (Figure 8B).

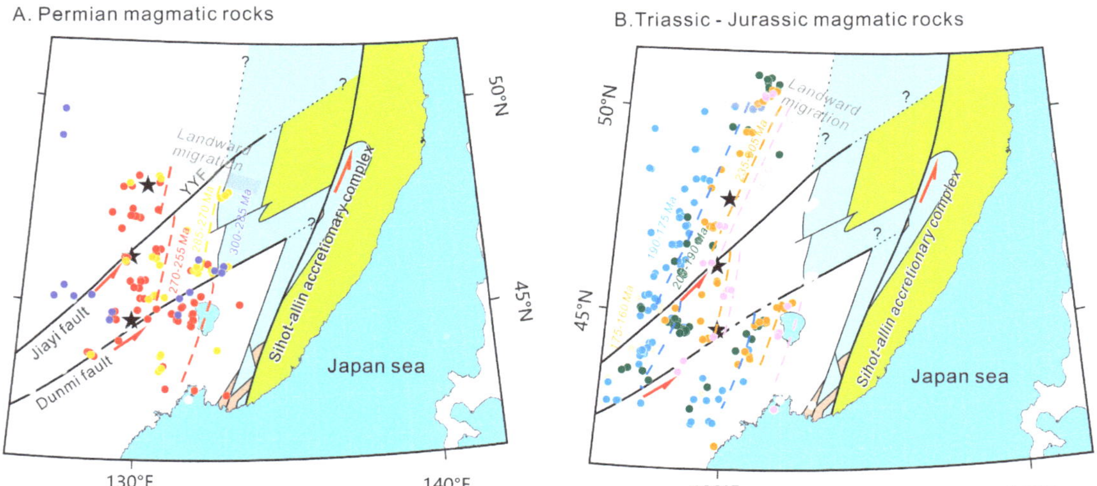

Figure 8. Spatial distribution of regional magmatic rocks. (**A**,**B**) show the westward younger trends of the Permian and Triassic-Jurassic calc-alkaline magmatic rocks (continental arc), respectively. Black stars show the positions of Luobei, Yilan, and Mudanjiang from north to south. The supposed extensions of the major faults are shown by dashed lines with question marks.

However, there are significant differences between the cases in NE China and the American Cordillera. The first is the nature of the arc magmatic migration. In NE China, the two migrations were independent from each other, that is, during the magmatic lull the magmas did not move back to the trench (east) but consistently migrated landward (westward) (Figure 8). However, in the American Cordillera, continental arc fronts tended to migrate towards the trench during the lulls and moved landwards again when the magmatic flare-ups were onset [3,5]. This difference indicates that the twice orogenic events in NE China, marked by the two magmatic flare-ups and crustal thickening events, were not two orogenic episodes within a single "cyclic orogen". The second difference is that regional extension was active in NE China during the orogenic events, evident from the opening of the Mudanjiang Ocean and regional A-type and bimodal volcanism [10]. In the Heilongjiang Complex, metamorphic oceanic crustal fragments were preserved as exhumed blocks within the strongly deformed matrix, such as MORB and OIB-like blueschist and greenschist, representing the opening of the Mudanjiang Ocean and island arc within it [15]. The protolith ages of these remnants of the oceanic crust range from c. 210 Ma to 290 Ma. In the Late Triassic, A-type and bimodal volcanic rocks were reported in the Lesser Xing'an-Zhangguangcai Range and Khanka block [10], indicating the occurrence of intra-continental extension during that period.

4.4. Tectonic Implications in NE China: Two Orogenic Events during the Closure of the Mudanjiang Ocean

According to the similarities and differences between magmatism in NE China and the American Cordillera, we propose a new model in our study area in order to explain the coupling between magmatic flare-up, landward migration, and crustal thickening (Figure 9).

During the early Permian, the eastern Jiamusi margin experienced a transition from a passive continental margin to an active one [26]. This transition produced a series of arc-related volcanic rocks with sedimentary strata (Figure 9A), aligning from north to south near Baoqing. Since then, the oceanic plate within the Panthalassa Ocean began to subduct beneath the Jiamusi block and triggered the first orogenic event in NE China. The slab dip angle became shallower as the oceanic plate subducted and thus resulted in the landward migration of the calc-alkaline magmatism in the Jiamusi block. The coupling between the overriding and subducting plates was strengthened, and thus the crust was thickened (Figure 9B). Crustal thickening caused warmer lower crust and deeper Moho in the region, facilitating deep crustal melting and resulted in more voluminous magmas in the Permian, namely the first magmatic flare-up. However, the asthenospheric mantle wedge could be largely expelled by the shallowing slab and thickening arc crust, and thus the arc magmatic source weakened after it migrated to the western margin of the Jiamusi block (Figure 9C). Deep source melting also vanished, causing the descending trend in the whole-rock Sr/Y, $(La/Yb)_N$, and zircon Eu/Eu*, Th/U, and $(Dy/Yb)_N$ ratios.

From the Early Triassic, the Mudanjiang Oceanic plate probably began to subduct westwards and produced calc-alkaline magmatism in the Lesser Xing'an-Zhangguangcai Range, the second continental arc flare-up (Figure 9D). Magmatism also moved westwards during the Middle Triassic to Middle Jurassic, probably related to the shallowing subducting slab dip, or associated with the removal of the fore-arc crust by subduction erosion. Regional extension may have existed in the arc or back-arc region during the Middle-Late Triassic, making the regional tectonics complicated than our schematic tectonic model. The closure of the Mudanjiang Ocean occurred after the Early Jurassic (the maximum depositional ages of most samples in this study), which should be the reason why the Early Jurassic magmatic flare-up ended abruptly at c. 170 Ma (Figure 9E) without remarkable decrease in the values of the whole-rock Sr/Y, $(La/Yb)_N$, and zircon Eu/Eu* ratios.

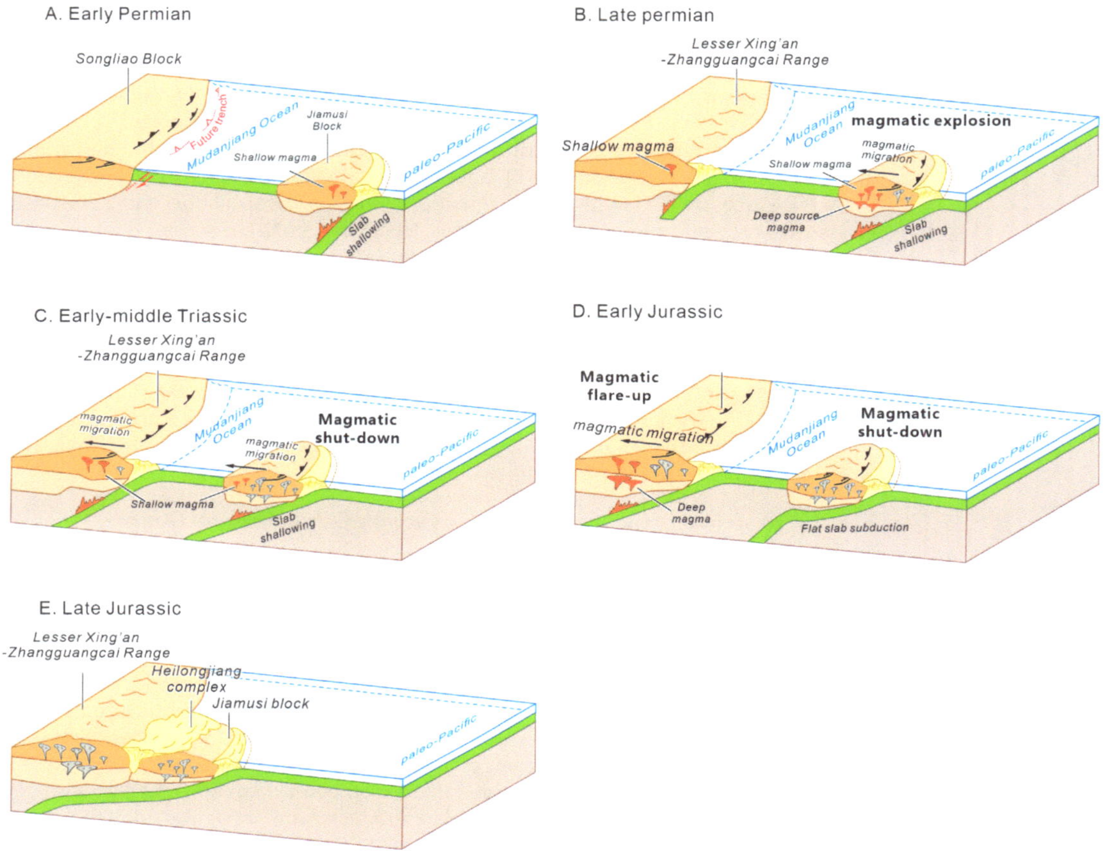

Figure 9. Tectono-magmatic evolution in the eastern Heilongjiang Province. It illustrates the two episodes of continental arc magmatic flare-ups and crustal thickening history.

5. Conclusions

The Heilongjiang Complex mainly received sediments from the both sides, including the Songliao block (the Lesser Xing'an-Zhangguangcai Range) and Jiamusi block, recording abundant geochemical information of the Permian to Middle Jurassic magmatism in NE China.

Trace elements from the detrital zircons (Eu/Eu*, Th/U, and $(Dy/Yb)_N$ ratios) and regional magmatic rocks (whole-rock Sr/Y, $(La/Yb)_N$ both indicate two crustal thickening events in the Permian and Early Jurassic, respectively. Crustal thickening was accompanied by magmatic flare-up and landward migration.

The two crustal thickening events and flare-ups are independent of each other, forming two orogenic events in NE China. The Permian crustal thickening event was caused by shallowing subducting slab angle and thus triggered magmatic flare-ups. The shallow slab subduction enhanced plate coupling and caused the westward subduction of the Mudanjiang Oceanic plate, generating the second flare-up in this region and the crustal thickening event.

Closure of the back-arc basin or marginal sea (the Mudanjiang Ocean) by the subduction of the Paleo-Pacific oceanic plate (Izanagi plate) could lead to several arc magmatic flare-ups and crustal thickening events, as revealed by our study. This is probably a common process in the accretionary orogens and contributed to the continental growth along active continental margins.

Supplementary Materials: The following supporting information can be downloaded at: https://www.mdpi.com/article/10.3390/min13091121/s1, Table S1. Detrital zircon U-Pb dating and trace elements results; Table S2. Data for Figure 7d,e; Table S3. Analytical methods. References [43,44] are cited in the Supplementary Materials.

Author Contributions: Conceptualization, K.L. and M.X.; methodology, K.L. and Y.P.; software, K.L. and Y.P.; validation, K.L. and Y.P.; formal analysis, M.X.; investigation, K.L., M.W. and Y.P.; resources, K.L.; data curation, K.L. and Y.P.; writing—original draft preparation, K.L. and Y.P.; writing—review and editing, K.L.; visualization, K.L. and Y.P.; supervision, K.L. and M.W.; project administration, K.L.; funding acquisition, K.L. All authors have read and agreed to the published version of the manuscript.

Funding: This research was funded by the National Natural Science Foundation of China (42272252, 41902233).

Data Availability Statement: The authors confirm that the data generated or analyzed during this study are provided in full within the published article. Zircon U-Pb and trace elements data from reference [13] are also compiled to generate the figures.

Acknowledgments: We deeply appreciate the discussions with Wenliang Xu and Jianbo Zhou on the tectonics in NE China.

Conflicts of Interest: The authors declare no conflict of interest.

References

1. Chapman, J.B.; Shields, J.E.; Ducea, M.N.; Paterson, S.R.; Attia, S.; Ardill, K.E. The causes of continental arc flare ups and drivers of episodic magmatic activity in Cordilleran orogenic systems. *Lithos* **2021**, *398–399*, 106307. [CrossRef]
2. Paterson, S.R.; Ducea, M.N. Arc Magmatic Tempos: Gathering the Evidence. *Elements* **2015**, *11*, 91–98. [CrossRef]
3. DeCelles, P.G.; Ducea, M.N.; Kapp, P.; Zandt, G. Cyclicity in Cordilleran orogenic systems. *Nat. Geosci.* **2009**, *2*, 251–257. [CrossRef]
4. Lee, C.-T.A.; Thurner, S.; Paterson, S.; Cao, W. The rise and fall of continental arcs: Interplays between magmatism, uplift, weathering, and climate. *Earth Planet. Sci. Lett.* **2015**, *425*, 105–119. [CrossRef]
5. Ducea, M. The California Arc: Thick Granitic Batholiths, Eclogitic Residues, Lithospheric-Scale Thrusting, and Magmatic Flare-Ups. *GSA Today* **2001**, *11*, 4. [CrossRef]
6. Chapman, A.D.; Saleeby, J.B.; Eiler, J. Slab Fattening Trigger for Isotopic Disturbance and Magmatic Flare-up in the Southernmost Sierra Nevada Batholith, California. *Geology* **2013**, *41*, 1007–1010. [CrossRef]
7. Karlstrom, L.; Lee, C.-T.A.; Manga, M. The role of magmatically driven lithospheric thickening on arc front migration. *Geochem. Geophys. Geosystems* **2014**, *15*, 2655–2675. [CrossRef]
8. Chapman, J.B.; Ducea, M.N. The role of arc migration in Cordilleran orogenic cyclicity. *Geology* **2019**, *47*, 627–631. [CrossRef]
9. Attia, S.; Cottle, J.M.; Paterson, S.R. Erupted zircon record of continental crust formation during mantle driven arc flare-ups. *Geology* **2020**, *48*, 446–451. [CrossRef]
10. Xu, W.-L.; Pei, F.-P.; Wang, F.; Meng, E.; Ji, W.-Q.; Yang, D.-B.; Wang, W. Spatial–temporal relationships of Mesozoic volcanic rocks in NE China: Constraints on tectonic overprinting and transformations between multiple tectonic regimes. *J. Asian Earth Sci.* **2013**, *74*, 167–193. [CrossRef]
11. Yang, H.; Ge, W.; Dong, Y.; Bi, J.; Wang, Z.; Ji, Z. Record of Permian–Early Triassic continental arc magmatism in the western margin of the Jiamusi Block, NE China: Petrogenesis and implications for Paleo-Pacific subduction. *Int. J. Earth Sci.* **2016**, *106*, 1919–1942. [CrossRef]
12. Sun, M.; Chen, H.; Milan, L.A.; Wilde, S.A.; Jourdan, F.; Xu, Y. Continental Arc and Back-Arc Migration in Eastern NE China: New Constraints on Cretaceous Paleo-Pacific Subduction and Rollback. *Tectonics* **2018**, *37*, 3893–3915. [CrossRef]
13. Dong, Y.; Ge, W.-C.; Yang, H.; Ji, Z.; He, Y.; Zhao, D.; Xu, W. Convergence history of the Jiamusi and Songnen-Zhangguangcai Range massifs: Insights from detrital zircon U-Pb geochronology of the Yilan Heilongjiang Complex, NE China. *Gondwana Res.* **2018**, *56*, 51–68. [CrossRef]
14. Liu, K.; Wilde, S.A.; Zhang, J.; Xiao, W.; Wang, M.; Ge, M. Zircon U–Pb dating and whole-rock geochemistry of volcanic rocks in eastern Heilongjiang Province, NE China: Implications for the tectonic evolution of the Mudanjiang and Paleo-Pacific oceans from the Jurassic to Cretaceous. *Geol. J.* **2019**, *55*, 1866–1889. [CrossRef]

15. Zhou, J.-B.; Wilde, S.A.; Zhang, X.-Z.; Zhao, G.-C.; Zheng, C.-Q.; Wang, Y.-J.; Zhang, X.-H. The onset of Pacific margin accretion in NE China: Evidence from the Heilongjiang high-pressure metamorphic belt. *Tectonophysics* **2009**, *478*, 230–246. [CrossRef]
16. Sundell, K.E.; George, S.W.; Carrapa, B.; Gehrels, G.E.; Ducea, M.N.; Saylor, J.E.; Pepper, M. Crustal Thickening of the Northern Central Andean Plateau Inferred from Trace Elements in Zircon. *Geophys. Res. Lett.* **2022**, *49*, e2021GL096443. [CrossRef]
17. Wilde, S.A. Final amalgamation of the Central Asian Orogenic Belt in NE China: Paleo-Asian Ocean closure versus Paleo-Pacific plate subduction—A review of the evidence. *Tectonophysics* **2015**, *662*, 345–362. [CrossRef]
18. Pei, F.; Xu, W.; Yang, D.; Zhao, Q.; Liu, X.; Hu, Z. Zircon U-Pb geochronology of basement metamorphic rocks in the Songliao Basin. *Chin. Sci. Bull.* **2007**, *52*, 942–948. [CrossRef]
19. Song, Y.; Stepashko, A.; Liu, K.; He, Q.; Shen, C.; Shi, B.; Ren, J. Post-rift Tectonic History of the Songliao Basin, NE China: Cooling Events and Post-rift Unconformities Driven by Orogenic Pulses from Plate Boundaries. *J. Geophys. Res. Solid Earth* **2018**, *123*, 2363–2395. [CrossRef]
20. Song, Y.; Ren, J.; Liu, K.; Lyu, D.; Feng, X.; Liu, Y.; Stepashko, A. Syn-rift to post-rift tectonic transition and drainage reorganization in continental rifting basins: Detrital zircon analysis from the Songliao Basin, NE China. *Geosci. Front.* **2022**, *13*, 101377. [CrossRef]
21. Wang, P.-J.; Mattern, F.; Didenko, N.A.; Zhu, D.-F.; Singer, B.; Sun, X.-M. Tectonics and cycle system of the Cretaceous Songliao Basin: An inverted active continental margin basin. *Earth-Sci. Rev.* **2016**, *159*, 82–102. [CrossRef]
22. Wu, F.Y.; Sun, D.Y.; Ge, W.C.; Zhang, Y.B.; Grant, M.L.; Wilde, S.A.; Jahn, B.M. Geochronology of the Phanerozoic granitoids in northeastern China. *J. Asian Earth Sci.* **2011**, *41*, 1–30. [CrossRef]
23. Yang, H.; Ge, W.-C.; Zhao, G.-C.; Bi, J.-H.; Wang, Z.-H.; Dong, Y.; Xu, W.-L. Zircon U–Pb ages and geochemistry of newly discovered Neoproterozoic orthogneisses in the Mishan region, NE China: Constraints on the high-grade metamorphism and tectonic affinity of the Jiamusi–Khanka Block. *Lithos* **2017**, *268–271*, 16–31. [CrossRef]
24. Sorokin, A.A.; Ovchinnikov, R.O.; Kudryashov, N.M.; Sorokina, A.P. An Early Neoproterozoic gabbro–granite association in the Bureya Continental Massif (Central Asian fold belt): First geochemical and geochronological data. *Dokl. Earth Sci.* **2016**, *471*, 1307–1311. [CrossRef]
25. Zhou, J.-B.; Wilde, S.A.; Zhang, X.-Z.; Zhao, G.-C.; Liu, F.-L.; Qiao, D.-W.; Ren, S.-M.; Liu, J.-H. A > 1300 km late Pan-African metamorphic belt in NE China: New evidence from the Xing'an block and its tectonic implications. *Tectonophysics* **2011**, *509*, 280–292. [CrossRef]
26. Li, G.-Y.; Zhou, J.-B.; Li, L. The Jiamusi Block: A hinge of the tectonic transition from the Paleo-Asian Ocean to the Paleo-Pacific Ocean regimes. *Earth-Sci. Rev.* **2022**, *236*, 104279. [CrossRef]
27. Liu, K.; Zhang, J.; Wilde, S.A.; Zhou, J.; Wang, M.; Ge, M.; Wang, J.; Ling, Y. Initial subduction of the Paleo-Pacific Oceanic plate in NE China: Constraints from whole-rock geochemistry and zircon U–Pb and Lu–Hf isotopes of the Khanka Lake granitoids. *Lithos* **2017**, *274–275*, 254–270. [CrossRef]
28. Zhu, C.Y.; Zhao, G.; Sun, M.; Liu, Q.; Han, Y.; Hou, W.; Zhang, X.; Eizenhofer, P.R. Geochronology and geochemistry of the Yilan blueschists in the Heilongjiang Complex, northeastern China and tectonic implications. *Lithos* **2015**, *216–217*, 241–253. [CrossRef]
29. Wang, F.; Xu, W.-L.; Gao, F.-H.; Meng, E.; Cao, H.-H.; Zhao, L.; Yang, Y. Tectonic history of the Zhangguangcailing Group in eastern Heilongjiang Province, NE China: Constraints from U–Pb geochronology of detrital and magmatic zircons. *Tectonophysics* **2012**, *566–567*, 105–122. [CrossRef]
30. Rubatto, D. Zircon: The Metamorphic Mineral. *Rev. Mineral. Geochem.* **2017**, *83*, 261–295. [CrossRef]
31. Vermeesch, P. IsoplotR: A free and open toolbox for geochronology. *Geosci. Front.* **2018**, *9*, 1479–1493. [CrossRef]
32. Fornelli, A.; Micheletti, F.; Langone, A.; Perrone, V. First U–Pb detrital zircon ages from Numidian sandstones in Southern Apennines (Italy): Evidences of African provenance. *Sediment. Geol.* **2015**, *320*, 19–29. [CrossRef]
33. Fornelli, A.; Micheletti, F.; Gallicchio, S.; Tursi, F.; Criniti, S.; Critelli, S. Detrital zircon ages of Oligocene to Miocene sandstone suites of the Southern Apennines foreland basin system, Italy. *J. Palaeogeogr.* **2022**, *11*, 222–237. [CrossRef]
34. Cawood, P.; Hawkesworth, C.; Dhuime, B. Detrital zircon record and tectonic setting. *Geology* **2012**, *40*, 875–878. [CrossRef]
35. Tang, M.; Ji, W.-Q.; Chu, X.; Wu, A.; Chen, C. Reconstructing crustal thickness evolution from europium anomalies in detrital zircons. *Geology* **2020**, *49*, 76–80. [CrossRef]
36. Profeta, L.; Ducea, M.N.; Chapman, J.B.; Paterson, S.R.; Gonzales, S.M.H.; Kirsch, M.; Petrescu, L.; DeCelles, P.G. Quantifying crustal thickness over time in magmatic arcs. *Sci. Rep.* **2015**, *5*, 17786. [CrossRef]
37. Farner, M.J.; Lee, C.-T.A. Effects of crustal thickness on magmatic differentiation in subduction zone volcanism: A global study. *Earth Planet. Sci. Lett.* **2017**, *470*, 96–107. [CrossRef]
38. Yakymchuk, C.; Kirkland, C.L.; Clark, C. Th/U ratios in metamorphic zircon. *J. Metamorph. Geol.* **2018**, *36*, 715–737. [CrossRef]
39. Harrison, T.M.; Watson, E.B.; Aikman, A.B. Temperature spectra of zircon crystallization in plutonic rocks. *Geology* **2007**, *35*, 635. [CrossRef]
40. Wu, J.T.-J.; Wu, J.; Alexandrov, I.; Lapen, T.; Lee, H.-Y.; Ivin, V. Continental growth during migrating arc magmatism and terrane accretion at Sikhote-Alin (Russian Far East) and adjacent northeast Asia. *Lithos* **2022**, *432–433*, 106891. [CrossRef]
41. Jara, J.J.; Barra, F.; Reich, M.; Leisen, M.; Romero, R.; Morata, D. Episodic construction of the early Andean Cordillera unravelled by zircon petrochronology. *Nat. Commun.* **2021**, *12*, 4930. [CrossRef] [PubMed]
42. Lieu, W.K.; Stern, R.J. The robustness of Sr/Y and La/Yb as proxies for crust thickness in modern arcs. *Geosphere* **2019**, *15*, 621–641. [CrossRef]

43. Ji, W.-Q.; Wu, F.-Y.; Wang, J.-M.; Liu, X.-C.; Liu, Z.-C.; Zhang, Z.; Cao, W.; Wang, J.-G.; Zhang, C. Early Evolution of Himalayan Orogenic Belt and Generation of Middle Eocene Magmatism: Constraint From Haweng Granodiorite Porphyry in the Tethyan Himalaya. *Front. Earth Sci.* **2020**, *8*, 236. [CrossRef]
44. Paton, C.; Hellstrom, J.; Paul, B.; Woodhead, J.; Hergt, J. Iolite: Freeware for the Visualisation and Processing of Mass Spectrometric Data. *J. Anal. At. Spectrom.* **2011**, *26*, 2508–2518. [CrossRef]

minerals

MDPI

Article

Middle Silurian–Middle Devonian Magmatic Rocks in the Eastern Segment of the Northern Margin of the North China Craton: Implications for Regional Tectonics

Bin Li [1,2], Jingsheng Chen [1,2,*], Fan Yang [3], Miao Liu [1,2], Yanqing Zang [1,2] and Chao Zhang [1,2]

1 Shenyang Geological Survey Center of China Geological Survey, Shenyang 110034, China; libin@mail.cgs.gov.cn
2 Northeast Geological S&T Innovation Center of China Geological Survey, Shenyang 110034, China
3 College of Resources and Environment, Jilin Agricultural University, Changchun 130061, China
* Correspondence: chenjingsheng@mail.cgs.gov.cn

Abstract: This paper presents a detailed study including LA-ICP-MS zircon U-Pb dating, geochemical, zircon Hf isotope, and whole rock Sr-Nd isotope analysis of magmatic rocks from the Yitong County, Jilin Province, NE China. These data are used to better constrain the Middle Silurian–Middle Devonian tectonic evolution in the eastern segment of the northern margin of the North China Craton (NCC). Zircon U-Pb dating results show that the Ximangzhang tonalite formed in the Late Silurian (425 ± 6 Ma); the basalt, andesite, and metamorphic olivine-bearing basalt in the Fangniugou volcanic rocks formed in the Middle Silurian (428 ± 6.6 Ma) and Middle Devonian (388.4 ± 3.9 Ma, and 384.1 ± 4.9 Ma). The Late Silurian tonalites are characterized by high SiO_2 and Na_2O and low K_2O, MgO, FeOT, and TiO_2, with an A/CNK ratio of 0.91–1.00, characteristic of calc-alkaline I-type granite. They are enriched in Rb, Ba, Th, U, and K, and depleted in Nb, Sr, P, and Ti, with positive $\varepsilon_{Nd}(t)$ (+0.35) and $\varepsilon_{Hf}(t)$ (+0.44 to +6.31) values, suggesting that they mainly originated from the partial melting of Meso–Neoproterozoic accretionary lower crustal material (basalt). The Middle Silurian basalts are characterized by low SiO_2, P_2O_5, TiO_2, and Na_2O and high Al_2O_3, FeOT, and K_2O, enriched in Rb, Ba, Th, U, and K and depleted in Nb, Ta, Sr, P, and Ti, indicative of shoshonitic basalt. The Late Silurian tonalites have positive $\varepsilon_{Nd}(t)$ (+4.91 to +6.18) values and primarily originated from depleted mantle magmas metasomatized by subduction fluids, supplemented by a small amount of subducted sediments and crustal materials. The Middle Devonian volcanic rocks exhibit low SiO_2, TiO_2, and Na_2O and high K_2O, and MgO, enriched in Rb, K, and LREEs and depleted in Nb, Ta, Sr, and HREEs, characteristic of shoshonitic volcanic rocks. Their $\varepsilon_{Nd}(t)$ (+2.11 to +3.77) and $\varepsilon_{Hf}(t)$ (+5.90 to +11.73) values are positive. These characteristics indicate that the Middle Devonian volcanic rocks primarily originated from depleted mantle magmas metasomatized by subduction fluids, with the addition of crustal materials or subducted sediments during their formation. Based on regional geological data, it is believed that the study area underwent the following evolutionary stages during the Silurian–Devonian period: (1) active continental margin stage of southward subduction of the Paleo–Asian Ocean (PAO) (443–419 Ma); (2) arc-continent collision stage (419–405 Ma); (3) post-collision extension stage (404–375 Ma); (4) active continental margin stage, with the PAO plate subducting southward once again (375–360 Ma).

Keywords: Silurian; Devonian; magmatic rocks; geochronology; geochemistry; petrogenesis; the eastern segment of the northern margin of the North China Craton

Citation: Li, B.; Chen, J.; Yang, F.; Liu, M.; Zang, Y.; Zhang, C. Middle Silurian–Middle Devonian Magmatic Rocks in the Eastern Segment of the Northern Margin of the North China Craton: Implications for Regional Tectonics. *Minerals* **2024**, *14*, 641. https://doi.org/10.3390/min14070641

Academic Editor: Alexandre V. Andronikov

Received: 19 May 2024
Revised: 15 June 2024
Accepted: 17 June 2024
Published: 23 June 2024

1. Introduction

The Central Asian Orogenic Belt (CAOB), situated between the North China, Tarim, and Siberian plates [1–6], represents the largest accretionary orogenic belt of the Phanerozoic. The CAOB is noted for its extensive crustal growth and significant mineralization potential during the same period [7–10]. The formation of the CAOB involved the subduction of the Paleo–Asian Ocean (PAO) and the amalgamation of terranes, consisting of a series of microcontinents, island arcs, accretionary complexes, ophiolites, forearc/back-arc basins, oceanic plateaus, and seamounts [1,4,11]. With a highly complicated geological tectonic evolution [12], the CAOB serves as a critical area for studying the processes of collision and amalgamation of micro-continentals, continental accretion, and ocean-continent transitions [6,13–17]. There is a consensus among scholars that the PAO closed along the Solonker-Xar Moron-Changchun-Yanji Suture (SXCYS) on the southeastern margin of the CAOB (Figure 1a) [4,6,12,13,18–29].

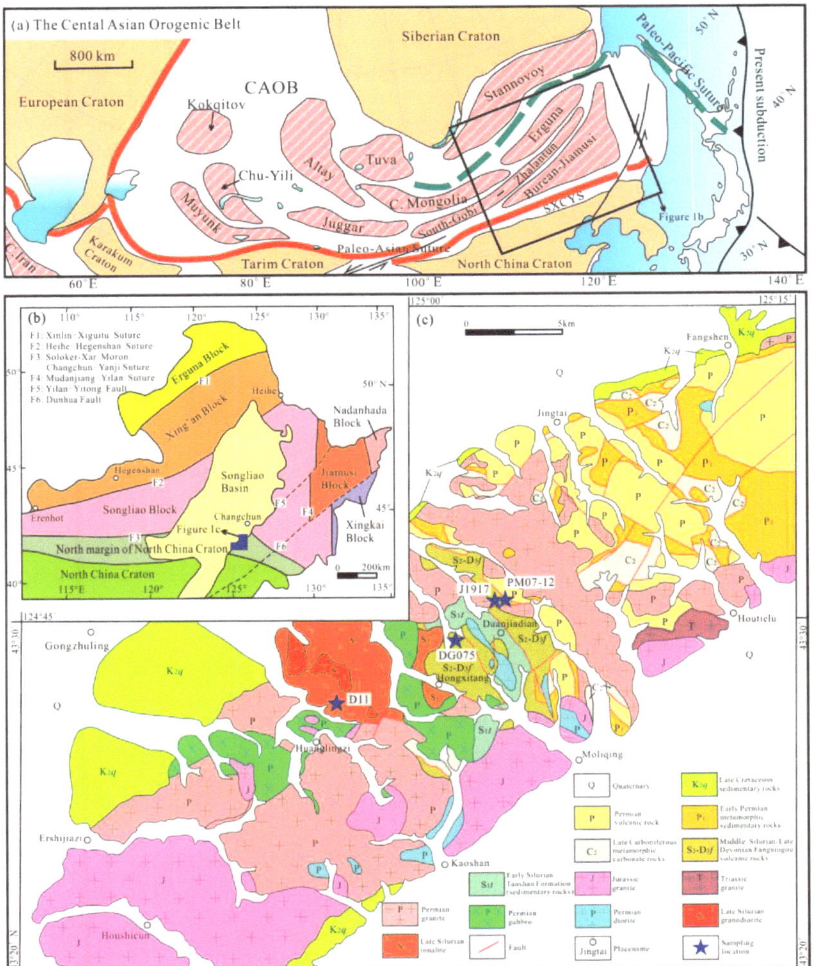

Figure 1. (**a**) Tectonic division of the Central Asian Orogenic Belt and adjacent regions (modified after [22]), and the position of Figure 1b (black color square); (**b**) Tectonic division of NE China (modified from [12]); (**c**) Geological sketch map of the Yitong area in Jilin Province (modified from [29]).

The northern margin of the North China Craton (NCC) is located at the southeastern margin of the CAOB (Figure 1a,b), north of the Bayan Obo-Chifeng-Kaiyuan fault and south of the SXCYS. This area primarily encompasses the Bainaimiao arc and the Ondor Sum subduction–accretionary complex [20,30,31]. The Bainaimiao arc extends approximately

1300 km from Baoerhantu to Bayan Obo, Bainaimiao, and Chifeng in Inner Mongolia to eastern Faku in Liaoning Province to northeastern Gongzhuling and Yanbian in Jilin Province [32]. At present, there are still different opinions on the basement properties, subduction polarity, and tectonic setting of the Bainaimiao arc. One suggests affinity of the arc basement with the NCC and the other proposes affinity with other continental blocks. It is also suggested that the Bainaimiao arc belt and the Ondor Sum subduction-accretionary complex constitute a trench-arc-basin system formed by the southward subduction of the PAO [15,20,33–41]; however, some argue for its formation through the northward subduction of the South Bainaimiao Ocean, accreting to the NCC via arc-continent collision [32,42–48]. Consequently, the Bainaimiao arc belt has been regarded as a Japan-style island arc [37,49,50], Andean-type continental margin arc [20,51], or exotic terrain [42]. In addition, there is controversy about the formation age of the Bainaimiao arc, suggested to be Middle Neoproterozoic [50,52,53] or Ordovician–Silurian [32,51,54,55].

Previous research was predominantly focused on the central and western regions of Inner Mongolia, while the northern margin of the NCC from northern Liaoning Province to central Jilin Province has received less attention [32,43,45,56,57]. Moreover, the northern Liaoning Province to central Jilin Province region, which was originally classified as Early Paleozoic, has now been classified as Late Paleozoic or Early Mesozoic [12,19,51,58–66]. Therefore, it remains uncertain whether the Bainaimiao arc belt can extend over the Songliao Basin into northeastern China [12,58,59,67]. With the deepening of research, some Ordovician–Devonian rocks have been discovered in recent years in northern Liaoning Province and central to eastern Jilin Province [43,44,48,56,64,68–71], providing valuable information for further understanding of tectonic evolution. The Silurian–Devonian rocks in Yitong County, Jilin Province, which document a continuous evolutionary history from the Silurian to the Permian, are particularly crucial for exploring the tectonic framework and accretionary processes along the eastern segment of the northern margin of the NCC. In this study, zircon U-Pb chronological, petrographical, geochemical, zircon Hf-isotope, and whole-rock Sr-Nd isotope analyses were conducted (Table 1), and the properties and tectonic setting of the source area of the Silurian–Devonian magmatic rocks in Yitong area are discussed. These data provide new evidence for the magmatic-tectonic evolution of the northern margin of the NCC in the Silurian–Devonian.

Table 1. The simplified list of the volcanic and intrusive rock samples in this study.

Age	Lithology	GPS	Samples No.
425 ± 6 Ma	Tonalite	E124°56′07″, N43°27′42″	Zircon U-Pb and Hf Isotope: D11 Whole-Rock: D11-1-1, D11-1-2, D11-1-3, D11-1-4 Sr-Nd: D11-1-1
428 ± 6.6 Ma	Basalt	E125°01′31″, N43°29′37″	Zircon U-Pb: DG075 Whole-Rock: DG075-1, DG075-2, DG075-3, DG075-4 Sr-Nd: DG075-1, DG075-2
388.4 ± 3.9 Ma	Andesite	E125°03′24″, N43°30′29″	Zircon U-Pb and Hf Isotope: J1917 Whole-Rock: J1917-2, J1917-5 Sr-Nd: J1917-2, J1917-5
384.1 ± 4.9 Ma	Metamorphic olivine-bearing basalt	E125°04′12″, N43°30′28″	Zircon U-Pb: PM07-12 Whole-Rock: PM07-9-2, PM07-9-4, PM07-9-5

2. Geological Setting and Samples

2.1. Geological Setting

The study area is located in the northwest of Yitong County, Jilin Province, Northeast China (Figure 1c). It is adjacent to the Yitong Basin to the east and the Songliao Basin to the west, belonging to the southeast margin of the CAOB and the eastern margin of the

NCC. This area mainly contains the Lower Silurian Taoshan Formation, Upper Silurian-Upper Devonian Fangniugou volcanic rocks, Upper Carboniferous metamorphic carbonate rocks, Lower Permian metamorphic clastic rocks, and Permian volcanic rocks. The main lithology of the Taoshan Formation consists of sandstone, siltstone, silty mudstone, and silty slate, interbedded with marlstone, and contains gracilite. Intrusive rocks include the Late Ordovician–Late Silurian tonalite and granodiorite; Middle–Late Permian granodiorite, monzonite, quartz monzonite, and gabbro; Early Triassic monzogranite; and Early–Middle Jurassic syenite granite, granodiorite, monzogranite and quartz diorite.

Fangniugou volcanic rocks, established as an informal mapping unit [72], primarily consist of intermediate to acidic volcanic rocks considered to have formed during the Ordovician. In recent years, scholars have obtained zircon U-Pb ages of 426 ± 4.8 Ma [29], 402 ± 4 Ma [68], 419 ± 3 Ma [43], 425 ± 4 Ma, 390 ± 4 Ma [44], and 375.5 ± 4.6Ma [29] from the Fangniugou volcanic rocks, suggesting that they were formed during the Late Silurian to Late Devonian period.

In the northeast of the Taoshan Formation, the Fangniugou volcanic rocks are exposed with a suite of metamorphic carbonates, clastic rocks, and volcanic rocks, which formed during the Carboniferous to Permian [32,45] (Figure 1c). The Paleozoic geological units in the study area are composed of a series of tectonic slices, which were formed during the Silurian to Permian, and gradually became younger from southwest to northeast. Reflected in the process of oceanic crust subduction, various geological bodies of different ages and environments were, in turn, collaged laterally, and then transformed by later tectonic events.

2.2. Sample Descriptions

The tonalite (D11) is gray in color, with a mylonitic structure and a fine-grained granitic texture (Figure 2a,b). The sample shows a typical mineral assemblage of plagioclase (~55 vol.%), quartz (~25 vol.%), biotite (~20 vol.%), and minor accessory minerals (~5 vol.%; e.g., allanite, epidote).

The basalt (DG075) is dark gray in color and shows a porphyritic texture with 27% phenocrysts (Figure 2c,d). The phenocrysts are mainly alkali feldspar (15%) and hornblende (12%). The matrix (73%) is predominantly microcrystalline and includes hornblende, alkali feldspar, and plagioclase.

The andesite (J1917) is gray in color and shows a porphyritic texture with 8% phenocrysts (Figure 2e,f). The phenocrysts are composed of plagioclase (5%) and hornblende (3%). Plagioclase undergoes alteration such as sericitization and chloritization. The matrix (92%) is predominantly microcrystalline and includes biotite, chlorite, sericite, felsic minerals, etc.

The metamorphic olivine-bearing basalt (PM07-12) is greenish–gray in color, with a porphyroblastic texture and massive structure (Figure 2g,h). The porphyroblasts are zoisite (6%), garnet (3%), and olivine (3%). The matrix (88%) is predominantly microcrystalline and includes zoisite, sericite, and plagioclase.

Figure 2. Field photographs and microphotographs of the Middle Silurian–Middle Devonian magmatic rocks from the Yitong area. (**a**,**b**) Tonalite (Sample D11); (**c**,**d**) Basalt (Sample DG075); (**e**,**f**) Andesite (Sample J1917); (**g**,**h**) Metamorphic olivine-bearing basalt (Sample PM07-12. In the figure, 12 represents the layer number of profile 07). Abbreviations: Qz, Quartz; Pl, Plagioclase; Kfs, K-feldspar; Srt, Sericite; Bi, Biotite; Hbl, hornblende; Ol, olivine; Ep, epidote; Zo, Zoisite; Grt, Garnet.

3. Analytical Methods

3.1. Whole-Rock Composition Analyses

Geochemical analyses of the rock samples were conducted at the Northeast China Supervision Inspection Center of Mineral Resources, Ministry of Natural Resources, Shenyang Center of Geological Survey, China Geological Survey, Shenyang, China, using contamination-free equipment. Major elements were quantified via X-ray fluorescence spectrometry (XRF), with errors in precision and accuracy of +/- 5%. Trace elements were measured using Inductively Coupled Plasma Mass Spectrometry (ICP-MS), with errors in precision and accuracy of less than 10%.

3.2. LA-ICP-MS Zircon U-Pb Dating

LA-ICP-MS Zircon U-Pb dating was performed at the Key Laboratory of Mineral Resources Evaluation in Northeast Asia, Ministry of Natural Resources, Jilin University, Changchun, China. The process utilized a COMPEx Pro type ArF excimer laser ablation system from the Coherent Company (Saxonburg, PA, USA) and an Agilent 7900 type ICP-MS (Agilent Technologies, Inc., Santa Clara, CA, USA). The laser spot diameter was set at 32 μm. High-purity helium served as the carrier gas, and argon was used as the auxiliary gas. Isotopic ratio corrections employed standard zircons 91500 and PLE, while the international standard NIST610 corrected the contents of Th, U, Pb, and other elements [73]. The procedures followed for these analyses are elaborated in Yuan et al. [74]. Age determinations were conducted using Isoplot software (version 3.0).

3.3. Zircon Lu-Hf Isotopic Analyses

Zircon Lu-Hf isotopic analyses were carried out at the Tianjin Institute of Geology and Mineral Resources. This process utilized a Thermo Fisher Neptune type Multi-Collector Inductively Coupled Plasma Mass Spectrometer (MC-ICPMS) and a 193 nm laser ablation system (NEW WAVE193nm FX), with a laser spot size of 50 μm (Thermo Fisher Scientific, Waltham, MA, USA). The methodologies applied, including isotopic fractionation corrections, are based on the techniques described by Geng Jianzhen et al. [75] and Wu Fuyuan et al. [17].

3.4. Whole-Rock Sr-Nd Isotopic Analyses

Whole-rock Sr-Nd isotopic analyses were conducted at the Analytical and Testing Research Center of the Beijing Research Institute of Uranium Geology. These analyses utilized an ISOPROBE-T thermal ionization mass spectrometer (Isotopx LTD., Cheshire, UK). The isotopic ratios for Sr and Nd were standardized using the values $^{86}Sr/^{88}Sr = 0.1194$ and $^{146}Nd/^{144}Nd = 0.7219$.

4. Analysis Results

4.1. Zircon U-Pb Geochronology

The zircon U-Pb dating results are shown in Supplementary Table S1.

Sample D11 is a tonalite from the Ximangzhang pluton. Twenty-five zircon grains from this sample were dated, with most results lying on or near the concordia line. These zircons are mostly granular and short columnar, with grain sizes of 100–200 μm and length/width ratios of 1:1 to 3:2. The cathodoluminescence (CL) images (Figure 3a) show that these zircon grains have typical zonal textures and Th/U ratios of 0.36–0.77, suggesting a magmatic origin. Of these, 17 zircon grains have $^{206}Pb/^{238}U$ ages of 420–437 Ma, with a weighted mean $^{206}Pb/^{238}U$ age of 425 ± 6 Ma (mean squared weighted deviation (MSWD) = 0.12) (Figure 4a), representing the best estimate for emplacement time of the tonalite. Another 8 zircon grains have $^{206}Pb/^{238}U$ ages of 445–459Ma, with a weighted mean $^{206}Pb/^{238}U$ age of 451 ± 10 Ma (MSWD = 0.08).

Sample DG075 is a basalt from the Fangniugou volcanic rocks. These zircons are short columnar, with grain sizes of 50–100 μm and length/width ratios of 2:1 to 3:2. The CL images (Figure 3b) show that these zircon grains have typical zonal textures and Th/U ratios of 0.41–0.75, suggesting magmatic zircons. Twenty zircon grains were analyzed in this study. Most results deviated from the concordia line, but all data could fit a line, showing a lower intercept age of 428 ± 6.6 Ma (MSWD = 0.77) (Figure 4b), which represents the best estimate for the eruption time of the basalt.

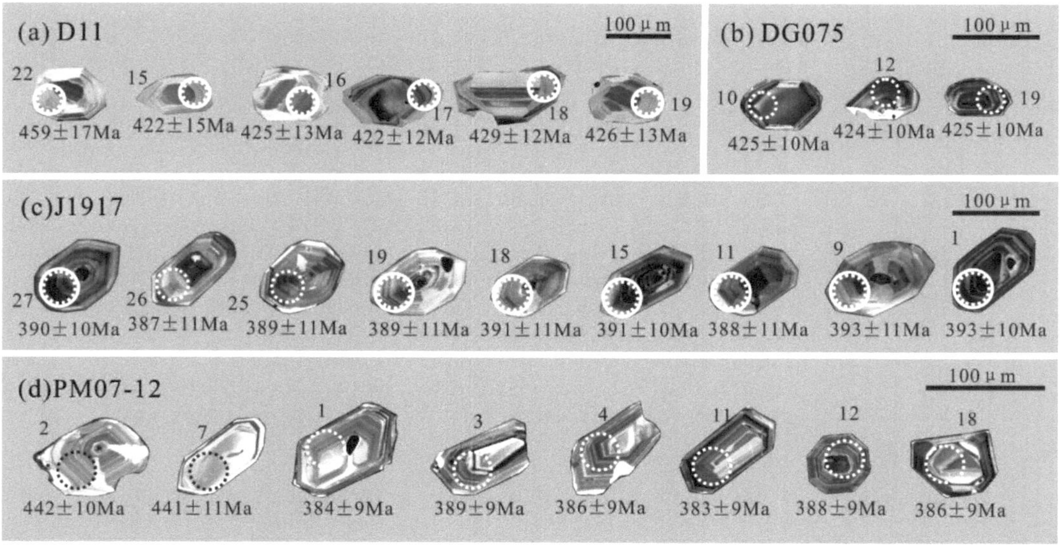

Figure 3. Representative CL images of zircons from the Middle Silurian–Middle Devonian magmatic rocks in the Yitong area. (**a**) CL image of zircons from sample D11. (**b**) CL image of zircons from sample DG075. (**c**) CL image of zircons from sample J1917. (**d**) CL image of zircons from sample PM07-12. The solid line circle is the Hf isotope analysis point; the dotted line circle is the U-Pb age analysis point. The numbers show the ages of the zircons (Ma) and the $\varepsilon_{Hf}(t)$ values.

Sample J1917 is an andesite from the Fangniugou volcanic rocks. These zircons are tabular and short columnar, with grain sizes of 50–100 μm and length/width ratios of 1:1 to 2:1. The CL images (Figure 3c) show that these zircon grains have typical zonal textures and Th/U ratios of 0.56–1.47, suggesting magmatic zircons. Twenty-nine zircon grains in this sample were dated, with all results lying on or near the concordia line. The analyses give a weighted mean $^{206}Pb/^{238}U$ age of 388.4 ± 3.9 Ma (MSWD = 0.23) (Figure 4c), representing the best estimate for the eruption time of the andesite.

Sample PM07-12 is a metamorphic olivine-bearing basalt, from the Fangniugou volcanic rocks. These zircons are tabular and short columnar (Figure 3d), with grain sizes of 50–120 μm and length/width ratios of 1:1 to 3:1. The CL images show that these zircon grains have typical zonal textures and Th/U ratios of 0.45–1.35, suggesting magmatic zircons. Eighteen zircon grains in this sample were dated, with all results lying on or near the concordia line. Of these, 13 zircon grains have $^{206}Pb/^{238}U$ ages of 372–389Ma, with a weighted mean $^{206}Pb/^{238}U$ age of 384.1 ± 4.9 Ma (MSWD = 0.33) (Figure 4d), representing the eruption time of the basalt. Another 4 zircon grains have $^{206}Pb/^{238}U$ ages of 446 ± 10 Ma, 442 ± 10 Ma, 441 ± 11 Ma, and 418 ± 11 Ma, most likely representing xenocrystic zircon ages.

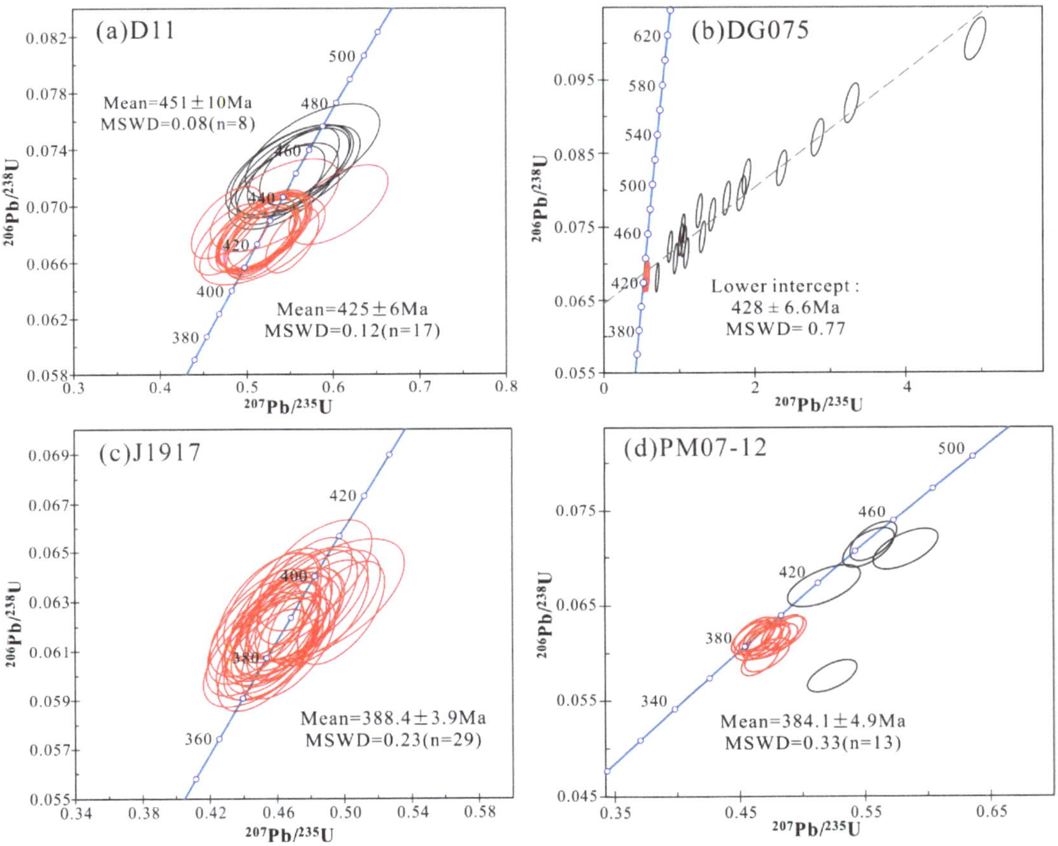

Figure 4. Zircon U-Pb concordia diagrams from the Middle Silurian–Middle Devonian magmatic rocks in the Yitong area. (**a**) Zircon U-Pb concordia diagrams from sample D11. (**b**) Zircon U-Pb concordia diagrams from sample DG075. (**c**) Zircon U-Pb concordia diagrams from sample J1917. (**d**) Zircon U-Pb concordia diagrams from sample PM07-12. Data-point error ellipses are 1 sigma. Age errors are 95% confidence.

4.2. Whole-Rock Major and Trace Elements Analysis

The whole-rock major and trace elements analysis results are shown in Supplementary Table S2.

The K_2O/Na_2O ratios of Late Silurian tonalite (D11) samples are 0.44–0.59, indicating that they are relatively enriched in Na and depleted in K. Their A/CNK ratios range from 0.91 to 1.00, showing metaluminous signatures. In the An-Ab-Or classification diagram (Figure 5a), data for the samples are located within the field of tonalite. As revealed by the rock series discrimination diagram (Figure 5d), the samples all fall into the calc-alkaline zone. The samples have relatively higher ΣREE contents (167.38–180.81 ppm), exhibiting LREE enrichment (La_N/Yb_N = 2.40–3.32) with slightly negative Eu anomalies (Eu/Eu* = 0.77–0.84) (Figure 6a). According to the parameters proposed by Irber [76] and Monecke [77], Ce_N/Ce^* and Pr_N/Pr^* of the tonalites are slightly less than 1, T1 is slightly greater than 0, and t1 is slightly less than 1 (Table S2), indicating that the first group of rare earth elements (La, Ce, Pr, and Nd) show a weak W-type tetrad effect. Tb_N/Tb^*, Dy_N/Dy^*, Tm_N/Tm^*, and Yb_N/Yb^* are significantly greater than 1, T3 is greater than 0, t3 and t4 are greater than 1 (Table S2), all indicating that the third group (Ga, Tb, Dy, and Ho) and the fourth group (Er, Tm, Yb, and Lu) of rare earth elements show a significant M-type tetrad effect. The trace element spider plot illustrates that the samples are relatively enriched in Rb, Ba, Th, U, and K (LILEs), and depleted in Nb, Sr, P, and Ti (HFSEs) (Figure 6b).

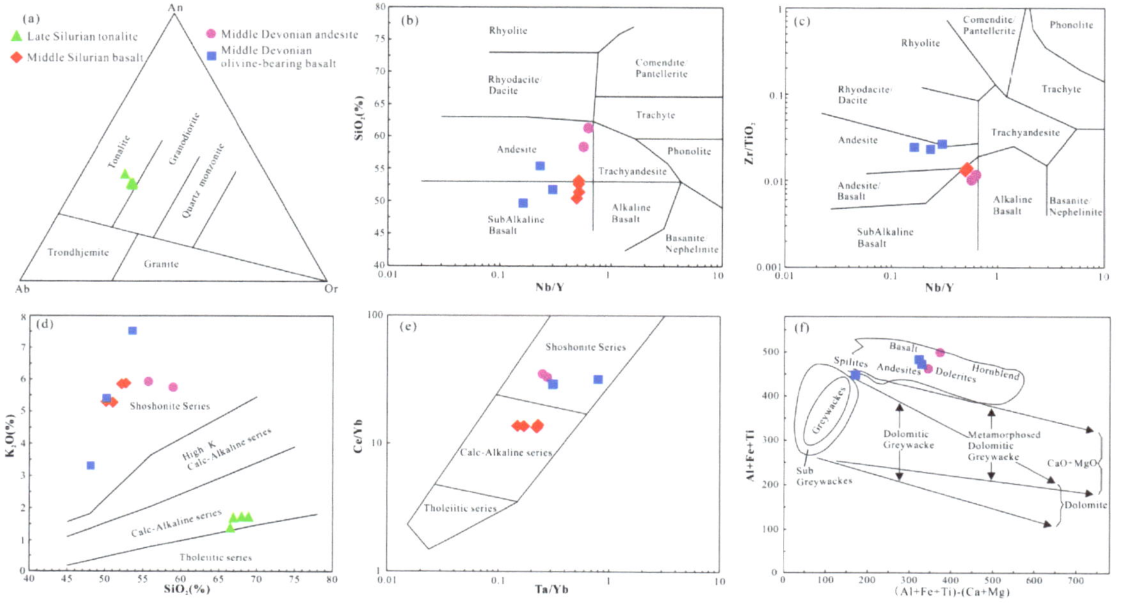

Figure 5. (**a**) An-Ab-Or [78], (**b**) SiO_2 versus Nb/Y [79], (**c**) Zr/TiO_2 versus Nb/Y [79], (**d**) K_2O versus SiO_2 [80], (**e**) Ce/Yb versus Ta/Yb [81], and (**f**) Al + Fe + Ti versus (Al + Fe + Ti) + (Ca-Mg) [82] diagrams for the Middle Silurian–Middle Devonian magmatic rocks in the Yitong area.

The K_2O/Na_2O ratios of Middle Silurian basalt (DG075) samples are 1.24–1.57, indicating that they are relatively enriched in K and depleted in Na. Their A/CNK ratios range from 0.58 to 0.64, showing metaluminous signatures. Their Rittmann index σ (9.70–11.75) suggests peralkaline characteristics. The basalt samples dominantly plot in the shoshonite series field (Figure 5d). In the SiO_2 vs. Nb/Y and Zr/TiO_2 vs. Nb/Y diagrams (Figure 5b,c), the data for the samples are located within the field of sub-alkaline basalt. They fall within the calc-alkaline series on the Ce/Yb vs. Ta/Yb diagram (Figure 5e) due to their low Ce content. The samples have relatively higher REE contents (145.15–152.30 ppm), exhibiting LREE enrichment (La_N/Yb_N = 5.89–6.16) with positive Eu anomalies (Eu/Eu* = 1.10–1.29) and negative Ce anomalies (Ce/Ce* = 0.68–0.74) (Figure 6a). The trace element spider plot illustrates that the samples are relatively enriched in Rb, Ba, Th, U, and K (LILEs), and depleted in Nb, Ta, Sr, P, and Ti (HFSEs) (Figure 6b).

The K_2O/Na_2O ratios of Middle Devonian andesite (J1917) samples are 57.74–150.66, indicating that they are enriched in K and depleted in Na. Their A/CNK ratios range from 2.03 to 2.10, showing peraluminous signatures. The andesite samples plot in the shoshonite series field (Figure 5d,e) and in the andesite and sub-alkaline basalt regions on the SiO_2 vs. Nb/Y and Zr/TiO_2 vs. Nb/Y diagrams (Figure 5b,c). The samples have relatively higher REE contents (171.29–188.49 ppm), exhibiting LREE enrichment (La_N/Yb_N = 12.15–14.54) with no Eu anomalies (Eu/Eu* = 0.91–0.95) and negative Ce anomalies (Ce/Ce* = 0.78–0.87) (Figure 6c). The trace element spider plot illustrates that the samples are relatively enriched in Rb, Th, U, and K (LILEs) and depleted in Nb, Ta, Sr, Zr, and HREEs (Figure 6d).

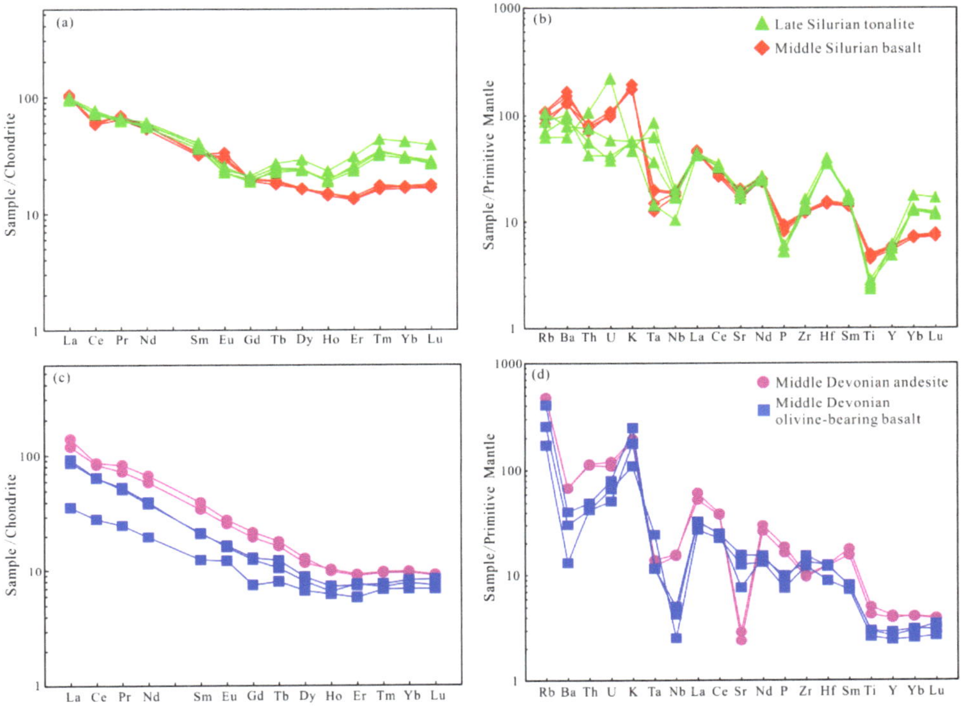

Figure 6. Chondrite–normalized REE patterns (**a**,**c**) (normalizing values after Boynton [83]), and primitive-mantle-normalized trace element spider diagrams (**b**,**d**) (normalizing values after Sun and Mc Donough [84]) for the Middle Silurian–Middle Devonian magmatic rocks in the Yitong area.

The K_2O/Na_2O ratios of the Middle Devonian metamorphic olivine-bearing basalt (PM07-12) samples are 13.25–24.14, indicating that they are enriched in K and depleted in Na. According to the (Al + Fe + Ti) vs. (Ca + Mg) diagram (Figure 5f), the samples predominantly classify within the regions of basalt and basaltic andesite volcanic rocks, suggesting that the protolith was an intermediate to mafic volcanic rock. Their A/CNK ratios range from 0.84 to 1.07, showing metaluminous–weakly peraluminous signatures. The andesite samples plot in the shoshonite series field (Figure 5d) and in the sub-alkaline basalt and andesite regions on the SiO_2 vs. Nb/Y and Zr/TiO_2 vs. Nb/Y diagrams (Figure 5b,c). The samples have relatively lower REE contents (93.52–106.89 ppm), exhibiting LREE enrichment (La_N/Yb_N = 9.77–10.23) with positive Eu anomalies (Eu/Eu* = 1.12–2.12) and no Ce anomalies (Ce/Ce* = 0.92–1.02) (Figure 6c). The trace element spider plot illustrates that the samples are relatively enriched in Rb and K and depleted in Ba, Th, U, Nb, Ta, Sr, P, Ti, and HREEs (Figure 6d).

4.3. Zircon Lu-Hf Isotopes

Based on LA-ICP-MS zircon U-Pb dating of Late Silurian tonalite and Middle Devonian andesite in the Yitong area, this study conducted in situ zircon Hf isotope analysis of some zircons from the dated samples (Figure 7a). The analytical results are shown in Supplementary Table S3.

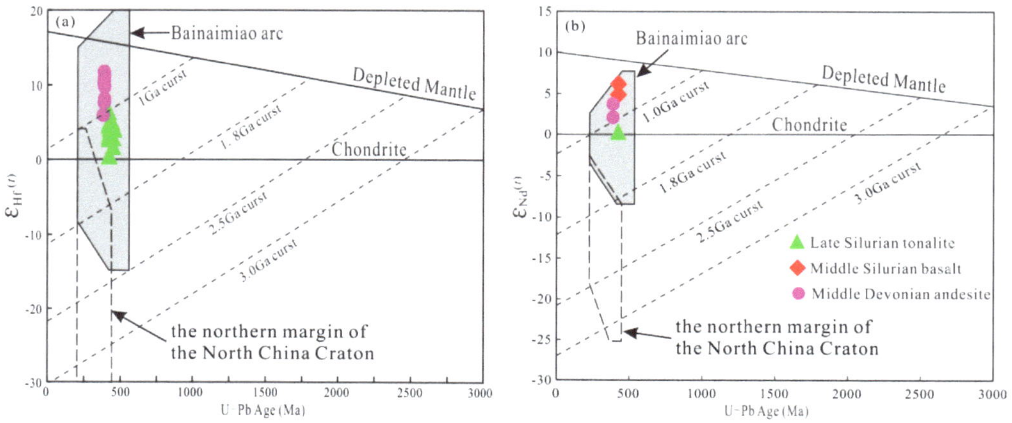

Figure 7. (**a**) $\varepsilon_{Hf}(t)$ versus U-Pb age for zircons [85], and (**b**) $\varepsilon_{Nd}(t)$ versus U-Pb age for zircons [86] of the Middle Silurian–Middle Devonian magmatic rocks in the Yitong area.

Sixteen zircons from Late Silurian tonalite sample D11 that represented the younger group gave $\varepsilon_{Hf}(t)$ values of +0.44–+6.31. They show single-stage Hf model ages (T_{DM1}) and two-stage Hf model ages (T_{DM2}) of 790–1027 Ma and 1005–1379 Ma, respectively. Ten zircons from Middle Devonian andesite sample J1917 gave $\varepsilon_{Hf}(t)$ values of +5.90–+11.73. They show single-stage Hf model ages (T_{DM1}) and two-stage Hf model ages (T_{DM2}) of 552–796 Ma and 636–1001 Ma, respectively.

4.4. Whole-Rock Sr-Nd Isotopes

Measured and age-corrected initial isotopic ratios of Sr and Nd are presented in Supplementary Table S4. The Sm, Nd, Rb, and Sr contents of these rocks are given in Supplementary Table S2.

The Late Silurian tonalite sample has initial $^{87}Sr/^{86}Sr$ ratios (I_{Sr}) of 0.70646 and $\varepsilon_{Nd}(t)$ values of +0.35. In contrast, the Middle Silurian basalt samples possess lower initial $^{87}Sr/^{86}Sr$ ratios (I_{Sr} = 0.70499–0.70508) and distinctly higher positive $\varepsilon_{Nd}(t)$ values (+4.91 to +6.18) (Figure 7b and Figure 9d). The Middle Devonian andesites have positive $\varepsilon_{Nd}(t)$ values of +2.11 to +3.77, but initial $^{87}Sr/^{86}Sr$ ratios (I_{Sr} = 0.66800–0.67546) are too low to have any genetic significance. All samples exhibit positive $\varepsilon_{Nd}(t)$ values, which is consistent with the Bainaimiao arc magmatic rocks in the CAOB (Figure 7b). The Middle Silurian basalts and Middle Devonian andesites have Neoproterozoic Nd model ages (T_{DM1} = 647–956 Ma; T_{DM2} = 663–964 Ma), whereas the Late Silurian tonalite has Mesoproterozoic Nd model age (T_{DM1} = 1119 Ma; T_{DM2} = 1138 Ma).

5. Discussions

5.1. Silurian to Devonian Magmatism in the Eastern Section of the Northern Margin of the NCC

Previous studies [29,32,43,44,48,68], indicate that the 424 ± 6 Ma Ximangzhang pluton underwent multiple episodes of magmatism during the Late Ordovician to Early Silurian (448–438 Ma), Late Silurian (425–419 Ma), and Early Devonian (414 Ma), forming a magmatic complex. The Fangniugou volcanic rocks emerged in the Middle Silurian to the Late Devonian, which was marked by multiple volcanic episodes at 428–425 Ma, 419 Ma, 404–400 Ma, 390–384 Ma, and 375 Ma.

In recent years, Cambrian–Devonian geological bodies have been reported successively from the eastern section of the northern margin of the NCC in Faku-Kaiyuan, Liaoning Province, and Gongzhuling-Liaoyuan-Zhangjiatun-Yanbian, Jilin Province (Figure 8) [29,43,44,48,56,64,68–70]. The Early Paleozoic magmatism in this region extended from the Late Cambrian to the Late Silurian, with peaks observed at 494 Ma, 486 Ma, 475 Ma, 467 Ma,

459 Ma, 438 Ma, and 422 Ma. These peaks resemble those of the Early Paleozoic Bainaimiao arc belt magmatic rocks [32], suggesting that the subduction-accretionary complex belt in the Faku-Yanji area is an extension of the Bainaimiao arc belt in the east. Combined with previous data, the Silurian magmatism in the northern margin of the NCC ranges from 443 to 419 Ma, and the Devonian magmatism ranges from 414 to 375 Ma (Table 2). The Silurian (443~419Ma) magmatism is frequent and distributed along the north margin of the NCC in an east–west direction, mainly exposed in the Faku, Gongzhuling, Yitong, Zhangjiatun, and Yanbian areas. The lithology is mainly intermediate-acid magmatic rocks, including andesite, quartz diorite, tonalite, granodiorite, rhyodacite, granite, rhyolite, etc. [29,32,43,44,48,56,57,87,88]. Mafic rocks are rarely exposed, only found in the Yitong area, as shoshonitic basalt. The Devonian (414–375 Ma) magmatism is weaker, exposed in Faku, Changtu, Gongzhuling, and Liaoyuan, with lithology mainly including granodiorite, rhyolite, rhyolitic tuff, dacite, monzogranite, etc., as well as a small amount of basalt and basaltic andesite [29,43,68–71,89].

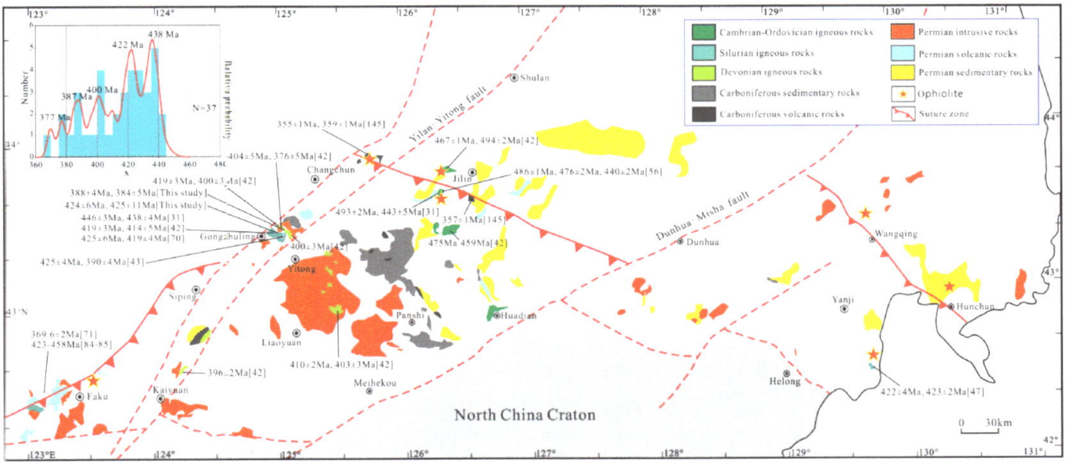

Figure 8. Distribution of Paleozoic magmatic rocks and strata from the Faku–Yanji area in the eastern segment of the northern margin of the NCC [17,29,31,42,43,47,56,70,71,84,85,89–91].

Table 2. Geochronological data for the Silurian–Devonian magmatic rocks in the eastern segment of the northern margin of the NCC.

Rocks	Sampling Location	Method	Ages (Ma)	References
Tonalite	E126°17′41″, N43°41′21″	LA-ICP-MS	443 ± 5	[56]
Tonalite	Zhangjiatun	LA-ICP-MS	440 ± 2	[57]
Quartz diorite	E124°56′15″, N43°27′38″	LA-ICP-MS	438 ± 4	[32]
Granodiorite	E123°14′51″, N42°32′59″	LA-ICP-MS	438 ± 4	[32]
Tonalite	E123°14′27″, N42°32′41″	SHRIMP	437 ± 3	[87]
Granite	E123°13′59″, N42°33′56″	LA-ICP-MS	436 ± 2	[32]
Granodiorite	E123°14′17″, N42°32′25″	LA-ICP-MS	436 ± 3	[32]
Granodiorite	E123°13′18″, N42°33′28″	LA-ICP-MS	434 ± 3	[32]
Granodiorite	E123°13′55″, N42°32′20″	SHRIMP	433 ± 3	[88]
Granodiorite (adakite)	E123°02′47″, N42°31′20″	SHRIMP	432 ± 3	[87]
Andesite	E125°01′51″, N43°28′00″	LA-ICP-MS	426 ± 4.8	[29]
Dacite	E125°08′51″, N43°28′04″	LA-ICP-MS	425 ± 4	[44]
Tonalitic mylonite	E124°58′29″, N43°29′40″	LA-ICP-MS	425 ± 6	[29]
Basalt	E125°01′31″, N43°29′37″	LA-ICP-MS	428 ± 6.6	This study
Tonalite	E124°56′07″, N43°27′42″	LA-ICP-MS	424 ± 6	This study

Table 2. *Cont.*

Rocks	Sampling Location	Method	Ages (Ma)	References
Felsic mylonite	E123°10′50″, N42°35′05″	SHRIMP	423 ± 5	[88]
Tonalite	E129°41′11″, N42°33′28″	LA-ICP-MS	423 ± 2	[48]
Tonalite	E129°40′37″, N42°33′12″	LA-ICP-MS	422 ± 4	[48]
Rhyolite	E125°02′37″, N43°29′38″	LA-ICP-MS	419 ± 3	[43]
Granodiorite	E124°56′03″, N43°27′44″	LA-ICP-MS	419 ± 3	[43]
Tonalite	E124°56′15″, N43°28′04″	LA-ICP-MS	419.5 ± 4	[29]
Granodiorite	E124°58′32″, N43°29′40″	LA-ICP-MS	414 ± 5	[43]
Rhyolitic tuff	E125°26′44″, N43°01′59″	LA-ICP-MS	410 ± 2	[43]
Rhyolitic tuff	E125°27′50″, N43°02′15″	LA-ICP-MS	403 ± 3	[43]
Basalt	Faku	LA-ICP-MS	405 ± 5	Shi et al. (unpublish)
Rhyolitic tuff rhyolitic crystal tuff rhyolitic crystal tuff	E125°00′51″, N43°31′52″	LA-ICP-MS	404 ± 4.5	[29]
Dacite	E125°01′40″, N43°53′11″	SHRIMP	402 ± 4	[68]
Monzogranite	E124°57′29″, N43°31′04″	LA-ICP-MS	400 ± 2	[43]
Monzogranite	E124°54′22″, N42°07′15″	LA-ICP-MS	396 ± 2	[43]
Rhyolite	E124°59′39″, N43°28′27″	LA-ICP-MS	390 ± 4	[44]
Biotite monzogranite	E124°16′44″, N42°48′08″	LA-ICP-MS	389 ± 2	[70]
Rhyolite crystal tuff	E124°27′13″, N42°54′15″	LA-ICP-MS	385 ± 2	[70]
Andesite	E125°03′24″, N43°30′29″	LA-ICP-MS	388 ± 4	This study
Metamorphic olivine-bearing basalt	E125°04′12″, N43°30′28″	LA-ICP-MS	384 ± 5	This study
Monzogranite	E124°15′47″, N42°49′01″	LA-ICP-MS	377 ± 2	[70]
Rhyolite	E125°01′20″, N43°32′00″	LA-ICP-MS	375.5 ± 5	[29]
High-Mg andesite	E123°33′19″, N43°39′13″	LA-ICP-MS	369.6 ± 2	[89]

5.2. Petrogenesis and Source Characteristics

5.2.1. Middle–Late Silurian Magmatic Rocks

The Middle–Late Silurian magmatic rocks display relatively weak metamorphism and deformation, with a low loss on ignition (LOI = 0.50–0.75), indicating minimal alteration effects. These geochemical characteristics make them well-suited for investigating their petrogenesis and source features.

The Late Silurian tonalites have high contents of SiO_2 and Na_2O, and low contents of K_2O, MgO, FeOT, and TiO_2, with an A/CNK ratio of 0.91–1.00, indicative of I-type granites. In the Na_2O vs. K_2O diagram (Figure 9a), the tonalite samples fall within the zone of the I-type granite. Additionally, in the petrogenetic discrimination diagrams (Figure 9b), all tonalite samples fall within the zone of unfractionated granitoids. These characteristics are consistent with those of metaluminous calc-alkaline I-type granitoids. They exhibit enrichment in Rb, Ba, Th, U, and K (LILEs) and depletion in Nb, Sr, P, and Ti (HFSEs), displaying magmatic arc features [19,92,93]. The explanations for the tetrad effect of rocks include mineral crystallization [94,95], late-stage magma-fluid interactions [76,96,97], or inheritance from the parental magma [76,97]. According to published papers [29,43], the Ximangzhang tonalites do not exhibit tetrad effects, indicating that this characteristic is not a feature of its primary magma. The low Zr/Hf (12.81–14.99) and Y/Ho (15.41–18.83) ratios in tonalites cannot be explained by mineral crystallization [98,99]. Crust-derived magmas that have undergone significant magma-hydrothermal exchange typically exhibit lower Zr/Hf ratios (<20) [76]. The Nb/Ta ratio (4.11–12.49, average 7.38) in tonalite is similar to the differentiated continental crust (11) [100]. Therefore, the REE anomalies may be caused by magma-fluid interactions. Increasing La concentrations correspond with a rise in the La/Sm ratio, indicative of partial melting processes during their formation. The absence of pronounced Sr and Eu anomalies further supports the partial melting of plagioclase in the source material. The tonalites have low Cr (24.70–30.82 ppm) and Ni (5.34–6.56 ppm) content and Mg# (39.26–43.79) values, indicating a crustal source. Their

La/Sm ratios between 3.72 and 4.34 (average 3.98) suggest a moderate degree of crustal contamination, given that ratios above 4.5 typically indicate strong contamination, while those below 2 suggest primary compositions [101]. Additionally, the tonalites display high Ba/Th (48.35–196.86, average 117.57) and Ba/La (14.27–24.49, average 19.79) ratios, but low Th content (3.63–9.13 ppm) and Th/Yb ratios (0.56–1.05), implying that slab-derived fluids primarily influenced the source of these Late Silurian magmatic rocks [102], with lesser contributions from sediments or melts. The Rb/Ba vs. Rb/Sr diagram (Figure 9c) shows that the source material for the tonalite is predominantly basalt. The $\varepsilon_{Nd}(t)$ vs. I_{sr} diagram (Figure 9d) indicates that the magmas are composed of approximately 70–75% basalt and 25–30% lower continental crust. The $\varepsilon_{Nd}(t)$ value of tonalite is +0.35, with a two-stage model age of 1138 Ma, and zircon $\varepsilon_{Hf}(t)$ values ranging from +0.44 to +6.31, with a two-stage model age ranging from 1005.0 to 1379.5 Ma. These characteristics indicate that the tonalite predominantly originated from partial melting of Mesoproterozoic accreted lower crustal material (basalt), and magma ascent was accompanied by contamination with lower crustal material.

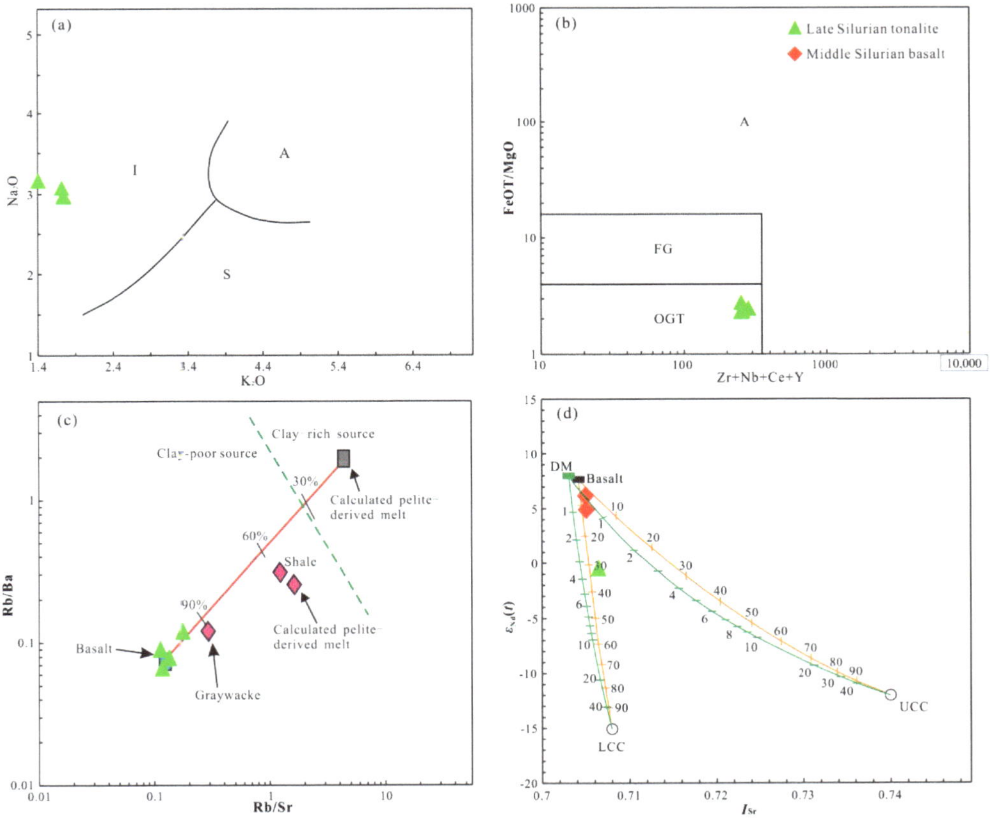

Figure 9. (**a**) Na_2O versus K_2O diagram [103], (**b**) FeOT/MgO versus (Zr + Nb + Ce + Y) diagram [104], (**c**) Rb/Ba versus Rb/Sr diagram (The red solid line represents the mixing curve between basalt- and pelite-derived melts; The green dashed line divides strongly peraluminous granites derived mostly from clay-poor sources from those formed mostly from clay-rich source rocks.) [105], and (**d**) $\varepsilon_{Nd}(t)$ versus I_{sr} diagram [105] for the Middle Silurian–Middle Devonian magmatic rocks in the Yitong area. FG: fractionated granitoid; OGT: unfractionated granitoid; I: I-type granitoid; S: S-type granitoid; A: A-type granitoid; DM: depleted mantle; LCC: lower continental crust; UCC: upper continental crust.

The Middle Silurian basalts have low contents of SiO_2, P_2O_5, TiO_2, and Na_2O and high contents of Al_2O_3, FeOT, CaO, and K_2O, with a K_2O/Na_2O ratio of 1.24–1.57. Notably, these basalts contain primarily K-feldspar phenocrysts, with σ values ranging from 9.70 to 11.75, and CIPW norm calculations indicating the presence of nepheline (Ne = 7.41–10.69), but alkaline minerals are absent in thin section observations. Positioned within the subalkaline basalt region in the SiO_2 vs. Nb/Y and Zr/TiO_2 vs. Nb/Y diagrams (Figure 5b,c), these samples are identified as shoshonitic rocks rather than typical alkaline rocks. The ratios of incompatible elements, largely unchanged by fractional crystallization, remain constant during the partial melting of the mantle, thereby serving as indicators of magma source characteristics [106]. Geochemical analysis reveals an average Lu/Yb ratio of 0.15–0.16, closely aligned with the typical mantle-derived magmatic value of 0.14–0.15 [107], and an average Nb/Ta ratio of 19.80, approaching the mantle standard of 17.5 [84,106], affirming their mantle-derived magmatic nature. Positive $\varepsilon_{Nd}(t)$ values ranging from +4.91 to +6.18 further affirm a mantle origin. Additionally, high ratios of La/Ta (37.94–60.22 > 22) and La/Nb (2.27–2.47 > 1.7), alongside low Zr/Ba (0.12–0.16) and La/Ba (0.03–0.04) ratios, point to a genesis within the lithospheric mantle, modified by subduction processes [108–112]. These basalts also exhibit high Ba/Th (128.08–177.84) and Ba/La (27.88–36.84) ratios and lower La/Sm (4.78–5.03) and Th/Yb (1.69–1.97) ratios, indicative of a mantle-derived magmatic source predominantly influenced by slab dehydration [102]. They are enriched in Rb, Ba, Th, U, K, and LREE, yet depleted in Nb, Ta, Sr, P, and Ti, characteristic of island arc magmatic rocks [19], suggesting an origin from the lithospheric mantle that has been metasomatized by subduction fluids [113]. Furthermore, relatively low Cr (67.60–70.51 ppm), Ni (38.95–40.13 ppm), and Co (17.96–19.97 ppm) contents and Mg# (32.15–34.95) values suggest a significant evolutionary process during its formation [114], such as fractional crystallization of olivine or pyroxene, or crustal contamination. Their average Rb/Sr ratio of 0.16 distinctly reflects a mixed mantle-crust source, consistent with the typical mantle-crust mixed source magmatic Rb/Sr ratio range of 0.05–0.50 [106]. Island arc volcanic rocks often exhibit a characteristic negative anomaly in Ce concentrations [115]. The samples have negative Ce anomaly (0.68–0.74) and low Ce/Th ratios (6.97–7.89), further suggesting the incorporation of subducted sediment melts into the source. The $\varepsilon_{Nd}(t)$ vs. I_{sr} diagram (Figure 9d) indicates that the magma mainly originated from a depleted mantle with some crustal material addition. The Nd isotope single-stage model age (T_{DM1}) ranges from 647 to 758 Ma, which represents the mixed age of juvenile crust and older crust, suggesting the presence of a Proterozoic crystalline basement in the study area. Collectively, these features indicate that the Middle Silurian basalts primarily originated from depleted mantle magmas metasomatized by subduction fluids, supplemented by a small amount of subducted sediments and crustal materials.

5.2.2. Middle Devonian Volcanic Rocks

The Middle Devonian volcanic rocks have undergone weak metamorphism and deformation with a slightly higher loss on ignition (LOI = 2.94–4.78). Therefore, to discern the rock origins and source characteristics, less mobile lithogenic elements along with rare earth elements and high field strength elements, which are less affected by metamorphism and alteration, are primarily utilized [116].

The Middle Devonian volcanic rocks have low SiO_2, TiO_2, and Na_2O content and high K_2O content, and a high K_2O/Na_2O ratio with a wide range of variation in Al_2O_3 content (ranging from 13.65 to 17.50 wt%). In the SiO_2 vs. Nb/Y and Zr/TiO_2 vs. Nb/Y diagrams (Figure 5b,c), these volcanic rocks primarily fall into the subalkaline basalt and andesite zones, indicating that the K-rich and Na-poor characteristics of the samples are a result of alteration. The samples are enriched in LILEs such as Rb, K, and LREEs and are depleted in Nb, Ta, Sr, and HREEs, showing geochemical characteristics similar to those of arc volcanics [117–121].

The Zr/Hf ratio of these volcanic rocks (28.74–49.78, average 38.59) is close to that of mantle-derived magmas (39) [122]; their average Lu/Yb ratio (0.14–0.17) is also close to the average mantle-derived magma value (0.14–0.15) [107], affirming their mantle-derived magmatic nature. The volcanic rocks have positive $\varepsilon_{Nd}(t)$ values (+2.11 to +3.77), also indicating mantle-derived magma. Additionally, high ratios of La/Ta (18.57–85.64, average 52.79 > 22) and La/Nb (3.29–10.22 > 1.7), alongside low Zr/Ba (0.22–1.50) and La/Ba (0.08–0.20) ratios, point to a genesis within the lithospheric mantle, modified by subduction processes [108–112].

It is generally believed that the fluids or melts formed during the subduction of oceanic crust and sediments metasomatize the overlying mantle wedge, resulting in partial melting and the formation of island arc magma [123,124]. The addition of subduction zone fluids significantly lowers the Ce/Pb and Nb/U ratios of magmas originating from depleted mantle [125]. The Middle Devonian volcanic rocks have Nb/U (1.27–4.86) and Ce/Pb (0.66–10.30) ratios lower than those of typical depleted mantle (Nb/U = 47 $\pm$ 10, Ce/Pb = 25 $\pm$ 5) [126], suggesting the addition of subduction zone fluids to the magma source. Sample J1917 has low Ce/Th ratios (7.07–7.13) and a Ce negative anomaly (0.78–0.87), suggesting the addition of subducted sediment melts to the source. Sample PM07 has higher Ce/Th ratios (10.36–11.98) and no Ce anomaly (0.92–1.02), suggesting no addition of reduced sediment melts in the source. The volcanic rocks have low Cr (2.35–10.09 ppm) and Ni (2.74–3.96 ppm) content and Mg# values (21.24–47.39), suggesting they may have undergone fractional crystallization of olivine and pyroxene, or crustal contamination [114]. Typically, the addition of crustal material leads to an increased abundance of LILEs (Rb, Ba, Th, etc.) and decreased content of HFSEs (Nb, Ta, Ti, etc.). Important indicators of crustal contamination include $(Th/Nb)_{PM} >> 1$ [127], La/Nb > 1 [128], and La/Sm values (>4.5) [101]. The Middle Devonian volcanic rocks exhibit $(Th/Nb)_{PM}$ = 7.12–17.02, La/Nb = 3.29–10.22, and La/Sm = 5.42–6.29, suggesting a limited degree of crustal contamination.

In the $\varepsilon_{Hf}(t)$ vs. t and $\varepsilon_{Nd}(t)$ vs. t diagrams (Figure 7a,b), the Middle Devonian volcanic rock samples fall within the $\varepsilon_{Hf}(t)$ and $\varepsilon_{Nd}(t)$ ranges of the Bainaimiao arc. The volcanic rocks have $\varepsilon_{Nd}(t)$ values of +2.11 to +3.77 and zircon $\varepsilon_{Hf}(t)$ values of +5.90 to +11.73, falling between the chondritic and depleted mantle evolution lines (Figure 7a,b), suggesting that their source mainly originated from depleted mantle; the decoupling of $\varepsilon_{Nd}(t)$ and $\varepsilon_{Hf}(t)$ indicates crustal contamination. The Nd isotope T_{DM2} ranges from 823 to 956 Ma, and the Hf isotope T_{DM2} ranges from 636.0 to 1001.4 Ma, representing the mixing ages of juvenile crust and Proterozoic crust, indicating the presence of a Proterozoic crystalline basement in the study area. These characteristics indicate that the Middle Devonian volcanic rocks primarily originated from depleted mantle magmas metasomatized by subduction fluids, with the addition of crustal materials or subducted sediments during their formation.

5.3. Tectonic Setting

In the study area, the Middle–Late Silurian magmatic rocks are mainly composed of basalt, andesite, tonalite, granodiorite, dacite, and rhyolite, which is similar to a typical island arc magmatic rocks assemblage. These rocks are enriched in Rb, Ba, Th, U, and K, while being depleted in Nb, Ta, Sr, P, and Ti, a signature suggesting formation in a subduction zone environment [92,129]. In the Sr/Y vs. Y and La/Yb vs. Th/Yb diagrams (Figure 10a,c), the tonalite and basalt samples fall within the zone of a typical island arc. Additionally, in the Rb vs. (Y + Nb) diagram (Figure 10d), the tonalite samples fall within the zone of a volcanic-arc granite. In the Th/Yb vs. Ta/Yb diagram (Figure 10d), the basalt samples fall within the active continental margin field. These Middle–Late Silurian magmatic rocks (428–419 Ma) share isotopic and geochemical traits with the earlier Late Ordovician to Early Silurian tonalite and quartz diorite (446 Ma, 438 Ma) [29,32]. Notably, the tonalite, characterized by low Sr and high Yb, formed at a depth similar to standard crustal thickness of about 30–40 km [130], suggesting that no arc-continent collision occurred during this period. Therefore, this study suggests that the Middle–Late Silurian magmatic rocks formed in an active continental margin environment.

The Middle Devonian volcanic rocks are marked by low SiO_2, Na_2O, MgO, Sr, and Y but are rich in K_2O. In the Sr/Y vs. Y diagram (Figure 10a), these samples fall within the zone of the typical island arc. In the Th/Yb vs. Ta/Yb diagram (Figure 10d), the samples fall within the active continental margin field, exhibiting characteristics of the calc-alkaline to shoshonitic series. The La/Yb vs. Th/Yb diagram (Figure 10c) further underscores their affinity with a mature continental arc. Furthermore, the Early Devonian rhyolitic tuff (404 ± 4.5 Ma) in the study area, showing the features of typical A2-type granite, formed in a post-collisional extension setting [29]. In the La/Sm vs. Sm/Yb diagram (Figure 11), the Middle Silurian and Middle Devonian volcanic rock samples can be derived from spinel-garnet lherzolite (Figure 11), with degrees of partial melting ranging from 1 to 5%, and magma source depths estimated between 60 and 80 km [131]. The Middle Silurian volcanic rocks can be derived from spinel lherzolite, whereas the Middle Devonian rocks are more likely derived from spinel-garnet lherzolite (spinel:garnet = 1:1), suggesting an increase in crustal thickness during the Middle Devonian. The collision of the Late Silurian Bainaimiao arc with the NCC led to the subsequent formation of a series of alkaline series magmatic rocks [31,86,132–138]. Combined with the Middle Devonian magmatic rocks in the northern Liaoning Province to central Jiling Province having characteristics of I- and A-type granites [43], this suggests that the study area was an extensional environment after the arc-continent collision during the Middle Devonian period.

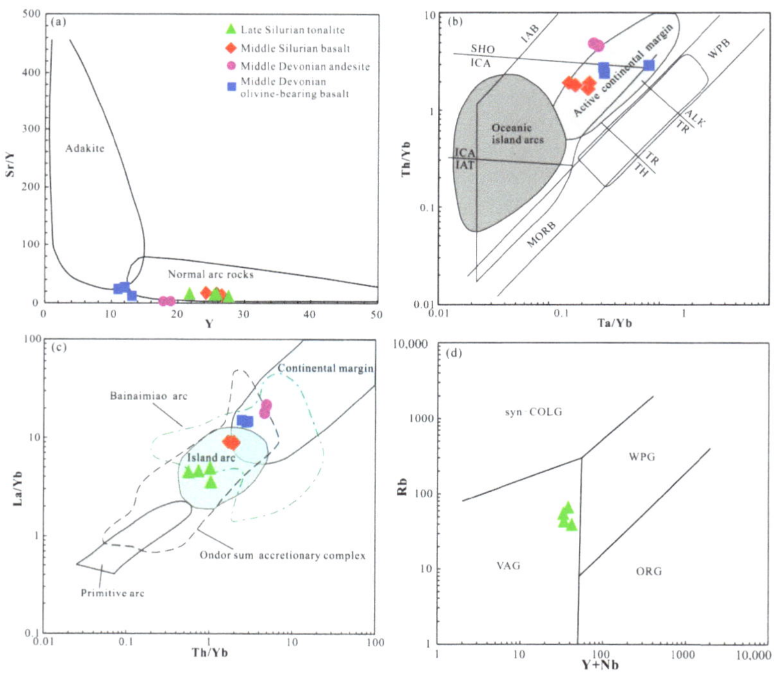

Figure 10. Tectonic environment discriminant diagrams for the Middle Silurian–Middle Devonian magmatic rocks in the Yitong area. (**a**) Sr/Y versus Sr diagram [139]; (**b**) Th/Yb versus Ta/Yb diagram [81]; (**c**) La/Yb versus Th/Yb diagram [140]; (**d**) Rb versus (Y + Nb) diagram [141]. MORB: mid-ocean ridge basalts; N-MORB: normal MORB; E-MORB: enriched MORB; WPB: within-plate basalt; OIB: ocean island basalts; OIT: ocean island tholeiitic; OIA: ocean island alkaline; SHO: shoshonite; CAB: calc–alkaline basalts; ICA: island-arc calc–alkaline basalts; IAT: island-arc tholeiitic; IAB: island-arc alkaline basalts; VAG: volcanic-arc granite; Syn-COLG: syn-collision granite; WPG: within plate granite; ORG: ocean-ridge granite.

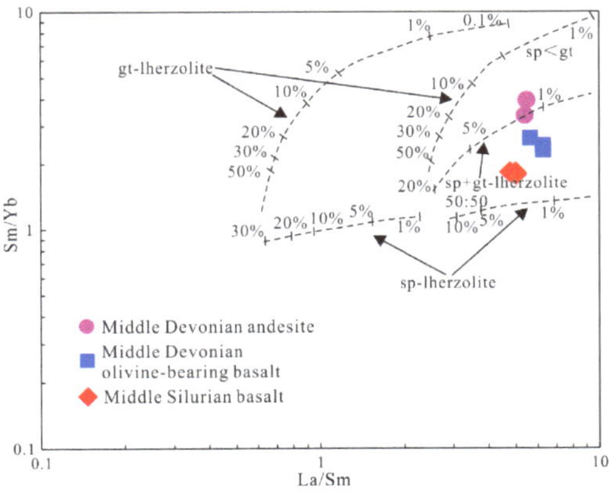

Figure 11. La/Sm-Sm/Yb diagram [141,142] for the Middle Silurian and Middle Devonian magmatic rocks in the Yitong area.

5.4. Silurian–Devonian Tectonic-Magmatic Evolution in the Eastern Section of the Northern Margin of the NCC

Based on previous studies and the division of tectonic evolution, the magmatic and tectonic progression of the eastern section of the northern margin of the NCC from the Silurian to Devonian can be delineated into four primary stages:

5.4.1. Active Continental Margin (443–419 Ma)

Early Paleozoic magmatism along the eastern segment of the northern margin of the NCC commenced in the Late Cambrian and persisted through the Late Silurian. The rock assemblages and peak ages of this Early Paleozoic magmatism align closely with those observed in the central-western segment of the northern margin of the NCC, indicating that the eastern magmatism is a continuation of the Bainaimiao arc [29,32]. Between the Late Cambrian and Middle Ordovician (493–458 Ma), the magmatic rocks are mainly composed of peridotite, pyroxenite, gabbro, high-Mg andesite, pyroxene andesite, quartz diorites, etc., representing the initial stages of subduction [43] within an intra-oceanic arc environment [57], and can be compared with the Ondor Sum subduction-accretionary complex. These early magmatic rocks are primarily located around areas such as Xiaosuhe, Zhangjiatun, Toudaogou, and Huadian, positioned northeastern of the study area, closer to the Changchun Yanji suture zone. Subsequent magmatism (446–419 Ma) is mainly composed of adakitic diorite, andesite, quartz diorites, granodiorites, tonalites, rhyolitic dacites, granites, and rhyolites. These rocks are similar to active continental margin rock assemblages and belong to the calc-alkaline to high-K calc-alkaline series, displaying mature island arc rock characteristics [43]. Silurian magmatic rocks are enriched in LREEs and LILEs, yet are depleted in HFSEs such as Nb, Ta, and Ti, mirroring the geochemistry of arc volcanic rocks. Discrimination diagrams suggest that these Silurian magmatic rocks were produced within an active continental margin setting. Due to subsequent strong modification, rocks from the Late Cambrian and Middle Ordovician are limited and scattered in exposure. Therefore, the position of the immature arc of the Late Cambrian and Middle Ordovician is not very clear. In general, the immature arc of the Late Cambrian and Middle Ordovician is located on the northeast side of the volcanic arc with an age of 446–419 Ma. They can be compared with the Bainaimiao Arc and Ondor Sum subduction-accretionary complex.

South of the Changchun-Yanji Suture zone, the study area records a continuous evolution process from the Late Ordovician to the Permian period. From southwest to northeast, Late Ordovician to Late Silurian intrusive rocks, Silurian to Devonian volcanic and clastic rocks, Late Carboniferous to Early Permian carbonate and clastic rocks, and

Permian metamorphic volcanic rocks are exposed sequentially. Based on rock assemblages, protolith formation, and geochemical characteristics, they can be divided into a Late Ordovician to Late Devonian arc-basin system and a Late Carboniferous to Late Permian arc-basin system. The Paleozoic geological bodies in the area are generally distributed in a northwest direction, and geological bodies of different ages are mostly in tectonic contact, reflecting the process of transpression during the southward subduction of the PAO. In summary, we suggest that the Silurian magmatic rocks in the area formed in an active continental margin environment during the southward subduction of the PAO.

5.4.2. Arc-Continent Collision (419–405 Ma)

The northern margin of the NCC and the Bainaimiao arc show different evolutionary histories and basement properties during the Early Paleozoic era [86]. Currently, many workers believe that the NCC and the Bainaimiao arc underwent an arc-continent collision at the end of the Early Paleozoic and the beginning of the Late Paleozoic [32,42–48]. The Xibiehe Formation and Zhangjiatun Formation, exposed in central Inner Mongolia to central Jilin Province, are considered to be post-collision molasse, unconformably overlying Early Paleozoic units [143]. The Xibiehe Formation in the Jiefangyingzi area of Chifeng City was deposited in the Early Devonian (with tuff interlayer ages of 405 $\pm$ 2 Ma); the Zhangjiatun Formation in central Jilin Province was deposited after 404 $\pm$ 14 Ma and 409 $\pm$ 8 Ma [43]. Detrital zircon geochronology from the Xibiehe Formation indicates that the arc-continent collision occurred during the Late Silurian to the Early Devonian [86]. Moreover, along the Chifeng-Kaiyuan fault, syn-collisional magmatic rocks have been dated to between 419 and 411 Ma [13,15,144,145]. Therefore, the arc-continent collision should have occurred in the Early Devonian (419–405 Ma).

5.4.3. Post-Collisional Extension (404–375 Ma)

During the Devonian period, the northern margin of the NCC and the central to the western section of the Bainaimiao arc show the development of an alkaline magmatic belt from 400 to 380 Ma [31], comprising mafic-ultramafic rocks, alkaline rocks, A-type granites, and bimodal volcanic rocks [31,62,133,135,137,146,147]. These features suggest that this region was in an extensional tectonic environment during the Devonian period. Devonian magmatism in the eastern section of the northern margin of the NCC (414–375 Ma) is mainly composed of basalt, basaltic andesite, andesite, granodiorite, monzonite, dacite, rhyolite, and rhyolite tuff, with notable magmatic peaks at 411 Ma, 400 Ma, 387 Ma, and 377 Ma (Figure 8). Age peaks correspond closely to those in the northern margin of NCC and the Bainaimiao arc (400 Ma, 392 Ma, 379 Ma) [31], suggesting a similar evolutionary history from the eastern to western sections of the northern margin of the NCC during the Devonian period.

In the study area, the Early Devonian rhyolitic tuff (404 Ma) shows A2-type granite geochemical characteristics and was formed in a post-collisional extensional tectonic setting [29]. The Early Devonian monzogranite (400 Ma) and Middle Devonian rhyolite (390 Ma) display adakitic geochemical characteristics, reflecting the thickening of the crust after collision [43,44]. In the Chifeng area, the Middle Devonian syenogranite (392 Ma), granite (385 Ma), and granitic porphyry (377 Ma) [132,148] are classified as A2-type granites, formed in a post-collisional or post-orogenic environment, indicating that the Middle Devonian setting was post-collisional. The intermediate and mafic volcanic rocks from the Middle Devonian (388–384 Ma) in the study area are rich in alkali elements and belong to the shoshonitic series, formed in an extensional setting. In summary, the northern margin of the NCC appears to have been in a post-collisional extensional environment during the Early to Middle Devonian (404–375 Ma).

5.4.4. Active Continental Margin (375–360 Ma)

Previous research suggests that extension during this period may have continued until the end of the Devonian period [70,147–149] or the Early Carboniferous period [150].

Studies on ophiolites and volcanic rocks indicate that southward subduction of the PAO commenced in the Late Devonian to Early Carboniferous. The discovery of Late Devonian calc-alkaline I-type rhyolite in the Yitong area (375.5 ± 4.6 Ma) [29] and high-K calc-alkaline I-type granite in the Changtu area (377 ± 2 Ma) [70] suggests that the PAO plate began to subduct southward during the Late Devonian. Late Devonian high-Mg andesite (370 Ma) in the Faku area, formed in a back-arc rift environment, supports that the PAO underwent southward subduction at this time, leading to the initiation of continental rifting in the Faku area and the development of a post-arc basin [71]. Song et al. [151] based on ophiolites exposed in the CAOB, proposed the existence of two major oceanic expansion/subduction cycles between 500–410 Ma and 360–220 Ma. Xie et al. [152] suggested that the Heishan mafic-ultramafic intrusive rock (357 ± 4 Ma) along the northern margin of the Tarim Craton formed in an active continental margin setting. In summary, this paper posits that the PAO began southward subduction during the Late Devonian (377–360 Ma), leading to the formation of a post-arc basin.

6. Conclusions

(1) Zircon U-Pb dating results show that the Ximangzhang tonalite formed in the Late Silurian (425 ± 6 Ma), while basalt, andesite, and metamorphic olivine-bearing basalt in the Fangniugou volcanic rocks formed in the Middle Silurian (428 ± 6.6 Ma) and Middle Devonian (388.4 ± 3.9 Ma, and 384.1 ± 4.9 Ma, respectively).

(2) The Late Silurian tonalites are characterized by high SiO_2 and Na_2O and low K_2O, MgO, FeOT, and TiO_2, with an A/CNK ratio of 0.91–1.00, and thus belong to the calc-alkaline series, identified as I-type granite. They are enriched in Rb, Ba, Th, U, and K and depleted in Nb, Sr, P, and Ti, with positive $\varepsilon_{Nd}(t)$ (+0.35) and $\varepsilon_{Hf}(t)$ (+0.44 to +6.31) values, suggesting that they mainly originated from partial melting of Meso–Neoproterozoic era accretionary lower crustal material (basalt). The Middle Silurian basalts are characterized by low SiO_2, P_2O_5, TiO_2, and Na_2O and high Al_2O_3, FeOT, and K_2O, enriched in Rb, Ba, Th, U, and K and depleted in Nb, Ta, Sr, P, and Ti, indicative of shoshonitic basalt. The Late Silurian tonalite has positive $\varepsilon_{Nd}(t)$ (+4.91 to +6.18) values and primarily originated from depleted mantle magmas metasomatized by subduction fluids, supplemented by a small amount of subducted sediments and crustal materials. The Middle–Late Silurian magmatic rocks formed in an active continental margin environment influenced by the southward subduction of the PAO.

(3) The Middle Devonian volcanic rocks exhibit low SiO_2, TiO_2, and Na_2O and high K_2O, and MgO and enriched in Rb, K, and LREEs and depleted in Nb, Ta, Sr, and HREEs, characteristic of shoshonitic volcanic rocks. $\varepsilon_{Nd}(t)$ (+2.11 to +3.77) and $\varepsilon_{Hf}(t)$ (+5.90 to +11.73) values are positive. These characteristics suggest that the Middle Devonian volcanic rocks primarily originated from depleted mantle magmas metasomatized by subduction fluids, with the addition of crustal materials or subducted sediments during their formation. The Middle Devonian magmatic rocks formed in an extensional environment after the arc-continent collision.

Supplementary Materials: The following supporting information can be downloaded at: https://www.mdpi.com/article/10.3390/min14070641/s1, Table S1. LA-ICP-MS zircon U-Pb data of samples from the Yitong Area; Table S2. Major element (wt.%), and trace elements (ppm) compositions of samples from the Yitong Area; Table S3. Lu-Hf isotopic compositions of zircons from the Yitong area; Table S4. Sr-Nd isotopic compositions of samples from the Yitong area.

Author Contributions: Methodology, B.L.; Investigation, B.L., F.Y., M.L., Y.Z. and C.Z.; Data curation, B.L.; Writing—original draft, B.L. and J.C., Writing—review & editing, B.L. and J.C. All authors have read and agreed to the published version of the manuscript.

Funding: This research was funded by the China Geological Survey (Grants DD20242929, DD20230210, and DD20190042-04) and the funding project of Northeast Geological S&T Innovation Center of China Geological Survey (NO. QCJJ2022–42).

Data Availability Statement: The authors confirm that the data generated or analyzed during this study are provided in full within the published article.

Conflicts of Interest: The authors declare no conflict of interest.

References

1. Badarch, G.; Cunningham, W.D.; Windley, B.F. A new terrane subdivision for Mongolia: Implications for the Phanerozoic crustal growth of central Asia. *J. Asian Earth Sci.* **2002**, *21*, 87–110. [CrossRef]
2. Buslov, M.M.; Saphonova, I.Y.; Watanabe, T.; Obut, O.T.; Fujiwara, Y.; Iwata, K.; Semakov, N.N.; Sugai, Y.; Smirnova, L.V. Kazansky, A.Y. Evolution of the Paleo-Asian Ocean (Altai-Sayan region, central Asia) and collision of possible Gondwana-derived terranes with the southern marginal part of the Siberian continent. *Geosci. J.* **2001**, *5*, 203–224. [CrossRef]
3. Xiao, W.J.; Huang, B.C.; Han, C.M.; Sun, S.; Li, J.L. A review of the western part of the Altaids: A key to understanding the architecture of accretionary orogens. *Gondwana Res.* **2010**, *18*, 253–273. [CrossRef]
4. Xiao, W.J.; Windley, B.F.; Sun, S.; Li, J.L.; Huang, B.C.; Han, C.M.; Yuan, C.; Sun, M.; Chen, H.L. A tale of amalgamation of three Permo-Triassic Collage Systems in Central Asia: Oroclines, sutures, and terminal accretion. *Annu. Rev. Earth Planet. Sci.* **2015**, *43*, 477–507. [CrossRef]
5. Xu, B.; Zhao, P.; Wang, Y.Y.; Liao, W.; Luo, Z.W.; Bao, Q.Z.; Zhou, Y.H. The Pre-Devonian tectonic framework of Xing'an-Mongolia orogenic belt (XMOB) in North China. *J. Asian Earth Sci.* **2015**, *97*, 183–196. [CrossRef]
6. Liu, Y.J.; Li, W.M.; Feng, Z.Q.; Wen, Q.B.; Neubauer, F.; Liang, C.Y. A review of the Paleozoic tectonics in the eastern part of Central Asian Orogenic Belt. *Gondwana Res.* **2017**, *43*, 123–148. [CrossRef]
7. Windley, B.F.; Allen, M.B.; Zhang, C.; Zhao, Z.Y.; Wang, G.R. Paleozoic accretion and Cenozoic redeformation of the Chinese Tien Shan Range. *Cent. Asia Geol.* **1990**, *18*, 128–131.
8. Sengör, A.M.C.; Natal'in, B.A.; Burtman, V.S. Evolution of the Altaid tectonic collage and Paleozoiccrustal growth in Eurasia. *Nature* **1993**, *364*, 299–307. [CrossRef]
9. Yakubchuk, A. Architecture and mineral deposit settings of the Altaid orogenic collage: A revised model. *J. Asian Earth Sci.* **2004**, *23*, 761–779. [CrossRef]
10. Jahn, B.M.; Wu, F.Y.; Hong, D.W. Important crustal growth in the Phanerozoic: Isotopic evidence of granitoids from East central Asia. *J. Earth Syst. Sci.* **2000**, *109*, 5–20. [CrossRef]
11. Liu, Y.J.; Li, W.M.; Ma, Y.F.; Feng, Z.Q.; Guan, Q.B.; Li, S.Z.; Chen, Z.X.; Liang, C.Y.; Wen, Q.B. An orocline in the eastern Central Asian Orogenic Belt. *Earth-Sci. Rev.* **2021**, *221*, 103808. [CrossRef]
12. Wu, F.Y.; De, Y.S.; Wen, C.G.; Zhang, Y.B.; Grant, M.L.; Wilde, S.A.; Jahn, B.M. Geochronology of the Phanerozoic granitoids in northeastern China. *J. Asian Earth Sci.* **2011**, *41*, 1–30. [CrossRef]
13. Jian, P.; Liu, D.Y.; Krner, A.; Windley, B.F.; Shi, Y.R.; Zhang, F.Q.; Shi, G.H.; Miao, L.C.; Zhang, W.; Zhang, Q.; et al. Time scale of an Early to Mid-Paleozoic orogenic cycle of the long-lived Central Asian Orogenic Belt, Inner Mongolia of China: Implications for continental growth. *Lithos* **2008**, *101*, 233–259. [CrossRef]
14. Xu, B.; Zhao, P.; Bao, Q.Z.; Zhou, Y.H.; Wang, Y.Y.; Luo, Z.W. Preliminary study on the Pre-Mesozoic tectonic unitdivision of the Xing-Meng Orogenic Belt(XMOB). *Acta Petrol. Sin.* **2014**, *30*, 1841–1857.
15. Chen, J.S.; Liu, M.; Li, B.; Li, W.; Li, W.W.; Yang, F.; Wang, Y. Zircon U-Pb chronology and geochemical characteristics of Late Silurian monzogranite in Ongniud Bannar, Inner Mongolia. *Geol. Bull. China* **2017**, *36*, 1359–1368. (In Chinese with English Abstract)
16. Feng, Z.Q.; Liu, Y.J.; Li, L.; Jin, W.; Jiang, L.W.; Li, W.M.; Wen, Q.B.; Zhao, Y.L. Geochemical and geochronological constraints on the tectonic setting of the xinlin ophiolite, northern Great Xing'an Range, NE china. *Lithos* **2019**, *326–327*, 213–229. [CrossRef]
17. Liu, Y.J.; Feng, Z.Q.; Jiang, L.W.; Jin, W.; Li, W.M.; Guan, Q.B.; Wen, Q.B.; Liang, C.Y. Ophiolite in the eastern Central Asian Orogenic Belt, NE China. *Acta Petrol. Sin.* **2019**, *35*, 3017–3047. (In Chinese with English Abstract)
18. Wu, F.Y.; Jahn, B.M.; Wilde, S.M.; Sun, D.Y. Phanerozoic crustal growth: U-Pb and Sr-Nd isotopic evidence from the granites in northeastern China. *Tectonophysics* **2000**, *328*, 89–113. [CrossRef]
19. Wu, F.Y.; Li, X.H.; Yang, J.H.; Zheng, Y.F. Discussions on the petrogenesis of granites. *Acta Petrol. Sin.* **2007**, *23*, 1217–1238. (In Chinese with English Abstract)
20. Xiao, W.J.; Windley, B.F.; Hao, J.; Zhai, M.G. Accretion leading to collision and the Permian Solonker suture, Inner Mongolia, China: Termination of the Central Asian Orogenic Belt. *Tectonics* **2003**, *22*, 1069. [CrossRef]
21. Xiao, W.J.; Krner, A.; Windley, B.F. Geodynamic evolution of Central Asia in the Paleozoic and Mesozoic. *Int. J. Earth Sci.* **2009**, *98*, 1185–1188. [CrossRef]
22. Li, J.Y. Permian geodynamic setting of Northeast China and adjacent regions: Closure of the Paleo-Asian Ocean and subduction of the Paleo-Pacific Plate. *J. Asian Earth Sci.* **2006**, *26*, 207–224. [CrossRef]
23. Jian, P.; Liu, D.Y.; Kröner, A.; Windley, B.F.; Shi, Y.R.; Zhang, W.; Zhang, F.Q.; Miao, L.C.; Zhang, L.Q.; Tomurhuu, D. Evolution of a Permian intraoceanic arc-trench system in the solonker suture zone, Central Asian Orogenic Belt, China and Mongolia. *Lithos* **2010**, *118*, 169–190. [CrossRef]
24. Wilde, S.A. Final amalgamation of the Central Asian Orogenic Belt in NE China: Paleo-Asian Ocean closure versus Paleo-Pacific plate subduction: A review of the evidence. *Tectonophysics* **2015**, *662*, 345–362. [CrossRef]

25. Chen, J.S.; Tian, D.X.; Yang, H.; Li, W.W.; Liu, M.; Li, B.; Yang, F.; Li, W.; Wu, Z. Triassic granitic magmatism at the northern margin of the North China Craton: Geochronology, geochemistry, and implications for the tectonic evolution of the Central Asian Orogenic Belt. *Acta Geol. Sin. (Engl. Ed.)* **2019**, *93*, 1325–1353. [CrossRef]
26. Zhang, C.; Neubauer, F.; Liu, Z.H.; Cui, F.H.; Guan, Q.B. Final-stage magmatic record of Paleo-Asian Oceanic subduction? Insights from Late Permian to Early Triassic intrusive rocks in the Yanbian area, easternmost Central Asian Orogenic Belt. *Minerals* **2020**, *10*, 799. [CrossRef]
27. Guan, Q.B.; Liu, Z.H.; Liu, Y.J.; Li, S.Z.; Wang, S.J.; Chen, Z.X.; Zhang, C. A tectonic transition from closure of the Paleo-Asian Ocean to subduction of the Paleo-Pacific plate: Insights from Early Mesozoic igneous rocks in eastern Jilin Province, NE China. *Gondwana Res.* **2022**, *102*, 332–353. [CrossRef]
28. Yang, H.; Ge, W.C.; Santosh, M.; Ji, Z.; Dong, Y.; Jing, Y.; Wu, H.R. The role of continental fragments in the formation of intra-oceanic arcs: Constraints from Sr-Nd-Hf-O isotopes of gabbro from the Jiamusi Block, NE China. *Gondwana Res.* **2022**, *103*, 297–313. [CrossRef]
29. Li, B.; Shi, Y.; Yang, F.; Liu, M.; Wu, Z. Middle Silurian-Late Devonian tectonic evolution in the eastern segment of the northern margin of the North China Plate: Evidence from magmatic rocks in Gongzhuling area. *Acta Petrol. Sin.* **2022**, *38*, 2364–2396. (In Chinese with English Abstract)
30. Shi, G.Z.; Faure, M.; Xu, B.; Zhao, P.; Chen, Y. Structural and kinematic analysis of the Early Paleozoic Ondor Sum-Hongqi mélange belt, eastern part of the Altaids (CAOB) in Inner Mongolia, China. *J. Asian Earth Sci.* **2013**, *66*, 123–139. [CrossRef]
31. Zhang, Q.Q.; Zhang, S.H.; Zhao, Y.; Liu, J.M. Devonian alkaline magmatic belt along the northern margin of the North China Block: Petrogenesis and tectonic implications. *Lithos* **2018**, *302*, 496–518. [CrossRef]
32. Zhang, S.H.; Zhao, Y.; Ye, H.; Liu, J.M.; Hu, Z.C. Origin and evolution of the Bainaimiao arc belt: Implications for crustal growth in the southern Central Asian Orogenic Belt. *Geol. Soc. Am. Bull.* **2014**, *126*, 1275–1300. [CrossRef]
33. Zhou, Z.H.; Mao, J.W.; Ma, X.H.; Che, H.W.; Ouyang, H.G.; Gao, X. Geochronological framework of the Early Paleozoic Bainaimiao Cu-Mo-Au deposit, NE China, and its tectonic implications. *J. Asian Earth Sci.* **2017**, *144*, 323–338. [CrossRef]
34. Qian, C.; Lu, L.; Wang, Y.; Guo, R.R.; Tang, Z. Age and geochemistry of amphibolite in Shuangsheng area, eastern Inner Mongolia: New evidence from the Paleoproterozoic basement of Bainaimiao island arc. *Geol. Bull. China* **2020**, *39*, 905–918. (In Chinese with English Abstract)
35. Chen, Y.; Zhang, Z.; Qian, X.; Li, J.; Ji, Z.; Wu, T. Early to Mid-Paleozoic magmatic and sedimentary records in the Bainaimiao Arc: An advancing subduction-induced terrane accretion along the northern margin of the North China Craton. *Gondwana Res.* **2020**, *79*, 263–282. [CrossRef]
36. Zhou, Z.G.; Zhang, D.; Gu, Y.C.; Wang, G.S.; Li, Y.H.; Yu, Y.S.; Liu, C.F.; Liu, W.C. Characteristics of Bainaimiao thrust belt along central Inner Mongolia in North China and its geological significance. *Geotecton. Metallog.* **2018**, *42*, 1–17. (In Chinese with English Abstract)
37. Gao, J.Y.; Wang, Y.X.; Qiu, Y.Z.; Zhang, Q. Islands-ocean structural evolution of mid-western continent in Inner Mongulia. *Geotecton. Metallog.* **2001**, *25*, 397–404. (In Chinese with English Abstract)
38. Shang, H.S.; Tao, J.X.; Baoyin, W.L.J.; Hao, X.Y. The arc-basin system and tectonic significance of Early Paleozoic in Baiyun'ebo area Inner Mongolia. *Geol. Surv. Res.* **2003**, *26*, 160–168 (In Chinese with English Abstract)
39. De Jong, K.; Xiao, W.J.; Windley, B.F.; Masago, H.; Lo, C.H. Ordovician $^{40}Ar/^{39}Ar$ phengite ages from the blueschist-facies OndorSum subduction-accretion complex (Inner Mongolia) and implications for the Early Paleozoic history of continental blocks in China and adjacent areas. *Am. J. Sci.* **2006**, *306*, 799–845. [CrossRef]
40. Xu, B.; Charvet, J.; Chen, Y.; Zhao, P.; Shi, G. Middle Paleozoic convergent orogenic belts in western Inner Mongolia (China): Framework, kinematics, geochronology and implications for tectonic evolution of the Central Asian Orogenic Belt. *Gondwana Res.* **2013**, *23*, 1342–1364. [CrossRef]
41. Wu. C.; Liu, C.F.; Zhu, Y.; Zhou, Z.G.; Jiang, T.; Liu, W.C.; Li, H.Y.; Wu, C.; Ye, B.Y. Early Paleozoic magmatic history of central Inner Mongolia, China: Implications for the tectonic evolution of the southeast Central Asian Orogenic Belt. *Int. J. Earth Sci.* **2016**, *105*, 1307–1327. [CrossRef]
42. Ma, S.X.; Wang, Z.; Zhang, Y.; Sun, J. Bainaimiao arc as an exotic terrane along the northern margin of the north china craton: Evidences from petrography, zircon U-Pb dating, and geochemistry of the Early Devonian deposits. *Tectonics* **2019**, *38*, 2606–2624. [CrossRef]
43. Pei, F.P.; Zhang, Y.; Wang, Z.W.; Cao, H.H.; Xu, W.L.; Wang, Z.J.; Wang, F.; Yang, C. Early-Middle Paleozoic subduction-collision history of the south-eastern Central Asian Orogenic Belt: Evidence from igneous and metasedimentary rocks of central Jilin Province, NE China. *Lithos* **2016**, *261*, 164–180. [CrossRef]
44. Han, Z.Z.; Song, Z.G.; Han, C.; Zhong, W.J.; Han, M.; Yan, J.L.; Liu, H.; Du, Q.X.; Gao, L.H.; Li, J.J. U-Pb ages and Hf isotopic composition of zircons and whole rock geochemistry of volcanic rocks from the Fangniugou area: Implications for Early–Middle Paleozoic tectonic evolution in Jilin Province, NE China. *J. Miner. Petrol. Sci.* **2018**, *113*, 10–23. [CrossRef]
45. Song, Z.G.; Han, C.; Liu, H.; Han, Z.Z.; Yan, J.L.; Zhong, W.J.; Gao, L.H.; Du, Q.X.; Han, M.; Li, J.J. Early-Middle Ordovician intermediate-mafic and ultramafic rocks in central Jilin Province, NE China: Geochronology, origin, and tectonic implications. *Min. Petrol.* **2019**, *113*, 393–415. [CrossRef]
46. Chen, Q.; Chou, G.L.; Xue, L.F.; Zhou, X.M.; Zhou, H.P.; Du, Y.S. Palaeoplate evolution in the southern part of the Inner Mongolian orogenic belt. *Geol. Rev.* **1993**, *39*, 477–483. (In Chinese with English Abstract)

47. Li, S.Q. Early Palaeozoic era of terrane pieced together and accreted in middle of Inner Mongolia. *Geol. Inn. Mong.* **1997**, *1*, 18–230. (In Chinese with English Abstract)
48. Wang, Z.W.; Pei, F.P.; Xu, W.L.; Cao, H.H.; Wang, Z.J.; Zhang, Y. Tectonic evolution of the eastern Central Asian Orogenic Belt Evidence from zircon U-Pb-Hf isotopes and geochemistry of Early Paleozoic rocks in Yanbian region, NE China. *Gondwana Res* **2016**, *38*, 334–350. [CrossRef]
49. Tang, K.D. *Tectonic Evolution and Minerogenetic Regularities of the Fold Belt along the Northern Margins of Sino-Korean Plate*; Peking University Press: Beijing, China, 1992; pp. 1–285. (In Chinese)
50. Hu, X.; Xu, C.S.; Niu, S.Y. *Continental Edge Evolution of Early Paleozoic Era in the Northern Margin of Huabei Platform*; Peking University Press: Beijing, China, 1990; pp. 1–273. (In Chinese with English Abstract)
51. Zhang, C.; Liu, Y.J.; Zhang, Z.L.; Cui, F.H.; Guan, Q.B.; Li, Y. Deformation and geochronological characteristics of Gudonghe ductile shear zone in Yanbian area. *Earth Sci.* **2019**, *44*, 3252–3264. (In Chinese with English Abstract)
52. Li, J.W.; Pei, R.F.; Zhang, D.Q.; Wang, C.X.; Po, Y.H. Metallogenic geological evolution of Bainaimiao copper deposit, Inner Mongolia. *Miner. Depos.* **2002**, *21*, 405–408. (In Chinese)
53. Nie, F.J.; Zhang, H.T.; Chen, Q.; Meng, L.Y.; Qiu, G.L.; Li, D.L. The zircon U-Pb age of metamorphosed basic volcanic rocks from the Bainaimiao Group in Inner Mongolia. *Chin. Sci. Bull.* **1991**, *36*, 738–742.
54. Li, J.J.; Dang, Z.C.; Zhao, Z.L.; Shi, Y.R.; Liu, D.Y.; Li, C.; Qu, W.J.; Wang, C.X.; Fu, C.; Tang, W.L.; et al. The metallogenic epochs of Bainaimiao copper deposit in Inner Mongolia. *Acta Geol. Sin.* **2015**, *89*, 1448–1457. (In Chinese with English Abstract)
55. Zhao, L.G.; Li, C.D.; Chang, Q.S.; Gao, X.S.; Xu, Y.W.; Fan, J.T.; Zhang, K.A. Tentative discussion on zircon U-Pb geochronologyand geochemistry of ore-bearing intermediate-acid intrusive rocks in the Bainaimiao copper ore district and the metallogenic epoch. *Geol. Bull. China* **2016**, *35*, 542–552. (In Chinese with English Abstract)
56. Pei, F.P.; Wang, Z.W.; Cao, H.H.; Xu, W.L.; Wang, F. Petrogenesis of the Early Paleozoic tonalite in the central Jilin Province: Evidence from zircon U-Pb chronology and geochemistry. *Acta Petrol. Sin.* **2014**, *30*, 2009–2019. (In Chinese with English Abstract)
57. Ma, H.T.; Ma, X.; Chen, J.F.; Chen, B.; Li, C.; Zhou, L.M.; Yang, H.Z. The Zhangjiatun igneous complex in the southeastern margin of the Central Asian Orogenic Belt, NE China: Evidence for an Early Paleozoic intra-oceanic arc. *J. Asian Earth Sci.* **2020**, *194*, 104182. [CrossRef]
58. Jia, D.C.; Hu, R.Z.; Lu, Y.; Qiu, X.L. Collision belt between the Khanka block and the North China block in the Yanbian region, Northeast China. *J. Asian Earth Sci.* **2004**, *23*, 211–219.
59. Zhang, Y.B.; Wu, F.; Wilde, S.A.; Zhai, M.; Lu, X.; Sun, D. Zircon U-Pb ages and tectonic implications of "Early Paleozoic" granitoids at Yanbian, Jilin Province, Northeast China. *Isl. Arc* **2004**, *13*, 484–505. [CrossRef]
60. Tang, K.D.; Shao, J.A.; Li, J.C.; Kang, Z. Nature of the Yanbian suture zone and structure of Northeast Asia. *Geol. Bull. China* **2004**, *23*, 885–891. (In Chinese with English Abstract)
61. Li, C.D.; Zhang, F.Q.; Miao, L.C.; Xie, H.Q.; Xu, Y.W. Zircon SHRIMP geochronology and geochemistry of Late Permian high-Mg andesites in Seluohe area, Jilin province, China. *Acta Petrol. Sin.* **2007**, *23*, 767–776. (In Chinese with English Abstract)
62. Zhang, C.Y. Tectonic Implication of Paleozoic Metamorphic Complexes in East Jilin. Ph.D. Thesis, Jilin University, Changchun, China, 2009. (In Chinese with English Abstract)
63. Zhang, C. The Mesozoic Tectonic Evolution of Yanbian Area in the Eastern Segment of Northern Margin of the North China Block. Ph.D. Thesis, Jilin University, Changchun, China, 2014. (In Chinese with English abstract)
64. Wang, Z.W.; Pei, F.P.; Xu, W.L.; Cao, H.H.; Wang, Z.J. Geochronology and geochemistry of Late Devonian and Early Carboniferous igneous rocks of central Jilin Province, NE China: Implications for the tectonic evolution of the eastern Central Asian Orogenic Belt. *J. Asian Earth Sci.* **2015**, *97*, 260–278. [CrossRef]
65. Zhao, C.; Gong, E.P.; Liu, J.; Song, S.E.; Peng, Y.B.; Cui, Y.S. Disinteration and reunderstanding of the "Kaiyuan rock group" in the east section of the northern margin of North China Craton. *Geol. Resour.* **2018**, *27*, 209–217. (In Chinese with English Abstract)
66. Zhou, J.B.; Cao, J.L.; Han, W.; Li, G.Y. The Changchun-Yanji suture zone: Nature and tectonic implications. *Acta Petrol. Sin.* **2020**, *36*, 636–643. (In Chinese with English Abstract)
67. Zhang, X.Z.; Ma, Y.X.; Chi, X.G.; Zhang, F.X.; Sun, Y.W.; Guo, Y.; Zeng, Z. Disscussion on Phanerozoic tectonic evolution in Northeastern China. *J. Jilin Univ. (Earth Sci. Ed.)* **2012**, *42*, 1269–1285. (In Chinese with English Abstract)
68. Jiang, Z.L.; Qiu, H.J.; Peng, Y.Q.; Zhang, W.M.; Liang, S. Zircon SHRIMP U-Pb dating for island arc volcanic rock of Fangniugou area in Yitong region of Jilin Province. *J. Cent. South Univ.* **2014**, *21*, 2877–2884. [CrossRef]
69. Han, Z.Z.; Song, Z.G.; Gao, L.H.; Han, M.; Guo, Z.P.; Zhong, W.J.; Li, J.J.; Du, Q.X.; Yan, J.L.; Liu, H. Late Paleozoic volcanic events in "Daheishan Horst" in Jilin, China and their tectonic significance. *Earth Sci. Front.* **2017**, *24*, 186–201. (In Chinese with English Abstract)
70. Zhang, C.; Quan, J.Y.; Zhang, L.; Ge, J.T.; Zhang, L.D. Devonian extension of the eastern section of the northern margin of the North China Plate: Evidence from post-collision magmatic rocks in Changtu area, northern Liaoning Province. *Acta Petrol. Sin.* **2022**, *38*, 2345–2363. (In Chinese with English Abstract)
71. Shi, Y.; Shi, S.S.; Liu, Z.H.; Wang, L.; Liu, J.; Chen, J.S.; Yang, F.; Zhang, C.; Li, B.; Zhang, L.D. Back-arc system formation and extinction in the southern Central Asian Orogenic Belt: New constraints from the Faku ophiolite in north Liaoning, NE China. *Gondwana Res.* **2022**, *103*, 64–83. [CrossRef]
72. Bureau of Geology and Mineral Resources of Jilin Province(BGMRJ). *Stratigraphy (Lithostratigraphy) of Jilin Province*; China University of Geosciences Press: Wuhan, China, 1997; pp. 1–324. (In Chinese)

73. Wiedenbeck, M.; Allé, P.; Corfu, F.; Griffin, W.L.; Meier, M.; Oberli, F.; Quadt, A.; Roddick, J.C.; Spiegel, W. Three natural zircon standards for U-Th-Pb, Lu-Hf, trace element and REE analyses. *Geostand. Geoanalytical Res.* **1995**, *19*, 1–23. [CrossRef]
74. Yuan, H.L.; Gao, S.; Liu, X.M.; Li, H.M.; Günther, D.; Wu, F.Y. Accurate U-Pb age and trace element determinations of zircon by laser ablation-inductively coupled plasma mass spectrometry. *Geostand. Geoanalytical Res.* **2004**, *28*, 353–370. [CrossRef]
75. Geng, J.Z.; Li, H.K.; Zhang, J.; Zhou, H.Y.; Li, H.M. Zircon Hf isotope analysis by means of LA-MC-ICP-MS. *Geol. Bull. China* **2011**, *30*, 1508–1513. (In Chinese with English Abstract)
76. Irber, W. The lanthanide tetrad effect and its correlation with K/Rb, Eu/Eu*, Sr/Eu, Y/Ho, and Zr/Hf of evolving peraluminous granitesuites. *Geochim. Cosmochim. Acta* **1999**, *63*, 489–508. [CrossRef]
77. Monecke, T.; Kempe, U.; Monecke, J.; Sala, M.; Wolf, D. Tetrad effect in rare earth element distribution patterns: A method of quantification with application to rock and mineral samples from granite-related rare metal deposits. *Geochim. Cosmochim. Acta* **2002**, *66*, 1185–1196. [CrossRef]
78. O'Connor, J. A classification for quartz-rich igneous rock based on feldspar ratios. *US Geol. Surv. Prof. Pap. B* **1965**, *525*, 79–84.
79. Winchester, J.A.; Floyd, P.A. Geochemical discrimination of different magma series and their differentiation products using immobile elements. *Chem. Geol.* **1977**, *20*, 325–343. [CrossRef]
80. Peccerillo, R.; Taylor, S.R. Geochemistry of eocene calc-alkaline volcanic rocks from the Kastamonu area, Northern Turkey. *Contrib. Miner. Pet.* **1976**, *58*, 63–81. [CrossRef]
81. Pearce, J.A. Trace elements characteristic of lavas from destructive plate boundaries. Andesites. In *Orogenic Andesites and Related Rocks*; Thorpe, R.S., Ed.; John Wiley & Sons: Chichester, UK, 1982; pp. 525–548.
82. Moine, B.; De La Roche, H. Nouvelle approche du problème de l'origine des amphibolites à partir de leur composition chimique. *C. R. Acad. Sci. Paris* **1968**, *267*, 2084–2087.
83. Boynton, W.V. Geochemistry of the rare earth elements: Meteorite studies. *Dev. Geochem.* **1984**, *2*, 63–114.
84. Sun, S.S.; McDonough, W.F. Chemical and isotopic systematics of oceanic basalts: Implications for mantle composition and processes. In *Magmatism in Ocean Basins*; Geological Society, London, Special Publication; Sanders, A.D., Norry, M.J., Eds.; Blackwell Scientific Publications: Oxford, UK, 1989; Volume 42, pp. 313–345.
85. Yang, J.H.; Wu, F.Y.; Shao, J.A.; Wilde, S.A.; Xie, L.W.; Liu, X.M. Constraints on the timing of uplift of the Yanshan fold and thrust belt, North China. *Earth Planet. Sci. Lett.* **2006**, *246*, 336–352. [CrossRef]
86. Zhang, Q.Q. Geological Records of the Late Silurian-Devonian Arc-Continent Collision in the Northern Margin of the North China Block and Craton-Orogen Boundary. Ph.D. Thesis, Chinese Academy of Geological Sciences, Beijing, China, 2021; pp. 1–203. (In Chinese with English Abstract)
87. Shi, Y.; Liu, Z.H.; Liu, Y.J.; Shi, S.S.; Wei, M.H.; Yang, J.J.; Gao, T. Late Paleozoic-Early Mesozoic southward subduction-closure of the Paleo-Asian Ocean: Proof from geochemistry and geochronology of Early Permian-Late Triassic felsic intrusive rocks from North Liaoning, NE China. *Lithos* **2019**, *346*, 105165.
88. Zhao, C.Q.; Zhang, Z.B.; Shi, Y.; Shi, S.S.; You, H.X.; Chen, C.; Zhao, J.Y. The deformation and metamorphic features of Faku metamorphic complex the composition and tectonic affinity in the northern Liaoning Province. *Acta Petrol. Sin.* **2021**, *37*, 1983–2006. (In Chinese with English Abstract)
89. Shi, Y.; Shi, S.S.; Chen, X.; Huan, F.M. The extinction process of the paleo-oceanic basin in the eastern part of the northern margin of the North China Block: Implications from the intermediate-mafic magmatic rocks in the northern Liaoning, NE China. *Acta Petrol. Sin.* **2022**, *38*, 2292–2322. (In Chinese with English Abstract)
90. Guan, Q.B. Permian-Early Jurassic Tectonic Evolution of Kaiyuan-Yanji Area in the Eastern Segment of the Northern Margin of the North China Block. Ph.D. Thesis, Jilin University, Changchun, China, 2018. (In Chinese with English Abstract)
91. Shi, Y.; Chen, J.S.; Wei, M.H.; Shi, S.S.; Zhang, C.; Zhang, L.D.; Hao, Y.J. Evolution of eastern segment of the Paleo-Asian Ocean in the Late Paleozoic: Geochronology and geochemistry constraints of granites in Faku area, North Liaoning, NE China. *Acta Petrol. Sin.* **2020**, *36*, 3287–3308. (In Chinese with English Abstract)
92. Kelemen, P.B.; Rilling, J.L.; Parmentier, E.M.; Mehl, L.; Hacker, B.R. Thermal structure due to solid-state flow in the mantle wedge beneath arcs. *Geophys. Monogr.-Am. Geophys. Union* **2003**, *138*, 293–311.
93. Liu, J.F. Late Paleozoic Magmatism and Its Constraints on Regional Tectonic Evolution in Linxi-Dongwuqi Area, Inner Mongolia. Ph.D. Thesis, Jilin University, Changchun, China, 2009. (In Chinese with English Abstract)
94. McLennan, S.M. Rare earth element geochemistry and the "tetrad" effect. *Geochim. Cosmochim. Acta* **1994**, *58*, 2025–2033. [CrossRef]
95. Zhao, J.X.; Cooper, J. Fractionation of monazite in the development of V-shaped REE patterns in leucogranite systems: Evidence from amuscovite leucogranite body in central Australia. *Lithos* **1993**, *30*, 23–32. [CrossRef]
96. Zhao, Z.H.; Musuda, A.; Shabani, M.B. Tetran effects of rare-earth elements in rare-metal granites. *Geochimica* **1992**, *3*, 221–233. (In Chinese with English Abstract)
97. Bau, M. Controls on the Fractionation of isovalent trace elements inmagmatic and aqueous systems: Evidence from Y/Ho, Zr/Hf, and Lanthanide tetrad effect. *Contrib. Miner. Petrol.* **1996**, *123*, 323–333. [CrossRef]
98. Ellison, A.J.; Hess, P.C. Solution behavior of 14 cations in high silicamelts: Petrologic and geochemical implications. *Contrib. Miner. Petrol.* **1986**, *94*, 343–351. [CrossRef]
99. Hu, G.Y.; Zeng, L.S.; Gao, L.E.; Xie, K.J. Lanthanide kinked shape, similar to tetrad effect, observed in sub-volcanic rocks from Qiaga, southern Tibet, China. *Geol. Bull. China* **2011**, *30*, 82–94. (In Chinese with English Abstract)

100. Rapp, R.P.; Watson, E.B. Monazite solubility and dissolution kinetics: Implication for the thorium and light rare earth geochemistry of felsic magmas. *Contrib. Miner. Petrol.* **1986**, *94*, 304–316. [CrossRef]
101. Lassiter, J.C.; DePaolo, D.J. Plume/lithosphere interaction in the generation of continental and oceanic flood basalts: Chemical and isotopic constraints. In *Large Igneous Provinces: Continental, Oceanic, and Planetary Flood Volcanism*; Mahoney, J.J., Coffin, M.F., Eds.; American Geophysical Union: Washington, DC, USA, 1997; Volume 100, pp. 335–355.
102. Genç, S.C.; Tüysüz, O. Tectonic setting of the Jurassic bimodal magmatism in the Sakarya Zone (Central and Western Pontides), Northern Turkey: A geochemical and isotopic approach. *Lithos* **2010**, *118*, 95–111. [CrossRef]
103. Collins, W.; Beams, S.; White, A.; Chappell, B.W. Nature and Origin of A-type Granites with Particular Reference to Southeastern Australia. *Contrib. Miner. Petrol.* **1982**, *80*, 189–200. [CrossRef]
104. Whalen, J.B.; Currie, K.L.; Chappell, B.W. A-type granites: Geochemical characteristics, discriminatuon and petrogenesis. *Contrib. Miner. Petrol.* **1987**, *95*, 407–419. [CrossRef]
105. Sylvester, P.J. Post-collisional strongly peraluminous granites. *Lithos* **1998**, *45*, 29–44. [CrossRef]
106. Taylor, S.R.; Mclennan, S. *The Continental Crust: Its Composition and Evolution*; Blackwell Scientific Publications: Boston, MA, USA, 1985; pp. 209–230.
107. Workman, R.K.; Hart, S.R. Major and trace element composition of the depleted MORB mantle (DMM). *Earth Planet. Sci. Lett.* **2005**, *231*, 1–328. [CrossRef]
108. Menzies, M.A.; Kyle, P.R.; Jones, M.; Ingram, G. Enriched and depleted source components for tholeiitic and alkaline lavas from Zuni-Bandera, New Mexico: Inferences about intraplate processes and stratified lithosphere. *J. Geophys. Res. Solid Earth* **1991**, *96*, 13645–13671. [CrossRef]
109. Saunders, A.D.; Norry, M.J.; Tarney, J. Fluid influence on the trace element compositions of subduction zone magmas. *Philos. Trans. R. Soc. A Math. Phys. Eng. Sci.* **1991**, *335*, 377–392.
110. Thompson, R.N.; Morrison, M.A. Asthenospheric and lower-lithospheric mantle contributions to continental extensional magmatism: An example from the British Tertiary Province. *Chem. Geol.* **1988**, *68*, 1–15. [CrossRef]
111. Yilmaz, Y.; Polat, A. Geology and evolution of the Thrace volcanism, Turkey. *Acta Vulcanol.* **1998**, *10*, 293–303.
112. Coish, R.A.; Sinton, C.W. Geochemistry of mafic dikes in the Adirondack mountains: Implications for Late Proterozoic continental rifting. *Contrib. Miner. Petrol.* **1992**, *110*, 500–514. [CrossRef]
113. Cao, H.H. Geochronology and Geochemistry of the Late Paleozoic-Early Mesozoic Igneous Rocks in the Eastern Segment of the Northern Margin of the North China Block. Ph.D. Thesis, Jilin University, Changchun, China, 2013. (In Chinese with English Abstract)
114. Frey, F.A. Integrated models of basalt petrogenesis: A study of quartz tholeiites to olivine melilitites from south eastern Australia. *J. Petrol.* **1978**, *19*, 463–513. [CrossRef]
115. White, W.M.; Patchett, J. Hf-Nd-Sr isotopes and incompatible element abundances in island arcs: Implications for magma origins and crust-mantle evolution. *Earth Planet. Sci. Lett.* **1984**, *67*, 167–185. [CrossRef]
116. Polat, A.; Hofmann, A.W.; Rosing, M.T. Boninite-like volcanic rocks in the 3.7–3.8 Ga Isua greenstone belt, West Greenland: Geochemical evidence for intra-oceanic subduction zone processes in the early earth. *Chem. Geol.* **2002**, *184*, 231–254. [CrossRef]
117. Morrison, G.W. Characteristics and tectonic setting of the shoshonite rock association. *Lithos* **1980**, *13*, 97–108. [CrossRef]
118. Müller, D.; Rock, N.M.S.; Groves, D.I. Geochemical discrimination between shoshonitic and potassic volcanic rocks in different tectonic settings: A pilot study. *Miner. Petrol.* **1992**, *46*, 259–289. [CrossRef]
119. Foley, S.F.; Peccerillo, A. Potassic and ultrapotassic magmas andtheir origin. *Lithos* **1992**, *28*, 181–185. [CrossRef]
120. Turner, S.; Arnaud, N.; Liu, J.; Rogers, N.; Hawkesworth, C.; Harris, N.; Kelley, S.; Van Calsteren, P.; Deng, W. Post-collision, shoshonitic volcanism on the Tibetan plateau: Implications for convective thinning of the lithosphere and the source of ocean island basalts. *J. Petrol.* **1996**, *37*, 45–71. [CrossRef]
121. Williams, H.M.; Turner, S.P.; Pearce, J.A.; Kelley, S.P.; Harris, N.B.W. Nature of the source regions for post-collisional, potassic magmatism in southern and northern Tibet from geochemical variations and inverse trace element modelling. *J. Petrol.* **2004**, *45*, 555–607. [CrossRef]
122. McDonough, W.F.; Sun, S.S. The composition of the earth. *Chem. Geol.* **1995**, *120*, 223–253. [CrossRef]
123. Sajona, F.C.; Maury, R.C.; Bellon, H.; Cotton, J.; Defant, M.J.; Pubellie, M. Initiation of subduction and the generation of slab melts in western and eastern Mindanao, Philippines. *Geology* **1993**, *21*, 1007–1010. [CrossRef]
124. Elliott, T.; Plank, T.; Zindler, A.; White, W.; Bourdon, B. Element transport from slab to volcanic front at the Mariana arc. *J. Geophys. Res.* **1997**, *102*, 14991–15019. [CrossRef]
125. Klein, E.M.; Karsten, J.L. Ocean ridge basalt with convergent margin geochemical affinities from the Chile Ridge. *Nature* **1995**, *374*, 52–57. [CrossRef]
126. Hofmann, A.W.; Jochum, K.P.; Seufert, M.; White, W.M. Nb and Pb in oceanic basalts: New constraints on mantle evolution. *Earth Planet. Sci. Lett.* **1986**, *79*, 33–45. [CrossRef]
127. Saunders, A.D.; Storey, M.; Kent, R.W.; Norry, M.J. Consequences of plume-lithosphere interactions. In *Magmatism and the Causes of Continental Break-Up*; Geological Society, London, Special Publications; Storey, B.C., Alabaster, T., Pankhurst, R.J., Eds.; The Geological Society: London, UK, 1992; Volume 68, pp. 41–60.
128. Kieffer, B.; Arndt, N.; Lapierre, H.; Lapierre, H.; Bastien, F.; Bosch, D.; Pecher, A.; Yirgu, G.; Ayalew, D.; Eeis, D.; et al. Flood and shield basalts from Ethiopia: Magmas from the African superwell. *J. Petrol.* **2004**, *45*, 793–834. [CrossRef]

129. Wu, C.Z. Geochronology, geochemistry and tectonic significances of the Hongyuntan granitoids in the Qoltag area, eastern Tianshan. *Acta Petrol. Sin.* **2006**, *22*, 1121–1134, (In Chinese with English Abstract).
130. Zhang, Q.; Jiao, S.T.; Liu, H.Y. Significance of Sr and Yb to granite theory: Philosophical thinking on granite research. *Gansu Geol.* **2021**, *30*, 1–15, (In Chinese with English Abstract).
131. Ellam, R.M. Lithospheric thickness as a control on basalt geochemistry. *Geology* **1992**, *20*, 153–156. [CrossRef]
132. Xu, B.W.; Xi, A.H.; Ge, Y.H.; Liu, Y.; Wang, M.Z.; Fang, C. Zircon U-Pb ages of the Late Paleozoic A-type granites in Chifeng. *Acta Geol. Sin.* **2015**, *89*, 58–69, (In Chinese with English Abstract).
133. Huang, D.L.; Hou, Q.Y. Devonian alkaline magmatism in the northern North China Craton: Geochemistry, SHRIMP zircon U-Pb geochronology and Sr-Nd-Hf isotopes. *Geosci. Front.* **2017**, *8*, 171–181. [CrossRef]
134. Zhang, S.H.; Zhao, Y.; Song, B.; Liu, D.Y. Petrogenesis of the Middle Devonian Gushan diorite pluton on the northern margin of the North China block and its tectonic implications. *Geol. Mag.* **2007**, *144*, 553–568. [CrossRef]
135. Zhang, S.H.; Zhao, Y.; Liu, J.M.; Hu, J.M.; Song, B.; Liu, J.; Wu, H. Geochronology, geochemistry and tectonic setting of the Late Paleozoic-Early Mesozoic magmatism in the northern margin of the North China Blcok: A preliminary review. *Acta Petrol. Miner.* **2010**, *29*, 824–842, (In Chinese with English Abstract).
136. Niu, X.L.; Liu, F.; Feng, G.Y.; Mao, X.H. Petrogenesis of the Late Silurian to Early Devonian potassic alkaline rocks on the northern margin of the North China Craton and their constraints on the tectonic evolution. *Acta Petrol. Miner.* **2021**, *40*, 835–858, (In Chinese with English Abstract).
137. Liu, J.F.; Li, J.Y.; Chi, X.G.; Feng, Q.W.; Hu, Z.C.; Zhou, K. Early Devonian felsic volcanic rocks related to the arc-continent collision on the northern margin of North China Craton—Evidences of zircon U-Pb dating and geochemical characteristics. *Geol. Bull. China* **2013**, *32*, 267–278, (In Chinese with English Abstract).
138. Wang, H.C.; Xiang, Z.Q.; Zhao, F.Q.; Li, H.M.; Yuan, G.B.; Chu, H. The alkaline plutons in eastern part of Guyang County, Inner Mongolia: Geochronology, petrogenesis and tectonic implications. *Acta Petrol. Sin.* **2012**, *28*, 2843–2854. (In Chinese with English Abstract)
139. Defant, M.J.; Drummond, M.S. Derivation of some modern arc magmas by melting of young subducted lithosphere. *Nature* **1990**, *347*, 662–665. [CrossRef]
140. Pearce, J.A.; Harris, N.B.; Tindle, A.G. Trace element discrimination diagrams for the tectonic interpretation of granitic rocks. *J. Petrol.* **1984**, *25*, 956–983. [CrossRef]
141. Kinzler, R.J. Melting of mantle peridotite at pressures approaching the spinel to garnet transition: Application to mid-ocean ridge basalt petrogenesis. *J. Geophys. Res. Solid Earth* **1997**, *102*, 853–874. [CrossRef]
142. Walter, M.J. Melting of garnet peridotite and the origin of komatiite and depleted lithosphere. *J. Petrol.* **1998**, *39*, 29–60. [CrossRef]
143. Zhang, Y.P.; Su, Y.Z.; Li, J.C. Regional tectonics significance of the Late Silurian Xibiehe Formation in central Inner Mongolia, China. *Geol. Bull. China* **2010**, *29*, 1599–1605. (In Chinese with English Abstract)
144. Zhang, W.; Jian, P. SHRIMP dating of Early Paleozoic granites from north Damaoqi, Inner Mongolia. *Acta Geol. Sin.* **2008**, *82*, 778–787. (In Chinese with English Abstract)
145. Zhang, W.; Jian, P.; Kröner, A.; Shi, Y.R. Magmatic and metamorphic development of an Early to Mid-Paleozoic continental margin arc in the southernmost Central Asian Orogenic Belt, Inner Mongolia, China. *J. Asian Earth Sci.* **2013**, *72*, 63–74, (In Chinese with English Abstract). [CrossRef]
146. Shi, Y.R.; Liu, D.Y.; Miao, L.C.; Zhang, H.F.; Jian, P.; Zhang, W.; Hou, K.J.; Xu, J.Y. Devonian A-type granitic magmatism on the northern margin of the North China Craton: SHRIMP U-Pb zircon dating and Hf-isotopes of the Hongshan granite at Chifeng, Inner Mongolia, China. *Gondwana Res.* **2010**, *17*, 632–641. [CrossRef]
147. Sun, L.X.; Ren, B.F.; Teng, F.; Zhang, Y.; Gu, Y.C.; Guo, H. LA-ICP-MS zircon U-Pb ages of the volcanoic rocks from the Chaotugou Formation in Aohan Banner, Inner Mongolia. *Geol. Bull. China* **2015**, *34*, 1493–1501. (In Chinese with English Abstract)
148. Wang, X.A. Tectonic Evolution in the Central Segment of the Northern Margin of the North China Plate from Early Paleozoic to Devonian. Ph.D. Thesis, Jilin University, Changchun, China, 2014. (In Chinese with English Abstract)
149. Chen, J.S.; Li, B.; Yang, H.; Liu, M.; Yang, F.; Li, W.; Wang, Y.; Cui, T.R. New zircon U-Pb age of granodiorite in Chifeng at the northern margin of North China Craton and constraints on plate tectonic evolution. *Acta Geol. Sin. (Engl. Ed.)* **2018**, *92*, 410–413. [CrossRef]
150. Wang, S.; Liang, H.J.; Wang, H.; Song, D.L.; Cheng, M.L.; Tan, C.L.; Ran, L.J.; Feng, Y.W.; Chen, Y.F.; Cui, Y.L.; et al. Discovery of the Late Devonian–Early Carboniferous bimodal volcanic rocks in the Aohanqi, Inner Mongolia and their geological significance. *Geochimica* **2020**, *49*, 150–167. (In Chinese with English Abstract)
151. Song, S.G.; Wang, M.M.; Xu, X.; Wang, C.; Niu, Y.L.; Allen, M.B.; Su, L. Ophiolites in the Xing'an-Inner Mongolia accretionary belt of the CAOB: Implications for two cycles of seafloor spreading and accretionary orogenic events. *Tectonics* **2015**, *34*, 2221–2248. [CrossRef]
152. Xie, W.; Song, X.Y.; Deng, Y.F.; Wang, Y.S.; Ba, D.H.; Zheng, W.Q.; Li, X.B. Geochemistry and petrogenetic implications of a Late Devonian mafic-ultramafic intrusion at the southern margin of the Central Asian Orogenic Belt. *Lithos* **2012**, *144–145*, 209–230. [CrossRef]

minerals

MDPI

Article

Permian Granitic Plutons from the Northern Margin of the North China Craton: Implications for the Tectonic Evolution of the Central Asian Orogenic Belt

Jingsheng Chen [1,2], Dexin Tian [3], Bin Li [1,2,*], Yi Shi [4,*], Zhonghui Gao [5], Yi Tian [5], Weiwei Li [5], Chao Zhang [1,2] and Yan Wang [1,2]

1 Shenyang Geological Survey Center of China Geological Survey, Shenyang 110034, China; jschen0712@126.com (J.C.); congray@163.com (C.Z.); wangy68413@163.com (Y.W.)
2 Northeast Geological S&T Innovation Center of China Geological Survey, Shenyang 110034, China
3 Non-Ferrous Metals Geological Exploration Bureau of Zhejiang Province, Shaoxing 312000, China; tiandx12@mails.jlu.edu.cn
4 School of Resource & Environment and Safety Engineering, University of South China, Hengyang 421001, China
5 Institute of Geology and Mineral Resources of Liaoning Co., Ltd., Shenyang 110029, China; 13897938707@163.com (Z.G.); lnsytianyi@163.com (Y.T.); 15040028747@163.com (W.L.)
* Correspondence: libin@mail.cgs.gov.cn (B.L.); aiwen9@126.com (Y.S.)

Abstract: As the world's largest accretionary orogen, the Central Asian Orogenic Belt (CAOB) underwent continuous juvenile crustal growth in the Phanerozoic. The northern margin of the North China Craton (NCC) and its adjacent area form the eastern segment of the CAOB, which is a key area for learning about the geological evolution of the Paleo-Asian Ocean (PAO). In the Permian, the west of the northern margin of the NCC was a post-collision extensional environment, while the east was in a subduction stage. As a connecting area, the Permian evolution of the PAO in the middle of the northern margin of the NCC has not been systematically studied. In order to fill the gap and understand the continuous temporal and spatial evolutionary process of the PAO, this paper focuses on the Permian granitic rocks in the Chifeng area. Zircon U-Pb dating and the geochemical analysis of whole-rock major and trace elements were conducted to build a granite chronological framework, and to discuss the genesis and tectonic background of the granitic rocks, along with tectono-magmatic evolutionary history in the Chifeng area. The respective LA-ICP-MS zircon U-Pb dating results from eight samples are 269 ± 1, 268 ± 3, 260 ± 4, 260 ± 1, 260 ± 1, 255 ± 2, 254 ± 2 and 256 ± 1 Ma, respectively. These results, combined with previous data, revealed that the Permian granitic rocks had undergone three events of magmatism: (1) monzogranitic-syenitic phase (294–284 Ma; Cisuralian); (2) monzogranitic phase (269–260 Ma; Guadalupian) and (3) late monzogranitic-syenitic phase (256–254 Ma; Lopingian). From the Early Permian (294–284 Ma) to the Middle Permian (269–260 Ma), granites with fine-medium-grained locally porphyritic texture and massive structure showed a high-potassium calc-alkaline series formed in a compressional setting, indicating a continuous collision between the Xing'an-Mongolian Orogenic Belt (XMOB) and the NCC. During the Late Permian-Early Triassic (256–248 Ma), granites with massive structure and medium-grained texture in the Chifeng area were magmatism dominated by A- and I-type granites of high-potassium calc-alkaline series, combined with the coeval basic rocks, which constituted a typical "bimodal" rock assemblage. This suggests that the Chifeng area was located in an extensional setting where the subducting slab broke off during the collision between the XMOB and NCC. These granitic plutons from the Permian are believed to have been generated by the subduction-collision of the Paleo-Asian oceanic crust beneath the NCC, according to emplacement time and occurrence location. Our findings provide strong evidence for Permian continuous temporal and spatial tectonic evolution and the characterization of the eventual closure of the PAO in Chifeng area at the northern margin of the NCC.

Citation: Chen, J.; Tian, D.; Li, B.; Shi, Y.; Gao, Z.; Tian, Y.; Li, W.; Zhang, C.; Wang, Y. Permian Granitic Plutons from the Northern Margin of the North China Craton: Implications for the Tectonic Evolution of the Central Asian Orogenic Belt. *Minerals* **2023**, *13*, 1554. https://doi.org/10.3390/min13121554

Academic Editor: Manuel Francisco Pereira

Received: 19 September 2023
Revised: 8 December 2023
Accepted: 11 December 2023
Published: 17 December 2023

Keywords: Permian granitic pluton; geochronology; geochemistry; tectonic evolution of the Paleo-Asian Ocean; northern margin of the North China Craton

1. Introduction

As the world's largest Phanerozoic subduction-accretion orogen, the Central Asian Orogenic Belt (CAOB) is located between the Siberian Plate (SP) and the North China Craton (NCC). Its formation is closely related to the evolution of the Paleo-Asian Ocean (PAO) [1–3]. The CAOB adjacent to the northern margin of the NCC is also known as the Xing'an-Mongolian Orogenic Belt (XMOB), which is the orogenic belt with the longest evolutionary history and the most complicated tectonic and magmatic activities in China (Figure 1a). Its formation is related to the closure of the PAO, continental crust accretion, and extensive long-lasting tectonic–magmatic activities throughout the Phanerozoic [4,5]. As a result, the XMOB was formed by the accretion of microcontinents and Phanerozoic orogenic belts [6]. After intense continental growth and multi-stage geological evolution, the XMOB had a complex and diverse structure and crustal composition, being rich in oil and gas and metal mineral resources. Therefore, it has become a natural laboratory for scholars around the world to study continental dynamic evolution, magmatism, and ore metals [3].

The PAO has undergone a long and complex evolutionary history, as concluded from various investigations. As the mid-ocean ridge expanded and the ocean reached its maximum width, the oceanic crust began to subduct toward the continental margins on both sides. Then, the PAO basins contracted and eventually closed, leading to the collision and amalgamation of the micro-continents between the two major plates [7,8]. After the collision, it entered the post-orogenic stage of stress relaxation, and a series of complex continental crust transformation events began to occur, including extension, structural collapse, and lithosphere delamination [9–11]. Increasing data show that the Solonker-Xar Moron-Changchun-Yanji Suture (SXCYS) is the final location where the two major plates collided and joined [7,12–16]. Currently, the focus of debate is the time of collision and suturing, including the following four main points of view:

(1) Based on the analysis of sedimentary petrography, mixed accumulation of ophiolite, formation of stratigraphic deposition, and contact between the Carboniferous strata and ophiolite, it has been proposed that the northern margin of the NCC and the XMOB collided and collaged in the early Late Paleozoic, resulting in the complete subduction of the PAO [17,18]. Since the Late Carboniferous and after the SP and NCC were united through the XMOB, the study area has been characterized by intracontinental rifting magmatic and sedimentary activities [17,18]. It is believed that the Permian basin is a new rift basin with extension-related volcanic activities [17]. Some scholars also believe that the PAO closed before the Early Permian based on magmatism [19,20].

(2) Based on the study on the Early-Middle Permian alkaline granites in the northern margin of the NCC, is assumed that the time limit for the final merging of the northern margin of the NCC and the XMOB to be the Early-Middle Permian [10,13,21,22].

(3) Some scholars consider that the subduction of the PAO occurred in the Late Paleozoic, and its closure took place in Late Permian to Early Triassic times based on (I) paleomagnetism, mixing of paleontological groups, and biological extinction events suggesting that the PAO closed at the end of the Permian [2,23,24]. (II) Geochronological and geochemical data of magmatism show that a post-collision extensional environment can be interpreted for the Early Triassic. Therefore, the closure of the PAO was delimited as Late Permian [4,8,12,25–29]. (III) Oceanic Carboniferous basalts and Permian fossil content were discovered in the north of the Xar Moron River [30,31]. The reefs and marine fossils of the late Permian and the P/T unconformity surface exposed in Balin Right Banner, Inner Mongolia, and Jiutai County, Jilin, proved that the PAO closed at the end of the Late Permian [32]. (IV) Ophiolite suites obducted in

the Late Permian-Early Triassic in the island arc/back-arc environment, indicating the two plates likely collided in this period [7,33–36]. (V) The zircon U-Pb age of the Permian clastic deposits in Balin Left Banner showed the final collision and merged time of the two plates were likely later than 266 Ma [37].

(4) Other researchers believe that the collision between the the NCC and SP occurred at the end of the Late Permian and lasted for a long time until the Late Triassic, based on the Middle Permian-Middle Triassic syn-collision granite [28,38] and Late Triassic post-orogenic granite [8,16,29,31,39,40].

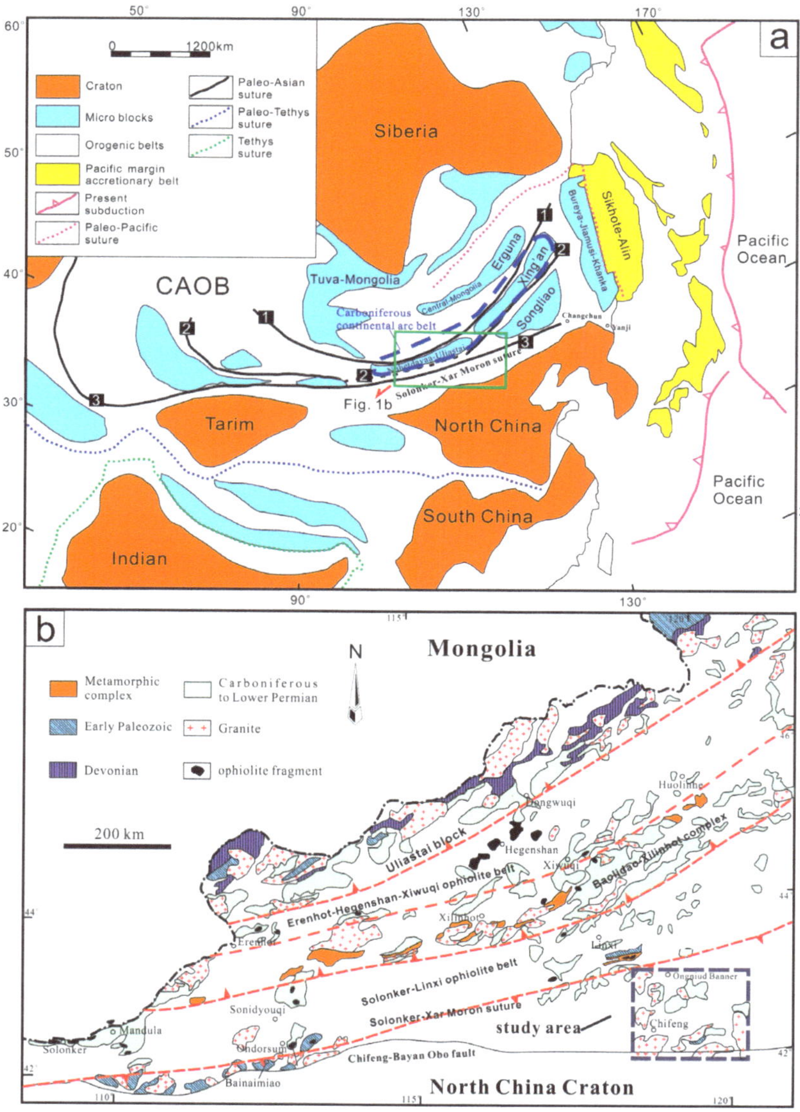

Figure 1. (**a**) Schematic tectonic map showing main tectonic subdivisions of central and eastern Asia and location of northeast China (modified from Li, 2006) [4]; (**b**) simplified geological map of central Inner Mongolia of the CAOB showing different tectonic units (modified from [7,8]).

The Chifeng area of Inner Mongolia, located at the junction of the XMOB and the northern margin of the NCC, is a key area to study the attributes, boundaries, collision and merging, and magmatic-tectonic evolution of the two tectonic units [17]. In recent years, geologists have identified multiple phases of granite assemblages from the Early Paleozoic to Mesozoic in southeastern Inner Mongolia: high-potassium, calc-alkaline series rock assemblages formed in an active continental margin environment [5,8,9,15,41,42]; potassium feldspar granites and monzogranite with syn-collision properties formed in a compressional environment [9,31,41]; and bimodal volcanic rocks and alkaline complex assemblages formed in a post-orogenic extensional environment [13,37,39]. These granites were formed in different magmatic-tectonic evolutionary stages during the PAO's closure and are controlled by different geodynamic backgrounds. Scholars have established the Triassic evolution model of the PAO by systematically studying the Triassic granitic rocks exposed in the Chifeng area [8]. However, the evolutionary process of the PAO in the Permian remains unclear. This study of the Permian granitic rocks exposed in the Chifeng area of the central-northern margin of the NCC, based on detailed field geological surveys, petrography, and geochemical and geochronological data, aims to shed light on the temporal and spatial tectonic–magmatic evolution of the PAO.

2. Geological Setting and Sample Descriptions

2.1. Geological Setting

The Chifeng area is located along the central-eastern part of the CAOB, between the Solonker-Xar Moron-Changchun-Yanji Suture (SXCYS) and Chifeng-Kaiyuan Fault (Figure 1b). Due to the quaternary covering, the bedrock mainly is composed of granitic rocks and the strata from different epochs [8,42]. The study area is bounded by the Chifeng-Kaiyuan Fault; the northern part belongs to the CAOB, showing Ordovician, Silurian, Carboniferous, and Permian strata (Figure 1b); and the southern part belongs to the NCC, showing Neoarchean metamorphic rocks and the Paleoproterozoic Baoyintu Group. During the Mesozoic, the Paleozoic basement was faulted, and the study area was transformed into an extensional basin. Continental molasse- and coal-bearing clastic rock formed in the early and Middle Jurassic; volcanic rocks formed in the Late Jurassic, and coal-bearing sediment was deposited in the Cretaceous. The graben basin formed in the Mesozoic era and changed over time due to uneven movements in the Cenozoic era. Sedimentary rocks were deposited in rivers and lakes, and basic volcanoes erupted [42].

2.2. Geological and Petrological Characteristics of Permian Granitoid Intrusions

There has been a lack of research on the granites in the Chifeng area. In recent years, in the process of regional geological surveys in Yuanbaoshan, Pingzhuang, Aohan Banner, and Beizifu in the east of Chifeng City, the author has disintegrated the plutons from 1:200,000 geological maps of Aohan Banner, Xiawa, and Chifeng City. Nine different Permian granitic intrusions have been identified in the Chifeng area (Figure 2), namely, the Mingshan, Jianshanzi, Aohan Banner, Daluobogou, Xiaodonghuang, Beizifu, Shangchaoyanggou, Qixieyingzi, and Erdaogou plutons. Eight samples were collected from each of these plutons for geochronological analyses (Table 1) and twenty-four samples for geochemical analyses (Figure 2).

There are only two Early Permian granitoid intrusions in the study area, the Mingshan and Jianshanzi plutons. The Jianshanzi pluton, dominated by syenogranite, located at west of the Wutonghua Town of Chifeng City (Figure 2), intrudes into the Carboniferous Jiujuzi Formation, Permian Sanmianjing and Elitu Formations, and is covered by the volcanic rocks of the Baiyingaolao and Hannuoba Formations [22]. The Mingshan pluton, dominated by porphyritic biotite monzogranite, located at south of Laofu Town in Chifeng City (Figure 2), intrudes into the Late Devonian granite and is covered by the Cretaceous volcanic rocks [22].

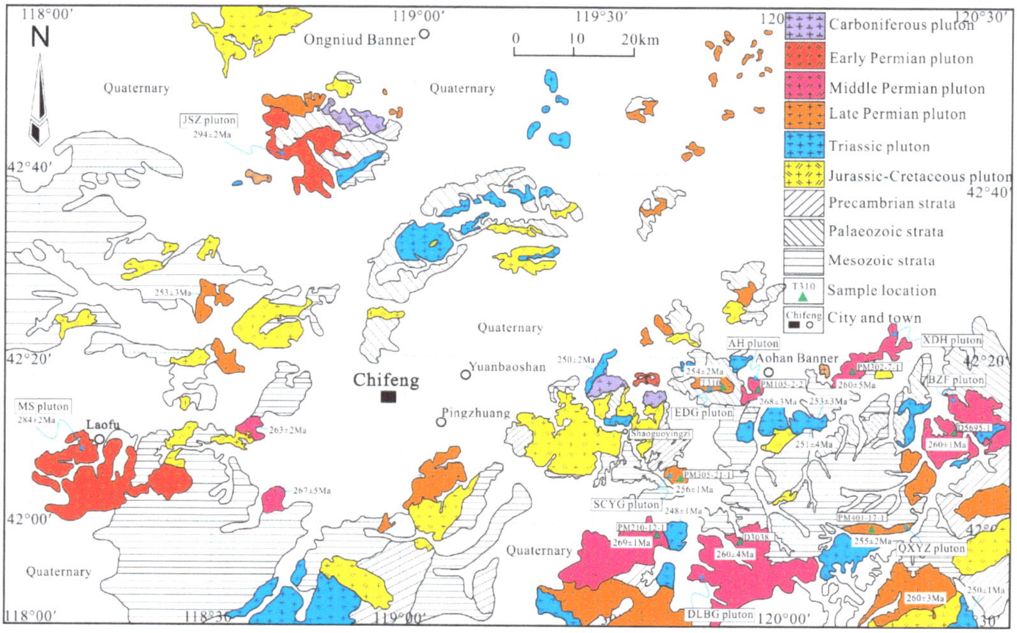

Figure 2. Simplified geological map of the study area in the north margin of the North China Craton showing sample locations, modified after [8]. AH–Aohan Banner, DLBG–Daluobogou, XDH–Xiaodonghuang, BZF–Beizifu, SCYG–Shangchaoyanggou, QXYZ–Qixieyingzi, EDG–Erdaogou.

Table 1. Location and lithology for the Permian granitic rocks of this study in the Chifeng area.

Sample No.	Name of Intrusion	Lithology	GPS Location	
PM210-12-1	Daluobogou pluton	monzogranite	119°41′28″	42°00′06″
PM105-2-2	Aohan Banner pluton	fine-grained biotite monzogranite	119°52′48″	42°17′58″
D3038	Daluobogou pluton	medium-grained monzogranite	119°51′53″	41°59′40″
PM302-7-1	Xiaodonghuang pluton	porphyry monzogranite	119°05′29″	42°19′21″
D5695	Beizifu pluton	medium-grained monzogranite	119°18′44″	42°07′51″
T310	Erdaogou pluton	medium-grained monzogranite	119°47′35″	42°18′50″
PM401-17-1	Qixieyingzi pluton	medium-grained syenogranite	119°08′56″	42°01′34″
PM305-21-1	Shangchaoyanggou pluton	monzogranite	119°42′24″	42°07′17″

The Aohan pluton, located west of Aohan Banner (Figure 2), is covered by the rhyolitic crystal tuff of the Upper Jurassic Manketouebo Formation and the andesitic tuff of the Lower Cretaceous Yixian Formation. This granitic pluton is dominated by monzogranites (samples PM105-2-1, PM105-2-2, PM105-2-3, PM105-2-4, PM105-2-4, and PM105-4-1) with a fine-grained granite texture and massive structure (Figure 3a). These samples are composed of potassium feldspar (34 wt%±), plagioclase (30 wt%±), quartz (30 wt%±), and biotite (5 wt%±), with accessory hematite, limonite, zircon, pyrite, and apatite (1 wt%±) (Figure 3b).

The Daluobogou pluton, located 80 km east of Chifeng City (Figure 2), is covered by the volcanic and sedimentary rocks of the Lower Cretaceous Yixian Formation. This pluton is dominated by monzogranites (samples PM210-12-1, D3038-1, D3038-2, D3038-3, D3047-1, D3048-1, and D3049-1) with fine-medium-grained texture and massive structure (Figure 3c,e). These samples consist of quartz (25 wt%±), plagioclase (28 wt%±), potassium feldspar (36 wt%±), and biotite (5 wt%±), with accessory limonite, zircon, and apatite (1 wt%±) (Figure 3d,f).

Figure 3. Outcrop pictures and microphotographs (cross-polarized light) of magmatic rocks from the Chifeng Area showing rock textures. (**a**,**b**) Aohan Banner pluton (fine-grained monzogranite); (**c**,**d**) Daluobogou pluton (fine-grained monzogranite); (**e**,**f**) Daluobogou pluton (medium-grained monzogranite) (**g**,**h**) Xiaodonghuang pluton (porphyritic biotite monzogranite); (**i**,**j**) Beizifu pluton (medium-grained monzogranite); (**k**,**l**) Shangchaoyanggou pluton(medium-grained monzogranite); (**m**,**n**) Qixieyingzi pluton (medium-grained syenogranite) and (**o**,**p**) Erdaogou pluton (medium-grained monzogranite); Qtz: quartz; Pl: plagioclase; Mic: microcline; Or: orthoclase; Bt: biotite.

The Xiaodonghuang pluton, located at Xiaodonghuang, Sijianfang, and Tengjiawopu in the northeast of Fengshou Town Aohan Banner (Figure 2), intrudes into the Upper Carboniferous Shizuizi Formation and is covered by the rhyolite and rhyolitic crystal tuff of the Upper Jurassic Manketouebo Formation. This pluton is dominated by biotite monzogranite (sample PM302-7-1) with porphyritic texture and massive structure (Figure 3g). The matrix is phanerocrystalline with medium-fine-grained granitic texture. These samples consist of quartz (25 wt%±), plagioclase (39wt%±), potassium feldspar (30 wt%±), and biotite (5 wt%±), with accessory zircon, magnetite, and titanite (1 wt%±) (Figure 3h).

The Bizifu pluton, located at Beizifu Town 30km east of the Aohan Banner (Figure 2), is intruded by the Mesozoic intrusions. This pluton is dominated by monzogranites (sample 5695-1) with massive structure and medium-fine-grained texture (Figure 3i). These samples consist of quartz (30 wt%±), plagioclase (32wt%±), potassium feldspar (34 wt%±), and biotite (3 wt%±), with accessory apatite, zircon, and titanite (1wt%±) (Figure 3j).

The Shangchaoyanggou pluton, located at Shangchaoyanggou, Yangjiaogou, and Weijiataizi in north of Jianping County (Figure 2), has fault contact with the Cretaceous Yixian Formation and the Sunjiawan Formation. This pluton is dominated by monzogranite (sample PM305-21-1) with massive structure and fine-grained texture (Figure 3k). These samples consist of quartz (25 wt%±), plagioclase (25wt%±), potassium feldspar (44 wt%±), and biotite (5 wt%±), with accessory hematite, titanite, and zircon (1 wt%±) (Figure 3l). The minerals in this monzogranite showed a weak certain orientation in the northeast direction locally because of the influence of Nenjiang–Balihan Fault.

The Qixieyingzi pluton, located at Qixieyingzi, Jiangjiagou, and Guluyingzi in the southeast of Aohan Banner (Figure 2), is intruded by Cretaceous monzogranite and covered by volcanic rocks from the Lower Cretaceous Yixian Formation. This pluton is dominated by syenogranite (samples PM403-2-1, PM403-2-2, PM403-2-3, PM403-2-4, PM403-2-5, PM403-2-6) with massive structure and medium-grained texture (Figure 3m). These samples consist of potassium feldspar (55 wt%±), quartz (25 wt%±), plagioclase (15wt%±), and biotite (4wt%±), with accessory apatite, zircon, and titanite (1 wt%±) (Figure 3n).

The Erdaohou pluton, located at Erdaogou, Gaojiayaozi, and Sanlamayaozi 15km northwest of the Aohan Banner (Figure 2), intrudes into the Permian strata (i.e., the Sanmianjing, Elitu, and Yujiabeigou Formations) and is covered by rhyolite and andesite from the Lower Cretaceous Yixian Formation. This pluton is dominated by monzogranites (samples T310, PM103-2-1, PM103-2-3, PM103-4-2, PM103-6-1, PM103-6-2, PM103-6-4,) with massive structure and medium-grained texture (Figure 3o). These samples consist of potassium feldspar (36 wt%±), quartz (30 wt%±), plagioclase (30wt%±), and biotite (3 wt%±), with accessory zircon, magnetite, hematite, and limonite (1 wt%±) (Figure 3p).

3. Analytical Methods

3.1. Sample Preparation

After being collected from the field, samples for geochronological analyses were cleaned, crushed, and ground. Next, zircon crystals were separated from these samples by means of conventional heavy liquid and magnetic techniques, and the separates were hand-picked and purified under a binocular microscope at Langfang Yuneng Mineral Separation Co., Ltd. (Langfang, China). Next, the zircons were selected and embedded in epoxy resin and polished, then imaged through a scanning electron microscope via cathodoluminescence (CL) to reveal their internal structures. CL images of five samples (PM105-2-2, D3038, PM302-7-1, PM401-17-1, and T310) were obtained from the Zhongnan Mineral Supervision and Testing Center of the Ministry of Land and Resources, and the others (PM210-12-1, PM305-21-1 and D5659-1) from the electron microprobe Laboratory of the Institute of Geology and Geophysics, Chinese Academy of Sciences.

3.2. Zircon LA-ICP-MS U-Pb Isotope Dating

The U-Pb zircon ages for the Permian granites in the Chifeng area were obtained using an Agilent 7500a inductively coupled plasma–mass spectrometry (ICP-MS) instrument, equipped with a 193 nm ArF Excimer laser-ablation (LA) system. The spot size for the analyses was 36 μm, with an energy density of 8.5 J/cm^2 and repetition rate of 10 Hz. Helium was used as the carrier gas to transport the ablated material from the standard LA cells. Zircon 91500 (^{206}Pb/^{238}U age of 1062.4 ± 0.4 Ma (2σ) and ^{206}Pb/^{207}Pb age of 1065.4 ± 0.3 Ma (2σ)) was used as the external standard for age calibration, while standard silicate NIST 610 glass was used to calibrate content calculations [43]. Zircon standards TEMORA (^{206}Pb/^{238}U age of 416.78 ± 0.33 Ma (2σ), [44]) and QH (160 ± 1 Ma) were also used as secondary standards to monitor any deviation in age measurement. Isotopic ratios and element contents were calculated using Glitter software (version 4.0; Macquarie University, Sydney, Australia), and the age calculation and concordant plots were obtained using Isoplot (v. 3.0) [45]. Common Pb was corrected using the method described by Andersen (2002) [46]. The analytical data are presented on U-Pb concordia diagrams with 1σ errors, and the mean ages are weighted means at the 95% confidence level [47]. Analyses of five samples (PM105-2-2, D3038, PM302-7-1, PM401-17-1, and T310) were conducted at the Zhongnan Mineral Supervision and Testing Center of the Ministry of

Land and Resources, and the others (PM210-12-1, PM305-21-1, and D5659-1) at the Geologic Laboratory Center, China University of Geosciences (Beijing, China). The analytical results are listed in Table S1.

3.3. Major- and Trace-Element Analyses

Next, major-, trace- and rare-earth-element (REE) analyses for 24 samples were conducted at the Northeast China Supervision and Inspection Center of Mineral Resources, Ministry of Land and Resources (Shenyang, China). After petrographic examination and removal of altered rock surfaces, the samples were crushed and ground to 74 μm in an agate mill. Then, whole-rock major-element contents were determined using X-ray fluorescence spectrometry (XRF), yielding analytical precisions exceeding 2%. The trace-element and REE contents were determined using ICP-MS and yielded analytical precisions exceeding 5% for elements with contents of >10 ppm, exceeding 8% for elements with contents of <10 ppm, and 10% for the transition metals. The results are listed in Table S2.

4. Analytical Results

4.1. Zircon U-Pb Ages

4.1.1. Early-Middle Permian

Sample PM210-12-1 is a monzogranite with fine-grained texture collected from the Daluobogou pluton in the Machang area. Zircons within this sample are generally colorless, transparent, and mostly subhedral to euhedral columnar in shape. They have length-to-width ratios of 3:2 to 4:1. In the cathodoluminescence (CL) images (Figure 4), most grains display fine-scale oscillatory growth zoning and clear internal structure. The zoning, combined with Th/U ratios of 0.42–1.57 (Table S1), indicates a magmatic origin [48,49]. A total of 24 isotopic analyses were conducted on 24 zircons grains from this sample. Most of the analyses are concordant and define a weighted mean $^{206}Pb/^{238}U$ age of 269 ± 1 Ma (MSDW = 0.90) (Figure 5a), which is interpreted as the crystallization age of the monzogranite.

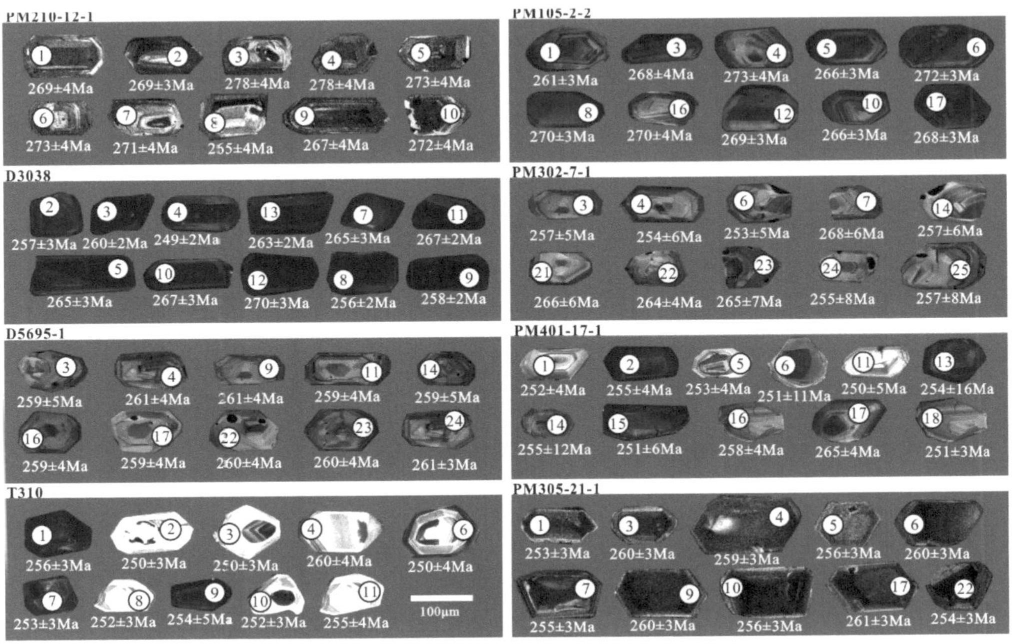

Figure 4. Cathodoluminescence (CL) images of selected zircon grains from the Permian granites in this study.

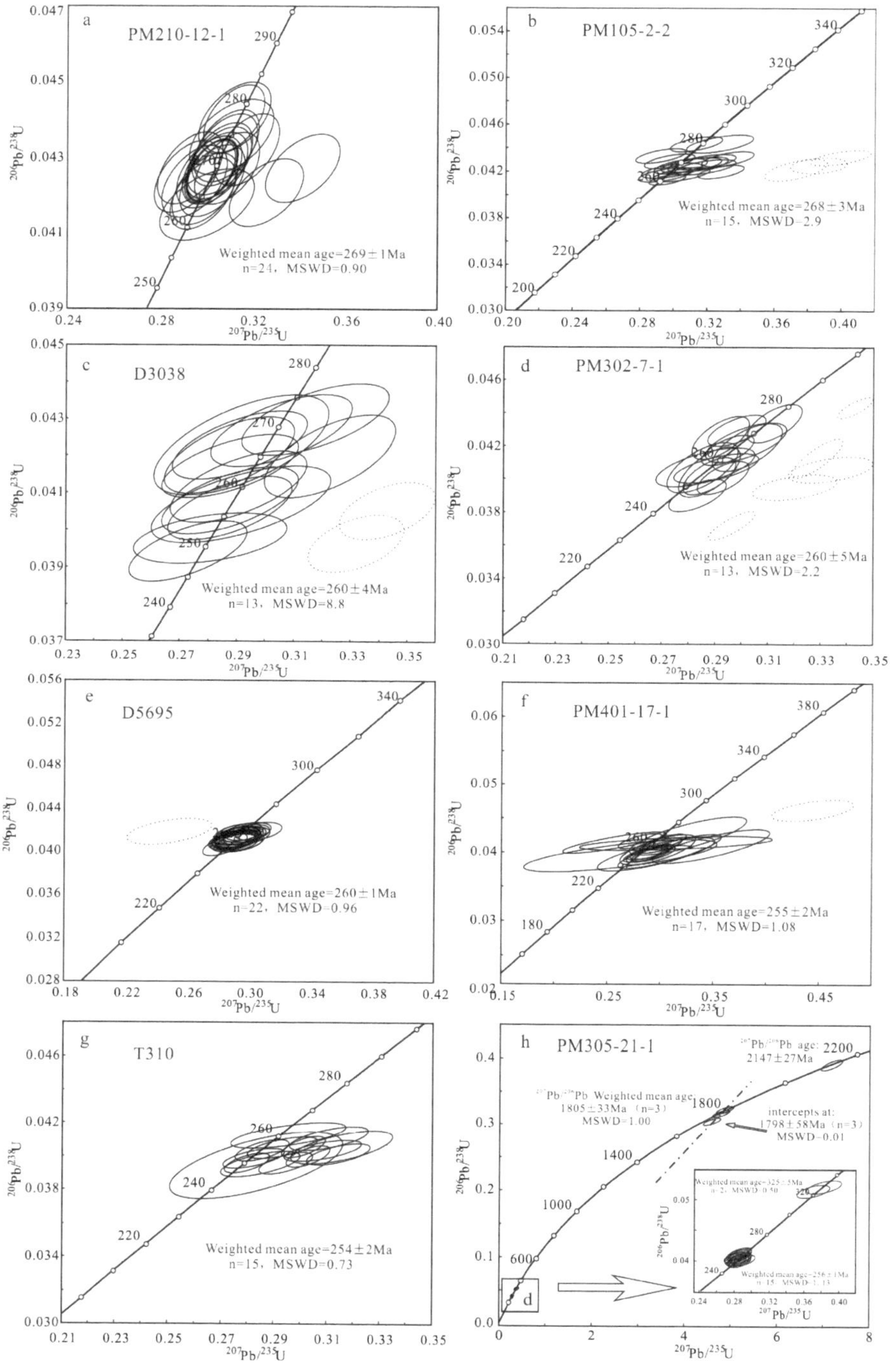

Figure 5. Zircon $^{207}Pb/^{235}U$-$^{206}Pb/^{238}U$ concordia diagrams of the Permian granites samples from this study. The dashed ellipse represents the discordant age. (**a**): PM210-12-1; (**b**): PM105-2-2); (**c**): D3038; (**d**): PM302-7-1; (**e**): D5695; (**f**): PM401-17-1; (**g**): T310; (**h**): PM305-21-1.

Sample PM105-2-2, a fine-grained biotite monzogranite, was collected from the Aohan pluton. Zircons from this sample are translucent and euhedral to subhedral or form short prisms to long columns in shape. They have length-to-width rations ranging from 2:1 to 3:1. CL imaging reveals that most zircons are bright with fine-scale oscillatory growth zoning (Figure 4), indicating a magmatic origin [48,49]. A total of 18 analyses were made on 18 zircons, which were shown to have high Th/U ratios (0.43–0.88) (Table S1). Three analysis points were not considered in the concordia diagram due to the low confidence (less than 90%) (Figure 5b). The remaining 15 analysis points were plotted on or near the concordia line. The concordant zircon data give a weighted mean $^{206}Pb/^{238}U$ age of 268 ± 3 Ma (MSDW = 2.90), interpreted as the crystallization age of the sample.

4.1.2. Late-Middle Permian

Sample D3038 is a medium-grained monzogranite from the Daluobogou pluton near the Luofugou area. The zircon grains from the sample are mostly euhedral elongate prisms with oscillatory zoning. They have length-to-width ratios of 2:1 to 4:1, and Th/U ratios of 0.52 to 1.31 (Figure 4), indicating magmatic origin [48,49]. A total of 15 analyses on 15 zircon grains were conducted. All but two of the analyses were concordant and yield a weighted mean $^{206}Pb/^{238}U$ age of 260 ± 4 Ma (MSDW = 8.8) (Figure 5c), which is regarded as the monzogranite's emplacement age.

Sample PM302-7-1, a porphyry monzogranite, was collected from the Xiaodonghuang pluton of Fengshou Village, Aohan Banner. Zircon grains from this sample are prismatic and stubby in shape, and the Th/U ratios mostly cluster around 0.43–1.38 (Table S1). A total of 23 analyses were conducted on 23 zircons; 10 analyses were excluded because of high discordancy, and the remaining 13 analysis spots yielded a weighted mean $^{206}Pb/^{238}U$ age of 260 ± 1 Ma (MSDW = 0.96) (Figure 5d), which is considered as representing the crystallization age of porphyritic monzogranite.

Sample D5695, a medium-grained monzogranite, was collected from the Beizifu pluton. Zircons from this sample are generally colorless, transparent, and euhedral in shape. In the CL images, most of them display fine-scale oscillatory zoning, with high Th/U ratios of 0.29–1.35, indicating magmatic origin [48,49]. A total of 25 spots were analyzed on 25 zircons from this sample; 3 analyses were excluded due to high discordancy, and the remaining 22 spots were concordant. The concordant zircon data give a weighted mean of $^{206}Pb/^{238}U$ age of 260 ± 1 Ma (MSDW = 0.96) (Figure 5e), interpreted as the sample's emplacement age.

4.1.3. Late Permian

Sample PM401-17-1, a medium-grained syenogranite, was collected from the Qixieyingzi pluton. Zircons within sample PM401-17-1 are generally euhedral and prismatic with length-to-width ratios of 2:1–4:1. They show variable degrees of oscillatory zoning visible during CL imaging (Figure 4) and have Th/U ratios of 0.43–2.83 (Table S1), characteristic of magmatic zircons of acid rocks [48,49]. Excluding three discordant data points likely caused by lead loss, the remaining 17 analysis spots yielded a weighted mean $^{206}Pb/^{238}U$ age of 255 ± 2 Ma (MSDW = 1.08) (Figure 5f), which is interpreted to be the crystallization age of the syenogranite.

Sample T310, a medium-grained monzogranite, was collected from the Erdaogou pluton. Zircon grains within this sample are generally colorless, transparent, and short columnar in shape, with a maximum length-to-width ratio of 2:1. The CL imaging reveals that most grains have fine oscillatory zones (Figure 4). These properties, together with their relatively high Th/U ratios (0.43–2.97) (Table S1), suggest the zircons are of magmatic origin [48,49]. The results of all 15 analyses are consistent and define a weighted mean $^{206}Pb/^{238}U$ age of 254 ± 2 Ma (MSDW = 0.73) (Figure 5g), which is interpreted as the time of crystallization of the monzogranite.

Sample PM305-21-1 is a monzogranite collected from the Shangchaoyanggou pluton. Zircons from this sample are generally euhedral and prismatic with length-to-width ratios of 2:1 to 3:1. In the CL images, most grains display fine-scale oscillatory growth zoning (Figure 4), with high Th/U values of (0.28–1.92) (Table S1). A total of 23 spots were analyzed on 23 zircons from this sample; 2 analyses were excluded due to high discordancy, and the remaining 21 analysis points were plotted on or near the concordant line, and gave four groups of ages (Figure 5h). The first group comprises one analysis (No. 16) with a $^{207}Pb/^{206}Pb$ age of 2147 ± 27 Ma, representing a middle Paleoproterozoic magmatic event. The second group consists of three analyses (Nos. 11, 18, 20), yielding a weighted mean $^{207}Pb/^{206}Pb$ age of 1,805 ± 33 Ma, which is consistent with their upper intercept age of 1798 ± 58 Ma, representing late Paleoproterozoic magmatic activity. The third group comprises two analyses (Nos. 14 and 15), yielding a weighted mean $^{206}Pb/^{238}U$ age of 352 ± 5 Ma (MSDW = 0.50) (Figure 5h). The fourth group comprises 15 analyses, yielding a weighted mean $^{206}Pb/^{238}U$ age of 256 ± 1 Ma (MSDW = 1.13). The youngest age of 256 ± 1 Ma is interpreted to be the emplacement age of the monzogranite, whereas the other three ages (ca. 2147, 1805, and 352 Ma) are interpreted to represent the crystallization ages of captured zircon entrained by the granitic magma.

4.2. *Major and Trace Element Geochemistry*

4.2.1. Early-Middle Permian Monzogranite

The monzogranites are high in SiO_2 (72.00–73.37 wt%), K_2O (4.32–4.70 wt%), and (Na_2O + K_2O) (14.17–14.62 wt%), with Na_2O/K_2O ratios of 0.88–1.01, yet low in TiO_2 (0.12–0.20 wt%), MnO (0.02–0.03 wt%), MgO (0.04–0.15 wt%), and P_2O_5 (0.04–0.07 wt%) (Table S2). They are classified in the alkalic series, and assigned to the high-K calc-alkaline igneous rocks (Figure 6a). Their A/CNK (molar $Al_2O_3/(CaO + Na_2O + K_2O)$) values range from 1.08 to 1.23, indicative of weakly peraluminous granites (Figure 6b). The total rare earth elements (REEs) range from 143.96 to 156.29 μg/g (with the exception of one datum lower than 66.45 μg/g). In the chondrite-normalized REE diagram (Figure 7a), the monzogranite samples have a pattern of slightly enriched LREEs but flat HREEs, with $(La/Yb)_N$ and ΣLREE/ΣHREE (total light rare-earth elements versus total heavy rare-earth elements) ratios of 3.54–19.62 and 5.29–10.06 and moderately negative Eu anomalies (Eu/Eu* = 0.42–0.57). In the primitive-mantle (PM) normalized trace element diagram (Figure 7b), all samples show negative Ba, Nb, Sr, P, and Ti anomalies and are all enriched in Rb, Th, K, La, Zr, and Hf.

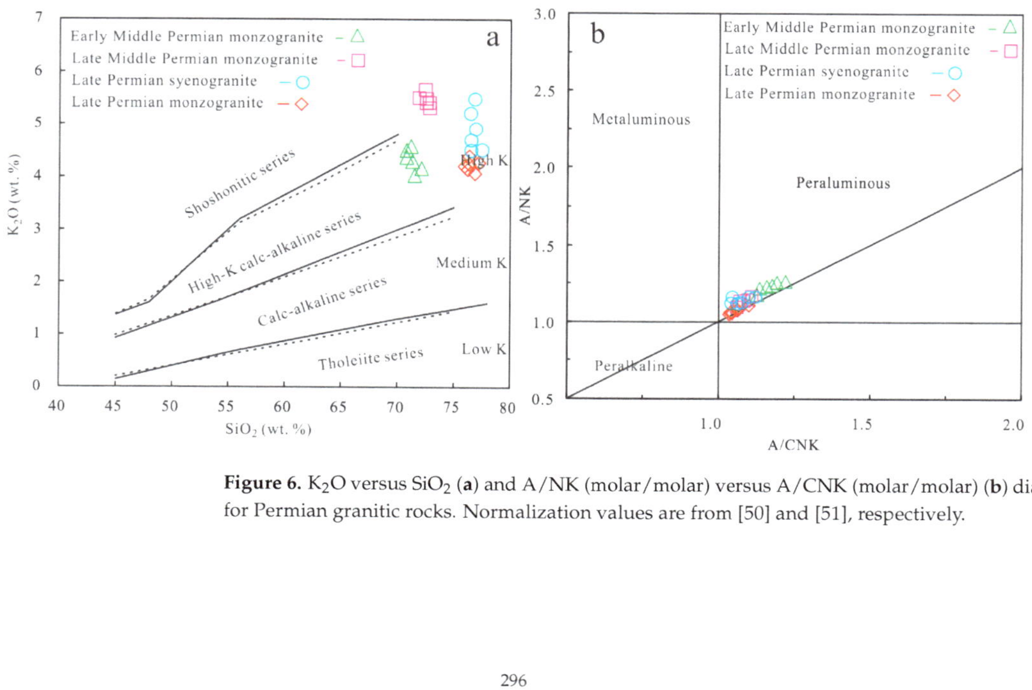

Figure 6. K_2O versus SiO_2 (**a**) and A/NK (molar/molar) versus A/CNK (molar/molar) (**b**) diagrams for Permian granitic rocks. Normalization values are from [50] and [51], respectively.

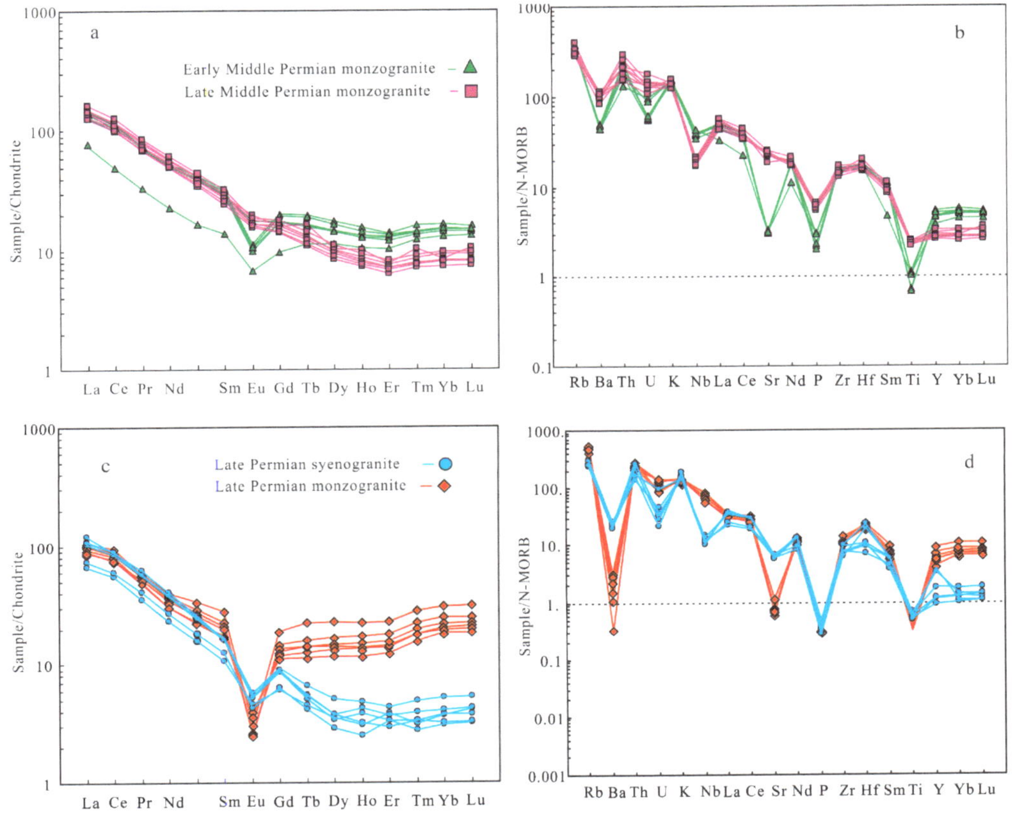

Figure 7. Chondrite-normalized REE patterns (**a**,**c**) and primitive mantle-normalized trace element spider diagrams (**b**,**d**) for these Permian granitic rocks in the Chifeng area. The values of chondrite and primitive mantle are from [52].

4.2.2. Late-Middle Permian Monzogranite

All of the monzogranite samples are alkalic- and silica-rich high-K calc-alkaline igneous rocks (Figure 6a). The whole-rock analysis results yield SiO_2 = 73.40–74.11 wt%, Na_2O/K_2O ratios = 0.65–070, and total $Na_2O + K_2O$ = 9.30–9.76. The samples are low in TiO_2 = 0.16–0.18 wt%, MnO = 0.03–0.04 wt%, MgO = 0.20 wt%, and P_2O_5 = 0.03 wt% (Table S2). They are classified as potassium basalt series and weakly peraluminous granites, due to their respective A/CNK, K_2O, and Al_2O_3 values of 1.07–1.11, 5.47–5.80 wt%, and 13.93–14.21 wt% (Figure 6b). The total REEs range from 272.5 to 340.64 μg/g. The chondrite-normalized REE diagrams invariably show a relative enrichment in LREEs, with $(La/Yb)_N$ and ΣLREE/ΣHREE ratios of 19.58–26.88 and 18.36–20.38, and moderate to significant negative Eu anomalies with Eu/Eu* ratios of 0.31–0.41 (Figure 7a). Furthermore, in the PM-normalized trace element diagram (Figure 7b), the rocks are enriched in large ion lithophile elements (LILEs; e.g., Rb, Th, K, La, and Ce) and depleted in high field-strength elements (HFSEs; e.g., Nb, P, and Ti). Finally, all samples show negative Ba anomalies.

4.2.3. Late Permian Monzogranite and Syenogranite

The Late Permian monzogranites have SiO_2 = 76.29–77.30 wt% and contain Al_2O_3 of 12.75–12.93 wt%, TiO_2 of 0.06–0.07 wt%, MnO of 0.01–0.08 wt%, MgO of 0.11 wt%, P_2O_5 of 0.01 wt%, and total $Na_2O + K_2O$ of 8.40–8.79 wt%, with Na_2O/K_2O ratios of 0.95–1.09 (Table S2). They are classified as the weakly peraluminous high-K calc-alkaline series on a plot of K_2O against SiO_2 (Figure 6a), and plot in the peraluminous field in the A/CNK vs. A/NK diagram (Figure 6b). The total REEs range from 101.75 to 128.91 μg/g. In the chondrite-normalized REE diagram, they are weakly fractionated, with $(La/Yb)_N$ and ΣLREE/ΣHREE ratios of 3.28–5.06 and 4.73–7.25 and strong negative Eu anomalies (Eu/Eu* = 0.13–0.28) (Figure 7c). In the PM-normalized trace-element diagram (Figure 7d), they show similar patterns, albeit with notable differences in the absolute element concentrations. All samples are enriched in Rb, Th, U, K, Nd, Zr, and Hf, but depleted in Ba, Sr, P, and Ti.

The Late Permian syenogranites are high in SiO_2 (76.22–77.58 wt%) and K_2O (4.35–5.62 wt%), but low in Al_2O_3 (12.05–12.63 wt%), TiO_2 (0.09–0.16 wt%), MnO (0.01–0.05 wt%), MgO (0.20–0.56 wt%), and P_2O_5 (0.01–0.02 wt%). They are high-K calc-alkaline, and weak peraluminous igneous rocks (Figure 6a), with total $Na_2O + K_2O$ of 0.49–0.86, and A/CNK ratios ranging from 1.05 to 1.16 (Figure 6b). All but two have high total REEs, with the total REE values ranging from 107.94 to 215.80 μg/g. The chondrite-normalized REE diagram shows that they are enriched in LREEs, with $(La/Yb)_N$ and ΣLREE/ΣHREE ratios of 18.05–32.16 and 16.77–25.86, respectively (Figure 7c). The granites also have moderately negative Eu anomalies (Eu/Eu* = 0.39–0.54). Finally, the PM-normalized trace-element diagram shows that they are enriched in LILEs (e.g., Rb, Th, K, La, and Ce) and LREEs (e.g., Nd, Zr, and Hf), and depleted in HFSEs (e.g., Ba, Nb, P, Ti, and Sr) (Figure 7d).

5. Discussion

5.1. Permian Magmatism in the Chifeng Area

Previous research on the Permian magmatism in the Chifeng area has been quite scarce, only covering the granodiorite located in the Late Permian in Mengguyingzi [15], diorite (256 ± 6 Ma) in Chaihuyingzi [53], gneissic monzogranite in the Liangjin mining area of Jinchanggou (249 ± 1 Ma) [40], Early Permian monzogranite in the Mingshan pluton, and syenite in the Jianshanzi pluton (284.4 ± 7.9 Ma, 294.7 ± 8.5 Ma) [22]. The coeval volcanic rocks include the andesite of the Permian Elitu Formation exposed in the Mingshan uplift belt in the west of Chifeng (273 Ma). Based on the available data, combined with those of this study, the Permian granitic chronology framework was established in the Chifeng area (Table 2).

Table 2. Geochronological data for the Permian granitic rocks in the Chifeng area.

Order	Sample	Pluton	Lithology	Age (Ma)	Method	References
1	14CH10	Jianshanzi	Monzogranite	294 ± 2	LA-ICPMS	[22]
2	14CH24	Mingshan	Granodiorite	284 ± 2	LA-ICPMS	[22]
3	PM210-12-1	Daluobogou	Monzogranite	269 ± 1	LA-ICPMS	This study
4	PM105-2-2	Aohan Banner	Monzogranite	268 ± 3	LA-ICPMS	This study
5	D3038	Daluobogou	Monzogranite	260 ± 4	LA-ICPMS	This study
6	PM302-7-1	Xiaodonghuang	Monzogranite	260 ± 5	LA-ICPMS	This study
7	D5695-1	Beizifu	Monzogranite	260 ± 1	LA-ICPMS	This study
8		Sidaogou	Syenogranite	267 ± 5	LA-ICPMS	Inner Mongolia 1:250,000 Chifeng regional geological survey
9		Jianshanzi	Monzogranite	263 ± 2	LA-ICPMS	
10		Majiadi	Granite	274 ± 4	LA-ICPMS	
11		Zhaojiawopu	Diorite	253 ± 3	LA-ICPMS	
12	PM305-21-1	Shangchaoyanggou	Granodiorite	256 ± 1	LA-ICPMS	This study
13	PM401-17-1	Qixieyingzi	Syenogranite	255 ± 2	LA-ICPMS	This study
14	T310	Erdaohou	Monzogranite	254 ± 2	LA-ICPMS	This study
15	T101	Mengguyingzi	Granodiorite	251 ± 4	LA-ICPMS	[8]
16	D2711-1	Xiaxinjing	Monzogranite	250 ± 2	LA-ICPMS	[8]
17	PM210-6-1	Daluobogou	Monzogranite	248 ± 1	LA-ICPMS	[8]
18	CH2	Chaihulanzi	Diorite	257 ± 6	SHRIMP	[53]
19	DJ10-18	Shuangjingzi	Granite	278 ± 1	LA-ICPMS	[54]
20	PM003-1	Mengguyingzi	Granodiorite	253 ± 3	LA-ICPMS	[15]
21		Dayingzi	Syenogranite	268 ± 3	LA-ICPMS	[55]
22	XG01	Jinchanggouliang	Monzogranite	250 ± 1	LA-ICPMS	[40]

5.1.1. Early Permian (294–284 Ma)

The magmatic activities are intense in the Permian in the northern margin of the NCC. However, these magmatic activities are mainly concentrated in the central region of Inner Mongolia and are rare in the Chifeng area. The Early Permian granite studied in this paper includes the syenite in the Jianshanzi pluton (294 Ma) and the monzogranite in the Mingshan pluton (284 Ma), indicating that the Early Permian granite of the Chifeng area formed during 294–284 Ma [22].

5.1.2. Middle Permian (269–260 Ma)

Late Paleozoic magmatic rocks are linearly distributed in the study area and its adjacent area, among which the Middle Permian magmatic rocks are the most widely distributed. They are exposed on the northern margin of the NCC and XMOB. In the Chifeng area, rocks that have been identified include syenogranite (269 ± 1 Ma) and monzogranite (260 ± 4 Ma) in the Daluobogou pluton, fine-grained monzogranite in Aohan Banner (268 ± 3 Ma), porphyritic medium-coarse monzogranite in Xiaodonghuang (260 ± 5 Ma), and medium-grained monzogranite in Beizifu (260 ± 1 Ma). The syenogranite at the margin of the Dayingzi pluton in the northwestern Chifeng has an age of 268 ± 3 Ma [55]. The geochronological results show that the Middle Permian intrusive rocks formed between 269 and 260 Ma, indicating that magmatism occurred in the Chifeng area in this period.

5.1.3. Late Permian-Early Triassic (256–248 Ma)

The Late Permian-Early Triassic granitic magmatism in the study area occurred frequently in the same tectonic background, and the volcanic activity in the Early Triassic was a continuation of that in the Late Permian [13,37]. The Late Permian granitic rocks identified in the research area include the weak gneissic monzogranite of the Shangchaoyanggou pluton (256 ± 1 Ma), the medium-grained syenogranite of the Qixieyingzi pluton (255 ± 2 Ma), and medium-grained monzogranite of the Erdaogou pluton (253 ± 2 Ma). There are also many other magmatic activities, such as the diorite hosting the granulite in Chaihuyingzi (257 ± 6 Ma) [53], fine-grained granodiorite in Mengguyingzi (252 ± 2 Ma) [15], gneissic monzonite in the Liangjin mining area of Jinchanggou (249 ± 1 Ma) [40], fine-grained granodiorite in the Mengguyingzi pluton (251 ± 2 Ma), medium-grained monzogranite in the Xiaxinjing pluton (250 ± 2 Ma), and medium-grained syenogranite in the Daluobogou pluton (247 ± 1 Ma) [8]. The zircon U-Pb ages of these granitic intrusions range from 256 to 247 Ma, indicating that the most intense magmatism occurred in the Late Permian-Early Triassic.

5.2. Genetic Types of the Middle-Late Permian Granitic Rocks in the Chifeng Area

5.2.1. Middle Permian Monzogranite

The early Middle Permian monzogranite (269 ± 1 Ma) is alkali-rich and high-potassium, calc-alkaline rock, and is rich in Si and Al. The A/CNK is between 1.08 and 1.23, indicating that it is peraluminous. The fine-grained monzogranite is aluminum-saturated, high-potassium, calc-alkaline rock. ΣREE is low, and close to the average level of the continental crust. Europium is moderately depleted. All of these granites fall in the range of classic island arc rocks in the $(La/Yb)_N$–Yb_N diagram and Y–Sr/Y diagram (Figure 8). By comparing the trace element ratio of fine-grained monzogranite to that of the continental crust average value and differentiated granite, it is concluded that the monzogranite of this phase is I-type granite.

The late Middle Permian monzogranite is alkali-rich, high-potassium, calcium-alkaline rock, rich in Si, with medium Al. The A/CNK value is between 1.07 and 1.11, indicating that it is slightly peraluminous. The medium-grained monzogranite is slightly aluminum-saturated, alkali-rich, high-potassium, calc-alkaline rock. The ΣREE content is much higher than the average level of the continental crust. Europium is moderately depleted, and these samples fall in the range of classic island arc rocks in the $(La/Yb)_N$–Yb_N and Y–Sr/Y diagrams (Figure 8). By comparing the trace-element ratio of medium-grained monzogranite to that of the continental crust average value and differentiated granite, it is concluded that this monzogranite in the Daluobogou pluton is I-type granite.

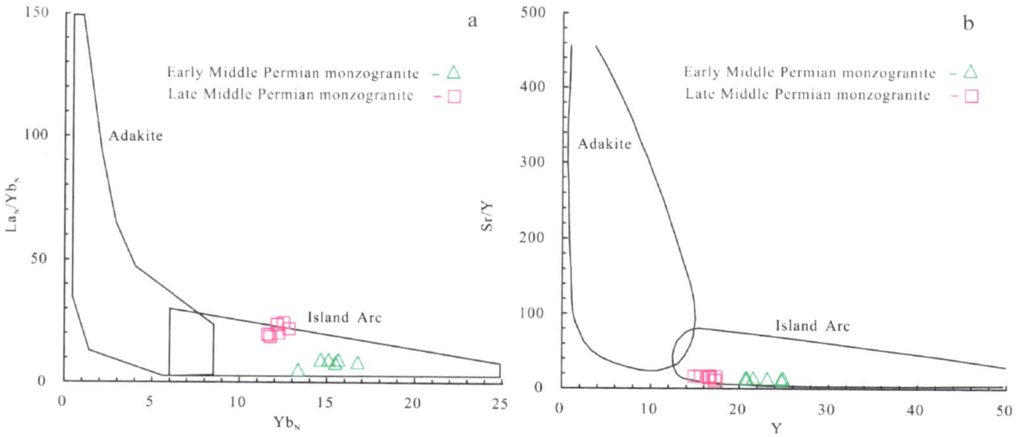

Figure 8. (**a**) La_N/Yb_N versus Yb_N; (**b**) Sr/Y versus Y diagram of the Middle Permian granites in Aohan area, modified after [56].

5.2.2. Genesis of the Late Permian Granite

The geochemistry characteristics show that the Late Permian monzogranite is alkali-rich, high-potassium, calc-alkaline rock, rich in Si and poor in Al. The A/CNK value is in the range of 0.95–1.05, indicating moderately peraluminous rock. The monzogranite belongs to alkali-rich, medium-Al, high-potassium, calc-alkaline rock. ΣREE is close to the average level of the continental crust. The trace elements have similar characteristics to those of A-type granite [57]. The light and heavy rare-earth elements are obviously fractionated, with strongly negative Eu anomalies. The rare-earth element distribution pattern forms a "V" shape. A-type granite has high heavy rare-earth elements and Zr, Nb, Ce, and Y content levels due to their rich alkalinity [57]. The observed enrichment of heavy rare-earth elements show high contents of high field-strength elements such as Nb, Zr, and Hf, yet low Sr and Ba contents (Figure 7), showing the characteristics of A-type granite. In the 10,000 Ga/Al-Zr, 10,000 Ga/Al-Nb, and 10,000 Ga/Al-$Na_2O + K_2O$ diagrams (Figure 9a, Figure 9b, and Figure 9c, respectively) of the Late Permian intrusive rocks, the monzogranite falls in the range of A-type granite. The (Zr + Nb + Ce + Y) content of the monzogranite is low (288.49–354.33 μg/g, less than 350 μg/g [57]), being slightly lower than the (Zr + Nb + Ce + Y) value of A-type granite (usually greater than 350 μg/g); however, there were some exceptions [58]. This puts the rock in the range of A-type granite in the (Zr + Nb + Ce + Y)–($K_2O + Na_2O$)/CaO diagram (Figure 9d). Zhang et al. (2012) suggested that A-type granite should be distinguished based on the comprehensive analyses of the (Zr + Nb + Ce + Y), rare-earth-element and trace-element diagrams rather than the (Zr + Nb + Ce + Y) diagram alone [57]. The most important geochemical characteristics of A-type granite are rich in SiO_2 (usually >70%, mostly >75%); rich in K_2O (4–6% or higher); depleted in Al_2O_3 (content of 12–13%), Sr, Ba, Eu, Ti, and P; and a "V"-shaped REE pattern with an obviously negative Eu anomaly [57]. The monzogranite has a SiO_2 content of 76.32–77.20%, a K_2O content of 4.15–4.47%, and an Al_2O_3 content of 12.77–12.93%, showing negative anomalies of Sr, Ba, Ti, and P. The monzogranite has a rare-earth element distribution pattern of a "V" shape with an obviously negative Eu anomaly, which is the same as the A-type granite distributed in northeastern China [6]. A-type granite is generally formed in high-temperature environments [57]. With the Zr saturation temperature of 807–828 °C, the monzogranite in the EDG pluton is not high-temperature granite. However, experimental petrology shows that when the pressure is less than 0.5 GPa and the water content is 1–6%, then the partial melting temperature of A-type granite can be reduced to 800–830 °C [59], or even lower [60–62]. In summary, it is concluded that the Late Permian monzonitic granite can be classified as A-type granite.

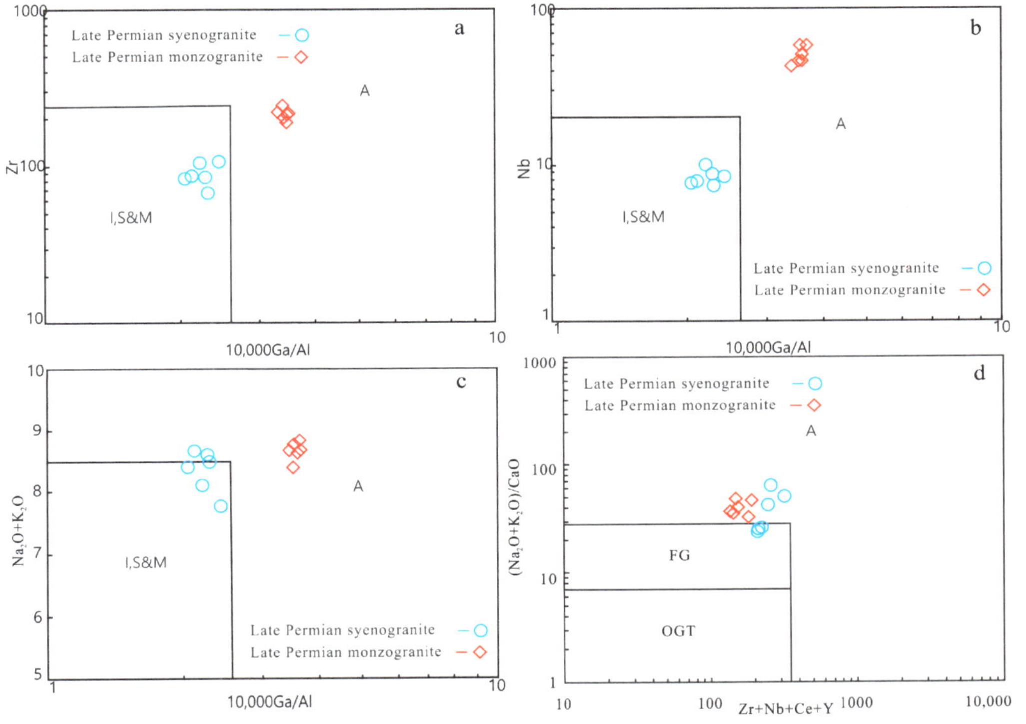

Figure 9. Diagram of 10,000 Ga/Al vs. Zr, Nb, $Na_2O + K_2O$ (**a**–**c**), and (Zr + Nb + Ce + Y) vs. ($Na_2O + K_2O/CaO$) (**d**) for the Late Permian granitic intrusions [56]. A, A-type granite; FG, highly fractionated I-type granite; I, S&M, OGT, unfractionated I, S, and M-type granite.

The syenogranite is high-potassium, calc-alkaline rock, rich in Si, and depleted in Al. The A/CNK value is in the range of 1.05–1.16, indicating slight peraluminum. The content of the ΣREE is close to the crust average level and the existing Eu moderately negative anomaly. The syenogranite does not fall in the range of A-type granite in the 10,000 Ga/Al–Zr and 10,000 Ga/Al–Nb diagrams (Figure 9). The syenogranite presented here shows the characteristics of A-type granite, with SiO_2 content of 76.22–77.58%, K_2O content of 4.35–4.87%, Al_2O_3 content of 12.05–12.56%, and negative anomalies of Sr, Ba, Eu, Ti, and P [57]. The samples fall in the transitional range of I-, S-, and A-type granite in the 10,000 Ga/Al–$Na_2O + K_2O$ diagram (Figure 9a–c), but in the range of A-type granite in the (Zr + Nb + Ce + Y)–($K_2O + Na_2O$)/CaO diagram (Figure 9d). In the standardized diagram of chondrite-normalized REE patterns (Figure 7c), the syenogranite does not show the "V"-shaped pattern or obviously negative Eu anomaly, which is inconsistent with the characteristics of A-type granite [6]. Combined with its relatively low content of (Zr + Nb + Ce + Y) (132.44–185.68 μg/g), it is considered that the syenogranite belongs to I-type rather than A-type granite.

5.2.3. Petrogenesis of High-K, I-Type Granitic Rocks of the Permian

As discussed above, both the Middle Permian monzogranite and Late Permian syenogranite in the Chifeng area are high-K calc-alkaline I-type granite, and the origins of such granite are still under debate [63,64]. Some geologists uphold the model of fractional crystallization and crustal assimilation of mantle-derived basaltic magma [65,66], while some insist on the model of mixing of crust-derived and mantle-derived magmas [67–70], and still others attribute the model of partial melting of hydrous medium- to high-K andesitic to basaltic meta-igneous rocks under crustal conditions [63,71,72].

The Permian granitic rocks are classic high-K, I-type granite. They have high silica and alkali contents, low MgO, TFe_2O_3, CaO, and transition-element contents, positive Th, Zr, and Hf anomalies, and negative Ba, Nb, Ta, P, and Ti anomalies in a primitive mantle-normalized diagram. These different characteristics indicate that the primary magma of these rocks originated from the partial melting of the continental crust, and may have been produced by the chemical differentiation of arc-derived magma [73,74]. The low Cao, Fe_2O_3, and MgO contents suggest that the source rock with this granitic component was likely formed under partial melting low-pressure conditions and that its source depths were relatively shallow. The negative anomalies of Nb were caused by the residual of rutile in the melt, since rutile is the main carrier mineral of Nb [75]. Therefore, the obvious depletions of Nb indicate residual rutile in the magma source of the Middle Permian monzogranite. Plagioclase, as the main carrier mineral of Ba, Sr, and Eu, becomes a residual mineral in the magma source in the event of negative anomalies in these elements. The negative Ba, Sr, and Eu anomalies of the granite sample indicate that plagioclase was retained in the residue of the source, or fractionated during magma evolution [76]. In conclusion, these Permian-age high-K, I-type granites were formed by partial melting of a lower crustal source under relatively low-pressure conditions.

5.3. *Tectonic Implications and Geological Significance*

5.3.1. Tectonic Setting

Tectonic Setting of the Middle Permian Monzogranite

In recent models, some scholars hold that the PAO closed in the Early Permian based on a late Early Permian-Middle Permian alkaline-peralkaline intrusion (276–259 Ma) distributed on the northern margin of the NCC, which indicates a post-collision environment in the Middle Permian [10,21,31]. All these intrusions studied by scholars are in the Jining area in the west section of the NCC's northern margin. However, the characteristics of the rock assemblages in the Middle Permian in the Changchun-Yanji area of the eastern section of the northern margin of the NCC show an active continental margin [74], while the widespread diorite, tonalite, and granodiorite indicate that the active continental marginal environment lasted until the Late Permian [38,77]. Therefore, the eastern section of the NCC's northern margin was still under the subduction background of the PAO in the Middle Permian [77]. In this case, then, what is the tectonic background of the Chifeng area in the central of the north margin of the NCC during the Middle Permian?

The Middle Permian granitic magma covers a large area in the Chifeng and is mainly dominated by monzogranite. Geochemical characteristics show that the monzogranite is the I-type granite originating from the lower crust. In the Y-Nb diagram, all samples fall into the volcanic arc and syn-collision area (Figure 10a). However, they fall into the transitional range between the volcanic arc and syn-collision and deviate to the syn-collision area in the Yb + Ta vs. Rb diagram (Figure 10b). In the R1-R2 structural environment discrimination diagram, they are in the group of syn-collision granite (Figure 11). Compared to the tectonic environment discriminant ratio of the orogenic granite proposed by Maniar and Piccoli (1989) [51], the monzogranite has characteristics like continental collision granite (CCG). Therefore, the Middle Permian monzogranite in the Chifeng area formed in a syn-collision environment.

Tectonic Setting of Late Permian Granite

Previous studies have shown that the Late Permian-Triassic magmatic rocks on the northern margin of the NCC exhibit post-collision/post-orogeny magmatism in terms of rock assemblage, mineral composition, high-potassium and calc-alkaline-based geochemical characteristics, magmatic evolution and isotope composition [8,28,29,37,53]. In the

R1-R2 diagram (Figure 11), all the Late Permian granites fall within the range of post-orogenic area, suggesting that the granite in the study area is related to the post-orogenic extension environment.

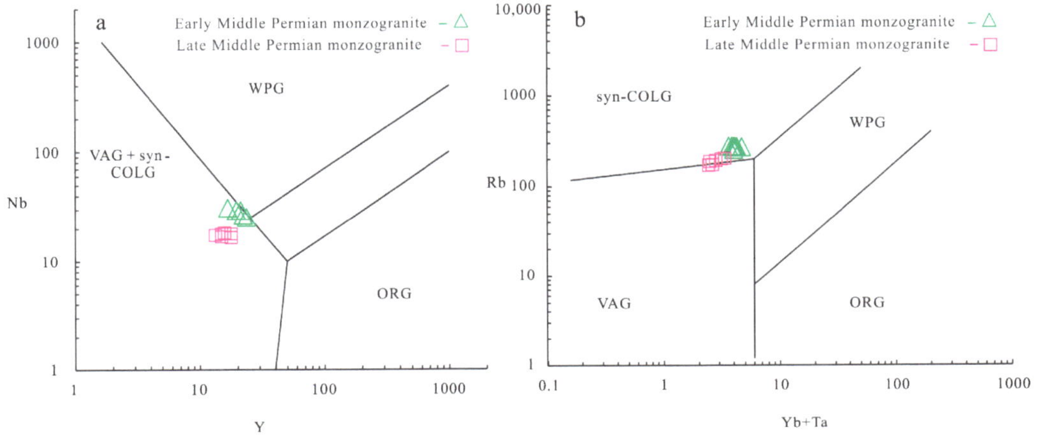

Figure 10. (**a**) Y versus Nb; (**b**) Yb + Ta versus Rb diagrams of the Middle Permian granites in Chifeng area (after [78]). The fields are ORG—oceanic ridge granites; syn-COLG—syn-collisional granites; VAG—volcanic arc granites; WPG—within-plate granites.

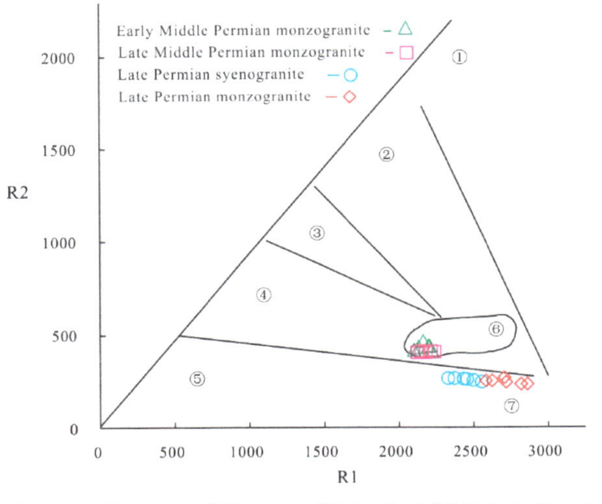

Figure 11. Diagram of R2 versus R1 for the Middle-Late Permian granites in the Chifeng area. ①—plagiogranite; ②—active continental margin granite; ③—collisional–orogenic granite; ④—late orogenic granite; ⑤—unorogenic A-type granite; ⑥—collisional granite (S-type); ⑦—post-orogenic granite (after [79]).

The granites that formed in the Chifeng area in the Late Permian and Early Triassic were of I- and A-type. Combined with the available data [8,15,40,53], these granites represent the acidic unit (256–248 Ma).The ophiolite formed from 256 to 246 Ma represents coeval basic members [7,33,35,36]. "Bimodal volcanic rocks" appeared between the northern margin of the NCC and XMOB in the Late Permian-Early Triassic. Previous studies have shown that bimodal volcanic rocks can be formed under the following extensional environments: continental rift, oceanic island extensional, plate margin extensional, post-orogenic extensional, oceanic island arc, mature island arc, and initial back-arc basin [80,81]. The Late Permian granitic rocks in the Chifeng area are dominated by middle-high potassium calc-alkaline I- and A-type granite, indicating a post-orogenic extensional environment [57,82–86].

The Late Permian granite is characterized by the enrichment of light rare-earth elements and large ion lithophile elements (LILEs), and depletion of high field-strength elements (HFSEs) and elements such as Nb, Ti, and P, showing the characteristics of the granite in the post-collision evolutionary stage of continental orogenic belts. The tectonic environment discriminant map (Figure 12a–c) shows that all granites fall in the range of post-collision granite. In the Y + Nb-Rb and Yb + Ta-Rb diagrams, all the samples fall in the post-collision range, too (Figure 12d,e).

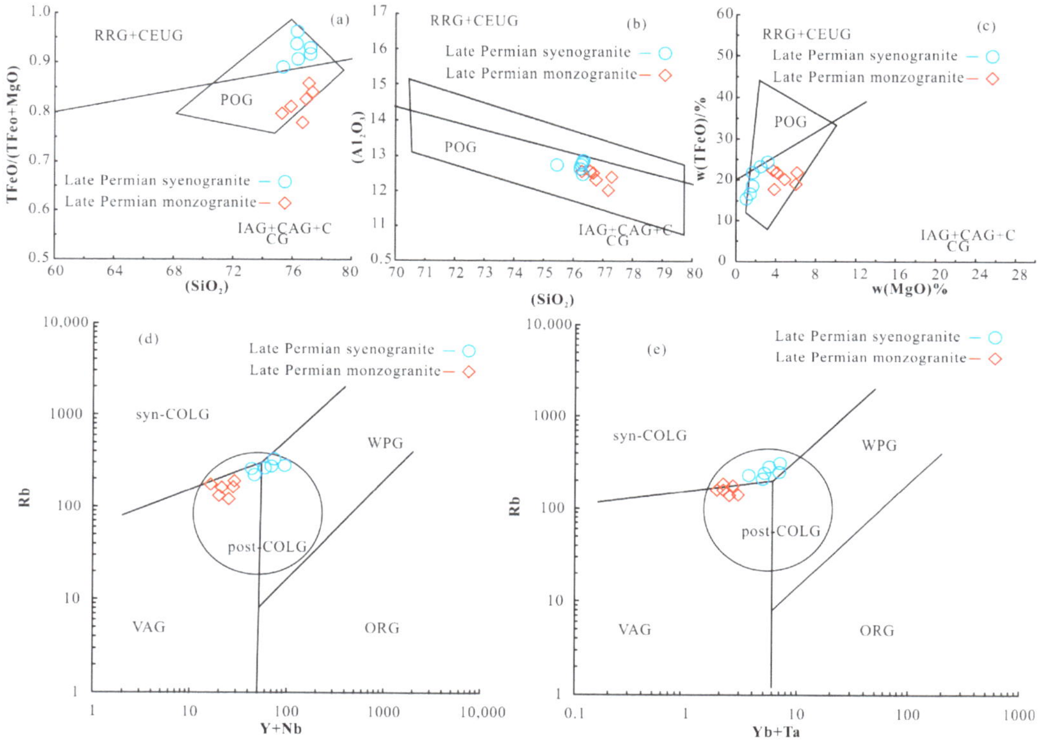

Figure 12. Identification diagram of tectonic setting for the Late Permian granites in the Chifeng area. (**a**) SiO_2 (wt%) versus TFeO/(TFeO + MgO); (**b**) SiO_2 (wt%) versus Al_2O_2 (wt%); (**c**) MgO (wt%) versus TFeO (wt%), IAG—island arc granite; CAG—continental arc granite; CCG—continental collision granite; POG—post-orogenic granite; RRG—rift related granite; CEUG—continental epeirogenic uplift granite (after [51]). (**d**) Yb + Nb versus Rb, (**e**) Yb + Ta versus Rb (after [78]), VAG—volcanic arc granites; syn-COLG—syn-collisional granites; WPG—within-plate granites; ORG—oceanic ridge granites; post-COLG—post-collisional granites.

The post-collisional, which is usually considered to begin after the closure of the main ocean, is a complex period when large-scale horizontal block movement still occurred along the giant shear zone [64]. Most post-collision granites in orogenic belts are dominated by medium-high potassium calc-alkaline I-type granite, along with some S-type and A-type granites [87]. The appearance of various types of rock assemblages shows the complexity and diversity of post-collision magmatism [37,87]. In a post collision environment, the pressure decreases under the action of extension, which facilitates rock melting. The extensional environment leads to lithospheric crust thinning and mantle-derived magma underplating. Sufficient heat then further drives the crust's partial melting, leading to the formation of many acidic rocks. Consequently, the appearance of these post-collision intermediate-acid igneous rocks may represent a transition environment from continental convergence to extension [64,88].

The following dynamic mechanisms may result in post-collision or post-orogenic magmatism: mantle convective thinning, gravity collapse, slab fragmentation, and delamination. As previously mentioned, the Late Permian-Early Triassic intrusive rocks with post-orogenic characteristics constitute a long and narrow magmatic rock belt, which spreads along both sides of the deep fault zone in the middle section of the northern margin of the NCC. The distribution pattern of Late Permian granites in the study area indicates that they were emplaced in an extensional environment caused by the slab breaking off after final closure of the PAO between the XMOB and NCC [8,89].

5.3.2. Permian Tectonic-Magmatic Evolution in the Chifeng Area

The chronological results show three phases of tectonic-magmatism in the Chifeng area during the Permian. The rock assemblage, geochemical characteristics, genesis, and tectonic setting of the granite from each phase are discussed here. Then, combined with previous research results, the Permian tectonic-magmatic evolution model is constructed (Figure 13), the specific dynamic process of which is as follows.

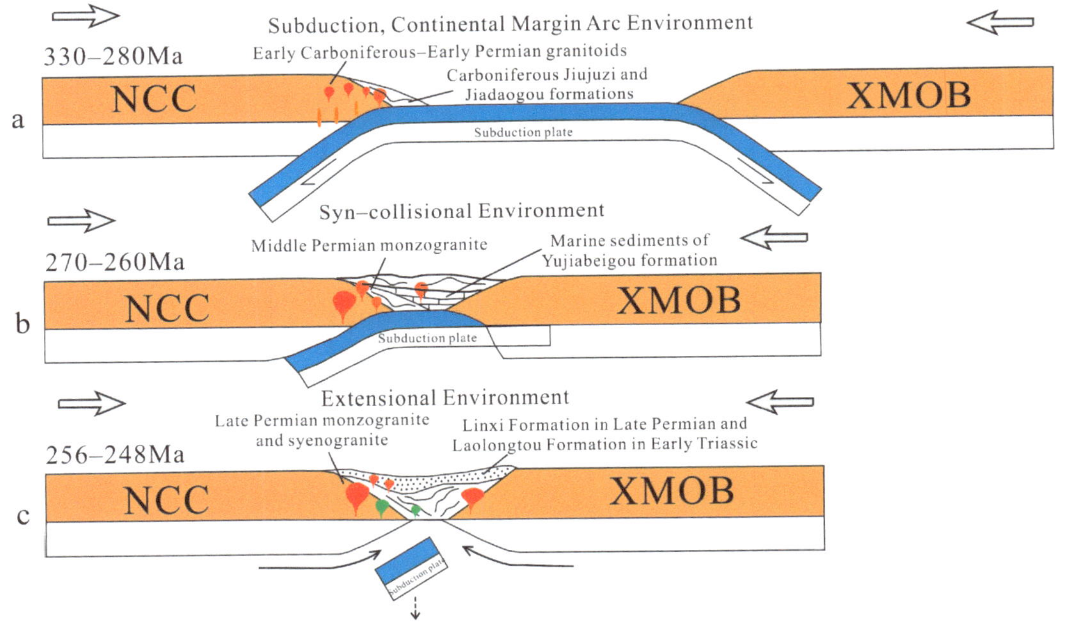

Figure 13. A tectonic evolution model from Carboniferous to Permian in the north margin of the NCC in the Chifeng-Aohan area. NCC, North China Craton; XMOB, Xing'an-Mongolian Orogenic Belt. (**a**): Subduction, Continental Margin Arc Environment; (**b**): Syn–collisional Environment; (**c**): Extensional Environment.

Early Permian–Active Continental Marginal Environment

The Early Permian granite in the Chifeng area formed in an active continental margin environment against the background of the PAO's southward subduction. The coeval volcanic rocks of the Permian Elitu Formation are also products of the subduction environment [22].

Regionally, the PAO began to subduct southward from the late Early Carboniferous, and magmatic rocks from the late Early Carboniferous to Middle Permian showed characteristics of active continental margins in terms of distribution, rock assemblages, mineral compositions, geochemical characteristics, and isotopes, also known as an Andean active continental marginal arc [12,21,25]. The results of the SHRIMP U-Pb ages and geochemistry of several granite intrusive rocks in the northern margin of the NCC show that the granite has the characteristics of calc-alkaline and subduction-related I-type granite, with an emplacement age of 300–320 Ma. This indicates that there was an Andean continental arc in the NCC's northern margin, and that the PAO still existed in the Late Carboniferous-Early Permian, then eventually closed after 290 Ma [25]. Continuous Late Carboniferous-Early Permian (324–274 Ma) rock assemblages were identified on the northern margin of the NCC, including hornblende gabbro, diorite, quartz diorite, granodiorite, and granite. According to the geochemical characteristics of this group of intrusions, it is believed that the continental arc in the northern margin of the NCC lasted at least 50 Ma [90]. By studying the Early Permian granite in Xilinhot, Li et al. (2018) also confirmed that the northern margin of the NCC was in a continental marginal arc environment from the Late Carboniferous to the end of the Early Permian [22]. By studying the granite in Siziwang Banner, Wang (2014) proposed that the Early Permian quartz diorite formed in a continental marginal arc environment [10]. Li et al. (2018) also held that the Early Permian monzogranite and syenogranite in the western Chifeng area formed in a continental marginal arc environment [22].

From the above studies, the middle section of the northern margin of the NCC was under the tectonic environment of the active continental margin, due to southward subduction of the PAO during the Early Carboniferous-Early Permian (Figure 13a). As a result, Andean-type active continental margin calc-alkaline rock assemblages of west-east trending formed along the northern margin of the NCC. Based on previous data, the existence of the Early Carboniferous-Early Permian intrusive rocks (330–280 Ma) with high-potassium and calc-alkaline characteristics related to subduction implies that the subduction of the PAO lasted for at least 50 Ma [10,13,19,22,54].

The evidence of magmatic events indicates a magmatic arc lasting about 80 Ma in the Jining-Chifeng area in the middle section of the northern margin of the NCC during the Early Carboniferous-Early Permian (330–280 Ma) (Figure 13a). The entire northern margin of the NCC was likely in an active continental margin environment under the subduction, indicating that the PAO was not closed at this time.

Middle Permian-Continent-Continent Collision Environment

Geochemical characteristics show that the Middle Permian granite in the study area is monzogranite formed in a syn-collisional environment, indicating that the XMOB collided with the NCC at this time (Figure 13b). Collisional granite with an age of 230–250 Ma was identified in Sonid Left Banner, thus indicating that the PAO closed between 310 and 230 Ma [9,41]; thus, the collision between the XMOB and the NCC lasted from the Middle Permian to the Middle-Late Triassic. Li et al. (2007) discovered granite with SHRIMP U-Pb ages of 229 and 237 Ma and considered it to be S-type granite formed by the crystallization of younger remelted crust. The geochemical characteristics show that the Shuangjingzi intrusions were like syn-collisional granite, implying they likely emplaced in the late-collisional orogenic period. It was thus concluded that the XMOB and NCC collided along the Xar Moron suture zone beginning at about 270 Ma, and ending at 230 Ma [31].

During the early Middle Permian, the plant fossils in the Yujiabeigou Formation (the zircon age of crystalline tuff of which was 270 Ma) all belong to Cathaysian flora, and there was no mixing growth phenomenon, indicating the study area was a Cathaysian flora distribution area. At that time, the Solonker-Xar Moron-Changchun-Yanji suture zone was not yet closed [91], and the PAO developed into an intercontinental residual ocean basin. Subsequently, the oceanic crust was subducted, and continent-continent collision occurred, in turn forming multiple Middle Permian syn-collisional granitic intrusions. The gray-black slate and shale of the Linxi Formation from the Late Permian distributed in the southern margin of Inner Mongolia are rich in freshwater bivalves, leaf limbs, and plant fossils [39].

The Linxi Formation, which has marine sediments at the bottom and delta-lacustrine-delta continental deposits in the middle-upper part, indicates that two large plates had collided and collaged together, then jointly assimilated the deposition during the Late Permian [14]. The above evidence shows that the northern margin of the NCC was in a co-collisional environment, and that the monzogranite formed in the Chifeng area.

During the Middle Permian, the Jining area in the western section of the northern margin of the NCC was in a post-collisional environment [10,21,31], yet the Yanbian area in the eastern section was in an active continental margin environment [14,74]. The study area covered in this paper was between these two areas; thus, in the Middle Permian, it was also in a co-collisional environment. This proves that the PAO closed in a "scissor closure" mode from west to east.

Late Permian-Early Triassic-PAO Final Closure and Subduction Slab Break-off

The granite of the Late Permian-Early Triassic consisted of A- and I-type granite, and it formed in a post-collisional extensional environment in response to slab fragmentation after the collisional orogeny between the XMOB and the NCC (Figure 13c).

Today, there is much evidence to support the conclusion that the PAO was closed from the Late Permian to the Early Triassic. The build-up types, biota, and tectonic activities on both sides of the Xar Moron fault were obviously different during the Carboniferous-Permian. The paleobiogeographical flora were in disarray until the Middle or Late Permian [92]. Moreover, the Linxi Formation from the Late Permian and the Xingfuzhilu Formation from the Early Triassic were in conformity contact, indicating that the study area was in the same tectonic environment from the Late Permian to the Early Triassic [39]. There were continental deposits and terrestrial flora and fauna fossils in the Late Permian [39], thus indicating that the two major plates had collided and assimilated sedimentation together, resulting in the formation of the Linxi Formation [14]. According to the age of the radiolarian in the siliceous rocks of the Xar Moron, the final mixed accumulation time of the ophiolite belt was proposed to be the Middle-Late Permian [10]. Both the biological extinction event and paleomagnetic evidence indicated that the PAO closed at the end of the Late Permian [24].

There were large quantities of intrusive rocks with continental margin arc magmatic character on the northern margin of the NCC, and the emplacement age ranged from 358 to 260 Ma [21,37,42,93]. The appearance of the post-collisional A-type granite and the bimodal volcanic rocks and the emplacement of the coeval ophiolite also indicated that the PAO completely disappeared during the Permian-Early Triassic, and finally closed along the Xar Moron Suture Zone [7,8,25,28,29,94].

In summary, immediately after the Middle Permian continental collisional orogeny, the subducted slabs remaining in the lithospheric mantle experienced eclogite facies metamorphism and fragmented under the influence of gravity from the Late Permian to Early Triassic (Figure 13c). After the collision between the XMOB and the NCC, the Chifeng area was in an extensional environment. The crust underwent partial melting in response to the underplating of mantle-derived magma, the upwelling of deep asthenospheric mantle materials, increasing heat, and decreasing pressure. As a result, the post-collisional A-type granite of west–east trending, bimodal volcanic rock, and ophiolite formed during the Late Permian-Early Triassic. The lacustrine Linxi Formation had abundant freshwater bivalves and gray-black slate and shale containing leaf limbs and plant fossils [14]. In addition, the paleobiogeographical divisions disappeared. All of these indicate that the PAO finally closed along the Solonker-Xar Moron-Changchun-Yanji Suture (SXCYS) during the Late Permian-Early Triassic, at which time the ocean basin completely disappeared [8,28,29].

6. Conclusions

In this study, on the Permian granitic rocks of the northern margin of the NCC, the following conclusions were obtained:

(1) According to the zircon U-Pb dating results, three events of Permian granitic magmatism were identified in the Chifeng area: (1) a suite of syenogranites and monzogranites dated at ca. 294–284 Ma (Cisuralian); (2) a suite of monzogranites dated at ca 269–260 Ma (Guadalupian); and (3) a suite of monzogranites and syenogranites dated at ca. 256–254 Ma (Lopingian).

(2) The two events related to the Middle Permian monzogranite represent I-type granite with Al-saturated, Si- and K-rich, negative Eu anomaly; enrichment of LILEs (Rb, Th, K, and La); and negative Sr, P, and Ti anomalies. The Late Permian syenogranite is rich in Si and K and poor in Al, classified as I-type granite with moderately negative Eu anomalies; enrichment of LILEs (Rb, Th, K, La, and Ce); minor enrichment in HFSEs (Nd, Zr, and Hf); and negative anomalies of Ba, P, Ti, and Sr. These Permian high-K I-type granites were formed by partial melting of the lower crustal source under relatively low-pressure conditions. The Late Permian monzogranite is A-type granite, displaying a "V" shape of rare-earth elements, with strongly negative Eu anomalies; enrichment of LILES (Rb, Th, U, and K); negative anomalies of Ba, P, and Ti; and obvious Sr negative anomalies, formed in a low-pressure environment caused by tension.

(3) The Permian tectono-magmatic evolution of the PAO can be divided into three events: (1) Early Permian granite formed in the setting of the PAO subducting to the NCC; (2) Middle Permian granite formed during the collision of the XMOB and the NCC; and (3) Late Permian-Early Triassic granite formed in an extensional setting induced by the slab breaking off after the collision between the XMOB and the NCC, responsible for the closing of the PAO.

Supplementary Materials: The following supporting information can be downloaded at https://www.mdpi.com/article/10.3390/min13121554/s1, Table S1. LA-ICP-MS zircon U-Pb data for the Permian granites in the Chifeng area; Table S2. Major (wt%) and trace (ppm) element data of Permian granites in the Chifeng area.

Author Contributions: Conceptualization, J.C.; formal analysis, J.C., D.T., B.L. and Y.S.; investigation, D.T., B.L., Y.S., Z.G., Y.T., W.L., C.Z. and Y.W.; writing—original draft preparation, J.C., B.L. and Y.S.; writing—review and editing, J.C., B.L. and Y.S.; projection administration, Y.W. All authors have read and agreed to the published version of the manuscript.

Funding: This research was funded by the China Geological Survey (Grants DD20230004, DD20190042 and DD20160048-05) and the National Natural Science Foundation of China (42272253, 41872203).

Data Availability Statement: The authors confirm that the data generated or analyzed during this study are provided in full within the published article.

Acknowledgments: We appreciate the editor and four anonymous reviewers for their critical reviews and excellent suggestions that helped to improve this manuscript. We thank the staff of the Geologic Lab Center, China University of Geosciences (Beijing), and the Zhongnan Mineral Supervision and Testing Center of the Ministry of Land and Resources, for their advice and assistance during the zircon U-Pb dating by LA-ICP-MS. We also appreciate the Northeast China Supervision and Inspection Center of Mineral Resources, Ministry of Land and Resources, Shenyang, China, for their assistance in the major- and trace-element analysis.

Conflicts of Interest: Authors Zhonghui Gao, Yi Tian, and Weiwei Li were employed by the company Institute of Geology and Mineral Resources of Liaoning Co., Ltd. The remaining authors declare that the research was conducted in the absence of any commercial or financial relationships that could be construed as a potential conflict of interest.

References

1. Sengör, A.M.C.; Natal'in, B.A. Paleotectonics of Asia: Fragments of a synthesis. In *The Tectonic Evolution of Asia*; Yin, A., Harrison, T.M., Eds.; Cambrige University Press: Cambridge, UK, 1996; pp. 486–641.
2. Sengör, A.M.C.; Natal'in, B.A.; Burtman, V.S. Evolution of the altaid tectonic collage and Palaeozoic crustal growth in Eurasia. *Nature* **1993**, *364*, 299–307. [CrossRef]

3. Jahn, B.M. The Central Asian Orogenic Belt and growth of the continental crust in the Phanerozoic. *Geol. Soc. Lond. Spec. Publ.* **2004**, *226*, 73–100. [CrossRef]
4. Li, J.Y. Permian geodynamic setting of northeast China and adjacent regions: Closure of the Paleo-Asian Ocean and subduction of the Paleo-Pacific Plate. *J. Asian Earth Sci.* **2006**, *26*, 207–224. [CrossRef]
5. Wu, F.Y.; Sun, D.Y.; Ge, W.C.; Zhang, Y.B.; Grant, M.L.; Wilde, S.A.; Jahn, B.M. Geochronology of the Phanerozoic granitoids in Northeastern China. *J. Asian Earth Sci.* **2011**, *41*, 1–30. [CrossRef]
6. Wu, F.Y.; Sun, D.Y.; Li, H.M.; Jahn, B.M.; Wilde, S.A. A-type granites in Northeastern China: Age and geochemical constraints on their petrogenesis. *Chem. Geol.* **2002**, *187*, 143–173. [CrossRef]
7. Song, S.G.; Wang, M.M.; Xu, X.; Wang, C.; Niu, Y.L.; Allen, M.B.; Su, L. Ophiolites in the Xing'an-Inner Mongolia accretionary belt of the CAOB: Implications for two cycles of seafloor spreading and accretionary orogenic events. *Tectonics* **2015**, *34*, 2221–2248. [CrossRef]
8. Chen, J.S.; Tian, D.X.; Yang, H.; Li, W.W.; Liu, M.; Li, B.; Yang, F.; Li, W.; Wu, Z. Triassic granitic magmatism at the northern margin of the North China Craton: Implications of geochronology and geochemistry for the tectonic evolution of the Central Asian Orogenic Belt. *Acta Geol. Sin.* **2019**, *93*, 1325–1353. [CrossRef]
9. Chen, B.; Jahn, B.M.; Tian, W. Evolution of the Solonker Suture Zone: Constraints from zircon U-Pb ages, Hf isotopic ratios and whole-rock Nd-Sr isotope compositions of subduction- and collision-related magmas and forearc sediments. *J. Asian Earth Sci.* **2009**, *34*, 245–257. [CrossRef]
10. Wang, W.Q. Late Paleozoic Tectonic Evolution of the Central-Northern Margin of the North China Craton: Constraints from Zircon U-Pb Ages and Geochemistry of Igneous Rocks in Ondor Sum-Jining Area. Ph.D. Thesis, Jilin University, Changchun, China, 2014; pp. 1–169. (In Chinese with English Abstract)
11. Chen, J.S.; Liu, Z.H.; Liu, Y.J.; Feng, Z.Q.; Zhang, L.D.; Wang, Y. Recent progress in the tectonic evolution of the eastern segment of Central Asia Orogenic Belt. *Acta Petrol. Sin.* **2022**, *38*, 2175–2180. (In Chinese with English Abstract)
12. Xiao, W.J.; Windley, B.F.; Hao, J.; Zhai, M.G. Accretion leading to collision and the Permian Solonker Suture, Inner Mongolia, China: Termination of the Central Asian Orogenic Belt. *Tectonics* **2003**, *22*, 1069–1090. [CrossRef]
13. Zhang, S.H.; Zhao, Y.; Song, B.; Hu, J.M.; Liu, S.W.; Yang, Y.H.; Chen, F.K.; Liu, X.M.; Liu, J. Contrasting Late Carboniferous and Late Permian-Middle Triassic intrusive suites from the northern margin of the North China Craton: Geochronology, petrogenesis, and tectonic implications. *Geol. Soc. Am. Bull.* **2009**, *121*, 181–200. [CrossRef]
14. Zhao, Y.L.; Li, W.M.; Wen, Q.B.; Liang, T.Y.; Feng, Z.Q.; Zhou, J.P.; Shen, L. Late Paleozoic tectonic framework of eastern Inner Mongolia: Evidence from the detrital zircon U-Pb ages of the Mid-Late Permian to Early Triassic sandstones. *Acta Petrol. Sin.* **2016**, *32*, 2807–2822. (In Chinese with English Abstract)
15. Liu, Y.; Xi, A.H.; Ge, Y.H.; Tang, X.Y.; Xu, B.W.; Wang, M.Z.; Ma, Y.J. LA-ICP-MS zircon U-Pb ages and petrogenesis of granodiorite in Mengguyingzi, Chifeng, Inner Mongolia. *Geol. Bull. China* **2015**, *34*, 437–446. (In Chinese with English Abstract)
16. Liu, Y.J.; Li, W.M.; Feng, Z.Q.; Wen, Q.B.; Neubauer, F.; Liang, C.Y. A review of the Paleozoic tectonics in the eastern part of Central Asian Orogenic Belt. *Gondwana Res.* **2017**, *43*, 123–148. [CrossRef]
17. Tang, K.D. Tectonic development of Paleozoic fold belts at the northern margin of the Sino-Korean Craton. *Tectonics* **1990**, *9*, 249–260. [CrossRef]
18. Tang, K.D.; Ju, N.; Zhang, D.Q.; Zhang, G.B.; Feng, Y.; Sun, J.G. Implication of the tectonic eveolution of Paleo-Asian Ocean. *Geol. Resour.* **2022**, *31*, 246–259. (In Chinese with English Abstract)
19. Zhang, S.H.; Zhao, Y.; Davis, G.A.; Ye, H.; Wu, F. Temporal and spatial variations of Mesozoic magmatism and deformation in the North China Craton: Implications for lithospheric thinning and decratonization. *Earth-Sci. Rev.* **2014**, *131*, 49–87. [CrossRef]
20. Zhao, P.; Jahn, B.; Xu, B.; Liao, W.; Wang, Y. Geochemistry, geochronology and zircon Hf isotopic study of peralkaline-alkaline intrusions along the northern margin of the North China Craton and its tectonic implication for the southeastern Central Asian Orogenic Belt. *Lithos* **2016**, *261*, 92–108. [CrossRef]
21. Zhang, S.H.; Zhao, Y.; Song, B.; Yang, Z.Y.; Hu, J.M.; Wu, H. Carboniferous granitic plutons from the northern margin of the North China Block: Implications for a Late Palaeozoic active continental margin. *J. Geol. Soc. Lond.* **2007**, *164*, 451–463. [CrossRef]
22. Li, B.; Chen, J.S.; Liu, M.; Yang, F.; Li, W.; Wu, Z.; Chen, M.H.; Wu, C.S. LA-ICP-MS zircon U-Pb geochronology and geochemical characteristics of the Early Permian granite in Chifeng area, Inner Mongolia. *Geol. Bull. China* **2018**, *37*, 1671–1681. (In Chinese with English Abstract)
23. Wang, Q.; Liu, X.Y. Paleoplate tectonics between Cathasia and Angaraland in Inner Mongolia of China. *Tectonics* **1986**, *5*, 1073–1088. [CrossRef]
24. Li, P.W.; Gao, R.; Guan, Y.; Li, Q.S. The closed times of the Paleo-Asian Ocean and the Paleo-Tethys Ocean: Implication for the tectonic cause of the End-Permian mass extinction. *J. Jilin Univ.* **2009**, *39*, 521–527. (In Chinese with English Abstract)
25. Zhang, S.H.; Zhao, Y.; Song, B. Hornblende thermobarometry of the Carboniferous granitoids from the Inner Mongolia Paleo-Uplift: Implications for the geotectonic evolution of the northern margin of North China Block. *Mineral. Petrol.* **2006**, *87*, 123–141. [CrossRef]
26. Windley, B.F.; Alexeiev, D.; Xiao, W.J.; Kroener, A.; Badarch, G. Tectonic models for accretion of the Central Asian Orogenic Belt. *J. Geol. Soc. Lond.* **2007**, *164*, 31–48. [CrossRef]

27. Guan, Q.B.; Liu, Z.H.; Liu, Y.J.; Liu, J.; Wang, S.J.; Tian, Y. Geochemistry and zircon U–Pb geochronology of mafic rocks in the Kaiyuan tectonic mélange of northern Liaoning Province, NE China: Constraints on the tectonic evolution of the Paleo-Asian Ocean. *Geol. J.* **2019**, *54*, 656–678. [CrossRef]
28. Shi, S.S.; Shi, Y.; Zhang, C.; Shi, J.M.; Jiang, S. Geochronology and geochemistry of the Triassic intrusive rocks in the Faku area, northern Liaoning, China: Constraints on the evolution of the Palaeo-Asian Ocean. *Geol. J.* **2022**, *57*, 1658–1681. [CrossRef]
29. Chen, J.S.; Li, W.W.; Shi, Y.; Li, B.; Zhao, C.Q.; Zhang, L.D. Evolution of the eastern segment of the the northern margin of the North China Craton in the Triassic: Evidence from the geochronology and geochemistry of magmatic rocks in Kaiyuan area, North Liaonign. *Acta Petrol. Sin.* **2022**, *38*, 2216–2248. (In Chinese with English Abstract)
30. Wang, Y.J. A review of the study of the Paleozoic radiation of the Paleozoic in China. *Acta Micropalaeontological Sin.* **2001**, *18*, 313–320. (In Chinese with English Abstract) [CrossRef]
31. Li, J.Y.; Gao, L.M.; Sun, G.H.; Li, Y.P.; Wang, Y.B. Shuangjingzi Middle Triassic syn-collisional crust-derived granite in the East Inner Mongolia and its constraint on the timing of collision between Siberian and Sino-Korean Paleo-plates. *Acta Petrol. Sin.* **2007**, *23*, 565–582. (In Chinese with English Abstract)
32. Tian, S.G.; Li, Z.S.; Zhang, Y.S.; Gong, Y.X.; Zhai, D.X.; Wang, M. Late Carboniferous-Permian tectono-geographical conditions and development in eastern Inner Mengolia and adjacent areas. *Acta Geol. Sin.* **2016**, *90*, 688–707. (In Chinese with English Abstract)
33. Miao, L.C.; Fan, W.M.; Liu, D.Y.; Zhang, F.Q.; Shi, Y.R.; Guo, F. Geochronology and geochemistry of the Hegenshan ophiolitic complex: Implications for late-stage tectonic evolution of the Inner Mongolia-Daxinganling Orogenic Belt, China. *J. Asian Earth Sci.* **2008**, *32*, 348–370. [CrossRef]
34. Robinson, P.T.; Zhou, M.F.; Hu, X.F.; Reynolds, P.; Bai, W.J.; Yang, J.S. Geochemical constraints on the oringin of the Hegenshan ophiolite, Inner Mongolia, China. *J. Asian Earth Sci.* **1999**, *17*, 423–442. [CrossRef]
35. Jian, P.; Liu, D.Y.; KröNer, A.; Windley, B.F.; Shi, Y.R.; Zhang, W.; Zhang, F.Q.; Miao, L.C.; Zhang, L.Q.; Tomurhuu, D. Evolution of a Permian intraoceanic arc-trench system in the Solonker Suture Zone, Central Asian Orogenic Belt, China and Mongolia. *Lithos* **2010**, *118*, 169–190. [CrossRef]
36. Chu, H.; Zhang, J.R.; Wei, C.J.; Wang, H.C.; Ren, Y.W. A new interpretation of the tectonic setting and age of meta-basic volcanics in the Ondor Sum Group, Inner Mongolia, China. *Sci. Bull.* **2013**, *58*, 3580–3587. [CrossRef]
37. Zhang, S.H.; Zhao, Y.; Liu, J.M.; Hu, J.M.; Song, B.; Liu, J.; Wu, H. Geochronology, geochemistry and tectonic setting of the Late Paleozoic-Early Mesozoic magmatism in the northern margin of the North China Block: A preliminary review. *Acta Petrol. Et Mineral.* **2010**, *29*, 824–842. (In Chinese with English Abstract)
38. Zhang, Y.B.; Wu, F.Y.; Wilde, S.A.; Zhai, M.G.; Lu, X.P.; Sun, D.Y. Zircon U-Pb ages and tectonic implications of "Early Paleozoic" granitoids at Yanbian, Jilin Province, Northeast China. *Isl. Arc.* **2004**, *13*, 484–505. [CrossRef]
39. Zhang, S.H.; Zhao, Y.; Ye, H.; Hou, K.J.; Li, C.F. Early Mesozoic alkaline complexes in the northern North China Craton: Implications for cratonic lithospheric destruction. *Lithos* **2012**, *155*, 1–18. [CrossRef]
40. Duan, P.X.; Li, C.M.; Liu, C.; Deng, J.F.; Zhao, G.C. Geochronology and geochemistry of the granites from the Jinchanggouliang gold deposit area in the Inner Mongolia and its geological significance. *Acta Petrol. Sin.* **2014**, *30*, 3189–3202. (In Chinese with English Abstract)
41. Chen, B.; Jahn, B.M.; Wilde, S.A.; Xu, B. Two contrasting Paleozoic magmatic belts in northern Inner Mongolia, China: Petrogenesis and tectonic implications. *Tectonophysics* **2000**, *328*, 157–182. [CrossRef]
42. Chen, J.S.; Li, B.; Yang, H.; Liu, M.; Yang, F.; Li, W.; Wang, Y.; Cui, T.R. New zircon U-Pb age of granodiorite in Chifeng at the northern margin of North China Craton and constraints on plate tectonic evolution. *Acta Geol. Sin.* **2018**, *92*, 410–413. [CrossRef]
43. Wiedenbeck, M.; Allé, P.; Corfu, F.; Griffin, W.L.; Meier, M.; Oberli, F.; Quadt, A.; Roddick, J.C.; Spiegel, W. Three natural zircon standards for U-Th-Pb, Lu-Hf, trace element and REE analyses. *Geostand. Geoanalytical Res.* **1995**, *19*, 1–23. [CrossRef]
44. Black, L.P.; Kamo, S.L.; Allen, C.M.; Davis, D.W.; Aleinikoff, J.N.; Valley, J.W.; Mundil, R.; Campbell, I.H.; Korsch, R.J.; Williams, I.S.; et al. Improved $^{206}Pb/^{238}U$ microprobe geochronology by the monitoring of a trace-element-related matrix effect; SHRIMP, ID–TIMS, ELA–ICP–MS and oxygen isotope documentation for a series of zircon standards. *Chem. Geol.* **2004**, *205*, 115–140. [CrossRef]
45. Ludwig, K.R. User's manual for isoplot 3.0: A geochronological toolkit for microsoft excel. *Berkeley Geochronol. Cent. Spec. Publ.* **2003**, *4*, 1–70.
46. Andersen, T. Correction of common Lead in U-Pb analyses that do not report ^{204}Pb. *Chem. Geol.* **2002**, *192*, 59–79. [CrossRef]
47. Horstwood, M.S.A.; Košler, J.; Gehrels, G.; Jackson, S.E.; McLean, N.M.; Paton, C.; Pearson, N.J.; Sircombe, K.; Sylvester, P.; Vermeesch, P. Community-derived standards for LA-ICP-MS U-(Th-) Pb geochronology–Uncertainty propagation, age interpretation and data reporting. *Geostand. Geoanalytical Res.* **2016**, *40*, 311–332. [CrossRef]
48. Belousova, E.; Griffin, W.; O'Reilly, S.Y.; Fisher, N. Igneous zircon: Trace element composition as an indicator of source rock type. *Contrib. Mineral. Petrol.* **2002**, *143*, 602–622. [CrossRef]
49. Hoskin, P.W.O.; Ireland, T.R. Rare earth element chemistry of zircon and its use as a provenance indicator. *Geology* **2000**, *28*, 627–630. [CrossRef]
50. Peccerillo, A.; Taylor, S.R. Geochemistry of Eocene calc-alkaline volcanic rocks from the Kastamonu area, northern Turkey. *Contrib. Mineral. Petrol.* **1976**, *58*, 63–81. [CrossRef]
51. Maniar, P.D.; Piccoli, P.M. Tectonic discrimination of granitoids. *Geol. Soc. Am. Bull.* **1989**, *101*, 635–643. [CrossRef]

52. Sun, S.S.; McDonough, W.F. Chemical and isotopic systematics of oceanic basalts: Implications for mantle composition and processes. *Geol. Soc. Lond. Spec. Publ.* **1989**, *42*, 313–345. [CrossRef]
53. Shao, J.A.; Zhang, Z.; She, H.Q.; Liu, D.S. The discovery of Phanerozoic granulite in Chifeng area of North Craton and its implication. *Earth Sci. Front.* **2012**, *19*, 187–198. (In Chinese with English Abstract)
54. Jiang, S.H.; Liang, Q.L.; Nie, F.J.; Liu, Y.F. A preliminary study of zircon LAMC-ICP-MS U-Pb ages of the Shuangjingzi complex in Linxi, Inner Mongolia. *Geol. China* **2014**, *41*, 1108–1123. (In Chinese with English Abstract)
55. Xi, A.H.; Ma, Y.J.; Ge, Y.H.; Tang, X.Y.; Liu, S.; Xu, B.W.; Liu, Y. Multi-stage intrusive evidence and geological significance of Dayingzi granite in Chifeng, Eastern Inner Mongolia. *J. Jilin Univ.* **2015**, *45*, 791–803. (In Chinese with English Abstract)
56. Whalen, J.B.; Currie, K.L.; Chappell, B.W. A-type granites: Geochemical characteristics, discrimination and petrogenesis. *Contrib. Mineral. Petrol.* **1987**, *95*, 407–419. [CrossRef]
57. Zhang, Q.; Ran, A.; Li, C.D. A-type granite: What is the essence? *Acta Petrol. Mineral.* **2012**, *31*, 621–626. (In Chinese with English Abstract)
58. Chappell, B.W.; White, A.J. I- and S-type granites in the Lachlan Fold Belt. *Trans. R. Soc. Edinb. Earth Sci.* **1992**, *83*, 1–26.
59. Clemens, J.D.; Holloway, J.R.; White, A.J. Origin of A-type granites: Experimental constraints. *Am. Miner.* **1986**, *71*, 317–324.
60. Anderson, J.L. Proterozoic anorogenic granite plutonism of North America. *Mem. Geol. Soc. Am.* **1983**, *161*, 133–154.
61. Scaillet, B.; Macdonald, R. Phase relations of peralkaline silicic magmas and petrogenetic implications. *J. Petrol.* **2001**, *42*, 825–845. [CrossRef]
62. Scaillet, B.; Macdonald, R. Experimental constraints on the relationships between peralkaline rhyolites of the Kenya Rift Valley. *J. Petrol.* **2003**, *44*, 1867–1894. [CrossRef]
63. Roberts, M.P.; Clemens, J.D. Origin of high-potassium, calc-alkaline, I-type granitoids. *Geology* **1993**, *21*, 825–828. [CrossRef]
64. Liegeois, J.P.; Navez, J.; Hertogen, J.; Black, R. Contrasting origin of postcollisional high-K calc-alkaline and shoshonitic versus alkaline and peralkaline granitoids: The use of sliding normalization. *Lithos* **1998**, *45*, 1–28. [CrossRef]
65. DePaolo, D.J. Trace element and isotopic effects of combined wallrock assimilation and fractional crystallization. *Earth Planet. Sci. Lett.* **1981**, *53*, 189–202. [CrossRef]
66. Moghazi, A.M. Geochemistry and petrogenesis of a high-K calc-alkaline Dokhan Volcanic suite, South Safaga area, Egypt: The role of late Neoproterozoic crustal extension. *Precambrian Res.* **2003**, *125*, 161–178. [CrossRef]
67. Dickinson, W.R. Potash-depth (K-h) relations in continental-margin and intraoceanic magmatic arcs. *Geology* **1975**, *3*, 53–56. [CrossRef]
68. Yang, H.; Ge, W.C.; Zhao, G.C.; Dong, Y.; Xu, W.L.; Wang, Z.H.; Ji, Z.; Yu, J.J. Late Triassic intrusive complex in the Jidong region, Jiamusi-Khanka Block, NE China: Geochemistry, zircon U-Pb ages, Lu-Hf isotopes, and implications for magma mingling and mixing. *Lithos* **2015**, *224–225*, 143–159. [CrossRef]
69. Yang, J.H.; Wu, F.Y.; Chung, S.L.; Wilde, S.A.; Chu, M.F. A hybrid origin for the Yianshan A-type granite, northeast China: Geochemical and Sr-Nd-Hf isotopic evidence. *Lithos* **2006**, *89*, 89–106. [CrossRef]
70. Clemens, J.D.; Darbyshire, D.P.F.; Flinders, J. Sources of post-orogenic calcalkaline magmas: The Arrochar and Garabal Hill-Glen Fyne complexes, Scotland. *Lithos* **2009**, *112*, 524–542. [CrossRef]
71. Sisson, T.W.; Ratajeski, K.; Hankins, W.B.; Glazner, A.F. Voluminous granitic magmas from common basaltic sources. *Contrib. Mineral. Petrol.* **2005**, *148*, 635–661. [CrossRef]
72. Topuz, G.; Altherr, R.; Siebel, W.; Schwarz, W.H.; Zack, T.; Hasözbek, A.; Barth, M.; Satır, M.; Şen, C. Carboniferous high-potassium I-type granitoid magmatism in the Eastern Pontides: The Gümüşhane pluton (NE Turkey). *Lithos* **2010**, *116*, 92–110. [CrossRef]
73. Hofmann, A.W. Chemical differentiation of the Earth: The relationship between mantle, continental crust, and oceanic crust. *Earth Planet. Sci. Lett.* **1988**, *90*, 297–314. [CrossRef]
74. Yu, J.J.; Wang, F.; Xu, W.L.; Gao, F.H.; Tang, J. Late Permian tectonic evolution at the southeastern margin of the Songnen-Zhangguangcai Range Massif, NE China: Constraints from geochronology and geochemistry of granitoids. *Gondwana Res.* **2013**, *24*, 635–647. [CrossRef]
75. Xiong, X.L.; Adam, J.; Green, T.H. Rutile stability and rutile/melt HFSE partitioning during partial melting of hydrous basalt: Implications for TTG genesis. *Chem. Geol.* **2005**, *218*, 339–359. [CrossRef]
76. Springer, W.; Seck, H.A. Partial fusion of basic granulites at 5-15 kbar: Implications for the origin of TTG magmas. *Contrib. Mineral. Petrol.* **1997**, *127*, 30–45. [CrossRef]
77. Cao, H.H.; Xu, W.L.; Pei, F.P.; Zhang, X.Z. Permian tectonic evolution in southwestern Khanka massif: Evidence from zircon U-Pb chronology, Hf isotope and geochemistry of gabbro and diorite. *Acta Geol. Sin.* **2011**, *85*, 1390–1402.
78. Pearce, J.A. A user's guide to basalt discrimination diagrams. In *Trace Element Geochemistry of Volcanic Rocks: Applications for Massive Sulphide Exploration*; Wyman, D.A., Ed.; Geological Association of Canada, Short Course Notes: St. John's, NF, Canada, 1996; Volume 12, pp. 79–113.
79. Batchelor, R.A.; Bowden, P. Petrogenetic interpretation of granitoid rock series using multicationic parameters. *Chem. Geol.* **1985**, *48*, 43–55. [CrossRef]
80. Ikeda, Y.; Yuasa, M. Volcanism in Nascent backarc basins behind the Schichito ridge and adjacent areas in the Izu-Ogasawara Arc, NW Pacific: Evidence for mixing between E-type MORB and island arc magmas at the initiation of back-arc rifing. *Contrib. Mineral. Petrol.* **1989**, *101*, 377–393. [CrossRef]

81. Pin, C.; Paquette, J.L. A mantle-derived bimodal suite in the Hercynian belt: Nd isotope and trace element evidence for a subduction-related rift origin of the Late Devonian brevenne metavolcanics, Massif Central (France). *Contrib. Mineral. Petrol.* **1997**, *129*, 222–238. [CrossRef]
82. Hildreth, W.E.; Halliday, A.N.; Christiansen, R.L. Isotopic and chemical evidence concerning the genesis and contamination of basaltic and rhyolitic magma beneath the Yellowstone Plateau Volcanic Field. *J. Petrol.* **1991**, *32*, 63–138. [CrossRef]
83. Eby, G.N. Chemical subdivision of the A-type granitoids: Petrogenetic and tectonic implications. *Geology* **1992**, *20*, 641–644. [CrossRef]
84. Turner, S.; Sandiford, M.; Foden, J. Some geodynamic and compositional constraints on "postorogenic" magmatism. *Geology* **1992**, *20*, 931–934. [CrossRef]
85. Frost, C.D.; Frost, B.R.; Chamberlain, K.R.; Edwards, B.R. Petrogenesis of the 1.43 Ga Sherman batholith, SE Wyoming, USA: A reduced, rapakivi-type anorogenic granite. *J. Petrol.* **1999**, *40*, 1771–1802. [CrossRef]
86. Bonin, B. Do coeval mafic and felsic magmas in post-collisional to within-plate regimes necessarily imply two contrasting, mantle and crustal, sources? A review. *Lithos* **2004**, *78*, 1–24. [CrossRef]
87. Han, B.F. Diverse post-collisional granitoids and their tectonic setting discrimination. *Earth Sci. Front.* **2007**, *14*, 64–72. (In Chinese with English Abstract)
88. Barbarin, B. A review of the relationships between granitoid types, their origins and their geodynamic environments. *Lithos* **1999**, *46*, 605–626. [CrossRef]
89. Von Blanckenburg, F.; Davies, J.H. Slab breakoff: A model for syncollisional magmatism and tectonics in the Alps. *Tectonics* **1995**, *14*, 120–131. [CrossRef]
90. Zhang, S.H.; Zhao, Y.; Kroner, A.; Liu, X.M.; Xie, L.W.; Chen, F.K. Early Permian plutons from the northern North China Block: Constraints on continental arc evolution and convergent margin magmatism related to the Central Asian Orogenic Belt. *Int. J. Earth Sci.* **2009**, *98*, 1441–1467. [CrossRef]
91. Sun, Y.W.; Ding, H.S.; Liu, H.; Zhang, D.J.; Gong, F.H.; Zheng, Y.J. Fossil plants from the Guadalupian Yujiabeigou Formation in the north margin of North China Plate and their tectonic implications. *J. Jilin Univ.* **2016**, *46*, 1268–1283. (In Chinese with English Abstract)
92. Huang, B.H.; Ding, Q.H. The angara flora from Northern China. *Acta Geosci. Sin.* **1998**, *19*, 97–104. (In Chinese with English Abstract)
93. Shi, Y.; Shi, S.S.; Liu, Z.H.; Liu, J.; Ju, N.; You, H.X.; Zhang, Z.B.; Zhao, C.Q. Petrogenesis of the late Early Palaeozoic adakitic granitoids in the southern margin of the Songliao Basin, NE China: Implications for the subduction of the Palaeo-Asian Ocean. *Geol. J.* **2019**, *54*, 3821–3839. [CrossRef]
94. Miao, L.C.; Zhang, F.Q.; Fan, W.M.; Liu, D.Y. Phanerozoic evolution of the Inner Mongolia-Daxinganling Orogenic Belt in North China: Constraints from geochronology of ophiolites and associated formations. *Geol. Soc. Lond. Spec. Publ.* **2007**, *280*, 223–237. [CrossRef]

minerals

MDPI

Article

In Situ Trace Element and Sulfur Isotope Composition of Pyrite from the Beiwagou Pb-Zn Deposit, Liaodong Peninsula, Northeast China: Implications for Ore Genesis

Qi Yu [1,2], Zhigao Wang [1,3,*], Qingfei Sun [1] and Keyong Wang [1]

1 College of Earth Sciences, Jilin University, Changchun 130061, China
2 Geological Survey Institute of Jilin Province, Changchun 130102, China
3 Northeast Geological S&T Innovation Center of China Geological Survey, Shenyang 110034, China
* Correspondence: wangzg19@jlu.edu.cn

Abstract: The Beiwagou Pb-Zn deposit, located in the western part of the Liaodong Peninsula, is a carbonate-hosted stratiform deposit with a Pb + Zn reserve of 0.08 Mt @ 4.14% (Pb + Zn). The orebodies occur as conformable layers and lenses and are strictly controlled by strata (the Paleoproterozoic Gaojiayu and Dashiqiao Formations) and lithology (plagioclase amphibolite and dolomitic marble). Given that previous studies have focused only on the mineralization features and mineralogy of deposits, herein, we report in situ trace element analyses of pyrite using LA-ICP-MS, together with in situ sulfur isotopes of pyrite, to constrain the composition, substitution mechanisms, source of sulfur, and sulfate reduction pathways of pyrite in the Beiwagou deposit. Based on pyrite morphology, texture, and chemistry, four pyrite types were identified: subhedral, porous-to-massive pyrite (Py1) related to chalcopyrite; subhedral, porous crushed pyrite (Py2) associated with fine-grained sphalerite; rounded and porous pyrite (Py3) related to the Zn-rich part of the laminated ore; and anhedral, porous-to-massive pyrite (Py4) associated with pyrrhotite, arsenopyrite, sphalerite, and galena. Py1 is characterized by high As, Ag, Cd, In, Au, Cu, and Zn concentrations and low Te, Bi, and Mo concentrations, whereas Py2 has high concentrations of Co and Ni and low concentrations of other trace elements, such as Cu, Zn, Bi, and Te. Py3 is characterized by elevated As concentrations, low Co, Ni, In, W, Te, and Tl concentrations, and varying Pb concentrations, whereas Py4 has low Ag, Cd, In, Zn, Cu, and Mn concentrations and varying W, Co, Ni, Pb, Sb, and As concentrations. Significant correlations between some elements in each pyrite type suggest substitution mechanisms, such as $(Zn^{2+} + Cu^{2+} + Mn^{2+} + Cd^{2+}) \leftrightarrow 2Fe^{2+}$, $Ag^{+} + (Sb)^{3+} \leftrightarrow 2Fe^{2+}$, and $(Te^{+} + Ag^{+}) + Sb^{3+} \leftrightarrow 2Fe^{2+}$, and the existence of a negative correlation between Co and Ni implies competition between both elements. The strongly positive $\delta^{34}S$ values (12.11‰–23.54‰) are similar to that of seawater sulfates and likely result from thermochemical sulfate reduction (TSR). In conclusion, the Beiwagou Pb-Zn deposit is a typical SEDEX deposit and mineralization likely occurred during diagenesis.

Keywords: Paleoproterozoic; SEDEX deposit; trace elements; sulfur isotope

Citation: Yu, Q.; Wang, Z.; Sun, Q.; Wang, K. In Situ Trace Element and Sulfur Isotope Composition of Pyrite from the Beiwagou Pb-Zn Deposit, Liaodong Peninsula, Northeast China: Implications for Ore Genesis. *Minerals* **2023**, *13*, 1176. https://doi.org/10.3390/min13091176

Academic Editor: Maria Boni

Received: 26 July 2023
Revised: 29 August 2023
Accepted: 5 September 2023
Published: 7 September 2023

1. Introduction

The Liaodong Peninsula is located in the southern part of Liaoning Province in Northeast China and hosts numerous medium- and large-scale lead-zinc, gold, and silver deposits. These deposits can be divided into Paleoproterozoic strata-bound Pb-Zn mineralization (e.g., Zhenzigou, Dongsheng, and Beiwagou Pb-Zn deposits [1,2]) and Mesozoic magmatism-related vein-type Pb-Zn-Au-Ag mineralization (e.g., Baiyun and Wulong Au and Gaojiapuzi Ag deposits [3–5]). Previous studies have shown that later magmatism-related Pb-Zn veins were spatially superimposed onto the former stratiform Pb-Zn orebodies (e.g., the Zhenzigou Pb-Zn deposit [2,6]), which is also documented in many ore districts worldwide, e.g., the Yenefrito Pb-Zn mine in the Pyrenees, Spain [7], the Macmillan Pass Pb-Zn ± Ba deposit in Yukou, Canada [8], and the Red Dog Zn-Pb-Ag district, Northern Alaska [9]. Due to essential base metal elements, studies on the mineralization process of lead-zinc deposits have been pursued by scholars around the globe, and the type of genesis can be classified into SEDEX [2], skarn [10,11], Mississippi Valley-type (MVT) [12], and hydrothermal vein-type deposits [13]. In this context, the ore genesis and metal sources of the Pb-Zn deposits in the Liaodong Peninsula are still debated, as indicated by the Proterozoic seafloor exhalation model [2,14] and the sedimentation metamorphism-hydrothermal reworking model [15,16]. In addition, hypotheses about these strata-bound sedimentary-dominated Pb-Zn deposits formed by either true exhalative processes or subseafloor replacement during sedimentation or early diagenesis are contentious [17–19].

As a ubiquitous mineral in various ore systems, pyrite incorporates numerous trace elements (e.g., Ag, As, Au, Bi, Co, Cu, Mn, Mo, Ni, Pb, Sb, Se, Tl, and Zn), which are either substituted within the crystal lattice or exist as nanoparticles or micro-inclusions [20,21]. Previous studies have shown that variations in pyrite textures and compositions can reflect changes in the fluid signatures and/or precipitation conditions during the evolution of the hydrothermal system [22–24]. For instance, pyrite has a homogeneous distribution of trace elements in sub-seafloor replacement processes but this is not the case in exhalation processes that involve a significant loss of metals through dispersion on the seafloor [25,26]. Detailed pyrite studies (e.g., those implementing laser ablation inductively coupled plasma mass spectrometry (LA-ICP-MS)) can detect these subtle variations during ore precipitation and have been commonly applied to porphyry [23], skarn [27], VMS [28,29], and SEDEX [19,30] deposits.

Previous sulfur isotope analyses of sulfide separates have some obvious limitations, especially when considering fine-grained sulfide minerals (e.g., pyrite) with complex textures and multiple sources in superimposing deposits. In situ sulfur isotope analysis using high-spatial-resolution ion microprobe instruments (e.g., those used in secondary ion mass spectrometry (SIMS)) can measure the variability in the sulfur isotope composition of fine-grained intergrown sulfide minerals [31,32]. This method is useful for distinguishing between syngenetic, diagenetic, and epigenetic pyrite in strata-bound sedimentary-dominated Pb-Zn deposits [33,34].

The Beiwagou deposit, a typical strata-bound Pb-Zn deposit in the Liaodong Peninsula, was formed in the Paleoproterozoic Gaojiayu and Dashiqiao Formations and has a Pb + Zn reserve of 0.08 Mt @ 4.14% (Pb + Zn) [35]. Previous studies have focused only on the mineralization features [35] and mineralogy [36] of deposits. The Beiwagou Pb-Zn deposit was proposed to be a sedimentary metamorphic deposit that was modified by later hydrothermal fluids [35]. Nevertheless, there is no convincing evidence on the mineralization process. Here, we present detailed in situ pyrite trace element and sulfur isotope composition measurements that can provide critical information about the pyrite's growth history and conditions and processes of ore formation and can distinguish origins. These results are useful for investigating the relationship between pyrite formation and base metal sulfide mineralization as well as for developing a robust deposit model for the formation of the Beiwagou deposit.

2. Geological Setting

The Liaodong Peninsula (the Liaodong Rift Zone, ca. 700 km in the NNE) is bordered by the Tanlu fault in the west and the Japan Sea in the east and records the main geological evolution of the North China Craton (NCC) from the Archean to the Cenozoic (Figure 1). Regionally exposed strata comprise (i) the Archean metamorphic Anshan Group, comprising mainly amphibolite, magnetite quartzite, plagiogneiss, and granulite [37], and (ii) the Paleoproterozoic metamorphic Liaohe Group (including Langzishan, Lieryu, Gaojiayu, Dashiqiao, and Gaixian Formations) comprising a suite of greenschist-epidote-amphibolite-facies metamorphosed clastic and volcano-sedimentary rocks [38], and (iii) Paleozoic-to-Mesozoic clastic and volcanic sequences, for example, Cambrian,

Ordovician, Carboniferous, and Permian-to-Cretaceous rocks [39]. The Anshan Group, a product of the recycling of ancient crustal material, is an Anshan-type BIF iron-ore host that includes the Waitoushan, Qidashan, Xi'anshan, Baoguolao, Xiaolaihe, and Yanglin iron deposits [37,40]. The Liaohe Group hosts strata-bound sedimentary-dominated deposits (including Pb-Zn, Cu-Co, boron, talc, and magnesite deposits) such as the Zhenzigou Pb-Zn [2], Dahenglu Cu-Co [41], Houxianyu boron [42], Yangjiadian talc [43], and Qingshanhuai magnesite [44] deposits. The protoliths of the metavolcanic rocks from the Liaohe Group were formed at ca. 2.19 Ga and were modified by a metamorphic event around 1.90 Ga. The evolutionary mechanism may be related to arc-continent collision [38,45]. These Paleoproterozoic strata and deposits are partly hosted in the core and limbs of the NWW/NEE-trending anticlines/synclines (e.g., Dashiqiao-Caohekou, Hupiyu-Kuandian, and Gaixian-Xiuyan). Paleozoic-to-Mesozoic volcano-sedimentary units are generally bounded by regional EW/NE-trending faults and their secondary structures (e.g., Fuzhou-Daziying, Yalvjiang, and Xiongyue-Anshan).

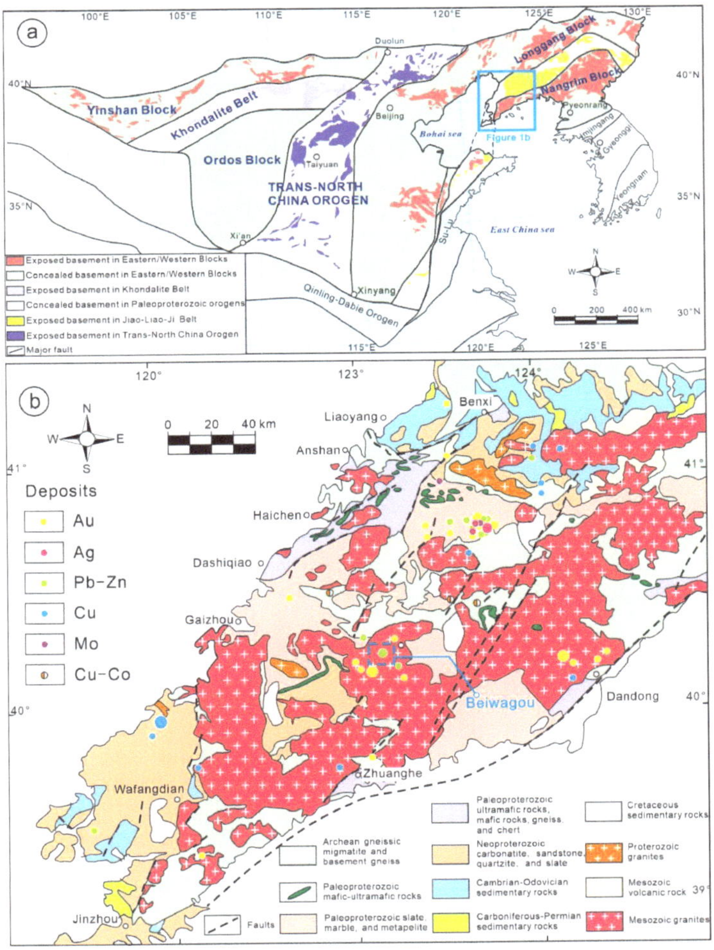

Figure 1. (**a**) Geologic map of the NCC (after [46]). (**b**) Geological map of the Liaodong Peninsula (after [3]).

Magmatism in the Liaodong Peninsula occurred during four episodes: (1) the Paleoproterozoic (ca. 2332–1756 Ma), (2) the Triassic (ca. 243–210 Ma), (3) the Jurassic (ca. 196–144 Ma), and (4) the Early Cretaceous (ca. 130–120 Ma) [2]. The Paleoproterozoic intrusive rocks comprise mainly gneissic granite, monzogranite, syenogranite, and granitic pegmatite [45,47], and their formation ages are concentrated at approximately 2332–2132 Ma and 1904–1756 Ma, respectively, which are related to collisions and extensions of the Liaonan block to the south and the Longgang block to the north [1,43,45]. Magmatism corresponds to three major orogenic/mobile belts during Mesozoic, namely the Central Asian orogenic belt on its northern margin, the Dabie–Sulu orogenic belt on its southern segment, and the subduction of the Pacific Plate on its eastern margin [2]. The Mesozoic plutons, including monzogranite, granodiorite, biotite granite, quartz diorite, etc. [39,47–51], show a close spatial-temporal-genetic link with the Au-Ag and local vein-type Pb-Zn mineralization (e.g., the Wulong Au, Baiyun Au, and Gaojiapuzi Ag deposits and the Xinling Ag-Pb-Zn deposits [2,52,53]).

3. Ore Deposit Geology

The Beiwagou Pb-Zn deposit (40°27′02″ N, 123°02′08″ E) is located in the western part of the Liaodong Peninsula (Figure 1b). Local strata mainly include the Gaojiayu and Dashiqiao Formations of the Paleoproterozoic Liaohe Group and Quaternary sediments (Figure 2). The Gaojiayu Formation (strike: 290–310°, NE dipping at 50–60°) is mainly distributed in the northern part of the ore district and is comprised of plagioclase hornblende, graphite-bearing marble, and biotite-tremolite-graphite granulite. The Dashiqiao Formation, the chief ore-bearing strata in the ore district, conformably overlies the Gaojiayu Formation and is distributed in the central part of the ore district. It can be divided into three units (e.g., D_1, D_2, and D_3) based on different lithological assemblages. Unit D_1 consists primarily of graphite-bearing marble, whereas unit D_2 comprises garnet/sillimanite-mica schist and dolomitic marble. Unit D_3 contains banded dolomitic and tremolite-bearing marble. Local structures include mainly NW/NE-trending faults and secondary EW-trending faults (Figure 2) and were formed after lead-zinc mineralization, which cut some stratiform orebodies but were limited in scale. Magmatic rocks are mainly discovered in the northern and southern parts of the ore district (Figure 2) and belong to the Paleoproterozoic Liaoji granite that was intruded upon by late Mesozoic magmatism (ca. 157.4–156.3 Ma [54]).

The orebodies of the Beiwagou deposit occur as conformable layers and lenses and are strictly controlled by strata (Paleoproterozoic Gaojiayu and Dashiqiao Formations) and lithology (plagioclase amphibolite and dolomitic marble) (Figure 3). Twelve NW-trending Pb-Zn orebodies (e.g., No. I, I-1, II, III, III-1, and V) have been delineated in the ore district, striking at 280–310° and dipping 35–60° NE. These orebodies are 46–305 m long and 0.61–13.03 m thick, with ore grades of 0.3%–0.79% Pb and 2.58%–9.74% Zn. Beiwagou Pb-Zn mineralization occurs mostly as layers, lenses, and veins and minorly as breccias and dissemination (Figure 4). The late Mesozoic intrusive bodies or dikes and vein-type Pb-Zn mineralization are significantly governed by the EW-trending and NW-trending faults, implying a close connection among them. The scales of vein-type orebodies are finite despite the grades achieving industrial requirements. The ore minerals of the Beiwagou Pb-Zn deposit are predominantly sphalerite, galena, and pyrite, along with minor chalcopyrite, arsenopyrite, marcasite, and pyrrhotite (Figure 5). The gangue minerals include dolomite, tremolite, diopside, talc, taxonite, and quartz. Wallrock alterations are predominantly serpentinization, dolomitization, and silicification, which are closely related to Pb-Zn mineralization.

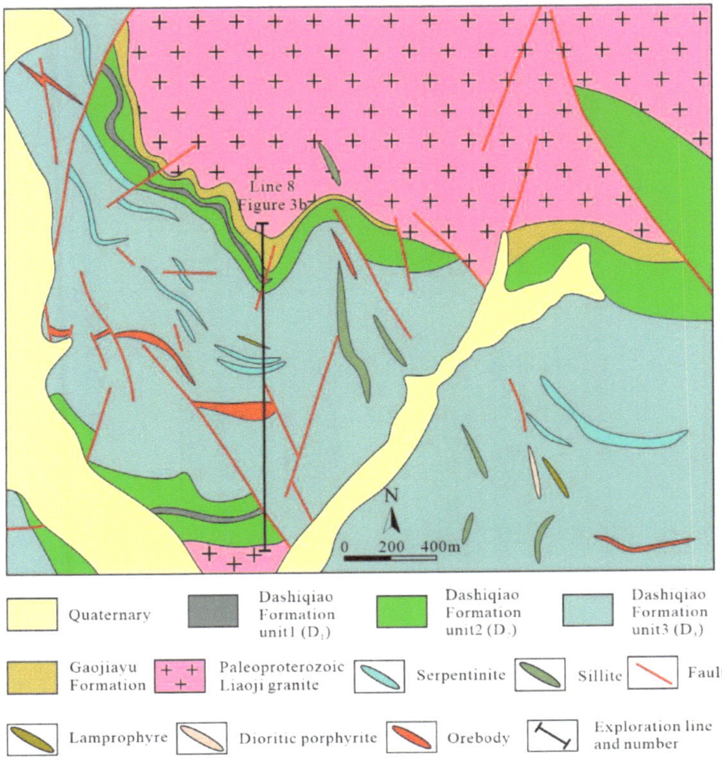

Figure 2. Geologic map of the Beiwagou Pb-Zn deposit (modified from [35]).

The delicately laminated nature of the Pb-Zn mineralization at Beiwagou and the surrounding carbonaceous host rocks (e.g., dolomitic marble) of the Gaojiayu and Dashiqiao Formations indicate that deposition occurred in a relatively static tectonic setting (Figures 3 and 4). Localized laminated orebodies show soft-sediment deformation features, such as slumping, folding, and dewatering pipes (Figure 4c,d). These soft-sediment deformation features may be related to the late Paleoproterozoic compressional tectonism (e.g., ca. 1800 Ma, during the "Lvliang Movement") [53,55,56].

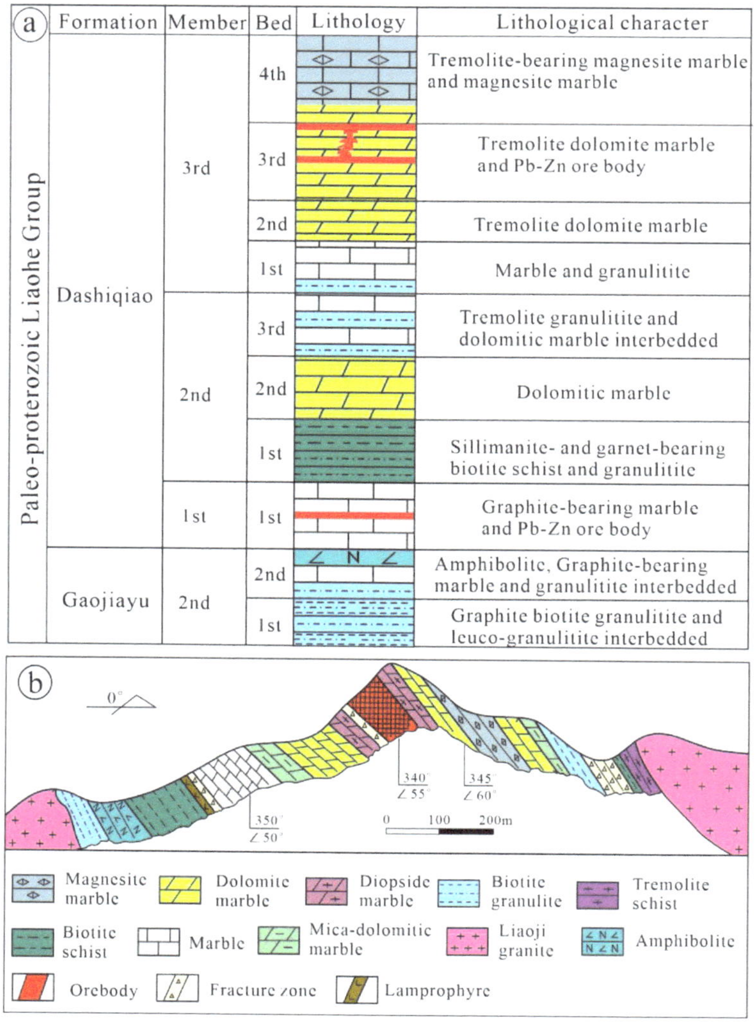

Figure 3. (**a**) Stratigraphic column of the Beiwagou Pb-Zn deposit; (**b**) The No. 8 exploration section map in the Beiwagou Pb-Zn deposit (modified from [35]).

Figure 4. Photographs of strata-bound mineralization in tremolite dolomite marble (**a**) and dolomitic marble (**b**) of the Dashiqiao Formation; (**c**,**d**) Deformation orebodies in the Beiwagou deposit.

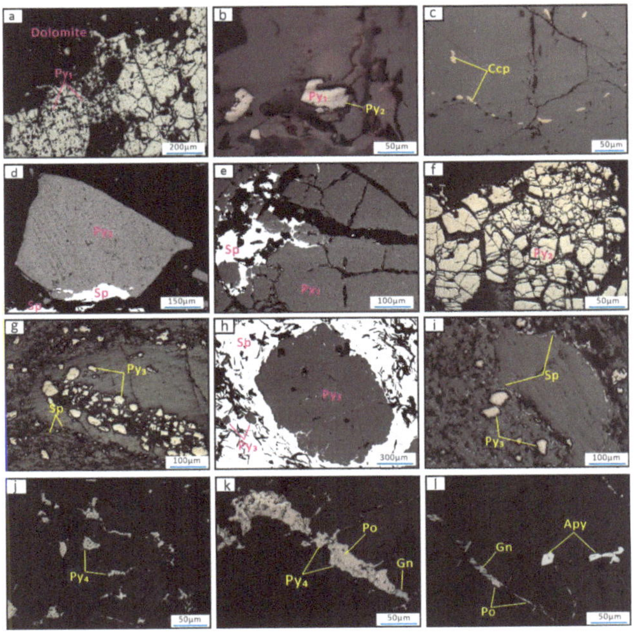

Figure 5. Photomicrographs of the Beiwagou Pb-Zn ore textures: (**a**) Subhedral, porous-to-massive pyrite (Py1); (**b**) Subhedral, porous-to-massive pyrite (Py1) and subhedral, porous crushed pyrite (Py2) developed in dolomitic marble; (**c**) Fine-grain chalcopyrite in the Py1; (**d**,**e**) Py2 replaced with sphalerite (BSE image); (**f**) Subhedral, porous crushed pyrite (Py2); (**g**) Clusters of rounded and porous pyrite (Py3); (**h**) Py3 replaced with sphalerite (BSE image); (**i**) Py3 replaced with sphalerite; (**j**) Anhedral, porous-to-massive pyrite (Py4); (**k**) Py4 replaced with pyrrhotite and galena; (**l**) Py4 replaced with pyrrhotite, galena, and arsenopyrite. Abbreviations: Apy = Arsenopyrite; Ccp = Chalcopyrite; Gn = Galena; Po = Pyrrhotite; Py = Pyrite; Sp = Sphalerite.

4. Sampling and Analytical Methods

4.1. Sampling Strategy

The samples were collected from underground mine working sites (e.g., PD3, 5, and 330) and drill cores (e.g., ZK8604, ZK8605, ZK8611, and ZK9006) from the Beiwagou deposit. Based on microscopic observations under transmitted and reflected light, a representative suite of samples were selected for trace element and sulfur isotopic analyses.

4.2. In Situ LA-ICP-MS Analyses

In situ LA-ICP-MS analyses were performed using an Analytik Jena PlasmaQuant MS Elite ICP-MS with a RESOlution SE 193 nm excimer laser denudation system at Beijing Createch Testing Technology Co., Ltd. (Beijing, China) [57]. Quantitative pyrite analyses were conducted by ablating a series of trenches in a high-purity He atmosphere to minimize beam attenuation and obtain an even ablation profile. The laser denudation was performed with a beam diameter of 30 μm, a laser frequency of 6 Hz, and an energy density of about 6 J/cm^2. The analysis time for each sample was 85 s, including a 40 s background with the laser off and 45 s analysis with the laser on. External standards, including synthetic glass standards NIST 611 and 612, as well as MASS-3, BHVO-2G, BCR-2G, and BIR-1G, were used at intervals of one every 10 denudation points for quantitative analyses [58]. Sample and blank signal selection, instrument sensitivity drift correction, and elemental content calculation were performed using ICPMSData Cal (v7.8) software.

4.3. In Situ Sulfur Isotope Analyses

In situ sulfur isotope analysis was performed on the same samples analyzed via LA-ICP-MS at Beijing Createch Testing Technology Co., Ltd (Beijing, China). The analysis was performed using a RESOlution SE 193 nm LA system coupled with a Neptune Plus Multi-Collector (MC)-ICP-MS. Pyrite and reference material samples were analyzed using 20 to 30 μm spot sizes at a frequency of 5 Hz with an energy density of 3 J/cm^2. High-purity helium was used as a carrier gas at a flow rate of 320 mL/min. ^{32}S and ^{34}S were statically received simultaneously with a Faraday cup detector array, and the integration time was 0.131 s. Each analysis took approximately 27 s to collect 200 sets of data. Standard WS-1 was utilized as external standard for sulfur isotope fractionation correction. Results are reported as $\delta^{34}S$ relative to Vienna Canyon Diablo Troilite (V-CDT) with a precision of ±0.2‰.

5. Results

5.1. Pyrite Types

Based on pyrite morphology, texture, and chemistry, four pyrite types were identified in the Beiwagou Pb-Zn deposit, including subhedral, porous-to-massive pyrite (Py1) (Figure 5a–c), subhedral, porous crushed pyrite (Py2) (Figure 5d–f), rounded and porous pyrite (Py3) (Figure 5g–i), and anhedral, porous-to-massive pyrite (Py4) (Figure 5j–l). In accordance with the formation sequence of the minerals and the associations of pyrite with the mineral assemblages, the four pyrite species were generated in the order of formation from Py1 to Py4. The detailed characteristics of each type of pyrite are described below.

Py1 was composed of subhedral-to-euhedral pyrite related to the Cu-rich part of the ore. The pyrite crystallites had a narrow range of diameters (40–100 μm, 65 μm on average) and locally occurred with chalcopyrite, commonly <10 μm in diameter.

Py2 was identified by its subhedral and crushed crystals, probably due to deformation from late Paleoproterozoic compressional tectonism. The grains were generally coarser (100–1100 μm, average of 400 μm) than in any other part of the deposit and locally filled with sphalerite (<10 μm) in the fracture space.

Py3 consisted of irregular and rounded pyrite crystallites related to the Zn-rich part of the laminated ore. These pyrite layers followed the bedding in the dolomitic marble independently or in groups. The grains had a wide range of diameters (<10–800 μm, 40 μm on average) and occurred with sphalerite crystallites, commonly 5 to 20 μm in diameter.

Py4 was defined by its association with pyrrhotite, arsenopyrite, sphalerite, and galena. Numerous macroscopic and petrographic observations have shown that pyrrhotite and sphalerite locally replace anhedral pyrite. The pyrite crystallites also had a wide range of diameters (20–500 μm, with an average of 150 μm).

5.2. Trace Element Characteristics of Pyrite

Seventeen trace elements were measured in all the pyrite varieties by LA-ICP-MS analysis (Table S1). Elements such as Mn, As, Co, Ni, Cu, Cd, Zn, In, W, Te, and Au show distinct trace element characteristics for each of the four pyrite types, as illustrated in Figure 6.

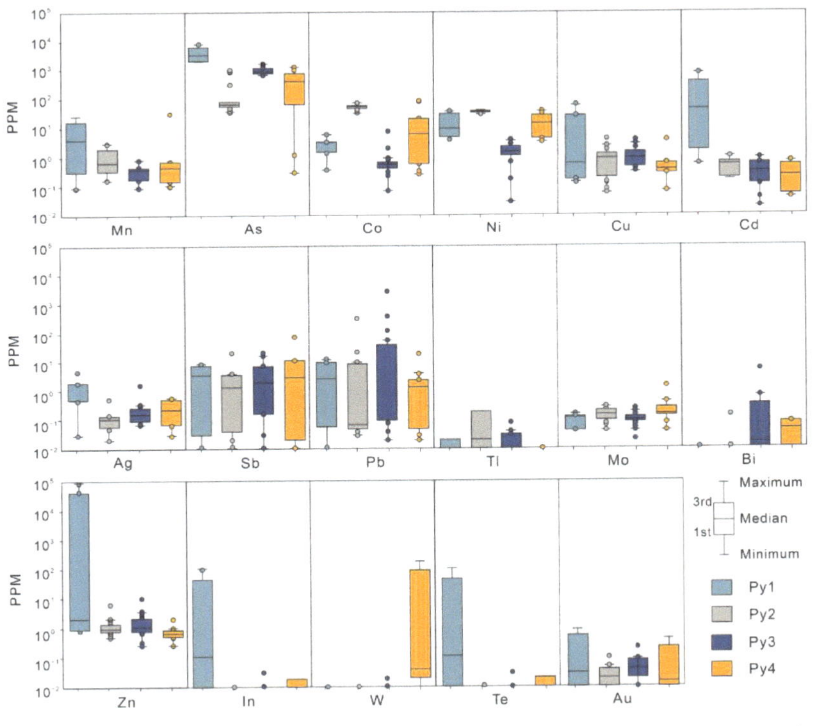

Figure 6. Simple box and whisker plots of trace element contents of each pyrite type (including Py1, Py2, Py3, Py4).

Py1: The trace element composition of the pyrite-chalcopyrite assemblage (Py1) is distinct in that it is characterized by elevated As (1980.75–8044.62 ppm), Ag (0.03–4.75 ppm), Cd (0.59–776.26 ppm), In (0.01–109.93 ppm), Au (0.01–0.88 ppm), Cu (0.14–59.60 ppm), and Zn (0.88–93,853.23 ppm) and low Te (<0.54 ppm), Bi (<0.01 ppm), and Mo (0.03–0.15 ppm) concentrations (Figure 6; Table S1). The data for Py1 formed the largest cluster in the box and whisker plots of trace element content when compared to the other pyrite types (Figure 6). The enriched Zn concentrations of Py1 were scattered within single grains and showed a strong association with Cd, In, Cu, Ag, and Mn and a minor association with Sb and Pb (Figure 7A).

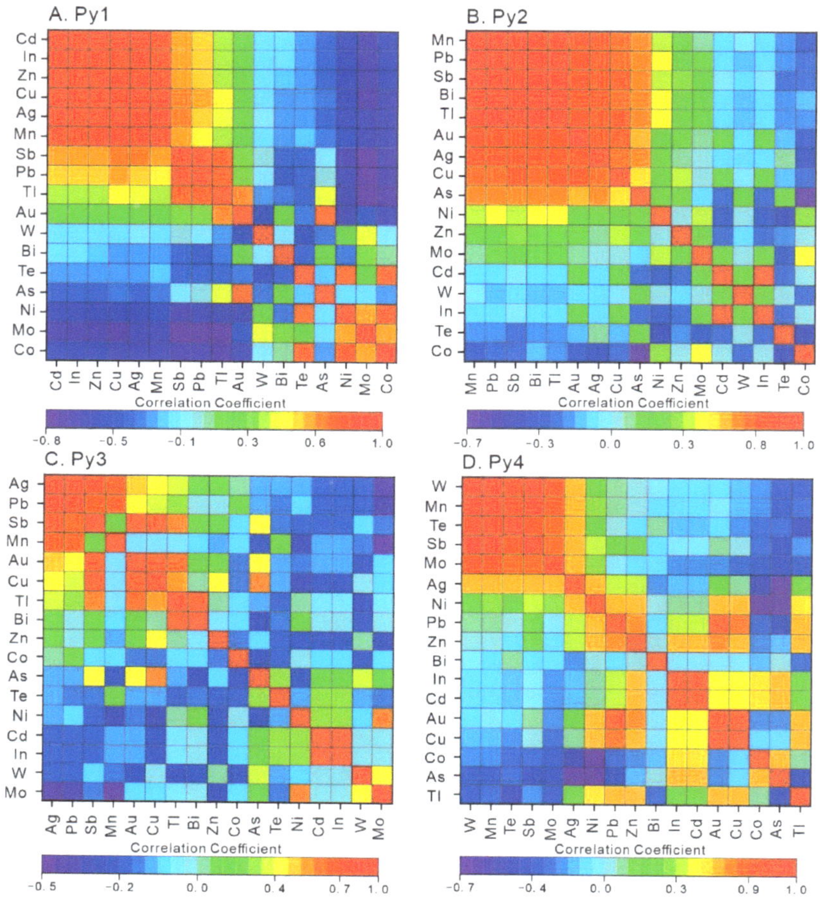

Figure 7. Correlation matrices showing element compositions of each pyrite type (including Py1, Py2, Py3, Py4).

Py2: Eighteen spots in four samples containing recrystallized and overgrown pyrite were analyzed. Py2 has higher Co (31.90–72.27 ppm) and Ni (29.52–41.14 ppm) concentrations than the other pyrite types (average = 1.64 ppm (Py3), 15.12 ppm (Py4), and 13.92 ppm (Py1)) (Figure 6; Table S1). But contents of some trace elements are low in Py2 grains, such as Cu (0.06–4.21 ppm), Zn (0.49–6.03 ppm), Bi (0.01–0.14 ppm), and Te (0.11–0.17 ppm). For Py2, positive correlations were observed between Pb and elements including Mn, Sb, Bi, Tl, Au, Ag, Cu, and As, together with the In vs. Cd correlation (Figure 7B).

Py3: Seventeen spots in three samples were observed on the rounded and porous pyrite. The result indicates very little chemical spread that is characterized by elevated As (665.86–1647.51 ppm), low Co (0.07–7.58 ppm), Ni (0.03–3.76 ppm), In (0.01–0.03 ppm), W (0.01–0.02 ppm), Te (0.01–0.89 ppm), and Tl (0.01–0.08 ppm), and varying Pb (0.02–2955.15 ppm) concentrations (Figure 6; Table S1). The scatter of Pb concentrations may be due to small inclusions in the porous pyrite. Positive correlations were observed between Pb and Ag, Sb, Mn, and Au, together with Au and Cu, Au and Tl, Au and Sb, In and Cd, and Bi and Tl (Figure 7C).

Py4: Chemically, the pyrite related to the pyrite-pyrrhotite-arsenopyrite assemblage (Py4) is clearly identified by low Ag (0.03–0.56 ppm), Cd (0.04–0.69 ppm), In (0.01–0.02 ppm), Zn (0.26–1.98 ppm), Cu (0.07–3.92 ppm), and Mn (0.11–25.45 ppm) and varying W (0.03–

180.75 ppm), Co (0.26–83.87 ppm), Ni (3.47–37.53 ppm), Pb (0.02–19.88 ppm), Sb (0.01–73.94 ppm), and As (0.29–1223.66 ppm) concentrations (Figure 6; Table S1). Py4 displays a broad positive correlation between W and elements such as Mn, Te, Sb, and Mo, together with Au vs. Cu and Cd vs. In (Figure 7D).

5.3. Sulfur Isotope Compositions

The sulfur isotope compositions of Py1, Py2, Py3, and Py4 had slightly narrow and highly positive ranges of $\delta^{34}S$ values (Table 1; Figure 8). The sulfide $\delta^{34}S_{V\text{-}CDT}$ values are 12.11‰–12.33‰ (mean: 12.23‰; Py1), 18.01‰–18.50‰ (mean: 18.25‰; Py2), 16.66‰–23.54‰ (mean: 18.89‰; Py3), and 14.24‰–15.00‰ (mean: 14.62‰; Py4).

Table 1. LA-MC-ICP-MS in situ S isotope compositions of each pyrite type from the Beiwagou Pb-Zn deposit.

Sample No.	$\delta^{34}S_{V\text{-}CDT}$ (‰)	Generations
BWG-4-1	12.33	Py1
BWG-4-2	12.11	Py1
BWG-4-3	12.25	Py1
BWG-4-4	18.01	Py2
BWG-7-1	18.23	Py2
BWG-7-2	18.50	Py2
BWG-3-1	14.24	Py4
BWG-3-2	15.00	Py4
BWG-6-1	16.66	Py3
BWG-6-2	17.38	Py3
BWG-6-3	19.65	Py3
BWG-6-4	23.54	Py3
BWG-6-5	17.24	Py3

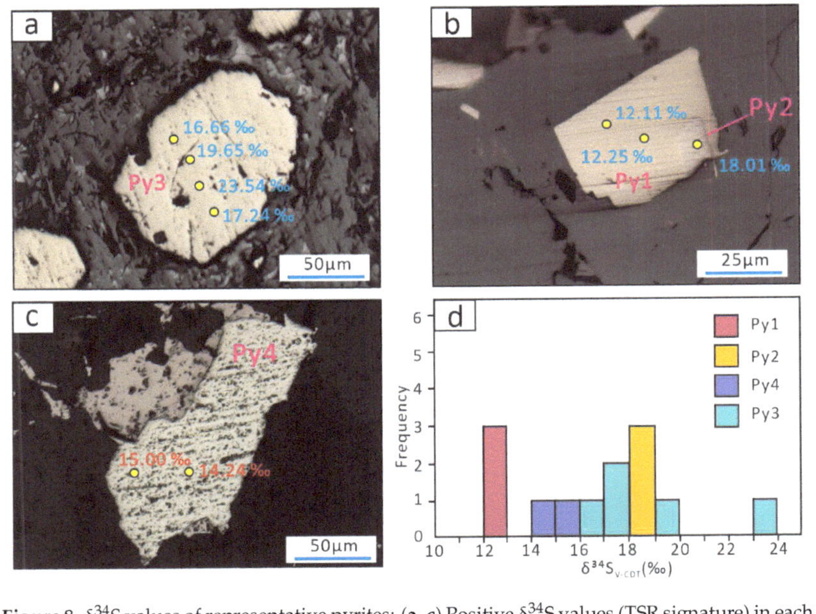

Figure 8. $\delta^{34}S$ values of representative pyrites: (**a–c**) Positive $\delta^{34}S$ values (TSR signature) in each pyrite type (including Py1, Py2, Py3, Py4); (**d**) Sulfur isotopic composition of pyrites in Beiwagou deposit.

6. Discussion

6.1. Partitioning of Trace Elements

Previous studies have demonstrated that many elements (e.g., Ag, Co, Ni, Sn, Zn, Au, Cu, As, Bi, Mn, Te, and Mo) have similar ionic radii to that of Fe^{2+} and are easily incorporated into the pyrite lattice via isomorphous substitution [59,60]. Huston et al. [20] suggested that these elements occur as either nonstoichiometric lattice substitutions (e.g., Cu, Mo, and Bi) or stoichiometric lattice substitutions (e.g., Co and Ni) within pyrite crystals. Therefore, the preferential concentration in pyrite can record not only changes in the physicochemical parameters of the parental fluid but also the subsequent deformation, dissolution-remobilization, and recrystallization processes.

The box and whisker plots of the trace element contents show that Py3 is more depleted in Co and Ni than Py4, Py1, and Py2 (Figure 6). This suggests that the fluids are barren in Co and Ni in nature or that little Co and Ni are partitioned to the pyrite via the isomorphous replacement of Fe. In addition, average Co and Ni ratios (1.38) are commonly accepted to indicate pyrite of sedimentary origin, as pyrite related to hydrothermal fluids has Co/Ni ratios of >>1 [61,62]. The abundance of As in all pyrite types may be associated with the significant substitution of As for S in the pyrite lattice. The Cd and In contents of all pyrite types showed a positive correlation (Figure 7), indicating that Cd and In entered the pyrite lattice via isomorphism. The low Mo content in the Beiwagou pyrite (up to 1.35 ppm) may reflect widespread oxidizing seawater conditions [63].

Py1 has higher Mn, Cd, As, In, and Te contents than other pyrite types, which is likely caused by surface defects in the pyrite lattice that facilitate elemental residence [64]. Metals with ionic sizes close to that of Fe^{2+}, such as Zn^{2+}, Cu^{2+}, Mn^{2+}, and Cd^{2+}, are generally incorporated into pyrite structures in the form $(Zn^{2+} + Cu^{2+} + Mn^{2+} + Cd^{2+}) \leftrightarrow 2Fe^{2+}$. The median values of As, Cd, In, and Te in Py1 were higher than those in the other pyrite types (Figure 6), indicating that the bulk replacement of S by As can promote a higher uptake of Cd, In, and Te in a more reduced environment. Py2 has higher Co (31.90–72.27 ppm) and Ni (29.52–41.14 ppm) concentrations than the other pyrite types (average = 13.92 ppm (Py1), 1.64 ppm (Py3), and 15.12 ppm (Py4)) (Figure 6; Table S1). However, a low correlation between Co and Ni was observed for each pyrite type, suggesting that Co and Ni are not direct substitutes for Fe in the pyrite lattice, forming Co- and Ni-bearing sulfides such as Co-pentlandite and pentlandite ($(Fe, Ni)_9S_8$). In addition, Sb shows positive correlations with Pb, Cu, and As in Py2 (Figure 7), which likely implies that some Sb may present in the pyrite lattice in the form of tetrahedrite ($(Cu, Fe)_{12}Sb_4S_{13}$) and freieslebenite ($AgPbSbS_3$) inclusions and behaves similarly to As when incorporated into the pyrite structure. The high correlation of the Ag, Pb, and Sb contents in Py3 suggests that Sb is incorporated into pyrite as a trivalent cation through the coupled substitution of $Ag^+ + (Sb)^{3+} \leftrightarrow 2Fe^{2+}$. However, due to its large ionic size, Pb barely enters the pyrite lattice. Most Pb occurs as a solid solution in pyrite and as a minor direct substitution of $Fe^{2+} \leftrightarrow Pb^{2+}$ in the pyrite structure [65]. Py4 and Py1 mostly have Co/Ni ratios < 1, which indicates that Co^{2+} and Ni^{2+} directly substitute for Fe^{2+} in pyrite [66,67]. The strong correlation between W, Mn, and Mo in Py4 may be attributed to the incorporation of them into the pyrite structure via the simple substitution equation: $M^{2+} \leftrightarrow Fe^{2+}$, while the strong correlation between monovalent and trivalent cations, such as Te, Ag, and Sb, may likely be caused by coupled substitutions in the pyrite lattice: $(Te^+ + Ag^+) + Sb^{3+} \leftrightarrow 2Fe^{2+}$.

6.2. Source of Sulfur

The $\delta^{34}S$ values for each pyrite type are strongly positive (12.11‰–23.54‰; Figure 8; Table 1) and similar to those of seawater sulfates (20‰ [68]), which is the predominant source of sulfur in stratiform, sediment-hosted Zn-Pb polymetallic deposits in most cases (e.g., [34,69]). Owing to the interference of the physicochemical properties of the ore fluid, such as the oxygen fugacity (f_{O2}), temperature, pH, and ionic strength [70], the sulfur isotopic composition ($\delta^{34}S_i$) of sulfides generally does not represent the sulfur content ($\delta^{34}S_{\Sigma S}$) of the ore fluids. In order to reliably estimate the sulfur content of the ore fluids from the pyrite data, it is necessary to know the temperature at which the pyrite was deposited. However, we could not obtain the formation temperature of the Beiwagou deposit because of a lack of microthermometric data for fluid inclusions and geothermal data for suitable mineral pairs. In most cases, the SEDEX-type Pb-Zn deposits were formed in a medium-low-temperature environment (e.g., 150–270 °C [71]). For this temperature range, the $\delta^{34}S$ value of pyrite corresponding to that of H_2S is 2‰ at 150 °C and 1‰ at 270 °C. Thus, we calculated that the $\delta^{34}S$ value of H_2S in the ore-forming fluid was 21‰–22‰ of the maximum $\delta^{34}S$ value analyzed for pyrite (~23‰) at Beiwagou and the above fractionation between pyrite and H_2S. Obviously, this is consistent with the $\delta^{34}S$ value of seawater for the Paleoproterozoic (~20‰ [72–74]). Therefore, we believe that the sulfur in the Beiwagou Pb-Zn deposit was sourced from coeval seawater sulfate.

6.3. Sulfate Reduction Pathways

There are several possible mechanisms for generating reduced sulfur (e.g., bacterial sulfate reduction (BSR) and thermochemical sulfate reduction (TSR)); the pathways by which sulfate is reduced to sulfide are still debated and remain contentious. BSR leads to the reduction of sulfate through bacterial respiration in an anaerobic environment, triggering a reaction of $2CH_2O + SO_4^{2-} \rightarrow 2HCO_3^- + H_2S$ [75]. TSR, which is an abiogenic process, reduces sulfate to sulfide by oxidizing labile organic carbon during diagenesis [76]. Although these two sulfate reduction pathways occur in different environments and processes, the end products of BSR and TSR are similar. Machel [75] suggested that the sulfur isotope composition of sulfide minerals could be effectively used to distinguish these processes. The presence of non-framboidal pyrite and high positive $\delta^{34}S$ values for each pyrite type from the Beiwagou deposit imply that the contribution of BSR sulfur to stratiform sediment-hosted Zn-Pb mineralization is not significant. This is because BSR can shrink the sulfate reservoir and subsequently increase the $\delta^{34}S$ value of sulfide as a sulfate-limited closed system in the ambient water column, which requires lower temperatures (<100 °C) to sustain microbial life [77]. BSR commonly leads to very negative $\delta^{34}S$ values and synsedimentary framboidal pyrite, which are shown in some SEDEX-type deposits elsewhere, such as the Howard's Pass SEDEX-type Zn-Pb deposit, Canada (−26 to −4.6‰ [33]), the Dajiangping pyrite deposit, China (−30 to −16‰ [34]), and the Navan Zn-Pb deposit, Ireland (−25 to −5‰ [78]).

As an important role in the formation of either vent-proximal or vent-distal SEDEX deposits, TSR by hydrothermal fluids generally occurs in higher-temperature environments (>100–140 °C) and significantly increase the $\delta^{34}S$ value of sulfide (TSR of seawater: 18‰–19‰ [79]). This is consistent with the highly positive $\delta^{34}S$ value of each pyrite type at Beiwagou from 12.11 to 23.54‰. Previous studies have experimentally determined that the kinetic fractionation between parent sulfate and daughter sulfide, expressed as $\Delta^{34}S_{SO4\text{-}H2S}$, is 20, 15, and 10‰ at 100, 150, and 200°C, respectively [80]. Assuming that the initial $\delta^{34}S$ sulfate value was approximately 30‰, the $\delta^{34}S$ values (12.11‰–23.54‰) of each pyrite type at Beiwagou are consistent with TSR at 150 to 200 °C ($\Delta^{34}S_{SO4\text{-}H2S}$ = 10‰–15‰). The temperature at which TSR occurs is consistent with the formation temperature of the SEDEX-type Pb-Zn deposit mentioned above.

6.4. Integrated Model for the Genesis of the Beiwagou Deposit

The stratiform nature of the orebodies and sulfide lamination are essential features of the Beiwagou deposit and are generally considered evidence of the SEDEX origin of deposits of this type (e.g., [19,34,81]). Recent studies on the genesis of stratiform sediment-hosted Pb-Zn deposits have shown that the diagenetic model is superior to the original synsedimentary-exhalative model. For instance, Gadd et al. [33] concluded that most of the mineralizing events in the Howard's Pass SEDEX Zn-Pb district took place in the pores of sediments during diagenesis, as is also the case for the HYC Zn-Pb-Ag and Tom-Jason Pb-Zn-Ba deposits [82,83]. Pyrite is a moderately refractory mineral that can retain its texture and trace composition during low-grade metamorphism but undergoes extensive recrystallization and trace composition changes in the presence of hydrothermal fluids [84]. Therefore, the texture and compositional changes of pyrite in different generations can effectively reflect the genesis of stratiform sediment-hosted Pb-Zn mineralization in Beiwagou.

As mentioned above, much of the Py3 in the stratified ores of the Beiwagou deposit occurs as rounded and anhedral crystals with a wide range of diameters (<10–800 μm, 40 μm on average), suggesting the replacement of the dolomitic marble during diagenesis. This is an important process in the formation of laminated sulfide ores in SEDEX deposits, such as the McArthur River (HYC), Mount Isa, and Hilton deposits [85,86]. Py4 crystals with anhedral shapes mostly occur in microfractures within the sediment layers (e.g., dolomitic marble), indicating the compaction during diagenesis. In contrast to Py3 and Py4, Py1, on the basis of its euhedral-to-subhedral pyrite cubes, clearly rules out the replacement model and likely precipitated from a mixture of vent fluid and seawater [87]. Owing to the subhedral and crushed crystals in Py2, the deformation from the late Paleoproterozoic compressional tectonism may have occurred within the Beiwagou deposit. Significantly, the trace element and sulfur isotope compositions in each pyrite type in the Beiwagou deposit suggest that Pb-Zn mineralization likely occurred within the sediments during diagenesis rather than only within the water column.

Mixing the exhalative metalliferous fluids with anoxic seawater caused the crystallization of sulfide particles in the water column along the interface between these two fluids. Sulfide aggregates gravitationally settle at the bottom of the basin, forming typical laminated sulfide textures and euhedral-to-subhedral pyrite cubes (e.g., Py1 and Py2) [88]. After deposition, the compaction and lithification during diagenesis could have induced the precipitated SEDEX-forming fluid to enter the unconsolidated carbonaceous sediments by displacing the less dense sediment pore water, as experimentally modeled by Sangster [89]. Hydrothermal chemical deposition near the sediment-water interface caused sulfate to undergo thermochemical sulfate reduction (TSR), thus turning it into sulfide and producing particulate pyrite and sphalerite. Continued compaction and lithification promoted the loss of permeability and vertical dewatering. Diagenetic fluid transport below the sediment-water interface may have deposited galena, sphalerite, and anhedral pyrite by replacing dolomitic marble during diagenesis.

The Liaodong Peninsula experienced an integrated introversion tectonic process that was initiated at ca. 2.2–2.0 Ga during an intra-continental rifting event and transitioned to an active continental margin at ca. 2.0–1.9 Ga, followed by an arc-continent collision at ca. 1.9 Ga and post-collisional extension at ca. 1.9–1.8 Ga [90]. LA-ICP-MS zircon U-Pb chronology data suggest that the Dashiqiao Formation was formed at 2.17–2.13 Ga while, at ca. 1866 $\pm$ 9 Ma, it underwent regional metamorphism [91], indicating that the stratiform Pb-Zn mineralization (SEDEX mineralization) in the Beiwagou ore district occurred during the tectonic setting of intra-continental rifting. The Dashiqiao Formation, as the chief ore-bearing stratum in the Beiwagou Pb-Zn district, is characterized by clastic deposition in the lower section, interbedded clastic deposition and carbonate in the middle section, and carbonate in the upper section, implying a mineralization environment of a transgression sedimentary cycle [35].

7. Conclusions

The Beiwagou sediment-hosted Pb-Zn deposit comprises stratiform orebodies hosted by carbonaceous rocks (e.g., dolomitic marble) of the Paleoproterozoic Gaojiayu and Dashiqiao Formations, similar to those of many SEDEX deposits. Based on the textual and compositional features of pyrite in the Beiwagou deposit, four pyrite generations (Py1, Py2, Py3, and Py4) were identified. Py1 is characterized by high As, Ag, Cd, In, Au, Cu, and Zn concentrations and low Te, Bi, and Mo concentrations, whereas Py2 is characterized by high concentrations of Co and Ni and low concentrations of other trace elements, such as Cu, Zn, Bi, and Te. Py3 was characterized by elevated As, low Co, Ni, In, W, Te, and Tl, and varying Pb concentrations, whereas Py4 was chemically identified by low Ag, Cd, In, Zn, Cu, and Mn and varying W, Co, Ni, Pb, Sb, and As concentrations. Significant correlations between some elements in each pyrite type suggest substitution mechanisms, such as (Zn^{2+} + Cu^{2+} + Mn^{2+} + Cd^{2+}) ↔ $2Fe^{2+}$, Ag^{+} + $(Sb)^{3+}$ ↔ $2Fe^{2+}$, and (Te^{+} + Ag^{+}) + Sb^{3+} ↔ $2Fe^{2+}$, and the existence of a negative correlation between Co and Ni implies competition between both elements. The average Co and Ni ratios of 1.38 may indicate pyrite of sedimentary origin, whereas the low Mo content (up to 1.35 ppm) may reflect seawater conditions as widespread oxidizing conditions. Therefore, each pyrite type in the Beiwagou deposit demonstrates the contribution of Paleoproterozoic seawater sulfate to mineralization. The ores have strongly positive $\delta^{34}S$ values, reflecting thermochemical sulfate reduction at 150 to 200 °C by hydrothermal fluids during diagenesis. Our data suggest that most mineralization events occurred due to the compaction and lithification during diagenesis, which induced sulfate to undergo TSR, loss of permeability, and vertical dewatering.

Supplementary Materials: The following supporting information can be downloaded at: https://www.mdpi.com/article/10.3390/min13091176/s1, Table S1. Trace element compositions of each pyrite type in the Beiwagou Pb-Zn deposit.xls.

Author Contributions: Conceptualization, Q.Y. and Z.W.; investigation, Q.Y., Z.W., Q.S. and K.W.; funding acquisition, Z.W. and K.W.; project administration, Z.W. and Q.S.; writing—original draft preparation, Q.Y.; writing—review and editing, Q.Y., Z.W. and Q.S. All authors have read and agreed to the published version of the manuscript.

Funding: This research was financially supported by the National Natural Science Foundation of China (Grants 42102074 and U2244201) and the funding project of Northeast Geological S&T Innovation Center of China Geological Survey (No. QCJJ2022-32).

Data Availability Statement: The experimental data used to support the conclusions of this study are included within the article and supplementary materials.

Acknowledgments: We are grateful to the staff from Beijing Createch Testing Technology Co., Ltd. for their advice and assistance in the experimental analysis. In addition, we wish to thank Xuebing Zhang (Xinjiang University) for providing valuable advice on the manuscript and the anonymous reviewers and editors for their insightful comments.

Conflicts of Interest: The authors declare no conflict of interest.

References

1. Duan, X.X.; Zeng, Q.D.; Wang, Y.B.; Zhou, L.L.; Chen, B. Genesis of the Pb-Zn deposits of the Qingchengzi ore field, eastern Liaoning, China; constraints from carbonate LA-ICPMS trace element analysis and C-O-S-Pb isotopes. *Ore Geol. Rev.* **2017**, *89*, 752–771. [CrossRef]
2. Li, J.; Cai, W.Y.; Li, B.; Wang, K.Y.; Liu, H.L.; Yassa, K.; Qian, Y.; Lee, G.J.; Yoo, B.C. Paleoproterozoic SEDEX-type stratiform mineralization overprinted by Mesozoic vein-type mineralization in the Qingchengzi Pb-Zn deposit, Northeastern China. *J. Asian Earth Sci.* **2019**, *184*, 104009. [CrossRef]
3. Liu, J.; Liu, F.X.; Li, S.H.; Lai, C.K. Formation of the Baiyun gold deposit, Liaodong Gold Province, NE China: Constraints from zircon U-Pb age, fluid inclusion, and C-H-O-Pb-He isotopes. *Ore Geol. Rev.* **2019**, *104*, 686–706. [CrossRef]
4. Li, J.; Wang, K.Y.; Cai, W.Y.; Sun, Q.F.; Liu, H.L.; Fu, L.J.; Qian, Y.; Lai, C.K. Triassic gold-silver metallogenesis in Qingchengzi orefield, North China Craton: Perspective from fluid inclusions, REE and H-O-S-Pb isotope systematics. *Ore Geol. Rev.* **2020**, *121*, 103567. [CrossRef]
5. Liu, Y.; Wei, J.; Tan, J.; Ulrich, T.; Zhang, D.; Lahaye, Y.; Zhang, X.; Chen, J. Sulfide micro-textures and in-situ compositions from the Wulong gold deposit, northeastern North China Craton: Implication for ore-forming processes. *Ore Geol. Rev.* **2022**, *148*, 105028. [CrossRef]
6. Ma, Y.B.; Bagas, L.; Xing, S.W.; Zhang, S.T.; Wang, R.J.; Li, N.; Zhang, Z.J.; Zou, Y.F.; Yang, X.Q.; Wang, Y.; et al. Genesis of the stratiform Zhenzigou Pb-Zn deposit in the North China Craton: Rb/Sr and C-O-S-Pb isotope constraints. *Ore Geol. Rev.* **2016**, *79*, 88–104. [CrossRef]
7. Subias, I.; Fanlo, I.; Yuste, A.; Fernandez-Nieto, C. The Yenefrito Pb-Zn mine (Spanish Central Pyrenees); an example of superimposed metallogenetic events. *Miner. Depos.* **1999**, *34*, 220–223. [CrossRef]
8. Magnall, J.M.; Gleeson, S.A.; Paradis, S. A new subseafloor replacement model for the macmillan pass clastic-dominant Zn-Pb ± Ba deposits (Yukon, Canada). *Econ. Geol.* **2020**, *115*, 953–959. [CrossRef]
9. Slack, J.F.; Kelley, K.D.; Anderson, V.M.; Clark, J.L.; Ayuso, R.A. Multistage hydrothermal silicification and Fe-Tl-As-Sb-Ge-REE enrichment in the Red Dog Zn-Pb-Ag district, northern Alaska: Geochemistry, origin, and exploration applications. *Econ. Geol.* **2004**, *99*, 1481–1508. [CrossRef]

10. Shu, Q.; Chang, Z.; Mavrogenes, J. Fluid compositions reveal fluid nature, metal deposition mechanisms, and mineralization potential: An example at the Haobugao Zn-Pb skarn, China. *Geology* **2021**, *49*, 473–477. [CrossRef]
11. Shu, Q.; Chang, Z.; Hammerli, J.; Lai, Y.; Huizenga, J.-M. Composition and evolution of fluids forming the Baiyinnuo'er skarn Zn-Pb deposit, northeastern China: Insights from laser ablation ICP-MS study of fluid inclusions. *Econ. Geol.* **2017**, *112*, 1441–1460. [CrossRef]
12. Liu, Q.; Mei, Y.; Etschmann, B.; Glenn, M.; MacRae, C.M.; Spinks, S.C.; Ryan, C.G.; Brugger, J.; Paterson, D.J. Germanium speciation in experimental and natural sphalerite: Implications for critical metal enrichment in hydrothermal Zn-Pb ores. *Geochim. Cosmochim. Acta* **2023**, *342*, 198–214. [CrossRef]
13. Esra, Ü.Ç.; Ahmet, G.; Chris, H. Genesis of Tertiary Akçakışla vein-type Pb-Zn-Cu mineralisation (Central Anatolia, Turkey): Evidence from fluid inclusion and O, H, S, Pb-isotope compositions. *J. Afr. Earth Sci.* **2023**, *205*, 105008.
14. Wang, K.Y.; Zhao, Y.M.; Cao, X.L. *The Proterozoic Type Lead-Zinc Deposits in Northern Border of North China Platform*; Geological Publishing House: Beijing, China, 1994; p. 190. (In Chinese)
15. Jiang, S.Y. Pb-Isotope Composition at Qingchengzi Lead-Zinc Deposit and Its Geological Application. *J. Peking Univ. (Nat. Sci.)* **1987**, *23*, 112–119, (In Chinese with English Abstract).
16. Ding, T.P.; Jiang, S.Y.; Wan, D.F.; Li, J.C.; Song, B.; Zhao, D.M. *Stable Isotope Studies on the Proterozoic Pb-Zn Mineral Belt of Northern China*; Beijing Publishing House of Science and Technology: Beijing, China, 1992; pp. 36–60. (In Chinese)
17. Leach, D.L.; Marsh, E.; Emsbo, P.; Rombach, C.S.; Kelley, K.D.; Anthony, M. Nature of Hydrothermal Fluids at the Shale-Hosted Red Dog Zn-Pb-Ag Deposits, Brooks Range, Alaska. *Econ. Geol.* **2004**, *99*, 1449–1480. [CrossRef]
18. Leach, D.L.; Bradley, D.C.; Huston, D.; Pisarevsky, S.A.; Taylor, R.D.; Gardoll, S.J. Sediment-Hosted Lead-Zinc Deposits in Earth History. *Econ. Geol.* **2010**, *105*, 593–625. [CrossRef]
19. Gadd, M.G.; Layton-Matthews, D.; Peter, J.M.; Paradis, S.J. The world-class Howard's Pass SEDEX Zn-Pb district, Selwyn Basin, Yukon; Part I, trace element compositions of pyrite record input of hydrothermal, diagenetic, and metamorphic fluids to mineralization. *Miner. Depos.* **2016**, *51*, 319–342. [CrossRef]
20. Huston, D.L.; Sie, S.H.; Suter, G.F.; Cooke, D.R.; Both, R.A. Trace elements in sulfide minerals from eastern Australian volcanic-hosted massive sulfide deposits: Part I. Proton microprobe analyses of pyrite, chalcopyrite, and sphalerite, and Part II. Selenium levels in pyrite: Comparison with $\delta^{34}S$ values and implications for the source of sulfur in volcanogenic hydrothermal systems. *Econ. Geol.* **1995**, *90*, 1167–1196.
21. Deditius, A.P.; Utsunomiya, S.; Reich, M.; Kesler, S.E.; Ewing, R.C.; Hough, R.; Walshe, J. Trace metal nanoparticles in pyrite. *Ore Geol. Rev.* **2011**, *42*, 32–46. [CrossRef]
22. Large, R.R.; Danyushevsky, L.; Hollit, C.; Maslennikov, V.; Meffre, S.; Gilbert, S.; Bull, S.; Scott, R.; Emsbo, P.; Thomas, H.; et al. Gold and Trace Element Zonation in Pyrite Using a Laser Imaging Technique: Implications for the Timing of Gold in Orogenic and Carlin-style Sediment-Hosted Deposits. *Econ. Geol.* **2009**, *104*, 635–668. [CrossRef]
23. Reich, M.; Deditius, A.; Chryssoulis, S.; Li, J.W.; Ma, C.Q.; Parada, M.A.; Barra, F.; Mittermayr, F. Pyrite as a record of hydrothermal fluid evolution in a porphyry copper system: A SIMS/EMPA trace element study. *Geochim. Cosmochim. Acta* **2013**, *104*, 42–62. [CrossRef]
24. Gregory, D.; Meffre, S.; Large, R. Comparison of metal enrichment in pyrite framboids from a metal-enriched and metal-poor estuary. *Am. Mineral.* **2014**, *99*, 633–644. [CrossRef]
25. Doyle, M.G.; Allen, R.L. Subsea-floor replacement in volcanic-hosted massive sulfide deposits. *Ore Geol. Rev.* **2003**, *23*, 183–222. [CrossRef]
26. Rona, P.A.; Bostrom, K.; Laubier, L.; Smith, K.L. Hydrothermal mineralization at seafloor spreading centers. *Earth-Sci. Rev.* **1984**, *20*, 1–104. [CrossRef]
27. Chen, F.; Deng, J.; Wang, Q.; Huizenga, J.M.; Li, G.; Gu, Y. LA-ICP-MS trace element analysis of magnetite and pyrite from the Hetaoping Fe-Zn-Pb skarn deposit in Baoshan block, SW China: Implications for ore-forming processes. *Ore Geol. Rev.* **2020**, *117*, 103309. [CrossRef]
28. Wagner, T.; Klemd, R.; Wenzel, T.; Mattsson, B. Gold upgrading in metamorphosed massive sulfide ore deposits: Direct evidence from laser-ablation-inductively coupled plasma-mass spectrometry analysis of invisible gold. *Geology* **2007**, *35*, 775–778. [CrossRef]
29. Genna, D.; Gaboury, D. Deciphering the Hydrothermal Evolution of a VMS System by LA-ICP-MS Using Trace Elements in Pyrite: An Example from the Bracemac-McLeod Deposits, Abitibi, Canada, and Implications for Exploration. *Econ. Geol.* **2015**, *110*, 2087–2108. [CrossRef]
30. Hazarika, P.; Mishra, B.; Saravanan Chinnasamy, S.; Bernhardt, H.J. Multi-stage growth and invisible gold distribution in pyrite from the Kundarkocha sediment-hosted gold deposit, eastern India. *Ore Geol. Rev.* **2013**, *55*, 134–145. [CrossRef]
31. Mason, P.R.D.; Košler, J.; de Hoog, J.C.M.; Sylvester, P.J.; Meffan-Main, S. In situ determination of sulfur isotopes in sulfur-rich materials by laser ablation-multiple collector-inductively coupled plasma-mass spectrometry (LA-MC-ICP-MS). *J. Anal. At. Spectrom.* **2006**, *21*, 177–186. [CrossRef]
32. Ulrich, T.; Long, D.G.F.; Kamber, B.S.; Whitehouse, M.J. In situ trace element and sulfur isotope analysis of pyrite in a Paleoproterozoic gold placer deposit, Pardo and Clement Townships, Ontario, Canada. *Econ. Geol.* **2011**, *106*, 667–686. [CrossRef]

33. Gadd, M.G.; Layton-Matthews, D.; Peter, J.M.; Paradis, S.; Jonasson, I.R. The world-class Howard's Pass SEDEX Zn-Pb district, Selwyn Basin, Yukon. Part II: The roles of thermochemical and bacterial sulfate reduction in metal fixation. *Miner. Depos.* **2017**, *52*, 405–419. [CrossRef]
34. Qiu, W.J.; Zhou, M.F.; Li, X.C.; Williams-Jones, A.E.; Yuan, H.L. The Ggenesis of the Giant Dajiangping SEDEX-Type Pyrite Deposit, South China. *Econ. Geol.* **2018**, *113*, 1419–1446. [CrossRef]
35. Gao, C.Y.; Wang, J.G.; Zhao, Y. *Liaoning Xiuyan Manchu Autonomous County North Tile Pb-Zn Deposit Geological Report Detailed Survey*; National Geological Library: Beijing, China, 1990; pp. 1–160. (In Chinese)
36. Jiang, Y.; Liu, X.L. The Discovery of Epiboulangerite and Its Relation with Gold mineralization in Beiwagou Lead-Zinc mine. *Miner. Resour. Geol.* **1998**, *5*, 26–30. (In Chinese)
37. Song, B ; Nutman, A.P.; Liu, D.Y.; Wu, J. 3800 to 2500 Ma crustal evolution in the Anshan area of Liaoning Province, Northeastern China. *Precambr. Res.* **1996**, *78*, 79–94. [CrossRef]
38. Zhang, Q.S.; Yang, Z.S.; Wang, Y.J. *Early Crust and Mineral Deposits of Liaodong Peninsula*; Geological Publishing House: Beijing, China, 1988. (In Chinese)
39. Zhang, P. Metallogenic Characteristics, Ore Genesis, Diagenesis and Metallogenic Geodynamic Process of Precious and Non Ferrous Metals Ore Deposits in Southeast Liaoning Province. Ph.D. Thesis, Jilin University, Changchun, China, 2018. (In Chinese with English Abstract).
40. Li, H.; Zhang, Z.; Li, L.; Zhang, Z.; Chen, J.; Yao, T. Types and general characteristics of the BIF-related iron deposits in China. *Ore Geol. Rev.* **2014**, *57*, 264–287. [CrossRef]
41. Tian, F.; Wang, K.Y.; Liang, Y.H.; Li, S.H.; Dong, G.H. Characteristics of Hydrothermal Superimposed Mineralization and Source of Ore-Forming Fluids in the Dahenglu Co-Cu Deposit. *Northwest. Geol.* **2017**, *50*, 167–177. (In Chinese)
42. Yan, X.L.; Chen, B. Chemical and boron isotopic compositions of tourmaline from the Paleoproterozoic Houxianyu borate deposit, NE China: Implications for the origin of borate deposit. *J. Asian Earth Sci.* **2014**, *94*, 252–266. [CrossRef]
43. Li, J. Metallogenesis of Lead-Zinc-Gold-Silver in the Qingchengzi Orefield, Liaoning Province. Ph.D. Thesis, Jilin University, Changchun, China, 2020. (In Chinese with English Abstract).
44. Luo, Y.X.; Zhu, J.R.; Wang, Y.K. A preliminary study on the genesis of the Dahiqiao sparry magnesite deposit in the light of some geological feature. *Miner. Depos.* **1989**, *1*, 71–84, (In Chinese with English Abstract).
45. Meng, E.; Wang, C.Y.; Liu, C.H.; Shi, J.R.; Li, Y.G. Geochronology and Petrogenesis and Constraints on Regional Tectonic Evolution of the Meta-Volcanic Rocks in Southeastern Liaodong Peninsula. *J. Jilin Univ. (Earth Sci. Ed.)* **2017**, *47*, 1589–1619, (In Chinese with English Abstract).
46. Yang, C.W.; Zhang, J.; Zhao, L.F.; Zhang, C.; Yang, H.X.; Lu, T.J.; Liu, J.L. Detrital zircon constraints on tectonic evolution of the Liaodong Paleoproterozoic orogenic belt, North China Craton. *Precambr. Res.* **2021**, *362*, 106152. [CrossRef]
47. Yang, J.H.; Wu, F.Y.; Liu, X.M.; Xie, L.W.; Yang, Y.H. Petrogenesis and geolgoical significance of the Jurassic Xiaoheishan pluton in the Liaodong Peninsula, East China: In-site zircon U-Pb dating and Hf isotope analysis. *Bull. Mineral. Petrol. Geochem.* **2007**, *26*, 197–211, (In Chinese with English Abstract).
48. Yu, G.; Chen, J.F.; Xue, C.J.; Chen, Y.C.; Chen, F.K.; Du, X.Y. Geochronological framework and Pb, Sr isotope geochemistry of the Qingchengzi Pb-Zn–Ag–Au orefield, Northeastern China. *Ore Geol. Rev.* **2009**, *35*, 367–382. [CrossRef]
49. Duan, X.X.; Zeng, Q.D.; Yang, J.H.; Liu, J.M.; Wang, Y.B.; Zhou, L.L. Geochronology, geochemistry and Hf isotope of Late Triassic magmatic rocks of Qingchengzi district in Liaodong peninsula, Northeast China. *J. Asian Earth Sci.* **2014**, *91*, 107–124. [CrossRef]
50. Zhou, G.C.; Wang, Y.W.; Li, D.D.; Shi, Y.; Xie, H.J. LA-ICP-MS Zircon U-Pb Dating of Dykes from the Baiyun Gold Deposit in Eastern Liaoning. *Bull. Mineral. Petrol. Geochem.* **2017**, *36*, 620–627, (In Chinese with English Abstract).
51. Wu, F.Y.; Yang, J.H.; Liu, X.M. Geochronological Framework of the Mesozoic Granitic Magmatism in the Liaodong Peninsula, Northeast China. *Geol. J. China Univ.* **2005**, *11*, 305–317, (In Chinese with English Abstract).
52. Yu, B.; Zeng, Q.D.; Frimmel, H.E.; Qiu, H.C.; Li, Q.L.; Yang, J.H.; Wang, Y.B.; Zhou, L.L.; Chen, P.W.; Li, J.P. The 127 Ma gold mineralization in the Wulong deposit, Liaodong Peninsula, China: Constraints from molybdenite Re-Os, monazite U-Th-Pb, and zircon U-Pb geochronology. *Ore Geol. Rev.* **2020**, *121*, 103542. [CrossRef]
53. Sun, G.T.; Zeng, Q.D.; Zhou, J.X.; Zhou, L.L.; Chen, P.W. Genesis of the Xinling vein-type Ag-Pb-Zn deposit, Liaodong Peninsula, China: Evidence from texture, composition and in situ S-Pb isotopes. *Ore Geol. Rev.* **2021**, *133*, 104120. [CrossRef]
54. Li, S.Z.; Zhao, G.C.; Sun, M.; Wu, F.Y.; Liu, J.Z.; Hao, D.F.; Han, Z.Z.; Luo, Y. Mesozoic, not paleoproterozoic SHRIMP U-Pb Zircon Ages of Two Liaoji Granites, Eastern Block, North China Craton. *Int. Geol. Rev.* **2004**, *46*, 162–176. [CrossRef]
55. Chang, R.H.; Liu, Y.J.; Li, W.M.; Gu, D.X. Deformation and rheological properties of Dashiqiao Formation of Liaohe Group in Hupiyu area, eastern Liaoning Province. *Geol. J.* **2019**, *54*, 804–818. [CrossRef]
56. Wang, F.; Liu, F.L.; Schertl, H.P.; Wang, X.; Liu, P.H.; Tian, Z.H. Detrital zircon U-Pb geochronology and Hf isotopes of the Liaohe Group, Jiao-Liao-Ji Belt: Implications for the Paleoproterozoic tectonic evolution. *Precambr. Res.* **2020**, *340*, 105633. [CrossRef]
57. Liu, Y.S.; Hu, Z.C.; Gao, S.; Günther, D.; Xu, J.; Gao, C.G.; Chen, H.H. In-situ Analysis of Major and Trace Elements of Anhydrous Minerals by LA-ICP-MS without Applying an Internal Standard. *Chem. Geol.* **2008**, *257*, 34–43. [CrossRef]
58. Pearce, N.J.G.; Perkins, W.T.; Westgate, J.A.; Gorton, M.P.; Chenery, S.P. A Compilation of new and published major and trace element data for NIST SRM 610 and NIST SRM 612 glass reference materials. Geostandards Newsletter. *Geostand. Geoanal. Res.* **1997**, *21*, 115–144. [CrossRef]

59. Price, B.J. Minor Elements in Pyrites from the Smithers Map Area, B.C. and Exploration Applications of Minor Element Studies. Master's Thesis, The University of British Columbia, Vancouver, BC, Canada, 1972.
60. Tribovillard, N.; Algeo, T.J.; Lyons, T.; Riboulleau, A. Trace metals as paleoredox and paleoproductivity proxies: An update. *Chem. Geol.* **2006**, *232*, 12–32. [CrossRef]
61. Thomas, H.V.; Large, R.R.; Bull, S.W.; Maslennikov, V.; Berry, R.F.; Fraser, R.; Froud, S.; Moye, R. Pyrite and Pyrrhotite Textures and Composition in Sediments, Laminated Quartz Veins, and Reefs at Bendigo Gold Mine, Australia: Insights for Ore Genesis. *Econ. Geol.* **2011**, *106*, 1–31. [CrossRef]
62. Meng, Y.M.; Hu, R.Z.; Huang, X.W.; Gao, J.F.; Sasseville, C. The origin of the carbonate-hosted Huize Zn–Pb–Ag deposit, Yunnan province, SW China: Constraints from the trace element and sulfur isotopic compositions of pyrite. *Mineral. Petrol.* **2019**, *113*, 369–391. [CrossRef]
63. Algeo, T.J.; Lyons, T.W. Mo-total organic carbon covariation in modern anoxic marine environments: Implications for analysis of paleoredox and paleohydrographic conditions. *Paleoceanography* **2006**, *21*, 1–23. [CrossRef]
64. Reich, M.; Kesler, S.E.; Utsunomiya, S.; Palenik, C.S.; Chryssoulis, S.L.; Ewing, R.C. Solubility of gold in arsenian pyrite. *Geochim. Cosmochim. Acta* **2005**, *69*, 2781–2796. [CrossRef]
65. Morse, J.W.; Luther, G.W. Chemical influences on trace metal-sulfide interactions in anoxic sediments. *Geochim. Cosmochim. Acta* **1999**, *63*, 3373–3378. [CrossRef]
66. Maslennikov, V.V.; Maslennikova, S.P.; Large, R.R.; Danyushevsky, L.V. Study of Trace Element Zonation in Vent Chimneys from the Silurian Yaman-Kasy Volcanic-Hosted Massive Sulfide deposit (Southern Urals, Russia) Using Laser Ablation-Inductively Coupled Plasma Mass Spectrometry (LA-ICPMS). *Econ. Geol.* **2009**, *104*, 1111–1141. [CrossRef]
67. Grant, H.L.J.; Hannington, M.D.; Petersen, S.; Frische, M.; Fuchs, S.H. Constraints on the behavior of trace elements in the actively-forming TAG deposit, Mid-Atlantic Ridge, based on LA-ICP-MS analyses of pyrite. *Chem. Geol.* **2018**, *498*, 45–71. [CrossRef]
68. Rollinson, H.R. A terrane interpretation of the Archaean Limpopo Belt. *Geol. Mag.* **1993**, *130*, 755–765. [CrossRef]
69. Leach, D.L.; Sangster, D.F.; Kelley, K.D.; Large, R.R.; Garven, G.; Allen, C.R.; Gutzmer, J.; Walters, S.G. Sediment-hosted lead-zinc deposits: A global perspective. In *Economic Geology 100th Anniversary Volume*; Society of Economic Geologist: Littleton, CO, USA, 2005; pp. 561–607.
70. Ohmoto, H. Systematics of Sulfur and Carbon isotopes in Hydrothermal Ore Deposits. *Econ. Geol.* **1972**, *67*, 551–578. [CrossRef]
71. Cooke, D.R.; Bull, S.W.; Large, R.R.; McGoldrick, P.J. The Importance of Oxidized Brines for the Formation of Australian Proterozoic stratiform Sediment-Hosted Pb-Zn (SEDEX) Deposits. *Econ. Geol.* **2000**, *95*, 1–18. [CrossRef]
72. Paiste, K.; Lepland, A.; Zerkle, A.L.; Kirsimäe, K.; Kreitsmann, T.; Mänd, K.; Romashkin, A.E.; Rychanchik, D.V.; Prave, A.R. Identifying global vs. basinal controls on Paleoproterozoic organic carbon and sulfur isotope records. *Earth-Sci. Rev.* **2020**, *207*, 103230. [CrossRef]
73. Scott, C.; Wing, B.A.; Bekker, A.; Planavsky, N.J.; Medvedev, P.; Bates, S.M.; Yun, M.; Lyons, T.W. Pyrite multiple-sulfur isotope evidence for rapid expansion and contraction of the early Paleoproterozoic seawater sulfate reservoir. *Earth Planet. Sci. Lett.* **2014**, *389*, 95–104. [CrossRef]
74. Williford, K.H.; Van Kranendonk, M.J.; Ushikubo, T.; Kozdon, R.; Valley, J.W. Constraining atmospheric oxygen and seawater sulfate concentrations during Paleoproterozoic glaciation: In situ sulfur three-isotope microanalysis of pyrite from the Turee Creek Group, Western Australia. *Geochim. Cosmochim. Acta* **2011**, *75*, 5686–5705. [CrossRef]
75. Froelich, P.N.; Klinkhammer, G.P.; Bender, M.L.; Luedtke, N.A.; Heath, G.R.; Cullen, D.; Dauphin, P.; Hammond, D.; Hartman, B.; Maynard, V. Early oxidation of organic matter in pelagic sediments of the eastern equatorial Atlantic: Suboxic diagenesis. *Geochim. Cosmochim. Acta* **1979**, *43*, 1075–1090. [CrossRef]
76. Machel, H.G. Bacterial and thermochemical sulfate reduction in diagenetic settings—Old and new insights. *Sediment. Geol.* **2001**, *140*, 143–175. [CrossRef]
77. Leavitt, W.D.; Halevy, I.; Bradley, A.S.; Johnston, D.T. Influence of sulfate reduction rates on the Phanerozoic sulfur isotope record. *Proc. Natl. Acad. Sci. USA* **2013**, *110*, 11244–11249. [CrossRef]
78. Anderson, I.K.; Ashton, J.H.; Boyce, A.J.; Fallick, A.E.; Russell, M.J. Ore depositional processes in the Navan Zn-Pb deposit, Ireland. *Econ. Geol.* **1998**, *93*, 535–563. [CrossRef]
79. Martin, A.J.; McDonald, I.; Jamieson, J.W.; Jenkin, G.R.T.; McFall, K.A.; Piercey, G.; MacLeod, C.J.; Layne, G.D. Mineral-scale variation in the trace metal and sulfur isotope composition of pyrite: Implications for metal and sulfur sources in mafic VMS deposits. *Miner. Depos.* **2022**, *57*, 911–933. [CrossRef]
80. Machel, H.G.; Krouse, H.R.; Sassen, R. Products and distinguishing criteria of bacterial and thermochemical sulfate reduction. *Appl. Geochem.* **1995**, *10*, 373–389. [CrossRef]
81. Rivai, T.A.; Yonezu, K.; Tindell, T.; Boyce, A.J.; Sanematsu, K.; Satori, S.; Watanabe, K. The Dairi SEDEX Zn + Pb + Ag deposit (North Sumatra, Indonesia): Insights from mineralogy and sulfur isotope systematics. *Ore Geol. Rev.* **2020**, *122*, 103510. [CrossRef]
82. Ireland, T.; Large, R.R.; McGoldrick, P.; Blake, M. Spatial Distribution Patterns of Sulfur Isotopes, Nodular Carbonate, and Ore Textures in the McArthur River (HYC) Zn-Pb-Ag Deposit, Northern Territory, Australia. *Econ. Geol.* **2004**, *99*, 1687–1709. [CrossRef]

83. Magnall, J.M.; Gleeson, S.A.; Stern, R.A.; Newton, R.J.; Poulton, S.W.; Paradis, S. Open system sulphate reduction in a diagenetic environment—Isotopic analysis of barite ($\delta^{34}S$ and $\delta^{18}O$) and pyrite ($\delta^{34}S$) from the Tom and Jason Late Devonian Zn-Pb-Ba deposits, Selwyn Basin, Canada. *Geochim. Cosmochim. Acta* **2016**, *180*, 146–163. [CrossRef]
84. Large, R.R.; Halpin, J.A.; Danyushevsky, L.V.; Maslennikov, V.V.; Bull, S.W.; Long, J.A.; Gregory, D.D.; Lounejeva, E.; Lyons, T.W.; Sack, P.J.; et al. Trace element content of sedimentary pyrite as a new proxy for deep-time ocean–atmosphere evolution. *Earth Planet. Sci. Lett.* **2014**, *389*, 209–220. [CrossRef]
85. Perkins, W.G.; Bell, T.H. Stratiform Replacement Lead-Zinc Deposits: A Comparison between Mount Isa, Hilton, and McArthur River. *Econ. Geol.* **1998**, *93*, 1190–1212. [CrossRef]
86. Eldridge, C.S.; Williams, N.; Walshe, J.L. Sulfur Isotope Variability in Sediment-Hosted Massive Sulfide Deposits as Determined Using the Ion Microprobe SHRIMP: II. A Study of the H.Y.C Deposit at McArthur River, Northern Territory, Australia. *Econ. Geol.* **1993**, *88*, 1–26. [CrossRef]
87. Feel, R.A.; Geiselman, T.L.; Baker, E.T.; Massoth, G.J.; Hammond, S.R. Distribution and composition of hydrothermal plume particles from the ASHES Vent Field at Axial Volcano, Juan de Fuca Ridge. *J. Geophys. Res.* **1990**, *95*, 12855–12873. [CrossRef]
88. Large, R.R.; Bull, S.W.; McGoldrick, P.J.; Walters, S.; Derrick, G.M.; Carr, G.R. Stratiform and strata-bound Zn-Pb-Ag deposits in Proterozoic sedimentary basins, northern Australia. In *Economic Geology 100th Anniversary Volume*; Society of Economic Geologist: Littleton, CO, USA, 2005; pp. 931–963.
89. Sangster, D.F. The role of dense brines in the formation of vent-distal sedimentary-exhalative (SEDEX) lead-zinc deposits: Field and laboratory evidence. *Miner. Depos.* **2002**, *37*, 149–157. [CrossRef]
90. Liu, J.; Zhang, J.; Liu, Z.H.; Yin, C.Q.; Zhao, C.; Li, Z.; Yang, Z.J.; Dou, S.Y. Geochemical and geochronological study on the Paleoproterozoic rock assemblage of the Xiuyan region: New constraints on an integrated rift-and-collision tectonic process involving the evolution of the Jiao-Liao-Ji Belt, North China Craton. *Precambr. Res.* **2018**, *310*, 179–197. [CrossRef]
91. Dong, Y.; Bi, J.H.; Xing, D.H.; Ge, W.C.; Jing, Y. Geochronology and geochemistry of Liaohe Group and Liaoji granitoid in the Jiao-Liao-Ji Belt, North China Craton: Implications for petrogenesis and tectonic evolution. *Precambr. Res.* **2019**, *332*, 105399. [CrossRef]

minerals

MDPI

Article

Late Triassic Tectonic Setting in Northeastern Margin of North China Craton: Insight into Sedimentary and Apatite Fission Tracks

Shuang Tang [1], Shichao Li [1,2,*], Xinmei Zhang [1], Daixin Zhang [1], Hongtao Wang [1], Lijun Nie [3], Xiaodong Zhou [3] and Mengqi Li [1]

1 College of Earth Sciences, Jilin University, Changchun 130061, China; shuangtang21@mails.jlu.edu.cn (S.T.)
2 Key Laboratory of Mineral Resources Evaluation in Northeast Asia, Ministry of Natural Resources, Changchun 130061, China
3 Survey of Regional Geological and Mineral Resource of Jilin Province, Changchun 130022, China
* Correspondence: lsc@jlu.edu.cn

Abstract: The closure timing of the Paleo-Asian Ocean and the terminal stage of the Central Asian Orogenic Belt have been widely debated in the geological community. It's known that the gradual scissor-like closure of the Paleo-Asian Ocean occurred from west to east during the Paleozoic period. However, it was not until the Triassic period that the complete closure of the ocean occurred at the northeastern margin of the North China Craton. Nevertheless, there is still much uncertainty regarding the Late Triassic tectonic setting in Northeast China. In this study, we focused on the Upper Triassic Dajianggang Formation, located in the Shuangyang area of central Jilin Province, which is situated on the northeastern margin of the North China Plate. Our aim was to determine the formation age of the Dajianggang Formation by analyzing the detrital particle composition, petrogeochemistry, detrital zircon U-Pb isotope dating, and apatite fission track thermochronology. Our results indicated that the primary sandstone provenance area of the Dajianggang Formation in the Shuangyang area is the island arc orogenic belt. The tectonic background of the sandstone provenance area is mainly a continental island arc environment. The provenance area is mostly composed of felsic rocks with sedimentary tendencies, and some of its material may have originated from the northern margin of the North China Craton or the eroded recycle orogenic belt. LA-ICP-MS U-Pb dating of detrital zircons shows that the Dajianggang Formation formed after 226.8 ± 5 Ma. Moreover, analysis of the thermal evolution history modelling shows that the Dajianggang Formation in the Shuangyang area continued to be deposited and heated in the early stage, and then experienced rapid exhumation around 30 Ma. This suggests that the study area underwent an orogenic process during the early stage of formation, but then transitioned into a post-orogenic extension period, which constrained the final closure of the Paleo-Asian Ocean prior to the Late Triassic period. In addition, our study indicates that the remote effect of the Pacific subduction did not reach the study area until 30 Ma. The central age of the detrital apatite fission track of sample 19DJ-1 is 94.2 ± 8.3 Ma, which is younger than its corresponding stratigraphic age. The two peak ages of the fission track analysis are 62.9 ± 5.4 Ma and 126 ± 11 Ma. These findings provide new evidence for the tectonic evolution of Northeast China and shed light on the Late Triassic tectonic setting, as well as the influence time of subsequent tectonic domains in the southern part of Northeast China.

Keywords: northern margin of North China Craton; Dajianggang Formation; sediment provenance; zircon U-Pb dating; thermochronology

Citation: Tang, S.; Li, S.; Zhang, X.; Zhang, D.; Wang, H.; Nie, L.; Zhou, X.; Li, M. Late Triassic Tectonic Setting in Northeastern Margin of North China Craton: Insight into Sedimentary and Apatite Fission Tracks. *Minerals* **2023**, *13*, 975. https://doi.org/10.3390/min13070975

Academic Editors: Jiafu Chen, Nan Ju, Zhonghai Zhao and Luca Aldega

Received: 16 May 2023
Revised: 3 July 2023
Accepted: 19 July 2023
Published: 22 July 2023

1. Introduction

Northeast China is situated at the convergence of the Siberian Plate and the North China Plate, characterized by the superposition of the tectonic domains of the Paleo-Asian, Mongolia–Okhotsk, and Pacific Oceans [1,2]. Its complex tectonic evolution is marked by the early subduction of the Paleo-Asian Ocean, the intercontinental accretion of the Siberian and North China plates, and the final collage in the Middle Triassic period [3–6]. Furthermore, the geological history of Northeast China has been shaped by the convergence and subduction of two major tectonic plates, the Mongolia–Okhotsk plate and the Paleo-Pacific plate, towards the Eurasian continent since the Late Jurassic period. These geological events have played a crucial role in the formation and evolution of the region, leading to the development of complex geological structures and the sediment of diverse rock types [7–14]. Northeast China has a long and complex tectonic evolution, and thus plays a key role in studying the history of the three tectonic domains present in the region. However, the Late Triassic tectonic setting connecting the two periods has been debatable. This period is particularly significant, as it marks the final closure of the Paleo-Asian Ocean. The precise timing of the Paleo-Asian and Pacific tectonic domain influences on Northeast China remains unclear. Therefore, further geological investigations are necessary in order to shed light on this issue. There is still ongoing debate among geologists regarding the tectonic setting of Northeast China during the Late Triassic period. Some scholars argue that the subduction of the Paleo-Pacific plate to the eastern margin of the northeast Asian continent began during this time, due to the presence of a high-pressure metamorphic belt in the region [15–17]. On the other hand, other researchers propose that the northeastern region experienced an extensional tectonic background in the Late Triassic period, which was related to the extension of the lithosphere after the closure of the Paleo-Asian Ocean, rather than to the subduction of the Paleo-Pacific plate. This view is supported by the presence of alkaline rocks and bimodal volcanic rocks in the area [9,18,19]. Despite the ongoing debate, it is widely acknowledged that the Late Triassic is a crucial period for understanding the tectonic evolution of Northeast China. This period witnesses the final closure of the Paleo-Asian Ocean and marks an important transition in the region's tectonic history. However, while much research has focused on Late Triassic magmatic activity, there is a scarcity of information regarding the tectonic setting reflected by sedimentary events during this time.

Considering the ongoing debate surrounding the Late Triassic tectonic setting in Northeast China, this study focuses on the Upper Triassic Dajianggang Formation, located in the northeastern margin of the North China Craton. Through the utilization of detrital zircon U-Pb isotope dating, major and trace element analysis, detrital particle composition analysis, apatite fission track thermochronology, and other techniques, the goal of this research is to elucidate the Late Triassic tectonic setting and the timing of subsequent tectonic domains in the southern part of Northeast China. By doing so, this paper aims to provide new insights into the overall tectonic evolution of this region.

2. Geological Setting and Sample Description

The study area is situated in the central part of Jilin Province, which is geotectonically located in the eastern portion of the accretionary belt on the northern margin of the North China Craton (as depicted in Figure 1a). It is bounded by the Songnen Block and Jiamusi-Xingkai block to the north, the Songliao Basin to the west, and the North China Plate to the south. In the north of the study area lies the Solon–Xar Moron–Changchun–Yanji suture zone, while the Yitong–Yilan fault can be found to the west. Furthermore, the Dunhua–Mishan fault, trending in the northeast direction, cuts through the area to the south.

The geological characteristics of the study area are dominated by Carboniferous, Triassic, Jurassic, and Quaternary strata (Figure 1b). The Dajianggang Formation, which is the focus of this study, is mainly distributed along the northeast fault basin in the syncline axis of Jiangjunling-Dajianggang village, on the southern margin of the Mesozoic basin in the Shuangyang area of central Jilin Province [20]. The Dajianggang Formation is unconformably overlain by the basal conglomerate of the Lower Jurassic Taiyangling Formation, and it unconformably overlies the Permian–Carboniferous Mopanshan Formation and the Upper Triassic Sihetun Formation (Figure 1b) [21]. The Dajianggang Formation is a set of clastic rocks with a positive sequence of fluvial, lake, and marsh sedimentary facies. The lithological assemblage consists of yellow-green conglomerate, yellow-green sandstone, siltstone, clayslate, and coal-bearing strata, with a total thickness of approximately 1439 m [21,22]. The study area is characterized by the presence of coarse terrigenous clastic rocks with molasse characteristics, which are located at the base of the Changchun–Yanji suture zone complex [16,23,24]. The accumulated thickness of these rocks is over 400 m, with the upper part consisting of coal-bearing sedimentary strata. Fossil evidence suggests that the Late Triassic Drepanzamites–Glocsophtllum assemblage of terrestrial plants existed within these strata [21,24], including species such as *Drepanozamites*, *Neocalamites carrerei*, *N. hoerensis*, *Todites williamsoni*, *Taeniopteris* sp., cf. *Glossophyllum florin*, and *Pityophyllum* sp. [21,22,25]. A strata column section is shown in Supplementary Figure S1.

For the purpose of laboratory research, five representative samples were collected (19DJ-1, 19DJ-2, 19DJ-3, 19DJ-4, and 19DJ-5; sample location shown as Figure 1b). Petrographic observations were carried out on all five samples, while three samples (19DJ-1, 19DJ-4, and 19DJ-5) were subjected to bulk rock geochemical analysis. Additionally, sample 19DJ-1 was chosen for detrital zircon U-Pb dating and apatite fission track testing.

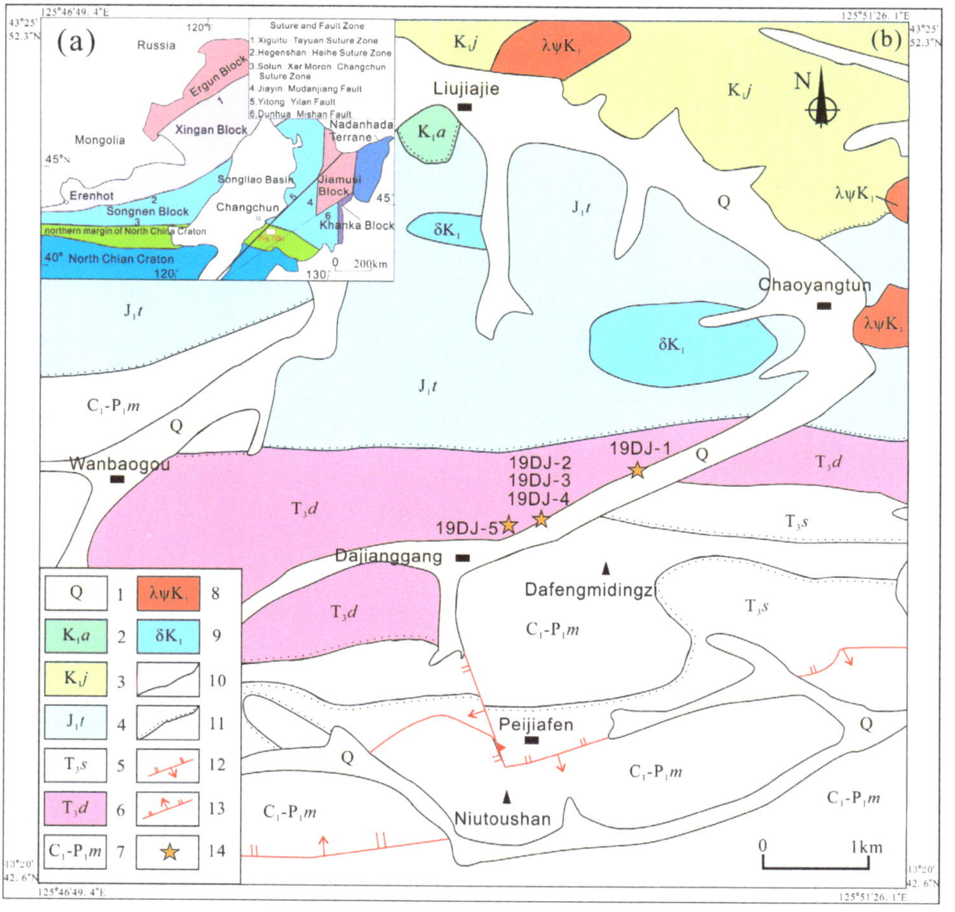

Figure 1. (**a**) Geological sketch map (modified after [23]). (**b**) Sampling location of study area. 1. Quaternary flood plain: alluvial sand and gravel; 2. Lower Cretaceous Anmin Formation: basalt, andesite, rhyolite, rhyolite containing agglomerate breccia, tuff breccia; 3. Lower Cretaceous Jiuda

Formation: grayish yellow pebbled coarse sandstone and fine sandstone; 4. Lower Jurassic Taiyangling Formation: grayish yellow conglomerate, medium fine and grained sandstone with coarse sandstone, thin conglomerate, upper black-gray siltstone, fine sandstone, mudstone, and coal seam; 5. Upper Triassic Sihetun Formation: Andesite, Andesitic litho-crystalline tuff, dacite, rhyolite litho-crystalline tuff, Dacitic breccia-bearing litho-crystalline tuff; 6. Upper Triassic Dajianggang Formation: black slate, sandstone, conglomerate mixed with acid, volcanic rock, and recoverable coal seam; 7. Permian–Carboniferous Mopanshan Formation: micrite limestone, micrite bioclastic limestone, chert-nodule and chert-banded limestone, dolomitic limestone, dolomite, off-white marble, siliceous rock; 8. Early Cretaceous quartz albite porphyry; 9. Early Cretaceous diorite; 10. Geological boundary; 11. Angular unconformity boundary; 12. Thrust fault; 13. Normal fault; 14. Sampling location.

3. Methodology

3.1. Analysis of Geochemical Characteristics

After observation under a microscope, the weathered and altered samples were removed, and three samples from the Dajianggang group were selected for research and analysis. Geochemical analysis of the major, minor, and rare earth elements of the samples was carried out at the Key Laboratory of Mineral Resources Evaluation in Northeast Asia, Jilin University. The fresh sample was pulverized to a particle size of <200 mesh (0.174 mm) using an agate ball mill. The major elements were determined using XRF spectroscopy, and the firing loss (LOI) was determined after the sample powder was ignited at 1000 °C for 1 h. The calcined sample (0.9 g) was added to 9.0 g lithium borate reagent ($Li_2B_4O_7$-$LiBO_2$), thoroughly mixed, and fused in a fusion machine between 1050 and 1100 °C. Glass melt samples were prepared from the obtained melt and measured using an X-ray fluorescence spectrometer (XRF). The determination of trace elements, including rare earth elements, was conducted using the inductively coupled plasma mass spectrometry (ICP-MS) method after the melting sample bomb mixed with high purity HF and HNO_3 was baked in a closed oven for 2 days. The detection limits of some key elements (integration time, 3 s) are as follows ($\times 10^{-6}$): Th (0.05), Nb (0.2), Hf (0.2), Zr (2), La (0.5), and Ce (0.5). Repeated analyses based on the analysis results and international standard reference materials (SRMS) show that the precision and accuracy of the data are better than 5% for major elements and better than 10% for trace elements.

3.2. Zircon U-Pb Isotope Dating

In this paper, sandstone sample 19DJ-1, from the Dajianggang Formation in the Shuangyang area, was chosen for LA-ICP-MS zircon analysis. The zircon dating was carried out at the Key Laboratory of Mineral Resources Evaluation in Northeast Asia, Ministry of Land and Resources, Jilin University, Changchun, China. The use of helium as a carrier gas provided efficient aerosol transport to the ICP and minimized aerosol deposition around the ablation site and within the transport tube [26,27]. Argon was used as the make-up gas and was mixed with the carrier gas via a T-connector before entering the ICP. The analysis spots were 32 μm in diameter, and U, Th, and Pb concentrations were calibrated using ^{29}Si as an internal standard. The standard zircon 91500 was used as an external standard in order to normalize isotopic fractionation during analysis. The analytical procedures used follow those described by Yuan, Gao [28]. The raw data were processed using the GLITTER program, and uncertainties of individual analyses were reported with 1σ error. Weighted mean ages were calculated at 1σ, and the data were processed using the IsoplotR program [29]. The zircon Plešovice was dated as an unknown sample and yielded a weighted mean $^{206}Pb/^{238}U$ age of 337.5 ± 2.4 Ma (n = 8, 2σ), which is in good agreement with the suggested age of 337.13 ± 0.37 Ma [30]. These methods and procedures ensured the accuracy and reliability of the zircon dating results.

3.3. Apatite Fission Track

Apatite fission track thermochronology is a method that utilizes the damage caused by the spontaneous fission of radioactive ^{238}U in a mineral lattice [31]. Many previous studies have been conducted on the annealing of apatite and the characteristics of partial annealing zones. The fission tracks in apatite provide information on the exhumation time, amplitude, rate, and mode of orogenic belts, as well as on the low-temperature domain of the thermal history of mountain exhumation [32,33]. The partial annealing zone of apatite is situated at a temperature range of 60–120 °C, which means that apatite fission track (AFT) data can indicate paleotemperature information of up to 120 °C, and fission tracks above 120 °C cannot be preserved on geological time scales. The accumulation of apatite tracks occurs with time at the temperature of the partial annealing zone or lower, and the age of a fission track can be determined by measuring its density and uranium content [34,35].

This study employed apatite fission track thermochronology in order to investigate the thermal history of the Dajianggang Formation in the Shuangyang area. The first step in the analysis was the collection of pure apatite grains in sandstone sample 19DJ-1. The sample was subjected to various preparatory procedures, including fragmentation, screening, magnetic separation, and heavy liquid separation in order to obtain pure apatite grains. Following this, apatite mounts in epoxy resin were ground and polished to expose maximum grain internal surfaces, which were then etched with 5.5 N HNO_3 at 20 °C for 20 s to reveal the spontaneous tracks. Low-uranium muscovite was then used as the external detector, while standard uranium glass IRRM540 was used as the dosimeter with which to detect irradiation neutron fluence. Subsequently, the pure apatite grains were irradiated with thermal neutrons at the Radiation Center in Oregon State University, USA. The fission track age of detrital apatite was then measured and analyzed in order to infer the thermal history of the Dajianggang Formation.

Fission track analysis was performed at the Fission Track Dating Laboratory, Northwest Institute of Eco-Environment and Resources, Chinese Academy of Sciences, using an automatic fission track counting system, consisting of a Zeiss AxioIMager Z2m microscope, ES16 platform, and Fission Track Studio software. Generally, about 80 particles per sample are analyzed to produce reliable statistical results. The fission track's age was determined by the ξ calibration method [36], and the personal ξ value was 257.56 ± 17.67. The χ^2 test was used to quantify age homogeneity. The value of $P(\chi^2) > 5\%$ represents a homogeneous age population [37]. The ability to retain fission tracks depends on the mineral and its composition, with apatite being a well-investigated mineral in this regard [34,38–41]. There was no obvious failure in the analyses. Moreover, the variation in Dpar is small and less likely to cause dispersion of the grain ages. For samples with a $P(\chi^2)$ value less than 5%, reflecting a wide distribution of individual grain ages, the distribution of the observed grain age should be decomposed to correspond to different grain age components of different clastic sources [42]. Track length distributions, combined with track density for a known uranium content, offer valuable insights into the thermal history of geological processes.

4. Petrographic Characteristics and Analysis of Detrital Particle Composition

In this study, the Dajianggang Formation in the Shuangyang area of central Jilin Province was systematically sampled to investigate the petrographic characteristics of the rock. In order to achieve this, five representative samples, namely 19DJ-1, 19DJ-2, 19DJ-3, 19DJ-4, and 19DJ-5, were selected for thin-section analysis. The Dajianggang Formation sandstones exhibit a medium–fine clastic texture and massive structure. They primarily consist of heterogenous matrix support, basement cement, and floating contacts between particles. The detrital particles are predominantly angular to subangular, with a wide range of particle sizes, varying from 0.1 to 2 mm, and the average content of heterogenous matrix is about 46%. Iron cement is the primary cementing agent in the sandstones. The clastic composition is dominated by feldspar, quartz, and lithic fragments, with an average content of feldspar at 29%, quartz at 30%, and lithic fragments at 41%. The rock type can be classified as feldspar lithic sandstone. The feldspar type includes plagioclase and alkaline feldspar, with plagioclase showing obvious polysynthetic twinning (Figure 2c). The alkaline feldspar type comprises potassium feldspar and microcline, with potassium feldspar showing Carlsbad twinning and microcline exhibiting grid twinning. The content of both types of alkaline feldspar is minimal. Monocrystalline quartz grains are small, irregular, and granular, with wavy extinction and no secondary enlarging. Polycrystalline quartz grains are large and often appear as aggregates or even gravel (Figure 2e,f). The lithic fragments vary in size, shape, and distribution and are dominated by volcanic and sedimentary lithic fragments, with rare metamorphic lithic fragments. Some volcanic lithic fragments show an obvious lattice made up of plagioclase (Figure 2d,g). Euhedral limonite is also observed (Figure 2e). The detritus exhibits poor sorting and roundness, uneven particle size, and a low component maturity, indicating that it originates from a nearby provenance area and has not been transported for a long distance or time.

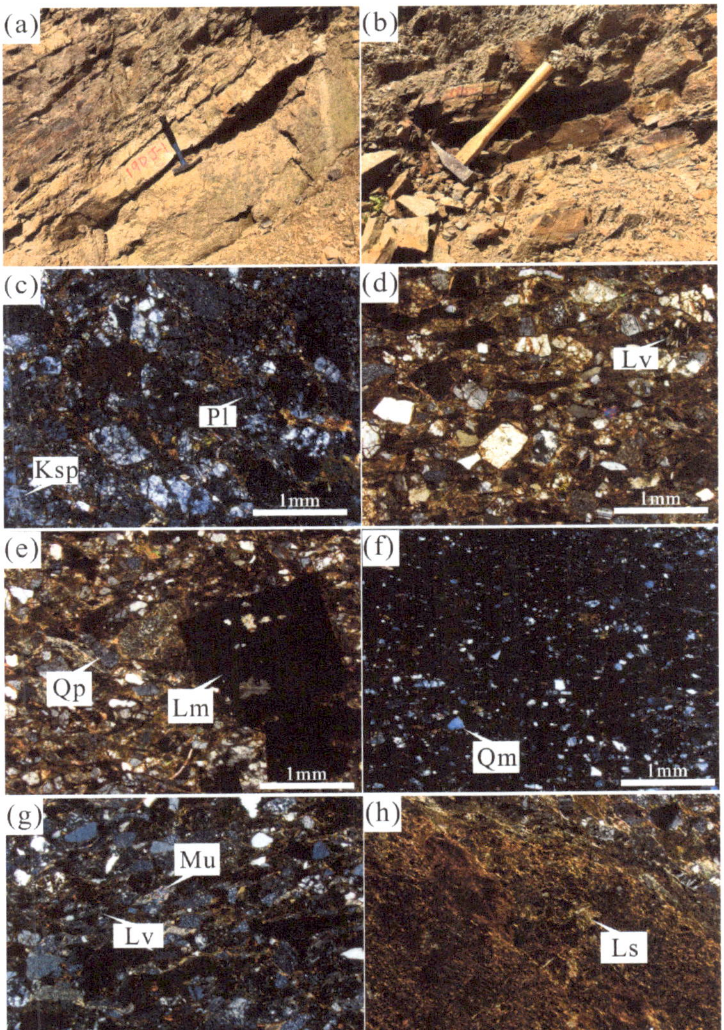

Figure 2. (**a**,**b**) Field photos of sandstone in the Dajianggang Formation; (**c**) microscopic photos of 19DJ-1; (**d**,**e**) microscopic photos of 19DJ-2; (**f**) microscopic photos of 19DJ-3; (**g**) microscopic photos of 19DJ-4; and (**h**) microscopic photos of 19DJ-5. Pl—Plagioclase; Ksp—K-feldspar; Qp—Polycrystalline quartz; Qm—Monocrystalline quartz; Ls—Lithic fragment of sedimentary rock; Lv—Lithic fragment of volcanic rock; Mu—Muscovite; Lm—Limonite.

The Gazzi–Dickinson point method was used to perform microscopic thin section analysis and Dickinson sandstone model analysis. The compositional analysis of the samples revealed the presence of monocrystalline quartz, polycrystalline quartz, total quartz, plagioclase, potassium feldspar, sedimentary rock debris, volcanic rock debris, total debris, muscovite, biotite, cement, or matrix. Such detailed petrographic analysis of the rock can provide valuable insights into the sedimentary environment and provenance of the rock [43,44]. The Gazzi–Dickinson point-counting method was used to conduct microscopic analysis of thin sections of five samples (19DJ-1, 19DJ-2, 19DJ-3, 19DJ-4, and 19DJ-5) selected from the Dajianggang Formation in the Shuangyang area of central Jilin Province. Each component's node count, obtained from pattern analysis, was converted into a percentage in order to determine its content in the sample. The statistical results of both the point-counting method and pattern analysis are presented in Table 1 and Figure 3, respectively. It should be noted that the section of sandstone sample 19DJ-5 contains gravels with diameters ranging from 3 to 12 mm and a content of 40%–50%; therefore, fewer than 300 particles met the statistical requirements.

Table 1. Analysis results of sandstone clastic particles in the Dajianggang Formation.

Group	Sample	Qm	Qp	Pl	Ksp	F	Ls	Lv	Lm	Mu	Bt	M	Total	Qt	L	Lt
Dajianggang Formation	19DJ-1	39	14	70	15	85	33	21	0	0	11	192	395	53	54	68
	19DJ-2	54	28	82	20	102	45	44	0	7	0	141	421	82	89	117
	19DJ-3	23	11	32	11	43	9	47	0	0	0	235	368	34	56	67
	19DJ-4	65	54	43	15	58	18	29	0	6	0	183	413	119	47	101
	19DJ-5	11	27	11	8	19	27	25	0	0	0	59	168	38	52	79

The values in the table are the numbers of various mineral particles counted using the Dickinson point method, the number of skeleton mineral components counted using Total, and the percentage of cement or matrix calculated using percentage. Qm—monocrystalline quartz; Qp—polycrystalline quartz; Qt—total quartz (Qt = Qm + Qp); F—feldspar (F = Pl + Ksp; Pl—plagioclase; Ksp—potassium feldspar); Ls—lithic fragment of sedimentary rock; Lv—lithic fragment of volcanic rock; Lm—lithic fragment of metamorphic rock; L—lithic fragments (L = Ls + Lv + Lm); Lt—total lithic fragments (Lt = L + Qp); Mu—Muscovite; Bt—biotite; M—cement or matrix.

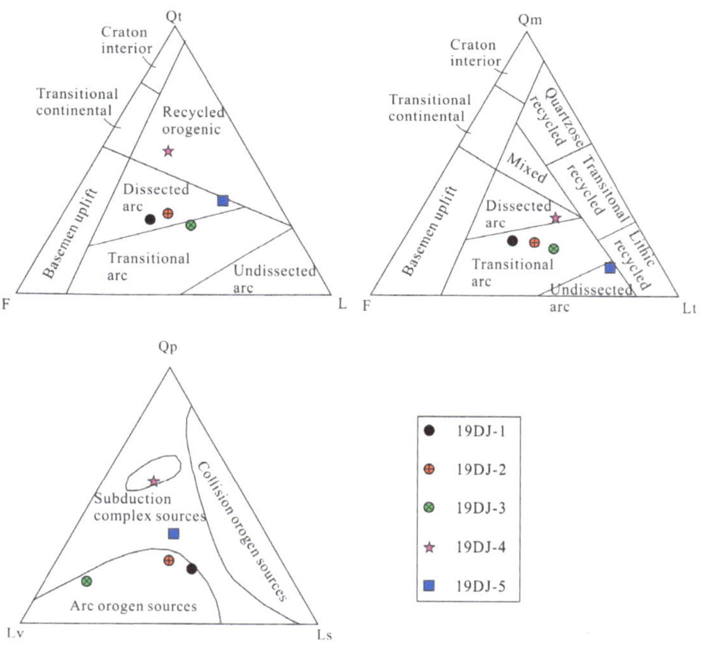

Figure 3. Structural diagram of sandstone skeleton triangle of the Dajianggang Formation (modified after [43]).

According to the statistical results of the detrital composition of sandstone and the Dickinson sandstone model analysis theory, Qt-F-L, Qm-F-Lt and Qp-Lv-Ls triangular diagrams were illustrated (Figure 3).

As shown in Figure 3, the diagrams reveal that some of the samples of the Dajianggang Formation belong to the side close to the dissected arc region within the recycled orogenic belt in the Qt-F-L diagram, while others fall on the dissected arc region, and the rest fall on the side close to the dissected arc region within the transitional arc region. In the Qm-F-Lt diagram, some samples are situated in the dissected arc area close to the transitional arc region, while most of them fall in the transitional arc area, and the remaining samples are located in the undissected arc area close to the transitional arc side. Regarding the Qp-Lv-Ls diagram, most of the Dajianggang Formation samples are distributed in and around the provenance area of the island arc orogenic belt, and some of them are found in the subduction complex belt. Based on the above analysis, it can be inferred that the sandstone provenance area of the Dajianggang Formation in the Shuangyang area is complex, mainly consisting of the provenance of the island arc orogenic belt.

5. Analysis of Geochemical Characteristics

The geochemical characteristics of clastic sedimentary rocks play a crucial role in understanding their origins, tectonic settings, and evolutionary processes. The composition of sedimentary rocks is closely linked to their provenance, tectonic environment, and depositional history. Studies have demonstrated that the texture of sandstones and their major element geochemistry can provide valuable information for evaluating provenance and tectonic settings. However, inactive trace elements and rare earth elements are even more effective in providing information on provenance. Hence, this study aims to examine the provenance of sandstones through the analysis of major elements and trace elements.

5.1. Geochemical Characteristics of Major Elements

The geochemical analysis results of the major elements of three sandstone samples from the Dajianggang Formation are shown in Table 2. The SiO_2 content of the sandstone is high, ranging from 63.01% to 72.92%, with an average of 66.97%. Meanwhile, the Al_2O_3 content is between 13.67% and 16.79%, with a mean of 15.69%, and the TFe_2O_3 content is between 3.82% and 8.37%, with a mean of 6.14%. The contents of CaO and MgO, which are indicators of carbonate rocks, are relatively low, ranging from 0.36% to 0.63% and 0.92% to 2.26%, respectively (the average contents are 0.49% and 1.38%, respectively). These results suggest that the proportion of carbonate rocks in the sample is minimal. The K_2O and Na_2O contents are average, ranging from 1.69% to 1.75% and 2.68% to 4.30% (average contents are 1.75% and 2.68%, respectively). The TiO_2 content ranges from 0.67 to 0.89, with an average content of 0.78%. The sample contains small amounts of P_2O_5 and MnO (ranging from 0.13% to 0.26% and 0.07% to 0.11%, with average contents of 0.20% and 0.09%, respectively), suggesting the presence of accessory minerals, such as apatite and epidote.

Table 2. Major elements of the Dajianggang Formation sandstone in the Shuangyang area (wt%).

Sample	Lithology	SiO_2	Al_2O_3	TFe_2O_3	CaO	MgO	K_2O	Na_2O	TiO_2	P_2O_5	MnO	LOI	Total	CIA	CIW	ICV
19DJ-1	conglomeratic sandstone	64.98	16.6	6.23	0.63	0.96	1.71	4.3	0.89	0.26	0.07	3.19	99.82	62.23	66.88	0.89
19DJ-4	sandstone	63.01	16.79	8.37	0.36	2.26	1.69	3.04	0.77	0.13	0.11	3.12	99.66	69.15	74.8	0.99
19DJ-5	greywacke	72.92	13.67	3.82	0.48	0.92	1.75	2.68	0.67	0.22	0.08	2.53	99.73	65.56	72.12	0.76

In order to evaluate the intensity and nature of weathering of the source rock in the provenance area, the difference in stability of major element oxides can be utilized. The chemical index of alteration (CIA) and chemical index of weathering (CIW) have been proposed for this purpose, allowing for the quantitative analysis of weathering intensity.

The chemical index of alteration (CIA), first introduced by Nesbitt and Young [45], is calculated based on the mole number of oxides CaO, Na_2O, K_2O, and Al_2O_3. Its formula is widely used in geochemical studies and can provide insight into the degree of chemical alteration of the source rock. The specific formula is as follows:

$$CIA = [(Al_2O_3)/(CaO^* + Na_2O + K_2O + Al_2O_3)] \times 100$$

Under natural conditions, the CIA value of rocks is usually greater than 50. The mole number of CaO, Na_2O, and K_2O in rocks decreases gradually with increasing weathering intensity, while the proportion of stable oxide Al_2O_3 increases accordingly, resulting in an increase in the CIA value. A range of 50–60 in the CIA value indicates initial weathering, a range of 60–80 indicates moderate weathering, and a CIA value greater than 80 indicates intense weathering [46]. The formula for the CIA calculation involves the mole number of each oxide, with CaO* representing the mole number of CaO in silicate minerals, excluding phosphate and carbonate minerals. Nesbitt and Young [45] first introduced the CIA as a quantitative parameter for evaluating the intensity of weathering of source rocks in the provenance area, allowing for the determination of the intensity and nature of weathering.

In evaluating weathering intensity using the chemical index of alteration (CIA), it is important to consider the potential influence of potassium metasomatism, which can cause the illitization of kaolinite and result in a lower calculated CIA value than the actual weathering intensity. In order to address this issue, Harnois [47] introduced the chemical index of weathering (CIW) by removing K_2O from the CIA formula. The CIW formula is as follows:

$$CIW = [(Al_2O_3)/(CaO^* + Na_2O + Al_2O_3)] \times 100$$

On the A-CN-K weathering triangle diagram, the CIW value of a rock can be determined by projecting the line connecting the K endpoint and the sample point onto the edge of A-CN. This diagram is widely used in the field of geochemistry to assess the degree of weathering and chemical alteration of rocks, particularly in the study of soils and sediments. The A-CN-K diagram plots the percentage of Al_2O_3 (A), CaO + Na_2O (CN), and K_2O(K) on a ternary plot, which enables geochemists to assess the extent of weathering and alteration based on the position of a given sample in the plot. The CIW value calculated from the A-CN-K diagram provides a more accurate estimate of weathering intensity than does the CIA, as it eliminates the potential bias introduced by potassium metasomatism.

The above methods for evaluating weathering intensity were applied to the sandstone of the Dajianggang Formation in the Shuangyang area, yielding an average CIA value of 65.65, which indicates a moderate level of weathering intensity (as shown in Table 2). In the A-CN-K diagram (Figure 4), in order to correct for potassium metasomatism, the average CIW was obtained by projecting a line connecting the K endpoint and the sample point, resulting in a value of 71.27, which is approximately 6 higher than the CIA value. Taken together, both the CIA and CIW values suggest that the sandstone of the Dajianggang Formation underwent moderate weathering during diagenesis. Additionally, the ICV, which is the indicator of compositional variation introduced by Cox, Lowe [48], can also be utilized to estimate the composition of clastic sandstone, with the following calculation formula:

$$ICV = (Fe_2O_3 + Na_2O + CaO + K_2O + MgO + MnO + TiO_2)/Al_2O_3$$

The ICV value is a useful indicator for estimating the composition of clastic sandstone, and it varies depending on the mineralogical composition. Clastic source rocks composed mainly of non-clay minerals typically have higher ICV values than those composed mainly of clay minerals. Sandstones composed primarily of biotite, amphibolite, or pyroxene, with low clay mineral content, usually have ICV values greater than 1 and are often associated with sediments from the first cycle in tectonically active areas [48]. In contrast, first-cycle terrigenous sediments that have undergone intense weathering and have been in the clay for a long time may have ICV values less than 1 due to further weathering. The ICV value of the sandstone from the Dajianggang Formation ranges from 0.76 to 0.99, with an average value of 0.88, which is less than 1, indicating that it may be the result of recycled sediments or sediments that have undergone intense weathering. Therefore, it can be concluded that the sandstone debris of the Dajianggang Formation has undergone the process of terrigenous recycling.

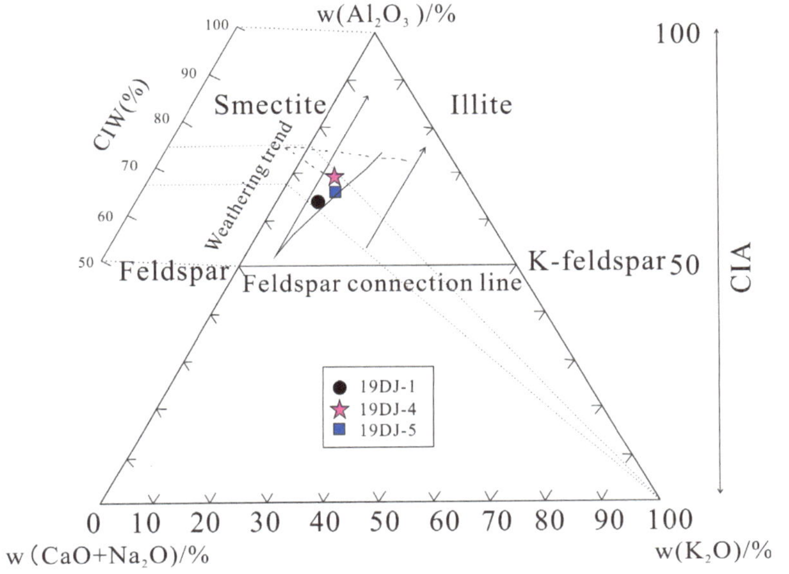

Figure 4. Weathering diagram of the major element A-CN-K of Dajianggang Formation sandstone in the Shuangyang area (Modified after [45]).

The geochemical characteristics of clastic sedimentary rocks are influenced by the composition of their source rocks. These characteristics not only reflect the type of source rock from a geochemical perspective, but also provide information about the tectonic setting of the formation. This is because plate tectonic activities are crucial in controlling the composition of source rocks, the sedimentary topography of provenance areas, the diagenesis of sedimentary rocks, the deposition rate, and the weathering intensity. The changes in geochemical characteristics can thus reflect the different tectonic settings of the sedimentary rocks. While it can be challenging to determine the relationship between geochemical characteristics and the tectonic setting of detrital sedimentary rocks throughout their long sedimentary history, researchers have established the relationship between them by analyzing the geochemical characteristics of sedimentary rocks with similar tectonic settings and studying their mechanisms from the perspective of diagenesis. Therefore, based on previous studies, we can discuss the tectonic background of the sandstone formation in the Dajianggang Formation in the Shuangyang area.

Roser and Korsch [49] developed a bivariate diagram based on the geochemical characteristics of sandstones and mudstones from modern tectonic settings to determine the tectonic background of sedimentary provenance. The diagram is based on the SiO_2-K_2O/Na_2O and K_2O/Na_2O-SiO_2/Al_2O_3 ratios, containing three categories of tectonic settings: active continental margin, passive continental margin, and island arc provenance. By projecting the geochemical data of sandstones from the Dajianggang Formation in the Shuangyang area onto this diagram (Figure 5), except for the graywacke sample 19DJ-5, the other two samples form in the island arc tectonic background. These results suggest that the provenance area of the Dajianggang Formation in the Shuangyang area was an island provenance area undergoing feldspar intrusive rock evolution. These findings are consistent with previous studies on the tectonic setting of the study area and provide further support for the geological evolution of the region.

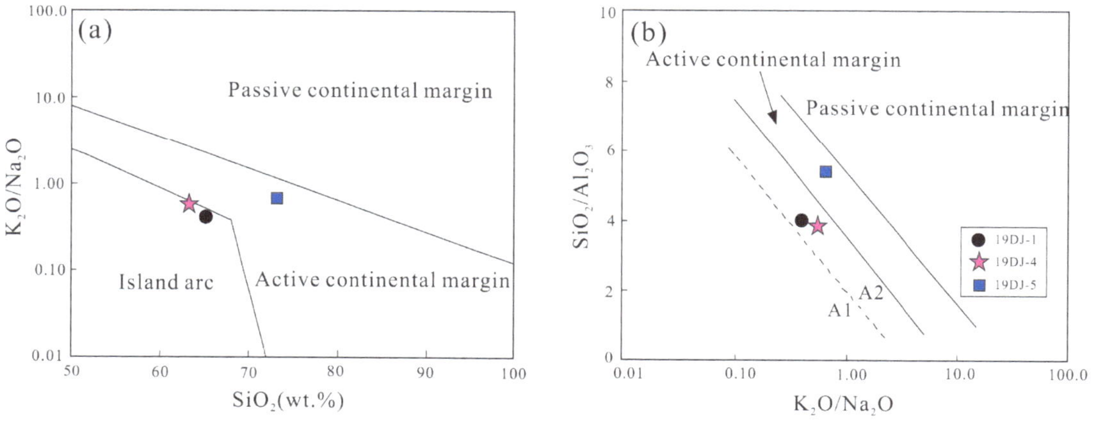

Figure 5. (**a**) SiO_2-K_2O/Na_2O and (**b**) K_2O/Na_2O-SiO_2/Al_2O_3 (Cited in [49]). A1—basaltic and andesitic island arc sources; A2—felsic intrusive rock evolution island arc source.

5.2. Geochemical Characteristics of Trace Elements

The results of trace and REE geochemical analysis of sandstone samples from the Dajianggang Formation are shown in Table 3. The large ion lithophile elements Rb, Sr, Ba, and Cs have concentrations ranging from 17.06 to 43.76 $\times 10^{-6}$, 126.60 to 185.60 $\times 10^{-6}$, 279.40 to 377.40 $\times 10^{-6}$, and 2.90 to 3.83 $\times 10^{-6}$, respectively. Notably, Rb, Sr, and Ba are found to be deficient relative to the upper continental crust component. Similarly, the high field intensity elements Nb, Ta, Th, and Ce exhibit deficits, with concentrations ranging from 11.61 to 14.49 $\times 10^{-6}$ (average of 12.51 $\times 10^{-6}$) for Nb, 0.88 to 1.02 $\times 10^{-6}$ (average of 0.95 $\times 10^{-6}$) for Ta, 5.06 to 8.69 $\times 10^{-6}$ (average of 7.40 $\times 10^{-6}$) for Th, and 30.99 to 58.42 $\times 10^{-6}$ for Ce. On the other hand, the sandstone samples from the Dajianggang Formation in the study area exhibit slight enrichment of Ti and Zr, with Ti content ranging from 4016 to 5335 $\times 10^{-6}$, with an average of 4655.33 $\times 10^{-6}$, and Zr content ranging from 241.70 to 284.70 $\times 10^{-6}$, with an average of 258.00 $\times 10^{-6}$. Meanwhile, other elements, such as U and Hf, show comparable contents to the average values of the upper continental crust. The U content ranges from 2.56 to 2.85 $\times 10^{-6}$, with an average of 2.74 $\times 10^{-6}$, while the Hf content ranges from 5.60 to 6.44 $\times 10^{-6}$, with an average of 6.00 $\times 10^{-6}$. The trace element distribution pattern of the sandstone samples in the study area is similar to the average component distribution of the upper crust, with a gentle curve on the spider diagram (Figure 6a). Floyd and Leveridge [50] introduced the Hf-La/Th discrimination diagram (Figure 7a) in order to investigate the relationship between different provenance components. The diagram suggests that the content of Hf elements in sandstones would increase due to the decomposition of zircons, as a result of the gradual erosion of island arcs or continental basement. Based on this diagram, the sandstone samples from the Dajianggang Formation in the study area are mainly characterized by a felsic provenance. In addition, the Ni-TiO_2 discrimination diagram (Figure 7b) shows that the samples are mainly distributed in the sandstone area with a sedimentary tendency. This result is consistent with the two-dimensional bivariate diagram of SiO_2-K_2O/Na_2O and K_2O/Na_2O-SiO_2/Al_2O_3, which further validates the conclusion. In summary, the provenance of the sandstones from the Dajianggang Formation in the study area can be attributed to felsic rocks with sedimentary tendency, and there might be a small amount of mixed source.

Table 3. Trace elements of the Dajianggang Formation sandstone in the Shuangyang area ($\times 10^{-6}$).

Sample	19DJ-1	19DJ-4	19DJ-5
Lithology	Conglomeratic Sandstone	Sandstone	Greywacke
trace elements ($\times 10^{-6}$)			
Li	26.66	36.37	14.15
Be	1.74	1.79	1.75
B	14.94	13.53	16.66
Sc	5.76	7.77	6.85
V	104.3	104.4	77.27
Cr	74.4	227	157.9
Co	8.75	17.45	10.46
Ni	53.73	105.2	62.63
Cu	20.49	22.16	19.58
Zn	83.64	113.4	60.55
Ga	19.95	21.02	18.22
Ge	6.94	9.54	7.83
As	22.85	7.63	7.42
Se	1.37	1.98	1.35
Rb	17.06	25.82	43.76
Sr	142.7	185.6	126.6
Y	9.83	17.73	19.22
Zr	284.7	247.6	241.7
Nb	14.49	11.42	11.61
Mo	10.89	10.57	11.74
Ag	0.42	0.42	0.36
Cd	2.5	2.17	2.23
Sn	1.63	2.05	1.95
Sb	3.12	2.79	3.18
Cs	2.95	2.9	3.83
Ba	279.4	310.1	377.4
La	15.11	27.49	27.51
Ce	30.99	58.42	46.72
Pr	4.438	7.412	6.91
Nd	16.8	27.77	25.57
Sm	3.25	5.06	4.81
Eu	0.89	1.16	1.28
Gd	2.89	4.59	4.44
Tb	0.44	0.67	0.67
Dy	2.37	3.67	3.55
Ho	0.5	0.74	0.75
Er	1.47	2.06	2.16
Tm	0.23	0.32	0.34
Yb	1.45	1.93	2.09
Lu	0.25	0.3	0.35
Hf	6.44	5.96	5.6
Ta	1.02	0.95	0.88
W	1.95	2.48	2.1
Au	0.09	0.08	0.09
Tl	0.4	0.54	0.47
Pb	18.93	19.59	14.69
Bi	0.15	0.29	0.19
Th	5.06	8.44	8.69
U	2.56	2.82	2.85
ΣREE	81.07	141.58	127.14
ΣLREE	71.48	127.3	112.8
ΣHREE	9.6	14.28	14.34
$(La/Yb)_N$	0.77	1.05	0.97
$(La/Sm)_N$	0.68	0.79	0.83
$(Eu/Yb)_N$	1.61	1.56	1.6

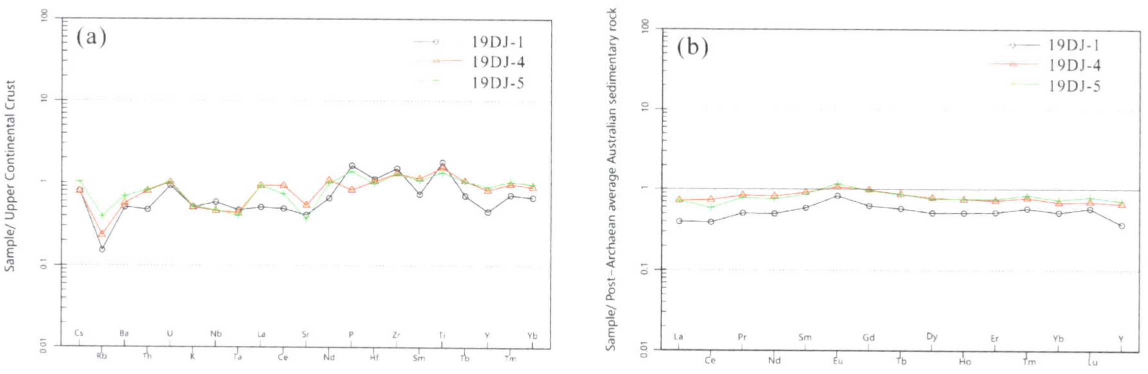

Figure 6. (**a**) A standardized spider diagram of trace elements in the upper crust of sandstone in the Dajianggang Formation (Modified after [51]) and (**b**) standard distribution model of rare earth elements in sandstone of the Dajianggang Formation (Modified after [52]).

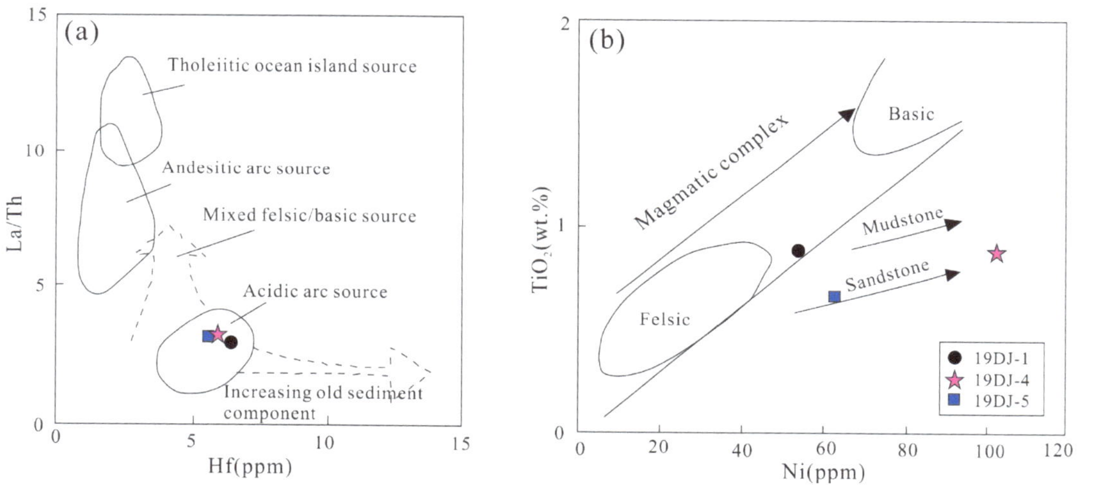

Figure 7. (**a**) Hf-La/Th source discrimination diagram (Modified after [50]); (**b**) Ni-TiO_2 diagram (Modified after [53]).

Trace elements, such as La, Ce, Y, Th, Zr, Nd, Hf, and Sc, are highly useful in identifying the provenance tectonic background of sedimentary rocks, due to their low mobility during the sedimentary process and high stability in seawater. These elements are abundant in clastic sedimentary rocks and are not prone to weathering and erosion, making them an ideal tool for understanding the characteristics of source rocks. Based on this premise, Bhatia and Crook [54] conducted a study on the tectonic significance of trace elements in provenance analysis, and categorized active continental margins (ACM), passive continental margins (PCM), oceanic island arcs (OIA), and continental island arcs (CIA) as the four major geotectonic source locations. In order to further understand the tectonic setting and provenance of the sandstones in the study area, several structure diagrams have been used. The Th-Co-Zr and Th-Sc-Zr triangle diagrams (Figure 8) were employed in order to analyze the compositional characteristics of the sandstones. The majority of the samples were found to fall within the continental island arc region. However, in the La/Sc-Ti/Zr diagram (Figure 9), they were mainly classified as active continental margin in source, with only one sample (19DJ-1) located at the boundary between the continental island arc and active continental margin, indicating that the tectonic setting of the Dajianggang Formation in the Shuangyang area is relatively complex. These findings provide valuable insights into the provenance and tectonic history of the study area. The sandstones of the Dajianggang Formation in the Shuangyang area were found to have a slightly scattered distribution of total rare earth elements (ΣREE), with values ranging from 81.07 to 141.58 × 10^{-6}, and an average value of 116.60 × 10^{-6}. The content of light rare earth elements (ΣLREE) ranges from 71.48 to 127.30 × 10^{-6}, with an average value of 103.86 × 10^{-6}. The ratio of $(La/Sm)_N$ is between 0.68 and 0.83. The content of heavy rare earth elements (ΣHREE) ranges from 9.60 to 14.34 × 10^{-6}, with a mean value of 12.74 × 10^{-6}. The $(Eu/Yb)_N$ ratio is between 1.56 and 1.61. The ratio of light rare earth to heavy rare earth elements (ΣLREE/ΣHREE) ranges from 7.45 to 8.91, with an average of 8.08. The $(La/Yb)_N$ ratio ranges from 0.77 to 1.05, with an average value of 0.93. Overall, the standard distribution pattern of rare earth elements in the sandstone samples of the Dajianggang Formation is very similar to that of post-Archaean Australia sedimentary rocks (Figure 6b), indicating a gentle trend.

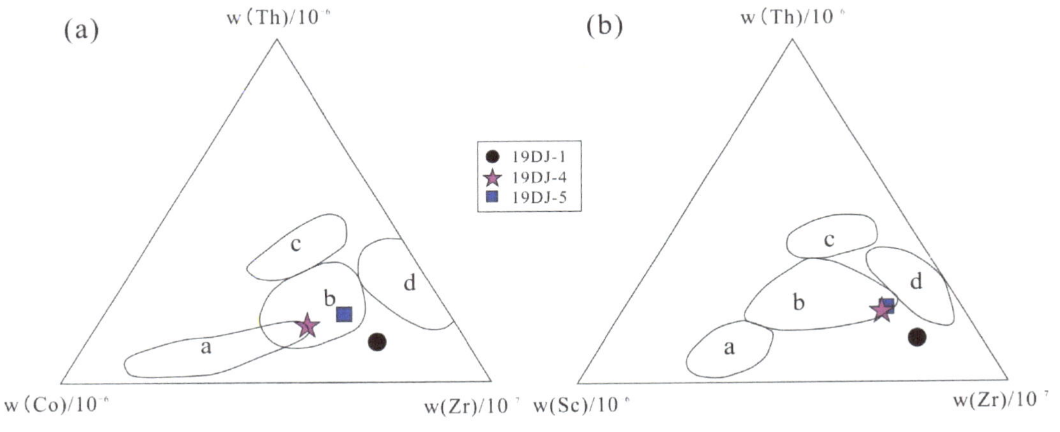

Figure 8. (**a**) Th-Co-Zr diagram of sandstone triangle structure of the Dajianggang Formation in the Shuangyang area and (**b**) Th-Sc-Zr diagram of sandstone triangle structure of the Dajianggang Formation in the Shuangyang area (Modified after [53]). a—oceanic island arc; b—continental island arc; c—active continental margin; d—passive continental margin.

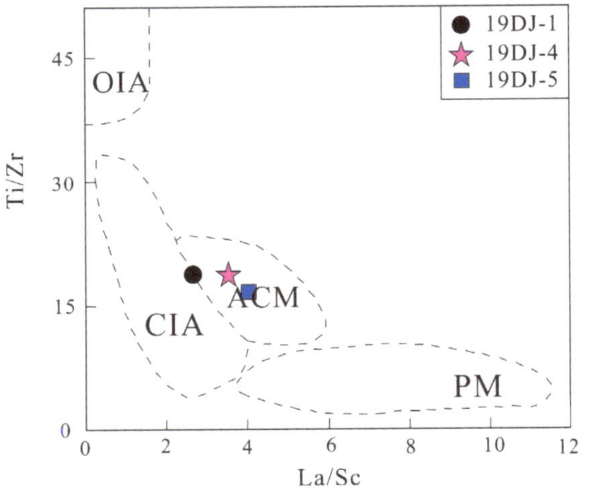

Figure 9. La/Sc-Ti/Zr tectonic diagram. PM—passive continental margin; ACM—active continental margin; CIA—continental island arc; OIA—oceanic island arc.

To summarize, the application of various structural background discrimination diagrams indicates that the sandstone of the Dajianggang Formation has been derived from the continental island arc environment, although there may be some contributions from mixed sources. By considering the geological time scale, this suggests that the tectonic activity in the source area was relatively unstable during the Late Triassic period, characterized by a geological setting with active tectonic movement.

6. Analysis of LA-ICP-MS Zircon Dating Results

In this study, LA-ICP-MS detrital zircon U-Pb dating was utilized in order to determine the age of sandstone sample 19DJ-1 from the Late Triassic Dajianggang Formation in the Shuangyang area, located in central Jilin Province. Detailed observations of the crystal morphology using CL images revealed that the majority of the detrital zircons are semi-euhedral and heteromorphic crystals with typical magmatic oscillating growth zones (Figure 10a). The presence of high Th/U values (0.31–1.61) further indicates their magmatic origin, which is in accordance with the geological context [55]. The age distribution of detrital zircons and the U-Pb Concordia diagram, as shown in the Figure 10, were used to analyze the provenance and geological history of the Dajianggang Formation in the Shuangyang area.

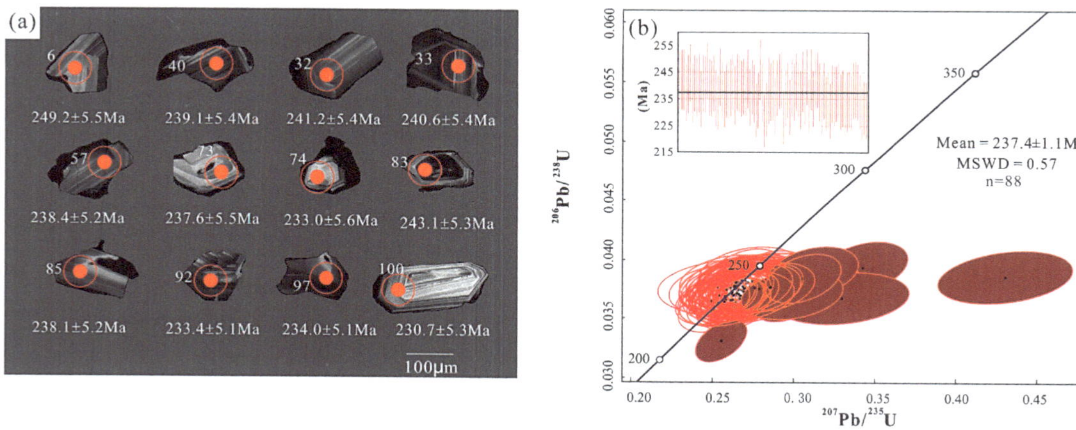

Figure 10. (**a**) Cathodoluminescence (CL) images of selected zircons from samples 19DJ-1. Solid circles show positions of LA-ICP-MS analytical sites with their ages in Ma. (**b**) Concordia diagram of conglomeratic sandstone in the Dajianggang Formation. Filled ellipses are discordant data.

The $^{206}Pb/^{238}U$ age range of the 100 analyzed zircons ranges from 210 to 363 Ma (Table 4), the majority of the 88 concordant zircons falling within the range of 226.8 to 246.5 Ma. The weighted mean $^{206}Pb/^{238}U$ age of these zircons is 237.4 $\pm$ 1.1 Ma (MSWD = 0.57), indicating that they were likely crystalized around the Middle to Late Triassic boundary (Figure 10b). The CL images of the detrital zircons show that most zircons are semi-euhedral and heteromorphic crystals, further supporting their magmatic origin.

Table 4. LA-ICP-MS Zircon U-Pb dating data of conglomeratic sandstone of the Dajianggang Formation.

Point Number	Content ($\times 10^{-6}$)			Th/U	$^{207}Pb/^{235}U$		$^{206}Pb/^{238}U$		$^{207}Pb/^{235}U$		$^{206}Pb/^{238}U$		Concordance (%)
	U	Th	Pb		Ratio	1σ	Ratio	1σ	Age (Ma)	1σ	Age (Ma)	1σ	
~~1 *~~	~~386~~	~~997~~	~~10,379~~	~~2.58~~	~~799.2065~~	~~20.6331~~	~~6.8468~~	~~0.1741~~	~~6787.7~~	~~26.2~~	~~*****~~	~~143~~	~~*****~~
2 *	564	713	31	1.26	0.3146	0.0098	0.0371	0.0008	277.8	7.6	234.9	5.2	83
3	399	421	21	1.06	0.2979	0.0103	0.0382	0.0009	264.8	8.1	241.8	5.4	91
4	200	97	9	0.49	0.2702	0.0123	0.0383	0.0009	242.8	9.8	242.2	5.6	100
5 *	400	350	21	0.88	0.3424	0.0114	0.0394	0.0009	299	8.6	249.2	5.5	82
6	162	103	8	0.64	0.2746	0.0137	0.0383	0.0009	246.4	11	242.1	5.7	98
7	272	326	15	1.2	0.2757	0.0111	0.0375	0.0009	247.3	8.9	237.3	5.4	96
8	240	125	11	0.52	0.2706	0.0114	0.038	0.0009	243.2	9.1	240.6	5.5	99
9	378	325	20	0.86	0.2845	0.0101	0.0387	0.0009	254.2	8	244.7	5.4	96
10	162	144	8	0.89	0.255	0.013	0.0375	0.0009	230.6	10.5	237.1	5.6	97
11	594	440	29	0.74	0.2715	0.0086	0.0381	0.0009	243.9	6.9	240.9	5.3	99
12	174	141	9	0.81	0.2798	0.0133	0.0378	0.0009	250.5	10.5	239.3	5.6	95
13	184	114	9	0.62	0.2642	0.0125	0.0381	0.0009	238	10	240.9	5.6	99
14	301	254	15	0.84	0.279	0.0108	0.0371	0.0009	249.9	8.5	234.8	5.3	94
15	489	358	24	0.73	0.2846	0.0096	0.0381	0.0009	254.3	7.6	240.8	5.3	95
16	526	518	28	0.98	0.267	0.0087	0.038	0.0009	240.3	7	240.5	5.3	100
17	388	475	22	1.23	0.2704	0.0097	0.0388	0.0009	243	7.8	245.5	5.4	99
18	309	196	15	0.63	0.2648	0.0103	0.0378	0.0009	238.6	8.3	239.1	5.4	100
19	469	217	21	0.46	0.2655	0.009	0.0377	0.0009	239.1	7.3	238.8	5.3	100
20	539	561	29	1.04	0.2719	0.0088	0.0381	0.0009	244.2	7.1	240.9	5.3	99
21	232	154	11	0.66	0.2761	0.0118	0.0372	0.0009	247.6	9.4	235.3	5.4	95
22	332	280	17	0.84	0.2747	0.0104	0.0369	0.0008	246.4	8.3	233.6	5.2	95
23	298	164	14	0.55	0.2842	0.0109	0.0385	0.0009	254	8.6	243.2	5.5	96
~~24 *~~	~~1382~~	~~2794~~	~~609~~	~~2.02~~	~~11.4984~~	~~0.3994~~	~~0.1275~~	~~0.004~~	~~2564.4~~	~~32.5~~	~~773.8~~	~~22.6~~	~~−7~~
25	457	409	23	0.89	0.2703	0.0091	0.0376	0.0008	242.9	7.3	238.2	5.2	98
26	319	351	17	1.1	0.2719	0.0102	0.0371	0.0008	244.2	8.1	234.7	5.3	96
27	1000	864	51	0.86	0.2608	0.0073	0.0376	0.0008	235.3	5.9	238.2	5.1	99
28	273	149	13	0.55	0.2605	0.0105	0.0377	0.0009	235.1	8.5	238.4	5.4	99
29	288	275	15	0.96	0.2632	0.0103	0.0378	0.0009	237.2	8.3	239.3	5.4	99
30	667	569	34	0.85	0.2785	0.0085	0.0382	0.0009	249.4	6.7	241.6	5.3	97
31	150	116	7	0.77	0.2757	0.0141	0.0374	0.0009	247.3	11.2	236.5	5.6	96
32	310	254	16	0.82	0.2673	0.0103	0.0381	0.0009	240.6	8.2	241.2	5.4	100
33	314	265	16	0.85	0.2669	0.0101	0.038	0.0009	240.2	8.1	240.6	5.4	100
34	416	298	20	0.72	0.2807	0.0097	0.0378	0.0009	251.2	7.7	238.9	5.3	95
35	394	558	23	1.42	0.2806	0.0098	0.0379	0.0009	251.2	7.8	239.6	5.3	95

Table 4. *Cont.*

Point Number	Content ($\times 10^{-6}$)			Th/U	$^{207}Pb/^{235}U$		$^{206}Pb/^{238}U$		$^{207}Pb/^{235}U$		$^{206}Pb/^{238}U$		Concordance (%)
	U	Th	Pb		Ratio	1σ	Ratio	1σ	Age (Ma)	1σ	Age (Ma)	1σ	
36	531	488	27	0.92	0.2685	0.0087	0.0377	0.0008	241.5	7	238.8	5.2	99
37	252	196	13	0.78	0.2757	0.0113	0.0384	0.0009	247.3	9	242.6	5.5	98
38	368	328	19	0.89	0.2803	0.0102	0.0372	0.0008	250.9	8.1	235.7	5.3	94
39	1095	960	56	0.88	0.275	0.0076	0.0379	0.0008	246.7	6	239.6	5.2	97
40	276	178	13	0.65	0.2675	0.0108	0.0375	0.0009	240.7	8.7	237.1	5.4	98
41	366	380	19	1.04	0.2527	0.0094	0.0364	0.0008	228.8	7.6	230.6	5.1	99
42	339	217	16	0.64	0.2716	0.0102	0.037	0.0008	244	8.2	234.4	5.2	96
43	170	108	8	0.63	0.2734	0.0131	0.0389	0.0009	245.4	10.5	245.9	5.7	100
44	198	157	10	0.79	0.2637	0.0123	0.0375	0.0009	237.6	9.9	237.5	5.5	100
45	435	444	22	1.02	0.2507	0.009	0.0358	0.0008	227.2	7.3	226.8	5	100
46 *	560	262	27	0.47	0.3162	0.0097	0.0379	0.0008	279	7.5	239.9	5.3	85
47	408	537	22	1.32	0.2494	0.0091	0.0365	0.0008	226.1	7.4	231	5.1	98
48	69	47	3	0.69	0.265	0.0199	0.0376	0.001	238.7	16	238.2	6.3	100
49	254	153	12	0.6	0.2787	0.0116	0.0376	0.0009	249.6	9.2	237.7	5.4	95
50	146	102	7	0.7	0.2633	0.0139	0.0377	0.0009	237.3	11.1	238.4	5.7	100
51	245	143	11	0.58	0.2667	0.0113	0.0367	0.0009	240	9	232.4	5.3	97
52	511	557	28	1.09	0.2768	0.009	0.0378	0.0009	248.1	7.2	239.3	5.3	96
53	757	693	40	0.92	0.2746	0.0081	0.0382	0.0008	246.3	6.5	241.7	5.2	98
54	702	1133	42	1.61	0.2783	0.0083	0.0383	0.0009	249.3	6.6	242.2	5.3	97
55	466	352	22	0.76	0.2631	0.0095	0.036	0.0008	237.2	7.6	228.2	5.1	96
56	291	291	15	1	0.2745	0.0107	0.0372	0.0009	246.3	8.5	235.6	5.3	96
57	652	657	34	1.01	0.2763	0.0085	0.0377	0.0008	247.7	6.8	238.4	5.2	96
58	378	406	20	1.08	0.2584	0.0095	0.0367	0.0008	233.4	7.6	232	5.2	99
59	167	139	9	0.83	0.2908	0.0138	0.0379	0.0009	259.2	10.9	239.7	5.6	92
60	254	197	13	0.77	0.2692	0.0112	0.0376	0.0009	242.1	9	237.7	5.4	98
61 *	124	102	6	0.82	0.3296	0.017	0.0368	0.0009	289.3	13	232.9	5.6	78
62	374	319	19	0.85	0.265	0.0096	0.0382	0.0009	238.7	7.7	241.3	5.4	99
63	546	448	28	0.82	0.273	0.0088	0.0377	0.0008	245.1	7	238.7	5.2	97
64	183	147	9	0.8	0.266	0.0127	0.0371	0.0009	239.5	10.2	234.7	5.5	98
65	153	101	7	0.66	0.2625	0.0134	0.0382	0.0009	236.7	10.8	241.7	5.7	98
66	263	264	14	1	0.262	0.0109	0.0373	0.0009	236.2	8.8	235.9	5.3	100
67	582	489	31	0.84	0.2978	0.0092	0.0385	0.0009	264.7	7.2	243.4	5.3	92
68*	793	749	41	0.94	0.297	0.0086	0.0367	0.0008	264	6.7	232.5	5.1	87
69	134	83	7	0.62	0.284	0.0148	0.0383	0.0009	253.9	11.7	242.5	5.8	95
70	69	23	14	0.33	0.2737	0.0087	0.0372	0.0008	245.7	6.9	235.2	5.2	96

Table 4. *Cont.*

Point Number	Content (×10^{-6})			Th/U	$^{207}Pb/^{235}U$		$^{206}Pb/^{238}U$		$^{207}Pb/^{235}U$		$^{206}Pb/^{238}U$		Concordance (%)
	U	Th	Pb		Ratio	1σ	Ratio	1σ	Age (Ma)	1σ	Age (Ma)	1σ	
71	960	887	51	0.92	0.2936	0.0082	0.0382	0.0008	261.4	6.4	241.6	5.2	92
72	838	675	43	0.81	0.268	0.0078	0.0383	0.0008	241.1	6.2	242.1	5.2	100
73	219	137	11	0.63	0.2796	0.0121	0.0376	0.0009	250.3	9.6	237.6	5.5	95
74	144	111	7	0.77	0.27	0.014	0.0368	0.0009	242.7	11.2	233	5.6	96
75	313	252	16	0.8	0.2534	0.0099	0.0376	0.0009	229.4	8	238.1	5.3	96
76	580	580	30	1	0.2703	0.0085	0.0369	0.0008	243	6.8	233.3	5.1	96
77 *	301	174	15	0.58	0.3115	0.0115	0.039	0.0009	275.4	8.9	246.5	5.5	89
78	438	417	22	0.95	0.2641	0.0092	0.037	0.0008	237.9	7.4	234.5	5.2	99
79	297	242	15	0.82	0.2644	0.0104	0.0364	0.0008	238.2	8.4	230.6	5.2	97
80	222	160	11	0.72	0.2586	0.0114	0.0375	0.0009	233.6	9.2	237.1	5.4	99
81	158	94	8	0.59	0.2689	0.0134	0.0373	0.0009	241.8	10.7	236.1	5.6	98
82 *	206	157	10	0.76	0.2989	0.0129	0.0371	0.0009	265.6	10.1	234.5	5.4	88
83 *	528	278	26	0.53	0.3056	0.0096	0.0384	0.0009	270.8	7.5	243.1	5.3	89
84	476	431	24	0.91	0.2657	0.009	0.0375	0.0008	239.2	7.2	237	5.2	99
85	689	676	36	0.98	0.2848	0.0085	0.0376	0.0008	254.5	6.8	238.1	5.2	93
86	409	445	22	1.09	0.2539	0.0092	0.037	0.0008	229.8	7.4	234	5.2	98
87	285	334	16	1.17	0.2649	0.0106	0.0372	0.0009	238.6	8.5	235.3	5.3	99
88	460	408	23	0.89	0.2609	0.009	0.0366	0.0008	235.4	7.2	231.5	5.1	98
89	147	99	7	0.67	0.267	0.0137	0.037	0.0009	240.3	10.9	233.9	5.6	97
90	341	368	18	1.08	0.2617	0.0099	0.0371	0.0009	236.1	8	235	5.3	100
91 *	2011	2044	96	1.02	0.2554	0.0065	0.0331	0.0007	231	5.2	210	4.5	90
92	613	531	31	0.87	0.2576	0.0082	0.0369	0.0008	232.8	6.6	233.4	5.1	100
93	429	239	20	0.56	0.2759	0.0096	0.0376	0.0009	247.4	7.6	238.2	5.3	96
94	296	220	15	0.74	0.2859	0.0111	0.0377	0.0009	255.4	8.7	238.6	5.4	93
95	330	222	16	0.67	0.2788	0.0105	0.0376	0.0009	249.7	8.3	237.6	5.3	95
~~96 *~~	~~190~~	~~141~~	~~11~~	~~0.74~~	~~0.4304~~	~~0.017~~	~~0.0386~~	~~0.0009~~	~~363.4~~	~~12.1~~	~~244.4~~	~~5.7~~	~~61~~
97	611	500	30	0.82	0.2628	0.0083	0.037	0.0008	236.9	6.7	234	5.1	99
98	666	576	33	0.87	0.2535	0.0079	0.0363	0.0008	229.4	6.4	229.8	5	100
99	786	758	40	0.96	0.2622	0.0079	0.0366	0.0008	236.5	6.3	231.7	5	98
100	211	101	10	0.48	0.2573	0.0117	0.0364	0.0009	232.4	9.5	230.7	5.3	99

Note: Data with * were discordant. Data with strikethrough were not shown in the figure due to huge discordance. ***** were the data with huge error that couldn't be calculated.

The conglomeratic sandstone samples from the Dajianggang Formation were subjected to statistical analysis of their ages, and a frequency distribution histogram of crystallization age–stratigraphic age was constructed in order to visualize the zircon age data (Figure 11a). The results show that the zircon ages are highly concentrated within a narrow range of 220–250 Ma, indicating a uniform provenance. The youngest single grain age of 226.8 $\pm$ 5 Ma in 88 concordant zircons is interpreted as the maximum depositional age of the strata. The cumulative probability graph demonstrates a steep line almost perpendicular to the horizontal line, indicating a sharp increase in the sedimentation rate during the period of zircon crystallization. Based on the empirical curve proposed by Cawood, Hawkesworth [56] for the relationship between detrital zircon ages and tectonic settings, and the cumulative probability distribution diagram of the difference between zircon crystallization age and the maximum deposition age (Figure 11b), it can be inferred that the zircon ages in the conglomeratic sandstone samples are closely linked to the maximum depositional age (227 Ma) and indicative of a convergent tectonic setting.

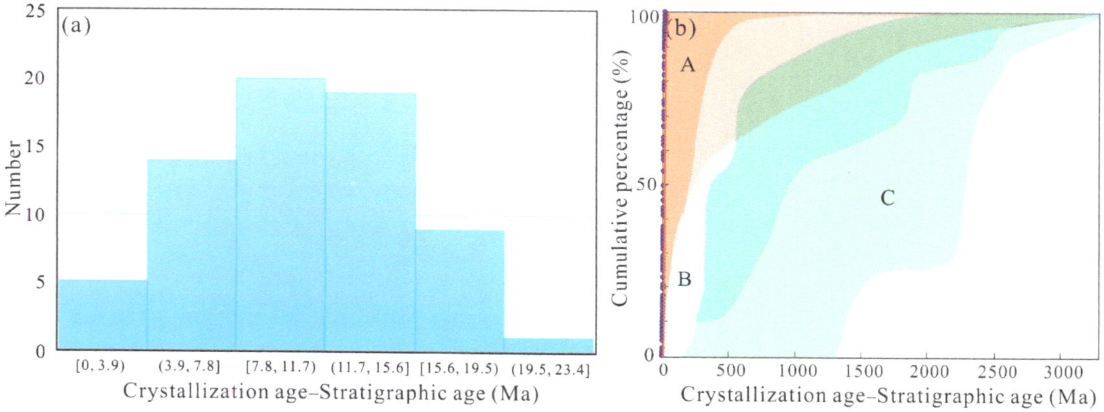

Figure 11. (**a**) Histogram of crystallization age–stratigraphic age frequency distribution of sandstone in the Dajianggang Formation and (**b**) cumulative probability graph. Summative drop point region: A—convergence plate edge background; B—syncollisional tectonic setting; C—Extensional basin setting. Purple dot—plot of the difference between the measured crystallization age and the depositional age of the succession in this study. Red line—variation of the difference between the measured crystallization age and the depositional age of the succession in which it occurs in this study.

7. Analysis of Apatite Fission Track Results

This study employed apatite fission track analysis in order to investigate the thermal history of sandstone sample 19DJ-1 from the Late Triassic Dajianggang Formation. A total of 73 apatite grains were analyzed, and the results are presented in Table 5. In order to further interpret the data, the results were processed using RadialPlotter [57], which produced a radar radiogram and age histogram with which to visualize the age components of the apatite grains in the sample. The results are presented in Figure 12.

Table 5. Apatite fission track analysis results of sample 19DJ-1 of the Dajianggang Formation.

Sample	Longitude (°E)	Latitude (°N)	Elevation (m)	Lithology/ Strata	Nc	ρ_s (10^5/cm^2) (Ns)	ρ_i (10^5/cm^2) (Ni)	ρ_d (10^5/cm^2) (Nd)	U ($\times10^{-6}$)	P(χ^2) /%	Central Age (Ma) 1σ	Dpar (μm)	MTL (μm) 1σ	N
19DJ-1	125°49′25″	43°22′53″	309	pebbly sandstone/T_3d	73	7.922 (1777)	19.759 (4432)	17 (15,655)	15.77	0	94.2 ± 8.3	2.15	11.53 ± 1.47	89

Abbreviations: Nc—number of grains; ρ_s and Ns—spontaneous track density of the sample ($\times 10^6$ tracks cm^{-2}) and number of spontaneous tracks counted; ρ_i and Ni—induced track density in external detector (muscovite) ($\times 10^6$ tracks cm^{-2}) and number of induced tracks counted; ρ_d and Nd—induced track density of dosimeter glass ($\times 10^6$ tracks cm^{-2}) and number of tracks counted; MTL and N—mean confined fission track length and number of lengths measured.

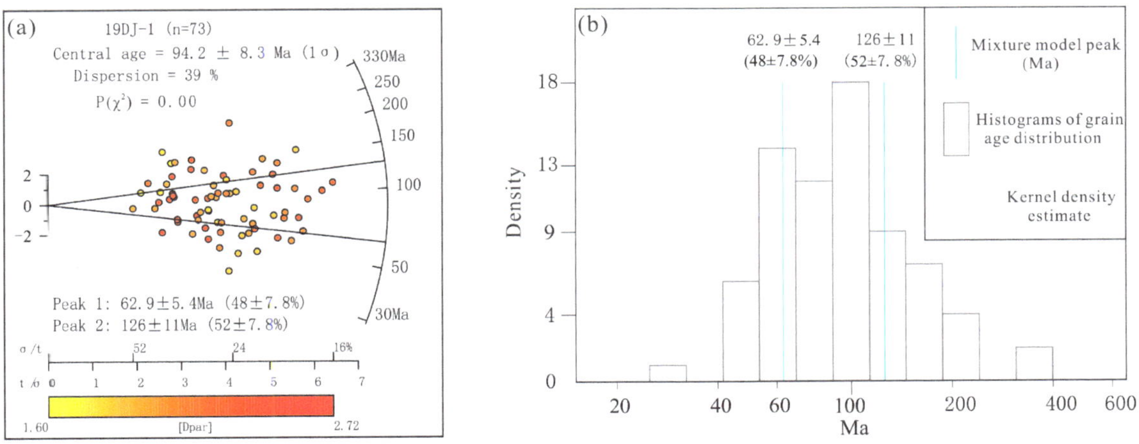

Figure 12. (**a**) Age distribution radar map of AFT particles from 19DJ-1 clastic rock of the Triassic Dajianggang Formation. (**b**) Age distribution density map of AFT particles from 19DJ-1 clastic rock of the Triassic Dajianggang Formation.

The sample did not pass the Chi-square test, as $P(\chi^2) < 5\%$, indicating that the age of sample particles was dispersed. This may be attributed to several factors, including changes in the chemical composition of apatite grains in a single sample [37,58], and difficulties in accurately identifying fission tracks and lengths due to a large number of fission tracks being shortened by the annealing process [34,59,60].

Using the age of apatite grains and the length of confined fission track obtained in this study, we carried out thermal history inversion modelling in order to infer the time–temperature evolution history of the sample [61,62]. We used HeFTy v1.9.1 thermal history simulation software in order to conduct thermal history inversion modelling [42,61,63], considering apatite fission track data such as grain age, confined fission track length, Dpar value, and C-axis Angle. The modelling results are shown in Figure 13 using two criteria, including (1) the first box constrained the deposition time of the strata, indicating that the surface temperature of the strata was 0–40 °C during the diagenesis; (2) the second box is the search box that allows the model to search for potential paths of the sample when it was deposited to the present time between the temperatures of 40 and 140 °C. We set the model to continuously run until it generated 100 good paths. The thermal history for the sample indicates the subsequent to deposition between ~230 and 30 Ma. After deposition, the sample showed rapid continuous cooling of 0.67–7.29 °C/Ma since 30 Ma.

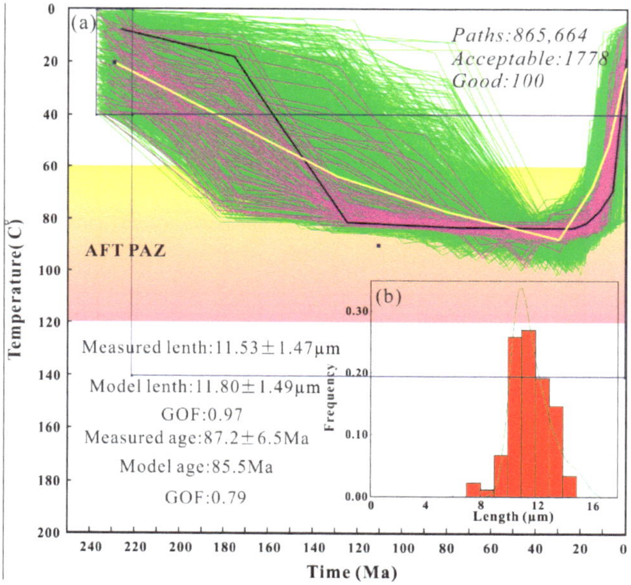

Figure 13. (**a**) Modelling result of AFT thermal history of clastic rock 19DJ-1 of Triassic Dajianggang Formation and (**b**) length distribution of fission tracks. (**a**) shows the results of the thermal history simulation based on grain age and track length. (**b**) shows the length distribution of the confined fission track and the comparison with the simulation results in (**a**); the purple lines represent simulated paths with good fit (GOF > 0.5), and the green lines represent simulated paths with acceptable fit (0.5 > GOF > 0.05). The blue boxes are used to constrain the Time–Temperature paths. The black solid line represents the optimal solution in the simulation result. The yellow solid line represents the weighted mean path. Abbreviation: AFT PAZ—partial annealing zone of apatite fission track.

8. Discussion

8.1. Provenance and Tectonic Setting Reflected by Clastic Components of Sandstone

The petrographic analysis of sandstone from the Dajianggang Formation in the study area reveals that it predominantly consists of feldspar lithic sandstone with low structural and compositional maturity. This indicates intense tectonic activity in the provenance area, where detrital material was rapidly eroded and deposited after being transported for a short distance. The Dickinson diagram analysis of the sandstone detrital components suggests that the primary provenance of the Dajianggang Formation sandstone in the Shuangyang area is an island arc orogenic belt. It is possible that the clastic material from the volcanic island arc mixed with the subduction complex material in the forearc basin, resulting in some sandstone samples showing the characteristics of recycled orogenic and subduction complex belts.

8.2. Provenance and Tectonic Setting Reflected by Geochemical Characteristics of Sandstone

Based on a comprehensive analysis of the geochemical characteristics of the sandstone samples, it has been determined that the tectonic setting and provenance of the Dajianggang Formation in the Shuangyang area is predominantly from a continental island arc environment, with a provenance type of feldspar island arc. In some samples, the structural diagram of La/Sc-Ti/Zr suggests an active continental margin provenance, indicating that materials from the active continental margin may have been mixed in the provenance. Along with receiving materials from the Late Triassic Changchun–Yanji suture zone island arc, the Dajianggang Formation in the Shuangyang area may also have received materials from the northern margin of the North China Craton or exhumed recycled orogenic belt, suggesting a complex provenance history for the formation.

8.3. Provenance and Tectonic Setting Reflected by LA-ICP-MS Analysis of Detrital Zircons from Sandstone

LA-ICP-MS analysis of detrital zircons from sandstone samples of the Dajianggang Formation in the study area provides valuable information about the age and origin of the formation. The peak $^{206}Pb/^{238}U$ age range of 226.8–246.5 Ma and the weighted average $^{206}Pb/^{238}U$ age of 237.4 ± 1.1 Ma suggest that the formation of the Dajianggang Formation occurred later than 226.8 ± 5 Ma, corresponding to the Carnian in the early Late Triassic period, based on the revised International Geological time scale of 2023. The high Th/U values of all detrital zircons (>0.3) indicate that most of the zircons are magmatic in origin, which is supported by the characteristics of the CL image. This suggests that the provenance area experienced magmatic events that resulted in the crystallization of zircons. These findings provide significant insights into the geological history of the study area and contribute to our understanding of Late Triassic sedimentation and tectonic evolution in this region.

Through statistical analysis of the age distribution of conglomeratic sandstone samples from the Dajianggang Formation, it is evident that the detrital zircon ages are predominantly clustered within a narrow range of 200–300 Ma. This high degree of concentration is further supported by the steep slope of the cumulative probability curve, which suggests a relatively uniform and consistent provenance for the sedimentary material. The nearly vertical trend of the curve near the age of the sedimentary strata indicates that the tectonic setting of the provenance area was likely convergent in nature.

Wang, Zhou [64] identified a zircon with an age of 2516 ± 18 Ma from the 238 zircons they dated from the Dajianggang Formation, indicating that the North China Craton was also a source of detrital materials. Consistent with previous studies, the Changchun–Yanji suture zone in central Jilin was believed to be the main source of the Dajianggang Formation, which is a subduction complex [7,16,23,65], while the Changchun–Yanji suture belt is not only an eastward extension of the Soloker–Xilamulon–Changchun suture belt [1,2,66–69] towards the east, but also the southern margin of the Jihei high pressure belt, which formed via the subduction of the Jiamusi–Xingkai Block to the northern margin of the North China Craton during the Late Triassic and Early Jurassic periods [16,65]. Based on the LA-ICP-MS results of detrital zircons from sandstone samples, it can be concluded that the provenance of the Dajianggang Formation may be mainly from the Late Triassic magmatic arc of the Changchun–Yanji suture zone, with possible contributions from the northern margin of the North China Craton, indicating a convergent tectonic setting during that time. Combining the previous report in the vicinity of the study area, the Paleo-Asian Ocean finished its closure before the Late Triassic period [70,71].

8.4. Tectonic Setting and Evolution History Reflected by Apatite Fission Track Results in Sandstone

The detritus apatite fission track (AFT) age analysis of sample 19DJ-1 shows a relatively dispersed age distribution, as indicated by a low P (χ^2) value of less than 5%. The central age of the detrital AFT in sample 19DJ-1 is 94.2 ± 8.3 Ma (Figure 12), which is younger than the corresponding stratigraphic age. The thermal evolution history of sample 19DJ-1 indicates that the early thermal events (240–30 Ma) may have been caused by burial, while the rapid cooling in the late stage reflects regional exhumation events (since 30 Ma). This indicates that the orogenic process of the Central Asian Orogenic Belt had ceased, and the study area had not been uplifted since the Late Triassic period. The rapid uplift of 30 Ma may correspond to the temporary squeezing of the northeast Asian continental margin and a series of inversion structures in basins caused by the subduction of the Pacific plate during the Middle Himalayan movement, reflecting a regional uplift event [72–74]. These findings suggest that the study area has experienced complex tectonic activities, including burial, subduction, and uplift events, and their effects on the region have been felt at different times throughout geological history [73,74].

9. Conclusions

(1) The Dajianggang Formation sandstone in the Shuangyang area is mainly feldspar lithic sandstone, and the lithic type is mainly volcanic lithic. Petrographic studies suggest that the clastic grains are angular, indicating near-source sedimentary characteristics. Dickinson diagram analysis reveals that the sandstone's main provenance is from island arcs, with some samples showing mixed characteristics from recycled orogenic belts and subduction complex belts.

(2) The geochemical analysis of sandstone samples from the Shuangyang area suggests a continental island arc environment, with felsic rocks dominating the provenance, mixed with active continental margin materials.

(3) The U-Pb chronology of detrital zircon LA-ICP-MS shows that the Dajianggang Formation has a maximum depositional age of 226.8 $\pm$ 5 Ma, indicating that it formed in the Carnian of the middle Late Triassic period, at the earliest. The distribution of zircon ages suggests that the main source of the Dajianggang Formation was the Late Triassic magmatic arc of the Changchun–Yanji suture zone, possibly mixed with materials from the northern margin of the North China Craton, in a convergent tectonic setting.

(4) The apatite fission track analysis reveals that the Dajianggang Formation underwent a thermal history that started around 240 Ma and experienced rapid exhumation to the surface since 30 Ma. This implies that the orogenic process in the study area ceased during the Late Triassic period, and the remote effect of Pacific subduction did not impact the study area until 30 Ma.

Supplementary Materials: The following supporting information can be downloaded at: https://www.mdpi.com/article/10.3390/min13070975/s1. Figure S1: The strata column section of Dajianggang Formation in Dajianggang area.

Author Contributions: Conceptualization, S.L.; methodology, S.L.; software, S.T. and H.W.; validation, X.Z. (Xinmei Zhang) and D.Z.; formal analysis, S.L. and H.W.; investigation, L.N. and X.Z. (Xiaodong Zhou); resources, S.L.; data curation, S.T., X.Z. (Xinmei Zhang), D.Z. and M.L.; writing—original draft preparation, S.T.; writing—review and editing, S.L.; visualization, S.T.; supervision, S.L.; project administration, S.T.; funding acquisition, S.L. All authors have read and agreed to the published version of the manuscript.

Funding: This research was funded by the National Natural Science Foundation of China (Grant No. 41872234). And the APC was funded by S.L.

Data Availability Statement: Not applicable.

Acknowledgments: We thank L.Y. Zhang and Z.R. Zhao for their assistance in the fieldwork and processing of U-Pb data, and H.T. Zhang for her technical support with the software. We also thank three reviewers for detailed reviews, as well as the editor.

Conflicts of Interest: The authors declare no conflict of interest.

References

1. Şengör, A.M.C.; Natal'in, B.A. Paleotectonics of Asia: Fragments of a synthesis. *Tecton. Evol. Asia* **1996**, *1*, 486–640.
2. Li, J.-Y. Permian geodynamic setting of Northeast China and adjacent regions: Closure of the Paleo-Asian Ocean and subduction of the Paleo-Pacific Plate. *J. Asian Earth Sci.* **2006**, *26*, 207–224. [CrossRef]
3. Eizenhöfer, P.R.; Zhao, G.; Zhang, J.; Sun, M. Final closure of the Paleo-Asian Ocean along the Solonker Suture Zone: Constraints from geochronological and geochemical data of Permian volcanic and sedimentary rocks. *Tectonics* **2014**, *33*, 441–463. [CrossRef]
4. Li, S.; Zhang, L.; Liu, Z.; Xu, Z. New Discovery of the Late Triassic Terrigenous Sediments in the Great Xing'an Range Region, NE China and its Geological Significance. *Acta Geol. Sin.* **2017**, *91*, 1928–1929. [CrossRef]
5. Wilde, S.A. Final amalgamation of the Central Asian Orogenic Belt in NE China: Paleo-Asian Ocean closure versus Paleo-Pacific plate subduction—A review of the evidence. *Tectonophysics* **2015**, *662*, 345–362. [CrossRef]

6. Zhou, J.-B.; Wilde, S.A.; Zhao, G.-C.; Han, J. Nature and assembly of microcontinental blocks within the Paleo-Asian Ocean. *Earth-Sci. Rev.* **2018**, *186*, 76–93. [CrossRef]
7. Wu, F.Y.; Sun, D.Y.; Ge, W.C.; Zhang, Y.B.; Grant, M.L.; Wilde, S.A.; Jahn, B.M. Geochronology of the Phanerozoic granitoids in northeastern China. *J. Asian Earth Sci.* **2011**, *41*, 1–30. [CrossRef]
8. Donskaya, T.; Gladkochub, D.; Mazukabzov, A.; Ivanov, A. Late Paleozoic—Mesozoic subduction-related magmatism at the southern margin of the Siberian continent and the 150 million-year history of the Mongol-Okhotsk Ocean. *J. Asian Earth Sci.* **2012**, *62*, 79–97. [CrossRef]
9. Xu, W.-L.; Pei, F.-P.; Wang, F.; Meng, E.; Ji, W.-Q.; Yang, D.-B.; Wang, W. Spatial–temporal relationships of Mesozoic volcanic rocks in NE China: Constraints on tectonic overprinting and trans-formations between multiple tectonic regimes. *J. Asian Earth Sci.* **2013**, *74*, 167–193. [CrossRef]
10. Safonova, I.; Santosh, M. Accretionary complexes in the Asia-Pacific region: Tracing archives of ocean plate stratigraphy and tracking mantle plumes. *Gondwana Res.* **2012**, *25*, 126–158. [CrossRef]
11. Tang, J.; Xu, W.-L.; Wang, F.; Zhao, S.; Wang, W. Early Mesozoic southward subduction history of the Mongol–Okhotsk oceanic plate: Evidence from geochronology and geochemistry of Early Mesozoic intrusive rocks in the Erguna Massif, NE China. *Gondwana Res.* **2016**, *31*, 218–240. [CrossRef]
12. Chen, Y.-J.; Zhang, C.; Wang, P.; Pirajno, F.; Li, N. The Mo deposits of Northeast China: A powerful indicator of tectonic settings and associated evolutionary trends. *Ore Geol. Rev.* **2017**, *81*, 602–640. [CrossRef]
13. Liu, Y.; Li, W.; Feng, Z.; Wen, Q.; Neubauer, F.; Liang, C. A review of the Paleozoic tectonics in the eastern part of Central Asian Orogenic Belt. *Gondwana Res.* **2017**, *43*, 123–148. [CrossRef]
14. Zhang, L.; Li, S.; Chu, X.; Shang, Y. Early Cretaceous Adakitic Rocks in the Northern Great Xing'an Range, NE China: Implications for the Final Closure of Mongol-Okhotsk Ocean and Regional Extensional Setting. *Acta Geol. Sin.* **2019**, *93*, 1544–1558. [CrossRef]
15. Wu, F.-Y.; Yang, J.-H.; Lo, C.-H.; Wilde, S.A.; Sun, D.-Y.; Jahn, B.-M. The Heilongjiang Group: A Jurassic accretionary complex in the Jiamusi Massif at the western Pacific margin of north-eastern China. *Isl. Arc* **2007**, *16*, 156–172. [CrossRef]
16. Zhou, J.-B.; Wilde, S.A. The crustal accretion history and tectonic evolution of the NE China segment of the Central Asian Orogenic Belt. *Gondwana Res.* **2013**, *23*, 1365–1377. [CrossRef]
17. Yang, H.; Ge, W.; Zhao, G.; Yu, J.; Zhang, Y. Early Permian–Late Triassic granitic magmatism in the Jiamusi–Khanka Massif, eastern segment of the Central Asian Orogenic Belt and its implications. *Gondwana Res.* **2015**, *27*, 1509–1533. [CrossRef]
18. Zhang, X.Z.; Guo, Y.; Zhou, J.B.; Zeng, Z.; Bin Pu, J.; Fu, Q.L. Late Paleozoic-Early Mesozoic tectonic evolution in the east margin of the Jiamusi massif, eastern northeastern China. *Russ. J. Pac. Geol.* **2015**, *9*, 103908. [CrossRef]
19. Tang, J.; Xu, W.; Wang, F.; Ge, W. Subduction history of the Paleo-Pacific slab beneath Eurasian continent: Mesozoic-Paleogene magmatic records in North-east Asia. *Sci. China Earth Sci.* **2018**, *61*, 527–559. [CrossRef]
20. RGSB; Wang, Z.Z. Discovery and Geological Significance of Late Triassic Dajianggang Formation in Shuangyang area, Jilin Province. *Jilin Geol.* **1979**, *3*, 22–27.
21. JBGMR. *Stratigraphy (Lithostatic) of Jilin Province*; China University of Geosciences Press: Wuhan, China, 1997.
22. Sun, G.; Zhao, Y.H.; Li, C.T. Late Triassic plants from Dajianggang of Shuangyang County, Jilin Province. *Acta Palaeontol. Sin.* **1983**, *22*, 447–459.
23. Wu, F.-Y.; Zhao, G.-C.; Sun, D.-Y.; Wilde, S.A.; Yang, J.-H. The Hulan Group: Its role in the evolution of the Central Asian Orogenic Belt of NE China. *J. Asian Earth Sci.* **2007**, *30*, 542–556. [CrossRef]
24. Xin, Y.L.; Ren, J.L.; Peng, Y.J.; Sun, X.Q. Ending of the mountain-building movement of Xing'an-Mongolian-Ji-Hei orogenic belt in northeast China: Evidence from Late Triassic molasse (geotectonic phase). *Geol. Resour.* **2011**, *20*, 413–419.
25. Li, D.J.; Zhao, Y.H.; Li, C.T. The finding of Late Triassic plant fossils and its significance in prospecting of Bamianshi in Shuanguang county, Jilin province. *Jilin Geol.* **1985**, *2*, 43–46.
26. Eggins, S.M.; Kinsley, L.P.J.; Shelley, J.M.M. Deposition elemental fractionation processes during atmospheric pressure laser sam-pling for analysis by ICP–MS. *Appl. Surf. Sci.* **1998**, *127–129*, 278–286. [CrossRef]
27. Jackson, S.E.; Pearson, N.J.; Griffin, W.L.; Belousova, E.A. The application of laser ablation–inductively coupled plasma–mass spectrometry to in situ U–Pb zircon geochronology. *Chem. Geol.* **2004**, *211*, 47–69. [CrossRef]
28. Yuan, H.; Gao, S.; Liu, X.; Li, H.; Günther, D.; Wu, F. Accurate U-Pb Age and Trace Element Determinations of Zircon by Laser Ablation-Inductively Coupled Plasma-Mass Spectrometry. *Geostand. Geoanalytical Res.* **2004**, *28*, 353–370. [CrossRef]
29. Vermeesch, P. IsoplotR: A free and open toolbox for geochronology. *Geosci. Front.* **2018**, *9*, 1479–1493. [CrossRef]
30. Sláma, J.; Košler, J.; Condon, D.J.; Crowley, J.L.; Gerdes, A.; Hanchar, J.M.; Horstwood, M.S.A.; Morris, G.A.; Nasdala, L.; Norberg, N.; et al. Plešovice zircon—A new natural reference material for U–Pb and Hf isotopic microanalysis. *Chem. Geol.* **2008**, *249*, 1–35. [CrossRef]
31. Gleadow, A.J.; Belton, D.X.; Kohn, B.P.; Brown, R.W. Fission Track Dating of Phosphate Minerals and the Thermochronology of Apatite. *Rev. Miner. Geochem.* **2002**, *48*, 579–630. [CrossRef]
32. Green, P.; Duddy, I.; Gleadow, A.; Tingate, P.; Laslett, G. Thermal annealing of fission tracks in apatite: 1. A qualitative description. *Chem. Geol. Isot. Geosci. Sect.* **1986**, *59*, 237–253. [CrossRef]
33. Hurford, A.J. Cooling and uplift patterns in the Lepontine Alps South Central Switzerland and an age of vertical movement on the In-subric fault line. *Contrib. Mineral. Petrol.* **1986**, *92*, 413–427. [CrossRef]

34. Gleadow, A.J.W.; Duddy, I.R.; Green, P.F.; Lovering, J.F. Confined fission track lengths in apatite: A diagnostic tool for thermal history analysis. *Contrib. Miner. Pet.* **1986**, *94*, 405–415. [CrossRef]
35. Zhou, Z.Y. *Low Temperature Thermochronology: Principles & Applications*; Science Press: Beijing, China, 2014.
36. Hurford, A.J.; Green, P.F. The zeta age calibration of fission-track dating. *Chem. Geol.* **1983**, *41*, 285–317. [CrossRef]
37. Galbraith, R.; Laslett, G. Statistical models for mixed fission track ages. *Nucl. Tracks Radiat. Meas.* **1993**, *21*, 459–470. [CrossRef]
38. Carlson, W.D.; Donelick, R.A.; Ketcham, R. Variability of apatite fission-track annealing kinetics; I, Experimental results. *Am. Miner.* **1999**, *84*, 1213–1223. [CrossRef]
39. Crowley, K.; Cameron, M.; Schaefer, R. Experimental studies of annealing of etched fission tracks in fluorapatite. *Geochim. Cosmochim. Acta* **1991**, *55*, 1449–1465. [CrossRef]
40. Donelick, R.A.; Ketcham, R.; Carlson, W.D. Variability of apatite fission-track annealing kinetics; II, Crystallographic orientation effects. *Am. Miner.* **1999**, *84*, 1224–1234. [CrossRef]
41. Duddy, I.; Green, P.; Laslett, G. Thermal annealing of fission tracks in apatite 3. Variable temperature behaviour. *Chem. Geol. Isot. Geosci. Sect.* **1988**, *73*, 25–38. [CrossRef]
42. Stewart, R.J.; Brandon, M.T. Detrital-zircon fission-track ages for the "Hoh Formation": Implications for late Cenozoic evolution of the Cascadia subduction wedge. *GSA Bull.* **2004**, *116*, 60–75. [CrossRef]
43. Dickinson, W.R.; Suczek, C.A. Plate Tectonics and Sandstone Compositions. *AAPG Bull.* **1979**, *63*, 2164–2182.
44. Dickinson, W.R.; Sue beard, L.; Brakenridge, G.R.; Erjavec, J.L.; Ferguson, R.C.; Inman, K.F.; Knepp, R.A.; Lindberg, F.A.; Ryberg, P.T. Provenance of North American Phanerozoic sandstones in relation to tectonic setting. *Geol. Soc. Am. Bull.* **1983**, *94*, 222. [CrossRef]
45. Nesbitt, H.W.; Young, G.M. Early Proterozoic climates and plate motions inferred from major element chemistry of lutites. *Nature* **1982**, *299*, 715–717. [CrossRef]
46. Fedo, C.M.; Nesbitt, H.W.; Young, G.M. Unravelling the effects of potassium metasomatism in sedimentary rocks and paleosols, with implications for paleoweathering conditions and provenance. *Geology* **1995**, *23*, 921–924. [CrossRef]
47. Harnois, L. The CIW index: A new chemical index of weathering. *Sediment. Geol.* **1988**, *55*, 319–322. [CrossRef]
48. Cox, R.; Lowe, D.R.; Cullers, R. The influence of sediment recycling and basement composition on evolution of mudrock chemistry in the southwestern United States. *Geochim. Cosmochim. Acta* **1995**, *59*, 2919–2940. [CrossRef]
49. Roser, B.P.; Korsch, R.J. Determination of Tectonic Setting of Sandstone-Mudstone Suites Using SiO_2 Content and K_2O/Na_2O Ratio. *J. Geol.* **1986**, *94*, 635–650. [CrossRef]
50. Floyd, P.A.; Leveridge, B.E. Tectonic environment of the Devonian Gramscatho basin, south Cornwall: Framework mode and geo-chemical evidence from turbiditic sandstones. *J. Geol. Soc.* **1987**, *144*, 531–542. [CrossRef]
51. Taylor, S.R.; Mclennan, S.M. The continental crust: Its composition and evolution. *The Journal of Geology.* **1985**, 57–72.
52. Mclennan, S.M. Rare earth elements in sedimentary rocks: Influence of provenance and sedimentary processes. *Mineral. Geochem.* **1989**, *21*, 169–200.
53. Floyd, P.A.; Winchester, J.A.; Park, R.G. Geochemistry and tectonic setting of Lewisian clastic metasediments from the Early Proterozoic Loch Maree Group of Gairloch, NW Scotland. *Precambrian Res.* **1989**, *45*, 203–214. [CrossRef]
54. Bhatia, M.R.; Crook, K.A.W. Trace element characteristics of graywackes and tectonic setting discrimination of sedimentary basins. *Contrib. Miner. Pet.* **1986**, *92*, 181–193. [CrossRef]
55. Koschek, G. Origin and significance of the SEM cathodoluminescence from zircon. *J. Microsc.* **1993**, *171*, 223–232. [CrossRef]
56. Cawood, P.; Hawkesworth, C.; Dhuime, B. Detrital zircon record and tectonic setting. *Geology* **2012**, *40*, 875–878. [CrossRef]
57. Vermeesch, P. RadialPlotter: A Java application for fission track, luminescence and other radial plots. *Radiat. Meas.* **2009**, *44*, 409–410. [CrossRef]
58. O'Sullivan, P.B.; Parrish, R.R. The importance of apatite composition and single-grain ages when interpreting fission track data from plutonic rocks: A case study from the Coast Ranges, British Columbia. Earth Planet. *Sci. Lett.* **1995**, *132*, 213–224. [CrossRef]
59. Green, P.F. The relationship between track shortening and fission track age reduction in apatite: Combined influences of inherent instability, annealing anisotropy, length bias and system calibration. *Earth Planet. Sci. Lett.* **1988**, *89*, 335–352. [CrossRef]
60. Lin, X.; Chen, H.; Wyrwoll, K.-H.; Batt, G.E.; Liao, L.; Xiao, J. The Uplift History of the Haiyuan-Liupan Shan Region Northeast of the Present Tibetan Plateau: Integrated Constraint from Stratigraphy and Thermochronology. *J. Geol.* **2011**, *119*, 372–393. [CrossRef]
61. Ketcham, R.A. Forward and Inverse Modeling of Low-Temperature Thermochronometry Data. *Rev. Miner. Geochem.* **2005**, *58*, 275–314. [CrossRef]
62. Gallagher, K. Transdimensional inverse thermal history modeling for quantitative thermochronology. *J. Geophys. Res. Solid Earth* **2012**, *117*, B02408. [CrossRef]
63. Ketcham, R.A.; Donelick, R.A.; Balestrieri, M.L.; Zattin, M. Reproducibility of apatite fission-track length data and thermal history reconstruction. *Earth Planet. Sci. Lett.* **2009**, *284*, 504–515. [CrossRef]
64. Wang, B.; Zhou, J.-B.; Wilde, S.A.; Zhang, X.-Z.; Ren, S.-M. The timing of final closure along the Changchun–Yanji suture zone: Constraints from detrital zircon U–Pb dating of the Triassic Dajianggang Formation, NE China. *Lithos* **2016**, *261*, 216–231. [CrossRef]
65. Zhou, J.B.; Han, J.; Wilde, S.; Guo, X.; Zeng, W.; Cao, J. A primary study of the Jilin-Heilongjiang high-pressure metamorphic belt: Evidence and tectonic implications. *Acta Petrol. Sin.* **2013**, *29*, 386–398.

66. Shao, J.A. *Crustal Evolution in the Middle Part of the Northern Margin of Sino–Korean Plate*; Beijing University Publication House: Beijing, China, 1991; p. 136.
67. Tang, K.D. Tectonic development of Paleozoic foldbelts at the north margin of the Sino-Korean Craton. *Tectonics* **1990**, *9*, 249–260. [CrossRef]
68. Xiao, W.; Windley, B.F.; Hao, J.; Zhai, M. Accretion leading to collision and the Permian Solonker suture, Inner Mongolia, China: Termination of the central Asian orogenic belt. *Tectonics* **2003**, *22*, 1069. [CrossRef]
69. Xu, G.Y. Some basic geological problems of Jilin Province. *Reg. Geol. China* **1995**, *1*, 6–14.
70. Han, Z.; Li, J.; Zhu, C.; Zhong, W.; Song, Z. The Late Triassic Molasse Deposits in Central Jilin Province, NE China: Constraints on the Paleo-Asian Ocean Closure. *Minerals* **2021**, *11*, 223. [CrossRef]
71. Yang, X.; Li, S.; Luo, W.; Wang, X.; Wang, H.; Wei, H. Final closure of the Paleo-Asian Ocean and initiation of the Paleo-Pacific subduction: Geophysical evidence from the Nadanhada Terrane. *Tectonophysics* **2023**, *859*, 229915. [CrossRef]
72. Yu, F.; Koyi, H. Cenozoic tectonic model of the Bohai Bay Basin in China. *Geol. Mag.* **2016**, *153*, 866–886. [CrossRef]
73. Li. S.Z.; Suo, Y.; Li, X.; Wang, Y.; Cao, X.; Wang, P.; Guo, L.; Yu, S.; Lan, H.; Li, S.; et al. Mesozoic plate subduction in West Pacific and tectono-magmatic response in the East Asian ocean-continent connection zone. *Chin. Sci. Bull.* **2018**, *63*, 1550–1593. [CrossRef]
74. Suo, Y.; Li, S.; Cao, X.; Li, X.; Liu, X.; Cao, H. Mesozoic-Cenozoic inversion tecvtonics of East China and its implications for the subduction process of the oceanic plate. *Earth Sci. Front.* **2017**, *24*, 249–267.

MDPI AG
Grosspeteranlage 5
4052 Basel
Switzerland
Tel.: +41 61 683 77 34

Minerals Editorial Office
E-mail: minerals@mdpi.com
www.mdpi.com/journal/minerals

www.ingramcontent.com/pod-product-compliance
Lightning Source LLC
Chambersburg PA
CBHW041427210926
51850CB00013B/3368
9783725825615